MOLECULAR CARCINOGENESIS AND THE MOLECULAR BIOLOGY OF HUMAN CANCER

MOLECULAR CARCINOGENESIS AND THE MOLECULAR BIOLOGY OF HUMAN CANCER

EDITED BY

DAVID WARSHAWSKY, PH.D.

Professor of Environmental Health
Department of Environmental Health
University of Cincinnati College of Medicine
Cincinnati, OH, USA

JOSEPH R. LANDOLPH JR., PH.D.

Associate Professor of Molecular Microbiology and Immunology,
Pathology, and Molecular Pharmacology and Toxicology
USC/Norris Comprehensive Cancer Center
Keck School of Medicine and School of Pharmacy
University of Southern California
Los Angeles, CA, USA

Taylor & Francis
Taylor & Francis Group
Boca Raton London New York

A CRC title, part of the Taylor & Francis imprint, a member of the
Taylor & Francis Group, the academic division of T&F Informa plc.

Published in 2006 by
CRC Press
Taylor & Francis Group
6000 Broken Sound Parkway NW, Suite 300
Boca Raton, FL 33487-2742

International Standard Book Number-10: 0-8493-1167-5 (Hardcover)
International Standard Book Number-13: 978-0-8493-1167-3 (Hardcover)
Library of Congress Card Number 2005050218

Library of Congress Cataloging-in-Publication Data

Molecular carcinogenesis and the molecular biology of human cancer / edited by David Warshawsky, Joseph R. Landolph.
    p. ; cm.
Includes bibliographical references and index.
ISBN 0-8493-1167-5 (alk. paper)
    1. Chemical carcinogenesis. 2. Cancer--Molecular aspects. I. Warshawsky, David. II. Landolph, Joseph R.
    [DNLM: 1. Neoplasms--chemically induced. 2. Carcinogens--toxicity. 3. Molecular Biology. QZ 200 M7183305 2005]

RC268.6.M65 2005
616.99'4071--dc22   2005050218

Taylor & Francis Group
is the Academic Division of T&F Informa plc.

Visit the Taylor & Francis Web site at
http://www.taylorandfrancis.com

and the CRC Press Web site at
http://www.crcpress.com

Preface

There is a strong need for a basic graduate textbook in molecular carcinogenesis. Dr. David Warshawsky has taught a comprehensive course in chemical carcinogenesis at the College of Medicine in the University of Cincinnati (UC) for many years. The colleagues of Dr. Warshawsky from UC who contributed to this textbook include Drs. Glenn Talaska (carcinogen biomarkers), George Leikauf and Jay Tichelaar (lung cancer), Mario Medvedovic (bioinformatics), Susan Heffelfinger (breast cancer), Zalfa Abdel-Malik (skin cancer) and Eula Bingham (carcinogen regulation). Hence, this book is the result of a strong USC–UC collaboration of Dr. Joseph Landolph, Jr. and Dr. Warshawsky and the colleagues of Professor Melvin Calvin at the University of California at Berkeley (Dr. Andrew Salmon) and in the laboratory of Professor Charles Heidelberger at the University of Wisconsin, Madison (Dr. Steven Nesnow). Colleagues of Dr. Landolph recruited from USC who have also contributed to this textbook are Drs. James Ou (Hepatitis B virus), Colin Hill (radiation carcinogenesis), and Nouri Neamati (proteomics). We have, in addition, recruited many experts and their collaborators from many other universities, research institutes, and regulatory bodies: Dr. Ole Laerum and Johan R. Lillehaug (cancer of the brain), University of Bergen; Dr. Dan Djakiew (prostate cancer), Georgetown University; Dr. Helmut Zarbl (oncogenes), Fred Hutchinson, Cancer Research Center; Dr. Bernard Weissman (tumor suppressor genes), University of North Carolina; Dr. Wynshaw-Boris (transgenic and knockout mice in carcinogenesis), University of California at San Diego; Dr. Jeff Ross (carcinogen DNA adducts), USEPA; and Dr. Gary Stoner (chemoprevention), Ohio State University.

Dr. Landolph has taught a similar course on chemical carcinogenesis at the Keck School of Medicine of the University of Southern California (USC) since 1982. Part I covers the basic science of cancer and includes lectures on cancer pathology, the epidemiology of cancer, stress responses, DNA methylation and cancer, oncogenes, tumor suppressor genes, DNA repair, chemical mutation, chemical carcinogenesis, viral carcinogenesis, and radiation carcinogenesis. Part II covers cancers of various organ systems and some treatments for these cancers, including breast cancer, prostate cancer, lung cancer, bladder cancer, colon cancer, leukemia, cancer immunology and immunotherapy, and gene therapy of cancer.

The collaborations between Dr. Landolph and Dr. Warshawsky go back to 1973, and have led directly to the writing of this book. They first met when Dr. Landolph was a graduate student (1971–1976), and Dr. Warshawsky was a postdoctoral fellow (1973–1976), in the chemical biodynamics laboratory of Professor Melvin Calvin at the University of California at Berkeley, California. Dr. Warshawsky received training in chemical carcinogenesis from Professors Calvin, Orchin, and Bingham. Similarly, Dr. Landolph received training from Professors Charles Heidelberger and Calvin.

This book is divided into two parts, similar to the organization of the two courses at USC and UC in the molecular biology of cancer. Part I covers the basic science of cancer, including an historical overview of cancer and chemical carcinogenesis, chemical carcinogenesis, viral carcinogenesis, and radiation carcinogenesis, oncogenes, tumor suppressor genes, and genomics and proteomics approaches to understanding the molecular nature of cancer. Part II covers an overview of human cancer induction and human exposure to carcinogens, complex mixtures of chemical carcinogens, and tumors of various human organs, including breast cancer, prostate cancer, skin cancer, cancer of the brain, and cancer of the thyroid. Part II continues with chemoprevention of chemical carcinogenesis and human cancer, exposure assessment and biomarkers, carcinogen risk assessment, and the regulation of carcinogens. Written by authors who are all experts in their chosen fields, this book should give the reader an overview of chemical, viral and radiation carcinogenesis, the carcinogenicity of complex mixtures and chemicals that cause human cancer, the proportional causes of human cancer, and the cell and molecular biology of specific important human tumors. This textbook is most appropriate for first and second year Ph.D. or M.D./Ph.D. students and postdoctoral fellows and is suitable for either a one or two semester course. However, this book should also be useful to advanced undergraduate students, to scientists moving into cancer research from other areas, to scientists teaching graduate courses in cancer biology and molecular carcinogenesis, to regulatory scientists, and to attorneys practicing law in the area of toxic torts.

We would like to thank two people without whom this book would not have been possible: Veronica Ratliff and Sireesha Kandula. Their collective skills were necessary to complete the book in a timely manner.

Dr. Warshawsky would like to thank his wife, Susan, and children Lisa and Bart, and Daniel and Deborah for their patience and understanding and insistence that this book be written to fill a need in molecular carcinogenesis. Dr. Landolph would like to thank his family, Alice and sons Joe III and Lewis, for their compassion and understanding in the writing of this book.

Editors

Dr. David Warshawsky is a professor of environmental health in the Division of Environmental Genetics and Molecular Toxicology in the College of Medicine at the University of Cincinnati. He has published over 120 papers in journals such as *Cancer Research, Chemical Research in Toxicology, Carcinogenesis, Molecular Carcinogenesis, Proceedings of the National Academy of Sciences, and Applied Environmental Microbiology.* Dr. Warshawsky has been funded by the NIH, USEPA, and US DOE. He received his B.S. with Robert Coates in chemistry from the University of Illinois, his M.S. in biochemistry from Rutgers University, his Ph.D. in organic chemistry with Milton Orchin from the University of Cincinnati, and did his postdoctoral training at University of California, Berkeley with Melvin Calvin, and research collaborations with Eula Bingham, Glenn Talaska, and Roy Albert at the University of Cincinnati. He has completed sabbaticals with Mina Bissell at Lawrence Berkeley Laboratory at Berkeley California, Steve Nesnow at Research Triangle Park at North Carolina, and Gary Stoner at Ohio State University. He is on the editorial board of *Toxicological Sciences* and has been on numerous review panels and has reviewed numerous manuscripts and proposals. He is a member of the Society of Toxicology, the American Association for Cancer Research, the American Chemical Society, and the American Association for the Advancement of Science.

Dr. Warshawsky has been a University of Cincinnati and State Representative on environmental issues for the Ohio Advisory Board from 1993 to 1996. He received an Outstanding Achievement Recognition Award from the department of environmental health, University of Cincinnati in 1995. He was also elected as Fellow of the Graduate School of the University of Cincinnati in 1998. Recently, he organized the 18th International Symposium on Polycyclic Aromatic Compounds that was held in Cincinnati in 2001 and he is the chair of the Oversight Committee for the University Mass Spectrometry Facility. He has been a technical consultant for OSHA. He was the Elected Councillor for the Society of Toxicology, Carcinogenesis Specialty Section, 1999 to 2000. He organized, with Dr. Landolph, a symposium in March 1998 at the Society of Toxicology meeting on the subject of molecular and cellular biology of chemical carcinogenesis.

Dr. Warshawsky's research focuses on the early events in the carcinogenic process for polyaromatic compounds and mixtures thereof including metals, the development of biomarkers of exposure, and methods for the microbial biodegradation of these recalcitrant compounds. He has investigated mechanisms of chemical carcinogenesis in target tissues, liver, skin, lung, and breast. This research involves the metabolism, binding to DNA and protein, gene expression, and growth factor regulation. The relationship of DNA adducts with specific mutations in the activated *RAS* oncogene and *p*53 tumor suppressor gene in target tissues is being assessed. His laboratory is studying mammary tumor prevention using angiogenic inhibitors. The role of angiogenesis in noninvasive tumors is largely unexplored. Although chemoprophylaxis in women at increased risk for developing invasive breast disease is a topic of national attention; all current approaches utilize hormone manipulation. This laboratory is the first to be able to delineate and characterize the degradation pathway of benzo[*a*]pyrene, a potent carcinogen. Work is underway to characterize the genes involved in the pathway. Based on the long history of this department on studying complex mixtures, studies are focused on the effects of a binary mixture on lung tumors, K-ras mutations, and DNA adducts. Dr. Warshawsky served on the National Academy of Sciences Committee on Complex Mixtures from 1986–1988.

Dr. Joseph R. Landolph, Jr., is currently associate professor of molecular microbiology and immunology and pathology and a member of the University Southern California (USC)/Norris Comprehensive Cancer Center, in the Keck School of Medicine, and associate professor of molecular pharmacology and toxicology, in the School of Pharmacy, with tenure, at the University Southern California in Los Angeles, California. Dr. Landolph received a B.S. degree in chemistry from Drexel University in 1971 and a commission as a 2nd Lieutenant in the U.S. Army through ROTC. He received a Ph.D. in chemistry from the University of California at Berkeley in 1976. At the University of California Berkeley, Dr. Landolph studied under the guidance of professor Melvin Calvin, and conducted research on the metabolism of the chemical carcinogen, benzo[*a*]pyrene (BaP), and its ability to induce cytotoxicity in cultured mouse liver epithelial cells and mechanisms of resistance to the cytotoxicity of BaP. He also studied BaP-induced cytotoxicity and morphological transformation in Balb/c3T3 mouse fibroblasts. Dr. Landolph did his postdoctoral study in chemically induced morphological and neoplastic cell transformation, chemically induced mutagenesis, and chemical carcinogenesis at the USC/Norris Comprehensive Cancer Center at the School of Medicine in the University of Southern California, under Professor Charles Heidelberger, from 1977 to 1980.

Dr. Landolph was appointed assistant professor of pathology in 1980, and associate professor of microbiology, pathology, and toxicology at USC in 1987. At USC, Dr. Landolph lectures to medical and Ph.D. students. These lectures include medical microbiology, prokaryotic molecular genetics, the molecular biology of cancer, the cytochrome P450 gene family and xenobiotic metabolism, and the role of oxygen radical generation in chemical carcinogenesis and tumor promotion. Dr. Landolph's research interests and activities include studies of the genetic toxicology and carcinogenicity of carcinogenic insoluble nickel compounds, carcinogenic chromium compounds, carcinogenic arsenic compounds, and carcinogenic polycyclic aromatic hydrocarbons. His laboratory studies the ability of these carcinogens to induce morphological and neoplastic transformation of C3H/10T1/2 mouse embryo cells and the cellular and molecular biology of the transformation process. His laboratory currently studies the ability of carcinogenic nickel compounds to induce activation of expression of oncogenes and inactivation of expression to tumor suppressor genes by mRNA differential display in cell lines transformed by insoluble carcinogenic nickel compounds, such as nickel subsulfide, crystalline nickel monosulfide, and green (high temperature) and black (low temperature) nickel oxides. His laboratory is also studying the molecular biology of chromium compound-induced cell transformation. Dr. Landolph is an expert in chemically induced morphological and neoplastic transformation and chemically induced mutation in murine and human fibroblasts. He is the author of 56 scientific publications and has held peer-reviewed research grant support from the U.S. Environmental Protection Agency (USEPA), the U.S. National Cancer Institute, and U.S. National Institute of Environmental Health Sciences.

Dr. Landolph has served as grant reviewer for the USEPA Health Effects Panel, for special Requests for Applications (RFAs) for the National Institute of Environmental Health Sciences (NIEHS), and as an ad hoc member of the Chemical Pathology Study Section and the Al-Tox-4 Study Section of the National Institutes of Health (NIH). Dr. Landolph is also a member of the Carcinogen Identification Committee of the Scientific Advisory Committee of the Office of Environmental Health Hazard Assessment of the California Environmental Protection Agency (EPA) (1994 to present). Dr. Landolph has served as a member of the Human Health Strategies Review Committee of the USEPA and is currently a member of the Scientific Review Panel for Toxic Air Contaminants of the California EPA, and is a member of the Science Advisory Board, the Drinking Water Committee, and the Science and Technology Achievement Award Committee of the USEPA. He is the recipient of numerous awards, including the Merck Award in Chemistry and the Superior Cadet Award in Reserve Officer Training Corps (ROTC) from Drexel University in 1971, an American Cancer Society Postdoctoral Fellowship from 1977 to 1979, the Edmundson Teaching Award in the Department of Pathology at USC in 1985, and a Traveling Lectureship Award from the U.S. Society of Toxicology in 1990.

Contributors

Zalfa A. Abdel-Malek, Ph.D.
Department of Dermatology
University of Cincinnati
Cincinnati, Ohio

Eula Bingham, Ph.D.
Department of Environmental Health
College of Medicine
University of Cincinnati
Cincinnati, Ohio

Christoph Borchers, Ph.D.
Department of Biochemistry and Biophysics
University of North Carolina
Chapel Hill, North Carolina

Ting Chen, Ph.D.
Departments of Biological Sciences,
Computer Science, and Mathematics
College of Letters, Arts,
and Sciences
University of Southern California
Los Angeles, California

Daniel Djakiew, Ph.D.
Department of Cell Biology
Georgetown University
School of Medicine
Washington, D.C.

Laura R. Erker, B.S.
Departments of Pediatrics and Medicine
University of California, San Diego
School of Medicine
La Jolla, California

Øystein Fluge, Ph.D.
Department of Molecular Biology
Molecular Biology Institute
University of Bergen
Bergen, Norway

Sue C. Heffelfinger, M.D., Ph.D.
Department of Pathology and Laboratory
Medicine
College of Medicine
University of Cincinnati
Cincinnati, Ohio

Colin K. Hill, Ph.D.
Cancer Research Laboratory
USC/Norris Comprehensive Cancer Center
Keck School of Medicine
University of Southern California
Los Angeles, California

Ana Luisa Kadekaro, Ph.D.
Deptartment of Dermatology
University of Cincinnati
Cincinnati, Ohio

Øle Didrik Laerum, M.D., Ph.D.
Department of Pathology
Gade Institute
Haukeland Hospital
University of Bergen
Bergen, Norway

Joseph R. Landolph, Jr., Ph.D.
Cancer Research Laboratory
USC/Norris Comprehensive Cancer Center
Keck School of Medicine
University of Southern California
Los Angeles, California

George D. Leikauf, Ph.D.
Department of Environmental Health
Kettering Laboratory
University of Cincinnati
Cincinnati, Ohio

Johan R. Lillehaug, Ph.D.
Department of Molecular Biology
Molecular Biology Institute
University of Bergen
Bergen, Norway

Mario Medvedovic, Ph.D.
Department of Environmental Health
College of Medicine
University of Cincinnati
Cincinnati, Ohio

Mark A. Morse, Ph.D.
Springborn Laboratories, Inc.
Spencerville, Ohio

Diya F. Mutasim, M.D.
Department of Dermatology
University of Cincinnati
Cincinnati, Ohio

Nouri Neamati, Ph.D.
Department of Pharmaceutical Sciences
School of Pharmacy
University of Southern California
Los Angeles, California

Stephen Nesnow, Ph.D.
United States Environmental Protection
Agency
ORD/NHEEL/ECD
Research Triangle Park, North Carolina

Jing-hsiung Ou, Ph.D.
Department of Molecular Microbiology and
Immunology
Keck School of Medicine
Health Sciences Campus
University of Southern California
Los Angeles, California

Michelle A. Pipitone, M.D.
Department of Dermatology
University of Cincinnati
Cincinnati, Ohio

Jon Reid, Ph.D.
Department of Environmental Health
Kettering Laboratory
University of Cincinatti
Cincinnati, Ohio

Jeffrey A. Ross, Ph.D.
United States Environmental Protection Agency
Research Triangle Park, North Carolina

Andrew G. Salmon, Ph.D.
Air Toxicology and Risk Assessment Unit
Office of Environmental Health Hazard
Assessment
California Environmental Protection Agency
Oakland, California

Gary D. Stoner, Ph.D.
Division of Environmental Health Sciences
The Ohio State University School of Public
Health
College of Medicine
Columbus, Ohio

Glenn Talaska, Ph.D.
Department of Environmental Health
College of Medicine
University of Cincinnati
Cincinnati, Ohio

Jay W. Tichelaar, Ph.D.
Department of Environmental Health
Kettering Laboratory
University of Cincinnati
Cincinnati, Ohio

Jan Erik Varhaug, M.D., Ph.D.
Department of Surgery
Haukeland University Hospital
University of Bergen
Bergen, Norway

David Warshawsky, Ph.D.
Department of Environmental Health
Kettering Laboratory
University of Cincinnati
Cincinnati, Ohio

Bernard E. Weissman, Ph.D.
Department of Pathology and Laboratory
Medicine
Comprehensive Cancer Center
University of North Carolina
Chapel Hill, North Carolina

Jonathan S. Wiest, Ph.D.
Office of Training and Education
National Cancer Institute
Bethesda, Maryland

Contributors

Donna M. Williams-Hill, Ph.D.
Pacific Regional Laboratory Southwest
U.S. Food and Drug Administration
Irvine, California

Helmut Zarbl
Fred Hutchinson Cancer Research Center
Human Biology and Public Services
Seattle, Washington

Anthony Wynshaw-Boris, M.D., Ph.D.
Departments of Pediatrics and Medicine
Center for Human Genetics and Genomics
University of California, San Diego
La Jolla, California

Weiling Xue, M.S.
Deptartment of Environmental Health
Kettering Laboratory
University of Cincinnati
Cincinnati, Ohio

Contents

1 Carcinogens and Mutagens

David Warshawsky

CONTENTS

SUMMARY

For residents in the United States, the probability of developing cancer at some point during the course of their lifetime is approximately one out of two among men and one out of three among women [1]. Nearly everyone's life has been directly or indirectly affected by cancer. Most cancer researchers believe that many of the cancers are associated with the environment in which we live and work. According to well-documented reports on association between occupational exposures and cancer, an estimated 40,000 new cancer cases and 20,000 deaths due to cancer in the United States each year are attributable to exposure to carcinogens in the occupational environment [2].

1.1   THE ENVIRONMENT, WORKPLACE, AND CANCER

1.1.1   HISTORICAL PERSPECTIVE

Until the late Middle Ages the environment in the workplace did not receive much attention from the medical profession. Due to the increased demands for gold, silver, iron, copper, and lead during the 15th century, miners and metal workers were among the earliest groups to be studied for occupationally related diseases. Ulrich Ellenberg, a German physician from Augsburg, published a pamphlet in 1472 entitled, "On the Poisons, Evil Vapors, and Fumes of Metals" detailed the irritating effects of the fumes of lead and mercury on goldsmiths. *De Re Metallica* was published by the Georgium Agricola in 1556 on the accidents and diseases among miners and smelters of gold and silver. He recommended the use of facial masks and ventilation for their chronic lung disease attributed to inhalation of dusts [3].

In 1700, Bernardino Ramazzini published the first edition of *Discourses on the Disease of Workers* (*DeMorbis Artificum Diatriba*). This work established the field of occupational medicine. This publication contained a survey of the existing knowledge of the nature of diseases thought to be associated with a particular profession or workplace environment [3,4].

TABLE 1.1
Early Landmarks — Physical, Chemical, and Infectious Carcinogens in Environmental and Occupational Settings [4,5]

Principal investigator(s)	Causative agent	Date	Type of cancer
Pott	Soot	1775	Scrotum
Ayrton	Arsenic containing material	1822	Skin
Thiersch	Sunlight	1875	Skin
Manourriez	Coal tar	1876	Skin
Harting and Hesse	Some fractions from distillation of crude petroleum	1879	Skin
Unna	Solar radiation	1893	Skin
Rehn	Manufacture of aniline dyes	1895	Bladder
Frieben	X-rays	1902	Skin
Ferguson	Egyptian peasants infected with parasite *Schistosoma haematobrium*	1911	Bladder
Davis	Pipe smokers and betel nut chewers	1915	Lip and mouth
Leitch and Seguina	Radium radiation	1920	Skin
Delore and Bergamo	Benzene	1928	Leukemia
Stephens	Nickel	1932	Lung
Alwens	Chromium compounds	1932	Lung
Wood and Gloyne	Arsenicals, beryllium, and asbestos	1934	Lung
Neitzel	Mineral oil mists and radiation	1934	Lung
Kawahata	Coal tar fumes	1936	Lung

In 1775, Percival Pott of London described the increased incidence of cancer of the scrotum among chimney sweeps and attributed this to their contact with soot. This was the first clinical report of occupational chemical carcinogenesis. Since then a number of physical, chemical, and infectious carcinogens in environmental and occupational settings have been reported (Table 1.1). Increased incidences of skin cancer are associated with exposure to chemicals generated from materials containing coal, petroleum, shale, and arsenic as well as from radiation. Bladder cancer in the workplace was associated with the manufacture of aniline dyes, such as magenta and auramine. A number of compounds containing metals, such as nickel, chromium, beryllium, and arsenic, coal and petroleum products, and asbestos, were thought to be responsible for the higher incidence of lung cancer in occupational environments. Similarly, leukemia was associated with benzene and ionizing radiation generated from radon and radium [3–5].

Early chemical carcinogenesis experiments were performed in the beginning of the 20th century (Table 1.2). In 1915, skin cancer was induced on rabbit ears by painting them with coal tar. Similar results were obtained with other animal species using a variety of mixtures from coal and petroleum. Benzo(a)pyrene (BaP), a strongly carcinogenic compound that has been used as an indicator of carcinogenic potency, was isolated from coal tar in 1933. A number of experiments were carried out in the 1930s and 1940s using polycyclic aromatic hydrocarbons (PAHs) and aromatic amines, which produced cancer in a variety of animal species. In 1941, Berenblum et al. proposed the two-stage mechanism in skin cancer using coal tar and BaP (Table 1.2) [6].

TABLE 1.2
Early Landmarks in Experimental Chemical Carcinogenesis [4,6]

Date	Early landmarks	Principal investigators
1915–1918	Induction of skin cancer in rabbits and mice by coal tar	Yamagiqa, Ichikawa, and Tsutsui
1930	Tumor induction by the first pure chemical carcinogen dibenz[a,h]anthracene	Kennaway and Hieger
1933	Isolation of the carcinogen benzo[a]pyrene from coal tar	Cook, Hewett, and Hieger
1933–1936	Induction of liver cancer in rats by o-aminoazotoluene and by p-dimethylaminoazobenzene	Yoshida and Kinosita
1937	Induction of urinary bladder cancer in dogs by 2-naphthylamine	Hueper, Wiley, and Wolfe
1941	Initiation and promotion stages in skin carcinogenesis with tar and benzo[a]pyrene	Berenblum, Rous, MacKenzie, and Kidd

1.1.2 PRESENT DAY KNOWLEDGE

Since these early observations on environmental carcinogenesis and experimental animal carcinogenesis, the number of identifiable carcinogens has increased. However, the number of new synthetic chemicals has been increasing even more rapidly. In the United States, there are more than 80,000 commercial chemicals registered and an estimated 2,000 new ones are introduced annually for use in everyday items such as foods, personal care products, prescription drugs, household cleaners, lawn care chemicals etc. [7]. These large number of chemicals are deposited into and distributed within the environment during their manufacture, distribution, use, and disposal. Reportedly, only <2% of chemicals in commerce have been tested for carcinogenicity [2].

Although relatively few chemicals are thought to pose significant risk to human health, it is important to identify the toxicological/carcinogenic effects of these chemicals as well as the levels of exposure at which they become hazardous to humans. People worldwide became more and more concerned about the relationship between their environment and cancer and they wanted additional information and knowledge on important public health issues. Since 1960, research on cancer and carcinogenesis has substantially increased, due to support and organization from governments, academia, and industrial firms. As a part of the World Health Organization (WHO), the International Agency for Research on Cancer (IARC) in Lyon, France, was created. From 1972 to 2002, it has evaluated 885 agents (chemicals, group of chemicals, complex mixtures, occupational exposures, cultural habits, biological, or physical agents) in the first 82 volumes of IARC Monographs series [8]. In accordance with the procedures adopted as standard IARC practice, the agents, mixtures, and exposures as evaluated are classified into four groups: (1) carcinogenic to humans, (2)(a) probably carcinogenic to humans; (b) possibly carcinogenic to humans, (3) not classifiable as carcinogenic to humans, and (4) probably not carcinogenic to humans. Of the total 885 agents, mixtures, and exposures in the last update [9] on December 4, 2002, 88 items (63 agents and groups of agents, 12 mixtures, 13 exposure circumstances) are listed in Group 1, 64 (55 + 5 + 4 in each category) in Group 2(a), 236 (220 + 12 + 4 in each category) in Group 2(b), 496 (477 + 12 + 7) in Group 3, and 1 item in Group 4. From earlier IARC Monographs [10], 34 industrial chemicals or processes subdivided into three groups are reported here in Table 1.3. It should be noted here that some of the compounds are grouped together, such as arsenic and certain arsenic compounds, chromium and certain chromium compounds, and soot, tars, and mineral oils and others by the process, manufacture of, refining and mining. This compilation of information is very useful in the legislative process, as well as to those monitoring or working in an occupational setting.

In the United States, the National Institute for Occupational Safety and Health (NIOSH) was created in the Department of Health and Human Services in the 1970s as a federal agency responsible

---

TABLE 1.3

Evaluation of the Carcinogenic Risk of Industrial Chemical Process to Humans [10]

1. Carcinogenic to humans
   4-Aminobiphenyl
   Arsenic and certain arsenic compounds
   Asbestos
   Manufacture of auramin
   Benzene
   Benzidine
   bis-Chloromethylether
   Chloromethyl methyl ether
   Chromium and certain chromium compounds
   Underground hematite mining
   Manufacture of isopropyl alcohol by strong acid process
   Mustard gas
   2-Naphthylamine
   Nickel refining
   Soots, tars, and mineral oils
   Vinyl chloride

2. Probably carcinogenic for humans
   Acrylonitrile
   Amitrole
   Auramine
   Beryllium compounds
   Carbon tetrachloride
   Cadmium and certain cadmium compounds
   Dimethyl sulfate
   Ethylene oxide
   Nickel and certain nickel compounds
   Polychlorinated biphenyls

3. Not classifiable as to carcinogenicity to humans
   Chlordane-heptachlor
   Chloroprene
   DDT
   Dieldrin
   Hematite
   Isopropyl oils
   Lead and certain lead compounds
   Trichloroethylene

---

for conducting research and making recommendations for the prevention of work-related injury and illness. Identifying occupational carcinogens is one of the priorities of NIOSH. NIOSH published a list of chemicals identified as Suspected Carcinogens. A sub-file of Toxic Substances was first published in 1974 and updated periodically. The latest list [11] of 132 substances that NIOSH considers to be potential occupational carcinogens was made public in 2005. This information on the workplace carcinogens can be further assessed by listing the carcinogens, suspected or confirmed by target organ (Table 1.4); particular target organs, that is, liver, nasal cavity and sinuses, lung, bladder, and bone marrow are associated with not only the carcinogens but also the occupations (Table 1.5).

TABLE 1.4

Confirmed and Suspected Occupational Carcinogens by Target Organ [12]

Target organ/tissue	Confirmed carcinogen	Suspected carcinogen
Bone		Beryllium
Brain	Vinyl chloride	
Gastroenteric tract	Asbestos	
Hematopoietic tissue (leukemia)	Benzene, ionizing radiation styrene butadiene and other rubber manufacture substances	
Kidney	Coke oven emissions	Lead
Larynx	Asbestos, chromium	
Liver	Vinyl chloride	Aldrin, carbon tetrachloride chloroform, DDT dieldrin, heptachlor PCB's, trichloroethylene
Lung	Arsenic, asbestos, *bis*-(chloromethyl)ether, chloromethyl methyl ether, chromates, coke oven emissions, mustard gas, nickel, soots and tars, ionizing radiation, vinyl chloride	Beryllium, cadmium, chloroprene, lead, hematite
Lymphatic tissue		Arsenic, benzene
Nasal cavity	Chromium, isopropyl oil, nickel, wood dusts	Leather dusts
Pancreas		Benzidine, PCBs
Pleural cavity	Asbestos	
Prostate		Cadmium
Scrotum	Soots and tars	
Skin	Arsenic, Coke oven emissions, cutting oils, soots and tars	Chloroprene
Urinary bladder	4-Aminobiphenyl, benzidine, β-naphthylamine	Auramine, 4-nitrodiphenyl, magenta

The pertinent information for each of the carcinogens listed in Tables 1.3 to 1.5 are described in detail in Table 1.6. Key animal studies are noted, particularly where these tests rather than human data are the primary basis for the carcinogenic potential of a compound. In some cases, the animal data were so convincing that widespread human exposure was averted. In some instances the work environment and activities in which exposure occurs is widespread while for other agents the exposure is questionable.

The route of absorption of an agent is dependent on both its physical and chemical properties and the route of exposure. Inhalation is the most frequent route of exposure to vapors and fumes while skin is the route of absorption for both liquids and gaseous agents. Ingestion can also occur, particularly with agents in the solid phase. It should be noted here that exposure to a carcinogen is confined to a limited number of target organs; in different species the same organs may not be affected by the same agent.

Among those chemicals and mixtures, a total of 32 listed in Table 1.7 are federally regulated under the Occupational Safety and Health Act [16]. Additionally, in 1978 the Department of Health and Human Services established the National Toxicology Program (NTP) consisting of relevant toxicology activities of federal health research and regulatory agencies to evaluate agents of public health concern. With assistances from other agencies and nongovernmental institutions, the NTP prepares the *Annual Report on Carcinogens* (RoC), which has been changed to a biennial report since 1993. The most recent RoC, the Eleventh Edition [17], was released publicly during January 2005. The Tenth RoC lists 49 substances as Known to be a Human Carcinogen and 174 substances as

TABLE 1.5
Occupations Associated with an Excess Risk of Cancer [13]

Occupations	Confirmed and suspected carcinogen	Site of cancer
Tanners, smelters; vineyard workers; plastic workers	Vinyl chloride	Liver
Glass, pottery, and linoleum workers; nickel smelters, mixers and roasters; electrolysis workers; wood, leather, and shoe workers	Chromium, isopropyl oil, nickel, wood, and leather dusts	Nasal cavity and sinuses
Vintners; miners; asbestos users; textile users; insulation workers; tanners; smelters; glass and pottery workers; coal tar and pitch workers; iron foundry workers; electrolysis workers; retort workers; radiologists; radium dial painters; chemical workers	Arsenic, asbestos, chromium, coal products, dusts, iron oxide, mustard gas, nickel, petroleum, ionizing radiation, *bis*-chloromethylether	Lung
Asphalt, coal tar, and pitch workers; gas stokers; still cleaners; dyestuffs users; rubber workers; textile dyers; paint manufacturers; leather and shoe workers	Coal products, aromatic amines	Bladder
Benzene, explosives, and rubber, cement workers; distillers; dye users; painters; radiologists	Benzene, ionizing radiation	Bone marrow (leukemia)

Reasonably Anticipated to be Human Carcinogen (Table 1.8) and gives a detailed profile for each substances listed. Referring to the latest list of Potential Occupational Carcinogens completed by NIOSH [11], 79 of total 132 substances are listed in the Tenth RoC, and asterisked in Table 1.8. Among them, 26 substances are in the Known to be Human Carcinogen category and 53 items are in the Reasonably Anticipated to be Human Carcinogen category.

For many years, researchers have studied various substances in order to identify those that may cause cancer. Much of this information on specific chemicals or occupational exposures has been published in the scientific literature or in publicly available and peer-reviewed technical reports. It provides meaningful and useful data on (1) the carcinogenicity, genotoxicity, and biologic mechanisms of these substances in people and in animals, (2) the potential for human exposure to them, and (3) Federal regulation to limit exposure.

1.2 CARCINOGENS AND MUTAGENS

1.2.1 DEFINITIONS

Cancer can be defined as an unregulated growth of cells arising from one cell. The scientific or medical term for cancer is malignant neoplasm, which is defined as a relatively autonomous growth of tissue not subject to the rules and regulations of normal growing cells. Tumor is a general term indicating any abnormal mass or growth of tissue. Therefore, a neoplasm is a tumor. Major features of benign tumors are encapsulation, slow growth, and non-invasion of surrounding tissue; that is, lack of metastasizing ability. Malignant tumors grow rapidly, are not encapsulated and invade surrounding tissue and metastasize. Benign growths generally have a normal complement of chromosomes, exhibit good differentiation, and have rare cell division. The opposite is characteristic of malignant neoplasms [18].

At the turn of the 20th century, cancer was the eighth leading cause of death due to diseases in the United States and heart disease was ranked fourth. Cancer caused 16% of the total deaths in the United States. Since 1990, about 16 million cancer cases have been diagnosed in the United States. In 2002, 1,284,900 new cancer cases were diagnosed and 555,500 Americans died of cancer, which contributed to a quarter of the total disease-associated deaths [1]. Ranking only behind the heart

TABLE 1.6
Occupational Carcinogens [3,10,13–15]

Agent	Chemical structure	Animal studies			Human data			Occupational exposure
		Species	Route of exposure	Target organ	Route of exposure	Target organ	Latency period (years)	
A. Organic agents								
1. Aromatic hydrocarbons								
Benzene		Mouse	Topical, S.C. injection	Inadequate	Inhalation, skin	Hemopoietic system, bone marrow leukemia	6–14	Benzene or rubber cement workers, distillers, dye users painters
Coke oven emissions (Benzo[a]pyrene)		Multiple	Topical, intratracheal	Lung	Inhalation, skin	Lung, kidney, skin	9–23	Coke oven workers
Soot, tars, oils-coal tars and pitch creosote oils, shale oils, cutting oils, coal tar fumes, mineral oils, coal soot		Mouse, rabbit	Topical, S.C. injection	Skin	Inhalation, skin	Skin, lung bladder, gastrointestinal tract, scrotum	9–23 or 12–30	Coal gas, coke and petroleum industry workers, miners, chimney sweeps, coal tar and pitch workers, textile weavers, rubber fillers
2. Amines								
Aliphatic amines, ethyleneimine	H N H_2C—CH_2	Rat, mouse	S.C. injection, P.O.	Sarcoma, kidney, Liver, lung	Inhalation, skin			Effluent treaters, organic chemical synthesizers, paper makers, textile workers; no documented human cases
N-nitrosodimethyl-amine	H_3C N—N=O H_3C	Rat	Inhalation P.O.	Liver, kidney Lung	Inhalation, skin			Dimethyl hydrazine makers

TABLE 1.6
(Continued)

Agent	Chemical structure	Animal studies			Human data			Occupational exposure
		Species	Route of exposure	Target organ	Route of exposure	Target organ	Latency period (years)	
3. Aromatic amines								
2-Acetylaminofluorene		Rat / Dog / Guinea Pig	P.O. / P.O. / P.O.	Liver / Bladder, liver / Negative	Inhalation, skin			Carcinogenic data averted commercial production
4-Aminobiphenyl		Mouse, rabbit, dog	P.O.	Bladder	Inhalation, skin	Bladder	15–35	Dye stuffs manufacturers and users, rubber workers, textile dyers
		Newborn mouse	S.C. injection	Liver				
		Rat	S.C. injection	Mammary gland Intestine				
1-Aminonaphthalene		Dog, mouse	P.O.	Inconclusive or negative	Inhalation, skin	Bladder	22	Dye stuffs manufacturers and users, rubber workers, activity has been attributed to contamination with 2-aminonaphthalene
2-Aminonaphthalene		Dog, hamster, monkey	P.O.	Bladder	Inhalation, skin	Bladder	16	Dye stuffs manufacturers and users, rubber workers
		Mouse	P.O.	Liver				
Auramine		Rat, mouse	P.O.	Liver	Inhalation, skin	Bladder	13–30	Dye stuffs manufacturers and users; auramine manufacture has been associated with increase in bladder cancer — the actual compound has not been identified

(continued)

Compound	Species	Route	Tumor site	Human route	Human site		Comments
Benzidine	Rat, hamster; Mouse, rat	P.O.; S.C. injection	Liver; Liver, ear duct in rat; Bladder	Inhalation, skin	Bladder	16	Dye stuff manufacturers and users, rubber workers; medical laboratory personnel and researchers use benzidine and derivatives
Dog	P.O.						
3,3′-Dichlorobenzidine	Rat, hamster	P.O.	Bladder	Inhalation, skin			Dye stuff manufacturers and users, cases have occurred with simultaneous exposure to benzidine
4-Dimethylaminoazobenzene	Rat, mouse	P.O.	Liver	Inhalation, skin			Carcinogenic animal data averted commercial production, no documented human cases
Magenta	Rat	S.C. injection	Local sarcomas	Inhalation, skin	Bladder		Dye stuff manufacturers and users, in 1895 Rehn's description of aniline tumors involved magenta manufacturers
4,4′-Methylene-*bis*-(2-chloroaniline) MOCA	Rat	P.O.	Liver, lung	Inhalation, skin			Elastomer makers, polyurethane foam workers no documented human cases
4. Biphenyls							
4-Nitrobiphenyl	Dog	P.O.	Bladder	Inhalation, skin			Research workers, carcinogenic animal data averted commercial production, exposure to compound occurred concurrently with 4-aminobiphenyl

TABLE 1.6
(Continued)

Agent	Chemical structure	Animal studies			Human data			Occupational exposure
		Species	Route of exposure	Target organ	Route of exposure	Target organ	Latency period (years)	
Polychlorinated biphenyls PCB		Mouse, rat	P.O.	Liver	Skin			Electric equipment makers and associated industries — capacitor producers and transformer workers. Dye stuffs manufacturers and users, plasticizer and resin makers, and wood preservers; slight increase in the incidence of cancer particularly melonoma of the skin reported in small groups of men exposed to mixture of PCB, Arochlor 1254, in occupational setting
5. Chlorinated hydrocarbons Aldrin/dieldrin		Dieldrin — mouse	P.O.	Liver	Inhalations, skin			Agricultural workers and insecticide manufacturers; Aldrin is metabolized to Dieldrin — involved too few subjects and insufficient follow up time to draw conclusion
		Dieldrin — rat, dog, monkey	P.O.	Negative or inconclusive				
		Aldrin — rat, mouse	P.O.	Negative or inconclusive				

(continued)

Compound	Structure	Species	Route	Target organ/Result	Exposure route	Comments
Carbon tetrachloride	CCl$_4$	Mouse, rat, hamster	P.O., inhalation	Liver	Inhalation, skin	Chemists, fluorocarbon skin makers, rubber workers, insecticide, refrigerant, lacquer and propellant makers, metal cleaners, three cases reported with liver tumors associated with cirrhosis following exposure
Chlordane/heptachlor		Mouse / Rat	P.O. / P.O.	Liver / Inconclusive	Inhalation, skin	Agricultural workers and insecticide manufacturers; these compounds are considered together because similar in structure and often contaminated with one another; 5 out of 14 children with neuroblastoma had prenatal and postnatal exposure to chlordane, three persons with acute leukemia had been exposed to chlordane (3 to 7% heptachlor)
Chloroform	HCCl$_3$	Mouse	P.O.	Liver	Inhalation, skin	Chemists, fluorocarbon makers, solvent workers, lacquer workers; human data too limited

TABLE 1.6
(Continued)

Agent	Chemical structure	Animal studies			Human data			
		Species	Route of exposure	Target organ	Route of exposure	Target organ	Latency period (years)	Occupational exposure
Chloroprene	$H_2C = CH - \overset{\displaystyle Cl}{\overset{\displaystyle |}{C}} = CH_2$	Mouse	Skin	Negative	Inhalations, skin			Synthetic rubber makers; epidemiological studies inconclusive, one case report of angiosarcoma of liver in worker exposed
		Rat	S.C. injection, ingestion	Negative				
DDT	(structure: CCl_3–CH with two chlorophenyl groups)	Mouse	P.O.	Liver	Inhalations, skin			Agricultural workers, insecticide manufacturers; epidemiological studies inadequate
		Rat	Ingestion	Liver (nonmetastasizing tumors)				
		Rat, guinea pig, dog, monkey	P.O.	Negative or inconclusive				
Ethylene dichloride	$ClCH_2 - CH_2Cl$	Rat	P.O.	Stomach, skin, breast	Inhalations, skin			Chemical makers, adhesive makers, plastic and solvent workers, degreasers, dry cleaner and insecticide makers; human data not available
		Mouse	P.O.	Breast, uterus, lung				
Mustard Gas	$ClCH_2 - \overset{\displaystyle H_2}{C}$ and $ClCH_2 - \overset{\displaystyle }{C}H_2$ joined by S	Mouse	I.V. injection or inhalation, S.C. injection	Lung, Local sarcoma	Inhalation	Respiratory tract	10–25	Mustard gas workers associated with chronic exposure

(continued)

Agent	Structure	Species	Route (animal)	Tumor site (animal)	Route (human)	Tumor site (human)	Latency (years)	Comments
Trichloroethylene	$ClCH = CCl_2$	Mouse	P.O.	Liver, lung	Inhalations, skin			Anesthetic workers and users, cleaners, degreaser, dry cleaners, printers, resin workers, rubber cementers and solvent workers; no human data available
Vinyl chloride	$CH_2 = CHCl$	Mouse, rat hamster	P.O., inhalation	Several sites including angiosarcoma of liver	Inhalation, skin	Angiosarcoma of liver and lung, brain, hemo-lymphopoietic system	20–30	Polyvinyl resin makers and rubber workers
6. Ethers								
Bichloromethyl ether (BCME)	$ClCH_2OCH_2Cl$	Mouse	Inhalation, topical, S.C. injection	Lung, site of application	Inhalation, skin	Lung	5+, 10–15	Ion exchange resin workers, small cell carcinoma
		Rat	S.C. injection	Site of application				
Chloromethyl methyl ether (CMME)	$ClCH_2OCH_3$	Mouse	Inhalation, topical, S.C. injection	Lung, site of application	Inhalation, skin	Lung	5+, 10–15	Chemists, ion exchange resin workers, small cell carcinomas, exposures generally involve CMME contaminated with BCME
		Rat	S.C. injection	Site of application				
7. Miscellaneous								
Acrylonitrile	$CH_2 = CHCN$	Rat	P.O. injection	Brain, forestomach, zymbal gland	Inhalation	Colon, lung	20–25	Acrylic fiber and resin lung makers — epidemiological studies lack smoking history, exposure to other chemicals, and incomplete follow-up
Ethylene oxide	ethylene oxide (epoxide ring)	Mouse	Topical	Inadequate	Inhalation	Leukemia, gastric cancer		Ethylene oxide workers exposed to other chemicals as well
		Rat	S.C. injection	Inadequate				

TABLE 1.6
(Continued)

Agent	Chemical structure	Animal studies			Human data			Occupational exposure
		Species	Route of exposure	Target organ	Route of exposure	Target organ	Latency period (years)	
Isopropyl alcohol	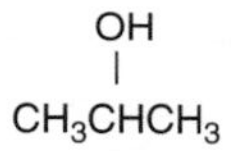 OH │ CH₃CHCH₃	Mouse	Inhalation, topical application, S.C. injection	Lung, inadequate	Inhalation	Nasal cavity, larynx	10+	Isopropyl alcohol makers isopropyl oils forms as by-product in manufacture of isopropyl alcohol by strong acid process
β-Propiolactone	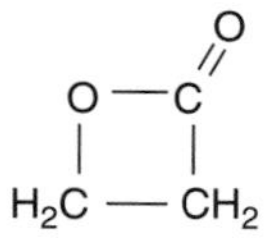	Mouse Rat	Topical application, P.O.	Site of application Stomach Liver	Inhalation, skin			Chemists, plastic and resin workers; no skin cancer documented
B. Inorganic agents **1. Metal and metal compounds** Arsenic and arsenic compounds	As, As₂O₃	Mouse Rat	P.O.	Negative/ inadequate	Inhalation, skin Oral	Lung Skin	10+ 15–35	Miners, smelters, dye stuff manufacturers and users, semiconductor compound makers; skin cancer associated with exposure to inorganic arsenic compounds; lung cancer increased in smelter workers exposed to arsenic trioxide

(continued)

Compound	Formula	Species	Route	Result	Route (human)	Organ (human)	Latency	Comments
Beryllium and beryllium compounds								
Beryl ore, beryllium sulfate, bertrandite, beryllium oxide	$3BeO \cdot Al_2O_3 \cdot 6SiO_2$, $BeSO_4$ $4BeO \cdot 2SiO_2 \cdot H_2O$, BeO	Rat Monkey	Inhalation Inhalation, intratracheal implantation	Lung Lung	Inhalation	Lung	10–15	Miners, smelters, refining alloy workers and users, metallurgists, electronic tube makers, nuclear reactor workers, rocket and aerosol research workers; other factors ruled out
Zinc beryllium silicate, beryllium, beryllium phosphate	$Zn\text{-}Be(SiO_2)_2$, Be, $Be_3(PO_4)_2$	Rabbit	I.V. injection	Bone tumors				
Chromium and chromium compounds								
Calcium chromate	$CaCrO_4$	Rat	Intratracheal implantation	Lung	Inhalation, skin	Nasal cavity	15–25	Producers, processors and users, metal workers and battery makers, increased incidence of lung cancer among chromate-producing industry and possibly chromate platers and chromium alloy workers; chromium compound(s) responsible — not known
Calcium chromate, strontium chromate and zinc chromate	$CaCrO_4$, $SrCrO_4$, $ZnCrO_4$	Rat	Topical application	Site of application		Sinuses, lung	10–30	
Barium chromate, lead chromate, chromic acetate, sodium chromate, and chromium carbonyl	$BaCrO_4$, $PbCrO_4$, $Cr(C_2H_3O_2)_3$, Na_2CrO_4, $Cr(CO)_6$	Rat, mouse	Inadequate					
Iron oxide and haemitite	Fe_2O_3	Hamster, mouse, guinea pig	Inhalation	Negative	Inhalation	Respiratory tract		Iron ore miners, metal grinders, and iron foundry workers; underground haemitite miners have high incidence of lung cancer, surface workers do not; excessing may be due to haemitite, radon, inhalation of ferric oxide or silica or combination of these

TABLE 1.6
(Continued)

Agent	Chemical structure	Animal studies			Human data			Occupational exposure
		Species	Route of exposure	Target organ	Route of exposure	Target organ	Latency period (years)	
Cadmium and cadmium compounds								
Cadmium chloride, cadmium oxide, cadmium sulfate, and cadmium sulfide	$CdCl_2$, CdO, $CdSO_4$, CdS	Rat	S.C. injection	Local sarcoma	Inhalation, skin			Miners, smelters, electroplaters, plastics alloy, solder, battery and insecticide workers, dye stuff manufacturers; increased risk to prostate, respiratory tract, and renal cancer due to Cd exposure perhaps CdO. The renal cancer risk doubled when cigarette smoking included in study
Cadmium powder, cadmium sulfide	Cd, CdS	Rat	I.M. injection	Local sarcoma				
Cadmium chloride, cadmium sulfate	$CdCl_2$, $CdSO_4$	Rat	S.C. injection	Testicular tumors				
Lead and lead compounds								
Lead and lead acetate	Pb, $Pb(C_2H_3O_2)_2$	Rat, mouse	P.O.	Renal tumors	Inhalation			Battery workers, gasoline additive workers, glass makers, insecticide makers, painters, plumbers, solderers, storage tank cleaners; no evidence of cancer in human
Lead acetate, lead subacetate, lead phosphate	$Pb(C_2H_3O_2)_2$, $Pb(C_2H_3O_2)_2 \cdot 2Pb(OH)_2$, $Pb_3(PO_4)_2$	Rat	P.O., S.C., or I.P. injection	Renal tumors				

(continued)

Compound	Formula	Species	Route	Organ	Route	Organ, latency	Remarks
Nickel and Nickel compounds							
Nickel sulfide	NiS	Rat mouse, rat, hamster	Inhalation I.M.	Lung Local			Refiners, founders, smelters, dye stuffs manufacturers and users, battery, ceramic makers; increased incidence of cancer of nasal cavity, lung and larynx in nickel refiners — cannot tell which specific nickel compound is carcinogenic
Nickel powder, subsulfide, nickel oxide, carbonate, and nickelocene	Ni, Ni_3S_2, NiO, $NiCO_3$, $Ni(C_5H_5)$						
Nickel carbonyl	$Ni(CO)_4$	Rat	Inhalation	Lung	Inhalation	Nasal sinuses, lung, larynx 3–30	
2. Fiber							
Asbestos — all forms of commercial asbestos	$Mg_3Si_2O_5(OH)_4$, $(Fe^{++}Mg)_7Si_8O_{22}(OH)_2$, $(MgFe^{++})_7Si_8O_{22}(OH)_2$, $NaFe^{++}Fe^{+++}Si_8O_{22}(OH)_2$	Mouse, rat, hamster, rabbit	Inhalation, intrapleural, Intratracheal and I.P. administration	Mesothelioma, and lung	Inhalation, ingestion	Nasal sinuses, lung 4–50	Miners, millers, textile, insulation and shipyard workers; exposure to chrysotile, amosite, anthophyllite and mixtures containing crocidolite result in high incidence of lung cancer; tremolitic material mixed with anthophyllite and small amounts of chrysotile increase incidence of lung cancer, pleural and peritoneal mesotheliomas observed after exposure to crocidolite, amosite, chrysotile; gastrointestinal tract and larynx cancer increase for groups exposed to amosite chrysotile or mixed fibers containing crocidolite; mesotheliomas occur in individuals living near asbestos factories and crocidolite mine and in persons living with asbestos workers; smoking and asbestos work act multiplicatively

TABLE 1.6
(Continued)

Agent	Chemical structure	Animal studies			Human data			
		Species	Route of exposure	Target organ	Route of exposure	Target organ	Latency period (years)	Occupational exposure
3. Dusts								
Wood	—	Rodent	S.C.	Local sarcomas	Inhalation	Nasal cavity and sinuses	10–40	Wood workers; hardwood dusts
Leather	—				Inhalation	Nasal cavity and sinuses	40–50	Leather and shoe workers; carcinogen unknown
C. Physical agents								
1. Nonionizing radiation								
Ultraviolet rays	—	Mouse	Irradiation	Skin	Skin	Skin (varies with skin pigment and texture)	5–50	Farmers, sailors, researchers, chemists, welders, optical equipment workers
2. Ionizing radiation								
X-rays	—	Multiple	Irradiation	Skin, breast, thyroid, bone, lung	Skin	Skin, bone marrow and leukemia	10–25	Radiologists, medical personnel, high-voltage, vacuum tube makers and users
Uranium/radon	U/Rn	Rat	Inhalation	Lung	Skin	Small cell carcinoma of lung, skin, leukemia	10–15	Underground miners, inhalation of radon daughters with irradiation of bronchial tree
Radium	Ra				Ingestion	Bone sarcoma, leukemia, and lymphoma	10–15 10–50	Radiologists, miners, radium dial painters, radium chemists

TABLE 1.7
Federally Regulated Workplace Carcinogens under Occupational Safety and Health Act [16]

Name of carcinogen	Effective date	Latest amended date
1. Asbestos	June 2, 1972	January 8, 1998
2. 13 Carcinogens		
4-Nitrobiphenyl	June 27, 1974	April 23, 1998
α-Naphthylamine	June 27, 1974	March 7, 1996
Methyl chloromethylether	June 27, 1974	March 7, 1996
3,3'-Dichlorobenzidine	—	March 7, 1996
bis-Chloromethyl ether	—	—
β-Naphthylamine	—	—
Benzidine	—	—
4-Aminodiphenyl	—	—
Ethyleneimine	—	—
β-Propiolactone	—	—
2-Acetylaminofluorene	—	—
4-Dimethylaminoazobenzene	—	—
N-Nitrosodimethylamine	—	—
3. Vinyl chloride	April 1, 1975	June 18, 1998
4. Inorganic arsenic	August 1, 1978	June 18, 1998
5. Cadmium	December 14, 1992	January 8, 1998
6. Benzene	December 10, 1987	April 23, 1998
7. Coke oven emissions	January 20, 1977	June 18, 1998
8. 1,2-Dibromo-3-chloropropane	June 30, 1993	January 8, 1998
9. Acrylonitrile	November 2, 1978	April 23, 1998
10. Ethyleneoxide	August 21, 1984	November 7, 2002
11. Formaldehyde	February 2, 1988	April 23, 1998
12. Methylenedianiline	September 9, 1992	November 7, 2002
13. 1,3-Butadiene	November 4, 1996	November 7, 2002
14. Methylene chloride	April 10, 1997	March 22, 1999
Other compliance related safety and health topics		**Revised date**
15. Diesel exhaust		November 13, 2002
16. Isocyanates		January 21, 2003
17. Silica (crystalline)		December 16, 2002
18. Lead		February 4, 2003
19. Metalworking fluids		January 13, 2003
20. Synthetic mineral fibers		April 2, 2003

diseases, cancer is presently the second major cause of death. About 5 to 10% of cancers are hereditary. The remaining cancers result from damage to genes that occur throughout our lifetime either due to internal factors or external factors. According to late statistics, probably 80% of all human cancer deaths are related to external factors that can be controlled or prevented [1,19,20]. This includes use of tobacco and tobacco products, occupationally induced cancers, virally induced cancers, iatrogenically induced (medically induced) cancers, and cancers that arise from diet and lifestyle considerations.

A carcinogen is an agent that causes cancer. A carcinogen is also defined as a physical, chemical, or biological agent or a combination of agents that produces cancers in an organism, that is, it produces neoplastic or metastatic tumors in the animal host. Physical agents include ionizing radiation

TABLE 1.8
Carcinogens Listed in the Tenth Report on Carcinogens [17]

Part A. Known to be a human carcinogen

Aflatoxins
Alcoholic beverage consumption
4-Aminodiphenyl*
Analgesic mixtures containing phenacetin
Arsenic compounds, inorganic*
Asbestos*
Azathioprine
Benzene*
Benzidine*
Beryllium and compounds*
1,3-Butadiene*
1,4-Butanediol dimethylsulfonate
Cadmium and compounds*
Chlorambucil
1-(2-Chloroethyl)-3-(4-methyl cyclohexyl)-1-nitrosourea
bis-(Chloromethyl) ether and technical-grade
  chloromethyl methyl ether*
Chromium hexavalent* compounds
Coal tar pitch*
Coke tars
Coke oven emissions*
Cyclophosphamide
Cyclosporin A
Diethylstilbestrol
Dues metabolized to benzidine*
Environmental tobacco smoke*
Erionite
Estrogen, steroidal
Ethylene oxide*
Melphalan
Methoxsalen with ultraviolet A therapy
Mineral oils
Mustard gas
2-Naphthylamine*
Nickel compounds*
Radon*
Silica, crystalline*
Smokeless tobacco
Solar radiation
Soot
Strong inorganic acid mists containing sulfuric acid
Sunlamps and sun beds exposure to Thiotepa
Tamoxifen
2,3,7,8-Tetrachlorodibenzo-*p*-Dioxin (TCDD)*
Thorium dioxide
Tobacco smoking*
Vinyl chloride*
Ultraviolet Radiation, Broad spectrum
UV radiation
Wood dust*

Part B. Reasonably anticipated to be a human carcinogen

Acetaldehyde*
2-Acetylaminofluorene*
Acrylamide*
Acrylonitrile*
Adriamycin
2-Aminoanthraquinone
o-Aminoazotoluene
1-Amino-3-methylanthraquinone
2-Amino-3-methylimidazole
  [4,5-*f*]quinoline
Amitrole*
o-Anisidine Hydrochloride*
Azacitidine
Benz[a]anthracene
Benzo[b]fluoranthene
Benzo[j]fluoranthene
Benzo[a]pyrene
Benzotrichloride
Bromodichloromethane
2,2-*bis*-(Bromoethyl)-1,3-propanediol (technical grade)
Butylated hydroxyanisole
Carbon tetrachloride*
Ceramic fibers
Chloramphenicol
Chlorendic acid
Chlorinated paraffins (C_{12}, 60% chlorine)
1-(2-Chloroethyl)-3-cyclo-hexyl-1-nitrosourea
bis-(Chloroethyl)-nitrosourea
Chloroform*
3-Chloro-2-methylpropene
4-Chloro-*o*-phenylenediamine
Chloroprene*
p-Chloro-*o*-toluidine
Chlorozotocin
C.I. basic red 9 hydrochloride
Cisplatin
p-Cresidine
CupferronDacarbazineDanthron2,4-Diaminoanisoleo
  sulfate*
2,4-Diaminotoluene
Dibenz[a,h]acridine
Dibenz[a,j]acridine
Dibenz[a,h]anthracene
7H-Dibenzo[c,g]carbazole
Dibenzo[a,e]pyrene
Dibenzo[a,h]pyrene
Dibenzo[a,i]pyrene
Dibenzo[a,l]pyrene
1,2-Dibromo-3-chloropropane*
1,2-Dibromoethane*

(continued)

TABLE 1.8
(Continued)

2,3-Dibromo-1-propanol
tris-(2,3-Dibromopropyl) phosphate
1,4-Dichlorobenzene*
3,3′-Dichlorobenzidine*
Dichlorodiphenyltrichloroethane (DDT)*
1,2-Dichloroethane*
1,3-Dichloropropene*
Diepoxybutane
Diesel exhaust particulate*
Diethyl sulfate
Diglycidyl ether (DGE)*; class, glycidyl ethers
3,3′-Dimethoxybenzidine
4-Dimethylaminobenzene*
3,3′-Dimethylbenzidine
Dimethylcarbomoyl chloride*
1,1-Dimethylhydrazine*
Dimethyl sulfate*
Dimethylvinyl chloride
1,6-Dinitropyrene
1,8-Dinitropyrene
1,4-Dioxane*
Disperse blue 1
Dyes metabolized to 3,3′-Dimethoxybenzidine
Dyes metabolized to 3,3′-Dimethylbenzidine
Epichlorohydrin*
Ethylene thiourea*
di(2-Ethylhexyl) phthalate
Ethyl methanesulfonate
Formaldehyde*
Furan
Glass wool (respirable size)
Glycidol
Hexachlorobenzene
Hexachlorocyclohexane isomers
Hexachloroethane*
Hexamethyl phosphoric triamide*
Hydrazine*
Hydrazobenzene
Indeno[1,2,3-cd]pyrene
Iron dextran complex
Isoprene
Kepone*
Lead acetate
Lead phosphate
Lindan
2-Methylaziridine
5-Methylcrysene
4,4′-Methylene-bis-(2-chloroaniline)
4,4′-Methylenebis(N,N-dimethyl)benzenamine
4,4-Methylenedianiline*
Methyleugenol

Methyl methanesulfonate
N-Methyl-N-nitro-N-nitroso-guanidine
Metronidazole
Michler's ketone
Mirex
Nickel and nickel compounds*
Nitrilotriacetic acid
o-Nitroanisole
6-Nitrochrysene
Nitrofen
Nitrogen mustard hydrochloride
2-Nitropropane*
1-Nitropyrene
4-Nitropyrene
N-Nitrosodi-n-butylamine
N-Nitrosodiethanolamine
N-Nitrosodiethylamine
N-Nitrosodimethylamine*
N-Nitrosodi-n-propylamine
N-Nitroso-N-ethylurea
4-(N-Nitrosomethylamino)-1-(3-pyridyl)-1-butanone
N-Nitroso-N-methylurea
N-Nitrosomethylvinylamine
N-Nitrosomorpholine
N-Nitrosonomicotine
N-Nitrosopiperidine
N-Nitrosopyrrolidine
N-Nitrososarcosine
Norethisterone
Ochratoxin A
4,4′-Oxydianiline
Oxymetholone
Phenacetin
Phenazopyridine hydrochloride
Phenolphthalein
Phenoxybenzamine hydrochloride
Phenytoin
Polybrominated biphenyl
Polychlorinated biphenyl*
Polycyclic aromatic hydrocarbons
Procarbazine hydrochloride
Progesterone
1,3-Propane sulfone*
β-Propiolactone*
Propylene oxide*
Propylthiouracil
Reserpine
Safrole
Selenium sulfate
Streptozotocin
Styrene-7,8-oxide

(continued)

TABLE 1.8
(Continued)

Sulfallate	2,4,6-Trichlorophenol
Tetrachloroethylene*	1,2,3-Trichloropropane*
Tetrafluoroethylene	**Ultraviolet A radiation**
Tetranitromethane	**Ultraviolet B radiation**
Thioacetamide	**Ultraviolet C radiation**
Thiourea	Urethane
Toluene diisocyanate*	**Vinyl bromide***
o-Toluidine*	Vinyl cyclohexene diepoxide*
Toxaphene	**Vinyl fluoride**
Trichloroethylene*	

Bold entries indicate new or changed listing in the RoC, 10th edition.
*Substance are listed in the latest NIOSH Carcinogen List (2002).

(x-rays), nonionizing radiation (UV light), and particles (asbestos). Chemical agents include calcium chromate, benzidine, and complex mixtures such as cigarette smoke and coal by-products. Biological agents included viruses and parasites. Each of these types of agents may be found in the environment. It is often difficult, however, to ascribe a specific cancer to a particular agent beyond a certain point in the development of cancer. The probability of cancer due to association increases with the potency of the carcinogen, the degree and duration of exposure, and the length of time elapsed since the exposure. For example, mesothelioma of the pleura and peritoneum in humans can result from exposure to asbestos.

A mutagen is an agent that produces a genetic event resulting in a heritable change. The majority of known chemical carcinogens are also mutagens; for example polycyclic aromatic hydrocarbons, aromatic amines, and N-nitrosamines. Therefore, mutagenicity has been considered a reasonable endpoint to use to detect mutagens and mutagenic carcinogens by virtue of their mutagenicity, and to assess relative hazards in the environment. Besides the association with carcinogens, mutations may have other harmful effects. The integrity of the human gene pool can be compromised by any increase in the mutation frequency of the population. Apart from the direct impact to the quality of human life, one must consider the possible damage to the ecological balance of other species caused by increases in the mutation frequency. Thus, mutagenesis is a useful indicator of deleterious biological effects.

1.2.2 CONCEPTS BASED ON DEFINITIONS

The following factors must be considered when assessing a carcinogenic agent [21]:

1. *Latency period* — The time between exposure and the onset of cancer can be many years in human (sometimes 20 to 30 years or more).
2. *Metabolism of chemical agent* — The metabolism of chemical carcinogens is not necessarily the same in humans and lower animals and will differ for organs in the same animal. This will result in differences in target organs affected in different species of animals.
3. *Dose–response relationship* — The increase in dose or duration of exposure to a carcinogen will result in an increase in the probability of production of cancer. Therefore, it is prudent to limit exposure to carcinogens to either no exposure, or to that dose that the regulatory agencies permit based on risk assessment calculations. (See Chapter 23 by Dr. Salmon.)

4. *Benign tumor* — There is not always a sharp distinction between benign and malignant tumors. This makes it difficult to assess cancer formation. Therefore, in some instances benign tumors are used as an indication of potential hazard.

5. *Cocarcinogenesis* — Cocarcinogens are agents administered together with carcinogens. Cocarcinogens have no carcinogenic activity of their own, but enhance the carcinogenic potency of a carcinogen. This can be accomplished by shortening the latency period or increasing the number of tumors produced by the carcinogen alone. Promoters can also enhance the potency of a carcinogen but are administered after the application of a carcinogen. Ferric oxide and dodecane have been found to be cocarcinogens while phorbol esters and tobacco smoke condensate have been found to be promoters. These types of compounds are important in the study of and exposure to complex mixtures derived from fossil fuels.

6. *Susceptibility* — The response to a substance will vary depending on the species, strain, age, sex, diet, health, etc. of the experimental animals or humans.

1.2.3 FACTORS TO BE CONSIDERED IN HANDLING CARCINOGENS

Most carcinogens have unique chemical and physical properties. For instance, carcinogens can be solids, liquids, or gases with low or high vapor pressures; photochemically reactive; direct acting (does not need to be metabolically activated to observe biological effects), or indirect acting compounds. Odors need not be present and particles or mists, if generated, may or may not be visible to the naked eye. Lastly, all of the compounds in the working environment need to be identified because of the synergistic effects of carcinogens with cocarginogenic or promoting agents.

In addition to the above considerations, the route of exposure is extremely important since carcinogens can be absorbed through the skin, eye, inhaled through the respiratory tract, or ingested through the gastrointestinal tract. The route of exposure could determine the target organ while the dose and length of exposure could affect the latency period.

REFERENCES

1. ACS, American Cancer Society, Cancer Facts and Figures 2002: Basic Facts, http://cancer.org/statistics/
2. NIOSH, National Institute for Occupational Safety and Health, Occupational Cancer, 2003, http://www.cdc.gov/niosh/cancer/
3. Schottenfeld, D. and Haas, H.F., Carcinogens in the workplace, *CA Cancer J. Clin.*, *29*: 144–168, 1978 and references therein.
4. Harris, C.C., Chemical and physical carcinogenesis: Advances and perspectives for the 1990s, *Cancer Res. (Suppl.)*, *51*: 5023s–5044s, 1991.
5. Harris, R.J.C., Occupational and environmental cancer in man, in *Cancer*, Penguin Books Ltd., Harmondsworth, Middlesex, England, 1976, pp. 57–75.
6. Miller, J.A., Carcinogenesis by chemicals: An overview, *Cancer Res.*, *30*: 559–576, 1970.
7. NTP, National Toxicology Program, Department of Health and Human Service, NTP Mission, 1981, http://ehp.niehs.nih.gov/
8. IARC, International Agency for Research on Cancer, IARC Monographs Programme on the Evaluation of Carcinogenic Risks to Humans, 2003, http://monographs.iarc.fr/
9. IRAC, List of IRAC Evaluation, 2002, http://monographs.iarc.fr/monoeval/grlist.html
10. IARC Working Group, An evaluation of chemicals and industrial processes associated with cancer in humans based on human and animal data: IARC monographs, Vol. 1–20, *Cancer Res.*, *40*: 1–12, 1980.
11. NIOSH, Carcinogen List, 2002, http://www.cdc.gov/niosh/npotocca.html
12. Bridbord, K., Wagoner, J.K., and Blejer H.P., Chemical carcinogens, in *Occupational Diseases, A Guide to their Recognition*, DHEW and NIOSH Key, M.M., Henschel, A.F., Butler, J., Ligo', R.N., Tabershaw, I.R., and Ede, L., Eds., Publication, 77–181, pp. 443–450, 1977.

13. Cole, P. and Goldman, M.B., Occupation, in *Persons at High Risk of Cancer*, Fraumeni, J.F., Jr., Ed., Academic Press, New York, 1975, pp. 167–184.
14. IARC Working Group, Evaluation of the carcinogenicity of chemicals: A review of the monograph program of the International Agency for Research on Cancer, *Cancer Res.*, *38*: 877–885, 1978.
15. National Toxicology Program, *Second Annual Report on Carcinogens*, U.S. Department of Health and Human Services Publication, NTP 81-43, December 1981.
16. OSHA, US Department of Labor, Occupational Safety & Health Administration, Safety and Health Topics: Carcinogens, Compliance, 2003. http://www.osha.gov/stlc/carcinogens/index,html
17. NTP, Report on Carcinogens, 11th ed., 2005, http://ehp.niehs.nih.gov/roc/toc10.html
18. Fidler, I.J. and Kripke, M.L., Tumor growth and spread, in *Cancer the Outlaw Cell*, LaFond, R.E., Ed., American Chemical Society Press, Washington, D.C., 1978, pp. 19–34.
19. WHO, Prevention of cancer, report of WHO expert committee, *WHO Tech. Rep. Ser.*, *276*: 1–53, 1964.
20. Tomatis, L., Fuff, J., Hertz-Picciotto, I., Dandler, D., Bucher, J., Boffeta, P., Axelson, O., Blaie, A., Taylor, J., Stayner, L., and Barrett, J.C., Avoided and avoidable risks of cancer, *Carcinogenesis*, *18*: 97–105, 1997.
21. Informing Workers and Employers about Occupational Cancer, U.S. Department of Labor Publication, 1978, pp. 5–6.

2 Whole Animal Carcinogenicity Bioassays

Joseph R. Landolph, Jr., Weiling Xue, and David Warshawsky

CONTENTS

2.1 INTRODUCTION

Epidemiologists, physicians, and laboratory scientists all play a role in identifying agents that contribute to causation of cancer in humans. Two hundred and twenty-eight years ago, the British physician, Percivall Pott, determined that there was an association between occupational exposure of chimney sweeps to soot in the chimneys they cleaned and the induction of scrotal cancer in these workers. Since that time, epidemiological studies have shown associations between exposure to cigarette smoking and lung cancer, exposure to asbestos and mesothelioma, exposure to β-naphthylamine and several other aromatic amines and urinary cancer, and exposure to aflatoxin B_1 and hepatocellular carcinoma [1–6]. There are therefore a variety of chemicals in various occupational settings and in the environment that are human and animal carcinogens. To determine whether an agent is a potential human carcinogen, the strongest tools to use are epidemiological studies. However, these studies are difficult, since they must rely on natural, not experimental, conditions [1]. Further, epidemiological studies are somewhat insensitive. It took many decades for epidemiologists to determine that cigarette smoke, a very strong carcinogenic mixture, was a cause of human lung cancer.

A second major group of tools in the area used to determine whether specific chemicals are carcinogens in humans are clinical observations, experimental long-term carcinogenicity bioassays in animals, initiation and promotion assays in animals, and short-term genotoxicity assays in the laboratory. The short-term genotoxicity assays include assays to detect chemically induced mutation in bacteria (*Salmonella typhimurium*, *Escherichia coli*), yeast, cultured mammalian cells, Drosophila, and in whole animals. Short-term genotoxicity assays also include assays to detect chemically induced DNA repair and chemical induction of chromosome aberrations, including micronucleus induction, in mammalian cells.

Among the laboratory studies that can be employed to detect carcinogens, the most important is the animal carcinogenesis bioassay. In many cases, specific agents were first found to cause cancer in animals, and then these agents were later found to induce cancer in humans. Experimental cancer useful to determine causes of human cancer research is based on the scientific assumption that substances causing cancer in animals will also cause cancer in humans [7]. Therefore, long-term carcinogenesis bioassay in animals, usually in two laboratory rodent species, mice and rats, over a range of doses for nearly the complete lifetime of the animals, are important assays that are used to detect chemical carcinogens. All experimental conditions are carefully chosen to maximize the likelihood of identifying any carcinogenic effects [8]. Although the responses of laboratory animals to agents may not always be exactly the same as for humans, animal studies remain the most valuable tool for determining the potential carcinogenicity of specific chemicals to humans [9,10]. In this chapter, we discuss the procedures used to conduct animal carcinogenicity bioassays, the variables that can affect the results of these assays, the interpretation of the data from these assays, and the utility of the animal carcinogenesis bioassay to detect chemical carcinogens.

2.2 EPIDEMIOLOGICAL STUDIES

Cancer epidemiology is very important because this approach provides direct evidence concerning factors involved in human cancer causation [11]. Epidemiological studies have shown that arsenic compounds are human carcinogens, even though arsenic compounds have not been shown to be carcinogenic alone in experimental studies [12].

Four basic epidemiological approaches have been used [11,13–15]. The informed hunch is used as a first step to develop a hypothesis concerning which type of exposure might be associated with a given cancer, for example, cigarette smoking and lung cancer. A second approach is the follow-up or prospective cohort study in which a large number of healthy people are followed for several years to determine which of them contract a specific cancer and die from that cancer. The logistical problems of prospective cohort studies, which require large numbers of healthy people to observe a statistically meaningful number of cases of cancer, prompted the development of case control or retrospective studies. Retrospective studies depend on locating cases with a particular disease as well as healthy or unrelated disease controls and asking them the same type of questions. The fourth approach, known as historical follow-up or retrospective cohort studies, is used to detect occupational causes of cancer. These studies exploit the follow-up approach, but here, the initial population of exposed workers, known as the cohort, is identified many years after the initial exposure.

When properly controlled retrospective or prospective studies indicate that an exposure to an agent increases the risk of cancer, it is accepted that the agent is carcinogenic. It should be noted that even though a working population develops cancer from exposure to an industrial carcinogen, the overall increase in worker death rate may not be great enough to exceed the death rate of a similar age group in the general population. Therefore, for any workforce under study, the control group of workers from other industries must be from industries where there is no exposure to the potential carcinogen. Negative results may not establish the noncarcinogenicity of agents. This may be because either the sample size of the number of people involved or the number of cases is insufficient, or because the exposure may be too brief or too small to produce a carcinogenic effect.

Problems encountered with epidemiological studies are inadequate follow-ups due to the loss of a number of people in the studies and the long latency period of 5 to 30 years for the expression of cancer in humans following exposure to carcinogens. This presents problems in the development of good data bases and adequate registries. Due to the length of the latency period for cancer development in humans, it is difficult to determine retrospectively the chemical involved, the dose and duration of exposure, and the extent to which social and personal habits as well as diet is related to cancer [11,13–15].

In general, epidemiological studies are not predictive but rather depend on the occurrence of human cancer related to chemical exposures. These types of studies provide important evidence for associations rather than identification of specific causative factors. It is also important, however, to prevent the possibility of human exposure to potential carcinogens. This preventive medicine approach requires that both experimental animal bioassays and short-term tests be used to determine the carcinogenicity of a variety of compounds before usage.

2.3 ANIMAL CARCINOGENESIS BIOASSAYS

2.3.1 GENERAL AND THEORETICAL CONSIDERATIONS

Long-term administration of chemical compounds to laboratory animals, usually rats or mice, is the best laboratory method for determining whether a compound is carcinogenic to lower animals, and therefore whether this specific compound may be carcinogenic to humans [7,11,15–19]. Bioassays of chemical carcinogens in whole animals involve administration of the candidate carcinogen by skin painting, inhalation, feeding, gavage, the drinking water route, and other modes of administration. For a critical appraisal of bioassay methodologies, the International Agency for Research on Cancer (IARC) book is an excellent reference [12].

Of course, human carcinogenesis testing is unethical and illegal. Futhermore, it is very difficult and inordinately expensive to induce tumors in primates with chemical carcinogens. Rodents are more sensitive to chemical carcinogens than primates and humans, and they also have a substantially shorter life span (2 years). Hence, rodents are substantially less expensive than primates as test animals. Rodents are therefore the animals of choice in most animal carcinogenesis bioassays. The primary bioassay for chemical carcinogens is currently conducted in both mice and rats.

Repeated applications of carcinogens to mouse skin give rise to papillomas and then carcinomas. It has never been proved whether papillomas progress to carcinomas or carcinomas arise independently (carcinomas are epithelial in origin). Locally, subcutaneous injection of carcinogens into mice and rats usually gives sarcomas (tumors of connective tissue). Systemic administration is usually accomplished through prolonged feeding experiments and usually results in liver tumors, but can also give many other types of tumors. Transplacental administration of the compounds to the pregnant mice results in tumors in the offspring. Rodents treated with carcinogens occasionally contract tumors in the F2 generation. Choriocarcinoma is an example of a tumor that arises in the fetus in women. Transplacental administration of carcinogens often results in tumors of the nervous system.

Carcinogenesis bioassays provide accurate information about dose and duration of exposure and are less affected than epidemiology studies by possible interactions of the substance with other chemicals or modifiers, since one compound can be tested at a time in a controlled fashion [8]. Thus, of known human carcinogens, only arsenic compounds have been found to be negative in classical rodent bioassays. However, recent studies from the laboratory of Dr. Michael Waalkes at the National Institute of Environmental Health Sciences (NIEHS) have shown that arsenic compounds are carcinogenic in transplacental studies. Occupational exposure of humans to benzene has resulted in induction of acute myelogenous leukemia (AML) as shown by epidemiological studies. However, the induction of leukemia in lower animals by benzene has not been reported [20], but recent studies have shown that benzene induces solid tumors of other organs in humans [21]. Hence, there is often but not always a strict concordance between tumor sites in lower animals and humans exposed to the same carcinogen.

Although it has long been known that there are significant interspecies differences in the metabolism of xenobiotics and hence in the toxicities of xenobiotics, the metabolic activation pathways of carcinogens and the resultant carcinogen–DNA adducts are qualitatively similar among various animal species, including humans [22]. This scientific evidence thus supports the quantitative extrapolation of carcinogenesis data from laboratory animals to humans, when the carcinogenicity data in rodents is scaled for body surface area and metabolic rates in humans.

The carcinogenicity of an agent is established when the route of administration to animals is adequately designed and the experiments conducted result in an increased incidence of one or more treatment-related types of cancer. Increased incidences of tumors are regarded with greater confidence if positive results are observed in both sexes of one species of animals, in different species of animals, and also by different laboratories. Bioassay treatments are usually repeated and given for a long period of time. The time for appearance of tumors in 50% of the animals treated with a carcinogen following initiation of application of a carcinogen is defined as the latent period. The latent period for mice can take 10 weeks (Figure 2.1), whereas for humans, the latent period can be decades, as for induction of mesothelioma induction following asbestos exposure (5 to 50 years). Tumor incidences are very difficult to analyze theoretically. The incidence of tumors is variable. Strong carcinogens at high doses can induce a 100% incidence of tumors. Weak carcinogens can induce incidences of tumors <50% at the conclusion of the assay (Figure 2.1). For a strong carcinogen giving a 100% incidence of tumors at the end of the assay, the latent period is defined at the time at which 50% of the tumors appear (Figure 2.1). For a weak carcinogen, giving <100% incidence of tumors, the latent period is defined as the time at which half of the final tumor incidence occurs (Figure 2.1). In the typical carcinogen bioassay, the number of tumors per animal is recorded, as is the percentage of animals with tumors. To have a complete carcinogen bioassay, it is necessary to have a complete autopsy of the animals. The investigators then conduct histopathological analysis of all tissues, determine a histological diagnosis of all tumors, and record all types of tumors that appear — papillomas, sarcomas, leukemias, and carcinomas — and the organs in which they occur. The Iball index has been used to relate the dose, latent period, and incidence of tumors (Figure 2.1 and Figure 2.2). This is not universally accepted, nor is any other relationship. The TD50 or dose at which 50% of the tumors occur, is more commonly reported.

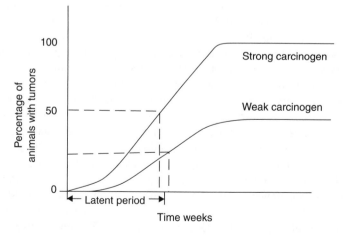

FIGURE 2.1   Idealized curves.

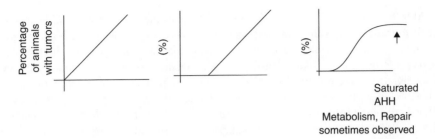

FIGURE 2.2   Dose linear curve (a) without threshold, (b) with threshold, and (c) linear above threshold saturating at high concentrator.

The evidence of carcinogenicity is strengthened if there is a dose response relationship, that is, the incidence of cancer increases monotonically with an increase in the dose of the compound being tested in the carcinogen bioassay. There are three general types of dose–response curves for carcinogenicity in animal experiments. The first curve is the linear, no threshold curve. This curve is observed with mutagenic or genotoxic carcinogens (Figure 2.2[a]). The second type of dose–response curve has a threshold concentration that must be exceeded before carcinogenicity manifests. Once the threshold is exceeded, the dose–response is then linear beyond the threshold concentration (Figure 2.2[b]). A third more complicated type of dose–response is a combination of the two threshold curves, which is linear above the threshold and then saturates at high concentrations (Figure 2.2[c]). The saturation in this curve may be due to a saturation of carcinogen-metabolizing enzyme activities beyond a specific dose of carcinogen.

A positive result in an animal carcinogen bioassay generally supersedes negative findings in epidemiological studies, due to the insensitivity of the epidemiological methods [11,13]. A negative result obtained in an animal bioassay conversely does not preclude the potential carcinogenicity of a compound in humans. In the animal bioassay, a negative result could be due to an inappropriate species having been chosen, too few animals having been tested, or too short an observation period having been used. It has been shown that there are significant variations between species of animals and strains within species in terms of their response to specific carcinogens. For example, 2-naphthylamine, an aromatic amine, does not induce bladder tumors in mice and rats, but does induce bladder tumors in hamsters, monkeys, and dogs, and also induces bladder tumors in humans. Extrapolation of carcinogenicity data in lower animals to humans is further strengthened when positive carcinogenicity results for a chemical are obtained in more than one species.

Note also that negative results in carcinogenicity bioassays obtained in one species do not detract from the significance of clearly positive results obtained in another species. Also, a lack of tumors in a carcinogenesis bioassay could easily result due to the utilization of too few animals. Therefore, to maximize the sensitivity of these bioassays to detect carcinogens, high doses of carcinogens, which are often many times the human exposure, are commonly employed to increase the sensitivity of the bioassay. In using such high doses, it is very important to utilize a dilution series, to determine whether a dose response for carcinogenicity exists for a specific chemical. If a dose–response can be found, then the results at high doses are considered valid. Genotoxic carcinogens yield a linear dose–response curve. This dose–response curve can be extrapolated from high doses to lower doses by a linear extrapolation. The carcinogenicity results can also be extrapolated to humans by use of appropriate scaling factors for surface area and metabolic rate [1,15–19]. Some nongenotoxic carcinogens may also follow a linear dose–response curve, while others that bind to specific biological receptors, such as hormone receptors, may exhibit a threshold in their dose–response curve for carcinogenicity (Figure 2.2[b]).

Extrapolation of carcinogenicity results from lower animals to humans for evaluating human risks entails uncertainties. An unresolved issue remains concerning whether carcinogens display a threshold or no-effect dose level and the precise shape of the curve at low and high doses. Until substantial information is available that demonstrates the existence of a threshold, the position of many scientists and most regulatory agencies is that one must assume that no threshold exists and that the shape of the dose–response curve for carcinogenesis is linear for genotoxic carcinogens and for nongenotoxic carcinogens that do not bind to specific biological receptors (Figure 2.2). The official position of most regulatory agencies is that unless there has been found strong evidence for the existence of a threshold, a threshold is presumed not to exist (Office of Environmental Health Hazard Assessment, California Environmental Protection Agency).

2.3.2 THE GENERAL DESIGN OF STANDARD ANIMAL CARCINOGENESIS BIOASSAYS

Animal bioassays must be well designed and each chemical agent evaluated individually [16–19]. Some of the important variables to be considered in the animal carcinogenesis bioassay are the route of administration, the species and strain of animal used, whether a dose–response relationship can be observed, the duration of dosing, metabolism of the carcinogen, the effect of modifying agents, and the dose of the agent to which humans are or will be exposed. Animal husbandry, diet, safety measures, and written protocols are important and are occasionally overlooked in design considerations. Data acquisition, data reduction, and statistical analysis of tumor induction data of the bioassay are important in estimating the potential human risk of carcinogens. There are some practical considerations that must also be taken into account in the bioassay testing, including the total number of animals, the length of the bioassay, its cost, dose levels, and the specific animal species used in the bioassay.

The standard rodent bioassays conducted by the National Cancer Institute (NCI) or by the National Toxicology Program (NTP) each require 600 animals per compound and require more than 2 years to complete. Each standard bioassay costs from \$250,000 to \$500,000 for skin studies and from \$500,000 to \$5 million for inhalation studies. This standard bioassay usually uses both B6C3F1 mice and F344 rats. This standard bioassay usually provides data at usually three doses, the maximum tolerated dose (MTD), $\frac{1}{2}$ the MTD, and $\frac{1}{10}$ the MTD (or some other fraction of the MTD) to determine whether a dose–response exists and to obtain statistically significant results if a dose–response does exist. At the end of the bioassay, a complete autopsy of the animal is conducted, with histopathology studies conducted on all tissues.

A number of practical considerations should also be used when selecting chemicals for carcinogenicity testing, particularly because these studies are very expensive and lengthy to conduct. The levels of production of a specific chemical, usage of this chemical, and occurrence of this specific chemical in the workplace, in commerce, and in the environment should be considered. Those chemicals with the highest production, use, and occurrence should be considered as high priority for bioassay testing. Those chemicals for which there is either the highest actual human exposure or the highest anticipated human exposure should also be considered a high priority for animal bioassay. Where there is epidemiological evidence to associate exposure to a specific chemical with a high incidence of cancer in humans, then this chemical should be considered a high priority for animal carcinogenicity bioassay. In addition, if a specific chemical is suspected to be a carcinogen due to its having a structural relationship with known carcinogens, or physical and chemical properties that suggest a potential carcinogenicity, then this chemical should also be considered a high priority for testing in animal carcinogenesis bioassays. Also, if a specific chemical has induced important deleterious biological effects in short-term assays, such as bacterial or mammalian cell mutagenesis assays, clastogenesis assays, or DNA repair assays, or has shown activity in other animal bioassays, then this chemical could also be considered a high priority for testing in animal carcinogenesis bioassays. If the specific chemical has shown biological interaction with tumor promoters or cocarcinogens, it could also be considered a priority for testing in animal carcinogenesis bioassays.

2.3.3 SPECIFIC PRACTICAL FACTORS THAT ARE IMPORTANT IN THE DESIGN OF ANIMAL CARCINOGENESIS BIOASSAYS

2.3.3.1 Personnel

Planning of a long-term carcinogenicity study involves animal science, pathology, toxicology, and biostatistics. It may also involve analytical and clinical chemistry, biochemistry, epidemiology, pharmacokinetics, hematology, microbiology, nutrition, and computer science. Hence, personnel skilled in all these areas are important to the proper conduct of animal carcinogenesis bioassays.

2.3.3.2 Properties of Chemicals Tested for Carcinogenicity

Information on the chemical and physical properties of the test agent, in addition to its structure, molecular weight, and chemical formula should be obtained before initiating the long-term carcinogenicity bioassay. The purity of the compound should be determined as well as the identification of the impurities present. Small amounts of carcinogenic impurities can give an incorrect answer in a carcinogenesis bioassay, and this has been a problem in the past. In addition, the source, batch, date received, method of storage, probable daily exposure level to this potential carcinogen for humans, and any biochemical information are also important variables and must be reported [16–19].

2.3.3.3 Species and Strains of Test Animals

There is no ideal animal species known for carcinogenicity testing, one in which the biological response to the test agent is always identical to that of humans. Therefore, at least two different species are used, usually rats and mice [15–19]. Selection of a particular animal species may be used based on metabolic and pharmacokinetic data or their susceptibility to a particular class of carcinogens. The preferred species are the rat and mouse, and to a lesser extent the hamster. These species are used based on practical considerations of short life span (2 years), small size, availability, relatively small cost, and considerable biological and genetic knowledge of these species. Sex differences have been documented with respect to carcinogen susceptibility. Therefore, equal numbers of male and female animals are included in the carcinogenesis bioassays. Furthermore, when a given species of animals is chosen, the strains of that species of animals used in a particular laboratory are based upon the experience and background knowledge of the colony of animals over the years. The strain usually used is one with little or no incidence of spontaneous tumors and a high and specific susceptibility to many carcinogens. Certain strains will produce a biological response to a test carcinogen and others may not. For example, a single administration of 7,12-dimethylbenz[a]anthracene (DMBA) induces mammary tumors in Sprague-Dawley rats but not in Long-Evans rats [23]. Similarly, aflatoxin B_1 induces a high incidence of liver tumors in both Wistar and Fisher rats and a high incidence of kidney tumors in Wistar rats but not in Fisher rats [24].

2.3.3.4 Route of Administration of Test Chemicals

To ensure that the carcinogenicity bioassay results are as relevant to humans as possible, the test agent should be administered by a relevant route of administration, that is, one that approximates as close as possible to the route by which humans are exposed to the chemical in question [15–19]. The three main routes of exposure are oral, cutaneous, and inhalation or intratracheal instillation. Examples of oral administration of chemicals are mixing the chemical in the diet and feeding it to animals, or dissolving the chemical in the drinking water of the animals, or administering the chemical to the animal by gastric intubation. Gastric intubation, which is usually performed over a 1 day, 5 days, or 1 week period, is less desirable because high concentrations may result in severe local damage to the upper digestive tract. If the specific chemical is administered in the diet or in the drinking water the exposure is continuous but there is the possibility of contamination of the laboratory facilities and exposure of laboratory personnel to a potential carcinogen. The ultimate choice of route of administration of the test chemical depends upon the volatility, solubility, stability, and palatability of the agent under test.

Cutaneous exposure is used to simulate dermal exposure, which is one major route of human exposure. Cutaneous exposure may also be used as a model system in animals to study induction of skin tumors. Using the cutaneous route, animals are usually dosed weekly twice until appearance of tumors or for the lifetime of the animals. A basic requirement for cutaneous studies includes clipping the hair to allow maximal skin absorption. As a skin tumor model, these assays may also be useful in initiation and promotion studies.

Inhalation through the respiratory tract is one of the main modes of human exposure for many substances, and therefore, is an important route of testing despite the large cost of inhalation carcinogenesis experiments. Properly designed exposure chambers and equipments for generation, sampling, and monitoring the test chambers are needed for studies involving vapors and aerosols [25]. Such inhalation carcinogenesis studies usually are carried out 4 to 6 h a day, 5 days a week or 22 h a day, 7 days a week. Aerosol exposure is generally not satisfactory for long-term exposures because of contamination of animal fur, resulting in undesired oral intake, or due to filtering of the air by the animals when they hide their noses in their own or in each other's fur. Individual housing and head and nose exposure chambers will reduce problems associated with aerosol exposure. Intratracheal instillation, which has been used in experimental respiratory tract carcinogenesis studies, is an alternative to inhalation for studying aerosols. Since humans are more likely to inhale air through the mouth than the rodents are, this type of instillation may mimic human exposure more closely. Intratracheal instillation of candidate carcinogens is usually administered to animals once or twice per week.

Subcutaneous, intraperitoneal, intramuscular, and intravenous injections are not used very often for testing potential carcinogens. Subcutaneous injection may be used when the agent is poorly absorbed after oral administration or to minimize contamination of the laboratory. The other types of injections may be used for testing for compounds readily absorbed by the digestive tract in humans and not mice or rats. The frequency of injections may be once or twice a week.

2.3.3.5 Dose Selection

An assay to determine carcinogenic activity may require only a high dose and a low dose [15–19]. However, generation of an accurate dose–response curve and use of carcinogen bioassay data in carcinogen risk assessment evaluations usually requires several dose levels with dose–response analyses, knowledge of the metabolism of the chemical being studied, and pharmacokinetic data on the specific chemical being studied. The high dose in many instances is the maximum tolerated dose (MTD), determined in a 90-day subchronic study. The MTD is used to maximize the sensitivity of the animal model to the chemical in order to determine whether a carcinogenic effect does exist. If the candidate chemical does exert carcinogenic effects, then lower dose levels are used to determine whether a dose–response exists (Figure 2.2), and to facilitate risk assessment evaluations. A major difficulty in these experiments is uncertainty about whether a threshold dose exists in these experiments. For genotoxic (mutagenic) carcinogens, and for many nongenotoxic carcinogens, it is widely believed that no threshold for carcinogenesis exists, and a linear, nonthreshold relationship is presumed to exist for the dose–response curve. For some nongenotoxic carcinogens, a threshold dose might exist, particularly if they exert their carcinogenicity by binding to biological receptors, such as hormone receptors.

To simply determine the potential carcinogenicity of a test agent, two dose levels plus the control treatment condition usually are sufficient. The high dose or MTD and two lower doses, usually $\frac{1}{2}$ the MTD and $\frac{1}{10}$ the MTD, are used to ensure that at least one group of animals can be compared meaningfully with the controls even if there was an error in the selection of the high dose, that is severe mortality or weight loss, and to evaluate whether a dose–response for carcinogenesis exists. Pharmacokinetic and metabolism data should be taken into consideration in selecting the lower dose(s). When it is important to determine whether a dose–response relationship exists for a known or candidate carcinogen, three or more dose levels are used, scaled down by factors of 3 to 5 and possibly 10 from the highest dose (MTD, $\frac{1}{2}$ MTD, $\frac{1}{10}$ MTD). This type of study becomes important when evaluating the human risk to a carcinogenic exposure.

2.3.3.6 Inception and Duration of Exposure

Exposure of animals to the test agent begins a few weeks after weaning the animals [16–19]. The animals are normally shipped right after weaning (3–4 weeks for most rodents) and quarantined for 1–2 weeks. The animals are then treated with the candidate carcinogen from 5–6 weeks of age through the major portion of the animal's life span. Studies with low doses of carcinogen are usually terminated between 104 to 130 weeks of age for rats, 99–120 weeks of age for mice, and 80–100 weeks of age for hamsters. For those chemicals that may be carcinogenic to fetal tissues and possibly to newborn and adults, the pregnant mother is treated and the exposure of the offspring is continued during infancy and adult life. An example of this type of chemical is diethylstilbestrol (DES), which resulted in an increased risk of vaginal cancer in daughters of those pregnant women administered DES [6]. This same effect was produced in hamsters treated with DES in utero [13].

2.3.3.7 Numbers of Animals, Methods of Randomization, and Statistics of the Carcinogenic Response

Animals used in a carcinogenesis bioassay must be from the same colony and shipment. Randomization procedures are used to allocate animals to different groups, including the positive and negative control groups. The birth date and weight of each animal are recorded, and a number is assigned to each animal. The animals are then randomized and weight distribution is considered in this randomization.

To allow for accurate statistical and biological evaluation, 50 animals are used per group. With a group size of 50, in which 10% of the control animals possess the same tumor, 28% of the animals in the experimental group must have tumors in order to achieve a significance of $P = .05$. In cases where no spontaneous tumors occur, only 12% incidence of tumors is needed in order to achieve a significance at the $P = .05$ level [17]. Another important factor to be considered is the relative potency of a carcinogen. With a group size of 50 animals only certain strongly carcinogenic compounds may be detected and weaker carcinogens may be detected only by a group size of 100. An effect that occurs in only 1% of the test animals will be entirely missed 37% of the time even with 100 animals and a 0% incidence in the controls. Therefore, carcinogens that are responsible for perhaps a 1% incidence of cancer in human populations could escape detection in animal bioassays due to a lack of sensitivity of these assays in detecting such weak carcinogens. An example of this relative potency is illustrated by the following: 1 μg of aflatoxin B_1 given daily to animals for their life span will produce cancers in 50% of the animals, whereas it would require 10 g of saccharin to produce the same tumor effect [26].

2.3.3.8 Animal Husbandry and Animal Diet

Housing conditions, animal care facilities, diseases in the test animals, purity of animals' diets, purity of air and water, composition of the animals' diets, composition of the animals' bedding, occurrence of cannibalism in the experiment, and pesticide use during the experiment can all alter the outcome of animal carcinogenesis bioassays [16–19]. Animal rooms should be well ventilated, with controlled lighting, temperature, and humidity. Animals received from outside a facility must be quarantined up to 2 weeks to prevent the introduction of unwanted pathogens and to allow the animals to recover from the stress of shipment. Infection of the animals with hepatitis viruses or other viruses could affect the outcome of carcinogenesis experiments.

Cages, racks, and other equipment should be constructed of materials that allow convenient and frequent cleaning, minimum stress to animals, and convenient and accurate feeding. Solid bottom caging should be fabricated from stainless steel, polycarbonate, or polypropylene and can be used with filter tops, which decreases exposure to microorganisms and fugitive contamination but increases levels of humidity, ammonia, CO_2, and temperature. Stainless-steel wire-mesh cages are durable and provide for animals' visibility, but the animal is not protected from microorganisms, nor does this method maximize environmental contamination of the test chemical. These types of cages are used in inhalation chambers. Laminar flow caging systems provide unidirectional airflow after passage through a high efficiency particulate aerosol (HEPA) filter, reduces the variability in environment but does protect animals from volatile chemicals in the room. Animal bedding should be absorbent, dust free, sterilizable, and not contaminated with pesticides or fungal products that would produce physiological changes in animals during carcinogenesis experiments.

Single housing facilitates husbandry and avoids cannibalism and autolysis. Single housing of animals also eliminates positive and negative correlations between animals in the same cage. Both single and multiple housing produce stresses, that is, isolation versus aggressive behavior, respectively. Attending the animals on a daily basis will reduce the stress due to isolation. At present, it is not known whether the cost of single housing would be better spent in purchasing more animals for each group. Most laboratories at present, normally engage in multiple housing, with control and treated groups being housed in separate racks [16–19].

Nutritional factors can also play an important role in carcinogenesis [16–19]. Therefore, irrespective of the source of the diet, for example, commercial, natural ingredients, laboratory designed, or semi-synthetic purified diets, the concentrations of common dietary constituents and possible dietary contaminants, should be determined. Diets used in animal carcinogenesis bioassays are of constant quality and contain standard amounts of essential nutrients. Animal feed older than 3 months is not used due to loss of labile nutrients. Finally, concentrations of dietary constituents should be determined on a periodic basis and the results retained and included in the final report [16–19].

2.3.3.9 Written Protocols and the Safety Action Plan

For each bioassay conducted, a detailed written protocol should be prepared. This written protocol should delineate the responsibility of each individual researcher involved in the assay, and all the aspects of the design of the study described previously in this section. The protocol should also include procedures for data collection, such as time of sacrifice, body weight, feed consumption, necropsy procedures, and for data analysis.

In addition, each chemical compound tested is regarded as potentially carcinogenic. Therefore, precautions need to be taken to prevent inadvertent exposure of laboratory personnel and laboratory facilities to each test chemical. A safety action plan (Table 2.1) should be constructed, and should include a list of the personnel that are potentially exposed, and descriptions of work practices, monitoring procedures, decontamination procedures, and disposal and emergency procedures (Safety Action Plan presented in Table 2.1 is based on a plan designed for and presently used by The Department of Environmental Health in the College of Medicine of the University of Cincinnati since 1975. References 27 and 28). Each safety action plan is designed specifically for the particular compound being studied. There are no general overall procedures that can be used. Each compound has unique physical, chemical, and biological properties, which require different means of disposal and surveillance procedures.

Of all the measures that are required in a safety action plan, the monitoring, decontamination, disposal, and emergency procedures are the most difficult to initiate. Before an animal carcinogenesis study is undertaken, the literature is reviewed for procedures employed with similar compounds. A number of chemical companies make available to investigators the preferred methods of disposal for their products. Through the National Cancer Institute (NCI), safe handling procedures are being developed and made available for a variety of carcinogenic and potentially carcinogenic compounds.

TABLE 2.1

Important Factors to be Considered in the Safety Action Plan

1. Name of test chemical
2. Personnel potentially exposed to carcinogens
3. Structure of chemical
4. Physical and chemical properties of chemical
5. Toxicity data
6. Carcinogenicity data
7. Description of work practices
    a. Personal protection
    b. Type of use
    c. Location of use
    d. Transportation
8. Monitoring procedures
    a. Medical surveillance
    b. Personnel monitoring
    c. Surveillance procedures for environmental contamination
9. Decontamination and disposal procedures for chemical
10. Emergency procedures

For example, it has been found that chlorinated hydrocarbons and polycyclic aromatic hydrocarbons should be incinerated, while nitrosourea compounds can be decomposed in sodium hydroxide. Standard textbooks that deal with decontamination and disposal of chemical carcinogens are available [27,29,30].

Procedures to monitor the health of laboratory workers who conduct animal carcinogenesis bioassays are conducted on an individual basis. If surveillance and personnel monitoring procedures are not available, new methods involving gas or liquid chromatography may have to be developed to conduct such monitoring studies. Medical surveillance procedures too may be difficult to assess due to the lack of appropriate endpoints to observe changes, for example, in personnel urinalysis, sputum cytology, or hematological tests. The safety action plan is practical because each procedure can be conducted routinely. All procedures are tested on a trial basis before the actual carcinogenesis bioassays are conducted to ensure a safe plan of action.

2.3.3.10 Data Collection and Statistical Analysis of the Data

Routine observations are conducted on a daily basis to provide accurate indications of the health status of the animals and biochemical analyses are conducted to aid in diagnosing diseases in the test animals. At the completion of the carcinogenesis bioassay, a complete autopsy and histopathological examinations are performed on all of the tissues of all the animals. Gross pathology analyses are performed first and the gross lesions are sized in standard units. All lesions are recorded and autopsies are performed on the dead or killed animals. All organs and tissues are fixed in appropriate formaldehyde or formalin solutions. Daily worksheets are kept, and if possible, the data on induction of tumors stored on a computerized data storage and retrieval system.

There are several different types of cells in lower animals and in humans from which tumors can arise. Epithelial cells are those cells covering the organs and very often in contact with air. Connective tissues, muscles, and bones comprise mesenchymal cells. In humans, the most common general type of tumors are carcinomas, which are derived from epithelial cells and comprise 92% of human tumors. Leukemias, derived from white blood cells, plus sarcomas, derived from mesenchymal cells, together comprise 8% of the total human cancer burden. The observed high cancer frequency in epithelial cells, which leads to the genesis of carcinomas in animals and humans, is a line of evidence that suggests environmental carcinogen may contribute to the human cancer burden.

Benign tumors are tumors which are locally encapsulated and that remain within the capsule. For instance, an astrocytoma is a benign brain tumor, but astrocytomas can kill the host by occupying space within the brain. Malignant tumors, such as carcinomas and sarcomas, invade into other tissues and can metastasize from the point of origin and spread to other tissues.

Statistical analysis of survival curves is used to determine the adequacy of the carcinogenesis bioassay, since loss of animals before completion of the assay can be a factor in the total tumor incidences. Other factors that are considered in the statistical analysis are the spontaneous incidences of tumors in animal strains, the total number of tumors in experimental groups versus controls, the latency periods, dose–response relationships, and differences in longevity in different experimental groups. The statistical methods are described elsewhere and are beyond the scope of this text. However, once the data has been analyzed, an evaluation of the bioassay can be undertaken [31]. Depending on whether the results are significant ($P = .01$) or marginally significant ($.01 < P < .1$ to $.01$), conclusions can be drawn as to whether the compound being assayed is carcinogenic or not in experimental animals.

Once the carcinogenic potential of a compound has been established in animals, extrapolations from these data to relatively safe levels and the probability of tumor induction in humans exposed to this chemical can be calculated [15]. The greatest difficulty with the extrapolation lies with the low end of the dose range and the potential of a threshold for a nongenotoxic carcinogen for the exposed population.

A number of mathematical models for statistical analyses of whole animal data have been used, such as the one hit (linear), multikit (k-hit), multistage, extreme value, and the log-probit. These models have been used in data evaluation and data extrapolation to humans. One other result of the extrapolation of carcinogenicity data from animal bioassays to humans is the incorporation of safety factors. These safety factors are used to calculate dose levels of agents that may be allowed in the environment and are considered safe for humans. A number of safety factors have been suggested: (1) to divide the observed experimental no observed effect level (NOEL) by 100; (2) to divide the lowest experimental effect level by 5,000 and (3) identification of a safe dose level at which the risk calculated for exposure to the carcinogen would not exceed very small level of 1 in 100,000,000 or 10^{-8}.

However, it should be noted that there is no way to demonstrate absolute safety of an agent on the basis of statistical analyses of data [15]. Pharmacokinetic and pharmacodynamic models, including physiologically based pharmacokinetic (PBPK) models have also been employed to enhance the precision of risk assessment procedures [32]. The problem, associated with extrapolation of lifetime exposure of animals to the MTD of an agent to exposure of humans to lower doses for shorter periods of time has been addressed by the EPA through the use of the Weibull Model [33].

Finally, in extrapolating animal carcinogenesis bioassay data to human risk, there are a number of factors that need to be considered. These include reproducibility of experimental data; dose-dependence of tumor incidence; experimental dose that approximates that of human exposure; availability of biochemical, metabolic, and pharmacokinetic data; relative potency of compounds as determined by human or experimental assays; the nature and quality of human exposure, and individual susceptibility [15–19]. It has been well-known for many years that there are quantitative differences in carcinogen metabolism and DNA adduct formation among different cell types, tissue types, and outbred individuals of the same species. A new research field in carcinogenesis, molecular epidemiology, was named for combined laboratory–epidemiological studies of human cancers [34]. The primary goal of molecular epidemiology is to identify individuals in human populations who are at the highest risk for contracting cancer. A second goal of molecular epidemiology is to determine the specific molecular mutations that occur in specific genes of specific tumors, and to identify the agents that have caused these tumors [35].

TABLE 2.2
Two-Stage Carcinogenesis Systems [36–38]

Organ system	Initiator	Promoter
Mouse skin	Polycyclic aromatic hydrocarbons, urethane, direct-acting electrophiles	Croton oil, phorbol esters, (TPA) fatty acids and fatty acid esters, surface-active agents, linear long chain alkanes, benzoyl peroxide, tobacco smoke condensate and extracts of unburned tobacco
Rat and mouse liver	2-Acetaminofluorene, dimethylnitrosamine, 2-methyl-N,N'-dimethyl-4-aminoazobenzene	Phenobarbital, DDT, butylated hydroxy toluene (BHT), polychlorobiphenyls (PCB), tetrachlorodibenzodioxin (TCDD)
Mouse lung	Urethane, polycyclic aromatic hydrocarbons	BHT, Phorbol
Rat colon	Dimethylhydrazine	Bile acids, high fat diet, high cholesterol diet
Rat bladder	N-methyl-N-nitrosourea	Saccharin, cyclamate
Rat and mouse mammary gland	Polycyclic aromatic hydrocarbons	Hormones, high fat diet, phorbol
Rat stomach	N-methyl-N'-nitro-N-nitrosoguanidine	Surfactants
Rat esophagus	Diethylnitrosamine	Diet
Mouse cell culture system	Polycyclic aromatic hydrocarbons, radiation	Phorbol esters, saccharin
Rat cell culture system	N-methyl-N'-nitro-N-nitrosoguanidine	Phorbol esters

2.4 MODIFIERS OF CARCINOGENESIS — INITIATORS, PROMOTERS, COCARCINOGENS, AND INHIBITORS OF CHEMICAL CARCINOGENESIS

Initiators and promoters (Table 2.2), cocarcinogens (Table 2.3), and inhibitors will modify the strength of a carcinogen. Promoters and cocarcinogens can either shorten the latency period or increase the number of tumors per animal. On the other hand, inhibitors of carcinogenesis (Table 2.4) will increase the latency period or decrease the number of tumors caused by treatment of animals with a carcinogen [36–41]. The mechanisms of each of these processes and specific examples of each type of agent are discussed below.

The mechanisms of action of initiation and promotion, or two-stage carcinogenesis, were first studied by Peyton Rous and Isaac Berenblum, and later by Philip Shubik in mouse skin. These studies have since been extended to mouse liver, mouse lung, and mouse mammary gland. In addition, initiation and promotion have been demonstrated in rat liver, rat colon, rat bladder, rat mammary gland, rat stomach, and rat esophagus. Initiation and promotion of neoplastic cell transformation have also been demonstrated *in vitro* in mouse and rat cell culture systems (Table 2.2). Initiation occurs when a tissue is treated with low levels of a complete carcinogen or an initiator, and is an irreversible process (Figure 2.3). Initiators bind covalently to DNA. Many scientists believe that initiation is due to a mutation, likely in cellular proto-oncogenes. Treatment of initiated cells with

TABLE 2.3
Cocarcinogens in Various Organs/Tissues [36–38]

Skin
Catechol, pyrogallol, lauryl alcohol, decane, undecane, tetradecane, *n*-dodecane, pyrene,
benzo[e]pyrene, fluoranthene, benzo[g,h,i]perylene, anthralin, croton oil, phorbol esters, phenols
nicotine, surfactants, radiation, viruses
Lung
Asbestos, radiation, *n*-dodecane, ferric oxide, magnesium oxide, hypoxia, ethanol
Mammary gland
Hormones, estradiol, prolactin
Bladder
L-Tryptophan, saccharin, cyclamate
Cheek pouch
X-Radiation, croton oil
Liver
Cyclopropenoid, fatty acids, alcohol

TABLE 2.4
Examples of Compounds that Inhibit Carcinogenesis in Various
Tissues [36–38]

Lung
Butylated hydroxy anisole (BHA), ethoxyquin, retinoids
Liver
BHT
Forestomach
BHA, BHT, disulfiram, benzylisothiocyanate, phenethylisothiocyanate, ethoxyquin
Breast
BHA, BHT, ethoxyquin, disulfiram, benzylisothiocyanate, phenethylisothiocyanate,
phenylisothiocyanate, benzylthiocyanate, cysteamine hydrochloride, retinoids
Large intestine
Disulfiram, diethyldithiocarbamate
Skin
Cycloparaffins, esculin, quereetin, squalene, oleic acid, eugenol, resorcinol, hexadecane,
hydroquinone, limonene, phenol, retinoids
Bladder
Retinoids

a tumor promoter, such as tetradecanoyl phorbol acetate (TPA), causes promotion or the growth
of initiated cells. Promotion is reversible up to a point (Figure 2.4, step 1). If application of the
promoter is stopped at this point, then the rate of cell death will equal the rate of cell growth, and
the developing tumor will regress. After a certain number of repeated applications of promoter, the
carcinogenesis process becomes irreversible, and the tumors become fixed and permanent (Figure 2.3
and Figure 2.4). In these initiation and promotion experiments on mouse skin, 92% of the tumors that
appear are papillomas (benign tumors of the papillary glands) and 8% of the tumors are carcinomas

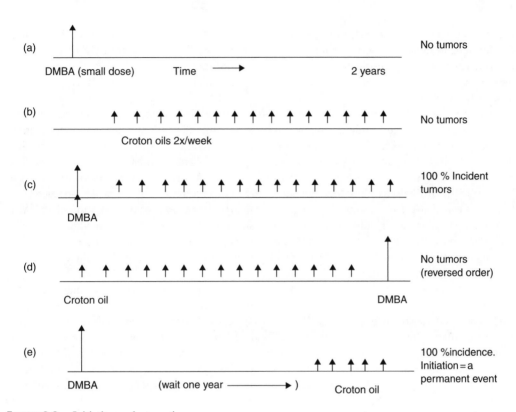

FIGURE 2.3   Initiation and promotion.

Initiation	Promotion	Progression
DMBA, low dose	Stage 1 TPA	Stage 2 Mezerein

FIGURE 2.4   Schematic of initiation, promotion, and progression.

(malignant tumors). Tumor promoters do not bind covalently to DNA, but bind to protein kinase C and stimulate the growth of initiated cells.

   Cigarette smoke condensate also contains tumor promoters. Environmental promoters may also be important in human cancer causation. TPA is one of the most powerful tumor promoters known. Four situations that have been clearly identified and well-studied where initiation and promotion occur are (1) in mouse skin treated with DMBA and TPA, (2) in rat liver treated with acetyl-aminofluorene (AAF) and phenobartital, (3) in rat bladder treated with N-methyl-N-nitrosurea and promoted with either saccharin or cyclamate, and (4) in rat colon treated with dimethylhydrazine and bile acids. Pure tumor promotes are not tumorigenic themselves, except at very high concentrations, where they are likely to promote already initiated cells. Cigarette smoke has very powerful promoters in it. There is synergism between initiators and promoters, such that the carcinogenic effect of initiators (or low doses of complete carcinogens) is enhanced by promoters to a greater than additive extent. TPA is a naturally occurring phorbol ester derived form the irritant croton oil, synthesized by the plant, *Euphorbia lathyris*. TPA was purified and identified by

Dr. Eric Hecker of the German Cancer Research Center in Heidelberg, Germany. TPA is widely used for mouse skin tumor promotion studies. 2,3,7,8-tetrachlorodibenzo-p-dioxin (TCDD), or dioxin, is formed as an impurity during the synthesis of trichlorophenol and also in fires. TCDD is one of the most effective promoter studies for rat liver carcinogenesis and is also effective as a tumor promoter in lung and skin [42].

Cocarcinogenesis is completely an operational definition. Cocarcinogenesis indicates that two agents are given together, simultaneously, and enhance the yield of carcinogenesis. In contrast, initiators and promoters are administered to animals at different times. An example of cocarcinogenesis is two carcinogens given together with a synergistic effect on carcinogenesis. Therefore, cigarette smoke contains cocarcinogens including catechols. A second example of cocarcinogenesis is intratracheal instillation of BaP together with ferric oxide in Syrian hamsters, which produced lung tumors, whereas the control, BaP, did not [43]. Similarly, cutaneous coadministration of BaP with n-dodecane to C3H mice resulted in an increase in skin tumors when compared to administration of BaP alone, providing another example of cocarcinogenesis [44]. Additional instances of cocarcinogenesis have been demonstrated to occur in skin, lung, mammary gland, bladder, the cheek pouch of hamsters, and in liver (Table 2.3).

Certain compounds have been shown to exert strong inhibitory effects on chemical carcinogenesis. For instance, BaP administered in the diet induces forestomach tumors in mice [45]. Addition of butylated hydroxyanisole (BHA) or butylated hydroxytoluene (BHT) together with BaP in the diet inhibits the formation of BaP-induced forestomach tumors in mice [45]. In the lung, BHA, ethoxyquin, and retinoids are inhibitors of chemical carcinogenesis. BHT inhibits chemical carcinogenesis in the liver and forestomach. In the breast, BHA, BHT, ethoxyquin, disulfiram, benzylisothiocyanate, phenethylisothiocyanate, phenylisothiocyanate, benzylthiocyanate, cysteamine hydrochloride, and retinoids inhibit chemical carcinogenesis. Disulfiram and diethyldithiocarbamate inhibit carcinogenesis in the large intestine. In the skin, cycloparaffins, esculin, quercetin, squalene, oleic acid, eugenol, resourcinol, hexadecane, hydroquinone, limonene, phenol, and retinoids inhibit chemical carcinogenesis. Retinoids also inhibit bladder carcinogenesis (summarized in Table 2.4).

Extensive examples of initiators, promoters, cocarcinogens, and inhibitors of carcinogenesis listed by target tissue are shown in Table 2.2 (initiators and promoters), Table 2.3 (cocarcinogens) and Table 2.4 (inhibitors of carcinogenesis) (reviewed in References 1, 36, and 46). As is indicated in Tables 2.2 to 2.4, these modifiers are ubiquitous in the environment and in particular in the occupational setting. Depending on the carcinogenesis bioassay protocol used, a compound may be a carcinogen or a promoter as indicated by skin painting studies, while phenol has been shown to be an inhibitor and a cocarcinogen on skin. It is clear that with the few exceptions where pure compounds have been implicated, mixtures containing carcinogens and modifying agents need to be examined more closely. It is unclear at this time what effects varying of the ratios of carcinogen to modifier will have on the carcinogenic response of mixtures. It may be that all that is necessary are very low doses of carcinogen in these mixtures of modifiers to produce detrimental effects in humans.

2.5 COMPLEX MIXTURES

The most common complex mixtures in the environment are those formed from the combustion of fossil fuels and tobacco smoke [47–52]. The conversion or processing of shale, coal, or petroleum involves elevated temperatures and elevated pressures. Certain compounds formed in these mixtures exhibit carcinogenic activities for a variety of organ sites in experimental animals, and epidemiological evidence implicates their role as carcinogens in humans. These mixtures contain carcinogenic and noncarcinogenic polycyclic aromatic hydrocarbons (PAHs), aromatic amines, nitrosamines, halogenated hydrocarbons, and metal compounds. In addition, these mixtures also contain cocarcinogens, tumor promoters, and inhibitors of carcinogenesis, such as straight aliphatic hydrocarbons and cycloparaffins. Interactions between chemicals in the mixture may be additive, inhibitory, antagonistic, or synergistic [53].

It is difficult to assess the carcinogenic potencies of these mixtures based solely on their chemical composition. Investigators have attempted to determine the relative potency of these mixtures by determining the concentrations of indicator compounds, such as BaP and benz(a)anthracene, with some success. It has been shown, however, that mixtures or fractions of mixtures are highly carcinogenic or mutagenic even without the presence of BaP in these mixtures [54] and without the PAHs in the basic fractions [55,56]. It also has been noted that the BaP content cannot account for the relative potency of mixtures [47,48,54,57]. An additional complicating factor is the effects of sunlight or near ultraviolet light on the carcinogenic potential of mixtures. Under appropriate conditions, there can be an enhancement of the carcinogenic activity of BaP [58], asphalt, and coal tar in the presence of ultraviolet light or sunlight [47,59,60]. It has been also shown that coal tar pitch fume is more than 90% aromatic in content, while asphalt is more than 99% aliphatic and less than 1% aromatic in content. The pitch fume materials showed effects consistent with high BaP content. The asphalt fume materials showed a higher activity than would be expected based on their BaP or total PAH content, suggesting the presence of other compounds contributing to the carcinogenic response, that is the aliphatic content [61].

Based on this discussion, it is apparent that it would be difficult to assess complex mixtures solely by use of long-term skin carcinogenesis bioassay due to the high cost and long duration of the experiments. Short-term assays have been developed, in particular the Salmonella/microsome mutation assay, to assess the potential activity of these mixtures and to prioritize the mixtures for further testing [31,62,63]. For a further in-depth discussion of mixtures, please see Chapter 14 by Dr. Stephen Nesnow.

REFERENCES

1. Pitot, H.C. *Fundamentals of Oncology,* 3rd edn. Marcel Dekker, New York, 1986.
2. Miller, J.A. Carcinogenesis by chemicals: An overview. *Cancer Res.*, *30*: 559–576, 1970.
3. Doll, R. and Peto, R. *The Causes of Cancer*, Oxford University Press, New York, 1981.
4. Miller, E.C. and Miller, J.A. Mechanisms of chemical carcinogenesis. *Cancer*, *47*: 1055–1065, 1981.
5. Cullen, M.R., Cherniack, M.G., and Rosenstock, L. Occupational medicine. *N. Engl. J. Med.*, *322*: 675–682, 1990.
6. Vainio, H., Coleman, M., and Wilbourn, J. Carcinogenicity evaluation and ongoing studies: The IARC database. *Environ. Health Perspect.*, *6*: 1653–1665, 1991.
7. Huff, J.E. Chemicals and cancer in humans: First evidence in experimental animals. *Environ. Health Perspect.*, *100*, 201–210, 1993.
8. Huff, J.E. Value, validity, and historical development of carcinogenesis studies for predicting and confirming carcinogenic risks to humans. In: *Carcinogenicity Testing, Predicting, and Interpreting Chemical Effects*. Kitchin, K.T., Ed., Marcel Dekker, New York, 1999, pp. 21–123.
9. Allen, D.G. Prediction of rodent carcinogenesis: An evaluation of prechronic liver lesions as forecasters of liver tumors in NTP carcinogenecity studies. *Toxicol. Pathol.*, *32*: 393–401, 2004.
10. Tomatis, L., Huff, J., Hertz-Picciotto, I., Dandler, D., Bucher, J., Boffeta, P., Axelson, O., Blaire, A., Taylor, J., Stayner, L., and Barrett, J.C. Avoided and avoidable risks of cancer. *Carcinogenesis*, *18*: 97–105, 1997.
11. Weinstein, I.B. Scientific basis for carcinogen detection and primary cancer prevention. *Cancer*, *47*: 1133–1141, 1981.
12. IARC Working Group. An evaluation of chemicals and industrial processes associated with cancer in humans based on human and animal data: IARC monographs, Vol. 1–20, *Cancer Res.*, *40*: 1–12, 1980.
13. Magos, L. Epidemiological and experimental aspects of metal carcinogenesis physiologicochemical properties. Kinetics and active species. *Environ. Health Perspect.*, *95*: 157–189, 1991.
14. Anderson, M. Chemical carcinogens: Whither epidemiology. *Br. Med. Bull.*, *36*: 95–100, 1980.

15. Pitot, H.C. Relationships of bioassay data on chemicals to their toxic and carcinogenic risk for humans. *J. Environ. Pathol. Toxicol.*, *3*: 431–450, 1980.
16. IARC Working Group. Basic requirements for long-term assays for carcinogenicity. In *Long-Term and Short-Term Screening Assays for Carcinogens: A Critical Appraisal*. IARC Monographs Supplement 2, IARC Publication, Lyon, France, 1980, pp. 23–83.
17. Kinnimulki, V. Smad3 regulates senescence and malignant conversion in a mouse multistage skin carcinogenesis model. *Cancer Res.*, *63*: 3447–3452, 2003.
18. Larcher, F. Modulation of the angiogenesis response through H-ras control, placenta growth factor, and angiopoitin expression in mouse skin carcinogenesis. *Mol. Carcinog.*, *37*: 83–90, 2003.
19. Munro, I.C. Considerations in chronic toxicity testing: The chemical, the dose, the design. *J. Environ. Pathol. Toxicol.*, *1*: 183–197, 1977.
20. Wallace, L.A. The exposure of the general population to benzene. *Cell Biol. Toxicol.*, *5*: 297–314, 1989.
21. Snyder, R. and Kalf., G.F. A perspective on benzene leukemogenesis. *Crit. Rev. Toxicol.*, *24*: 177–209, 1994.
22. Harris, C.C. Human tissues and cells in carcinogenesis research. *Cancer Res.*, *47*: 1–10, 1987.
23. Syndor, K.L., Butenandt, O. Brillanter, F.P., and Huggins. C. Race-strain factor related to hydrocarbon-induced mammary cancer in rats. *J. Natl. Cancer Inst.*, *29*: 805–814, 1962.
24. Wogan, G.N. Aflatoxin carcinogenesis. In: *Methods in Cancer Research.*, Vol. 7., Busch, H., Ed. Academic Press, New York., 1973, pp. 309–344.
25. Willeke, K., Ed. *Generation of Aerosols.* Ann Arbor Science Publishers, Inc., Ann Arbor, MI, 1980.
26. Maugh, T.H. Chemical carcinogens: How dangerous are low doses? *Science*, *202*: 37–41, 1978.
27. Montesano, R., Bartsch, H., Boyland, E., Porta, G.D., Fishbein, L., Griesemer, R.A., Swan, A.B., and Tomatis, L., Eds. *Handling Chemical Carcinogens in the Laboratory Problems of Safety*, IARC Scientific Publication No. 33, Lyon, France, 1979.
28. NIH Guidelines for Laboratory Use of Chemical Carcinogens, NIH publications No. 81–2385, May, 1981.
29. Walters, D.B., Ed. *Safe Handling of Chemical Carcinogens, Mutagens and Teratogens and Highly Toxic Chemicals*, 2 Vol., Ann Arbor Science, Ann Arbor, MI, 1980.
30. Slein, M.W. and Sansone, E.B. *Degradation of Chemical Carcinogens*, Van Nostrand-Reinhold, Co., New York, 1980.
31. Warshawsky, D., Schoeny, R., and Moore, G. Evaluation of coal liquefaction technologies by *Salmonella* mutagenesis. *Toxicol. Lett.*, *10*: 121–127, 1982.
32. Anderson, M.E., Tissue dosimetry, physiologically based pharmacokinetic modeling and cancer risk assessment. *Cell Biol. Toxicol.*, *5*: 405–415, 1989.
33. Hanes, B. and Wedel, T. A selected review of risk models: Onehit multihit, multistage, probit, weibull, and pharmacokinetic. *J. Am. Coll. Toxicol.*, *4*: 271–278, 1985.
34. Perera, F.P. and Weinstein, I.B. Molecular epidemiology and carcinogen-DNA adduct detection: New approaches to studies of human cancer causation. *J. Chronic Dis.*, *35*: 581–600, 1982.
35. Harris, C.C. Molecular epidemiology of human cancer in the 1990's. In: *Trends in Biological Dosimetry*, Gledhill, B.L. and Mauro, F., Eds. Wiley-Liss, New York, 1991.
36. Slaga, T.J., Sivak, A., and Boutwell, R.K. Ed. *Carcinogenesis — A Comprehensive Survey, Vol. 2, Mechanisms of Tumor Promotion and Cocarcinogenesis.* Raven Press, New York, 1978.
37. Slaga, T.J., Fischer, S.M., Triplett, L.L., and Nesnow, S. Comparison of complete carcinogenesis and tumor initiation and promotion in mouse skin: The induction of papillomas by tumor initiation promotion a reliable short-term assay. *J. Am. Coll. Toxicol.*, *1*: 83–99, 1982.
38. Slaga, T.J., Triplett, L.L., and Nesnow, S. Mutagenic and carcinogenic potency of extracts of diesel and related environmental emissions: Two-stage carcinogenesis in skin tumor sensitive mice. In: *Proceedings of the International Symposium on Health Effects of Diesel Engine Emissions*, 2 Vol., EPA 600/9-90-0576, 1980, pp. 874–897.
39. Boyd, J.A. and Barret, I.C. Genetic and cellular basis of multistep carcinogenesis. *Pharmacol. Ther.*, *46*: 469–486, 1990.
40. Harris, C.C. Chemical and physical carcinogenesis: Advances and perspectives for the 1990's. *Cancer Res.*, *51*: 5023–5044, 1991.

41. Pitot, H.C. and Dragan, Y.P. The instability of tumor production to human cancer risk. In: *Growth Factors and Tumor Promotion: Implication for Risk Assessment, Progress in Clinic and Biological Research.*, McElain, J., Slaga, T.J., LeBoeut, R., and Pitot, H.C., Eds., *Vol. 391.* Wiley-Liss, New York, 1995, pp. 21–38.

42. Wattenberg, L.W., Lam, L.K.T., Speier, J.L., Loub, W.D., and Borchert, P. Inhibition of chemical carcinogenesis. In: *Origins of Human Cancer Book B on Mechanisms of Carcinogenesis,* Hiatt, H.H., Watson, J.D., and Winsten, J.A., Eds. Cold Spring Harbor Laboratory, Woodbury, New York, 1977, pp. 785–799.

43. Saffiotti, D., Montesano, R., Sellakumar, A.R., Cefis, F., and Kaufman, D.G. Respiratory tract carcinogenesis in hamsters induced by benzo(a)pyrene and ferric oxide. *Cancer Res., 32*: 1073–1081, 1972.

44. Bingham, E. and Falk, H. Environmental carcinogenesis: Threshold concentrations of carcinogens and cocarcinogens. *Arch. Environ. Health, 19*: 779–783, 1969.

45. Van Den Bert, M., De Jongh, J., Poiger, H., and Olson, J.R. The toxicokinetics and metabolism of polychlorinated dibenzo-p-dioxin (PCDDs) and dibenzoturan (PCDFs) and their relevance for toxicity. *Crit. Rev. Toxicol., 24*: 1–74, 1994.

46. Bingham, E. and Nord, P.J. Cocarcinogenic effects of N-alkanes and ultraviolet light on mice. *J. Am. Chem. Soc., 58*: 1099–1101, 1977.

47. Bingham, E., Niemeier, R.W., and Reid, J.B. Multiple factors in carcinogenesis. *Ann. N. Y. Acad. Sci., 271*: 14–21, 1976.

48. Trosset, R., Warshawsky, D., Menefee, C., and Bingham, E. Investigation of Selected Potential Environmental Contaminants: Asphalts and Coal Tar Pitch. *EPA Publication* 56012–77–005, 1978.

49. Bingham, E., Trosset, R.P., and Warshawsky., D. Carcinogenic potential of petroleum hydrocarbons — A critical review. *J. Environ. Pathol. Toxicol., 3*: 483–563, 1979.

50. Nesnow, S., Triplett, L.L., and Slaga, T.J., Comparative tumor — initiating activity of complex mixtures from environmental particulate emissions on SENCAR mouse skin. *J. Natl. Cancer Inst.* 68: 829–834, 1982.

51. Albert, R.E. Comparative carcinogenic potencies of particulates from diesel engine exhausts, coke oven emissions, roofing tar aerosols and cigarette smoke. *Environ. Health Perspect., 47*: 335–341, 1983.

52. Berger, M.R. Synergism and antagonism between chemical carcinogens. In: *Chemical Induction of Cancer*, Arcos, J.C., Argus, M.F. and Woo, Y.T. Eds. Basel, Birkhauser, 1995, pp. 23–49.

53. Manderly, J.L., Toxicological approaches to complex mixtures. *Environ. Health Perspect., 101*: 155–165, 1993.

54. Bingham, E. and Barkley, W. Bioassay of complex mixtures derived from fossil fuels. *Environ. Health Perspect., 30*: 157–163, 1979.

55. Guerin, M.R., Ho, C.H., Rao, J.K., Clark, B.R., and Epler, J.L. Polycyclic aromatic primary amines as determinant chemical mutagens in petroleum substitutes. *Environ. Res., 23*: 42–53, 1980.

56. Wilson, B.W., Pelroy, R., and Cresto, J.T. Identification of primary aromatic amines in mutagenically active subfractions from coal liquefraction materials. *Mutat. Res., 79*: 193–202, 1980.

57. Warshawsky, D., Bakley, W. and Bingham, E. Factors affecting carcinogenic potential of mixtures. *Fundam. Appl. Toxicol., 20*: 376–382, 1993.

58. Cavalieri, E. and Calvin, M. Photochemical coupling of benzo(a)pyrene with 1-methylcytosinei photo enhancement of carcinogenicity. *Photochem. Photobiol., 14*: 641–653, 1971.

59. Davis, R.E. Interaction of Light and Chemicals in Carcinogenesis. NCI Monograph No. 50. International Conference on Ultraviolet Carcinogenesis DHEW Publication #78-1532, 1978, pp. 45–50.

60. Thayer, P.S., Menzies, K.T., and Von Thuna, P.C. Roofing Asphalts, Pitch, and UVL Carcinogenesis. NIOSH Contract 210-78-0035, by Arthur D. Little, Inc., Cambridge, MA, November, 1981.

61. Thayer, P.S., Harris, J.C., Menzies, K., and Niemeier, R.W. Integrated chemical and biological analysis of asphalt and pitch fumes. *Enviorn. Sci. Res., 27*: 351–366, 1983.

62. Whong, W.Z., Sorenson, W.G., Elliott, J.A., Stewart, J., Simpson, J., Piacitelli, L., McCawley, M., and Ong, T. Mutagenicity of oil-shale ash. *Mutat. Res., 103*: 5–12, 1982.

63. Nachtrnan, J.P., Xu Xiao-bai, Rappaport, S.M., Talcott, R.E., and Wei, E.T. Mutagenic activity in diesel exhaust particulates. *Bull. Environ. Contam. Toxicol., 27*: 463–466, 1981.

64. Informing Workers and Employers about Occupational Cancer. U.S. Department of Labor Publication, 1978, pp. 5–6.

65. Hueper, W.C., Wiley, F.H., and Wolte, H.D. Experimental production of bladder tumors in dogs by administration of beta-naphthylamine. *J. Ind. Hyg. Toxicol.*, *20*: 46–84, 1938.

66. Connolly, J.G. and White, E.P. Malignant cells in the urine of men exposed to beta-naphthylamine.*Can. Med. Assoc. J.*, *100*: 874–882, 1969.

67. Lida, M., Changes in the global gene and protein expression during early mouse liver carcinogenesis induced by nongenotoxic model carcinogens oxazepam and wyeth-14,643. *Carcinogenesis 24*: 757–770, 2003.

68. Slaga, T.J., Klein-Szanto, A.J.P., Triplett, L.L., Yotti, L.P., and Trosko, J.E. Skin tumor-promoting activity of benzoyl peroxide, a widely used free radical generating compound. *Science*, *213*: 1023–1025, 1981.

69. Van Duuren, B.L. and Goldschmidt, B.M. Cocarcinogenic and tumor-promoting agents in tobacco carcinogenesis. *J. Natl. Cancer Inst.*, *56*: 1237–1242, 1976.

70. Furstenberger, G., Berry, D.L., Sorg, B., and Marks, F. Skin tumor promoting by phorbol esters in a two-stage process. *Proc. Natl. Acad. Sci.*, *78*: 7722–7726, 1981.

71. Sporn, M.B. Prevention of epithelial cancer by vitamin A and its synthetic analogs (retinoids). In: *Origins of Human Cancer,* Hiatt, H.H., Watson, J.D., and Winston, J.A., Eds. Cold Spring Harbor Laboratory, Woodburg, New York, pp. 801–807, 1977.

72. Chu, K.C., Cueto, C. and Ward, J.M. Factors in the evaluation of 200 NCI carcinogen bioassays. *J. Toxicol. Environ. Health*, *8*: 251–280, 1981.

3 Molecular Mechanisms of Action of Selected Organic Carcinogens

Weiling Xue and David Warshawsky

CONTENTS

SUMMARY

After many studies, we still have a long way to go to completely understand the nature of carcinogenesis. We are much closer to achieving this goal since realizing that in living organisms chemical carcinogens are altered through metabolism [1–4].

3.1 HISTORICAL

Paleopathologists have found cancerous lesions in dinosaur bones, such as osteomas and hemangiomas. It has also been reported that cancers have been found in both plants and animals, such as planaria and fish, which suggests that cancer has existed for most of the evolutionary period of life on earth. History also shows evidence of schistosomiasis in Egyptian mummies of the fifth dynasty (about 3000 B.C.), which has been associated with bladder cancer in Egyptians of that period [5–9].

However, Egyptians were aware of cancer in human patients. Autopsies of mummies have shown bone lesions and other cancerous processes. By the fourth century B.C., tumors, such as stomach and uterine cancer were described. Hippocrates used the term carcinoma to refer to tumors that spread and killed the patient. Not until the 19th century did physicians and scientists begin to study cancer systemically; Bichat, Muller, Pasteur, Cohnheim, and Virchow were among the more prominent.

Many theories have been developed in the 19th century to explain the origin and development of cancer. The irritation theory dealt with the little known effects of chemicals and radiation as causes of cancer; there appeared to be a relationship between chronic ulcerations and cancer. The embryonal theory indicated that all cancers developed from cells originating in fetal life and remained alive in adults. The infectious theory resulted from scientific information obtained from infectious diseases in man and animal. Otto Warburg suggested that cancer development was due to abnormal cell respiration that resulted in an increase in the fermentation of glucose to lactic acid.

Present theories of cancer formation can be divided into two groups: genetic and epigenetic. The genetic theory states that tumor formation and development begins with changes in genetic information Deoxyribonucleic acid (DNA) either by addition, alteration, subtraction, or substitution. Loss, substitution, or alteration of the DNA results in a mutational event. Chemicals or radiation can cause mutations in DNA but the question is whether such mutations can be causally related to cancer or in the demethylation of promoters of proto-oncogenes, leading to expression of previously repressed proto-oncogenes, or in the methylation of promoters of tumor suppressor genes, leading to repression of their expression. Addition of DNA results in the integration of new genetic materials into the cell's DNA and cancer-causing viruses can add viral DNA to the genetic material. The epigenetic theory states that tumors result from alterations in Ribonucleic acid (RNA) and protein resulting in the persistent change in cell expression. This means that the DNA remains unaltered but that its messages are expressed incorrectly.

3.2 CELLULAR ASPECTS

In molecular biology advances have been made in understanding the mechanism of cancer at the cellular level [10]. Cancers are groups of cells that arise from a singly transformed cell. Normal cell proliferation is controlled or inhibited by contact of cell membranes with other normal cells. Cancerous cells lose contact inhibition and multiply indiscriminately. These transformed cells show an increase in anaerobic respiration and a decrease in aerobic respiration. In comparison with normal cells, they also show characteristics of fetal cells with less differentiation.

In normal cells, just below the surface membrane lies the cell's skeleton composed of large thick sheaths or cables of structural proteins called actin. These cables disappear in transformed cells. It appears that cancer-causing agents alter either the cell membrane or the cell skeleton or both. The cell loses its capacity to halt growth. Cancer cells that are less adhesive than normal can dislodge from a growing tumor. They can be transported to other sites in the body and form secondary tumors. This accounts for invasiveness and metastasizing ability of transformed cells. It should be noted,

---

TABLE 3.1
Chemical Carcinogens

A. Direct-acting carcinogens
Nitrogen or sulfur mustards
Propane sulfone
Methyl methane sulfonate
Ethyleneimine
B-Propiolactone
1,2,3,4-Diepoxybutane
Dimethyl sulfate
Bis-(Chloromethyl)ether
Dimethylcarbamyl chloride

B. Chemicals requiring activation
Polycyclic aromatic hydrocarbons and heterocyclic aromatics
Aromatic amines
N-Nitrosoamines
Azo dyes
Hydrazines
Cycasin
Safrole
Chlorinated hydrocarbons
Aflatoxin
Mycotoxin
Pyrrolizidine alkaloids
Bracken fern
Carbamates

---

however, that cell surface changes may be secondary effects and not directly responsible for the production of cancer.

3.3 CARCINOGENIC AGENTS

As indicated earlier, there appears to be three general classes of carcinogenic agents: physical agents (radiation), viruses, and chemicals [2,11,12]. Ultraviolet radiation (UV) has been shown to induce skin cancers in humans and experimental animals. Ionizing radiation causes cancer in both humans and animals. The possible role of viruses in the etiology of human cancers is not clear. There have been positive correlations between human hepatic cancer and exposure to hepatitis B virus, between cervical cancer and herpes-virus infections and between a herpes-virus and Burkitt's lymphoma in children in certain parts of Africa. Zur Hausen of the German Cancer Center in Heidelberg, Germany, postulates that 10% of human cancer may be due to viruses. These include HIV and AIDS and Human T Cell Leukemia viruses in Japan and the Carribean and Human T Cell Leukemia. Synthetic and naturally occurring chemicals in the environment, in cigarette smoke, in special occupational processes, in food, in chemically polluted air, and polluted water, also contribute to increased frequencies of human cancers. Therefore, most of our effort is directed in understanding the mechanisms of action of chemical carcinogens.

By mode of action, chemical carcinogens are classified into two groups: direct-acting alkylation agents and chemicals requiring activation (Table 3.1). Direct-acting carcinogens are chemically and biologically reactive by virtue of their structure and are electrophilic in character, that is, electron deficient. These chemicals can interact with the nucleophilic centers in nucleic acids and proteins and are termed ultimate carcinogens [12].

On the other hand, majority of chemical carcinogens are chemically or biologically inert and are termed procarcinogens. Specific spontaneous or biochemical activation reactions convert procarcinogens to either proximal or intermediate carcinogens and to ultimate electrophilic carcinogens. Procarcinogens that require chemical or spontaneous conversion exhibit a wide variety of activities in species and target organs. Procarcinogens that require biochemical conversion may exhibit organ specificity since the conversion depends on specific enzyme systems. Therefore, modifiers of enzymes systems or lack of these systems will dictate the carcinogenic effect of chemicals in a variety of animals as a function of age, sex, strain, and species and organ specificity.

It should also be noted that not all electrophiles are equally active. Spatial arrangements of atoms will determine whether or not a molecule will interact with macromolecules and cell receptors. Lipid solubility of a molecule will determine whether it will be able to pass through the cell membrane to reach targets within the cell. Reactivity of a molecule will determine whether an electrophile will react with water or targets in the cell.

3.4 GENERAL MECHANISMS OF CHEMICAL CARCINOGENS

An overriding theory at present is that a chemical carcinogen, electrophilic in character, will cause a change in DNA, RNA, or protein, resulting in an altered receptor (Figure 3.1) [11,12]. This is known as the initiation phase of carcinogenesis and is thought to be irreversible [4,11,14]. There appears to be substantial evidence on electrophilic interactions with DNA, on the mutagenicity of electrophilic reagents, and on the correlations between their mutagenicities and carcinogenicities of chemicals on their metabolites [15]. These data suggest that a mutation of DNA or other genetic information by an electrophilic form of the carcinogen is the primary step in the initiation phase. However, that does not mean that epigenetic changes of the RNA or protein, are not the primary events but at present the

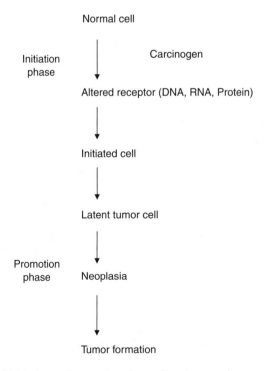

FIGURE 3.1   Schematic of initiation and promotion phase of carcinogenesis.

evidence does not strongly support this initial step. The second phase of carcinogenesis, termed the promotion phase, results in the formation of latent tumor cells, which leads to a neoplastic growth and eventual tumor formation. This phase is considered to be reversible up to a certain point.

There are a number of biological endpoints that result from a mutational event in the initiation phase following exposure to a chemical [16] (Figure 3.2). If the mutational event occurs in germinal cells either cell death or transmission of genetic information can occur. On the other hand, if the mutational event occurs in somatic cells, terata (malformation of the fetus) or neoplasia can result. This scheme does not take into account the excision–repair mechanisms of DNA that are present to correct these mutational events nor the changes in DNA, which have no biologic effect. Therefore, it should be noted that a mutational event would not necessarily produce a neoplastic response.

A more detailed scheme of the possible mechanisms of action is shown in Figure 3.3. The action of the carcinogen in terms of its effect on a biological system is dependent on the species, strain, sex, age, diet, intestinal flora, and health status of the organism [17]. All of these factors or modifiers can play a role as to whether or not a compound will be metabolized, and therefore to the extent to which it can interact with macromolecules in a genetic or epigenetic manner and its consequent ability to form tumors.

The carcinogen (procarcinogen) can undergo metabolism by a mixed function oxidase enzyme system (MFO) to form a series of metabolites through epoxidation, hydroxylation, glucuronidation, sulfation, epoxide hydration, and glutathionation [8,13,17,18]. The MFO system, also commonly called the cytochrome P450 enzyme system, is present in the smooth endoplasmic reticulum and in the nuclear envelope of the cell and is found in a variety of tissues and organs, in particular in liver, lung, and skin. Of these, the liver contains higher levels of the MFO system and therefore has the greater capacity to metabolize these compounds. The series of metabolites formed can be either activating or deactivating; activation of the compound leads to proximate or ultimate carcinogen formation while deactivation leads to detoxification of the compound. One of these metabolites formed, the ultimate carcinogen, is a strong electrophilic reagent that can undergo noncritical binding, spontaneous decomposition, or critical covalent binding with nucleophilic centers in cellular material, such as the nucleic acid bases and the amino acid residue of protein. Figure 3.4 shows the chemical structures and nucleophilic centers in representative nucleobases and amino acids. The binding to DNA, for example, can be eliminated by the DNA repair system if it occurs prior to DNA replication and the cell will retain its normal properties. On the other hand, the particular DNA binding may lead to changes in the genome responsible for transformation, and thereby, an initiated cell will give rise to latent tumor or transformed cells eventually leading to tumor formation.

The positions involved in binding to macromolecules depend on the carcinogen [19,20]; for example, acylating and alkylating agents bind at the exocyclic oxygen (O^6) and nitrogen seven (N^7)

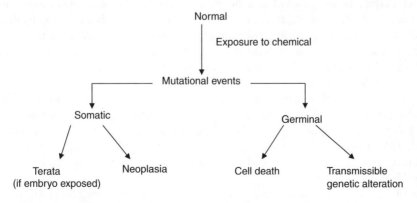

FIGURE 3.2 Schematic of mutational events.

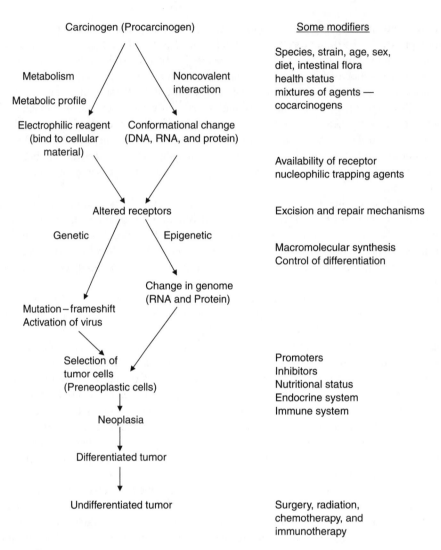

FIGURE 3.3   Schematic of mechanism of action of calcinogens.

positions of guanine while aromatic amines react at the carbon eight (C^8) position and benzo[a]pyrene (BaP) reacts at the 2-amino group of guanine. Binding also takes place at cytosine and adenine but to a lesser degree. Most of the electrophiles affect DNA and RNA in the same way by distorting the helical structure of the nucleic acids. For example, the bond between the guanine base and sugar can be distorted so that the guanine no longer pairs with the adjacent adenine. The adenine is now adjacent to the electrophile and the guanine base is out of helical structure. These chemical changes in nucleic acids affect their normal biological activities.

There are two types of noncovalent binding of carcinogens to macromolecules. Intercalation involves the insertion or stacking of molecules, planar aromatic rings, between base pairs in DNA or RNA. This process will cause the uncoiling of the helical structure and result in a conformational change. A second noncovalent binding involves the carcinogen being inserted along the nucleic acid axis and perpendicular to the nucleic acid bases. The significance of this type of binding is unknown [20].

The covalent binding of carcinogen to proteins has been known for over 50 years even before the nucleic acid structure was determined. However, little is known about the sites affected in proteins, the modified protein structure, and the consequences of such alterations [2,20]. Nevertheless it is

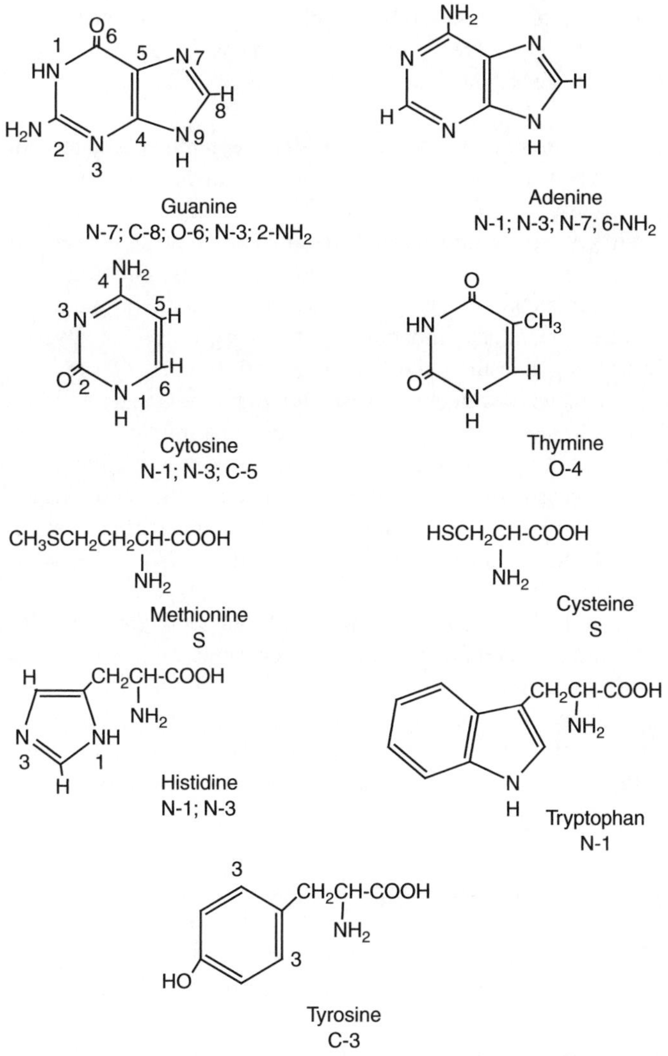

FIGURE 3.4 Examples of cellular nucleophiles and positions that can be involved in binding.

known that aromatic amines and azo dyes bind to proteins with the nucleophilic sulfur of methionine and that dimethylnitrosamine adds a methyl group to nitrogen in the imidazole ring of histidine.

Irrespective of whether it is by covalent or noncovalent binding, an altered receptor results from the interaction of a procarcinogen, proximate, or ultimate carcinogen with macromolecules. The altered receptors, in turn, can lead to genetic or epigenetic effects. Two examples of genetic effects are mutational events and virus activation. The mutational events have been described earlier (Figure 3.2). The activation of a virus as a result of the altered receptor has been suggested as occurring by the derepression of a gene controlling the viral genetic information that is present in normal cellular DNA [2,21]. Epigenetic effects can be manifested in cell expression, such as cellular differentiation, as a result of carcinogen interaction with RNA or protein.

All of the discussions based on the flow diagram in Figure 3.3 has been concerned with the initiation phase of carcinogenesis. Modifiers of this phase in addition to species, strain, age, etc. are cocarcinogens, receptor availability, nucleophilic trapping agents, excision and repair mechanisms, macromolecular synthesis, and control of differentiation. Cocarcinogens, which are noncarcinogenic agents, augment carcinogens through concomitant exposure with a carcinogen. Cocarcinogens primarily influence the initiation phase through an effect on the carcinogen (cell permeability, metabolism, or reparability of the altered receptor) or the target tissue [22]. Aliphatic and certain aromatic hydrocarbons, phenols, and certain hormones are examples of cocarcinogens (see Chapter 2 for more details).

The onset of latent tumor or preneoplastic cells is the beginning of the promotion phase of carcinogenesis. It is unclear when latent tumor cells are formed. All that can be stated is that following the genetic or epigenetic events, cells obtain the capability of being transformed into tumor cells. These preneoplastic cells will give rise to neoplasia, differentiated tumors, and eventually undifferentiated tumors. Modifiers of this phase of carcinogenesis are promoters, inhibitors, nutritional status, the endocrine system, the immune system, surgery, radiation, chemotherapy, and immunotherapy [12].

Promoting agents are those noncarcinogenic agents when given repeatedly following the exposure to a single dose of a carcinogen result in the increase of tumors and the decrease of the latency period of the carcinogen. Based on work with phorbol ester tumor promoters, it appears that these promoters may induce (1) changes in the phenotype of normal cells that mimic features of transformed cells, (2) dedifferentiation of adult epidermal cells, (3) increase protease activity, and (4) stimulation of phospholipid synthesis, macromolecular synthesis, and cell proliferation [22]. On the other hand, tumor inhibitors have the opposite effect of tumor promoters on the promotion phase. Cycloparaffins found in petroleum [23] and butylated hydroxy anisole (BHA) [24] are examples of inhibitors (see Chapter 2 for more details).

In summary, multiple factors are involved in the carcinogenic process [23]. Genetic determinants, such as immune defenses and hormonal imbalances, cultural influences such as smoking and diet and health status have an influence on cancer incidence. Furthermore, in the workplace environment, workers are exposed to single agents and multiple agents in sequence or in combination and complex mixtures. Table 3.2 summarizes the types of chemical agents that can be found in the workplace [25,26]. In various combinations these agents can have additive, synergistic, or inhibitory effects on carcinogenic agents present.

3.5 CLASSES OF ORGANIC CHEMICAL CARCINOGENIC AGENTS

As a class of important chemical carcinogens, organic carcinogens still vary widely in their chemical structures, their mode of action, their source, and the organs in which they cause tumor, etc. It is impossible to discuss these agents in a single group; in this chapter the organic chemical carcinogens are subclassified based mainly on the functional group in the compound: polycyclic aromatic hydrocarbons, aromatic amines, nitrosamines, halogenated compounds, naturally occurring agents, and alkylating agents. The six subclasses will be discussed briefly in terms of their sources, carcinogenicity data, and mechanisms of action.

3.5.1 POLYCYCLIC AROMATIC HYDROCARBONS

3.5.1.1 Source

Polycyclic aromatic hydrocarbons (PAHs) and their derivatives represent one of the major classes of organic molecules known. Chemically they consist of two or more fused benzo rings with theoretically a large number of isomers. It should also be noted that the carcinogenic PAHs are in the size range of four to six rings. PAHs are ubiquitous throughout the environment [27–32]. The primary sources of PAHs are from the combustion of fossil fuels, such as coal, shale, and petroleum, and to a somewhat lesser extent, forest fires, and cigarette smoke. During the pyrolysis process, free radicals are formed that polymerize to form stable aromatized compounds. The higher the temperature the faster these compounds are formed. As the temperature is raised, more unsubstituted PAHs are produced in relation to alkyl substituted PAHs.

TABLE 3.2
Classes of Agents in Workplace

Type	Mode of Action	Example
1. Direct Acting Carcinogen	Electrophilic organic compounds interacts with macromolecules, in particular, DNA	Methyl sulfone
2. Pro-carcinogen	Requires conversion through metabolic activation by type 1	Vinyl chloride, 4&5 membered ring PAHs, β-naphthylamine
3. Inorganic carcinogen	Not directly genetic, lead to changes in DNA by selective alteration in fidelity of DNA replication	Nickel, chromium
4. Solid-state carcinogen	Exact mechanism unknown; usually affects mesenchymal cells and tissues; physical form essential	Asbestos, metal foil
5. Co-carcinogen	Does not alter genetic material and is not carcinogenic; but enhances effect of type 1 or 2 agents when given at same time	Phorbolesters, pyrene catechol, n-dodecane, SO_2, ethanol, sulfur type compounds
6. Promoters	Does not alter genetic material and is not carcinogenic; but enhances effect of type 1 or 2 agent when given subsequently	Phorbolesters, phenols, saccharin, bile acids, anthralin, dodecane
7. Hormone	Mainly alters endocrine system balance and differentiation, often acts as promoter	Diethylstilbestrol estradiol
8. Inhibitor	Does not alter genetic material but suppresses effect of type 1 or 2 agent when given at same time or subsequently	BHT, BHA, cyclo-paraffins
9. Mixture	Additive, synergistic or inhibitory effect of carcinogen with co-carcinogen, tumor promoter and inhibitor	Shale oil, petroleum and coal conversion products

BaP and dibenz[a,h]anthracene were the first two PAHs isolated from coal tar in the 1930s. Since then hundreds of PAHs have been identified in the air, water, soil, and in the by-products from fossil fuel combustion. In addition to PAHs, which contain only carbon and hydrogen, heterocyclic analogs containing one or more nitrogens, sulfurs, or oxygens have also been identified in these mixtures [27,28,30,32]. Some of the compounds that have been identified include the following:

Anthracene
Benz[a]anthracene (WC, weak carcinogen)
Benz[c]acridine
Benzo[a]pyrene (strongly carcinogenic, SC)
Benzo[b]fluoranthene (carcinogenic, C)
Benzo[c]phenanthrene (SC)
Benzo[e]pyrene
Benzo[j]fluoranthene (C)
Benzo[j]fluoranthene (C)
Dibenzo[a,1]pyrene (SC)

Chrysene (WC)
Dibenz[a,h]acridine (C)
Dibenz[a,j]acridine (C)
Dibenz[a,h]anthracene (SC)
7H-Dibenzo[c,g]carbazole (SC)
7,12-Dimethylbenz[a]anthracene (SC)
Fluoranthene
Indeno[1,2,3-cd]pyrene (C)
Naphthalene
Pyrene

The chemical structures of selected carcinogenic PAHs are shown in Figure 3.5.

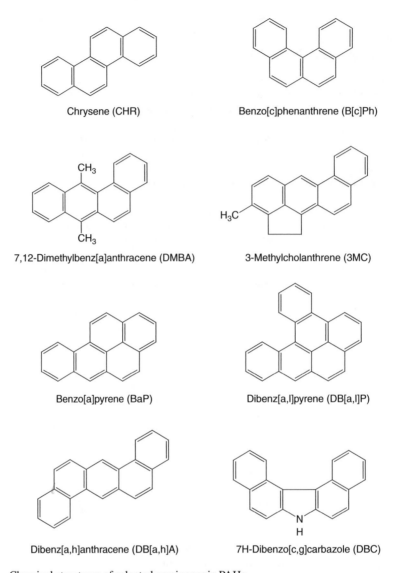

FIGURE 3.5   Chemical structures of selected carcinogenic PAHs.

A strongly carcinogenic BaP is used as the indicator of potential hazard due to its ease of detection in mixtures. In a 1972 report [33], it was stated that significant amounts of BaP in tons per year were produced and emitted into the environment from a variety of sources: cars — 20; heat and power generation — 500; refuse burning — 600; coke production — 200. The total concentration of PAHs in air has been measured to be as high as 100–150 ng/m³ especially in cities using coke ovens while cities not using coke ovens have levels of 1–2 ng/m³. BaP levels during the winter in various cities in the United States, Australia, South Africa, and Europe have varied from 0.6–104 ng/m³ to 4 ng/m³ in summer. The levels of BaP in drinking water have varied from 1.2–2.4 ng/m³ to as high as 35 mg/m³ in sewage water. Levels as high as 200 mg of PAHs per kilogram of dried soil have been found near an oil refinery plant whereas soil samples in rural areas have levels of 2 μg/kg dried soil. In particular, benz[a]anthracene, a weak carcinogen and a potential indicator compound, has been found in concentrations per kilogram soil near an industrial center, highly traversed highway, and in soil polluted with coal tar pitch at 390 μg, 1500 μg, and 2500 mg, respectively [27,28,30,32].

Finally, cigarette smoke contains approximately 150 PAHs in the gas phase and more than 2000 PAHs in the particulate phase. In one cigarette there are 10–50 ng of BaP and 12–140 ng of benz[a]anthracene. A person who smokes one pack of unfiltered cigarettes per day inhales 0.7 μg of BaP per day.

3.5.1.2 Carcinogenicity

Many PAHs are unique in their ability to produce tumors in the skin and in most of the epithelial tissues of practically all animal species tested and in particular in mice, rats, and hamsters [27,28,30,32]. Tumor formation can be induced by acute exposures to microgram quantities of PAHs. The latency periods can be very short, 8 to 10 weeks, and the tumors resemble human carcinomas. The major concern in carcinogenicity studies has revolved around the effects of the carcinogens on the skin or lungs. With perhaps one or two exceptions, PAHs have not been found to be carcinogenic to the liver. It should be noted that there are no documented cases of human cancer due to PAHs alone. However, there are numerous cases of human skin cancer associated with complex mixtures containing these PAHs. The PAHs present in these mixtures, such as coal tar and shale oil appear to be partly responsible for the carcinogenic activity of these materials [29,34–36].

In the first part of the 20th century, Japanese investigators (Yawagiwa and Ichikawa) were able to induce tumors on rabbit ears at the site of application with undiluted coal tar. This was the first study of its kind in experimental carcinogenesis. It was not long after when it was recognized that PAHs were present in coal tar. Benzo[a]pyrene and dibenz[a, h]anthracene were isolated from coal tar and these substances along with chemically synthesized PAHs were found to be carcinogenic [28,30]. Of the many PAHs studied, the following PAHs have been investigated in detail:

Benz[a]anthracene (WC)	Dibenz[a,h]anthracene (SC)
Benz[a]fluoranthene (C)	Dibenzo[c,g]carbazole (SC)
Benzo[a]pyrene (SC)	7,12-Dimethylbenz[a]anthracene (SC)
Benzo[c]phenanthrene (SC)	Indeno[1,2,3-cd]pyrene (C)
Benzo[j]fluoranthene (C)	3-Methylcholanthrene (SC)
Chrysene (WC)	Dibenzo[a,l]pyrene (SC)

Like the PAH levels, BaP has been used as the indicator compound in term of relative carcinogenicity [35].

One of the two major routes of exposure to PAHs has been by inhalation. It has been shown that in an environmental or occupational setting PAHs are usually absorbed on inert particulate, which slows the rate of decomposition of the PAHs in the presence of light [29,37]. Therefore, PAHs apparently can reach the lungs without decomposition. Particulate also appears to retain the PAHs in the lung for longer periods of time and enhances its carcinogenicity. Particulates, such as ferric oxide [38] and carbon black [39] have been shown to enhance the carcinogenicity of the respiratory tract carcinogen BaP in hamsters and rats, respectively. The only other compound studied under these conditions in the presence of ferric oxide is 7H-dibenzo[c,g]carbazole (DBC) that resulted in an even stronger carcinogenic response in the respiratory tract of hamsters than BaP [40].

The second major route of exposure to PAHs has been by absorption through the skin. In environmental and occupational settings, light appears to alter the biological effects of PAHs on skin [41]. It has been shown that light will enhance the carcinogenicity of BaP on the skin of mice but will have no effect on inducing a carcinogenic response in noncarcinogenic PAHs, such as pyrene and benzo[e]pyrene [42].

In summary, there are many PAHs that are potent carcinogens in laboratory animal studies. Therefore, even though there are no documented human cases of cancer directly attributed to PAHs on their own, there is overwhelming evidence that they contribute to the overall carcinogenic risk for occupational workers exposed to complex mixtures [27–30].

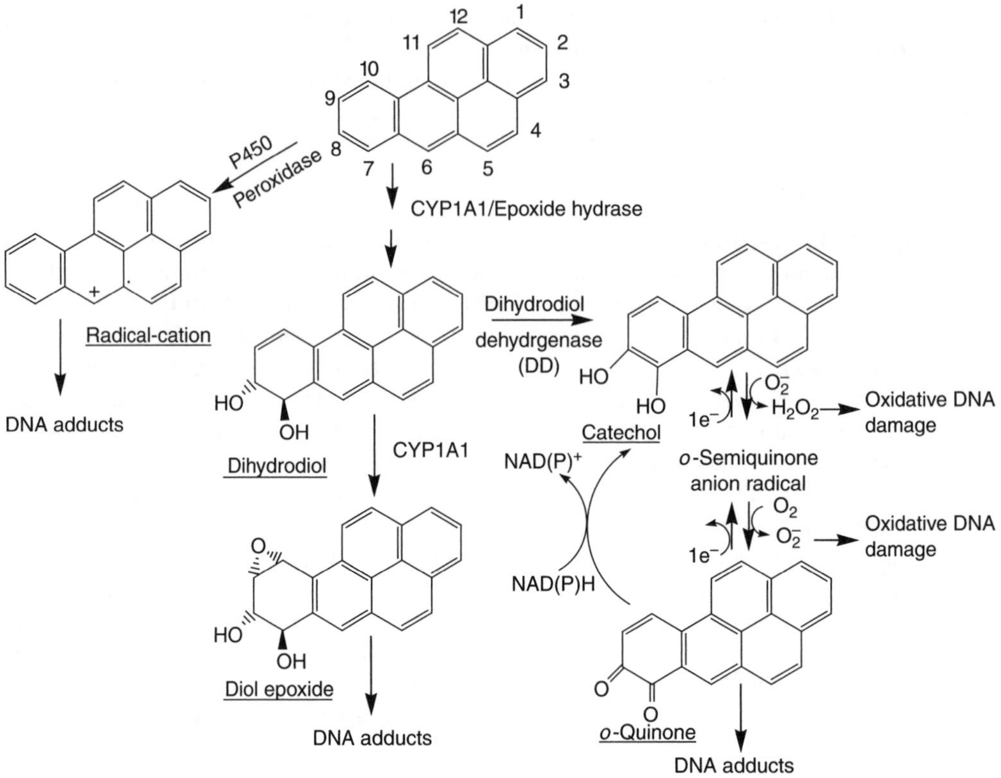

FIGURE 3.6   Metabolic activation pathways of PAHs (Bap as shown).

    PAHs are chemically and biologically inert. They require activation to form electrophilic moieties capable of binding to cellular nucleophilic target to initiate their tumorgenic process. This activation is achieved by complex metabolic routes involving several enzyme-mediated oxidative steps.

3.5.1.3 Mechanisms of Action

In all the PAHs studies conducted, BaP has been investigated in great detail. *In vivo* studies have shown that activated PAHs, and in particular BaP, bind to cellular components, that is, DNA, RNA, and protein.

    Three major metabolic activation pathways have been proposed. The first involves the formation of diol epoxides catalyzed by cytochrome P450s, where P450 1A1 plays a dominant role [43–46]. A second route of metabolic activation is the formation of radical cations catalyzed by P450 peroxidase [47,48]. A third mechanism of metabolic activation for PAH is the formation of reactive and redox active *o*-quinones catalyzed by dihydrodiol dehydrogenase (DD) [49,50] (Figure 3.6).

    1. Activation to diol epoxide was the earliest proposed and widely accepted pathway of PAH activation. In BaP, for example (Figure 3.7) initial epoxidation occurs mainly on the terminal benzene ring by P450s to form an epoxide that is then hydrolyzed to form the *trans*-7,8-dihydrodiol of BaP by epoxide hydrolase as the proximate metabolite [32,51,52]. Further epoxidation by P450 1A1 gives the ultimate metabolite diol epoxide. The antiisomer (+) (7R,8S)-dihydroxy-(9S,10R)-epoxy-7,8,9,10-tetrahydro BaP [(+) anti-BPDE] was found to be the most mutagenic metabolites of BaP [53–55] and the most tumorigenic in mice [56–58]. BPDE binds to DNA *in vitro* and *in vivo* and the DNA adducts formed are isolated and chemically elucidated; the products resulting from reaction of the exocyclic

amino group (N^2) of deoxyguanosine residue with C-10 of the BPDE, that is, BPDE-10-N^2-dG, are the primary adducts [44,59–61] (Figure 3.7). Furthermore, anti-BPDE reacts with the *p53* tumor suppressor gene leading to adduct formation in codons 157, 248, and 273 [62–64].

2. The pathway of PAH activation to form radical cation intermediates is catalyzed by P450 peroxide in 3-methylcholanthrene-induced rat liver microsomes through one-electron oxidation [47,48,65]. The most electrophilic site is the position with maximum charge density in a radical cation of PAH, for example, C6 on BaP, from where it binds to the nucleophilic centers in DNA to form covalent adducts [47,66]. It is reported that the adducts formed are unstable and undergo spontaneous cleavage of the glycosidic bond to produce depurinating adducts [47]. For BaP the major adducts are the N7 position of adenine (A) bound to C6 of BaP (BaP6N7A) and N7 and C8 positions of guanine (G) bound to C6 of BaP (BaP6N7G and BaP6C8G) [47,65,66]. In *H-ras* oncogene and *p53* tumor suppressor gene, G to T transversions are observed.

The value of ionization potential (IP) of the PAH is a factor that plays an important role in activation by one-electron oxidation. Above certain IP it is more difficult to remove one electron from the aromatic system to form a radical cation. The cutoff IP above which the activation by one electron oxidation is unlikely, is reportedly to be about 7.35 eV [48,67].

3. Activation to *o*-quinones is a newly proposed mechanism of PAH which involves the formation of reactive and redox active *o*-quinones by DD [49,50]. Rat and human DD successfully compete with P450 for the trans-dihydrodial intermediate of PAH [49,68]. These diols undergo $NADP^+$-dependent oxidation to produce catechols, which are unstable and undergo autoxidation in the air to give the fully oxidized *o*-quinones and superoxide anion (O_2^-) through the formation of an *o*-semiquinone anion radical (SQ) (Figure 3.5) [50,69]. The *o*-quinones of PAHs have been found to be highly reactive Michael acceptors that bind to DNA and have the potential to cause the G to T transversions observed in *ras* and *p53* genes [70,71]. The formation of reactive oxygen species (ROS) during the activation process of PAH is significant. ROS may cause oxidative damage on DNA to produce 8-hydroxy-deoxyguanosine [72–74]. In radiation-induced carcinogenesis, ROS are found to be a causative factor [75].

Besides homo-PAHs, the heterocyclic analogue, for example, 7H-dibenzo-[c,g]carbazole (DBC) reportedly undergoes the *o*-quinone pathway *in vitro* to produce a 3,4-dione derivative which interacts with nucleic acid bases giving adducts at the C-1 position of DBC through 1,3-Michael addtion [76].

In summary, it is likely that all the three pathways diol epoxides, radical cations and *o*-quinones contribute to the metabolic activation of PAHs to generate tumors. The relative significance of each mechanism depends on the nature of the individual PAH (e.g., chemical structure, IP value, etc.) and may also be related to the animal species, the tumor site, and the level of the activation enzyme expression.

3.5.2 AROMATIC AMINES

3.5.2.1 Source

The aromatic amines and amides (AA) represent a large family of compounds that have wide utilization in the industrial setting and in particular as intermediates in the manufacture of a wide variety of dyes that are used in the textile, leather, plastic, and paper products, and in permanent and semipermanent coloring products [77–79]. AA present in cigarette smoke appear to account for the majority of DNA adducts detected in the uroethelium of smokers and are therefore considered to be the primary etiological agents in human urinary bladder cancer [80]. Heterocyclic AA are formed in cooked foods and released into the environment. The exposure to these chemicals occurs primarily through skin contact and inhalation. The most important members of the series from a standpoint of industrial epidemiology are 2-naphthylamine, benzidine, 4-aminobiphenyl, 4,4′-methylenebis(2-chloroaniline), 4,4′-methyllenedianiline, 3,3′-dichlorobenzidine, diphenylamine, *o*-tolidine and *o*-dianisidine (Figure 3.8), as well as azo dyes derived from these amines. Among

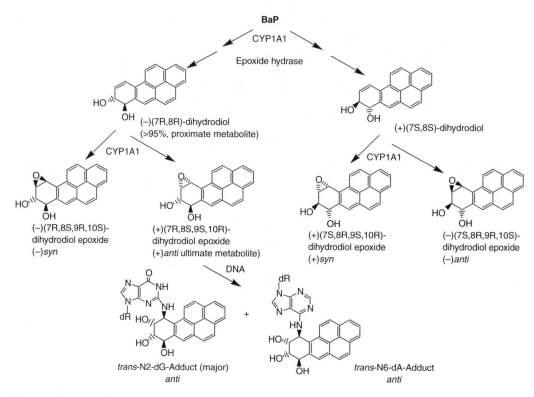

FIGURE 3.7   Configurational isomers of DNA adducts of BaP-diol epoxide.

these are anilines that have proven responsible for a majority of, if not almost all, the known incidence of industrial bladder cancer. In particular, 2-naphthylamine, benzidine, and 4-aminobiphenyl are proven human carcinogens while the others are highly suspect human carcinogens.

3.5.2.2 Carcinogenicity

Aromatic amines and amides were some of those early chemicals found to be carcinogenic in humans and in experimental animals. A correlation between aromatic amine exposure and human bladder cancer was reported first by Rehn in the late 19th century. He attributed the bladder cancer to aniline, which was used in the making of magenta [77,81–83]. Later the chemicals attributed to this industrial disease were benzidine, 2-napthylamine, and 1-napthylamine. Epidemiological evidence based on the British chemical industry between 1921 and 1950 showed that benzidine and 2-naphthylamine were directly related to the incidence of bladder cancer and that aniline induced excess bladder tumors only when used in the manufacture of auramine and magenta but not in other instances. It appeared that just a 6-month exposure to 2-naphthylamine increased the risk of bladder cancer while five years or more were necessary for 1-naphthylamine. The average latency period benzidine and 2-naphthylamine was 17–18 years and 22 years for 1-naphthylamine. Since 1-naphthylamine at the time of the study contained 4–10% of 2-naphthylamine, it is now believed that the effects of the 1-isomer are due to its contamination with the carcinogenic 2-isomer.

One other chemical, 4-aminobiphenyl shown to be a human carcinogen, was used in the United States between 1935 and 1954. Of a population of 315 men exposed to 4-aminobiphenyl, 53 of them developed bladder tumors [77,81–83].

In addition to the epidemiological data, experimental animal models for aromatic amines were developed in the 1930s. Hueper et al. [84] showed that 2-naphthylamine was a bladder carcinogen

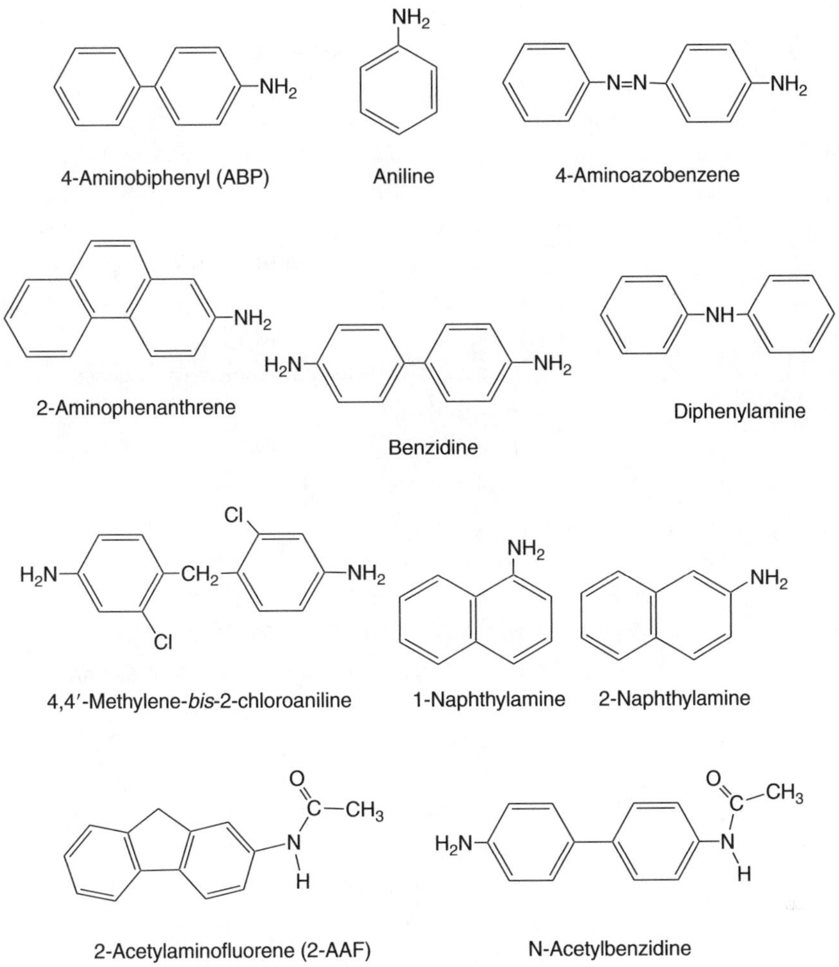

FIGURE 3.8 Structures of carcinogenic aromatic amines and guide.

in dogs and that aromatic amines did not produce tumors at the site of administration but in distant sites such as liver, intestine, or urinary bladder. It also appeared that aromatic amines were species specific. For example, 2-naphthylamine was carcinogenic in the bladder of man and dog but weakly carcinogenic in rat and rabbit [77].

The effects of chronic administration of primary amines on the bladder of dogs indicate that 4-aminobiphenyl and 2-naphthylamine are potent bladder carcinogens in dogs as they are in man [83]. Carcinogenicity of these and other compounds using other species are reported [77,81–83]. Finally, it should be noted that aromatic amines in coal and shale-derived petroleum substitutes make a major contribution to the overall mutagenicities of these mixtures, and therefore, may be responsible for increased biological activity of these substitutes relative to petroleum [85,86].

3.5.2.3 Mechanisms of Action

The different carcinogenecities of aromatic amines in various animal species implies that metabolic activation is required for this group of chemicals to exert their genotoxic effects [5]. Generally, evidence has demonstrated that N-hydroxylation is the first step [77,87,88]. This reaction may be followed by esterification of the hydroxyl group in some but not all cases. The acetoxy group is a

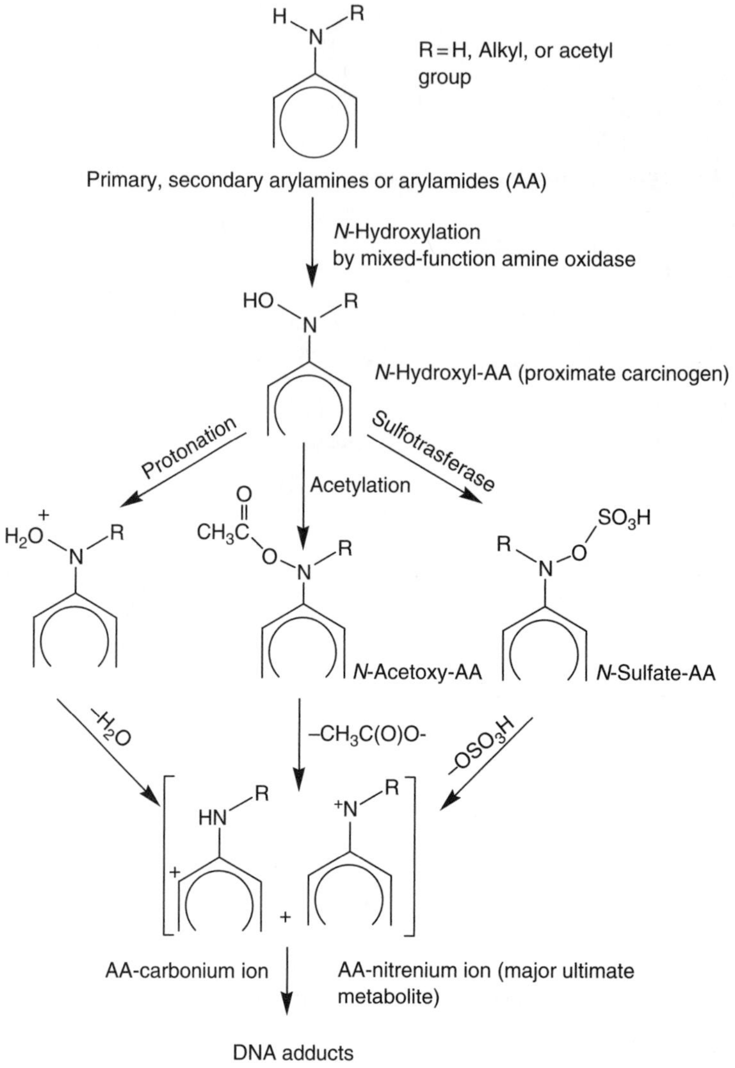

FIGURE 3.9   Metabolic activation pathways of aromatic amine and amide.

chemically good leaving group, leading the esterified N-hydroxyl amine intermediate to break down to form an electrophilic nitrenium ion that interacts with DNA (Figure 3.9). Oxidation of primary or secondary aromatic amines (AA) to N-hydroxy derivatives is the major route in animals occurring within the endoplasmic reticulum. These reactions are considered to be involved in the phase I metabolism [14], which are catalyzed by mixed-function amine oxidase. These metabolic activation reactions introduce molecular oxygen to a substrate that converts it to a more polar intermediate, N-hydroxylated AA, in this case. When the hydroxylation occur on aromatic ring, it is considered to be a detoxification route [2]. 2-Acetylaminofluorene (2-AAF), as an example of AA, has been reported to undergo both N- and ring hydroxylation [77]. Some amines cannot be enzymatically N-hydroxylated and hence are noncarcinogenic. Generally, N-hydroxy compounds are not usually the ultimate carcinogenic form of AA under physiological conditions but are proximate metabolites and further metabolism is needed. The N-hydroxylation of AA can be followed by esterification of the N-hydroxyl group to yield highly reactive intermediates capable of interaction with nucleophilic sites in cellular macromolecules [11].

A variety of esters of AA have been identified. Esterification by acetate is a common conjugation mechanism in the metabolism of AA. In the reaction of N-hydroxy-2-AAF with acetyl CoA, N-acetoxy-2-AAF is formed [77]. This compound is a potential carcinogen subcutaneously; tumors are induced at the injection site and it reacts *in vitro* with cellular macromolecules to yield covalently bound adducts [77,87]. However, N-acetoxy-2-AAF does not appear to be the one actually generated within the cell. It has been shown that N-hydroxy-2-AAF could be esterified with 3-phosphoadenosine-5-phosphosulfate by an enzyme in the soluble fraction of the liver cell i.e. sulfotransferase, to give N-sulfate-2-AAF [77,89,90]. The toxicity and reactivity of N-hydroxy-2-AAF are increased by simultaneous administration of sulfate ion. The fact that depletion of the intracellular sulfate level with acetanilide reduces the carcinogenicity of 2-AAF was discovered as a confirmatory evidence for the involvement of the N-sulfate in the mechanism of carcinogenesis [77]. Phosphate esters, glucuronate esters, and carbonate esters may also be involved in the activation mechanism of AA [77,89,90].

Based on experimental results with model system and reaction kinetics, the covalent binding of electrophilic AA metabolites to DNA *in vivo* can result in both N- and aromatic ring adduction at nucleophilic sites of deoxyguanosine (dG) and deoxyadenosine (dA) bases [89,91,92]. The predominant DNA adducts formed with majority of AA carcinogens are C8-dG-N-arylamine or arylamide derivatives, and minor include N^2-dG, O^6-dG, C8-dA, or N^6-dA derivatives [93].

3.5.3 N-NITROSO COMPOUNDS

3.5.3.1 Source

In 1962, Druckery et al. reported that from tobacco amines and nitrous gases, N-nitroso compounds were formed as a group of carcinogens [94]. N-nitrosamines are found widespread in the environment. Concentrations in the nanograms to micrograms per unit volume or mass range have been reported in air, water, soil, foodstuffs, plants, and cigarette smoke. For example, N-nitrosodimethylamine (DMN) has been found in soil samples at 1–8 μg/kg, wastewater at 3–4 μg/l, drinking water at 0.1–0.5 μg/l, food at 1–10 μg/kg, and mushrooms at 0.4–30 μg/kg. Nitrosamines, including N-nitrosonornicotine, are present in cigarette smoke up to 10 μg/cigarette. Compounds, such as DMN, N-nitrosodiethylamine (DEN), N-nitrosopyrrolidine, and N-nitrosopyridine have been detected in meat products at 1–80 ppb and nitrosamines have been found in fresh, salted, smoked, or fried fish at 1–26 ppb, and in fish meal at 0.12–0.45 ppm [82,94–99].

In the industrial settings prior to 1976, DMN was used as an intermediate in the production of 1,1-dimethylhydrazine, a storable liquid rocket fuel that contained 0.1% DMN as an impurity. DMN has also been used as a solvent in the fiber and plastics industry, as an antioxidant, a softener of copolymers, and as an additive for lubricants. Workers in dimethylamine and dimethylhydrazine manufacturing plants, leather, oil and gas extraction, and pesticide workers are likely to be exposed to DMN. 640 μg/l of DMN has been found in the pesticide 2,3,6-trichlorobenzoic acid that contains dimethylamine. DMN has been found in the air above an industrial plant producing dimethylamine ranging from 0.001–0.43 ppb. DEN is used in plastics as a lubricant additive and as a minor antioxidant has been found in wastewater from chemical plants up to 0.24 μg/l. Cutting fluids have been reported to contain up to 3% of N-nitrosodiethanol amine (NEA) while pesticides such as atrazine contain 0.5 μg/kg NEA [100].

Since the formation of N-nitroso compounds can be accomplished through reactions of secondary or tertiary amines and nitrous acid, there is increasing evidence of an etiologic role for endogenously formed N-nitroso compounds in the development of certain human cancers [101]. The nitrosating compound could be formed by the reaction of gastric juice with nitrite compounds that are present in foods and are high in some vegetables [8,93,100,102]. Nitrates that are more widely distributed in the environment may be reduced to nitrites under conditions of low acidity or during storage by nitrosoreductase containing bacteria. Nitrates are also present in the water supply and meat curing process and are converted to nitrites by bacteria or *in vivo* in the human body in the mouth, stomach, intestine, or esophagus [20,102]. It has also been suggested that nitrites can be formed *in vivo* via heterotrophic nitrification of ammonia and other reduced nitrogen compounds in the distal ileum, although this is debatable [99,103]. In the United States the average mean intake of nitrate plus nitrite on a daily basis is 120 mg [104].

Secondary and tertiary amines, quaternary ammonium compounds, ureas, carbamates, and guanidines are present in vegetables, beer, tea, fish, some drugs and pesticides, and tobacco and tobacco smoke. Following the ingestion of foodstuffs, 15.3 mg of dimethylamine and 5.7 mg piperidine and pyrrolidine have been measured in the urine over a 24-h period [100].

N-nitroso formations have been shown to occur from the reaction of nitrites and amines *in vitro* with animal and human gastric juices. It has been demonstrated with rather convincing evidence that the formation of nitrosamines or nitrosamides do occur *in vivo* [100,102]. Nitrosamines may also be formed from secondary amines and nitrites at neutral pH by nitrate reducing bacteria, including those found in the intestinal tract although more recent evidence would tend to indicate that *N*-nitroso compounds are not formed in the lower gastrointestinal tract [100,105]. On the other hand, bacteria have been shown to also degrade nitrosamines to the parent amines and nitrites. Ascorbic acid and vitamin E apparently inhibit nitrosamine formation by possibly reacting with nitrites whereas thiocyanate anions tend to catalyze nitrosamine formation [100,102].

3.5.3.2 Carcinogenicity

In the 1930s, two cases of severe liver toxicity were observed in chemists preparing DMN. Subsequently in the 1950s it was found that DMN was also a potent hepatocarcinogen in rats that were being studied for chronic DMN toxicity. These initial studies uncovered a new class of chemical carcinogens and are described in great detail in references [82,94,100].

The *N*-nitroso compounds can be divided into two major categories namely nitrosamines, such as DMN and DEN and nitrosamides such as *N*-methylnitrosourea (MNU) and *N*-ethylnitrosourea (ENU). Of 130 different nitrosamines that have been tested for carcinogenicy in rats, 106 induced tumors in 14 different organs; in hamsters, 41 nitrosamines generated carcinogenic responses in 8 different organs [106]. The major sites of cancer production are liver, stomach, intestine, bladder, esophagus, and to a lesser extent the respiratory tract, and the nasal cavities. A variety of species have been used in carcinogenicity studies, which include monkey, rodent, rabbit, bird, and rainbow trout [100]. It is apparent that there are a wide range in their carcinogenic specificity in terms of species and tissue susceptibility and the degree of carcinogenicity [100,107]. It appears that the carcinogenic potency and toxicity of *N*-alkyl nitroso compounds is inversely proportional to the size and is proportional to the number of alkyl substituents present. For example, DMN is more toxic than di-*n*-arylnitrosamine while methylpropylnitrosamine, an esophageal carcinogen in rat, is more potent than methyldecylnitrosamine, a bladder carcinogen in rat [100,107]. These variations probably relate to the relative chemical stability of compounds under physiological conditions and the inducibility of the drug metabolizing enzymes in the tissues being tested. Furthermore the apparent high degree of organ specificity of many nitroso compounds may be related to the relative susceptibility of a given tissue, which can be a function of the ability of specific enzymes to activate the compound for decomposition in the proximity of the target site [7,107]. Finally, it should be noted that single doses of potent nitroso compounds such as DMN, DEN, and *N*-methyl-*N*-nitro-*N*-nitrosoguanidine (MNNG) administered to a day-old mice, hamster, and rat respectively produced tumors in the adults [100].

So far there is no conclusive epidemiological evidence that a single or groups of nitrosamines are carcinogenic to humans. However, the overwhelming consistency of tumorigenic responses in all animal species tested suggests no reason that humans are exclusive [106,108]. In fact, certain nitrosamines have been suggested as a contributing factor in stomach and esophageal cancer in man in certain parts of China, Iran, and parts of eastern and southern Africa. It has been suggested that all of these regions have low intake of foods with vitamin C, which has been shown to inhibit nitrosation *in vitro* [100,102,108].

3.5.3.3 Mechanisms of Action

When considering mutagenicity, carcinogenicity, and other biological effects of N-nitroso compounds, clear distinction is demonstrated between the two categories of compounds, nitrosamines, and nitrosamides. N-nitrosamides give rise to mutagenic effects in almost all genetic indicators, while nitrosamines do not show such activities in bacteria, yeast, or fungal assay systems, but they are active in higher organisms [100]. Similarly, nitrosamines do not induce tumors at the site of application but rather produce tumors at a distant target site. Nitrosamines are stable chemically when not exposed to UV light. On the other hand, nitrosamides often produce tumors at the site of application. They are inherently chemically unstable in aqueous media, and therefore, decompose spontaneously or catalytically under physiological conditions to produce reactive intermediates [100]. However, N-nitrosamines require metabolic activation to generate the active moiety catalyzed by activating enzymes that the simple cells do not have. The activation pathways of N-nitrosamines presently considered to be most likely are shown in Figure 3.10. Early studies on DMN established that the first step in the metabolic activation is oxidation by the mixed function oxidase system to give α-hydroxylation [100,109]. The formation of α-hydroxyalkylnitrosamines mediated by cytochrome

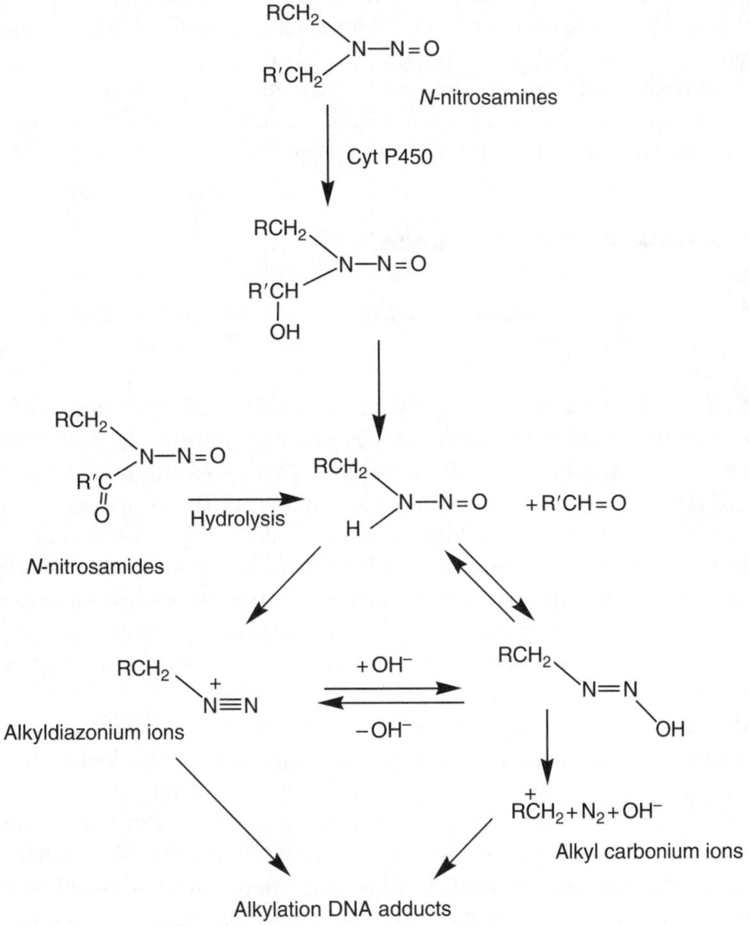

FIGURE 3.10 Metabolic activation pathways of N-nitroso compounds.

P450 was reported in animals [110] and in humans [111]. The α-hydroxyalkylnitrosamines would then be easily hydrolyzed to form the corresponding aldehyde and monoalkylnitrosamines and the tautomeric diazohydroxide. Loss of nitrogen leads to the formation of alkylcarbonium ions, the electrophilic alkylating species considered as the ultimate carcinogen [100,112].

Another proposed pathway suggested that α-hydroxyalkylnitrosamines decompose spontaneously to give aldehydes and alkyldiazonium ions which are considered to be the ultimate alkylating agent [113,114]. The actual nature of the interaction with DNA bases remains unclear. Evidence suggests that alkyldiazonium ions, but not carbonium ions are the reactive moieties to bind to DNA [112]. The chemical instability of N-nitrosamides is thus comparable with the α-hydroxynitrosamines that have been described earlier. An identical degradation scheme would follow the nonenzymatic hydrolysis of the labile nitrosamide bond resulting in the formation of alkyldiazonium ions and alkylcarbonium ions binding to nucleophilic centers in macromoleculars (see Figure 3.10).

It appears that the N-7 and O-6 of guanine and the N-3 of adenine are the predominant sites of alkylation by N-nitrosamines to form DNA adducts [87,100,112]. The apurinic sites produced by glycosylase action may lead to mutations [112]. It has been reported that DMN-induced kidney tumors in rats had a very high incidence (88%) of mutated $p53$ genes with GC–AT transversions at codons 204 and 213 [115] and MNU-induced rat mammary carcinomas had a high incidence of c-H-ras oncogenes with GC–AT transversions at the 12th codon [112]. Higher di-n-alkyl amines with more than two carbons per alkyl group and cyclic nitrosamines may involve β- or γ- hydroxylation. Relevant pathways might be minor in nature and major routes of metabolism may be detoxification pathways. Small changes in chemical structure might lead to large differences in the pharmacokinetics and the carcinogenic potency [107,116,117].

Dimethylhydrazine, a weak carcinogen, and cycasin, present in the cycad nut, a carcinogen in animal studies and a neurotoxin in humans are similar in structure to nitrosamines and appear to undergo similar metabolic activations [100].

3.5.4 HALOGENATED ORGANIC COMPOUNDS

3.5.4.1 Source

Halogenated hydrocarbons are ubiquitous in the environment [118–121]. Chloroform, first synthesized in 1831, was used as a general anesthetic in 1847 and along with other chlorinated hydrocarbons was used in the development of modern surgery. In the past 50 years, chlorinated hydrocarbons have been incorporated into medicines such as antiseptics, diuretics, sedatives, and antihistamines, and as anesthetics in routine medical surgery. With the advent of World War II, second generation pesticides were introduced into the agricultural arena, such as dieldrin, DDT, lindane, aldrin, chlordane, and heptachlor, and so on, and they are all halogenated organic compounds. These compounds led to the damage of nontarget organisms and persisted in the environment, while insects developed resistance to them. Along with these insecticides, halogenated hydrocarbons were synthesized as fungicides and herbicides. As a by-product in the preparation of the so-called orange agent (2,4,5-trichlorophenoxy acetic acid), 2,3,7,8-tetrachloro-dibenzodioxin (TCDD) has been found to be an extremely toxic compound causing cancer and reproduction problems [121–123].

The industrial and commercial applications of the halogenated hydrocarbons are pervasive. They are used as solvents, fire retardants, dry-cleaning fluids, degreasing agents, fuel additives, fumigants, and intermediates for the synthesis of a multitude of compounds. Other compounds, such as poly-chlorinated biphenyls (PCBs) are used as dielectric fluids in the construction of transformers and capacitors. These compounds are extremely stable and are not hydrolyzed by water, acid, or alkali and can withstand temperatures up to 650°C without decomposition and therefore are resistant to degradation in the environment. They are well suited for lubricants, hydraulic fluids, liquid seals, cutting oils, vacuum diffusion pump oils, and as vapor suppressants for insecticide formulations.

Similarly, chlorofluorocarbons are used as refrigerants and have also been used as aerosol propellants for the dispersal of household and industrial products. Some polymers containing halogens are used as fireproofing, while others such as neoprene are resistant to solvents, oils, and weathering. Polyvinyl chloride is most widely used in the plastics industry in which chlorinated hydrocarbons occur in films used as food wrapping, latex paint, and rubber. Due to wide usage of the halogenated hydrocarbons, a number of these compounds have been found in drinking water, such as vinyl chloride, vinylidene chloride, 1,2-dichloroethylene, 1,1,2-trichloroethylene, tetrachloroethylene, and hexachlorobutadiene.

3.5.4.2 Carcinogenicity

Since the time vinyl chloride was found to be carcinogenic, there has been a great effort to study a series of related saturated and unsaturated halogenated short-chain hydrocarbons [121,124]. The National Cancer Institute conducted a study that involved testing a series of these compounds using both mice and rats of both sexes [121,125]. Some of the more interesting results indicated that 1,2-dibromoethane and 1,2-dibromo-3-chloropropane produced gastric tumors in mice and rats, chloroform produced hepatocellular carcinomas in mice and kidney tumors in rats, 1,2-dichloroethane induced liver and lung tumors in mice, and 1,1-dichloroethane, 1,1,2-trichloroethane (TCE) produced hepatocellular carcinomas in mice [121,126]. Based on these results the coffee industry opted for a substitute for TCE that was used in their manufacturing process. Of the three or four compounds that were considered, dichloromethane was eventually substituted for TCE.

Due to the fact that vinyl chloride has been found to be a human carcinogen, with the production of angiosarcomas of the liver, a number of other compounds are under intensive investigation which are similar in structure. Among these are vinylidene chloride, tetrachloroethylene, allyl chloride, 1,3-dichloropropene, chloroprene, and hexachlorobutadiene [121,126–128].

Organochlorine pesticides such as chlordecone, heptachlor, hexachlorabenzene, hexachlorocyclohexane (lindan) and toxasphene and so on have been found to be carcinogenic in two or more species of rodent while aldrin, chlorine, dieldrin and DDT and so on are carcinogenic in mice [121,129]. Most of these pesticides are banned from use. Finally, it has been postulated that when in the upper atmosphere chlorofluorocarbons will photochemically decompose releasing chlorine atoms, which would then destroy the ozone layer. Depletion of the ozone layer would lead to an increased incidence of skin cancer since this layer shields the earth's surface from shortwave UV light, which is known to cause cancer. Based on this premise and other scientific data, the fluorocarbons were banned as an aerosol propellant [118].

3.5.4.3 Mechanisms of Action

Generally, the mechanism of action of halogenated hydrocarbons has not been intensively investigated as other classes of carcinogens, such as PAHs, AA, and nitrosamines. As a subgroup of the halogenated hydrocarbons, methyl[ethyl] halides and halomethyl-substituted aromatic hydrocarbons (arylalkylhalides) are direct acting electrophiles and therefore require no metabolic activation to bind to nuclei acid constituents. *In vitro*, these agents react predominately with ring nitrogens on adenine, guanine, and cytosine bases in DNA to form N-1-aralkylated adenine, N-7 guanine and N-3-cytosine [128]. From various organs of mice treated intravenously (i.v.) with benzyl chloride, N-7 benzylguanine was detected [130]. The short chain polyhalogenated hydrocarbons are also the most extensively studied compounds of this class [126]. An example of this is the mechanism of activation of 1,2-dihaloethane. It has been reported that the major metabolic pathway of 1,2-dihaloethanes for the generation of reactive, electrophilic species involves glutathion (GSH) conjugation which is responsible for DNA adduct formation [131,132]. Oxidation of 1,2-dihaloethanes produces a 2-haloacetaldehyde which can bind to DNA to give rise to ethano adducts [133]. In the liver, kidney, spleen, and stomach of rats treated with 1,2-difromoethane, the adduct S-[2-(N-7-guanyl)ethyl]GSH has been detected [131,134].

Vinyl halides also require metabolic activation through oxidation catalyzed by microsomal monoxygenase to form chloroethylene oxide which binds to DNA [135,136]. The major DNA adduct is N-7-(2-oxoethyl)guanine [135,136]. The reactive oxide metabolite can also be deactivated by a glutathion transferase but rearrange, in part, to 2-chloroacetaldehyde, then oxidized to chloroacetic acid [127]. Epoxide metabolites have also been postulated for structurally more complicated chlorine pesticides, DDT, and heptachlor [118].

3.5.5 NATURALLY OCCURRING ORGANIC AGENTS

3.5.5.1 Source and Carcinogenicity

In 1942, it was discovered that crude ergot was carcinogenic in animals when the compound was fed for at least 6 months. It has not been investigated further [137,138]. During the 1960s it was reported that safrole, which occurs in natural oils such as camphor or sassafras oils, caused liver tumors in animals fed with high doses of the substance. Safrole had been previously used as a synthetic flavoring agent in beverages and food [2,137,138]. From the 1970s there has been increased emphasis on the role of naturally occurring substances in the etiology of cancer [139]. Substantial evidences have been accumulated to show that a number of natural agents are potential human health hazards. A number of plants, some of which are edible, contain carcinogenic chemicals whose structures have been elucidated [140]. Table 3.3 shows some naturally occurring carcinogens.

Cycasin present in the nuts of the cycad plant was found to cause liver, kidney, and intestinal tumors in rats. It also apparently has neurological effects on the cattle grazing on cycads. The husk of the nut is chewed by people working in the fields; the nut is soaked in water to remove the toxin. Cycasin apparently caused sclerosis of the liver in people living in Guam [137,138]. *Senecio* plants are quite widespread and some forms known as "golden ragwort" are found in North America. These plants are quite toxic to grazing cattle and in some instances fatal. However, they are used as herbal medicines despite the toxicity. The Senecio plants contain pyrrolizidine alkaloids such as retrorsine, insatidine, and lasiocarpine. Honey made from these plants may contain these alkaloids some of which are liver carcinogens in rats [137,138]. Until recently one of the materials used to flavor vermouth, a constituent of drinks such as martinis or manhattans was calamus root of the *Acorus calamus*. In 1967 it was found that β-asarone a major flavoring material in calamus caused intestinal tumors in rats. Estrazole, a component of oil of tarragon obtained from the tarragon plant, *Artemisia dracunculus* was found to be a weak liver carcinogen in mice [137,138]. Ethyl carbamate or urethane causes lung tumors in mice or liver tumors in mice or rats. Foods and beverages processed by fermentation, such as beer, wine, bread, olives, and yogurt contain measurable levels of the compound [137,138].

In 1960, an outbreak in England of a fatal liver disease in young turkeys was traced to peanut meal in the feed, which had been infected with the fungus *Aspergillus flavus*. This fungus is capable of producing mycotoxins, such as aflatoxins B_1 (AFB$_1$, Figure 3.11) and B_2 and to a lesser extent aflatoxins G_1 and G_2 as secondary fungal metabolites. AFB$_1$ is one of the most potent carcinogens known to humans [141,142]. It causes liver tumors in rats, at doses as low as 1 μg. In certain strains of rats fed with 0.1 ppm of AFB$_1$ for 50–80 weeks a 50% incidence of liver cancer occurred. Rainbow trout were affected by levels of 0.1 ppb. However, adult mice are quite resistant to aflatoxins. The fungus can infect rice, corn, and cottonseed meal. Aflatoxin M_1 (Figure 3.11) is excreted in cow's milk when fed AFB$_1$, which suggests that M_1 is a detoxification product [8,137,138,141–145]. Aflatoxins have been attributed to human liver cancers in areas where growth is favorable for these mycotoxins. In the warm humid areas of the world where there are heavy rainfalls ingestion of contaminated food crops (as high as 222 ng/kg/day of AFB$_1$) can be correlated with liver cancer incidence in the population. Certain areas of the Far East and Africa have an inordinatly high incidence of liver

TABLE 3.3
Some Naturally Occurring Carcinogens

Toxin	Source	Species Affected	Organ Developing Tumors
Actinomycin D	*Streptomyces chrysomallus* or *S. antibioticus*	Mouse, Rat	Local reaction
Aflatoxin B$_1$	*Aspergillus flavus* from moldy grains	Man(?)	Liver
Asarone	Calamus root	Rat	Liver
Cycasin	Cycads	Guinea pig, Mouse, Rat	Intestine, Kidney, Liver
Elaiomycin	*Streptomyces hepaticus*	Rat	Various sites
Ergot (crude)	*Claviceps purpiirea* fungus on rye	Rat	Fibrous tissues
Estragole	Oil of tarragon	Mouse	Liver
Ethyl carbamate	Fermented products	Mouse	Lung
Griseofulvin	*Penicillium griseofulvum*	Mouse	Liver
Nitrosamines	Tobacco	Rat	Liver
Pyrrolizidine alkaloids	Senecio Plants	Rat	Liver
Safrole	Natural plant oils	Mouse, Rat	Liver
Sterigmatocystin	*Aspergillus versicolor*	Mouse, Rat	Liver
Streptozotocin	*Streptomyces achromogenes*	Rat	Kidney, Lung
Unknown	Bracken fern	Cattle, Mouse Rat	Bladder, Intestine, Lung

cancers and hepatitis associated with AFB$_1$ and related toxins due to contamination of many of the farm products such as grain and peanuts [137,141,145].

Sterigmatocystin, another member of mycotoxins, produced by *Asperigllus versicolor*, is carcinogenic in rats and mice but is used as a chemotherapeutic agent in the treatment of neoplastic diseases. Elaiomycin produced from *Streptomyces hepaticus* has a structure similar to cycasin and is a weak carcinogen in rats. Griseofulvin from *Penicillium griseofulvum,* used in the treatment of fungal skin infections, is a heptocarcinogen in mice. Finally, Streptozotocin from *Streptomyces achromogenes* containing N-nitrosourea structure induces kidney and lung tumors in rats but is also considered a chemotherapeutic agent [137,138].

In some instances the active component in plants has not been characterized. Bracken fern, common to open woods and waste places and the young shoots called fiddleheads, are eaten as a delicacy in some countries. Cattle that graze on bracken fern are found to have bladder and stomach cancer. This is particularly so in Turkey, Wales, and Scotland. Over the years many chemical components have been isolated from bracken by different workers, and in some cases they have been tested for carcinogenic activities [146]. Tobacco is another example where the active carcinogenic component has not been identified in a plant; the carcinogenic agents are products of combustion and include nitrosamines, hydrazine, vinyl chloride, and PAHs [137,138,146].

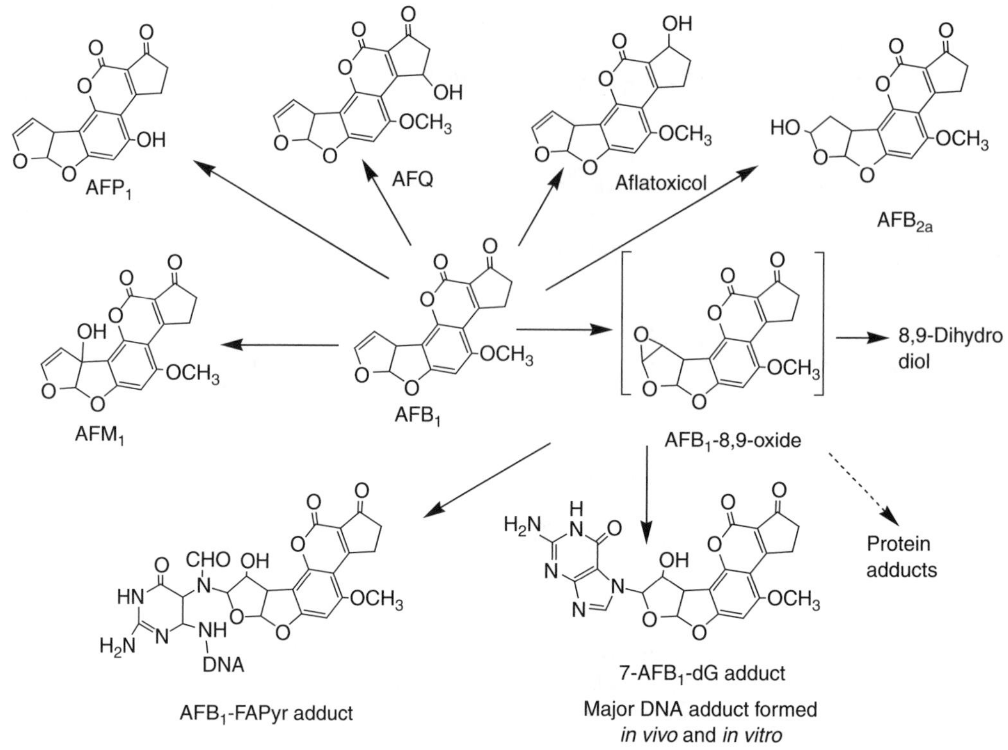

FIGURE 3.11   Metabolic activation of affatoxin B, by hepatic monogenase(s) (MFO).

3.5.5.2 Mechanisms of Action

In terms of mechanisms of action, of all the naturally occurring carcinogens discussed in this section, only two have been studied in depth namely AFB_1 and safrole. To indicate the complexity involved, the metabolism of AFB_1 is illustrated in Figure 3.11. Safrole is described in detail in the references [2,147].

Like a wide variety of other xenobiotic compounds, the aflatoxins are primarily metabolized in animal by the microsomal mixed-function oxigenase (MFO) system mainly localized on the endoplasmic reticulum of liver cells, as well as in lungs, kidney, skin, and other organs. These enzymes catalyze the oxidative metabolism of AFB_1, resulting in the formation of various hydroxylated derivatives or a highly reactive epoxide metabolite [137,142,143]. The reactive electrophilic AFB_1-8,9-epoxide is known to be the ultimate carcinogen that binds to nucleophilic cellular macromolecules including DNA. The main form of epoxide generated is the exo form but there is increasing evidence of a role of the endo form [148]. The epoxide is reported to bind preferentially to the N-7 position of guanine although there is evidence that it binds to other bases such as adenine and cytosine [149]. In addition to the AFB_1 epoxide, aflatoxin M_1 (AFM_1) was the first identified metabolite of AFB_1. It was detected in the milk of cows fed with aflatoxin-contaminated ground nut meal [141] and in the urine of sheep given a single dose of aflatoxins. AFM_1 has been structurally characterized as 10-hydroxy-AFB_1 [150]. AFM_1 was toxic and carcinogenic to rats and mutagenic to bacteria. It has been detected in urine and feces of varieties of animals and also in the urine of humans who consumed AFB-contaminated peanut butter [141]. Water soluble AFM_1 conjugates (glucoronide, sulfate) were found in animal urine [141]. AFB_{2a} is formed from AFB_1 by hydration of the 8,9-double bond resulting in hydroxylation at the 8-position [151]. It is essentially nontoxic and nonmutagenic in the Ames assay. After incubation of AFB_1 with rat or hamster liver microsome pre-parations, AFB_{2a} is not detected but the 8,9-dihydrodiol aflatoxicol (AFL) is formed by the reduction of the cyclopentenone carbonyl group of AFB_1 to a hydroxyl by an NADPH-dependent cytoplas-mic enzymatic fraction *in vitro* [141]. AFL has been reported to be highly mutagenic to bacteria and toxic and carcinogenic in rats that suggests that AFL is an active metabolite [144]. Aflatoxin P_1 (AFP_1), an *O*-demethylation product of AFB_1, is isolated from the urine of monkeys treated with AFB_1 and structurally elucidated [152]. It was the major identifiable excretory metabolite in the urine of AFB_1 treated monkeys and occurred mainly as glucuronide and sulfate conjug-ates [141,152,153]. AFP_1 is nonmutagenic in Ames assay, weakly toxic in comparison to AFB_1 in mice, and nontoxic to chicken embryos [141]. AFQ is formed from AFB_1, by ring hydroxylation at 3-position. It is slightly mutagenic in the Ames assay and weakly toxic to chicken embryos [141]. Pathways summarizing the metabolism and macromolecular binding of AFB_1 are shown in Figure 3.11.

3.5.6 Direct Acting Agents

As discussed earlier, direct acting agents are those chemicals that act as the ultimate carcinogen without metabolic activation.

3.5.6.1 Source

There are a large number of direct acting carcinogens that have been described in detail else-where [154]. Direct acting carcinogens can be grouped into alkylating and acylating agents [11]. Typical direct acting carcinogens are as follows: aralkyl halides (benzylchloride), *bis*-(chloromethyl) ether, chloromethyl methyl ether, dimethylcarbamyl chloride, dimethyl sulfate, ethyleneimine, epox-ides (ethylene oxide), melphalan, sulfonate, mustard gas (*bis*-[2-chloroethyl] sulfide), nitrogen mustard, *N*-acetylimidazole, sulfones, and lactones.

The usage of these chemicals is widespread in the industry. Benzylchloride has been identified as a major environmental pollutant in eastern United States [155]. Some of these chemicals such as ethylene oxide, ethyleneimine and β-propiolactone have been used as intermediates in chemical syntheses, and ethyleneimine has been used in the paper and textile industries as well [154,156,157].

3.5.6.2 Carcinogenicity

Because of the extreme reactivity of these chemicals they can induce tumors at the site of application, as well as sites distant from administration, such as liver, lung, stomach, and kidney. They have been administered subcutaneously, topically, orally, or intravenously to mice or rats.

The classical case of cancer caused by an industrial chemical was the induction of respiratory tract neoplasia by mustard gas as a result of long exposure. Three hundred and twenty-two male workers were exposed in a Japanese mustard gas plant between 1929 and 1945, which resulted in 33 deaths from such neoplasma traced between 1948 and 1966. Dimethylsulfate has been associated with the death of one worker from oat cell type carcinoma. The latency period was 11 years. Dimetylsulfate has also proven to be a carcinogen in the rat following inhalation exposure to this compound [154]. *Bis*-chloromethyl ether, chloromethyl ether, β-propiolactone, ethyleneimine, mustard gas, and ethyl-ene oxide have either been deemed as human carcinogens or probable human carcinogens and all are regulated under the Occupationl Safety and Health Act (see chapters 1 and 28).

3.5.6.3 Mechanisms of Action

These compounds do not need metabolic activation but rather undergo spontaneous breakdown to produce a very reactive electrophile [154]. They react with significant receptors of macromolecular material *in vivo*, in the way as *in vitro*, to initiate the carcinogenic events. The common site of alkylation in the nucleic acid has been the N-7 of guanine with minor sites in the N-1, N-3, and N-7 of adenine, the O-6 and N-3 of guanine, and the N-3 of cytosine [127,158–160].

It appears that the mustard gas and its analog nitrogen mustard alkylate both form cross-links of strands of DNA by a neighboring group effect mechanism [117,154,161]. This cross-linking of the DNA can lead to a modified template of DNA that will block DNA synthesis during the normal cell cycle or potentiate DNA synthesis with a changed base sequence. The blocked DNA synthesis can result in DNA excision–repair while the potentiation of DNA synthesis can result in potential incorporation of an incorrect base pair. In any case whether due to deletion, repair, or incorporation of base sequence, potential carcinogenic events may result.

REFERENCES

1. Miller, E.C. and Miller, J.A. Mechanisms of chemical carcinogenesis: Nature of proximate carcinogens and interactions with macromolecules. *Pharmacol. Rev.*, *18*: 805–838, 1966.
2. Miller, E.C. and Miller, J.A. Searches for ultimate chemical carcinogens and their reactions with cellular macromolecules. *Cancer*, *47*: 2327–2345, 1981.
3. Harris, C.C. Chemical and physical carcinogenesis: Advances and perspectives for the 1990s. *Cancer Res.*, *51 (Suppl.)*: 5023–5044, 1991.
4. Guengerich, F.P. Metabolism of chemical carcinogens. *Carcinogenesis*, *21*: 345–351, 2000.
5. Miller, J.A. Carcinogenesis by chemicals: An overview, *Cancer Res.*, *30*: 559–576, 1970.
6. Pitot, H.C. Cancer — an overview. In *Cancer the Outlaw Cell*, LeFond, R.E., Ed., American Chemical Society Press, Washington, D.C., 1978, pp. 1–18.
7. Braun, A.C. Cancer as a problem in development. In *Cancer the Outlaw Cell*, LeFond, R.E., Ed., American Chemical Society Press, Washington, D.C., 1978, pp. 45–59.
8. Selkirk, J.K. Chemical carcinogenesis: A brief overview of action of the mechanisms of poly-cyclic aromatic hydrocarbons, aromatic amines, nitrosamines and aflatoxins. In *Carcinogenesis, Vol. 5, Modifiers of Chemical Carcinogenesis*, Slaga, T.J., Ed., Raven Press, N.Y., 1980, pp. 1–31.
9. Lawly, P.D. Historical origin of current concept of carcinogenesis. *Adv. Cancer Res.*, *65*: 17–111, 1994.
10. Meyer, D.L. and Burger, M.M. Puzzling role of cell surfaces. In *Cancer the Outlaw Cell*, LeFond, R.E., Ed., American Chemical Society Press, Washington, D.C., 1978, pp. 61–72.
11. Miller, E.C. and Miller, J.A. Mechanisms of chemical carcinogenesis. *Cancer*, *47*: 1055–1065, 1981.
12. Weisburger, J.H. Bioassays and tests for chemical carcinogens. In *Chemical Carcinogens*, Searle, C.E., Ed., American Chemical Society Monograph 173, Washington, D.C., 1976, pp. 1–23.
13. Vainio, H.C., Coleman, M., and Wilbourn, J. Carcinogenesis evaluations and ongoing studies: The IARC database. *Environ. Health Perspect.*, *96*: 5–9, 1991.
14. Goldstein, J.A. and Faletto, M.B. Advances in mechanisms of activation and deactivation of environmental chemicals. *Environ. Health Perspect.*, *100*: 169–176, 1993.
15. Ames, B.N., Durston, W.E., Yamasaki, E., and Lee, F.D. Carcinogens and mutagens: A simple test system combining liver homogenates for activation and bacteria for detection. *Proc. Natl. Acad. Sci. USA*, *70*: 2281–2285, 1973.
16. Working Group of Subcommittee on Environmental Mutagenesis for DREW Committee to Coordinate Toxicology and Related Programs. Approaches to determining the mutagenic properties of chemicals: Risk to future generations. *J. Environ. Pathol. Toxicol.*, *1*: 301–352, 1977.
17. Lucier, G.W., Lui, E.M.K., and Lamartimiere, C.A. Metabolic activation/deactivation reactions during prenatal development. *Environ. Health Perspect.*, *29*: 7–16, 1979.
18. Fishbein, L. An overview of some metabolic and modulating factors in toxicity and chemical carcinogenesis. *J. Am. Coll. Toxicol.*, *2*: 63–89, 1983.
19. Farber, E. Chemical carcinogenesis, a biologic perspective. *Am. J. Pathol.*, *106*: 271–296, 1982.
20. Weisburger, E.K. Mechanisms of chemical carcinogenesis. *Annu. Rev. Pharmacol. Toxicol.*, *18*: 395–415, 1978.

21. Bishop, J.M. Oncogenes. *Sci. Am.*, *246*: 81–92, 1982.
22. Slaga, T.J., Triplett, L.L., and Nesnow, S. Mutagenic and carcinogenic potency of extracts of diesel and related environmental emissions: Two-stage carcinogenesis in skin tumor sensitive mice. In *Health Effects of Diesel Engine Emissions Proceedings of an International Symposium*, 2 Vol., EPA 600/9-90-0576, 1980, pp. 874–897.
23. Bingham, E., Niemeier, R.W., and Reid, J.B. Multiple factors in carcinogenesis. *Ann. N.Y. Acad. Sci.*, *271*: 14–21, 1976.
24. Slaga, T.J., Fischer, S.M., Triplett, L.L., and Nesnow, S. Comparison of complete carcinogenesis and tumor initiation and promotion in mouse skin: The induction of papillomas by tumor initiation promotion a reliable short-term assay. *J. Am. Coll. Toxicol.*, *1*: 83–99, 1982.
25. Weisburger, J.H. and Williams, G.M. Carcinogen testing: Current problems and new approaches. *Science*, *214*: 401–407, 1981.
26. Anttila, A., Sallmen, M., and Hemminki, K. Carcinogenic chemicals in the occupational environment. *Pharmacol. Toxicol.*, *72*: 69–76, 1993.
27. Neff, J.M. *Polycyclic Aromatic Hydrocarbons in the Aquatic Environment*. Applied Science Publishers, London, 1979.
28. Zedeck, M.S. Polycyclic aromatic hydrocarbons — a review. *J. Environ. Pathol. Toxicol.*, *3*: 537–567, 1980.
29. Perera, F. Carcinogenicity of airborne fine particulate BaP: An appraisal of the evidence and the need for control. *Environ. Health Perspect.*, *42*: 163–185, 1981.
30. IARC. Polynuclear aromatic compounds part 1: Chemical, environmental and experimental data. *IARC Monograph on the Evaluation of Carcinogenic Risk of Chemicals to Man*, Vol. 32, Lyon, France, 1983.
31. Dipple, A. Polynuclear aromatic carcinogens. In *Chemical Carcinogens*, Charles E. Searle, Ed., ACS Monograph 173, American Chemical Society Press, 1976, pp. 245–314.
32. Harvey, R.G. *Polycydic Aromatic Hydrocarbons, Chemistry and Carcinogenicity*, Cambridge University Press, Cambridge, UK, 1991, pp. 11–87.
33. National Academy of Sciences. *Polycyclic Particulate Organic Matter*, Washington, D.C., 1972.
34. Trosset R., Warshawsky, D., Menefee, C., and Bingham, E. *Investigation of Selected Potential Environmental Contaminants: Asphalts and Coal Tar Pitch*, EPA publication 56012-77-005, 1978.
35. Bingham, E., Trosset, R.P., and Warshawsky, D. Carcinogenic potential of petroleum hydrocarbons — a critical review. *J. Environ. Pathol. Toxicol.*, *3*: 483–563, 1979.
36. Syracuse Research Corporation, Toxicological Profile for Creosote, U.S. Department of Health and Human Services, 2000, pp. 17–293.
37. Butler, J.D. and Crossley, P. Reactivity of PAHs adsorbed on soot particles. *Atmos. Environ.*, *15*: 91–94, 1981.
38. Saffiotti, D., Montesano, R., Sellakumar, A.R., Cefis, F., and Kaufman, D.G. Respiratory tract carcinogenesis in hamsters induced by benzo[a]pyrene and ferric oxide. *Cancer Res.*, *32*: 1073–1081, 1972.
39. Shabad, L.M. Dose–response studies in experimentally induced lung tumors. *Environ. Res.*, *4*: 305–315, 1971.
40. Sellakumar, A. and Shubik, L. Carcinogenicity of 7H-dibenzo[c,g]carbazole in the respiratory tract of hamsters. *J. Natl. Cancer Inst.*, *48*: 1641–1646, 1972.
41. Davis, R.E. Interaction of light and chemicals in carcinogenesis. NCI Monograph No. 50. *International Conference on Ultraviolet Carcinogenesis*, DHEW publication #78-1532, 1978, pp. 45–50.
42. Cavalieri, E. and Calvin, M. Photochemical coupling of benzo[a]pyrene with 1-methylcytosine photo enhancement of carcinogenicity. *Photochem. Photobiol.*, *14*: 641–653, 1971.
43. Thakker, D.R., Yagi, H., Lu, A.Y.H., Levin, W., Comey, A.H., and Jerina, D.M. Metabolism of benzo[a]pyrene: Conversion of (±)trans-7,8-dihydroxy-7,8-dihydrobenzo[a]pyrene to highly mutagenic 7,8-diol-9,10-epoxide. *Proc. Natl. Acad. Sci. USA*, *73*: 3381–3385, 1976.
44. Jeffrey, A.M., Jennette, K.W., Blobstein, S.H., Weinstein, I.B., Beland, F.A., Harvey, R.G., Kasai, H., Miura, I., and Nakanishi, K. Benzo[a]pyrene-nucleic acid derivatives found in vivo: Structure of a benzo[a]pyrene-tetrahydrodiol epoxide-guanosine adduct. *J. Am. Chem. Soc.*, *98*: 5714–5715, 1976.

45. Yang, S.K., McCourt, D.W., Roller, P.P., and Gelboin, H.V. Enzymatic conversion of benzo[a]pyrene leading predominantly to the diol-epoxide r-7, t-8-dihydroxy-t-9,10-oxy-7,8,9,10-tetrahydrobenzo[a]pyrene. *Proc. Natl. Acad. Sci. USA*, *73*: 2594–2598, 1976.

46. Conney, A.H. Induction of microsomal enzymes by foreign chemicals and carcinogenesis by polycyclic aromatic hydrocarbons. G.H.A. Clowes Memorial Lecture. *Cancer Res.*, *42*: 4875–4917, 1982.

47. Cavalieri, E.L. and Rogan, E.G. Central role of radical cations in metabolic activation of polycyclic aromatic Hydrocarbons. *Xenobiotica*, *25*: 677–688, 1995.

48. Cavalieri, E.L. and Rogan, E.G. The approach to understanding aromatic hydrocarbon carcinogenesis. The central role of radical cations in metabolic activation. *Pharmacol. Ther.*, *55*: 183–199, 1992.

49. Shou, M., Harvey, R.G., and Penning, T.M. Contributions of dihydrodial-dehydrogenase to the metabolism of (±)*trans*-7,8-dihydroxy-7,8 dihydro-benzo[a]pyrene: A fortified rat liver subcellular fractions. *Carcinogenesis*, *13*: 1575–1582, 1992.

50. Smithgall, T.E., Harvey, R.G., and Penning, T.M. Spectroscopic identification of ortho-quinones as the products of polycyclic aromatic trans-dihydro-diol oxidation catalyzed by dihydrodial dehydrogenase. A potential route of proximate carcinogen metabolism. *J. Biol. Chem.*, *263*: 1814–1820, 1988.

51. Gelboin, H.V. Benzo[a]pyrene metabolism, activation and carcinogenesis: Role of regulation of mixed function oxidases and related enzymes. *Physiol. Rev.*, *60*: 1107–1166, 1980.

52. Yang, S.K., Roller, P.P., and Gelboin, H.V. Enzymatic mechanism of benzo[a]pyrene conversion to phenols and diols and an improved high-pressure liquid chromotographic separation of benzo[a]pyrene derivatives. *Biochemistry*, *16*: 3680–3687, 1977.

53. Newbold, R.F. and Brookes, P. Exceptional mutagenicity of a benzo[a]pyrene diol-epoxide in cultured mammalian cells. *Nature*, *261*: 52–54, 1976.

54. Nagao, N. and Sugrimira, T. Mutagenesis: Microbial systems. In *Polycyclic Hydrocarbons and Cancer*, Vol. 1, Ts'O, P.O.P. and Gelboin, H.V., Eds., Academic Press, New York, 1977, pp. 99–121.

55. Malaville, C., Karoki, T., Sims, P., Grover, P.L., and Bartsch, H. Mutagenicity of isomeric diol-epoxides of benzo[a]pyrene and benzo[a]anthracene in *S. typhimurium* TA98 and TA100 and in V79 Chinese hamster cells. *Mutat. Res.*, *44*: 313–326, 1977.

56. Buening, M.K., Wilsocki, P.G., Levin, W., Yagi, H., Thakker, D.R., Akagi, H., Koreeda, M., Jerina, D.M., and Conney, A.H. Tumorigenicity of the optical enantiomers of the diastereomeric benzo[a]pyrene-7,8-diol-9,10-epoxides in newborn mice: Exceptional activity of (±)-7β,8α-dihydroxy-9α,10α-epoxy-7,8,9,10-tetrahydrobenzo[a]pyrene. *Proc. Natl. Acad. Sci. USA*, *75*: 5358–5361, 1978.

57. Kapitulnik, J., Wislocki, P.G., Levin, W., Yagi, H., Jerina, D.M., and Conney, A.H. Tumorigenicity studies with diol-epoxides of benzo[a]pyrene which indicate that (±)-trans-7β,8α-dihydroxy-9α,10α-epoxy-7,8,9,10–tetrahydrobenzo[a]pyrene is an ultimate carcinogen in newborn mice. *Cancer Res.*, *38*: 354–358, 1978.

58. Change, R.L., Wood, A.W., Conney, A.H., Yagi, H., Sayer, J.M., Thakker, D.R., Jerina, D.M., and Levin, W. Role of diaxial versus diequatorial hydroxy groups in the tumorigenic activity of benzo[a]pyrene bay-region diol-epoxide. *Proc. Natl. Acad. Sci. USA*, *84*: 8633–8636, 1987.

59. Osborne, M.R., Bland, F.A., Harvey R.G., and Brookes, P. The reaction of (±)7α-8β-dihydroxy-9β,10β epoxy-7,8,9,10-tetrahydrobenzo[a]pyrene with DNA. *Int. J. Cancer*, *18*: 363–368, 1976.

60. Jennette, K.W., Jeffrey, A.M., Blobstein, S.H., Bland, F.A., Harvey, R.G., and Weinstein, I.B. Nucleoside adducts from the in vitro reaction of benzo[a]pyrene-4,5-oxide with nucleic acids. *Biochemistry*, *16*: 932–938, 1977.

61. Koreeda, M., Moore, P.D., Wislocki, P.G., Levin, W., Conney, A.H., Yagi, H., and Jerina, D.M. Binding of benzo[a]pyrene-7,8-diol-9,10-epoxide to DNA, RNA and protein of mouse skin occurs with high stereoselectivity. *Science*, *199*: 778–781, 1978.

62. Marshall, C.J., Vousden, K.H., and Phillips, D.H. Activation of c-Ha-ras-1 proto-oncogene by in vitro chemical modification with a chemical carcinogen benzo[a]pyrene diol-epoxide. *Nature*, *310*: 585–589, 1984.

63. Puisieux A., Lin, S., Groopman, J., and Ozturk, M. Selective targeting of *p53* gene mutational hotspots in human cancer by etiologically defined carcinogens. *Cancer Res.*, *51*: 6185–6189, 1991.
64. Denissenko, M.F., Pao, A., Tang, M.S., and Pfieifer, G.P. Preferential formation of benzo[a]pyrene adducts at lung cancer mutational hotspots in *p53*. *Science*, *274*: 430–432, 1996.
65. Chen, L., Devanesan, P.D., Higginbotham, S., Ariese, F., Jankowiak, R., Small, G.J., Rogan, E.G., and Cavalieri, E. Expanded analysis of benzo[a]pyrene-DNA adducts formed in vitro and in mouse skin: Their significance in tumor initiation. *Chem. Res. Toxicol.*, *9*: 897–903, 1996.
66. DeVanesan, P.D., Ramakrishna, N.V.S., Todorovic, R., Rogan, E.G., Cavalieri, E.L., Jeony, H., Jankowiak, R., and Small, G.J. Identification and quantitation of benzo[a]pyrene-DNA adducts formed by rat liver microsomes in vitro. *Chem. Res. Toxicol.*, *5*: 302–309, 1992.
67. Xue, W., Zapien, D., and Warshawsky, D. Ionization potentials and metabolic activation of carbazole and acridine derivatives. *Chem. Res. Toxicol.*, *12*: 234–1239, 1999.
68. Flowers-Geary, L., Harvey, R.G., and Penning, T.M. Identification of benzo[a]pyrene-7,8-dione as an authentic metabolite of (±)-*trans*-7,8-dihydroxy-7,8-dihydrobenzo[a]pyrene in isolated rat hepatocytes. *Carcinogenesis*, *16*: 2707–2715, 1996.
69. Penning, T.M., Ohnishi, S.T., Ohnishi, T., and Harvey, R.G. Generation of reactive oxygen species during the enzymatic oxidation of polycyclic aromatic hydrocarbon trans-dihydrodiols catalyzed by dihydrodiol dehydrogenase. *Chem. Res. Toxicol.*, *9*: 84–92, 1996.
70. Shou, M., Harvey, R.G., and Penning, T.M. Reactivity of benzo[a]pyrene-7,8-dione with DNA. Evidence for the formation of deoxyguanosine adducts. *Carcinogenesis*, *14*: 475–482, 1993.
71. McCoull, K.D., Rindgen, D., Blair, I.A., and Penning, T.M. Synthesis and characterization of polycyclic aromatic hydrocarbon *o*-quinones depurinating N7-guanine adducts. *Chem. Res. Toxicol. 12*: 237–242, 1999.
72. Kasai, H. and Nishimura, S. Hydroxylation of deoxyguanosine at the C-8 position by ascorbic acid and other reducing agents. *Nuclei Acid Res.*, *12*: 2137–2145, 1984.
73. Kasai, H., Crain, P.F., Kuchino, Y., Nishimura, S., Ootsuyama, A., and Tanooka, H. Formation of 8-hydroxyguanine moiety in cellular DNA by agents producing oxygen radicals and evidence for its repair. *Carcinogenesis*, *7*: 1847–1851, 1986.
74. Frenkel, K. Carcinogen-mediated oxidant formation and oxidative DNA damage. *Pharmacol. Ther.*, *53*: 127–166, 1992.
75. Halliwell, B. and Gutteridge, J.M.C. *Free Radicals in Biology and Medicine*, 2nd ed., Clarendon Press, Oxford, UK, 1989.
76. Xue, W., Siner, A., Rance, M., Jayasimuhula, K., Talaska, G., and Warshawsky, D. A metabolic activation mechanism of 7H-dibenzo[c,g]carbazole via *o*-quinone. Part 2 covalent adducts of 7H-dibenzo[c,g]carbazole-3,4-dione with nucleic acid bases and nucleosides. *Chem. Res. Toxicol.*, *15*: 915–921, 2002.
77. Garner, R.C., Martin, N., and Clayson, D.B. Carcinogenic aromatic amines and related compounds. In *Chemical Carcinogenesis*, Searle, C.E., Ed., ACS Monograph 182, American Chemical Society, Washington, D.C., 1984, pp. 175–276.
78. Parkes, H.G. and Evans, A.E.J. The epidemiology of aromatic amine cancers. In *Chemical Carcinogens*, Searle, C.E., Ed., ACS Monograph 182, American Chemical Society, Washington, D.C., 1984, pp. 277–301.
79. Kadlubar, F.F. Metabolism and DNA binding of carcinogenic aromatic amines. *ISI Atlas Sci. Pharmacol.*, *1*: 129–132, 1987.
80. Talaska, G., Schamer, M., Skipper, P., Tannebaum, S., Caporaso, N., Unruh, L., Kadlubar, F.F., Bartsch, H., Malaville, C., and Vineis, P. Detection of carcinogen DNA adducts in exfoliated urothelial cells of cigarette smokers: Association with smoking, hemoglobin and urinary mutagenicity. *Cancer Epidemiol. Biomarkers Prev.*, *1*: 61–66, 1991.
81. Kriek, E. Carcinogenesis by aromatic amines. *Biochem. Biophys. Acta*, *355*: 177–203, 1974.
82. IARC Monograph. Some aromatic amines, hydrazine and related substances, *N*-Nitroso compounds and miscellaneous alkylating agents. In *Monograph on the Evaluation of Carcinogenic Risk of Chemicals to Man*, Vol. 4, Lyon France, 1974 and Vol. 16, 1978, Vol. 27, 1982, Vol. 56, 1993, Vol. 57, 1993.

83. Radokmski, J.L. The primary aromatic amines. Their biological properties and structure–activity relationships. *Ann. Rev. Pharmacol. Toxicol.*, *19*: 129–157, 1979.

84. Hueper, W.C., Wiley, F.H., and Wolfe, H.D. Experimental production of bladder tumors in dogs by administration of betanaphthylamine. *J. Ind. Hyg. Toxicol.*, *20*: 46–84, 1938.

85. Guerin, M.R., Ho, C.H., Rao, J.K., Clark, B.R., and Epler, J.L. Polycyclic aromatic primary amines as determinant chemical mutagens in petroleum substitutes. *Environ. Res.*, *23*: 42–53, 1980.

86. Wilson, B.W., Pelroy, R., and Cresto, J.T. Identification of primary aromatic amines in mutagenically active subfractions from coal liquefaction materials. *Mutat. Res.*, *79*: 193–202, 1980.

87. Lotlikar, P.D. Metabolic activation of aromatic amines and dialkylnitrosamines. *J. Cancer Res. Clin. Oncol.*, *99*: 125–136, 1981.

88. Parris, G.E. Environmental and metabolic transformation of primary aromatic amines and related compounds. *Residue Rev.*, *76*: 1–30, 1980.

89. Kadlubar, F.F. and Belond, F.A. Chemical properties of ultimate carcinogenic metabolites of arylamines and arylamides. In *Polycyclic Hydrocarbons and Carcinogenesis*, Harvey, R.G., Ed., ACS Symposium Series 283, American Chemical Society, Washington, D.C., 1985, pp. 341–370.

90. Kadlubar, F.F., Butler, M.A., Kaderlik, K.R., Chou, H.-C., and Lang, N.P. Polymorphisms for aromatic amine metabolism in humans: Relevance for human carcinogenesis. *Environ. Health Perspect.*, *98*: 69–74, 1992.

91. Minchin, R.F., Ilett, K.F., Teitel, C.T., Reeves, P.T., and Kadlubar, F.F. Direct *O*-acetylations of *N*-hydroxy arylamines by acetylsalicylic acid to form carcinogen — DNA adducts. *Carcinogenesis*, *13*: 661–667, 1992.

92. Novak, M. and Rangappa, K.S. Nucleophilic substitution on the ultimate hepatocarcinogen *N*-(sulfonatoxy)-2-(acetylamino) fluorene by aromatic amines. *J. Org. Chem.*, *57*: 1285–1290, 1992.

93. Kadlubar, F.F. DNA adducts of carcinogenic aromatic amines. In *DNA Adducts: Identification and Biological Significance*, Hemminki, K. et al., Eds., IARC Scientific Publications No. 125, Lyon, France, 1994, pp. 199–216.

94. Preussmann, R. and Eisenbrand, G. *N*-Nitroso carcinogens in the environment. In *Chemical Carcinogens*, 2nd ed., Vol. 2, Searle C.E., Ed., ACS Monograph 182, American Chemical Society, Washington, D.C., 1984, pp. 828–868.

95. Ender, F. and Ceh, L. Occurrence of nitrosamines in foodstuffs for human and animal consumption. *Food Cosmet. Toxicol.*, *6*: 569–571, 1968.

96. Montesano, R. and Bartsch, H. Mutagenic and carcinogenic *N*-nitroso compounds: Possible environmental hazards. *Mutat. Res.*, *32*: 179–228, 1976.

97. IARC Monograph Publication 14. *N-Nitroso Compounds in Air and Water*, Lyon, France, 1976 .

98. Fine, D.H., Rounbehler, D.P., Rounbehler, A., and Silvergleid, A. Determination of dimethylnitrosamine in air, water, and soil by thermal energy analysis: Measurements in Baltimore, MD. *Environ. Sci. Technol.*, *11A*: 581–584, 1977.

99. Hoffmann, D., Adams, J.D., Brunjnemann, K.D., and Hecht, S.S., Formation, occurrence and carcinogenicity of *N*-nitrosamines in tobacco products. In *Nitroso Compounds*, Scanlan, R.A. and Tannebaum, S.R., Eds., ACS Symposium Series #174, American Chemical Society, Washington, D.C., 1981, pp. 247–273.

100. Preussmann, R. and Stewart, B.W. N-nitroso carcinogens. In *Chemical Carcinogens*, 2nd Ed., Vol. 2, Searle, C.E., Ed., ACS Monograph 182, American Chemical Society, Washington, D.C., 1984, pp. 643–828.

101. Bartsch, H., Ohbhima, H., and Shuker, D.E.G. Exposure of humans to endogenous *N*-nitroso compounds: Implications in cancer etiology. *Mutat. Res.*, *238*: 255–267, 1990.

102. Coulston, F. and Olajos, E.J. Comparative toxicology of *N*-nitroso compounds and their carcinogenic potential to man. *Potential Carcinogenicity of Nitrosable Drugs*, WHO Symposium, Geneva, June 1978, pp. 25–86.

103. Saul, R.L., Kabir, S.H., Cohen, Z., Bruce, W. R., and Archer, M.C. Re-evaluation of nitrate and nitrite levels in human intestine. *Cancer Res.*, *41*: 2280–2283, 1981.

104. White, J.W., Jr., Relative significance of dietary sources of nitrate and nitrite. *J. Agric. Food Chem.*, *23*: 886–891, 1975.

105. Lee, L., Acher, M.C., and Bruce, W.R. Absence of volatile nitrosamines in human feces. *Cancer Res.*, *41*: 3992–3994, 1981.
106. Lijinsky, W. *Chemistry and Biology of N-Nitroso Compounds*, Cambridge University Press, New York, 1992.
107. Lijinsky, W. Structure–activity relationships among *N*-nitroso compounds. In *N-Nitroso Compounds*, Scanlan, R.A. and Tannenbaum, S.R. Eds., ACS Symposium Series #174, American Chemical Society, Washington, D.C., 1981, pp. 88–99.
108. Bartsch, H. and Montesano, R. Relevance of nitrosamines to human cancer. *Carcinogenesis, 5*: 1381–1393, 1984.
109. Linjinsky, W., Loo, W., and Ross, A.E. Mechanism of alkylation of nucleic acids by nitrosodimethyl-amine. *Nature, 218*: 1174–1175, 1968.
110. Yang, C.S., Yoo, J.-S.H., Ishizaki, H., and Hong, J. Cytochrome P450 II El: Roles in nitrosamine metabolism and mechanism of regulation. *Drug Metab. Rev.*, *22*: 147–159, 1990.
111. Gruengerich, F.P. and Shimadal, T. Oxidation of toxic and carcinogenic chemicals by humans cytochrome P450 enzymes. *Chem. Res. Toxicol.*, *4*: 391–407, 1991.
112. Shuker, D.E.G. and Bartsch, H. DNA adducts of nitrosamines. In *DNA Adducts: Identification and Biological Significance*, Hemminki, K. et al., Eds., IARC Scientific Publications No. 125, Lyon, France, 1994, pp. 73–89.
113. Hecker, L.I., Saavedra, J.E., Farrelly, J.E., and Andrews, A.W. Mutagenecity of potassium alkane-diazotates and their use as model compounds for activated nitrosamines. *Cancer Res.*, *43*: 4078–4082, 1983.
114. Lown, J.W., Chauhan, S.M.S., Koganty, R.R., and Sapse, A.M. Alkyldinitrogen species implicated in the carcinogenic, mutagenic and anticancer activities of *N*-nitroso compounds: Characterization by nitrogen-15 NMR of nitrogen-15-enriched compounds and analysis of DNA base situ selectivity by ab initio calculations. *J. Am. Chem. Soc.*, *106*: 6401–6408, 1984.
115. Ohgaki, H., Hard, G.C., Hirota, N., Maekawa, A., Takahashi, M., and Kleihues, P. Selective mutation of codons 204 and 213 of the *p53* gene in rat tumors induced by alkylating *N*-nitroso compounds. *Cancer Res.*, *52*: 2995–2998, 1992.
116. Hecht, S.S., McCoy, G.D., Chen, C.H.B., and Hoffman, D. The metabolism of cyclic nitrosamines. In *N-Nitroso Compounds*, Scanlon, R.A. et al., Eds., ACS Symposium Series 174, American Chemical Society, Washington, D.C., 1981, pp. 49–75.
117. Loeppky, R.N., Outram, J.R., Tomasik, W., and McKinley, W., Chemical and biochemical transformation of β-oxidized nitrosamines. In *N-Nitroso Compounds*, Scanlon, R.A. et al., Eds., ACS Symposium Series 174, American Chemical Society, Washington, D.C., 1981, pp. 21–37.
118. Burchfield, H.P. and Storrs, E.E. Organohalogen carcinogens. In *Advances in Modern Toxicology*, Vol. 3, Kraybill, H.F. and Mehlman, M.A., Eds., Hemisphere Publishing Co., Washington, D.C., 1977, pp. 319–371.
119. Allen, J.R. and Norback, D.H. Carcinogenic potential of polychlorinated biphenyls. In *Origins of Human Cancer*, Hiatt, H.H., Watson, J.D., and Winsten, J.A., Eds., Cold Spring Harbor Laboratory, Cold Spring Harbor, NY, 1977, pp. 173–203.
120. Epstein, S.S. The carcinogenicity of organochlorine pesticides. In *Origins of Human Cancer*, Hiatt, H.H., Watson, J.D., and Winsten, J.A., Eds., Cold Spring Harbor Laboratory, Cold Spring Harbor, NY, 1977, pp. 243–265.
121. Greim, H. and Wolff, T. Carcinogenicity of organic halogenated compounds. In *Chemical Carcinogens*, 2nd ed., Searle, C.E., Ed., ACS Monograph 182, American Chemical Society, Washington, D.C., 1984, pp. 525–575.
122. Poland, A. and Knutson, J.C. 2,3,7,8-Tetrachlorodibenzo-p-dioxin and related halogenated aromatic hydrocarbons: Examinations of the mechanism of toxicity. *Annu. Rev. Pharmacol. Toxicol.*, *22*: 517–554, 1982.
123. Van den Berg, M., De Jongh, J., Poiger, H., and Olson, J.R. The toxicokinetics and metabolism of polychlorinated dibenzo-p-dioxin (PCDDs) and dibenzo/furan (PCDFs) and their relevance for toxicity. *Crit. Rev. Toxicol.*, *24*: 1–74, 1994.

124. VanDuuren, B.L. Carcinogenicity and metabolism of some halogenated olefinic and aliphatic hydro-carbons. *Banbury Report #5 Ethylene Dichloride: A Potential Risk*, Cold Spring Harbor Laboratory, Cold Spring Harbor, NY, 1980, pp. 189–205.

125. National Cancer Institute, Bioassay program reports beginning in 1975, Government Printing Office, Washington, D.C., DHEW Publications include pp. 76–802, 77–813, 78–828, 78–1323 on Tri-chloroethylene, Chloroform, Tetrachloroethylene, Epichlorohydrin, Dibromochloropropane and Allyl Chloride.

126. Guengerich, F.P., Min, K.-S., Persmarl, M., Kim, M.-S., Humphreys, W.G., Cmarick, J.M., and Thies, R. Dihaloalkanes and polyhaloakenes. In *DNA Adducts: Identification and Biological Significance*, Hemminki, K. et al., Eds., IARC Scientific Publications No. 125, Lyon, France, 1994, pp. 57–72.

127. Bolt, H.M. Vinyl halides, haloaldehydes and monohaloalkanes. In *DNA Adducts: Identification and Biological Significance*, Hemminki, K. et al., Eds., IARC Scientific Publications No. 125, Lyon, France, 1994, pp. 141–150.

128. Moschel, R.C. Reactions of aralkyl halides with nucleic acid components and DNA. In *DNA Adducts: Identification and Biological Significance*, Hemminki, K. et al., Eds., IARC Scientific Publication No. 125, Lyon, France, 1994, pp. 25–36.

129. Waters, M.D., Simmon, V.F., Mitchell, A.D., Jorgenson T.A., and Valencia, R. An overview of short term tests for Mutagenic and carcinogenic potential of pesticides. *J. Environ. Sci. Health, B15*: 867–906, 1980.

130. Walles, S.A.S. Reaction of benzyl chloride with haemoglobin and DNA in various organs of mice. *Toxicol. Lett., 9*: 379–387, 1981.

131. Kim, D.H. and Guengerich, F.P. Formation of DNA adduct S-[2-(N^7-guanyl)-ethyl]glutathione from ethylene dibromide: Effects of modulation of glutathione and glutathione S-transferase levels and the lack of a role for sulfation. *Carcinogenesis, 11*: 419–424, 1990.

132. Guengerich, F.P., Humphreys, W.G., Kim, D.H., Oida, T., and Cmarik, J.L. DNA-glutathione adducts derived from vic-dihaloalkanes: Mechanisms of mutagenesis. In *Xenobiotics of Cancer*, Ernster, L., Esumi, H., Fujii, Y. Gelboin, H.V., Kato R., and Sugimura, T., Eds., Scientific Societies Press, Tokyo, Japan, 1991, pp. 101–107.

133. Lenoard, N.J. Etheno-substituted nucleotides and coenzymes: Fluorescence and biological activity. *Crit. Rev. Biochem., 15*: 125–199, 1984.

134. Inskeep, P.B., Koga, N., Cmarik, J.L., and Guengerich, F.P. Covalent binding of 1,2 dihaloalkanes to DNA and stability of the major DNA adducts, s-[2-(N^7-guanyl) ethyl]glutathione. *Cancer Res., 46*: 2839–2844, 1986.

135. Guengerich, F.P., Mason, P.S., Stott, W.T., Fox, T.R., and Watanabe, P.G. Roles of 2-haloethylene oxides and 2-haloacetaldehydes derived from vinyl bromide and vinyl chloride in irreversible binding to protein and DNA. *Cancer Res., 41*: 4391–4398, 1981.

136. Guengerich, F.P. Roles of the vinyl chloride oxidation products 2-chlorooxirane and 2-chloroacetaldehyde in the in vitro formation of etheno adducts of DNA bases. *Chem. Res. Toxicol., 5*: 2–5, 1992.

137. Schoental, R. Carcinogens in plants and microorganisms. In *Chemical Carcinogens*, Searle C.E., Ed., American Chemical Society, Washington, D.C., 1976, pp. 626–689.

138. Weisburger, E.K. Natural carcinogenic products. *Environ. Sci. Technol., 13*: 278–281, 1979.

139. Gold, L.S., Stone, T.H., Stern, B.R., Manley, N.B., and Ames, B.N. Rodent carcinogens: Setting priorities. *Science, 258*: 261–265, 1992.

140. Hirono, I. Edible plants containing naturally occurring carcinogens in Japan. *Jpn. J. Cancer Res., 84*: 997–1006, 1993.

141. Busby, W.F., Jr. and Wogan, G.N. Aflatoxins. In *Chemical Carcinogens*, 2nd ed., Vol. 2, Searle, C.E., Ed., ACS Monograph 182, American Chemical Society, Washington, D.C., 1984, pp. 945–1136.

142. Dragan, Y.P. and Pitot, H.C. Aflatoxin carcinogenesis in the content of the multi-stage nature of cancer. In *The Toxicology of Aflatoxins: Human Health, Veterinary and Agricultural Significance*, Eaton, D.L. and Groopman, J.D., Eds., Academic Press, San Diego, 1994, pp. 179–206.

143. Garner, R.C. Carcinogenesis by fungal products. *Brit. Med. Bull., 36*: 47–52, 1980.

144. Nixon, J.E., Hendricks, J.D., Pawlosurki, N.E., Loveland, P.M., and Sinnhuber, R.O. Carcinogenicity of aflatoxicol in Fischer 344 rats. *J. Natl. Cancer Inst.*, *66*: 1159–1163, 1981.

145. Wogan, G.N. Aflatoxins as risk factors for hepatocellular carcinoma in humans. *Cancer Res.*, *52*: 2114s–2118s, 1992.

146. Evans, I.A. Bracken carcinogen. In *Chemical Carcinogens*, 2nd ed., Vol. 2, Searle, C.E., Eds., ACS Monograph 182, American Chemical Society, Washington, D.C., 1984, pp. 1171–1204.

147. Ioannides, C., De LaForge, M., and Parke, D.V., Safrole: Its metabolism carcinogenicity and interactions with cytochrome P-450. *Food Cosmet. Toxicol.*, *19*: 657–666, 1981.

148. Raney, K.D., Coles, B., Gruengerich, F.P., and Harris, T.M. The endo-8,9-epoxide of aflatoxin B1: A new metabolite. *Chem. Res. Toxicol.*, *5*: 333–335, 1992.

149. Yu, V.L., Huang, J.-X., Bender, W., Wu, Z., and Chang, J.C.S. Evidence for the covalent binding of aflatoxin B1-dichloride to cytosine in DNA. *Carcinogenesis*, *12*: 997–1002, 1991.

150. Holzapfel, C.W., Steyn, P.S., and Purchase, I.F.H. Isolation and structure of aflatoxin M_1 and M_2. *Tetrahedron Lett.*, 2799–2803, 1966.

151. Pohland, A.E., Cushmac, M.E., and Andrellos, P.J. Aflatoxin B1 Hemiacetal. *J. Assoc. Off. Anal. Chem.*, *51*: 907–910, 1968.

152. Dalezios, J., Wogan, G.N., and Weinkeb, S.M. Aflatoxin P1: A new aflatoxin metabolite in monkeys. *Science*, *171*: 584–585, 1971.

153. Dalezios, J. and Wogan, G.N. Metabolism of aflatoxin B1 in rhesus monkeys. *Cancer Res.*, *32*: 2297–2303, 1972.

154. Lawley, P.D. Carcinogenesis by alkylating agents. In *Chemical Carcinogens*, Searle, C.E., Ed., ACS Monograph #182, American Chemical Society, Washington, D.C., 1984, pp. 325–484.

155. Saxena, S. and Abdel-Rohman, M.S. Pharmacodynamics of benzyl chloride in rats. *Arch. Environ. Contam. Toxicol.*, *18*: 669–677, 1989.

156. Hemminki, K. Nucleic acid adducts of chemical carcinogens and mutagens. *Arch. Toxicol.*, *152*: 249–285, 1983.

157. National Toxicology Program (NTP). *Report on Carcinogens*, 10th ed., U.S. Department of Health and Human Services Publication, 2002.

158. Beranek, D.T., Weis, C.C., and Swenson, D.H. A comprehensive, quantitative analysis of methylated and ethylated DNA using high pressure liquid chromatography. *Carcinogenesis*, *1*: 595–606, 1980.

159. Segerback, D. DNA alkylation by ethylene oxide and some mono-substituted epoxides. In *DNA Adduces: Identification and Biological Significance*, Hemminki, K. et al., Eds., IARC Scientific Publications No. 125, Lyon, France, 1994, pp. 37–48.

160. Solomon, J.J. DNA adducts of lactones, sultones, acylating agents and acrylic compounds. In *DNA Adduces: Identification and Biological Significance*, Hemminki, K. et al., Eds., IARC Scientific Publication No. 125, Lyon, France, 1994, pp. 179–198.

161. Hemminki, K. DNA adducts of nitrogen mustards and ethylene imines. In *DNA Adduces: Identification and Biological Significance*, Hemminki, K. et al., Eds., IARC Scientific Publication No. 125, Lyon, France, 1994, pp. 313–322.

4 Binding of Carcinogens to DNA and Covalent Adducts and DNA Damage — PAH, Aromatic Amines, Nitro-aromatic Compounds, and Halogenated Compounds

Jeffrey A. Ross

CONTENTS

SUMMARY

The formation of covalent DNA adducts by the binding of reactive metabolites of chemical carcinogens to cellular DNA represents a key early event in the cascade of molecular alterations that leads to the transformation of a normal cell into a cancer cell. The specific structures of the adducts formed and the extent of adduction induced are key determinants of the relative potencies of many chemical carcinogens. This has led to much research to understand the biological processes leading to the formation and repair of DNA adducts, and to utilizing DNA adducts as biomarkers for both exposure and effect. This chapter introduces methodologies that have been developed for the detection and measurement of DNA adducts and the application of such measurements to molecular epidemiology. A brief survey of the types of DNA adducts induced by a variety of chemical carcinogens is also included.

4.1 INTRODUCTION

DNA adducts are the covalent addition products resulting from binding of reactive chemical species to DNA bases. The cancer initiating role of DNA adducts is well established, and is clearly reflected in the high cancer incidence observed in individuals with deficiencies in any of a variety of DNA repair enzymes [1–5]. As early as the mid-1960s, researchers in chemical carcinogenesis were reaching the consensus that there was a strong relationship between the extent of binding of carcinogens or their metabolites to DNA and their carcinogenic potency [6,7]. Research interest in DNA adducts derives from the central role of DNA damage in inducing heritable changes in DNA structure that can lead to a variety of adverse health effects. Figure 4.1 illustrates the context for a central role of DNA adducts in the process of genotoxic carcinogenesis.

An individual may be exposed to a certain extent depending on the concentration of a genotoxic carcinogen present in the environment (the external dose). Some fraction of the external dose is absorbed (internal dose) and some fraction of that internal dose is then transported to the target tissue (target tissue dose). Once delivered to the cells of the target tissue metabolizing enzymes may convert the parent chemical to a variety of metabolites, some of which may be chemically reactive. These chemically reactive species interact with cellular components including macromolecules such as protein, RNA, and DNA. Binding to protein and RNA can induce a number of generally cytotoxic effects, but it is the binding to DNA that is paramount in inducing mutations. The extent of binding

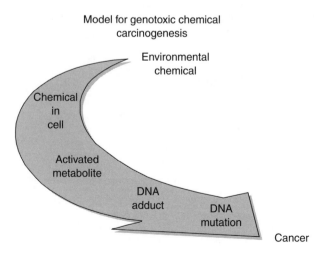

FIGURE 4.1 Overview of the process of genotoxic chemical carcinogenesis. The formation of DNA damage is an essential step in the induction of mutations in cellular control genes.

of reactive metabolites to form DNA adducts in the target cells of the target tissues can be considered to represent the effective molecular delivered dose of the carcinogen.

It is important to distinguish that DNA adducts are not DNA mutations, but can be direct inducers of mutations. The process of mutagenesis is generally an active cellular process. It results from the actions of cellular repair or replication machinery on the adducted DNA. This implies that the types of mutations induced by DNA adducts will be determined by not only the adduct structure, but also by the specific repair and replication enzyme systems acting upon those adducts.

4.1.1 GENERAL ASPECTS OF DNA ADDUCTION

The structure of a DNA adduct is dependent both on the structure of the reactive species that attacks the DNA as well as the specific DNA base and site of reaction on the DNA base. It is seldom the case that a genotoxic chemical induces a single type of DNA adduct. Depending on the type of chemical, multiple metabolizing enzymes may act upon it, and primary metabolites may be further metabolized. Each metabolic pathway may lead to an array of metabolites, more than one of which may react with DNA. Each reactive metabolite may have the ability to attack DNA at more than one base and at more than one position on a given base. Further, for some chemicals it is possible to form stereoisomers and conformational isomers of a given individual adduct. It is not uncommon for a single polycyclic aromatic hydrocarbon (PAH) to induce several distinct DNA adducts in a given tissue. For example, for benzo[a]pyrene, one of the primary reactive metabolites is the bay-region dihydrodiol epoxide that exists in two diasteriomeric forms, termed *syn* and *anti*, each of which exists in two optically active enantiomers. In most mammalian systems, it is primarily the (+)-*anti*-dihydrodiol epoxide that is the predominant adducting species, and it is the N^2-deoxyguanosine adduct that is the major DNA adduct formed. However, that is not the only DNA adduct formed. Additional adducts are formed as a consequence of formation of the phenolic 9-hydroxy-B[a]P followed by additional epoxidation, probably at the 4,5 bond. In some tissues additional adducts are also formed by one-electron oxidation [8], while yet others may be induced by binding of benzo[a]pyrene B[a]P quinones [9]. In other species, the patterns of metabolic activation, and hence the spectrum of adducts induced can be completely different. In some mollusks, the predominant pathway of B[a]P activation appears not to be via the diol epoxide, although other B[a]P-derived adducts are detectable following B[a]P administration [10].

Since the formation of DNA adducts is usually dependent on metabolic activation, the number and types of adducts observed following exposure to a parent chemical is dependent upon the available active metabolizing enzymes. These enzymes can vary from tissue to tissue, individual to individual, and species to species, and so the observed spectrum of adducts also varies among these categories [11,12]. Any internal or external factors that can influence the expression and activity of metabolic processes can influence the type and extent of DNA adducts produced. Similarly, differences in DNA repair processes among tissues, individuals, and species may have a critical role in influencing the efficiency with which DNA adducts are converted into mutations.

Specific atoms on the DNA bases are more prone to adduction than others. For example, the reaction of PAH metabolites with DNA demonstrates marked base and position specificity. PAH dihydrodiol epoxides, which are formed by monooxygenation mediated by cytochromes P450, react with DNA primarily at the exocyclic guanine nitrogen N^2, although N^6 of adenine is also a major site of binding for some PAH dihydrodiol epoxides. Whether a particular reactive PAH dihydrodiol epoxide binds to guanine, adenine, or both depends upon the structure and reactivity of the metabolite [13]. In contrast, PAH metabolites resulting from one-electron oxidation tend to bind to N-7 and C-8 of guanine, while reactive arylamine metabolite nitrenium ions tend to bind primarily to the C-8 of guanine, with only relatively little binding to guanine N^2.

4.1.2 CATEGORIZATION OF DNA ADDUCTS

DNA adducts can be categorized according to several criteria. One common characteristic is the molecular size of the DNA adduct, leading to a dichotomy between bulky and small DNA adducts.

These can differ in the degree to which they distort the DNA helical structure, and in the pathways by which they are repaired. Many types of small DNA adducts can result from both exogenous exposures to chemicals as well as via normal metabolic processes. Examples include 8-oxo-deoxyguanosine, and the small alkylated bases, for example, 7-methylguanine and 6-methyladenine. It should be noted that the 5-position on cytosine is a site that is frequently observed to be methylated in eukaryotic DNA. Methylations at this position are formed and removed enzymatically and serve an important role in regulating gene expression. Thus 5-methylcytosine is not usually considered to be a DNA adduct. Bulky DNA adducts include those formed by polycyclic aromatic hydrocarbons and arylamines, as well as those that can be formed endogenously by catechol estrogens. The differentiation of a particular adduct as either bulky or small is perhaps most usefully predicated depending upon the effect induced by the adduct upon the local structure of the DNA, and, consequently, the repair pathways utilized by the cell to remove the adducts. In general, nucleotide excision pathways dominate in the repair of bulky adducts, while base excision repair and alkyltransferases comprise the primary pathways for small alkyl adducts.

An additional dichotomy of adducts can be defined between stable and unstable adducts. Adduction at specific sites on the DNA bases, for example, the N-7 of guanine, results in an excess positive charge on the ring and leads to the labilization of the glycosyl bond of the adducted base, such that the adducted base spontaneously leaves the DNA phosphodiester backbone, resulting in the formation of abasic sites. These abasic sites may be mutagenic in their own right and may contribute to the observed carcinogenicity of those chemicals whose adducts are capable of forming such unstable adducts. However, abasic sites are also formed spontaneously in most cells, and the relative contribution of the additional abasic sites induced by chemical exposures relative to stable adduct contributions to the process of mutagenesis and carcinogenesis remains an ongoing area of active investigation.

4.1.3 RELATIONSHIPS BETWEEN DNA ADDUCTS AND MUTATIONS

Many studies have demonstrated that specific DNA adducts tend to induce reproducible spectra of mutations, and these may be similar across species. Thus, benzo[a]pyrene has been shown to induce a large proportion of G to T transversions in *Escherichia coli* [14,15], in transgene targets in mice [16], and in the Ki-*ras* oncogenes in benzo[a]pyrene-induced lung tumors [17]. The mutation spectra induced appear to be relatable to the structures of the DNA adducts formed. This is true at the level of the DNA base adducted, such that chemicals that bind to adenine more than guanine tend to induce more mutations at adenines [18]. This is also true at the level of adduct structure, such that cyclopenta-fused PAH induce characteristic G to C transversions whereas bay-region dihydrodiol epoxides tend to induce more G to T transversions in codon 12 of Ki-*ras* [17].

Different chemicals also exhibit different DNA sequence specific effects on binding efficiency. Within a given gene sequence, not all guanosines are equally likely to be adducted. "Hot spots" for binding, that is bases at which the level of binding is several-fold greater than the average level of binding, have been identified for a variety of bulky adducts. In many cases, these correlate well with hot spots for DNA mutation. For example, HELA cells treated *in vitro* with B[a]P diol epoxide exhibit preferential adduction to specific bases within the *p53* gene that correlate with frequently observed sites of mutations in *p53* observed in lung tumors from smokers [19].

4.2 TECHNIQUES FOR MEASURING DNA ADDUCTS

A variety of techniques have been developed to identify and quantitate the formation of DNA adducts, and have been recently reviewed [20,21].

4.2.1 RADIOLABELED CHEMICAL BINDING STUDIES

One of the oldest approaches relied on the synthesis of radiolabeled (usually with ^{3}H or ^{14}C) chemicals, and the identification of radioactivity of the covalently bound radiolabeled chemicals to DNA following administration to animals, or *in vitro* reaction with DNA in the presence or absence of metabolic activating systems. Further characterization was accomplished by chemically or enzymatically hydrolyzing the DNA to its constituent bases, nucleosides, or nucleotides, and then separating the adducts by liquid or thin layer chromatography with detection and quantitation based on the associated radioactivity of the adduct. While much of our early understanding of the processes of DNA adduct formation was gained from this approach, it is currently of limited application. One reason for this is that the available specific activities of radiolabeled carcinogens do not allow the sensitivity of detection that other techniques permit, and the difficulty and expense of obtaining the radiolabeled chemicals is an obstacle. Also, this approach is generally not applicable to studies of DNA adduction in human subjects.

4.2.2 MASS SPECTROSCOPIC TECHNIQUES

Mass spectroscopy approaches, particularly in combination with gas chromatography (GC-MS) or liquid chromatography (LC-MS), have been developed that permit sensitive detection and quantitation of DNA adducts, particularly in those instances where the structures of the expected adducts are known. One of the particular strengths of MS approaches is that the technique provides some degree of direct structural information about the adducts detected, aiding in identification of the precursor metabolite, the DNA base, and the position of binding.

4.2.3 SYNCHRONOUS FLUORESCENCE

Synchronous fluorescence has been effectively utilized to detect and quantitate specific DNA adducts that have intrinsic fluorescence. This has been applied to the quantitation of benzo[*a*]pyrene adducts from a variety of biological tissues. Fluorescence signals are measured with a fixed difference in wavelength settings for excitation and emission monochromators that is equal to the difference between the maximum of the lowest energy excitation maximum and the highest energy emission maximum for the fluorescent adduct. A sharp peak is observed in the fluorescence output when the target fluorophore is present. This technique can be highly specific and quite sensitive, but is limited to application on only fluorescent adducts. This technique has been applied to benzo[*a*]pyrene tetrols released from benzo[*a*]pyrene-adducted DNA by hydrolysis, and its sensitivity can be enhanced by preenrichment of adducts with immunoaffinity columns [22,23]. However, protein and RNA benzo[*a*]pyrene adducts are as capable of releasing tetrols under hydrolysis conditions as are DNA adducts, so adequate sample preparation and characterization is essential to avoid contributions from these sources.

4.2.4 IMMUNOASSAY APPROACHES

Immunoassay approaches to DNA adduct analysis capitalize on the ability of adducted DNA to elicit an immune response and the generation of antibodies specific to certain types of DNA damage [24]. Monoclonal antibodies raised against BPDE-modified DNA have found wide use in the assessment of benzo[*a*]pyrene exposure in humans, for example. Most, if not all, of the antibodies developed for DNA adduct analysis can recognize a variety of related types of DNA adducts in addition to the one used as an immunogen in their production. This broad range increases the utility of this approach in surveying the extent of DNA damage induced from unknown exposures. However, there may be significant differences in the efficiency of recognition of different types of damage, so caution must be used in interpreting quantitative results from this technique among samples that may represent different distributions of adducts. For example, assume that two PAH–DNA adducts bind with different affinities to an antibody. The higher affinity binding of one adduct will require less adduct in a sample to generate a certain amount of signal in an assay than will the lower affinity adduct. Thus a given result in an assay could be due to small amounts of high affinity adduct, larger amounts of

TABLE 4.1
Comparison of Adduct Detection Methods

Detection method	Typical amount of DNA required (μg)	Sensitivity (adducts/normal nucleotides)
Radiolabeled compound	10–100	10^{-7}–10^{-8}
Immunoassay	3–50	10^{-9}
^{32}P-postlabeling	1–50	10^{-10}
Synchronous fluorescence spectroscopy	100	10^{-8}
Mass spectroscopy	100	10^{-9}
Accelerator mass spectroscopy	200	10^{-12}

low affinity adduct, or some indeterminate mixture of the two. In typical human exposures to PAHs, individuals are exposed to complex mixtures of hundreds of potentially adducting species. Each of the many DNA adducts induced will bind with different affinities to an antibody, and will generate a response that is a function of the particular antibody used and the particular adducts present in that sample. The response from a different mixture of adducts may be rather different. Thus two analyses of different mixtures that generate the same response may not actually represent similar levels of adducts. Results of immunoassays for PAH adducts are frequently reported in terms of benzo[a]pyrene equivalents of binding, but it should be kept in mind that such values are directly comparable only for analyses conducted on similar mixtures.

4.2.5 POSTLABELING APPROACHES

Postlabeling is another technique that has come into prominence for measuring a variety of DNA adducts [25]. Typically, DNA is enzymatically hydrolyzed to 3′-mononucleotides which are then labeled at the 5′-hydroxyl of the deoxyribose with ^{32}P (occasionally ^{33}P or ^{35}S) using polynuc-leotide kinase and γ-^{32}P-ATP. The labeled *bis*phosphate adducts are then resolved from the normal nucleotides by either liquid chromatography or thin layer chromatography. For postlabeling bulky adducts, normal nucleotides are usually removed prior to the labeling step to enhance the efficiency of labeling the adducts, and thus markedly increasing the sensitivity of the assay. This preenrich-ment step can be accomplished enzymatically with nuclease P1 [26], by solvent extraction with butanol [27], or, especially for small adducts, by immunoaffinity chromatography [28]. Postlabeling can be extremely sensitive and requires relatively small amounts of DNA for analysis, which can be advantageous in studies where sample sizes are limited. However, postlabeling generally does not provide direct structural information about the adducts detected. Identification of adducts by postlabeling is generally inferred from cochromatography with adduct standards of known structure. However, particularly for *in vivo* human studies, the patterns of adducts observed are quite complex and many adducts may not be resolvable from each other, so that identification of individual adducts is not always possible.

4.3 CONSIDERATIONS FOR THE USE OF DNA ADDUCTS IN MOLECULAR EPIDEMIOLOGY

The central role for DNA adducts in the process of genotoxic chemical carcinogenesis makes them ideal candidates for use as biomarkers of both exposure and effect. DNA adducts, at least in the case of bulky adducts, retain enough chemical structural identity to be directly relatable back to the chemical exposure that induced them. While mutations in critical genes may be more directly relatable to tumorigenesis, they provide no direct information regarding the identity of the initiating chemical. For example, the observation of a G to T transversion in codon 12 of K-*ras* may indicate that a critical early event in tumorigenesis has occurred, but it says nothing about the nature of the DNA damage that induced the mutation.

While the conceptual use of adducts as biomarkers seems straightforward, there are important factors that should be considered in either designing or interpreting such studies. These include the selection of tissues for monitoring, the determination of when to measure adducts, and understanding the relationships between the observed adducts and adverse health endpoints.

In many instances, chemical carcinogens exhibit marked target tissue specificity. In such cases, DNA adducts formed in the target tissues are typically the most relevant to relate molecular dose to adverse health effect. In laboratory studies, this usually presents no difficulty. However, for the application of DNA adducts to biomonitoring studies in human populations, target tissues may frequently be inaccessible. As a consequence, much human adduct work has focused on the use of more easily obtainable surrogate tissues, such as peripheral blood lymphocytes, buccal cells, or exfoliated bladder cells, rather than the internal tissues that may be the actual targets for tumorigenesis. To fully utilize DNA adducts in surrogate tissues for such studies, it is necessary to understand the relationships between the types of adducts formed and the kinetics of adduct formation and removal in both the surrogate tissue and the target tissue. In laboratory animal studies, it can be demonstrated that administration of PAHs results in systemic distribution, and that DNA adducts can be detected in most tissues examined. Following single intraperitoneal (i.p.) doses of B[*a*]P or benzo[*b*]fluoranthene in rats, simple relationships can be observed between adduct levels in lung, liver, and peripheral blood lymphocytes [29,30].

When attempting to relate DNA adduction with the induction of cancer it is also important to consider the temporal relationships between the induction and removal of DNA damage and the induction of adverse health effects. Cancer in humans is a disease that has a latency period between initial induction and progression to a detectable tumor that is measured in years or decades. By the time of tumor formation, any adducts measured within that tumor or adjacent normal tissue are not the adducts that gave rise to that tumor, and have no association with the original adducts that induced the tumor. The cell integrates DNA adducts over a time span that is a function of cell turnover, DNA repair rates, and chemical stability of the adducts in the specific tissue measured. Cell turnover rates may be directly affected by the extent of DNA damage induced. The DNA adducts are only a measure of this damage. In many laboratory studies, levels of DNA adducts have been determined at a single arbitrary time point following exposure. This measured adduct level is then typically used as an index of DNA damage that is related to administered dose, adverse health endpoint, or both. This approach is only valid in those rare cases where steady-state levels of DNA damage are attained. In all other cases, it is unlikely that any useful conclusion may be drawn from such measurements of DNA adduct levels at a single arbitrary time point.

In general, after single administration of a genotoxic chemical, DNA adduct levels in a tissue increase relatively rapidly, reach a maximal level, and then decline at a more gradual rate (see Figure 4.2). Tumors induced by the DNA adducts only become apparent long after the DNA adducts which induced them are gone. The exact kinetics depend upon the chemical, the dose, the route of administration, the tissue, and the species. Thus, the instantaneous adduct level at any single point of time following an unknown external dose of a carcinogen is not going to necessarily be predictive of risk.

This is illustrated in Figure 4.3, which shows the hypothetical DNA adduct levels present in a tissue after four different arbitrary exposures to a carcinogen that take place at different times. The curve A and the curve B represent the adducts induced by identical doses of the carcinogen, while the curve D represents a higher exposure and the curve C represents a lower exposure. We wish to predict the tumor yield at some point far in the future as a consequence of these exposures, but have no a priori information concerning the magnitude or timing of the exposures. If we measure DNA adducts at time T1, we observe that one individual (curve A) has a significant adduct level, and thus

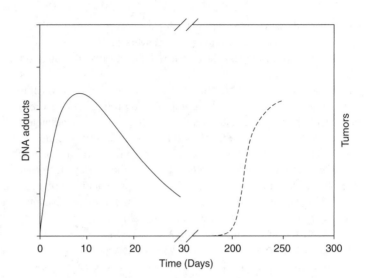

FIGURE 4.2 The temporal relationship between formation of DNA adducts (solid curve) and tumors (dashed curve) following a single exposure. Note that there is a significant latency period between DNA adduct induction and visible tumor formation. These curves and times are typical for polycyclic aromatic hydrocarbon adducts and induced tumors in the lungs of A/J mice. Similar patterns would be observed for other tissues and species, but with a different time scale. For humans, the latency period may be measured in years or decades.

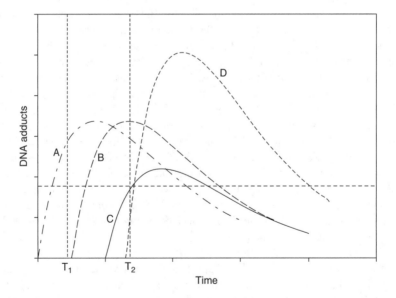

FIGURE 4.3 Hypothetical DNA adduct formation for four individuals exposed to an adducting chemical at four different times. The black and red curve represent DNA adducts following exposure to the same amount of chemical, while the blue curve represents adducts induced by a higher exposure and the magenta curve represents a lower exposure. While the total adduct burden, and thus risks of cancer should be blue > black = red > magenta, the actual relationship in adduct levels observed is strongly dependent upon when the adducts are measured.

some risk of tumorigenesis, but find no DNA adducts for the other three individuals. If we sample at time T2 we find that all of the individuals have detectable adducts, that is the greatest adduct level, and thus risk is in the individual represented by the curve B; with somewhat lower adduct levels in the individual represented by the curve A, and even fewer by the curve C, with the fewest observed in the individual represented by the curve D. It is clear that the greatest adduct burden is experienced by the individual represented by D, the least by C, and an intermediate and equal level by the A and B. Similarly, it is clear that a given adduct level, represented by the horizontal line, can be found in each of these individuals at one or more time points.

This relationship between DNA adduct levels in a target tissue following administration of a carcinogen and the subsequent induction of cancer has been examined in a various rodent models. In several studies, it appears that the total adduct burden experienced by the target tissue over time is predictive of the risk of tumorigenesis, whether as a result of chronic [31] or acute administration [32,33]. This suggests that the induction of cancer is a stochastic process at the level of DNA adduction. The probability of generating a necessary complement of oncogenic alterations in the DNA of a cell is determined by the number of times that cellular repair or replication enzymes encounter a promutagenic adduct. Interestingly, for a number of diverse PAHs, the probability of tumorigenesis is similar for similar levels of DNA adduction over time [32]. This suggests that while the molecular details of specific DNA adducts may be important in dictating what specific base misincorporation may occur, the probability of a base misincorporation occurring may be more a property of the replication and repair enzymes when they encounter an adduct. One final caveat to the interpretation of human biomonitoring studies using adducts is that, in most cases, exposures are neither single acute events nor chronic exposures, but are more often episodic exposures of uncertain magnitude and timing.

4.4 EXAMPLES OF DNA ADDUCTS

4.4.1 DNA ADDUCTS OF BAY-REGION POLYCYCLIC AROMATIC HYDROCARBONS

The polycyclic aromatic hydrocarbons (PAHs) are a large class of incomplete combustion by-products consisting of fused aromatic rings. In general, PAHs are incapable of binding to DNA in the absence of metabolic activation. Much of the focus of PAH carcinogenesis has been on the stable DNA adducts formed by the epoxides resulting from aromatic ring oxidation carried out by cytochromes P450. In particular, the bay-region diol epoxide adducts of a number of PAHs have been shown to be capable of forming a variety of stable covalent DNA adducts.

Benzo[a]pyrene has served as the archetypical bay-region PAH in studies of DNA adduction. Most of the research on B[a]P has centered on the role of the bay-region diol epoxide adducts. B[a]P-7,8-dihydrodiol-9,10-epoxide (BPDE) can exist in two diastereomeric forms, referred to as *syn-*, having the hydroxyl at position 7 on the same face of the molecule as the epoxide, and *anti-*, having the hydroxyl at position 7 on the opposite face of the molecule from the epoxide. Confusingly, the *anti-*diol epoxide has been referred to in the literature as BPDE I or BPDE 2, while the *syn-*diol epoxide has been referred to as BPDE II or BPDE 1. Note that two enantiomeric forms are possible for each diastereomer. To further complicate the issue, some sources refer to the 9,10-dihydrodiol-7,8-epoxide as BPDE II, while others refer to this metabolite as BPDE III or as reverse-diol epoxide. It is perhaps least ambiguous to refer to the various diol epoxides by the configurational assignments of the oxidized ring. Thus, the *anti-*B[a]P-7,8-dihydrodiol-9,10-epoxides are either (+)7R,8S-dihydroxy-9S, 10R-epoxy-7,8,9,10-tetrahydrobenzo[a]pyrene or (−)7S,8R-dihydroxy-9R, 10S-epoxy-7,8,9,10-tetrahydrobenzo[a]pyrene, while the *syn-*diol epoxides are either (−)7R,8S-dihydroxy-9R,10S-epoxy-7,8,9,10-tetrahydrobenzo[a]pyrene or (+)7S,8R-dihydroxy-9S,10R-epoxy-7,8,9,10-tetrahydrobenzo[a]pyrene (Figure 4.4). Each of these diol epoxides possesses differing reactivities with DNA, which have earlier been reviewed in detail [34]. The primary form in most mammalian tissues is (+)7R,8S-dihydroxy-9S,10R-epoxy-7,8,9,10-tetrahydrobenzo[a]pyrene [35], which binds primarily to deoxyguanosine to form 7R,8S,9S-trihydroxy-10S-N^2-deoxyguanosinyl)-7,8,9,10-tetrahydrobenzo[a]pyrene, also referred to simply as 10S(+)-*trans-anti*-benzo[a]pyrene-N^2-dG. The other enantiomers also bind to DNA,

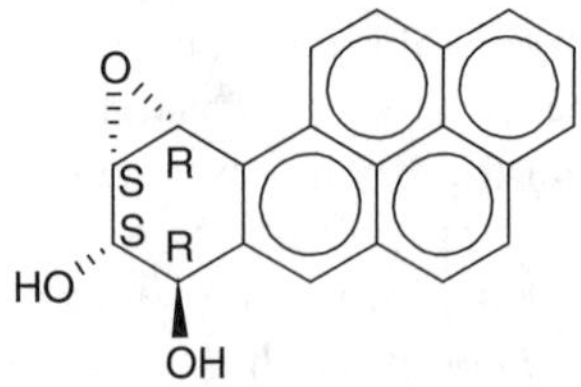

(+)7R,8S-dihydrodiol-9S,10R-epoxy-7,8,9,10-tetrahydrobenzo[a]pyrene

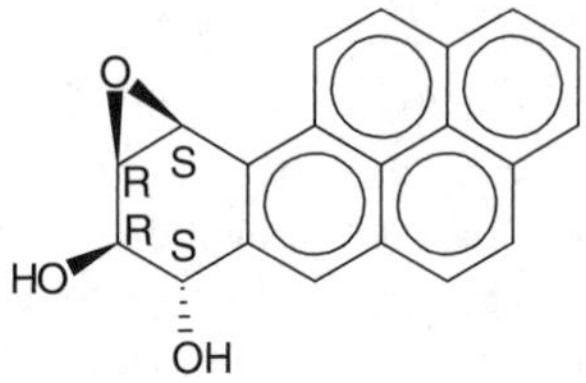

(−)7S,8R-dihydrodiol-9R,10S-epoxy-7,8,9,10-tetrahydrobenzo[a]pyrene

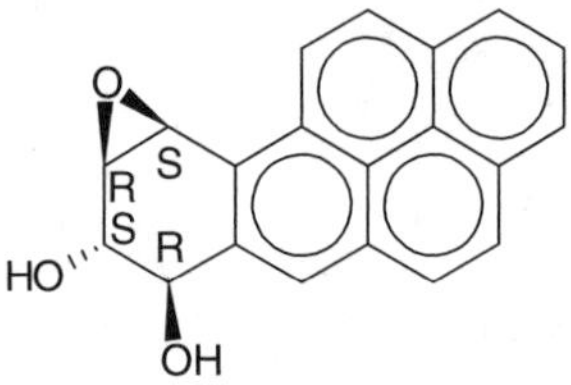

(−)7R,8S-dihydrodiol-9R,10S-epoxy-7,8,9,10-tetrahydrobenzo[a]pyrene

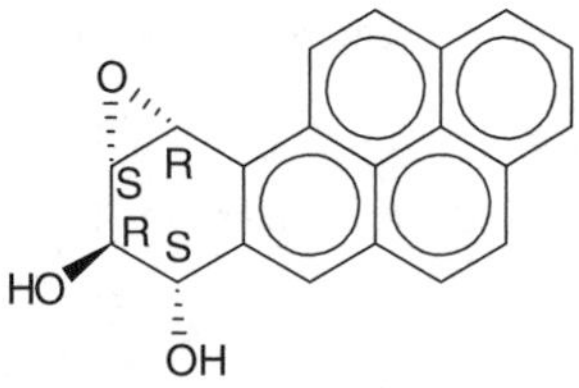

(+)7S,8R-dihydrodiol-9S,10R-epoxy-7,8,9,10-tetrahydrobenzo[a]pyrene

FIGURE 4.4   Reactive bay-region dihydrodiol epoxides of benzo[a]pyrene.

and the primary site of adduction (∼95%) is the N^2 of deoxyguanosine, with only minor binding to deoxyadenosine [36] (Figure 4.5). An additional adduct has been identified *in vivo* and results from the binding of 9-hydroxy-4,5-epoxy-benzo[a]pyrene to deoxyguanosine, although the absolute structure has not been elucidated.

In addition to the monooxygenation of B[a]P to diol epoxide route of metabolic activation, other pathways of B[a]P metabolism leading to DNA adducts have been identified. In particular, one-electron activation leads to the formation of a radical cation. This radical cation is capable of binding to DNA to form several covalent adducts (Figure 4.6), each of which is unstable, spontaneously depurinating to form abasic sites [8,37–39]. Analysis of the stable and unstable adducts formed in isolated rat liver nuclei showed that about 80% of the total B[a]P adducts formed were unstable adducts, and only about 10% could be identified as (+)7R,8S-dihydrodiol-9S, 10R-epoxy-7,8,9,10-tetrahydrobenzo[a]pyrene-N^2-deoxyguanosine. The depurinating adducts identified were

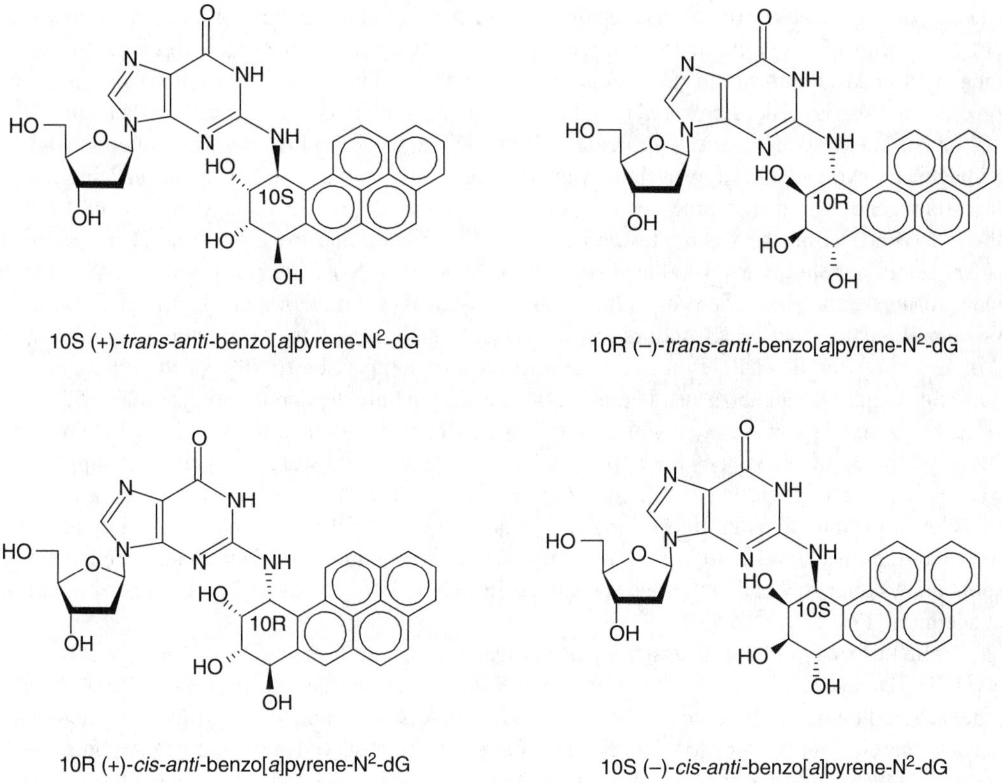

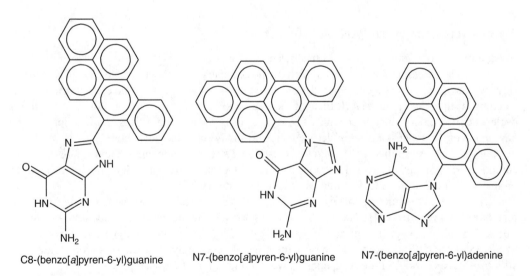

FIGURE 4.5 N^2-deoxyguanosine adducts of benzo[a]pyrene.

C8-(benzo[a]pyren-6-yl)guanine   N7-(benzo[a]pyren-6-yl)guanine   N7-(benzo[a]pyren-6-yl)adenine

FIGURE 4.6 Depurinating adducts of benzo[a]pyrene resulting from one-electron oxidation.

N7-(benzo[a]pyren-6-yl)guanine ($\sim$20%), N7-(benzo[a]pyren-6-yl)adenine ($\sim$60%), and C8-(benzo[a]pyren-6-yl)guanine ($\sim$5%).

There is ongoing debate as to the relative roles of depurinating and stable adducts in the etiology of PAH carcinogenesis. On one hand, abasic sites are promutagenic, and an excess production of abasic sites beyond the capacity of cellular repair mechanisms should be capable of inducing mutations in critical genes. Additionally, many of the most carcinogenic PAHs have ionization potentials in the range that makes them candidates for one-electron activation in addition to monooxygenation, and several have also been demonstrated to induce depurinating adducts *in vivo*. On the other hand, 10^3 to 10^4 abasic sites/day are generated in each cell [40,41], and efficient mechanisms to cope with this damage have evolved. One approach to evaluating the relative roles of stable versus unstable adducts in carcinogenesis is to examine the mutations in oncogenes present in tumors induced by these PAHs for correlations between sites and extents of adduction and mutation. The H-*ras* oncogene is frequently mutated in PAH-induced skin tumors, while the K-*ras* gene is often mutated in PAH-induced lung tumors in rodent models. In one study of H-*ras* mutations in B[*a*]P-induced mouse skin papillomas, 7 out of 13 tumors (54%) had detectable mutations at guanine positions, while 2 of 13 (15%) had detectable mutations at adenines, in comparison to 70% of the total adduction occurring at guanine and 30% at adenine [42]. Considering only the depurinating adducts, 59% were at guanines and 41% were at adenines. While the agreement between the fraction of adduction at adenines and at guanines does not appear to be robust, these data were interpreted as supporting a role for depurinating adducts as the major species contributing to H-*ras* activating mutations. In a study of K-*ras* mutations in mouse lung adenomas induced by B[*a*]P, 22 out of 22 tumors reported had mutations in codon 12, all at guanine residues, with no mutations at adenines observed. Among 9 spontaneous tumors, 7 had mutations at guanines in codons 12 or 13, while 2 had codon 61 mutations at adenines [17].

While this study did not measure depurinating adducts, there is no evidence for a contribution of any DNA damage at adenines to the spectrum of K-*ras* mutations recovered. In a similar study of the potent PAH dibenzo[*a*, *l*]pyrene in mouse lung adenomas, stable adducts were identified at guanines and adenines in equal proportion, and K-*ras* mutations were observed at similar levels at guanines in codon 12 (44%) and adenines in codon 61 (56%) [33]. While these studies do not exclude a role for unstable adducts in PAH carcinogenesis, most of the data are consistent with a major role for stable adducts.

4.4.2 CYCLOPENTA-FUSED PAH

In addition to bay-region activation, other structural features of PAHs provide sites for metabolic activation. The addition of an ethylene moiety to the periphery of a PAH to create a fused cylcopenta ring provides a ready substrate for epoxidation, and many such cyclopenta-fused PAH are more potent carcinogens than their parent PAH. Each of these are epoxidated *in vivo* to form the corresponding cyclopenta ring oxide which is capable of binding to DNA. The primary sites of adduction, as with bay-region diol epoxides, are N^2 of deoxyguanosine and N^6 of deoxyadenosine.

One of the best studied of the cyclopenta-fused PAH is cyclopenta[*c,d*]pyrene (CPP). CPP administered to Sprague Dawley rats induces DNA adducts that cochromatograph with the reaction products of CPP-3,4-epoxide with deoxyadenosine and deoxyguanosine, with deoxyadenosine predominating [43]. The structures of several CPP-deoxyguanosine adducts formed in strain A/J mouse lung have been reported [44], and are shown in Figure 4.7. The major adducts identified were *cis*-(3R,4S)N^2-CPP-deoxyguanosine and *trans*-(3S,4S)N^2-CPP-deoxyguanosine, with smaller amounts of *cis*-(3S,4R)N^2-CPP-deoxyguanosine also identified. Interestingly, no adduct cochromatographing with *trans*-(3R,4R)N^2-CPP-deoxyguanosine were found in the *in vivo* samples.

4.4.3 NITRO-PAH

Many environmental PAHs are nitrated at one or more positions, and many of these are potent carcinogens. For many nitro-PAHs, the primary activation pathway is via reduction to the corresponding amine, rather than monooxygenation. These amino-PAHs are then activated via the same *N*-hydroxylation/esterification pathways as are other arylamines, yielding predominately

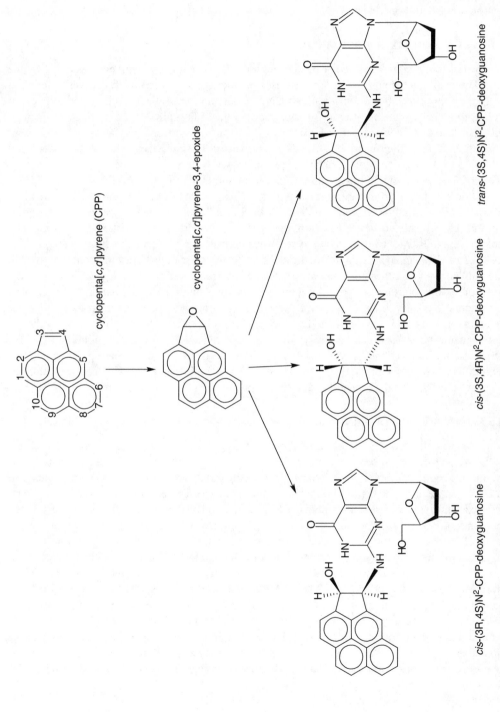

FIGURE 4.7 Cyclopenta ring activation and characterized adducts of cyclopenta[c,d]pyrene.

C-8-deoxyguanosine adducts. Thus 1,6-dinitropyrene yields N-(deoxy-guanosin-8-yl)-1-amino-6-nitropyrene [45], and 1-nitropyrene yields N-(deoxy-guanosin-8-yl)-1-aminopyrene [46]. However, some nitro-PAHs are activated via ring oxidation as well as nitroreduction. For example, 5-nitrobenzo[b]naphtho[2,1-d]thiophene is metabolized by both reduction of the nitrogen as well as oxidation of the 9,10-bond [47].

4.4.4 DNA ADDUCTS OF AROMATIC AMINES

The occurrence of a high incidence of bladder tumors among textile dye workers was noted by the 1890s. The causative agents were identified as aromatic amines. Some of the earliest studies to identify the role of metabolic activation and DNA adduction in chemical carcinogenesis involved elucidation of the mechanisms of action of arylamines by the Millers [48,49]. The primary pathway of metabolism leading to DNA adduction by aromatic amines involves the hydroxylation and subsequent esterification of the amino function. These N-hydroxyesters form reactive nitrenium ions that bind primarily to the C-8 of deoxyguanosine residues in DNA.

One of the best-studied aromatic amines is 2-acetylaminofluorene (2-AAF), and much of our current understanding of aromatic amines metabolic activation and DNA adduction derives from these studies. 2-AAF is initially metabolized to N-hydroxy-2-acetylaminofluorene by various cytochromes P450, and subsequently converted to the sulphate ester by the action of a sulfotransferase [50]. The sulphate ester typically decomposes rapidly to form a reactive nitrenium ion that binds preferentially to the C-8 of deoxyguanosine in DNA. In addition to the sulphotransferase-mediated activation to the sulphates ester, other enzymes may catalyze the formation of esters that also lead to the reactive nitrenium ion. One of the major alternate pathways involves the formation of N-acetoxy-2-acetylaminofluorene produced by O-acetylases. The nitrenium ion can also rearrange to form a carbonium ion at the C-3 of the fluorenyl moiety, which preferentially reacts with the exocyclic N^2 of deoxyguanosine. 2-AAF can be enzymatically deacetylated to form 2-aminofluorene (2-AF), which is also metabolized via N-hydroxylation and esterification to yield the corresponding reactive nitrenium ion. In mammalian systems, the major DNA adduct detected following administration of either 2-AAF or 2-AF is the deacetylated N-(deoxyguanosin-8-yl)-2-aminofluorene (Figure 4.8).

4.4.5 DNA ADDUCTS OF AFLATOXINS

The aflatoxins were identified in the early 1960s as the agents responsible for "Turkey X disease." The disease was traced to feed containing ground peanuts, which proved to be contaminated with the mold *Aspergillus flavus*. Two related *bis*furanocoumarins were identified and named aflatoxin (from *A. flavus* toxin) B and G (for blue and green fluorescence). These compounds were shown to be hepatotoxic and to induce hepatocarcinomas (HCC) in rats and other species. Aflatoxin exposure due to mold in foodstuffs is a major cause of liver cancer in human populations, especially in some areas of the world where food storage conditions favor mold growth. Aflatoxin exposure also appears to be synergistic with hepatitis B viral infection in inducing human HCC. The major adduct of Aflatoxin B_1 is shown in Figure 4.9, and results from binding to N7 of deoxyguanosine. As the adduction at N7 induces an excess positive charge on the N7, the adduct is somewhat unstable, and either hydrolyzes at the glycosyl bond, forming an apurinic site, or hydrolyzes the imidazole ring to form a ring-opened Aflatoxin B_1-formamidopyrimidine (FAPY) adduct, which exits as an equilibrium between two rotationally related forms. Analysis of the intrinsic mutagenicity of these adducts in *E. coli* demonstrates that both the AFB$_1$-N7-dG and FAPY adducts are highly mutagenic, and induce the same types of mutations as are seen in mutated *p53* genes in human HCC induced by aflatoxin B_1 [51].

4.4.6 SMALL ALKYL DNA ADDUCTS

Many potent carcinogens act as alkylating agents. These include dimethyl and diethyl nitrosamine, N-methyl-N-nitrosourea, N-ethyl-N-nitrosourea, and 1,2-dimethyl hydrazine. Each of these

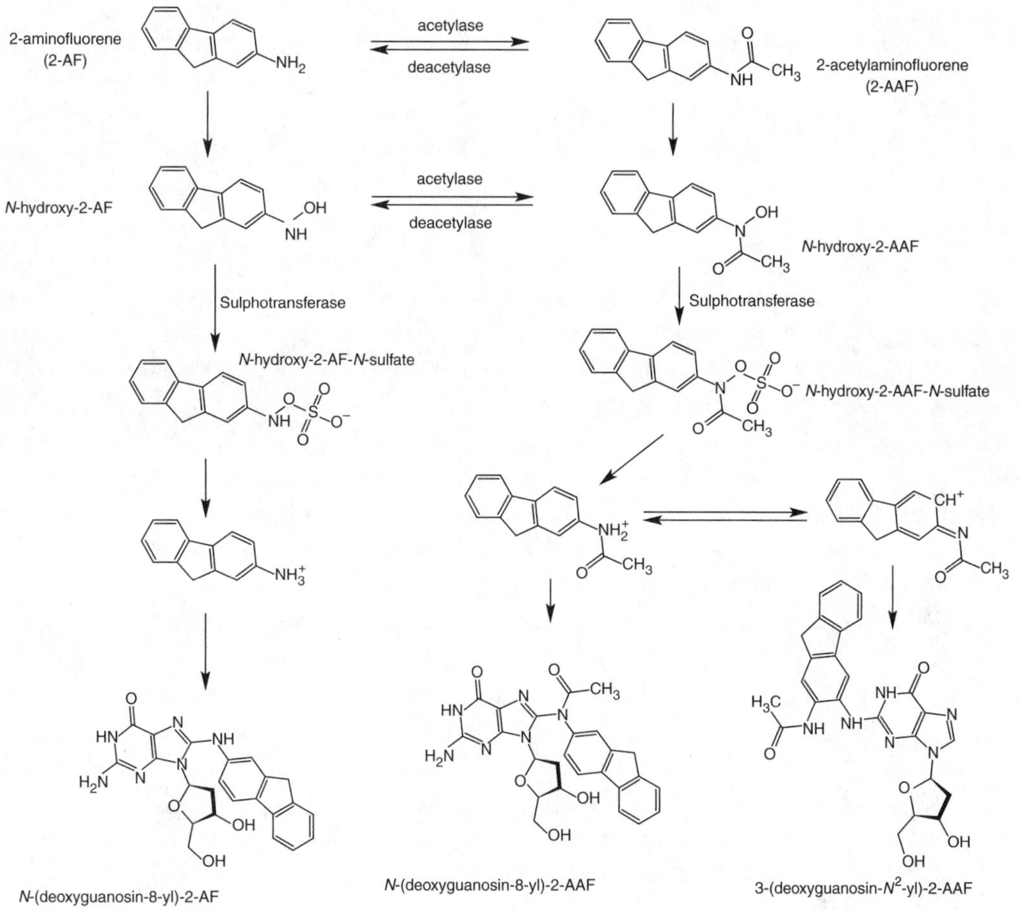

FIGURE 4.8 Metabolic activation and major DNA adducts of 2-acetylaminofluorene and 2-aminofluorene.

produces a major adduct by alkylation of the N7 of deoxyguanosine. The major adduct formed in liver following administration of dimethylnitrosamine is 7-methylguanine, and administration of diethylnitrosamine yields predominately 7-ethylguanine. Attack at the phosphate of the DNA backbone also occurs to a significant extent with these agents. In addition, each produces lesser amounts of 1-, 3-, or 7-alkyl adenine, 3- or O^6-alkyl guanine, 3-, O^{2-}, and O^4-alkylthymine, and 3- or O^2-alkylcytosine.

4.4.7 DNA ADDUCTS OF HALOGENATED COMPOUNDS

Vinyl chloride is a widely used industrial chemical, and is known to be carcinogenic in humans, with a highly significant occurrence of angiosarcoma of the liver associated with exposure. Vinyl chloride is metabolized primarily by CYP450 2E1 to form chloroethylene oxide, which reacts with DNA to form predominately (~98%) 7-(2-oxoethyl)-deoxyguanosine, as well as much smaller amounts of N^6,3-ethenodeoxyguanosine, 1,N^6-ethenodeoxyadenosine, and 3,N^4-ethenodeoxycytosine (Figure 4.10). However, 7-(2-oxoethyl)-deoxyguanosine has not been demonstrated to be highly promutagenic. It is the minor etheno-base adducts that appear to contribute most to the carcinogenicity of vinyl chloride. Cloroethylene oxide can also rearrange to form chloroacetaldehyde, which can also react with DNA to form the etheno-base adducts, but not 7-(2-oxoethyl)-deoxyguanosine.

Ethylene dibromide is activated via an altogether different mechanism involving conjugation to glutathione. Glutathione S-transferase theta catalyzes the formation of

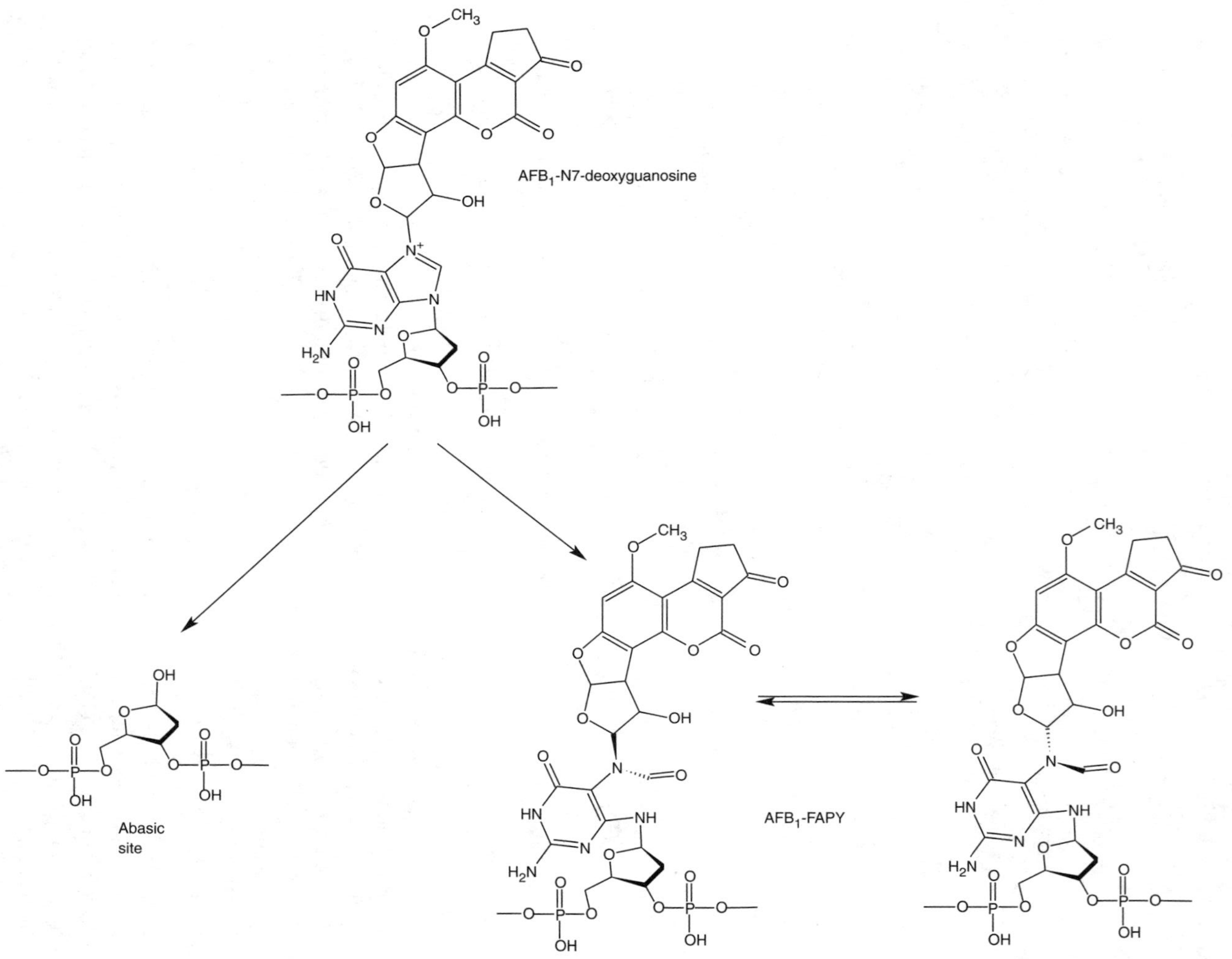

FIGURE 4.9 Structures of aflatoxin B_1-N7-deoxyguanosine adduct and its ring-opened formamidopyrimidine (FAPY) form.

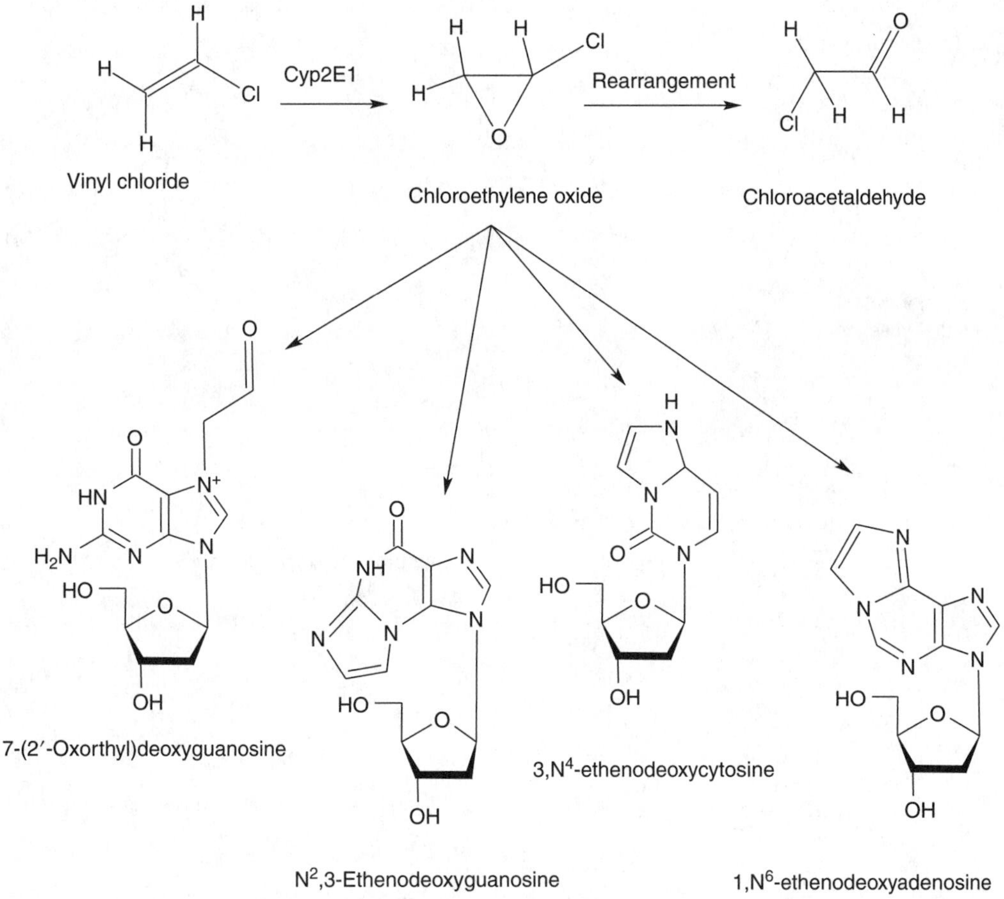

FIGURE 4.10 Activation and DNA binding products of vinyl chloride.

S-(2-bromoethyl)glutathione from the conjugation of ethylene dibromide with glutathione. This conjugate yields a reactive episulfonium ion that reacts with the N-7 of deoxyguanosine to form *S*-[2-(N7-deoxyguanosinyl)ethyl]glutathione [52–54] (Figure 4.11). The same pathway pertains to other dihaloethanes; 1,2-dichloroethane administered to rats induces the same adduct in the liver and kidney. An additional minor adduct formed via the same pathway but involving binding to deoxyadenosine has been identified as *S*-[2-(N1-adenyl)ethyl]glutathione [55].

4.4.8 ENDOGENOUS ADDUCTS

A variety of endogenous DNA adducts are also formed in most mammalian species studied. One of the major sources of endogenous adducts results from lipid peroxidation. Malondialdehyde is one of the major products of lipid peroxidation and binds to deoxyguanosine residues to form pyrimido[1,2-α]purin-10(3H)-one (M_1G) [56]. This adduct is highly promutagenic [57], and can also be formed via the reaction of base propenals with DNA [58]. Another abundant lipid peroxidation product is 4-hydroxynonenal, which is oxidized to form reactive 2,3-epoxy-4-hydroxynonenal. This can react with DNA bases to form 1,N^2-ethenodeoxyguanosine, N^2,3-ethenodeoxyguanosine, 3,N^4-ethenodeoxycytosine, and 1,N^6-ethenodeoxyadenosine. An additional deoxyadenosine adduct has also been reported, 1″-[3-(2′-deoxyribosyl)-3H-imidazole[2,1-i]purin-7-yl]heptane-2′-one [59] (Figure 4.12).

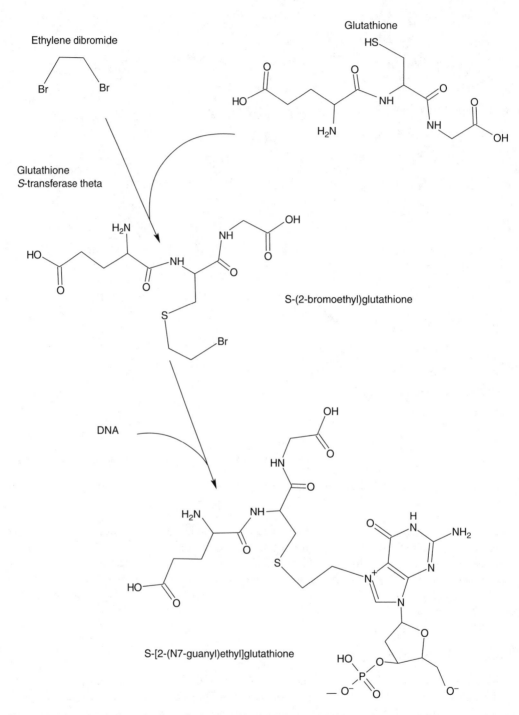

FIGURE 4.11 Metabolic activation of ethylene dibromide by glutathione conjugation and structure of major DNA adduct.

4.5 CONCLUDING REMARKS

The establishment of the specific linkages between the formation of DNA adducts and the subsequent induction of oncogene mutations and tumors in a variety of model systems has cemented the causative role of DNA adducts in the mode of action of many genotoxic chemicals. The development of molecular epidemiological approaches utilizing DNA adducts as both biomarkers of exposure and predictors

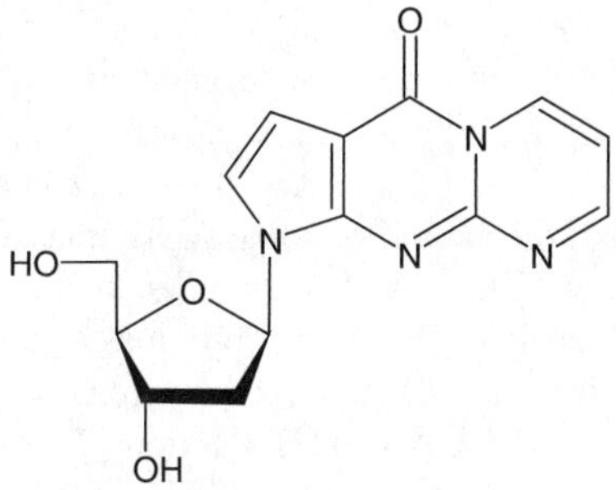

pyrimido[1,2-a]purin-10(3H)-one (M₁G)

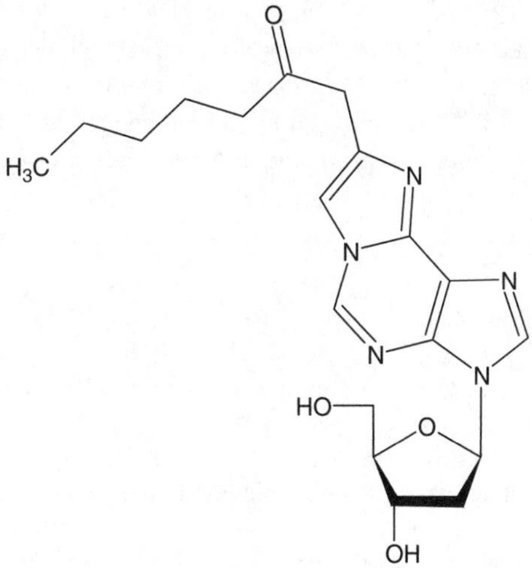

1″-[3-(2′-deoxyribosyl)-3H-imidazole[2,1-i]purin-7-yl]heptane-2″-one

FIGURE 4.12 Major DNA adducts induced by lipid peroxidation by-products.

of adverse health outcomes has been an active and fruitful area of research over the past two decades. Such approaches will benefit from further research to define the impact of individual polymorphisms of metabolism and repair genes in modulating the biological effects of DNA adduction, as well as further development of sensitive methods for adduct measurement and identification. Ultimately, DNA adducts may prove to serve as a useful metric for extrapolating cancer risk from model studies to exposed human populations.

N-Heterocyclic aromatics require metabolic activation to electrophilic species in order to bind covalently with cellular macromolecules and exert their carcinogenic effects [60–63]. Dibenz[*c,g*]carbazole is metabolized to a series of phenols, of which mutagenic 3-hydroxyDBC is considered to be the proximate metabolite in mouse liver [64], which further reacts to bind to liver and skin DNA [64,65] preferentially to polyguanylic acid *in vitro* [66]. DBC appears to be organotropic, in that there is a relationship between DBC–DNA adduct levels and target organ specificity [65,67].

The pattern of mutations in the *ras* gene appears to be compound-specific and consistent with the DNA adduct profile of the compound. For example, the majority of K-*ras* mutations found in BaP-induced lung tumors in strain A/J mice were GC to TA transversions at the second base of codon 12 [68,69]. The mutations occurring in BaP-induced lung tumors involving a guanine base are consistent with the major DNA adduct being formed with the bay-region diol epoxide of BaP, N^2-{10β-(+7β,8α,9α-trihydroxy-7,8,9,10-tetrahydrobenzo[*a*]pyrene)y1}deoxyguanosine [69]. The primary K-*ras* mutation observed with BJA and CPP is a GC to CG transversion at codon 2, which appear to be a result of cyclopenta ring-derived adducts [69,70].

Interestingly, other PAHs also induce mouse tumors with similar *ras* codon 61 mutations. In liver, only 7,12-dimethylbenz[*a*]anthracene given intraperitoneally has been shown to induce liver tumors with Ha- and Ki-*ras* gene mutations [71,72]. These tumors have Ha-*ras* codon 61 CAA to CTA mutations, much like the DBC-induced tumors, but also show Ki-*ras* codon 13 GGC to CGC mutations. Both 7,12-dimethylbenz[*a*]anthracene and dibenz[*c,h*]acridine papillomas and skin carcinomas have Ha-*ras* codon 61 CAA to CTA mutations [73], which are similar to those induced by DBC. Brown et al. [74] reported that 7,12-dimethylbenz[*a*]anthracene papillomas and skin carcinomas with the same Ha-*ras* codon 61 mutations. On the other hand, benzo[*a*]pyrene papillomas had mostly (70%) Ha-*ras* codon 12 and 13 mutations and 20% CAA to CTA codon 61 mutations [75]. Benzo[*c*]phenanthrene-3,4-diol-1,2-epoxide induces mouse skin tumors of which 75% have Ha-*ras* codon 61 CAA to CTA mutations at the 100-nmol dose and 34.5% at the 400-nmole dose [76].

This chapter has been reviewed for publication by the National Health and Environmental Effects Research Laboratory of the U.S. Environmental Protection Agency and approved for publication. Approval does not signify that the contents necessarily reflect the views of the agency, nor does the mention of trade names, products, or techniques constitute endorsement or recommendation for use.

REFERENCES

 1. Furuichi, Y. (2001) Premature aging and predisposition to cancers caused by mutations in RecQ family helicases. *Ann. N.Y. Acad. Sci.*, 928, 121–131.
 2. Lehmann, A.R. and Norris, P.G. (1990) DNA repair deficient photodermatoses. *Semin. Dermatol.*, 9, 55–62.
 3. (1998) Genetic susceptibility to cancer. ICRP publication 79. Approved by the Commission in May 1997. International Commission on Radiological Protection. *Ann. ICRP*, 28, 1–157.
 4. Benhamou, S. and Sarasin, A. (2000) Variability in nucleotide excision repair and cancer risk: A review. *Mutat. Res.*, 462, 149–158.
 5. Sekiguchi, M. and Sakumi, K. (1997) Roles of DNA repair methyltransferase in mutagenesis and carcinogenesis. *Jpn. J. Hum. Genet.*, 42, 389–399.
 6. Brookes, P. (1966) Quantitative aspects of the reaction of some carcinogens with nucleic acids and the possible significance of such reactions in the process of carcinogenesis. *Cancer Res.*, 26, 1994–2003.
 7. Miller, J.A. and Miller, E.C. (1965) Natural and synthetic chemical carcinogens in the etiology of cancer. *Cancer Res.*, 25, 1292–1304.
 8. Cavalieri, E. and Rogan, E. (1985) Role of radical cations in aromatic hydrocarbon carcinogenesis. *Environ. Health Perspect.*, 64, 69–84.
 9. Penning, T.M. (1993) Dihydrodiol dehydrogenase and its role in polycyclic aromatic hydrocarbon metabolism. *Chem. Biol. Interact.*, 89, 1–34.
10. Canova, S., Degan, P., Peters, L.D., Livingstone, D.R., Voltan, R., and Venier, P. (1998) Tissue dose, DNA adducts, oxidative DNA damage and CYP1A-immunopositive proteins in mussels exposed to waterborne benzo[a]pyrene. *Mutat. Res.*, 399, 17–30.
11. Kadlubar, F.F. and Badawi, A.F. (1995) Genetic susceptibility and carcinogen-DNA adduct formation in human urinary bladder carcinogenesis. *Toxicol. Lett.*, 82–83, 627–632.
12. Badawi, A.F., Stern, S.J., Lang, N.P., and Kadlubar, F.F. (1996) Cytochrome P-450 and acetyltransferase expression as biomarkers of carcinogen-DNA adduct levels and human cancer susceptibility. *Prog. Clin. Biol. Res.*, 395, 109–140.
13. Ford, G.P. and Scribner, J.D. (1990) Prediction of nucleoside-carcinogen reactivity. Alkylation of adenine, cytosine, guanine, and thymine and their deoxynucleosides by alkanediazonium ions. *Chem. Res. Toxicol.*, 3, 219–230.
14. Mackay, W., Benasutti, M., Drouin, E., and Loechler, E.L. (1992) Mutagenesis by (+)-anti-B[a]P-N2-Gua, the major adduct of activated benzo[a]pyrene, when studied in an Escherichia coli plasmid using site-directed methods. *Carcinogenesis*, 13, 1415–1425.
15. Eisenstadt, E., Warren, A.J., Porter, J., Atkins, D., and Miller, J.H. (1982) Carcinogenic epoxides of benzo[a]pyrene and cyclopenta[c,d]pyrene induce base substitutions via specific transversions. *Proc. Natl. Acad. Sci. USA*, 79, 1945–1949.

16. Kohler, S.W., Provost, G.S., Fieck, A., Kretz, P.L., Bullock, W.O., Sorge, J.A., Putman, D.L., and Short, J.M. (1991) Spectra of spontaneous and mutagen-induced mutations in the lacI gene in transgenic mice. *Proc. Natl. Acad. Sci. USA*, 88, 7958–7962.

17. Mass, M.J., Jeffers, A.J., Ross, J.A., Nelson, G., Galati, A.J., Stoner, G.D., and Nesnow, S. (1993) Ki-*ras* oncogene mutations in tumors and DNA adducts formed by benz[*j*]aceanthrylene and benzo[*a*]pyrene in the lungs of strain A/J mice. *Mol. Carcinog.*, 8, 186–192.

18. Gorelick, N.J., Andrews, J.L., Gu, M., and Glickman, B.W. (1995) Mutational spectra in the lacI gene in skin from 7,12-dimethylbenz[a]anthracene-treated and untreated transgenic mice. *Mol. Carcinog.*, 14, 53–62.

19. Denissenko, M.F., Pao, A., Tang, M., and Pfeifer, G.P. (1996) Preferential formation of benzo[*a*]pyrene adducts at lung cancer mutational hotspots in P53. *Science*, 274, 430–432.

20. de Kok, T.M., Moonen, H.J., van Delft, J., and van Schooten, F.J. (2002) Methodologies for bulky DNA adduct analysis and biomonitoring of environmental and occupational exposures. *J. Chromatogr. B Analyt. Technol. Biomed. Life Sci.*, 778, 345–355.

21. Poirier, M.C., Santella, R.M., and Weston, A. (2000) Carcinogen macromolecular adducts and their measurement. *Carcinogenesis*, 21, 353–359.

22. Weston, A., Rowe, M., Poirier, M., Trivers, G., Vahakangas, K., Newman, M., Haugen, A., Manchester, D., Mann, D., and Harris, C. (1988) The application of immunoassays and fluorometry to the detection of polycyclic hydrocarbon-macromolecular adducts and anti-adduct antibodies in humans. *Int. Arch. Occup. Environ. Health*, 60, 157–162.

23. Weston, A., Rowe, M.L., Manchester, D.K., Farmer, P.B., Mann, D.L., and Harris, C.C. (1989) Fluorescence and mass spectral evidence for the formation of benzo[a]pyrene anti-diol-epoxide-DNA and -hemoglobin adducts in humans. *Carcinogenesis*, 10, 251–257.

24. Poirier, M.C. (1984) The use of carcinogen-DNA adduct antisera for quantitation and localization of genomic damage in animal models and the human population. *Environ. Mutagen.*, 6, 879–887.

25. Randerath, K., Reddy, M.V., and Gupta, R.C. (1981). ^{32}P-labeling test for DNA damage. *Proc. Natl. Acad. Sci. USA* 78, 6126–6129.

26. Reddy, M.V. and Randerath, K. (1987) 32P-postlabeling assay for carcinogen-DNA adducts: Nuclease P1-mediated enhancement of its sensitivity and applications. *Environ. Health Perspect.*, 76, 41–47.

27. Gupta, R.C. (1985) Enhanced sensitivity of 32P-postlabeling analysis of aromatic carcinogen: DNA adducts. *Cancer Res.*, 45, 5656–5662.

28. Guichard, Y., Nair, J., Barbin, A., and Bartsch, H. (1993) Immunoaffinity clean-up combined with 32P-postlabelling analysis of 1,N6-ethenoadenine and 3,N4-ethenocytosine in DNA. In *Postlabelling Methods for Detection of DNA Adducts*, Vol. 124 (Phillips, D.H. et al., Eds.), pp. 263–269. IARC Scientific Publications, Lyon, France.

29. Ross, J., Nelson, G., Kligerman, A., Erexson, G., Bryant, M., Earley, K., Gupta, R., and Nesnow, S. (1990) Formation and persistence of novel benzo(a)pyrene adducts in rat lung, liver, and peripheral blood lymphocyte DNA. *Cancer Res.*, 50, 5088–5094.

30. Ross, J.A., Nelson, G.B., Holden, K.L., Kligerman, A.D., Erexson, G.L., Bryant, M.F., Earley, K., Beach, A.C., Gupta, R.C., and Nesnow, S. (1992) DNA adducts and induction of sister chromatid exchanges in the rat following benzo[b]fluoranthene administration. *Carcinogenesis*, 13, 1731–1734.

31. Poirier, M.C. and Beland, F.A. (1992) DNA adduct measurements and tumor incidence during chronic carcinogen exposure in animal models: Implications for DNA adduct-based human cancer risk assessment. *Chem. Res. Toxicol.*, 5, 749–755.

32. Ross, J.A., Nelson, G.B., Wilson, K.H., Rabinowitz, J.R., Galati, A., Stoner, G.D., Nesnow, S., and Mass, M.J. (1995) Adenomas induced by polycyclic aromatic hydrocarbons in strain A/J mouse lung correlate with time-integrated DNA adduct levels. *Cancer Res.*, 55, 1039–1044.

33. Prahalad, A.K., Ross, J.A., Nelson, G.B., Roop, B.C., King, L.C., Nesnow, S., and Mass, M.J. (1997) Dibenzo[*a*, *l*]pyrene-induced DNA adduction, tumorigenicity, and Ki-*ras* oncogene mutations in strain A/J mouse lung. *Carcinogenesis*, 18, 1955–1963.

34. Szeliga, J. and Dipple, A. (1998) DNA adduct formation by polycyclic aromatic hydrocarbon dihydrodiol epoxides. *Chem. Res. Toxicol.*, 11, 1–11.

35. Conney, A.H. (1982) Induction of microsomal enzymes by foreign chemicals and carcinogenesis by polycyclic aromatic hydrocarbons: G.H.A. Clowes Memorial Lecture. *Cancer Res.*, 42, 4875–4917.

36. Meehan, T., Straub, K., and Calvin, M. (1977) Benzo[alpha]pyrene diol epoxide covalently binds to deoxyguanosine and deoxyadenosine in DNA. *Nature*, 269, 725–727.

37. Cavalieri, E.L. and Rogan, E.G. (1992) The approach to understanding aromatic hydrocarbon carcinogenesis. The central role of radical cations in metabolic activation. *Pharmacol. Ther.*, 55, 183–199.

38. Cavalieri, E.L. and Rogan, E.G. (1995) Central role of radical cations in metabolic activation of polycyclic aromatic hydrocarbons. *Xenobiotica*, 25, 677–688.

39. Devanesan, P.D., Higginbotham, S., Ariese, F., Jankowiak, R., Suh, M., Small, G.J., Cavalieri, E.L., and Rogan, E.G. (1996) Depurinating and stable benzo[*a*]pyrene-DNA adducts formed in isolated rat liver nuclei. *Chem. Res. Toxicol.*, 9, 1113–1116.

40. Nakamura, J. and Swenberg, J.A. (1999) Endogenous apurinic/apyrimidinic sites in genomic DNA of mammalian tissues. *Cancer Res.*, 59, 2522–2526.

41. Atamna, H., Cheung, I., and Ames, B.N. (2000) A method for detecting abasic sites in living cells: Age-dependent changes in base excision repair. *Proc. Natl. Acad. Sci. USA*, 97, 686–691.

42. Chakravarti, D., Pelling, J.C., Cavalieri, E.L., and Rogan, E.G. (1995) Relating aromatic hydrocarbon-induced DNA adducts and c-H-*ras* mutations in mouse skin papillomas: The role of apurinic sites. *Proc. Natl. Acad. Sci. USA*, 92, 10422–10426.

43. Beach, A.C. and Gupta, R.C. (1994) DNA adducts of the ubiquitous environmental contaminant cyclopenta[*c,d*]pyrene. *Carcinogenesis*, 15, 1065–1072.

44. Nelson, G., Ross, J., Prusiewicz, C., Sangaiah, R., and Gold, A. (2002) Identification of stereochemical configurations of cyclopenta[*c,d*]pyrene DNA adducts in strain A/J mouse lung and C3H10T1/2CL8 cells. *Polycyclic Aromatic Compd.*, 22, 923–931.

45. Djuric, Z., Potter, D.W., Culp, S.J., Luongo, D.A., and Beland, F.A. (1993) Formation of DNA adducts and oxidative DNA damage in rats treated with 1,6-dinitropyrene. *Cancer Lett.*, 71, 51–56.

46. Zhou, L. and Cho, B.P. (1998) Synthesis, characterization, and comparative conformational analysis of *N*-(deoxyguanosin-8-yl)aminopyrene adducts derived from the isomeric carcinogens 1-, 2-, and 4-nitropyrene. *Chem. Res. Toxicol.*, 11, 35–43.

47. King, L.C., Kohan, M.J., Brooks, L., Nelson, G.B., Ross, J.A., Allison, J., Adams, L., Desai, D., Amin, S., Padgett, W., Lambert, G.R., Richard, A.M., and Nesnow, S. (2001) An evaluation of the mutagenicity, metabolism, and DNA adduct formation of 5-nitrobenzo[b]naphtho[2,1-d]thiophene. *Chem. Res. Toxicol.*, 14, 661–671.

48. Miller, E.C., Juhl, U., and Miller, J.A. (1966) Nucleic acid guanine: Reaction with the carcinogen *N*-acetoxy-2-acetylaminofluorene. *Science*, 153, 1125–1127.

49. Miller, E.C. and Miller, J.A. (1966) Mechanisms of chemical carcinogenesis: Nature of proximate carcinogens and interactions with macromolecules. *Pharmacol. Rev.*, 18, 805–838.

50. Lai, C.C., Miller, J.A., Miller, E.C., and Liem, A. (1985) *N*-sulfooxy-2-aminofluorene is the major ultimate electrophilic and carcinogenic metabolite of *N*-hydroxy-2-acetylaminofluorene in the livers of infant male C57BL/6J × C3H/HeJ F1 (B6C3F1) mice. *Carcinogenesis*, 6, 1037–1045.

51. Smela, M.E., Hamm, M.L., Henderson, P.T., Harris, C.M., Harris, T.M., and Essigmann, J.M. (2002) The aflatoxin B(1) formamidopyrimidine adduct plays a major role in causing the types of mutations observed in human hepatocellular carcinoma. *Proc. Natl. Acad. Sci. USA*, 99, 6655–6660.

52. Ozawa, N. and Guengerich, F.P. (1983) Evidence for formation of an S-[2-(N7-guanyl)ethyl]glutathione adduct in glutathione-mediated binding of the carcinogen 1,2-dibromoethane to DNA. *Proc. Natl. Acad. Sci. USA*, 80, 5266–5270.

53. Koga, N., Inskeep, P.B., Harris, T.M., and Guengerich, F.P. (1986) S-[2-(N7-guanyl)ethyl]glutathione, the major DNA adduct formed from 1,2-dibromoethane. *Biochemistry*, 25, 2192–2198.

54. Inskeep, P.B., Koga, N., Cmarik, J.L., and Guengerich, F.P. (1986) Covalent binding of 1,2-dihaloalkanes to DNA and stability of the major DNA adduct, S-[2-(N7-guanyl)ethyl]glutathione. *Cancer Res.*, 46, 2839–2844.

55. Kim, D.H., Humphreys, W.G., and Guengerich, F.P. (1990) Characterization of S-[2-(N1-adenyl)ethyl]glutathione as an adduct formed in RNA and DNA from 1,2-dibromoethane. *Chem. Res. Toxicol.*, 3, 587–594.

56. Basu, A.K., O'Hara, S.M., Valladier, P., Stone, K., Mols, O., and Marnett, L.J. (1988) Identification of adducts formed by reaction of guanine nucleosides with malondialdehyde and structurally related aldehydes. *Chem. Res. Toxicol.*, 1, 53–59.

57. Fink, S.P., Reddy, G.R., and Marnett, L.J. (1997) Mutagenicity in *Escherichia coli* of the major DNA adduct derived from the endogenous mutagen malondialdehyde. *Proc. Natl. Acad. Sci. USA*, 94, 8652–8657.

58. Dedon, P.C., Plastaras, J.P., Rouzer, C.A., and Marnett, L.J. (1998) Indirect mutagenesis by oxidative DNA damage: Formation of the pyrimidopurinone adduct of deoxyguanosine by base propenal. *Proc. Natl. Acad. Sci. USA*, 95, 11113–11116.

59. Lee, S.H., Rindgen, D., Bible, R.H., Jr., Hajdu, E., and Blair, I.A. (2000) Characterization of 2′-deoxyadenosine adducts derived from 4-oxo-2-nonenal, a novel product of lipid peroxidation. *Chem. Res. Toxicol.*, 13, 565–574.

60. International Agency for Research on Cancer (1983) *Polynuclear Aromatic Compounds, Part I*. Vol. 2. IARC Monographs on the Evaluation of the Carcinogenic Risk of Chemicals to Humans, IARC, Lyon, France.

61. Gill, J.H., Duke, C.C., Rosario, C.A., Ryan, A.J., and Holder, J.M. (1986) Dibenz[*a,j*]acridine metabolism: Identification of *in vitro* products formed by liver microsomes from 3-methylcholanthrene pretreated rats. *Carcinogenesis*, 7, 1371–1378.

62. Wan, L., Xue, W., Reilman, R., Radike, M., and Warshawsky, D. (1992) Comparative metabolism of 7H-dibenzo(c,g)carbazole and dibenz(a,j)acridine by rat and mouse liver microsomes. *Chem. Biol. Interact.*, 81, 131–147.

63. Perin, F., Dufour, M., Mispelter, J., Ekert, B., Kunneke, C., Oesch, F., and Zajdela, F. (1981) Heterocyclic polycyclic aromatic hydrocarbon carcinogenesis: 7H-dibenzo(c,g)carbazole metabolism by microsomal enzymes from mouse and rat liver. *Chem. Biol. Interact.*, 35, 267–284.

64. Schurdak, M.E., Stong, D.B., Warshawsky, D., and Randerath, K. (1987) ^{32}P-Postlabeling analysis of DNA adduction in mice by synthetic metabolites of the environmental carcinogen, 7H-dibenzo[c,g]carbazole: Chromatographic evidence for 3-hydroxy-7H-dibenzo[c,g]carbazole being a proximate genotoxicant in liver but not skin. *Carcinogenesis*, 8, 591–597.

65. Talaska, G., Reilman, R., Schamer, M., Roh, J.H., Xue, W., Fremont, S.L., and Warshawsky, D. (1994) Tissue distribution of DNA adducts of 7*H*-dibenzo(c,g)carbazole in mice following topical application of the parent compound and its major metabolites. *Chem. Res. Toxicol.*, 7, 374–379.

66. Lindquist, B. and Warshawsky, D. (1989) Binding of 7H-dibenzo[c,g]carbazole to polynucleotides and DNA *in vitro*. *Carcinogenesis*, 10, 2187–2195.

67. Warshawsky, D., Fremont, S., Xue, W., Schneider, J., Jaeger, M., Collins, T., Reilman, R., O'Connor, P., and Talaska, G. (1994) Target organ specificity for N-heterocyclic aromatics. *Polycyclic Aromatic Compd.*, 6, 27–34.

68. You, M., Candrian, U., Maronpot, R., Stoner, G., and Anderson, M. (1989) Activation of the *K-ras* protooncogene in spontaneously occurring and chemically induced lung tumors of the strain A mouse. *Proc. Natl. Acad. Sci. USA*, 86, 3070–3074.

69. Mass, M.J., Jeffers, A.J., Ross, J.A., Nelson, G., Galati, A.J., Stoner, G.D., and Nesnow, S. (1993) *Ki-ras* oncogene mutations in tumors and DNA adducts formed by benz(j)aceanthrylene and benzo(a)pyrene in the lungs of strain *A/J* mice. *Mol. Carcinog.*, 8, 186–192.

70. Nesnow, S., Ross, J.A., Nelson, G. et al. (1994) Cyclopenta[c,d]pyrene induced tumorigenicity, *Ki-ras* codon 12 mutations and DNA adducts in strain *A/J* mouse lung. *Carcinogenesis*, 15, 601–606.

71. Manam, S., Storer, R.D., Prahalada, S. et al. (1992) Activation of the Ha-, Ki and N-ras genes in chemically induced liver tumors from CD-1 mice. *Cancer Res.*, 52, 3347–3352.

72. Manam, S., Shinder, G.A., Joslyn, D.J. et al. (1995) Dose-related changes in the profile of ras mutations in chemically induced CD-1 mouse liver tumors. *Carcinogenesis*, 16, 1113–1119.

73. Bizub, D., Wood, A.W., and Skalka, A.M. (1986) Mutagenesis of the Ha-ras oncogene in mouse skin tumors induced by polycyclic aromatic hydrocarbons. *Proc. Natl. Acad. Sci. USA*, 83, 6048–6052.

74. Brown, K., Buchmann, A., and Balmain, A.W. (1990) Carcinogen-induced mutations in the mouse c-Ha-ras gene provide evidence for multiple pathways for tumor progression. *Proc. Natl. Acad. Sci. USA*, 87, 538–542.

75. Colapietro, A.-M., Goodell, A.L., and Smart, R.C. (1993) Characterization of benzo[a]pyrene-initiated mouse skin papillomas for Ha-ras mutations and protein kinase C levels. *Carcinogenesis*, 14, 2289–2295.

76. Ronai, l.A., Gradia, S., EI-Bayoumy, A.K., Amin 5, and Hecht 55 (1994) Contrasting incidence of ras mutations in rat mammary and mouse skin tumors induced by anti-benzo[c]phenanthrene-3,4-diol-1,2-epoxide. *Carcinogenesis*, 15, 2113–2116.

5 Cellular Oncogenes and Carcinogenesis

Helmut Zarbl

CONTENTS

5.1 INTRODUCTION

Cancer can be described as a process that leads to the abnormal spatial and temporal growth of cells and tissues within an organism. Cancers are almost always clonal in origin and result from the accumulation of genetic and epigenetic changes within a single somatic cell. As a result of these changes, cancer cells and their progeny fail to respond to normal growth regulatory signals from their environment, do not respect the normal physical boundaries and temporal limits spelled out by their normal developmental program, and acquire a selective growth advantage over their neighbors. It is therefore not surprising that cancer-specific mutations and epigenetic changes occur in genes that function in regulating normal cellular and organismal processes such as development, tissue homeostasis and repair, as well as cell growth and turnover.

To comprehend the complex molecular basis of cancer, it is helpful to consider the role of somatic cells within the context of a multicellular organism. Single celled organisms are generally characterized by unrestrained growth that is limited only by the availability of nutrients, space constraints, and the build-up of toxic metabolites. In contrast, somatic cells in multicellular organisms must be programmed for a life of altruism, wherein their primary mission is to support the sexual transmission of their genes through the germ cells. As a result, the developmental program of each somatic cell spells out specific rules governing a cell's growth, division, position, and migration within a tissue, as well as the time and conditions for programmed cell death. These rules must in turn be responsive to environmental stimuli including the microenvironment of the cell, including oxygen tension, nutrient status, injury, and build-up of toxins, as well as messages in the form of hormones or growth factors generated by neighboring cells and in some cases distant cells within the organism. Each of these biological processes is in turn regulated by multiple layers of positive and negative regulatory networks of signaling molecules.

As a result carcinogenesis is a multistep process resulting from the sequential perturbation of both positive and regulatory networks that normally allow a somatic cell to live a cooperative existence within the society of normal cells that comprise an organism. Normal cells are even programmed to give their own life for the good of the organism. Any genetic or epigenetic changes that allow a cell to escape these normal societal constraints represent a step toward cancer. Once on the road to cancer, the altruism of the initiated cells succumbs to the process of molecular evolution, where genetic and genomic instability that allow for survival are favored over the stability and apoptosis of abnormal cells. Survival of the fittest cells allows for the clonal expansion of progeny cells with ever increasing numbers of genetic or epigenetic changes that favor even greater antisocial and selfish behavior of the cancer cell within the organism.

Cancer specific mutations can be separated into those affecting positive, and those affecting negative growth/survival regulatory networks. *Oncogenes*, or dominant cancer genes, are defined as gain of function mutants in normal genes or *proto-oncogenes* that promote cell growth, division and survival. *Tumor suppressor genes* encode proteins that normally function to suppress cell growth or protect the genome from damage that could lead to uncontrolled growth. Inherited mutations in tumor suppressor genes are usually recessive and may be passed on through the germline in the heterozygous state, leading to increased susceptibility to cancer. However, in some tumor suppressor genes (e.g., the *p53* tumor suppressor), dominant negative mutations may appear as a genetically dominant genotype, sometimes leading to their misclassification as dominant oncogenes. Tumor suppressor genes and their mechanisms of action will be discussed in detail in the Chapter 6. The focus of this chapter will be the mechanisms of oncogene activation and the molecular mechanisms by which they contribute to cancerous cell growth. However, since more than a 100 oncogenes have been identified in human and experimental tumors to date, the discussion will of necessity be limited to specific examples that exemplify the regulatory pathways disrupted by oncogenic mutations. It is also impossible to cite all relevant primary literature. Thus, in this chapter, I cite primarily review articles that reference important primary literature pertaining to each topic (with apologies to authors of the primary studies).

5.2 IDENTIFICATION OF ONCOGENES AND MECHANISMS OF ACTIVATION

5.2.1 Retroviral Transduction

The oncogene hypothesis, first proposed by Huebner and Todaro [1] in 1969, predicted that within the genome of an organism, there exists a set of genes that can be activated into forms which promote the cancerous growth of cells. Research with acutely transforming retroviruses in animal models had clearly demonstrated that viral genomes carried transforming regions that were able to induce tumors in animals. In many cases, the acute tumorigenicity was eliminated by deletion of a single gene from the viral genome. In what can only be described as one of the most important discoveries in cancer research, Bishop and Varmus, in studies for which they were awarded the Nobel Prize in Medicine, demonstrated that the retroviral oncogenes had their origins in the host genome. Using Southern blot hybridization, these investigators demonstrated that the sarcoma inducing *src* gene, present in the genome of the Rous Sarcoma Virus, had significant DNA sequence similarity with an endogenous gene (*c-src*) present in the genomic sequences of the chicken [2]. Nonetheless, the *v-src* gene present in the transforming virus was significantly altered by mutations that resulted in the constitutive activation of its endogenous protein kinase activity. The Rous sarcoma virus was somewhat unique in that the integration of the *src* gene into the viral genome did not interfere with the ability of the retrovirus to replicate. As a result, the Rous sarcoma virus is a rare example of replication competent RNA tumor virus.

Given the propensity for retroviral genomes to recombine during their replicative integration into host genomic sequences, it seemed likely that other transforming retrovirus also carried oncogenes of a cellular origin. Indeed, analysis of the transforming sequences of other acute transforming retroviruses were shown to have extensive sequence similarity with host genomic sequences, providing further evidence for the oncogene hypothesis [3,4]. A significant number of oncogenes were identified by studying the acute transforming retroviruses including *mos*, *myc*, *fos*, *rel*, *myb*, *fms*, *fes*, *sis*, *erbB*, *ets*, *abl*, and *jun*. However, unlike the Rous sarcoma virus, the genomes of other acute transforming retroviruses were rendered replication incompetent by the rearrangements that lead to the integration of oncogene sequences from the host cell. As a result, most acute transforming retroviruses require coinfection with a compatible replication competent retrovirus. Retroviral transduction of cellular oncogenes remains a rare event in nature, and few previously unknown oncogenes were identified since the initial rush to characterize the transforming genes of all the known, acute transforming retroviruses.

5.2.2 Retroviral Insertional Mutagenesis

The acute transforming retroviruses carry activated versions of cellular oncogenes and therefore can induce tumors in animals, within weeks after infection. However, in many cases the nondefective parental retroviruses can also induce malignancies, primarily lymphomas, in infected animals albeit after a long latency period of months to years. These replication-competent viruses do not carry endogenous oncogenes and are not capable of transforming cultured cells. They do, however, induce persistent viremia in their animal hosts, suggesting that transformation by these retroviruses is the result of rare, insertional mutagenesis events that lead to proto-oncogene activation. Although the identification of putative oncogene sequences that are close to the sites of retroviral insertion in specific tumor types has led to the identification of several important oncogenes [5], this approach remains an arduous undertaking. The first step is to select a model system in which the virus consistently induces tumors of the same tumor type. A large number of tumors are isolated and assessed for retroviral integration by Southern blot analysis, using carefully chosen restriction enzyme digests and specific viral probes. Tumors with one or two integration sites are then selected. The proviral integration sites are then cloned from a bacteriophage library prepared with tumor DNA screened with the same hybridization probes. Sequences from each disrupted locus are then used to probe Southern blots of tumor DNAs to determine the frequency with which the locus is the target of retroviral insertion in the type of tumor induced by the virus. If a locus is a frequent target for integration in a given tumor type, then the locus is likely to harbor an oncogene that is activated by retroviral insertional mutagenesis.

This approach was used to successfully clone a number of oncogenes in animal tumor models. Among the genes identified were the *c-myc* gene in ALV-induced leukemias, and the *Wnt* and *int* oncogenes identified by studies of breast cancers induced in female mice by infecting them with Mouse Mammary Tumor Virus (MMTV) [6]. Several oncogenes were identified as repeated sites

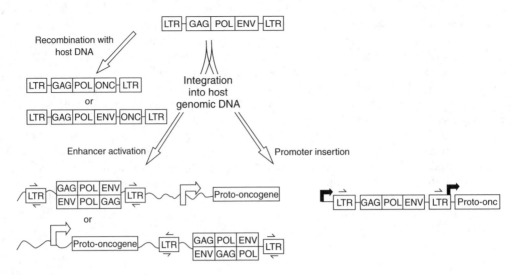

FIGURE 5.1   Mechanisms of proto-oncogene activation by retroviruses.

of retroviral insertion in rat lymphomas induced after prolonged infection with Moloney Murine Leukemia Virus [7]. Careful analysis of the disrupted oncogenes in tumors indicated that there are several mechanisms by which retroviral insertions can lead to oncogene activation Figure 5.1. The simplest mechanism, promoter activation, involves insertion of the retrovirus between upstream of the coding sequences so that the viral promoter within the 3' Long Terminal Repeat (LTR) drives transcription of the proto-oncogene. As a result the gene is transcribed expressed at greatly elevated levels. Another mechanism of oncogene activation, enhancer activation, is independent of position or orientation relative to the proto-oncogene coding sequences. Rather, the integration of the retroviral sequences places the endogenous promoter of the proto-oncogene under the control of the strong enhancer elements present in the viral LTR. Thus, insertion may occur upstream or downstream of the coding sequences and can occur at some distance away from the promoter. In many cases, however, the mechanism remains unclear. For example, many loci such as the *Mlvi-1/mis1/pvt-1*, locus identified in rat thymomas do not harbor oncogenes per se. Rather the targeted loci are located at ~100 kb upstream of the *c-myc* gene and effect its transcriptional regulation [8]. These finding raise the possibility that the retroviral insertions can also activate proto-oncogenes via epigenetic mechanisms such as chromosomal remodeling. Retroviral insertional mutagenesis has resurfaced in the postgenomic era as a tool for cancer gene identification using high throughput screens [9].

5.2.3   DNA TRANSFECTION

However, the studies with animal retroviruses could not provide direct evidence for the role of proto-oncogene mutation in human cancer, since there were no known acute transforming human retroviruses. Direct evidence for the role of oncogenes in human cancer came from the application of DNA transfection using the calcium phosphate coprecipitation technique [10]. DNA transfection was first used to demonstrate cell-transforming activity by ectopically expressing the genomes of the acute transforming retrovirus into tissue culture cells. The assay relied on the availability of immortalized rodent cell lines such as NIH/3T3, Rat-1, and 208F fibroblasts, that retained a flat, normal morphology when cultured *in vitro*, and when cultured to confluence, exhibited contact inhibited growth. When infected with an acute transforming retrovirus, or transfected with transforming genes from these viruses, these cells acquired a more rounded and refractile morphology, and their growth was no longer restricted by contact inhibition due to confluence. As a result, the cells continued to grow in a disorderly pattern and piled up in multiple layers to form *transformed foci* (Figure 5.2). The latter had

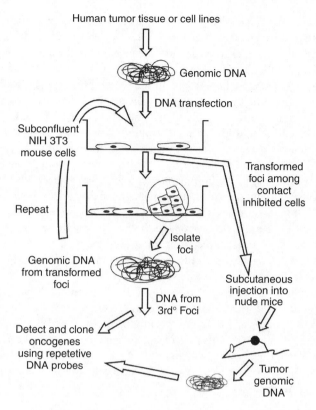

FIGURE 5.2　Detection and isolation of oncogenes from tumor cells using DNA transfection assays.

characteristics reminiscent of tumor cells, including the ability to form colonies in soft agar or methylcellulose (anchorage independent growth). Indeed cells comprising foci were capable of inducing rapidly growing tumors within weeks after subcutaneous injection into athymic nude mice. By contrast the "nontransformed" albeit immortalized parental cells only rarely produced nude mouse tumors, and then only after an extended latency period.

In the early 1980s, several groups began to use the DNA transfection assay to screen human tumors for transforming oncogenes. Since the retroviral *src* gene of chicken origin was able to transform mouse fibroblasts, the use of rodent cells to detect human oncogenes did not seem unreasonable, and indeed the gamble was rewarded [10]. Several groups almost simultaneously reported detection of cell transforming activity in genomic DNA isolated from human tumors, but not in DNA from normal human cells. Nonetheless, identification of the transforming sequences still presented a hurdle. Transfection of large genomic DNA fragments into fibroblast cells results in a complex series of events. Cells that take up DNA often accept large amounts of DNA, which is initially assembled into a replicating extrachromosomal complex, which is eventually stabilized by integration into the host genome. To detect the human sequences that integrated into the genome of the host rodent cells and presumably contributed to the development of transformed foci, researchers made use of DNA hybridization probes for species specific repetitive DNA sequences that are dispersed throughout the donor genome. To reduce the complexity of the transfected donor sequences, genomic DNA from the transformed foci were used in repeated rounds of transfection until the foci contained a limited number of human DNA, as detectable by hybridization with Alu repeat sequence probes. At this point, human DNA fragments were cloned and sequenced using standard techniques. Researchers were initially surprised that they did not detect the human *src* homolog. However, reservations about the significance of these finding were assuaged when it was determined that the putative human oncogene detected in several laboratories had extensive sequence similarity with the *ras* oncogene also transduced in the acute transforming Harvey and Kirsten Murine Sarcoma Viruses. DNA transfection technology quickly became the standard assay for the detection of oncogenes in human and experimental tumors, and led to the identification of dozens of oncogenes [11].

The use of DNA transfection can also lead to the serendipitous identification of human genes with transforming activities. Fragments of randomly sheared genomic DNA from the donor are often ligated together within the transfected cells to generate *in situ* DNA recombination events. Occasionally such fusion events would lead to the activation of a proto-oncogene that was not present in the DNA of the donor cancer cells. Thus, when using transfection to detect oncogenes, it is essential to go back and demonstrate the presence of the putative oncogene in the donor tumor DNA before inferring causality in the original tumor. The *de novo* generation of oncogenes during transfection also explained why transforming activity was occasionally detected during the transfection of control DNA from normal cells. Activation could occur as a result of several mechanims including gene fusions, gene truncations, promoter insertion, amplifications, or enhancer activation (see Sections 5.2.4 and 5.3). An example of an oncogene identified as a result of a *de novo* DNA recombination is the *trk* oncogene which fused to the tropomyocin gene to the amino terminus of the nerve growth factor receptor tyrosine kinase domain [12].

Despite the successes of DNA transfection, it was also apparent early on that not all oncogenes were detectable when using *in vitro* transformation as the endpoint. Thus, attempts were made to bypass the *in vitro* selection step and assay for transformation *in vitro*. The assay developed included the cotransfection of genomic DNA with a selectable marker gene, such as the *neo* gene. Transfected cells, which harbored both genomic DNA and the marker plasmid DNA, were then selected by *in vitro* culture in the presence of G418. A pool of resistant cells was then injected directly into nude mice to assay for tumorigenicity.

5.2.4 CHROMOSOMAL REARRANGEMENTS AND AMPLIFICATIONS

Detection of oncogenes by studying illegitimate DNA recombination is, however, not limited to *in vitro* accidents. In fact, one of the success stories in targeted molecular therapy of cancer has its roots in the study of chromosomal rearrangements. While studying the molecular basis of Chronic Myelogenous Leukemia (CML), researchers demonstrated that the Philadelphia Chromosome (Ph), the first cancer specific chromosomal rearrangement described, always resulted in the activation of the *ABL* specific oncogene. As a result of the reciprocal t(9;22) translocation that defines the Ph chromosome and is characteristic of most Chronic Myeloid Leukemias (CML), the *c-ABL* oncogene on Chromosome 9 is fused with the *BCR* gene on Chromosome 22. The fusion always occurs between the specific exons of the two reciprocal genes, resulting in the characteristic Bcr-Abl fusion proteins [13]. Within the fusion protein, the tyrosine kinase activity of the Abl protein is constitutively derepressed and accounts for its transforming activity. In recent years, this information was used in the development of Gleevec, a drug that specifically inhibits the Abl kinase activity, and has proved to be remarkably effective in treating *CML*, often curing this deadly disease [14].

Molecular analyses of chromosomal breakpoints have demonstrated that chromosomal translocations can activate proto-oncogenes by a variety of mechanisms including promoter insertion, enhancer activation, chromatin remodeling, and the generation of protein fusion or truncation (see Molecular Mechanisms of Proto-Oncogene Activation). Despite the prevalence of multiple chromosomal rearrangements in all cancers [15], the identification of oncogenes using this approach is limited due to the lack of specific chromosomal rearrangement that are reproducibly associated with a specific tumor type, especially solid tumors.

Another approach for detecting oncogenes has been to identify genes within chromosomal regions that are amplified in tumor cells. Chromosomal amplifications can be detected by cytogenetic analysis as Homogeneous Staining Regions (HSR), Double-Minute Chromosomes (DM), or aneuploidy. Amplified oncogenes associated have been detected in a wide variety of tumors, and include the

HER-2/neu oncogene frequently amplified in human breast cancers, members of the MYC family of proto-oncogenes (*c-MYC, N-MYC, and L-MYC*), the *RAS* family of genes and many others. The *mdm-2* oncogene, an important posttranslational regulator of the *p53* tumor suppressor gene, was first identified and cloned as a gene present on DM chromosomes replicating in a murine tumor cell line [16].

Regions of chromosomal amplification can also be detected and often more precisely delineated using molecular approaches such as Comparative Genome Hybridization (CGH) and DNA microarray technologies [17]. To perform CGH, genomic DNA from tumor cells and reference normal cells are differentially labeled with fluorochromes, and cohybridized to metaphase chromosomes. A reference DNA is typically generated by pooling equal amounts of genomic DNA from lymphocytes of at least ten healthy male or female donors. Metaphase chromosomal spreads are prepared from Phytohemagglutinin stimulated lymphocytes pooled from appropriate normal nonsmoking males and females. The reference and test DNAs are then differentially labeled with derivatized nucleotide triphosphates such as biotin-14-dATP or digoxigenin-11-dUTP using nick translation. The labeled probes are then pooled and hybridized to the metaphase chromosome spreads. Hybridizations are typically repeated with the chromofluores reversed to verify findings. Chromosomes are identified by inspection of inverted digital DAPI stained images. The mean ratios of fluorescence intensities for each test DNA relative to the appropriate reference DNA are determined for at least 25 metaphase spreads. Chromosomal losses or gains associated with a specific tumor are then identified as statistically significant deviations in the ratios of fluorescence intensities at specific chromosomal regions. CGH provides a comprehensive overview of chromosomal losses or gains in a single assay. The resolution of this technique is at the level of chromosome subregions. While CGH does not provide resolution at the locus or gene level, CGH can implicate chromosomal regions that merit further study. Using a combination using molecular and bioinformatics approaches, investigators have identified several putative oncogenes in subchromosomal regions identified by CGH in a variety of tumor types.

As an alternative, DNA microarrays can be used to identify chromosomal regions and genes that are amplified in tumors. CGH on metaphase chromosomes can provide information on chromosomal changes at the resolution of chromosomal bands. An adaptation of this approach is to hybridize differentially labeled genomic DNAs to cDNA microarrays to measure gene copy number, although the hybridization of genomic DNA to cDNA often yields low signal to noise ratios, making it difficult to detect subtle alterations in copy number. A more recent application of this technology is the use of Bacterial Arteficial Chromosomes (BAC) microarrays, which can provide resolution at the level of cytogenetic bands. The test and reference DNAs are labeled with either Cy3 or Cy5 dNTPs and cohybridized to the BAC array, washed, scanned and the fluorescence data processed using appropriate software. The Cy5/Cy3 ratios determined from multiple replicates for each BAC are averaged and subjected to statistical analyses to determine if changes in DNA copy number have occurred in the tumor cells within the region encompassed by any individual BACs. BAC arrays may be able to detect single copy number changes. The use of BAC arrays will no doubt shed more light on the issue of how proto-oncogenes can function as both tumor suppressors and oncogenes based on their copy number.

5.3 MOLECULAR MECHANISMS OF PROTO-ONCOGENE ACTIVATION

The mechanisms leading to the activation of proto-oncogenes fall into two categories: (1) gain of function mutations that alter the activity of the oncoprotein, and (2) mutations that lead to deregulation of or inappropriate expression of the proto-oncogene protein. Gain of function mutations can include specific point mutations that result in constitutive activation of signaling. Point mutations are most often associated with activation of the *ras* of proto-oncogenes [18]. However, mutations that result in the constitutive protein tyrosine kinase activity have been detected in a variety of other oncogenes, including growth factor receptors such as *HER2/neu* oncogene, and nonreceptor tyrosine kinases such as *SRC* [19]. Gain of function mutations can also result from chromosomal rearrangements that generate truncated proteins or novel fusion proteins with activated growth-promoting activities. Mutations that lead to deregulated expression of proto-oncogenes include promoter insertion, enhancer activation, gene amplification, and possibly epigenetic mechanisms that alter chromatin structure. However, in many cases oncogenes are activated by a combination of mutation and deregulation.

5.3.1 Proto-Oncogenes Activated by Point Mutations

The members of the *ras* family of genes are the prototypical proto-oncogenes that acquire their transforming potential as a result of specific point mutations [20]. Initially identified by their ability to transform NIH 3T3 cells, mutant versions of the *ras* oncogenes can be detected in approximately 25% of all human tumors, but the frequencies can be very high in specific tumor types [21,22]. For example, mutated Ki-*ras* oncogenes can be detected in more than 90% of all human pancreatic tumors [23], while the genomes of almost 30% of Acute Myeloid Leukemias harbor mutated *N-ras* oncogenes [24]. The vast majority of these activating mutations occur in codon 12 of the ras proteins, which invariably encodes a glycine (Gly) moiety. Mutations in the Gly residue at codon 13 have also been detected in some tumors. Mutational analyses of *ras* oncogenes from a variety of human and experimental tumors have detected many different amino acid substitutions at codons 12 [22]. *In vitro* studies using saturation mutagenesis have confirmed that substitution of the crucial Gly residue with any amino acid except proline (Pro) activates the transforming potential of the *ras* oncogenes [25]. Structure modeling predicted and x-ray crystallographic data subsequently confirmed that the Gly and Pro residues, but no other residues, allow for normal folding of the ras proteins [26]. All other amino acid substitutions induce a structural change that leads to constitutive inactivation of the proteins' intrinsic GTPase activity. As a result, the oncogene remains in the active GTP bound form and continually activates downstream effectors of signaling and cell growth (see Section 5.4.2).

The *ras* proto-oncogenes can also be activated by amino acid substitution at codon 61, and possibly codon 59. The latter was initially detected in the viral Ha-*ras* gene and was shown to have some transforming activity *in vitro*, but has seldom been detected in tumors. By contrast, a variety of mutations at codon 61 occur in both human and animal tumors. The invariable glutamine residue forms part of the active site for the intrinsic GTPase activity. Consequently most amino acid substitutions at this residue also lead to a constitutively activated form of the oncogene.

Recent studies have demonstrated that in conjunction with SV40 Large T antigen and human telomerase (*h-tert*), activated Ha-*ras* can initiate the transformation of normal human epithelial and fibroblastic cells [27]. Another study with transgenic animals harboring an inducible Ha-*ras* transgene demonstrated that continued expression of the oncogene is necessary for the genesis and maintenance of solid tumors *in vivo* [28]. Collectively these *in vitro* and *in vivo* findings are consistent with the notion that *ras* oncogenes activated by mutations are dominant oncogenes that play an important role in cell transformation and tumorigenesis.

Despite the wealth of supportive data, several investigators have questioned the dominance of oncogenes, including *ras* genes with activating point mutations [29,30,31]. In the case of the *ras* oncogenes, Duesberg and his colleagues noted that the mechanism of oncogene activation in the Harvey Rat Sarcoma virus (Ha-*ras*) and Kirsten Rat Sarcoma virus (Ki-*ras*) oncogenes, transduction always involved the truncation that removed the 5' regulatory regions. Hence, oncogene transcription was placed under the influence of the enhancer sequences in viral LTR. Some transfection studies using genomic DNA also seemed to indicate that in order for a *ras* oncogene to have transforming activity *in vitro*, its expression has to be deregulated. Further support for this supposition comes from *in vivo* observations indicating that most tumors carrying activated *ras* oncogenes also express these genes at elevated levels. As a direct test of the hypothesis that mutated Ha-*ras* oncogenes were dominant, Finney and Bishop [32] used homologous recombination to introduce a mutant codon 12 into the endogenous Ha-*ras* gene of a rat fibroblast cell line. Consistent with the hypothesis that transforming potential requires deregulation of expression, cells carrying a mutant Ha-*ras* allele under the control of the endogenous promoter did not have a transformed phenotype. When these cells were passaged *in vitro*, spontaneous foci of cells arose within the cultures. Significantly, in all cases the spontaneous transformants were found to express elevated levels of the mutant Ha-*ras* allele as a result of gene amplification or transcriptional deregulation. *In vivo* studies using the rat mammary tumor model also indicated that Ha-*ras-1* oncogenes in normal mammary tissue do not acquire a significant growth advantage in the absence of additional genetic or epigenetic stimuli changes [33]. In a typical transfection experiment the multiple copies of the transfected gene are incorporated into the transformant. Thus the ectopic gene is expressed at much higher levels than the endogene, whereas activating point mutations that occur in cells are usually haploid. Thus one possible explanation for the lack of dominance of mutated *ras* alleles *in vivo* may be that the presence of the normal *ras* allele can suppress the penetrance of the mutant allele by competing for downstream effectors. The latter hypothesis is supported by recent genetic demonstrating that transgenic mice heterozygous for a mutant alleles of Ki-*ras* appeared to be more resistant to carcinogenesis than homozygous mutants [34,35]. While these data indicate that wild type *ras* genes can reduce the transforming potential of activated *ras* oncogenes, expression of very high levels of wild-type *ras* can transform cells *in vitro* and are often detected in tumors. Although paradoxical at first glance, excessive expression of wild-type *ras* may activate downstream signaling by overwhelming negative regulators of *ras* signaling, for example RasGAP or Rev proteins (see Section 4). Thus the effect of wild-type and mutant *ras* genes on cell signaling may be competitive and dosage dependent.

Another interesting example of a proto-oncogene activated by a specific point mutation is the rat *Her-2/ErbB-2/neu* oncogene, first identified by transfecting DNA from a rat neuroblastoma cell lines into NIH 3T3 cells [36]. Unlike the *ras* genes, the *neu* proto-oncogene gene was activated by a single amino acid substitution in the transmembrane domain of the growth factor receptor encoded by this gene. Activation always involves a single base pair substitution that changes a valine residue to a glutamic acid residue. Interestingly, this specific mutation is not detected in *HER-2*, the human homolog of the *neu* oncogene, even though amplification and over expression of this proto-oncogene is detected in ~30% of all human breast cancers and is associated with a poor prognosis [37]. However, the lack of activating point mutations in the *Her-2* gene is easily explained on the basis of sequence polymorphisms between the transmembrane domains of the two genes. While a single base pair change can activate the rat *neu* gene, two specific point mutations are required to induce the amino acid change that can activate the human gene. Nonetheless, amplification is common in breast tumors and has led to the development of a gene-based therapeutic agent (Herceptin) that specifically targets the receptor molecule.

More recent studies have identified activating point mutation in the *BRAF* proto-oncogene [38]. *BRAF* is a member of the RAF family of serine–threonine kinases and was originally identified by transduction into acute transforming chicken and mouse retroviruses. *BRAF* activation of the RAS/RAF/MAPK signaling pathway MAPK kinase pathway is mediated by its binding to activated Ras proteins. Several studies have identified single, and occasionally double, activating point mutations in a variety of human cancers, including up to 80% of melanomas [38]. In about 80% of these, the mutations resulted in valine to glutamic acid substitutions at amino acid 599 (G599A), and all of these resulted from T to A transversion at base pair 1796 in exon 15. The G599A as well as all other transforming mutations detected to date occur within the kinase domain of Braf, and result in elevated kinase activity that is associated with its RAS independent transforming activity.

5.3.2 PROTO-ONCOGENES ACTIVATION BY CHROMOSOMAL REARRANGEMENTS

As demonstrated at the turn of the last century by pathologists, essentially all tumors are characterized by a loss of chromosome stability and aneuploidy, which at the cytogenetic levels are detected by a seemingly endless complexity of chromosomal gains, losses, amplifications, and rearrangements [39]. These random changes can lead to both loss of function mutation in suppressor genes, and gain of function mutations in proto-oncogenes [40]. Mutations that are not lethal to the cell are retained and may contribute to the clonal evolution of the cancer cell [41].

Chromosomal rearrangements in somatic cells are the result of illegitimate recombination events that arise during mitosis or as a result of error prone repair or nonhomologous end joining of damaged chromosomes with single or double DNA strand breaks. As already indicated above, analysis of genes comprising the breakpoints of specific chromosomal rearrangements that are reproducibly detected in tumors has led to the identification of many oncogenes. The mechanisms by which these rearrangements lead to the activation of proto-oncogenes include both gain of function mutations and deregulation of expression.

The best characterized among these gain of function mutants is the gene fusion associated with the Ph chromosome detected in over 90% of all CML, 30% of Acute Lymphocytic Leukemias (ALL) and 1% of Acute Myeloid Leukemias (Figure 5.3). The t(9;22) translocation that characterizes the Ph chromosome invariably occurs in breakpoint cluster of ∼6 kb, that includes the first in intron and an alternatively spliced exon of the ABL proto-oncogene. Likewise, the breakpoint cluster on chromosome 22 comprises a region between the fourth through the eighth exons of the *BCR* gene. As a result of these reciprocal translocations, most of the *ABL* proto-oncogene is fused to the 5′ end of the *BCR* gene to generate the chimeric *BCR–ABL* gene. The transcript of this fusion gene usually encodes a 210 KDa fusion protein whose expression is under the regulatory control of the *BCR* gene promoter, although other chimeric proteins can also be produced. The transforming protein kinase activity of the *Abl* oncoprotein is also derepressed in the fusion protein. Although some studies have indicated that the fusion protein generated on the reciprocal chromosome may also have transforming potential, though its contribution to the pathogenesis of CML remains unclear.

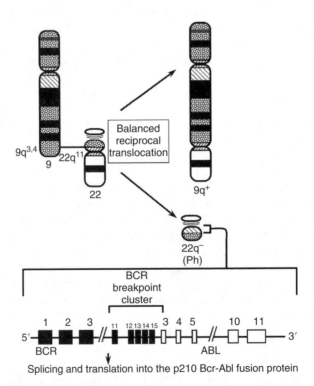

FIGURE 5.3   The t(9;22) rearrangement leading to the Philadelphia chromosome: Generation of the oncogenic Bcr-Abl fusion protein.

Studies of translocations in Burkitts Lymphoma (BL) first implicated the *c-MYC* proto-oncogene in this B cell malignancy. However unlike in CML, the chromosomal breakpoints are not clustered in BL. Breakpoint analyses demonstrated that the tumor cells often carried rearrangements between the *c-MYC* oncogene on Chromosome 8q24 and the immunoglobulin loci on chromosomes 2, 14, or 22, all of which result in the deregulation of *c-MYC* proto-oncogene expression. In the most common of these, the t(8;14) translocation juxtaposed the *c-MYC* proto-oncogene adjacent to a strong enhancer sequence within the IgH locus, suggesting that *c-MYC* was deregulated by enhancer activation. However, the mechanisms of *c-MYC* deregulation are not apparent in all BL-specific chromosomal translocation, and may in some cases involve long range or epigenetic effects similar to those detected after retroviral insertion. Other examples of important oncogenes detected by the investigation of tumor type specific chromosomal rearrangements include a variety of other genes, encoding basic-helix-loop-helix transcription factors, LIM proteins, and the antiapoptotic *BCL-2* gene [40].

5.3.3 PROTO-ONCOGENES ACTIVATION BY AMPLIFICATION

Oncogenes can also be activated as a result of an increase in copy number. Amplification can occur at the level of entire chromosomes as a result of nondisjunction during mitosis, or as a result of tandem gene replication. Although the mechanisms of gene amplification are still to be elucidated, it is clear that cells with increasing gene copy numbers can be selected if they provide a selective growth or survival advantage, and can play a major role in drug resistance [42]. At the cytogenetic level, gene amplifications are sometimes detected as homogeneously staining regions (HSR) on banded chromosome. Tandem gene amplifications can sometimes give rise to double-minute chromosomes which lack centromeres and replicate by as yet undetermined mechanisms [43]. Amplification of oncogenes plays an important role in many human cancers, and can be a major prognostic indicator of many cancer types, prompting their consideration as important targets for therapeutic interventions [43,44].

5.4 SYSTEMS BIOLOGY OF ONCOGENE INDUCED CELL TRANSFORMATION

Cells communicate with each other and respond to their environment via a series of specific receptor molecules that can detect extracellular molecules, including nutrients, toxicants, and secreted proteins such as growth factors, differentiation factors, apoptotic factors, extracellular matrix proteins, and antigens. The signals detected by these receptors are then amplified and relayed to the nucleus by proteins that comprise complex signal transduction cascades, many of which interact with and affect the activity of one another (cross-talk). Within the nucleus, the inputs from multiple signaling cascades are integrated to alter the activity of both negative and regulatory factors controlling gene expression. The resulting alterations in the pattern of gene expression ultimately determine the cell's response to its environment. Understanding how these complex systems integrate signals to generate an organism as complex as a human being remains a major challenge, and will require all the tools of modern genetics, functional genomics, proteomics, computational, and systems biology [45].

Despite the complexity of the signal integration required to maintain homeostasis, a great deal has been learned about the processes involved by studying oncogenes and the mechanisms through which they perturb normal cell growth, differentiation, and programmed death. Oncogenes alter normal signaling pathways at all levels. Proto-oncogenes encode extracellular growth factors, mutant versions of all major cellular growth factor receptors, as well as molecules involved in downstream signaling cascades, transcriptional regulators, cell cycle control proteins, and proteins involved in apoptosis. All are characterized by their ability to generate constitutive signal transduction in the absence of the appropriate stimulus.

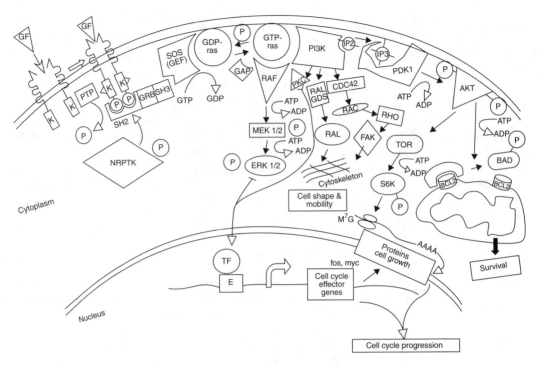

FIGURE 5.4   Schematic of the signaling pathways activated by binding of a growth factor to a generic cell surface receptor tyrosine kinase: Role of proto-oncogenes.

Early studies focused on the identification of individual oncogenes in specific tumor types. Although some tumor types were reproducibly associated with the activation of a specific oncogene (e.g., detection of the Ki-*ras* oncogene in about 90% of pancreatic cancers), this was not always the case. The failure to reproducibly detect the same activating mutations in all tumors simply reflects the fact that oncogenes are a single component of specific signaling pathways, which are in turn branches of complex interacting systems (Figure 5.4). Thus while specific oncogenes may be activated in only a fraction of any given tumor type, it is likely that major signaling pathways will play a role in almost all tumors. Pathways can be activated by mutations at many levels. For example, the RAS/RAF/MEK kinase pathway can be activated by a variety of mutations other than those in the *ras* oncogenes. Alternatively, the mutations in other signaling pathways that ultimately impact the same effector proteins as the RAS/RAF/MEK pathway may be present in other tumors. The positive signal generated by oncogenes must then be integrated with negative growth signals generated by tumor suppressor genes. Obviously understanding how the accumulation of cancer specific mutations contribute to deregulated cell growth and the progression of mutant cells toward malignancy will require the concepts and approaches of systems biology [46].

5.4.1 GROWTH FACTORS AS ONCOGENES

The idea that growth factors could contribute to malignant transformation of cells was based on the observation that many tumor cells produced elevated levels of Growth Factors (GF) as well as their cognate receptors. It was therefore reasonable to posit that autocrine stimulation via GFs could contribute to the mechanism of malignant transformation [47]. This hypothesis was validated when two groups independently cloned and partially sequenced Platelet Derived Growth Factor (PDGF), and found that the sequence were highly similar to *v-sis*, the oncogene of the acute transforming Simian Sarcoma Virus. The hypothesis was further substantiated by a plethora of studies demonstrating that a variety of growth factors isolated from different tissues had transforming potential. These included Epidermal Growth Factor (EGF), and the Transforming Growth Factors alpha and beta (TGF-α and TGF-β). EGF and TGF-α both bind to the EGF receptor [48], while TGF-β binds to its own receptors. TGF-β is somewhat an enigma in that it was originally discovered as a transforming GF, but later experiments demonstrated that it actually inhibits the transformation of many cell types including most epithelial cells, even fibroblasts under some conditions, by inducing the p27 cell-cycle inhibitor. More recent studies suggest that the role of TGF-β may depend on the genetic and epigenetic context playing a tumor suppressor role in normal cells and playing a growth promoting role during progression of malignant cells [49]. Other growth factors with transforming potential include the Fibroblast Growth Factor (FGF) family of proteins, the neurotrophins, the insulin family of growth factors (IGF), and the several classes of hematopoetic growth factors, including Interleukins (IL), Colony Stimulating Factors (CSF), and others [19]. The Wnt family of GFs were initially identified as sites of repeated retroviral integration in MMTV-induced mammary tumors in mice, and were subsequently shown to play a major role in developmental patterning in the embryo by binding to the Frizzled receptor [50]. The *in vitro* transforming potentials of many GFs have been confirmed *in vivo* using transgenic mouse models.

5.4.2 GROWTH FACTOR RECEPTOR AS ONCOGENES

Since elevated levels or inappropriate spatial/temporal expression of growth factors can contribute to cell transformation, it stands to reason that their cognate receptors also have oncogenic potential. Elevated Expression of the EGF Receptor (EGFR) as a result of gene amplification or transcriptional deregulation is a hallmark of many human cancers. The mechanisms by which elevated levels of the EGF receptor stimulate cell growth in the absence of EGF provides a paradigm for all GF Receptor Protein Tyrosine Kinases (RPTK) [19,48]. Although GF receptors with tyrosine kinase activity fall into several classes, all are characterized by an extracellular ligand binding domain, a transmembrane domain that plays a key role in transducing the signal, and an intracellular effector domain (Figure 5.4). The latter comprises of a number of effector protein binding domains and the crucial protein tyrosine kinase activity. The activities of RPTK molecules are tightly regulated so that they do not relay signals to the downstream effector molecules in the absence of appropriate stimuli.

The stringent autoregulation of RPTK activity is mediated by protein–protein interactions involving specific regulatory protein loops and the catalytic domains. These interactions either prevent interaction of the substrate tyrosine with the catalytic domain or prevent binding of ATP. Oligomerization of the receptor as a result of ligand binding or due to a high concentration of the receptor molecules in the cell membrane shifts the equilibrium of the regulatory protein–protein interactions toward a conformation that relieves these inhibitory constraints. As a result, specific tyrosine residues within the intracellular domains of each RPTK molecule are autophosphorylated, resulting in activation of the kinase activity. The phosphotyrosine residues within the RPTK molecules are ligands for effector proteins containing various PTB (phosphotyrosine binding) domains and *Src* Homology (SH) domains. SH domains are so named because they were initially identified as regions with sequence similarity to domains found within the *Src* oncogene. The PTB and SH domains bind to phosphotyrosine with a high degree of substrate specificity that is determined by flanking amino acid sequences. This specificity in binding effector molecules determines the downstream effects that result from the autophosphorylation of a given receptor in a particular cell type. The signal generated by RPTK autophosphorylation can also be attenuated by a series of regulated Protein Tyrosine Phosphatases (PTPase) that recognize specific phosphotyrosine residues via different SH domains [51].

Clearly, cellular homeostasis requires that the signal generated by a receptor is proportional to the amount of growth factor stimulation presented at the extracellular surface. The activity of the RPTK is thus under stringent autorepression by multiple layers of regulation. Mutations that convert

RPTK molecules into oncogenes invariably relieve these normal regulatory constraints and allow for uncontrolled signaling. One mechanism of activation is over expression of the RPTK molecule. Ligand binding to RPTK normally leads to receptor dimerization and conformation changes that relieve the regulatory constraints. Elevated expression of the RPTK as a result of gene amplification or transcription deregulation is thought to bring receptor molecules into closer proximity in the absence of ligand, and allow for conformational changes that mimic ligand binding. The conformational changes induced by ligand binding or concentration dependent dimerization of the receptors then activate receptor autophosphorylation by their intrinsic tyrosine kinase activity.

There are also numerous examples of RPTK that are activated by mutations that lead to constitutive activation of their tyrosine kinase activity. The prototypical GF receptor oncogene activated by truncation is the *v-erbB* oncogene. This oncogene, a member of the EGF family of RPTKs, was transduced by the Avian Erythroblastosis Virus as a highly mutated and truncation version of the receptor that has a constitutive PTK activity. As is the case for many other *RPTK* oncogenes, the activating mutations are truncations that remove crucial regulatory domains that normally repress the intrinsic tyrosine kinase activity. Some oncogenic *RPTK* mutations are truncations that remove part or all or the extracellular ligand domain that facilitate conformational changes that allow for receptor autophosphorylation. There is also evidence to suggest that mutations in transmembrane domain can play a critical role in transducing these conformational changes. For example, the *HER-2/neu* receptor can be transformed into an oncogene by a single, specific amino acid substitution within its transmembrane domain. Other *RPTK* oncogenes are activated by small deletions or specific point mutations; usually gain of function mutations result in constitutive activation of the tyrosine kinase activity. PTK receptor oncogenes can also be activated as a result of cancer associated chromosomal rearrangements, as is the case for the t(5;12) translocation that fuses the *ETS* related *TEL* gene to the kinase domain of the PDGFR beta subunit. However, *in situ* recombination during DNA-mediated gene transfer can also lead to the generation of novel fusion oncogenes with activated RPTK activities [12]. Other mutations that can contribute to the activation of receptor PTK activity are mutations within the intracellular domain that prevent signal attenuation by interfering with the activity of PTPases. Peptide Growth Factor receptors with oncogenic potential can be found among all the different receptors families, and are activated by a wide variety of mutational events, all of which lead to inappropriate or excess downstream signaling from activated RPTK.

Another class of cell surface receptors that are associated with cancer comprises the serpentine receptors that recognize small molecules such as hormones and ions. These are large proteins that traverse the lipid bilayer a total of seven times, and are coupled to heterotrimeric GTP-binding proteins that transduce signals to intracellular effort molecules. Mutations in either the serpentine receptor for Thyroid Stimulating Hormone (TSH), or the corresponding G proteins have been associated with oncogenic activity in thyroid tumors [52].

5.4.3 INTRACELLULAR SIGNALING MOLECULES AS ONCOGENES

5.4.3.1 The JAK/STAT Pathway

Not all growth factor receptors have intrinsic protein kinase activities. Some of these receptors, which includes many cytokine receptors and the prolactin receptor, have short intracellular domains, with constitutively associated, enzymatically latent Janus tyrosine kinase (JAK) proteins [53]. Upon ligand binding, the JAK kinase molecules are activated and phosphorylate the receptor molecules. The phosphotyrosine residues then serve as binding sites for the SH2 domains of the Signal Transducers and Activators of Transcription (STAT) family of proteins. The STAT proteins are then phosphorylated by the oligomerized JAKs. The STAT proteins then form heterodimers via reciprocal interactions between the phosphotyrosine residues and their SH2 domains. The STAT dimers then interact with the importin protein and translocate to the nucleus where they function as transcriptional regulators.

DELIVERY ADDRESS
Spatial House
Spatial House
Castle Donington
DE74 2TW,
GBR

Bookbarn International Ltd
Unit 1 Hallatrow Business Park, Wells
Road, Hallatrow,
Bristol
Somerset, BS39 6EX,
UNITED KINGDOM

Packing Slip / Invoice
Price: $82.81

Standard International

Order Date: 19/05/2021
Customer Contact: Spatial House

BBI Order Number: 1339372
Website Order ID: 68597983

NOS1
NOS2

SKU InvID	Locators	Item Information
3236965	C61-00-05	Molecular Carcinogenesis and the Molecular Biology of Human
10292803	318299	Cancer
		Blue
		HBDJ
		EXLib - N
		U:VG

albuk

If you wish to contact us regarding this order, please email us via Alibris quoting your order number.

Thanks for shopping with us

1339372

BOOKBARN INTERNATIONAL

Greetings from Bookbarn International,

Thank you so much for placing an order with us, in doing so you've helped us grow as an independent bookseller.

We hope you're happy with your purchase, however if you have any issues or queries please do not hesitate to contact us.

Contact Seller:
Log into the website you bought from, locate your order in your purchase history and

Our small and committed team provides a fast response to ensure you are fully satisfied.

Kind regards & happy reading!
The BBI Team

Salutations de Bookbarn International,

Merci beaucoup d'avoir passé une commande chez nous.
Vous nous avez aidé à grandir en tant que libraire indépendant.

Nous espérons que vous êtes satisfait de votre achat. Si vous avez des questions ou des questions, n'hésitez pas à nous contacter.

Connectez-vous au site Web où vous avez commandé votre livre, trouvez votre commande dans votre historique d'achat et contactez le vendeur.

Notre équipe petite et engagée fournit une réponse rapide pour s'assurer que vous êtes pleinement satisfait.

Bonne lecture!

Cordialement,
L'équipe BBI

Grüße von Bookbarn International,

Vielen Dank für Ihre Bestellung. Sie haben uns dabei geholfen, ein unabhängiger Buchhändler zu werden.

Wir hoffen, dass Sie mit Ihrem Kauf zufrieden sind. Wenn Sie jedoch Probleme oder Fragen haben, zögern Sie bitte nicht, uns zu kontaktieren.

Gehen Sie auf Meine Bestellungen, suchen Sie Ihre Bestellung in der Liste und klicken Sie auf Verkäufer kontaktieren.

Unser kleines und engagiertes Team bietet eine schnelle Antwort, um sicherzustellen, dass Sie voll zufrieden sind.

Mit freundlichen Grüßen und viel Spaß beim Lesen!
Das BBI-Team

Saludos desde Bookbarn International,

Muchas gracias por hacer un pedido con nosotros. Nos ha ayudado a crecer como librería independiente.

Esperamos que esté satisfecho con su compra. Si tiene algún problema o consulta, no dude en contactarnos: Inicie sesión en el sitio web que utilizó para efectuar su compra, busque su pedido en su historial de compras y comuníquese con el vendedor.

Somos un equipo pequeño, comprometido con brindar una respuesta rápida para garantizar su completa satisfacción.

¡Saludos cordiales y feliz lectura!

El equipo de BBI

Bookbarn Internationalからのご挨拶、

この度はご注文をしていただきありがとうございます。
あなたは私たちが独立系書店として成長するのを手伝ってくれました

ご購入いただいた商品に問題がある場合や質問がある場合は、遠慮なくお問い合わせください。

ご購入いただいた商品は、お客様が注文したウェブサイトにログインして、購入履歴から注文を見つけて販売者に連絡してください。

当社の小規模で献身的なチームは、お客様が完全にご満足いただけるように迅速な対応を行っています。

どうぞよろしくお願いいたします。

書籍を楽しんでいただけることを願っております！

BBI チーム

Although the canonical pathway for activation of STATs is via the receptor associated JAKs, STATs can also be activated directly by the kinase activity associated with RPTK such as EGF and PDGF. Alternatively, STATs can be activated by members of the src of Nonreceptor Protein Tyrosine Kinases (NRPTK) that are recruited and activated by RPTK or G protein coupled receptors. Oncogenic mutations that lead to constitutive activation of these types of JAK/STAT signaling have been primarily associated with leukemias and lymphomas. Since both JAK and STAT can activate other PI(3) Kinase and *Src*, their mechanisms of transformation probably includes cross-talk with other signaling pathways.

5.4.3.2 Nonreceptor Protein Tyrosine Kinases

Among the earliest discovered oncogenes are several members of intracellular protein tryosine kinases, which comprises 32 known genes, approximately half of which have been implicated in human cancer. These include the *c-Src* gene, the homolog of the first oncogene identified in a transforming retrovirus (Rous Sarcoma Virus), and the first oncogene to be identified by DNA mediated gene transfer. Other members of the cytoplasmic NRPTK class of oncogenes are *c-abl*, the cellular homolog of the Abelson Murine Sarcoma virus oncogene and many others including *yes*, *fgr*, *fyn*, *lyk*, *fps/fes*, and *trk*A [19]. The *c-ABL* gene is also involved in the Ph chromosome translocation associated chronic myelogenous leukemia (see Section 3). As is the case for the receptor protein tyrosine kinases, the activities of these signaling molecules are under stringent autoregulation.

Members of this family of kinases share a similar protein domain structure that includes a highly conserved kinase domain, or Src homology domain 1 (SH1), and a number of regulatory/effector binding domains. The effector binding sites include one or more SRC homology domains (SH2 and SH3) that recognize and bind to specific tyrosine residues that are the substrates for both phosphorylation by other tyrosine kinase, and dephosphorylation by protein tyrosine phosphatases. In the nonstimulated state, one of the SH2 domains of the prototypical SRC protein binds to a crucial phosphotyrosine residue near the carboxy terminus (Y527 in the murine *Src* gene). As a result, the SH3 domain interacts with a polyproline helix between the SH2 domain and the kinase domain, thereby preventing activation of the Src. The tyrosine kinase activity of *c-Src* can be activated by two separate mechanisms. The first involves Protein Tyrosine Phosphatase (PTP)-induced dephosphorylation of the carboxy terminal P-TYR (murine Y527). Accordingly, over expression of a wide variety of PTPs has been implicated with increased Src kinase activity. The Src kinase activity can also be activated by binding them to other effectors activated by phosphorylation on specific tyrosine residues. These P-TYR residues bind the SH2 or SH3 domains of inactive c-*Src*, displacing the intramolecular interactons and opening the conformation of the SRC protein. The active conformation of the c-Src protein can then be stabilized by autophosphorylation (*in trans*) of another crucial tyrosine residue (Y416 in murine Src).

The sequence context of each tyrosine determines the specificity for interactions with proteins that bind to the SH2 and SH3 homology domains. Unlike the receptor coupled tyrosine kinase proteins, cytoplasmic, NRPTK are not directly activated by ligand binding. Rather the proteins are normally in the inactive state due to the interaction of the kinase domain with a regulatory domain, typically one of the other SH domains. When an appropriate stimulus arrives in the cell, a specific tyrosine kinase enzyme is activated. When the latter phosphorylates the appropriate SH2 or SH3 domain, it activates the tyrosine kinase activity of the c-Src kinase. The phosphotyrosine residue then serves as a binding site for other specific SH2 or SH3 binding proteins, which are then phosphorylated and converted to the active form. The SH2 or SH3 binding proteins can be adapter molecules such as SOS or GRB that are phosphorylated [54] and then function as effectors of other signaling pathways, such as the RAS/RAF/MAPK pathways (Figure 5.4). The c-Src kinase substrates may themselves be protein kinases that then transmit the signal to downstream effectors. Examples of kinases activated by the c-Src kinase include Protein Kinase c (PKC), Focal Adhesion Kinase (FAK) and cdc2 kinase.

As a result, activation of c-Src related protein tyrosine kinases affect a variety of cellular functions including proliferation, motility, and survival.

The c-Src family of NRPTKs often acquires transforming activity as a result of mutations that lead to constitutive derepression of their tyrosine kinase activity. Activating mutations include truncations or point mutations affecting the autoregulatory domains. *NRPTK* oncogenes have been implicated in a number of chromosomal translocations in humans and gene fusions. Over expression of *c-SRC* has also been detected in a number of tumors. Elevated levels of c-Src kinase activity are generally associated with a poor prognosis in human colon cancer.

5.4.3.3 The RAS/RAF/MAPK Signaling Network

The Ras proteins are a closely related set of genes that encode membrane-associated proteins involved in cell proliferation and differentiation. In order for Ras proteins to be localized in the inner aspect of the cell membrane and mediate their effects, they must be prenylated and fatty acid acetylated at their carboxy termini [55]. More specifically, the specific cysteine residues are myristoylation, farnesylation, and geranylgeranylation. As a result a significant amount of research has focused on using inhibitors of farnesylation to block *ras* mediated transformation [56].

The Ras proteins belong to the family of small GDP/GTP-binding proteins that transduce signals, including those activated by cell surface tyrosine kinase receptors (Figure 5.4). Ras proteins receive and amplify signals from secondary messengers such as GRB, SOS, SRC, G proteins, that activate guanine nucleotide exchange factors (RASGEF). The RASGEFs then reactivate the inactive, GDP-bound form of Ras by catalyzing the exchange for GTP. As a result, mutations that result in the constitutive activation of RASGEF activity shift the equilibrium toward the activated form of Ras and are sufficient for transformation in human cells. The active GTP-bound form is normally reverted to the inactive GDP-bound form by an inherent GTPase activity. However, the inherent Ras GTPase activity is extremely low and is positively regulated by the interaction of Ras with GTPase Activating Proteins (GAPs). The Ras-GAP enzymes are themselves regulated by phosphorylation. As a result, the steady state levels of active GTP bound Ras can also be affected by mutations or environmental stimuli that regulated the activity of Ras-GAP. A clear example is the case of the *NF1* gene, which encodes the neurofibromin protein. Mutations in NF1 leads to the development of Neurofibromatosis, one of the most common autosomal dominant disorders in man [57]. Patients carrying the mutation are susceptible to a variety of skin lesions including "cafe-au-lait" spots, benign neurofibromas, and numerous malignancies including, neurosarcomas, gliomas, and pheochromocytomas. In NF1 patients, neurofibromin is unable to activate the intrinsic GTPase activity of Ras proteins.

Activated Ras proteins mediate their effects on normal cell growth by regulating the levels or activities of key regulators of cell cycle progression, including the induction of cyclin *D1* gene, suppression of CDK inhibitors, and subsequent Rb phosphorylation [58]. Ras signaling also induces changes in protein synthesis, cell growth, cellular adhesion and mobility, as well as enhancing cell survival and preventing apoptosis. When activated by oncogenic mutations that inhibit the ability to hydrolyze GTP, or as a result of mutations in key regulators of activity that affect and shift the equilibrium toward the active GTP bound form, ras proteins continue signaling to effectors in the absence of mitogenic stimuli. Ras proteins are able to transform a number of immortalized cell lines *in vitro*, as well as decrease tumor latency and increase tumor frequencies in transgenic animals. *Ras* oncogenes have been implicated in a wide variety of human and experimental tumors, and are present in $\sim$15% of all human tumors. In some human cancers, including pancreatic adenocarcinoma (25–87%), bronchioepithelial adenocarcinoma of the lung (25–48%), and colon adenocarcinoma (7–80%), myeloid leukemia (30%), activated *ras* oncogenes are detected at a much higher frequency. These findings suggest that the Ras proteins themselves represent a key step in the regulation of mitogenic signaling. The pivotal function of the Ras proteins is accentuated by the existence of regulatory G proteins that function as negative regulators of Ras signaling [59]. One of the Ki-*rev* was initially identified in a functional DNA mediated gene transfer experiment assay designed to detect suppressors of Ki-*ras* induced cell transformation. Ki-*rev* encodes a GTP binding protein closely related to the *ras* gene, which in its activated form competes with Ki-*ras* for binding to effector proteins. However, unlike Ki-Ras, the Ki-Rev protein inhibits rather than stimulates the activity of these effector proteins.

The molecular mechanisms of Ras mediated signal transduction has been extensively studied both *in vitro* and *in vivo*. It is now clear that activated Ras proteins serve as a vital point in the branch in signaling interaction, affecting three major classes of effector proteins — Rafs, PI(3)-kinase, and RalGefs [60]. The first of these comprise a small family of small proteins with serine/threonine protein kinase activities (Figure 5.4). The prototype of these enzymes, *v-raf* first encodes the acute transforming oncogene of several avian and murine retroviruses. Along with its cellular homolog, *c-raf-1*, this family of genes includes the *A-raf*, *B-raf*, *MEKK1*, and *MEKK2* genes, and are collectively referred to as Mitogen Activated Protein Kinase Kinases Kinases (MAPKKK) [61]. Acitvated MAPKKKs in turn phosphorylate and activate a family of MAP Kinase Kinases (MAPKK) enzymes that include MEK1/2, JNKK, and MEK3/6. These MAPKKs in turn phosphorylate and activate respectively the ERK1/2, JUNK/SAPK, and p38 classes of MAP Kinases (MAPK). Each of these signaling cassettes in turn mediates changes in cell proliferation, differentiation, cytoskeletal architecture, stress responses, and apoptosis. The ERK1/2 cassette, which alters gene expression, phospholipid metabolism, and protein synthesis, has been strongly implicated in oncogenesis. First of all, c-Raf-1 is a component of the ERK1/2 cassette. ERK1/2 can itself induce cell growth and interacts with a number of other oncoproteins. More recent studies have demonstrated that mutations in the *BRAF*, a homolog of *c-Raf-1*, can substitute for *Ras* oncogenes in a variety of human tumors. Although there is some overlap in the molecular responses evoked by the JNKK and p38 MAPKKK cassettes, and may contribute to the phenotype of the transformed cells, neither of these pathways has been shown to have transforming potential.

5.4.3.4 The PI(3) Kinase/AKT Pathway

The phosphoinositol-3-kinase/AKT mediated signaling pathway is another frequent source of oncogenic mutations [62]. The PI(3) kinase protein is activated by a number of other signaling pathways including RTPKs, including the insulin receptor, the Src family of protein tyrosine kinases, and the RAS/RAF/MAPK pathway. In addition to activating the MAPKKK networks, Ras signaling also activates the cell membrane associated phosphatidylinositol-3-kinase activity (PI(3)K), which phosphorylates the D3 position of membrane-associated phosphotidylinositol-4,5-*bis*phosphate (PIP$_2$) to generate phosphotidylinositol-3,4,5-triphosphate (PIP$_3$). The transforming potential of the PI(3)K gene was demonstrated by its transduction as the transforming GAG–P3K fusion protein encoded by Avian Sarcoma Virus 16 (ASV16). Membrane associated PIP3 then binds to the pleckstrin homology domain of 3-Phosphoinositide-dependent protein kinases PDK1 and PDK2, which in turn activate the AKT protein kinase. *AKT* was first identified as the oncogene of the transforming AKT8 murine retrovirus [63]. The AKT protein kinase then phosphorylates a large number of effector proteins that culminate in the inhibition of apoptosis and cell growth.

PIP2 is a substrate for the phosphatidylinositol specific phospholipase C (PLC) family of enzymes. Enzymatic cleavage of membrane bound PIP2 by PLC releases two important signal-transducing molecules. The first of these, diacylglycerol (DAG) is an allosteric activator of Calcium dependent protein kinases (PKC), which play a role in numerous cellular functions including growth and differentiation. Several tumor promoting chemicals, including phorbol esters mimic the effects of DAG on protein PKC, indicating that activation of PLC by *ras* signaling may also contribute to tumor progression. The second signaling molecule released by PLC mediated cleavage of PIP2 is soluble phosphatidylinositol-1,4,5-triphosphate, or PtdIns(1,4,5)P3. The latter diffuses through the cytoplasm and opens PtdIns(1,4,5)P3-gated calcium channels in the endoplasmic reticulum, leading to further activation of PKC and other Calcium dependent signaling processes that promote differentiation, and possibly prevent abnormal growth.

5.5 DEVELOPMENTAL REGULATORS AS ONCOGENES

Another important signal transduction pathway that is frequently a source of oncogenic mutations is the Wnt signaling pathway [64,65]. Although the roles of the Wingless (WG or Wnt) and Hedgehog (HH) signaling pathways in cell growth and patterning during embryogenesis were elucidated many years ago, the fact that these pathways were also deregulated in cancer was not appreciated until the last decade. Once this convergence was understood scientists had access to the tremendous wealth of genetic and mechanistic studies from the field of developmental biology. The *Wnt* gene (originally designated *Int-1*) was initially identified in mouse mammary tumors as a target for transcriptional activation, following insertional mutagenesis by the MMTV retrovirus [6]. The *Wnt* genes encode a family of soluble peptide growth factors that are characterized by their ability to induce the formation of a new embryonic axis in a variety of organisms. The interaction of the Wnt and HH signaling pathways is, for example, essential for dorsal/ventral patterning at the segment boundaries in embryo of the fruit fly (*Drosophila melanogaster*). A greatly simplified scheme of the signaling networks involved is presented in Figure 5.5 in order to illustrate overlap with genes involved in carcinogenesis. What is immediately apparent is that patterning involves the generation of reciprocal signals that allow each cell to determine and maintain its position in embryos. Thus, the posterior cell secretes HH, another soluble protein which the binds to the Smoothened receptor (SMO) on the anterior cell. The SMO protein is a member of the seven transmembrane receptor family that is closely related to the Wnt receptor proteins. The ability of the SMO receptor to transduce a signal is negatively regulated by the Patched (*PTCH*) gene, possibly by the formation of heteromeric complexes. Significantly, the human *PTCH* gene encodes the tumor suppressor gene mutated in familial nevoid basal cell carcinoma [66]. Binding of HH to SMO appears to uncouple the receptor from PTCH mediated repression, and activates a complex signaling network. Activation of the serine/threonine kinase encoded by the Fused (*FU*) gene allows for dissociation of the transcription factor encoded by the Cubitus Interuptus (*CI*) gene from a large complex with tubulin and allows for its translocation to the nuclears. The CI transcription factor then activates transcription of WG (*Wnt*) gene, as well as PTCH resulting in a negative feedback loop. CI also induces the expression of decapentaplegic (*DPP*), which encodes a putative tumor suppressor gene deleted in human pancreatic cancers. The WG (Wnt) protein is then secreted by the anterior cell and binds to the Wnt receptor on the posterior cell (Figure 5.5). Wnt receptors in mammalian cells consist of a receptor molecule encoded by the Frizzled gene and a coreceptor molecule encoded by a member of the low-density Lipoprotein Receptor Proteins (LRP). Upon binding, the product of the *Disheveled* gene mediates the dissociation of the *Armadillo* (*ARM*), which encodes the β-catenin protein, from a complex *Zeste White* (*ZW*), which encodes the Glycogen Synthase Kinase-3β (GSK-3β). In the absence of Wnt signaling, GSK-3β phosphorylates and recruits β-catenin to a complex containing Axin and the *APC* (Adenomatous Polyposis Coli tumor suppressor) gene, which then promotes degradation of β-catenin by the proteosome system. Thus in the presence of Wnt signaling, the β-catenin (*ARM*) protein is stabilized and transported to the nucleus where it induces the expression of genes, including HH, and engrailed. In mammalian cells, stabilized β-catenin functions as a transcriptional coactivator by associating with the Tcf/LEF family of transcription factors.

The importance of this signaling pathway in carcinogenesis is evinced by the fact that numerous genes comprising these pathways are involved in human cancer. The β-catenin, Wnt, Gli, Smo, and others can be activated into oncogenes, while several other genes including *APC*, *PTCH*, *DPP*, and others function as tumor suppressor genes. However, since all of these genes are involved in embryogenesis, how is it that mutations in these oncogenes and tumor suppressor genes are associated with both pediatric and adult cancer? Clearly the effects of the Wnt signaling pathways

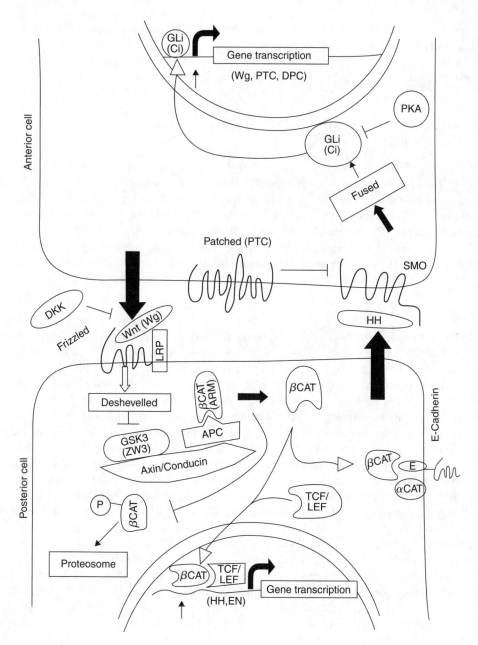

FIGURE 5.5   The Wnt signaling pathway in development and in cancer.

cannot be explained solely on the basis of growth stimulation. A reasonable supposition is that the Wnt signaling pathway also plays a role in tissue homeostasis [50]. In metazoans, cells are continuously lost due to programmed cells death as well as normal wear and tear. As a result some tissues, including the epidermis and lining of the intestinal tract, are continuously replaced by the process of tissue turnover. Under normal circumstances, this is a highly controlled process involving stem cell turnover units comprised of single stem cells, which undergoes asymmetrical cell division to yield a daughter stem cell and the first transition cell. The latter then divides for a specified number of additional cell divisions, eventually giving rise to terminally differentiated daughter cells. The Wnt signaling pathway probably plays a role in maintaining the identity of the stem cells and the transition cells so that the stem cell compartment does not change and organ size remains constant. Such a model explains why germ line mutations in the *APC* and *PTCH* tumor suppressor genes lead to the generation of multiple, premalignant lesions characterized by what appears to be an expansion in the number of stem cells after stochastic loss of the remaining tumor suppressor allele.

5.6 NUCLEAR TRANSCRIPTIONAL REGULATORS AS ONCOGENES

Ultimately all signaling pathways directly or indirectly impinge upon the nucleus to modulate gene expression patterns and effect phenotypic change, be it cell growth or arrest, differentiation, senescence, or apoptosis. Constitutive expression, or for that matter inappropriate spatial/temporal expression of effector genes can significantly impact normal cellular programs and contribute to malignant growth. It is therefore logical to assume that mutations that lead to altered function or levels of key transcriptional regulatory proteins can contribute to cell transformation. Indeed many important human oncogenes are mutated version of transcription factors that play roles in regulating the expression of genes control key cellular processes [67].

5.6.1 FOS/JUN/AP-1

Although not specifically detected as an activated oncogene in human cancers, the *c-fos* proto-oncogene plays a central role in regulating cell growth. Initially detected as the acute transforming component of the FBJ and FBR murine sarcoma viruses, the *c-fos* gene encodes the prototypical leucine zipper transcription factor protein [68]. The Fos Protein forms heterodimeric complexes with members of the JUN family of transcription factors, via interactions through their respective leucine zipper domains. A mutated version of the *c-jun* is responsible for the transforming activity of the transforming avian retrovirus [69]. The Jun family of proteins can also homodimerize and form heterodimeric complexes with other leucine zipper transcription factors such as those encoded by the FRA family of genes. The Fos and Jun protein dimers, also referred to as AP-1 complexes, bind to specific DNA sequences and transactivate gene expression. Different heterodimeric complexes show differences in binding affinities for specific, although sometimes closely related DNA binding sites. The proteins comprising the AP-1 complexes are usually induced at the transcriptional level shortly after exposure of cells to a variety of stimuli, including mitogens. This immediate early response is attenuated by the presence of RNA destabilizing sequences within the noncoding sequences of the mRNAs. As a result, the immediate early genes such as *Fos* have a short half-life in the cell. In order to function as an oncogene, the *Fos* proto-oncogene must not only be deregulated transcriptionally, but the destabilizing sequences must be removed from the mRNA sequence. The mechanisms by which Fos and Jun proteins induce cell transformation remain to be detailed. It is likely that stabilization of one member of a several families of interacting leucine zipper transcriptional regulators will have a profound effect on the cell's pattern of gene expression.

5.6.2 c-myc

Perhaps the most important of all the nuclear proto-oncogenes is *c-myc*, initially identified as the transforming component of the avian myelocytomatosis retrovirus [70]. As already indicated above, the *c-myc* gene is also activated in Burkitt's lymphoma as a result of chromosomal translocations that result in the deregulated expression of the proto-oncogene. However, elevated c-myc expression is also detected in a wide variety of human and experimental tumors, and is usually associated with more aggressive tumor phenotypes and poor patient prognosis. There are at least three mechanisms by which deregulated expression of c-myc modulates the cancerous growth of cells, including effects on cell growth and proliferation, inhibition of terminal differentiation, and sensitizing cells to apoptosis (Figure 5.6).

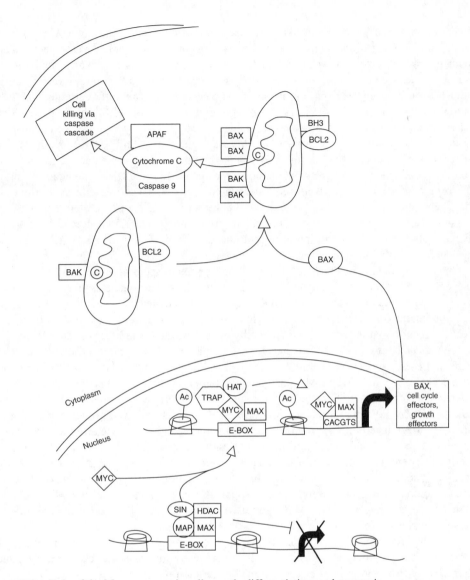

Figure 5.6   Role of the Myc oncogene in cell growth, differentiation, and apoptosis.

The proto-oncogene encodes a 439 amino acid protein with a carboxyterminal basis-Helix-Loop-Helix-Zipper (bHLHZ) domain. The c-MYC protein forms heterodimeric protein complexes with c-MAX, another bHLHZ transcription factor, which then bind to specific DNA sequences in the promoters of target genes (e.g., the CACGTG E-box sequence). The amino terminus of the protein contains two conserved Myc Box sequences (MB I and MB II) that are required for interactions with other transcriptional activators. Among the proteins that interact with the transactivating domain is the transformation/transcription-domain associated protein (TRAP), a component of the histone acetyltransferase complex. These findings suggest that c-myc may play a direct role in chromosomal remodeling. Mutations that abolish the ability of c-myc to heterodimerize or interact with transcriptional activators also abolish the effect of the proto-oncogene on cell growth, proliferation, differentiation, and apoptosis. Together, these findings indicate that the c-Myc protein mediates its pleiotrophic effects by altering the cell's pattern of gene expression.

The *c-myc* proto-oncogene plays a central role in promoting both cell growth (increase in cell size) and proliferation (increase in cell number). Expression of the Myc protein is essentially undetectable in nondividing cells, is quickly induced in response to mitogenic stimuli, and again tapers off as cells return to a nonproliferating state. Myc expression is high during embryonic development and tapers off with diminished cell growth and terminal differentiation. In adult tissues, expression is highest in cells from tissues with a high turnover rate, such as the colon and skin. It is therefore not surprising that ablation of the *c-myc* gene has dramatic effects on cell growth and differentiation both *in vitro* and *in vivo*. Although the role of c-MYC in promoting progression of cells through the G_1–S phase of the cell cycle activity has long been known, studies have only recently begun to shed light upon the mechanism by which elevated expression of c-Myc activates the cyclin E-CDK2 (cyclin dependent kinase) complex. Among the targets of c-Myc transcriptional regulation are genes encoding several key regulators of cell cycle progression, including the cyclin D2 and CDK4. The latter two proteins then form a complex that effectively sequesters the p27 (KIP1) CDK inhibitor away from cyclin E-CDK2 complexes. c-Myc may also promote the degradation of the p27 protein by inducing the expression of key regulators of degradation via the proteasome system. Recent studies suggest that c-Myc may also inhibit the transcription of gene encoding the INK4a (p15) and WAF1 (p21) cyclin dependent kinase inhibitors, resulting in increased levels of cyclin dependent kinase activities in the cell, which then promote the phosphorylation of the Rb protein and progression into S phase.

Several lines of evidence indicate that regulation of protein synthesis plays a key role in cancer [71]. The ability of c-Myc to alter the cell growth is mediated by its effects on the expression of rate limiting factors in protein synthesis. The c-Myc protein induces the expression of the eIF4E and eIF2α translational initiation factors, thereby increasing the overall rate of mRNA translation. Cells with ablated c-myc show defects in cell growth, which in turn limit the ability of cells to enter into the cell cycle, suggesting a link between cell size and division. These studies highlight the importance of c-Myc induced translational deregulation in cell growth and cancer.

Deregulated expression of c-Myc also appears to inhibit terminal differentiation, which may in some instances be sufficient for malignant growth. The stimulation of cell growth and proliferation by the MYC/MAX heterodimer must be overcome in order to induce growth arrest and terminal differentiation, although some exceptions may exist. Studies have identified a family of closely related transcriptional repressors that appear to counter the effects of c-Myc. The members of the *MAD* gene family encode proteins that have been linked to inhibition of cell cycle progression, terminal differentiation, and some may even be tumor suppressor genes. *MAD* genes encode bHLHZ proteins that heterodimerize with the c-Myc partner, MAD, and sequester the latter into transcriptional repressor complexes that compete for the same DNA binding sites. In addition the Max/Mad complexes bind directly to the SIN3 protein, which in turn recruits histone deacetylase complexes. Thus, the Max/Mad complex may also have the opposite effects of the Myc/Max complex, by inducing histone deacetylation and promoting chromatin condensation.

5.6.3 THE RETINOIC ACID RECEPTOR

Another interesting example of a transcription factor that functions as a transcriptional regulator and oncogene is the PML-RARα fusion gene detected in promyelocytic leukemia cells with the t(15;17) chromosomal translocation [72]. This translocation fuses the Promyelocytic Leukemia (*PML*) gene on chromosome 15 with the Retinoic Acid Receptor alpha (RARα) on chromosome 17. The RARα receptor, is a nuclear receptor that forms heterodimers with the retinoid X receptor. The heterodimer represses gene expression by binding to specific DNA sequences present in the promoter sequences of target genes and recruiting histone deacetylase (HDAC)/corepressor complexes. Normally, binding of the ligand to the receptor allows for the dissociation of HDAC/corepressor complexes and recruitment coactivator/histone acetyltransferase complexes. The target genes activated by Retinoic Acid (RA) induced depression are required for the normal differentiation of promyelocytes into erythrocytes, monocytes, and granulocytes [73,74]. By contrast the *PML-RARα* oncogene is potent repressor of gene expression that requires elevated levels of RA to induce dissociation of the corepressor complex. Studies have shown that the PML domain of the *PML-RARα* oncogene also recruits HDAC/corepressor complexes, which may contribute to its potency as a transcriptional repressor. Thus, retinoic acid can reverse the effects of the *PML-RARα* oncogene by inducing differentiation of the leukemic cells. As a result RA therapy has been used as a standard therapy for these leukemias with translocations that involve the *RARα* gene. Recently, arsenic compounds were shown to be the active ingredient in a traditional Chinese medicine shown to be highly effective in the treatment of promyelocytic leukemias. Subsequent studies showed that like RA, arsenic was able to derepress the expression of genes inhibited by *PML-RARα*. The mechanism by which arsenic compounds cause derepression appear to involve both the dissociation of HDAC complexes, as well as the sumoylation of the PML moiety, which leads to its targeted degradation by the proteosome system. As a result, arsenic appears to be a more potent treatment of promyelocytic leukemias, even in patients that do not respond to RA, and its therapeutic effects are additive with those of RA.

The *PML* gene, originally thought to be a transcription factor, also appears to play a role in the transforming potential of the oncogene. Knockout mice lacking *PML* show defects in myeloid differentiation, are resistant to apoptosis, have increased sensitivity to carcinogens, and have aberrant *PML* expression and have been detected in a variety of human tumors, while over expression of PML induces growth arrest, senescence, and apoptosis. Studies have shown the PML-RARα appears to interfere with PML nuclear body formation. Treatment with RA or arsenic compounds has been shown to induce proteolytic degradation of the PML-RARα oncoprotein. Since the nuclear bodies can be reformed after either RA or arsenic therapy, both of which can induce apoptosis, these finding suggest that the ability of PML-RARα to block PML induced apoptosis may also contribute to its mechanisms of carcinogenesis.

5.7 REGULATORS OF APOPTOSIS AS ONCOGENES

Apoptosis, or programmed cell death, plays a critical role in multicellular organisms ensuring a constant size as cells continue to turnover in the adult as well as removing cells that have acquired significant levels of irreparable DNA damage [75]. The process of apoptosis itself involves a complex and highly regulated network of signals that result in the activation of a series of proteases (Caspases) and nucleases that eventually lead to controlled disintegration of the nucleus and the cell. There are at least two distinct pathways to programmed cell death. Both involve the activation of caspases that are synthesized as inactive proenzymes that must undergo specific proteolytic cleavage in order to be activated and in turn activate other caspases in the cascade. The extrinsic pathways is mediated by the binding of the tumor necrosis factor family of ligands to specific death receptors on the surface of the target cell. Ligand binding recruits caspase 8 via the FADD adaptor protein and activates caspase-3. The intrinsic pathway is activated by the formation of pores that allow for the release of solutes, especially Cytochrome C from the mitochondria (Figure 5.6). The latter then activates the apoptosome, comprised of the activator *APAF1* gene and Caspase-9, which then sets in motion a proteolytic cascade that culminates in destruction of the cell by caspase-3 and caspase-7. The triggering event, namely increased mitochondrial permeability is a regulated fine balance between pro- and anti-apoptotic factors. The most important of these is the protein encoded by the *BLC2* gene [76], originally identified as an oncogene as a result of its transcriptional activation by specific chromosomal translocations in follicular B cell lymphomas. In the lymphocytes, chronic over expression of the Bcl2 protein presumably prevents normal cell turnover, thereby allowing for the uncontrolled expansion of the affected clones.

The mechanism by which Bcl2 and other antiapoptotic proteins (e.g., Bcl-X_L, Bcl-W) prevent cell death involves two additional classes of structurally related proapoptotic proteins. Bcl2 is normally integrated into the outer mitochondrial membrane, where it prevents the oligomerization of the members of the BAX family of proapoptotic genes (e.g., *Bax*, *Bak*, *Bok*). When the cell stress pathway is activated, members of the BH3 family of proapoptotic proteins, normally sequestered via a variety of mechanisms are released and bind to the BCL2 family of proteins. BH3 binding neutralizes the effect of Bcl2 on the Bak family of proapoptotic proteins, which then form homo-oligiomers that increase the permeability of the outer mitochondrial membrane. The consequent release of Cytochrome C then activates the lethal caspase cascade.

A number of different oncogenes, including *c-myc*, appear to play an important role in the regulation of programmed cell death or apoptosis. The initial observations made *in vitro* indicated that cells expressing ectopic *c-myc* were more sensitive to apoptosis under a variety of conditions including deprivation of growth/survival factor, exposure to tumor necrosis factor, hypoxia, DNA damage, or the accumulation of reactive oxygen species. Studies *in vitro* and in transgenic animals indicated that the *c-myc* induced sensitivity to apoptosis could be overcome by ectopic expression of antiapoptotic genes, such as *bcl-2* or *bcl-x$_L$*, or the loss of tumor suppressors such as *p53* and *ARF* genes. These findings suggested that in a society of cells, cell proliferation and apoptosis are coupled as part of an altruistic defense against clonal outgrowth. Thus cells that acquire mutations in growth promoting genes such as *c-myc* sacrifice themselves for the good of the organism. Although many details remain to be elucidated, apoptosis can be triggered by a variety of stimuli including the posttranslational stabilization of the *p53* tumor suppressor. The p53 protein can be stabilized by a number of pathways, including the activation of the DNA damage checkpoints cascade. Alternatively, p53 degradation can be inhibited by mitogenic signals leading to the increased expression of *ARF*. The latter then sequesters mdm2, a negative regulator of p53, that was initially identified as an oncogene present on double-minute chromosomes in a mouse plasmocytoma [16]. The *p53* gene can then promote apoptosis by increasing the activity of proapoptotic genes such as *BAX*, or by increasing negative regulators of the prosurvival members of the *bcl2* gene family (e.g., *Puma, Noxa*).

Although it is clear that deregulation of apoptosis by oncogenes can lead to preferential survival of tumor cells under conditions that would normally induce cellular suicide (e.g., lack of mitogenic signals, anoxia), not all tumors have lost the ability to undergo programmed cell death. This has important implications for therapy since many chemotherapeutic agents kill cells via their genotoxic effects, presumably by inducing apoptosis. This notion is supported by the observations that over expression of Bcl2 makes cells resistant to a variety of cancer therapies, and resistant clones that arise after therapy are often associated with elevated levels of Bcl2. Thus the genes comprising the apoptotic signaling networks remain interesting targets for directed cancer therapies.

REFERENCES

1. Huebner, R.J. and Todaro, G.J. Oncogenes of RNA tumor viruses as determinants of cancer. *Proc. Natl. Acad. Sci. USA, 64*: 1087–1094, 1969.
2. Stehelin, D., Varmus, H.E., Bishop, J.M., and Vogt, P.K. DNA related to the transforming gene(s) of avian sarcoma viruses is present in normal avian DNA. *Nature, 260*: 170–173, 1976.
3. Bishop, J.M. Viruses, genes, and cancer. II. Retroviruses and cancer genes. *Cancer, 55*: 2329–2333, 1985.
4. Varmus, H. Retroviruses. *Science, 240*: 1427–1435, 1988.
5. Jonkers, J. and Berns, A. Retroviral insertional mutagenesis as a strategy to identify cancer genes. *Biochim. Biophys. Acta, 1287*: 29–57, 1996.
6. Nusse, R. Insertional mutagenesis in mouse mammary tumorigenesis. *Curr. Top. Microbiol. Immunol., 171*: 43–65, 1991.
7. Tsichlis, P.N., Strauss, P.G., and Hu, L.F. A common region for proviral DNA integration in MoMuLV-induced rat thymic lymphomas. *Nature, 302*: 445–449, 1983.
8. Lazo, P.A., Lee, J.S., and Tsichlis, P.N. Long-distance activation of the Myc protooncogene by provirus insertion in Mlvi-1 or Mlvi-4 in rat T-cell lymphomas. *Proc. Natl. Acad. Sci. USA, 87*: 170–173, 1990.
9. Neil, J.C. and Cameron, E.R. Retroviral insertion sites and cancer: Fountain of all knowledge? *Cancer Cell, 2*: 253–255, 2002.
10. Weinberg, R.A. A molecular basis of cancer. *Sci. Am., 249*: 126–142, 1983.
11. Gibbs, W.W. Untangling the roots of cancer. *Sci. Am., 289*: 56–65, 2003.

12. Barbacid, M., Lamballe, F., Pulido, D., and Klein, R. The trk family of tyrosine protein kinase receptors. *Biochim. Biophys. Acta, 1072*: 115–127, 1991.

13. Pane, F., Intrieri, M., Quintarelli, C., Izzo, B., Muccioli, G.C., and Salvatore, F. BCR/ABL genes and leukemic phenotype: From molecular mechanisms to clinical correlations. *Oncogene, 21*: 8652–8667, 2002.

14. Druker, B.J. STI571 (Gleevec) as a paradigm for cancer therapy. *Trends Mol. Med., 8*: S14–S18, 2002.

15. Boveri, T. *Zur Frage der Enstehung maligner Tumoren.* Gustav Fisher, Jena, 1914.

16. Momand, J. and Zambetti, G.P. Mdm-2: "big brother" of p53. *J. Cell. Biochem., 64*: 343–352, 1997.

17. Albertson, D.G. and Pinkel, D. Genomic microarrays in human genetic disease and cancer. *Hum. Mol. Genet.*, 2003.

18. Balmain, A. and Brown, K. Oncogene activation in chemical carcinogenesis. *Adv. Cancer. Res., 51*: 147–182, 1988.

19. Blume-Jensen, P. and Hunter, T. Oncogenic kinase signalling. *Nature, 411*: 355–365, 2001.

20. Malumbres, M. and Barbacid, M. RAS oncogenes: The first 30 years. *Nat. Rev. Cancer, 3*: 459–465, 2003.

21. Barbacid, M. ras oncogenes: Their role in neoplasia. *Eur. J. Clin. Invest., 20*: 225–235, 1990.

22. Barbacid, M. ras genes. *Annu. Rev. Biochem., 56*: 779–827, 1987.

23. Almoguera, C., Shibata, D., Forrester, K., Martin, J., Arnheim, N., and Perucho, M. Most human carcinomas of the exocrine pancreas contain mutant c-K-ras genes. *Cell, 53*: 549–554, 1988.

24. Bartram, C.R. Mutations in ras genes in myelocytic leukemias and myelodysplastic syndromes. *Blood Cells, 14*: 533–538, 1988.

25. Fasano, O., Aldrich, T., Tamanoi, F., Taparowsky, E., Furth, M., and Wigler, M. Analysis of the transforming potential of the human H-ras gene by random mutagenesis. *Proc. Natl. Acad. Sci. USA, 81*: 4008–4012, 1984.

26. Shih, T.Y., Hattori, S., Clanton, D.J., Ulsh, L.S., Chen, Z.Q., Lautenberger, J.A., and Papas, T.S. Structure and function of p21 ras proteins. *Gene Amplif. Anal., 4*: 53–72, 1986.

27. Hahn, W.C., Counter, C.M., Lundberg, A.S., Beijersbergen, R.L., Brooks, M.W., and Weinberg, R.A. Creation of human tumour cells with defined genetic elements. *Nature, 400*: 464–468, 1999.

28. Chin, L., Tam, A., Pomerantz, J., Wong, M., Holash, J., Bardeesy, N., Shen, Q., O'Hagan, R., Pantginis, J., Zhou, H., Horner, J.W., 2nd, Cordon-Cardo, C., Yancopoulos, G.D., and DePinho, R.A. Essential role for oncogenic Ras in tumour maintenance. *Nature, 400*: 468–472, 1999.

29. Duesberg, P.H. Are cancers dependent on oncogenes or on aneuploidy? *Cancer Genet. Cytogenet., 143*: 89–91, 2003.

30. Duesberg, P.H. Oncogenes and cancer. *Science, 267*: 1407–1408, 1995.

31. Sonneschein, C. and Soto, A. *The Society of Cells: Cancer and Control of Cell Proliferation.* BIOS Scientific Publishers, Ltd, Oxford UK and Springer Verlag, New York, 1999.

32. Finney, R.E. and Bishop, J.M. Predisposition to neoplastic transformation caused by gene replacement of H-ras1. *Science, 260*: 1524–1527, 1993.

33. Cha, R.S., Thilly, W.G., and Zarbl, H. *N*-nitroso-*N*-methylurea-induced rat mammary tumors arise from cells with preexisting oncogenic *Hras1s* gene mutations. *Proc. Natl. Acad. Sci. USA, 91*: 3749–3753, 1994.

34. Zhang, Z., Wang, Y., Vikis, H.G., Johnson, L., Liu, G., Li, J., Anderson, M.W., Sills, R.C., Hong, H.L., Devereux, T.R., Jacks, T., Guan, K.L., and You, M. Wildtype Kras2 can inhibit lung carcinogenesis in mice. *Nat. Genet., 29*: 25–33, 2001.

35. Diaz, R., Ahn, D., Lopez-Barcons, L., Malumbres, M., Perez de Castro, I., Lue, J., Ferrer-Miralles, N., Mangues, R., Tsong, J., Garcia, R., Perez-Soler, R., and Pellicer, A. The N-ras proto-oncogene can suppress the malignant phenotype in the presence or absence of its oncogene. *Cancer Res., 62*: 4514–4518, 2002.

36. Bargmann, C.I., Hung, M.C., and Weinberg, R.A. Multiple independent activations of the neu oncogene by a point mutation altering the transmembrane domain of p185. *Cell, 45*: 649–657, 1986.

37. Yarden, Y. and Sliwkowski, M.X. Untangling the ErbB signalling network. *Nat. Rev. Mol. Cell Biol., 2*: 127–137, 2001.

38. Mercer, K.E. and Pritchard, C.A. Raf proteins and cancer: B-Raf is identified as a mutational target. *Biochim. Biophys. Acta, 1653*: 25–40, 2003.

39. Rowley, J.D. Chromosome translocations: Dangerous liaisons revisited. *Nat. Rev. Cancer*, *1*: 245–250, 2001.
40. Rabbitts, T.H. Chromosomal translocations in human cancer. *Nature*, *372*: 143–149, 1994.
41. Nowell, P.C. The clonal evolution of tumor cell populations. *Science*, *194*: 23–28, 1976.
42. Schimke, R.T. Gene amplification, drug resistance, and cancer. *Cancer Res.*, *44*: 1735–1742, 1984.
43. Hahn, P.J. Molecular biology of double-minute chromosomes. *Bioessays*, *15*: 477–484, 1993.
44. Brodeur, G.M. Neuroblastoma: Biological insights into a clinical enigma. *Nat. Rev. Cancer*, *3*: 203–216, 2003.
45. Ideker, T., Galitski, T., and Hood, L. A new approach to decoding life: Systems biology. *Annu. Rev. Genomics Hum. Genet.*, *2*: 343–372, 2001.
46. Pawson, T. and Nash, P. Assembly of cell regulatory systems through protein interaction domains. *Science*, *300*: 445–452, 2003.
47. Aaronson, S.A. Growth factors and cancer. *Science*, *254*: 1146–1153, 1991.
48. Wells, A. EGF receptor. *Int. J. Biochem. Cell Biol.*, *31*: 637–643, 1999.
49. Roberts, A.B. and Wakefield, L.M. The two faces of transforming growth factor beta in carcinogenesis. *Proc. Natl. Acad. Sci. USA*, *100*: 8621–8623, 2003.
50. Taipale, J. and Beachy, P.A. The Hedgehog and Wnt signalling pathways in cancer. *Nature*, *411*: 349–354, 2001.
51. Neel, B.G., Gu, H., and Pao, L. The 'Shp'ing news: SH2 domain-containing tyrosine phosphatases in cell signaling. *Trends Biochem. Sci.*, *28*: 284–293, 2003.
52. Rodien, P., Ho, S.C., Vlaeminck, V., Vassart, G., and Costagliola, S. Activating mutations of TSH receptor. *Ann. Endocrinol. (Paris)*, *64*: 12–16, 2003.
53. Boudny, V. and Kovarik, J. JAK/STAT signaling pathways and cancer. Janus kinases/signal transducers and activators of transcription. *Neoplasma*, *49*: 349–355, 2002.
54. Pawson, T. Regulation and targets of receptor tyrosine kinases. *Eur. J. Cancer*, *38 (Suppl 5)*: S3–S10, 2002.
55. Resh, M.D. Regulation of cellular signalling by fatty acid acylation and prenylation of signal transduction proteins. *Cell Signal.*, *8*: 403–412, 1996.
56. Dinsmore, C.J. and Bell, I.M. Inhibitors of farnesyltransferase and geranylgeranyltransferase-I for antitumor therapy: Substrate-based design, conformational constraint and biological activity. *Curr. Top. Med. Chem.*, *3*: 1075–1093, 2003.
57. Dasgupta, B. and Gutmann, D.H. Neurofibromatosis 1: Closing the GAP between mice and men. *Curr. Opin. Genet. Dev.*, *13*: 20–27, 2003.
58. Malumbres, M. and Barbacid, M. To cycle or not to cycle: A critical decision in cancer. *Nat. Rev. Cancer*, *1*: 222–231, 2001.
59. Noda, M. Structures and functions of the K rev-1 transformation suppressor gene and its relatives. *Biochim. Biophys. Acta*, *1155*: 97–109, 1993.
60. Chang, F., Steelman, L.S., Shelton, J.G., Lee, J.T., Navolanic, P.M., Blalock, W.L., Franklin, R., and McCubrey, J.A. Regulation of cell cycle progression and apoptosis by the Ras/Raf/MEK/ERK pathway (Review). *Int. J. Oncol.*, *22*: 469–480, 2003.
61. Su, B. and Karin, M. Mitogen-activated protein kinase cascades and regulation of gene expression. *Curr. Opin. Immunol.*, *8*: 402–411, 1996.
62. Vivanco, I. and Sawyers, C.L. The phosphatidylinositol 3-kinase AKT pathway in human cancer. *Nat. Rev. Cancer*, *2*: 489–501, 2002.
63. Staal, S.P. and Hartley, J.W. Thymic lymphoma induction by the AKT8 murine retrovirus. *J. Exp. Med.*, *167*: 1259–1264, 1988.
64. Giles, R.H., van Es, J.H., and Clevers, H. Caught up in a Wnt storm: Wnt signaling in cancer. *Biochim. Biophys. Acta*, *1653*: 1–24, 2003.
65. Polakis, P. Wnt signaling and cancer. *Genes Dev.*, *14*: 1837–1851, 2000.
66. Dean, M. Towards a unified model of tumor suppression: Lessons learned from the human patched gene. *Biochim. Biophys. Acta*, *1332*: M43–M52, 1997.
67. Darnell, J.E., Jr. Transcription factors as targets for cancer therapy. *Nat. Rev. Cancer*, *2*: 740–749, 2002.
68. Forrest, D. and Curran, T. Crossed signals: Oncogenic transcription factors. *Curr. Opin. Genet Dev.*, *2*: 19–27, 1992.

69. Vogt, P.K. Jun, the oncoprotein. *Oncogene, 20*: 2365–2377, 2001.

70. Pelengaris, S., Khan, M., and Evan, G. c-MYC: More than just a matter of life and death. *Nat. Rev. Cancer, 2*: 764–776, 2002.

71. Ruggero, D. and Pandolfi, P.P. Does the ribosome translate cancer? *Nat. Rev. Cancer, 3*: 179–192, 2003.

72. Zhu, J., Chen, Z., Lallemand-Breitenbach, V., and de The, H. How acute promyelocytic leukaemia revived arsenic. *Nat. Rev. Cancer, 2*: 705–713, 2002.

73. He, L.Z., Tolentino, T., Grayson, P., Zhong, S., Warrell, R.P., Jr., Rifkind, R.A., Marks, P.A., Richon, V.M., and Pandolfi, P.P. Histone deacetylase inhibitors induce remission in transgenic models of therapy-resistant acute promyelocytic leukemia. *J. Clin. Invest., 108*: 1321–1330, 2001.

74. Weston, A.D., Blumberg, B., and Underhill, T.M. Active repression by unliganded retinoid receptors in development: Less is sometimes more. *J. Cell Biol., 161*: 223–228, 2003.

75. Evan, G.I. and Vousden, K.H. Proliferation, cell cycle and apoptosis in cancer. *Nature, 411*: 342–348, 2001.

76. Cory, S. and Adams, J.M. The Bcl2 family: Regulators of the cellular life-or-death switch. *Nat. Rev. Cancer, 2*: 647–656, 2002.

6 Tumor Suppressor Genes

Bernard E. Weissman

CONTENTS

6.1 INTRODUCTION

Investigations into the causes of human cancer have increasingly focused on the molecular basis of its development. Although we now understand many facets of cellular transformation, the event(s) responsible for many of these changes still remain ill defined. Recent reports have delineated some important changes during malignant transformation, including aberrant growth factor signaling, defects in programmed cell death, altered interactions in the tumor microenvironment, and inappropriate differentiation patterns. Many of these changes can arise from altered proto-oncogene activity due to increased expression, mutations, or translocation. However, the loss or inactivation of specific genetic information, that is, tumor suppressor genes, during human neoplastic initiation and progression has also emerged as an equally potent mechanism for these changes. Indeed, without an understanding of these molecular events, the relative contributions of genetic susceptibility and environmental factors to the development of human cancer remain nebulous.

This chapter focuses on the role of tumor suppressor genes in the development of human cancers. It examines their identification and characterization within the theoretical and historical context of tumorigenesis. It then discusses specific examples of how the loss of these genes leads to neoplastic transformation and, finally, how this information can improve the treatment and prognosis for cancer patients.

6.2 DEFINITION OF TUMOR SUPPRESSOR GENES

While the number of known human oncogenes increased substantially since the demonstration of *ras* mutations in 1982, the identification of tumor suppressor genes has remained an arduous process. Isolation of novel oncogenes comes from multiple, straightforward assays, for example, focus formation in culture, tumor formation in animals, chromosome fusions, etc. However, identification of tumor suppressor genes has mainly relied upon arduous positional cloning. Despite the intense efforts of many laboratories, molecular studies have isolated only a limited number of putative tumor suppressor genes. The normal functions of these tumor suppressor genes and their effects upon introduction into tumor cells cover a broad range of cell activities. Therefore, to avoid confusion in this chapter, I define the term "tumor suppressor" gene as any gene lost or inactivated during the development of human cancer.

6.3 THEORETICAL FRAMEWORK FOR CANCER

The molecular basis of cancer has been the focus of intense investigation for over 100 years. Although we have catalogued many of the consequences of cellular transformation, we are just beginning to understand the complicated event(s) responsible for these changes. In order to explain the mechanism by which a normal cell transforms into a neoplastic cell, several general theories on the cause(s) of cancer have appeared. These include theories based on viruses as etiological agents [1–3] and those based on somatic mutation [4,5].

6.3.1 ONCOGENE HYPOTHESIS OF HUEBNER AND TODARO

In 1969, Huebner and Todaro proposed that malignant transformation results from the activation of a transforming gene called the oncogene, transmitted vertically in the germ line of each species (see Chapter 5) [2]. Several features of this theory made it an attractive model for cancer. The oncogene represented one part of a larger viral genome that coded for the production of infectious virus. Because only the oncogene fueled transformation, cells may undergo transformation without producing viral particles. Early studies showed that some transformed cell lines produced viruses while others did not [6]. Morever, in normal cells, oncogene expression would normally remain suppressed. Upon physical insult from radiation or mutagens, or as a result of the aging process, expression could be activated. Finally, the susceptibility of a cell to transformation would depend on the "strength" of the oncogene repressor system. This could account for the existence of strains of mice that varied in their incidence of leukemogenesis [7]. Why would such a deleterious gene be conserved in the germ line? Huebner and Todaro suggested that the oncogene might play a role in the normal development of an animal [2]. Thus, oncogenes might control different cells undergoing unlimited proliferation during tissue formation. After differentiation, cells would repress the expression of the oncogene. Therefore, a strong evolutionary selection exists for the preservation of oncogenes.

Howard Temin proposed a second viral theory, the protovirus theory [1]. According to this theory, malignancy resulted from a continuous process of genome rearrangement in somatic cells including RNA to DNA to RNA gene recombination. For example, a nontransforming RNA virus or protovirus could be transcribed into DNA by reverse transcriptase and inserted randomly into the host genome. Upon excision, the virus might integrate a host gene that confers upon it the ability to transform normal cells. Feline leukemia virus could have arisen by this mechanism as a result of recombination between the proviral DNA or a simian retrovirus and the feline host genome [8]. Carcinogens in this model would promote this program of recombination. The theory also suggested that the different rearrangement of somatic cell genomes caused by this process could lead to the normal variety of differentiated cell types. Unlike the oncogene theory, the protovirus theory proposed that reverse transcriptase would exist in all somatic cells instead of only cancer cells and that the germ line did not necessarily carry endogenous retroviruses.

Early evidence appeared to favor the position that viruses were intimately involved with human cancer. Numerous reports had established viral involvement in neoplastic transformation in animal systems. In 1908, avian leukemia was shown to be transmissible by a filterable agent [9,10]. Since that initial discovery, examples of small RNA virus-induced leukemias have appeared in many species including mice, rats, cats, and even apes [11]. In addition to type C viruses, which cause leukemias, some type C viruses are capable of transforming cells in culture and producing a variety of solid tumors *in vivo*. Thus, it seemed logical that type C viruses might also underlie all neoplasms, including those of humans. However, only a limited number of human cancers appear to arise from a viral etiology [11,12].

6.3.2 Chromosomal Changes or Genetic Balance

In contrast to theories resting upon a viral origin of cancer, several investigators proposed that cancer is a result of a somatic mutation event. As early as 1914, Boveri recognized the importance of mutation in genetic disorders and suggested that a kind of a "genetic imbalance" could result in the expression of cancer [4]. In 1971, Knudson proposed that cancer arose from two different genetic events, otherwise known as the "two-hit" hypothesis [13]. Although he did not explicitly propose that these two genetic alterations resulted in a loss of information, the statistical analysis of the rare pediatric cancer, retinoblastoma, provided the theoretical framework for the role of tumor suppressor genes in carcinogenesis. In 1973, Comings presented a theory of cancer based on somatic mutation events that incorporated both tumor suppressor genes and oncogenes [5]. In this theory, he proposed that all cells possess multiple genes called transforming genes that can code for factors that allow a cell to escape from normal growth control. In addition, diploid sets of genes exist that regulate the expression of the transforming genes. Therefore, in normal cells, the putative gene product of the regulatory gene suppresses the transforming gene and maintains normal cell cycle control. Neoplastic transformation results from the inactivation of the normal regulatory gene leading to the subsequent activation of the transforming genes. A variety of agents could cause inactivation of the regulatory genes including carcinogens, spontaneous mutations, or oncogenic viruses. Like in the oncogene hypothesis, Comings suggested that these transforming genes might become active at some point in development to allow for the cellular proliferation necessary for tissue formation. In this sense, each transforming gene may be specific for a type of somatic tissue. Finally, RNA or DNA tumor viruses could arise by a genetic recombination between a transforming gene and a virus. The similarity between the nucleotide sequences in mouse retroviral *src* genes and sequences in the genome of a variety of other species, including man, supports this notion [14–16].

Several lines of evidence support the notion that rare genetic events lead to the development of human neoplasia. The majority of the experimental and clinical evidence suggests a clonal origin for human cancers [17,18]. For example, tumors from female patients, whose normal tissues are heterozygous for the glucose-6-phosphate dehydrogenase isoenzymes, expressed only one isoenzyme [18]. Furthermore, as discussed later, several human cancers display an autosomal dominant pattern of inheritance including retinoblastoma, Wilms' tumor, and breast cancer. These findings suggest that in genetically inherited cancers, one regulatory gene undergoes inactivation in the germ line of the patient. Consequently, only one mutational event is required for neoplastic transformation. Other genetic diseases also exist with a predisposition for cancer such as Xeroderma pigmentosum and Ataxia telangiectasia [19]. The common feature of these diseases comes from a decreased capacity for repair of DNA damage [20]. Thus, an inability to correct random genetic damage leads to a predisposition for cancer.

Now, more than 30 years later, we know that parts of each theory appear correct. Genetic recombination often occurs during the development of human cancers that causes activation of proto-oncogenes or transforming genes. Both DNA and RNA tumor viruses play a role in the etiology of specific types of human cancers, a number that appears on the increase. Regulatory genes or tumor suppressor genes also exist that counteract the effects of DNA damage and oncogene expression. Indeed, in most cases, it appears that loss of both copies of these genes must precede cancer development. However, the notion that regulatory genes control the expression of oncogenes has proved more complicated than originally outlined.

6.4 EVIDENCE FOR EXISTENCE OF TUMOR SUPPRESSOR GENES

How do we know that tumor suppressor genes exist? The body of evidence supporting the notion that loss of genetic information contributes to tumor initiation and progression has accumulated over many years. The supporting data come primarily from two general types of studies — demonstration of physical loss of chromosomes or chromosome regions and functional studies using cell culture models. Together, these studies formed the conceptual framework that led to the isolation of tumor suppressor genes for human and rodent tumors.

6.4.1 CYTOGENETICS/CGH

One of the first lines of evidence supporting the loss of genetic information during tumor initiation and progression came from the development of chromosome banding techniques. These protocols allowed for the identification of individual chromosomes in normal and tumor cells on the patterns of bands formed after treatment with proteases and DNA binding proteins or staining with DNA-binding fluorescent agents [21–24]. Detailed analyses of primary tumor samples from human patients and rodent models revealed specific losses of chromosome regions. In some cases, many types of cancer showed similar deletions. For example, cancers of the lung, kidney, ovary, and breast display deletions in human chromosome 3p21 [25]. Many tumors also showed a common area of loss in chromosome 1p36 [26–29]. In other cases, these changes appeared primarily in one kind of tumor. Examples include deletions in chromosome 13q14 in retinoblastoma and 11p15.5 in Wilms tumor [30,31]. However, the size of these visible deletions (>20 Mb) made it difficult to identify the operative tumor suppressor gene in the region. The development of chromosomal and array comparative genomic hybridization (CGH) has greatly improved the specificity of identifying cytogenetic changes in tumor cells [32]. CGH locates regions of chromosomal loss or gain by hybridizing fluorescently tagged DNAs from normal and tumor cells together to either chromosomes or genomic DNA libraries. If both normal and tumor cells possess the same number of copies of a chromosome region, one color, derived from the equal amount of fluorescence signal from each chromosome complement, appears. However, losses or gains of a chromosomal region lead to an altered color. The array CGH approach can refine the chromosomal deletion to a size of 1 Mb or less [32].

6.4.2 MOLECULAR GENETICS/LOH

Somatic cells contain two copies of each chromosome, one set derived from the mother and the other set contributed by the father. While karyotypic analysis of tumor cells by G-banding can identify each individual chromosome, it cannot determine their parental origin. Therefore, if exchange of genetic information between two copies of the same chromosome occurred during tumor development, one could not identify this event by a cytogenetic approach. This could prove particularly disadvantageous if one was searching for tumor suppressor genes. Because of the need to lose both copies of this class of genes, mitotic recombination could provide an important mechanism for this task. In other words, if one chromosome possesses a defective copy of a tumor suppressor, a mitotic recombination event with its sister chromosome could lead to loss of the remaining wild-type allele.

The recognition of naturally occurring variations (polymorphisms) in the sequence of human DNA provided the first means to distinguish between copies of same chromosomes in tumor cells [33]. Because many DNA polymorphisms exist between different individuals, one can distinguish between paternally and maternally derived copies of chromosomes [34]. Therefore, based on these poly-morphic sequences, one can use several different approaches to distinguish between two copies of the same chromosome. Initially, restriction fragment length polymorphisms (RFLPs) were used to show that the chromosomal composition of tumor cells often differed from the normal tissue in patients. This technique exploited differences in restriction enzyme recognition sequences between copies of the same chromosomes using restriction enzyme digestion of normal and tumor DNA followed by Southern blotting [35]. More recent approaches depend on differences in the number of tandem repeat sequences between different chromosome copies using polymerase chain reaction (PCR) amp-lification followed by polyacrylamide gel electrophoresis. The development of high throughput array analyses can now allow an investigator to screen multiple tumor samples especially in the introns and in the region between genes.

In the initial studies by Cavenee et al. [35], mitotic recombination appeared frequently in tumors from patients with a rare pediatric cancer of the eye, retinoblastoma. This result suggested that one copy of a tumor suppressor gene had undergone inactivation. Mitotic recombination then provided the mechanism for the loss of the remaining wild-type allele of the tumor suppressor gene. At that point, the cells would become initiated for subsequent cellular transformation. These and numerous subsequent studies on other types of human cancer strongly supported the "two-hit" hypothesis of Alfred Knudson and gave impetus to identify the operative tumor suppressor genes for human cancers [13,36].

6.4.3 SOMATIC CELL GENETICS

Some of the earliest studies suggesting the existence of functional tumor suppressor genes came from somatic cell genetics. The majority of these reports, including our own, have shown that hybrids between tumor cells and their normal counterparts failed to form tumors upon inoculation into appropriate animals [37]. These results were consistent with the presence of tumor suppressor information in the normal parental cell line. Indeed, all these studies have demonstrated that loss of specific chromosomes from the normal parent correlated with the reexpression of tumorigenicity in the hybrid cells [38,39]. Somatic cell hybrid studies have also specified a second class of tumor suppressor genes that control cellular immortality [40]. Thus, the majority of hybrids between human tumor cells and normal fibroblasts will proliferate for a limited number of population doubling in culture [40]. A more recent report also demonstrated that human cell hybrids lose the gene amplification phenotype of the tumoregenic parent [41]. This phenotype segregates independently of the other two phenotypes [41].

6.5 ISOLATION OF TUMOR SUPPRESSOR GENES

The isolation of tumor suppressor genes has proven a difficult task due to the lack of facile assays. Identification of new oncogenes rests upon bestowing a transformed phenotype to normal cells. Many hallmarks of transformation provide excellent markers for finding a newly transformed cell in a background of normal cells, that is, loss of density-dependent inhibition of growth, ability to grow in semi-solid medium, and tumorigenic potential in animals. However, one can see the difficulty of the reverse operation — isolating a nontransformed cell from a population of transformed cells. The only assays that provide some selectivity use the same principle as chemotherapy, which uses reagents that preferentially kill rapidly proliferating cells. This has not proved an efficient way of isolating tumor suppressor genes. Thus, discovery of most tumor suppressor genes has relied mainly upon positional cloning techniques. This approach relies upon localizing the position of the tumor suppressor gene to a small region of the human genome (generally <1 Mb) and characterizing each gene in the area for genetic and epigenetic abnormalities. A few examples cited later in this chapter illustrate the isolation of several well-known tumor suppressor genes.

6.5.1 RETINOBLASTOMA — THE ORIGINAL TUMOR SUPPRESSOR GENE

As briefly described above, the etiology of retinoblastoma provided the first example of tumor suppressor gene contributing to human cancer development. Retinoblastoma, a cancer of retinal origin, occurs in children under the age of seven. The tumor may be present in only one eye, the unilateral form, or in both eyes, the bilateral form. The tumor could arise in a sporadic fashion or in an inherited form that displays an autosomal dominant pattern. The finding that the bilateral form occurred more frequently in the inherited form of the disease led Alfred Knudson to develop the "two-hit" hypothesis described earlier [13]. The clear implication from this hypothesis was that a loss or inactivation of genetic material must take place in order for retinoblastoma development. Cytogenetic evidence already suggested a location for this tumor suppressor gene. As early as 1962, reports described a recurrent chromosome deletion in the D group in retinoblastoma patients [42]. With the advent of chromosomal banding techniques, the site of deletion was localized to chromosome 13 band q14 [43]. However, only about 25% of the retinoblastomas showed a visible loss of chromosome 13q14. How do we know that a selection for loss of information in this region occurs during the onset of retinoblastoma?

6.5.1.1 Linkage Analysis Demonstrates Recessive Genetic Nature of the Disease

The discovery of a gene that mapped to the 13q14 region provided an answer to the above question. In 1979, Sparkes and his colleagues showed that esterase D maps to 13q14 [44,45]. The proximity of esterase D to a potential "*Rb*" tumor suppressor gene allowed these investigators to test whether the loss of information from the other apparently intact copy of chromosome 13 occurs in these tumors. Because one could measure esterase D activity in a biochemical assay, one could determine the number of copies of esterase D in cells. Indeed, tumor cells from patients with a cytogenetic deletion of 13q14 showed a 50% reduction in esterase D activity [45]. In 1983, Benedict and his collaborators used this information to firmly establish that loss of genetic information in chromosome 13q14 occurs even in the absence of a visible cytogenetic deletion. They found that normal cells from a retinoblastoma patient had one-half of the expected esterase D activity but no apparent deletion of 13q14 [46]. However, cells from two different retinoblastomas from this same patient retained only one copy of chromosome 13 and did not express detectable esterase D activity [46]. These data strongly support the scenario that this patient carried one copy of chromosome 13 with a submicroscopic deletion of the region carrying the esterase D and putative *Rb* tumor suppressor gene. Upon loss of the intact chromosome 13, possessing only the copies of the esterase D and *Rb* tumor suppressor genes, retinoblastoma could develop. Although the esterase D studies provided support for a loss of heterozygosity or LOH in the region of chromosome 13q14 in retinoblastoma, this approach was limited by the assay and the need to screen for tumors showing a reduction in enzyme activity. The subsequent development of DNA polymorphism markers as outlined earlier allowed researchers to screen any retinoblastoma sample where normal cells from the same patient were available. The pioneering studies of Cavenee, White, and others demonstrated that LOH for chromosome 13q14 appears in almost all retinoblastomas, both in sporadic and hereditary cases [35,47].

6.5.1.2 Positional Cloning of the *Rb* Gene

By 1985, several lines of evidence had mapped the first putative human tumor suppressor gene for retinoblastoma to human chromosome 13q14. However, this region appeared quite large containing over 10 Mb of genomic DNA. In order to further define the location of the *Rb* tumor suppressor, Dryja and coworkers isolated new DNA polymorphic markers from cloned DNA fragments of chromosome 13. They identified three RFLPs that appeared homozygous in a high percentage of human osteosarcomas, a secondary tumor associated with hereditary retinoblastoma patients. In order to identify potential coding sequences within this region, Drs. Steven Friend and Robert Weinberg hybridized chromosome 13 fragments to DNA from a variety of different species. If a portion of a gene resided on a fragment, it might recognize orthologous genes from other species due to conservation. In contrast, fragments without coding regions should hybridize poorly to other species. Using this approach, they identified a DNA fragment that recognized a DNA sequence in the mouse genome and also in human chromosome 13q14 [48]. They then determined whether this fragment could recognize a specific mRNA in human cell lines. By Northern blotting, they showed that it recognized a 4.7 kb RNA transcript in a cell line derived from normal retinal cells but not in four retinoblastoma cell lines. They also screened RNA samples from different tumor types and observed the presence of the 4.7 kb RNA in all tumors except retinoblastomas and retinoblastoma-associated osteosarcomas. Finally, they determined the genomic structure of this gene and showed that it ranged over 70 kb of genomic DNA. When they characterized the status of the gene in 50 retinoblastomas or associated osteosarcomas, they found gene rearrangements, heterozygous deletions, and homozygous deletions in about 30% of the samples. One tumor possessed a homozygous deletion within the genomic locus of the gene providing convincing evidence that the *Rb* tumor suppressor gene had been isolated. Subsequent studies by Lee, Fung, and others further supported this cDNA as the bonafide *Rb* tumor suppressor gene [49,50].

6.5.2 ISOLATION OF OTHER TUMOR SUPPRESSOR GENES

While cytogenetic evidence pointed researchers in the right direction for the eventual isolation of the *Rb* tumor suppressor gene, this advantage disappears when examining adult tumors. While most pediatric malignancies show limited cytogenetic changes, the more common epithelially derived tumors of adults undergo frequent and often complicated chromosomal rearrangements. Therefore, clear and consistent cytogenetic abnormalities in breast or lung cancer do not commonly appear. One way that researchers have overcome this obstacle is due to the availability of familial forms of human cancer. Like retinoblastoma, many adult malignancies can occur more frequently in some families. Following the paradigm of the two-hit hypothesis, affected family members must presumably carry one inactive or mutant copy of the operative tumor suppressor gene. Therefore, one can carry out a genetic linkage analysis of these families to pinpoint a common genetic locus associated with increased cancer risk. Once this region is mapped, one can apply standard positional cloning approaches to isolate the putative tumor suppressor gene.

This plan of attack has led to the identification of many important tumor suppressor genes. Mary-Claire King and colleagues established that breast cancer can occur in a familial form based upon an increased incidence in siblings if their mother developed this malignancy at an early age [51]. Using polymorphic markers, this group later mapped the susceptibility locus to human chromosome 17q21 [52]. Using this information, Miki, Futreal, and collaborators isolated the *BRCA1* gene in 1994 [53,54]. Following the same paradigm, Wooster et al. identified a second familial breast cancer gene, *BRCA2*, at chromosome 13q12.3 [55]. Other examples of familial tumor suppressor genes include the Von Hippel Lindau (*VHL*) gene at 3p26–p25, the *WT-1* gene at 11p13, the *NF1* gene at 17q11.2, and the ataxia telangiectasia mutated (*ATM*) gene at 11q22.3 [56–60].

The existence of several other tumor suppressor genes came to light from allelotyping of tumor samples. This screen involves a LOH analysis of each autosomal and sex chromosome arm with a limited number of polymorphic markers. One can subsequently use additional polymorphic markers to minimize the LOH regions on chromosome arms showing a high frequency of LOH for a particular tumor. In the case of pancreatic cancer, Kern, Hruban, and colleagues characterized a large number of pancreatic adenocarcinomas for LOH on all chromosomes [61]. Their analyses showed several hot spots for LOH including 9p, 13q, 17p, and 18q [61]. The sites of chromosomes 9, 13, and 17 correlated with known tumor suppressor genes $p16^{INK4A}$, BRCA2, and p53, respectively. However, the locus on chromosome 18 seemed novel. The only known tumor suppressor gene within the region, DCC (deleted in colorectal carcinoma), was retained in many of the tumors displaying 18q LOH [62]. Positional cloning efforts by this group led to the isolation of the DPC4 or Smad4 gene [62]. Interestingly, Howe et al. [63] later linked mutations in this gene to a familial cancer-prone syndrome, Juvenile Polyposis.

One caveat from position cloning studies lies in the lack of a functional assay for validation of putative tumor suppressor gene. If a region of the human genome frequently suffers mutations from exposure to environmental insults, then mutations in multiple genes in the region can occur. If this happens in a region containing a tumor suppressor gene, then relying upon the presence of mutations in a gene as the basis for a tumor suppressor gene can prove erroneous. Vogelstein and colleagues encountered this problem when searching for the familial adenomatous polyposis (FAP) gene underlying a form of familial colon cancer. Molecular genetics, cytogenetic, and functional studies had mapped the FAP gene to chromosome 5p21 [64–67]. By positional cloning, Kinzler et al. [68] had isolated the MCC (mutated in colorectal carcinoma) that showed mutations in some primary tumor samples and colorectal cancer cell lines. However, in a later report that same year, the same investigators as well as Groden et al., showed that mutations in a different gene, located next to MCC, actually caused FAP [69,70]. Therefore, validation of an operative tumor suppressor gene must rely upon more that the presence of mutations in tumor samples.

6.5.3 CURRENT APPROACHES

Even today, the basic approach of position cloning has proven the mainstay for identification of tumor suppressor genes. However, several advances have significantly streamlined this process. The completion of the human genome project has provided researchers with the identity of almost every human gene including its location in the genome and its genomic structure (exons and introns). Therefore, once a putative tumor suppressor gene has been localized to a specific region of the human genome, one can quickly ascertain the number of genes in the area and prioritize them for characterization. Further, the identification of numerous single nucleotide polymorphisms or SNPs has facilitated rapid screens for LOH in multiple tumor samples. Combined with DNA microarray platforms that can contain from 10,000 to 100,000 SNPs, researchers can easily determine novel areas of LOH. In a complementary fashion, cDNA microarrays can identify genes showing lower expression in tumor cells versus normal tissue. Therefore, one can find tumor suppressor genes that may undergo inactivation by epigenetic mechanisms, for example, DNA promoter methylation or loss of a transcription factor, during carcinogenesis. Furthermore, mouse geneticists have also applied these approaches to the identification of novel tumor suppressor genes from mouse models of carcinogenesis. These latest developments should significantly increase the number of tumor suppressor genes that will be identified over the next few years.

6.6 NORMAL FUNCTIONS OF TUMOR SUPPRESSOR GENES

Since the reports of the isolation of the Rb gene in 1986, a small number of other tumor suppressor genes have emerged each year. As with oncogenes, tumor suppressor genes normally participate in a variety of physiological processes, including signal transduction pathways for growth, differentiation,

---

TABLE 6.1
Known Tumor Suppressor Genes

Cell cycle control genes
$Rb, p53, p16^{INK4a}, p14^{ARF1}$

Cellular structure genes
$APC, NF-2, DCC$

Transcriptional regulation
WT-1, snf5/ini1, p53

Signal transduction
sMAD4, FHIT, NF-1, PTEN/MMAC1, PTC

DNA Replication/repair
BRCA1, BRCA2, MSH2, MLH1, ATM, MRE11, WRN, BLM, FA, XP

Protein degradation
VHL

---

and programmed cell death. These tumor suppressor genes have been grouped by their primary biological role in Table 6.1, although many of them perform multiple functions within cells. Both the p53 and Rb proteins play major roles in the control of cell division under normal conditions and in response to environmental damage [71,72]. Other tumor suppressor genes fall into the family of cyclin-dependent kinase inhibitors (CDKIs) that regulate the activity of the Rb protein [73,74]. The *ATM* gene, whose loss underlies the cancer-prone disease ataxia-telangiectasia, appears to help regulate the cell's response to DNA damage by a variety of agents [56]. This chapter, examines the roles of representative tumor suppressor genes in the context of normal cellular biological processes and discusses the concept of inactivation of tumor suppressor "pathways" as opposed to individual genes.

6.6.1 *Rb* TUMOR SUPPRESSOR GENE

The product of the *Rb* tumor suppressor gene participates in a variety of cellular activities including cell cycle regulation, gene repression apoptosis, and senescence. The traditional and best-studied interaction involves regulation of the E2F family of transcription factors. The E2F family constitutes the major regulators of cell cycle progression from G0 into G1 [75]. In 1989, several groups showed that the Rb protein undergoes phosphorylation/dephosphorylation in a cell cycle dependent fashion. At G0/G1 checkpoint, virtually the entire Rb protein appears unphosphorylated while it becomes phosphorylated during S and G2 phases (Figure 6.1) [76–78]. Moreover, the unphosphorylated form of Rb binds E2F1 and prevents it from activating transcription of cell cycle progression genes [79]. Phosphorylation of the Rb protein at the G0/G1 interface then releases E2F proteins and allows progression into G1.

Initial models proposed that the mere release of the bound E2F proteins from Rb accounted for the transition from G0 to G1. However, recent studies have indicated a more complicated mechanism. Zhang et al. [80] reported that Rb and E2F proteins form a complex that represses gene transcription. Other known repressors of transcription can also form repressive complexes with Rb including chromatin remodeling proteins such as HDAC and BRG1 [81,82]. Dahiya et al. [83] also found an association between Rb and Polycomb (PcG) group proteins that results in a repressor complex that blocks the progression of cells from G2 to mitosis. The PcG proteins play key roles in early mammalian development by repressing the transcription of specific pattern forming genes. Therefore, the existence of Rb-PcG complex provides a link between Rb-mediated growth arrest and the differentiation events leading to embryonic pattern formation. Several reports have indicated a

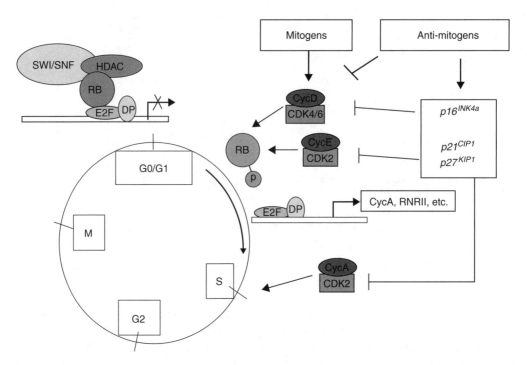

FIGURE 6.1   The *Rb* tumor suppressor gene pathway (Courtesy of Dr. E. Knudsen, University of Cincinnati, Cincinnati, OH).

role for Rb in the regulation of apoptosis. Hsieh et al. [84] showed that binding of Rb to MDM2 abrogates the antiapoptotic function of MDM2 and its ability to induce degradation of the *p53* tumor suppressor gene (see under *p53*). Pennaneach et al. [85] showed that Rb mediates both cell survival and cell cycle arrest after DNA damage. They showed that Rb directly interacts with the large subunit of the replication factor C (RFC) complex. Disruption of this binding inhibits RFC's ability to promote cell survival after DNA damage.

The *Rb* gene has also been shown to contribute to the regulation of cellular senescence. Normal cells proliferate for a limited number of passages in cell culture followed by a stable form of cell cycle arrest [86]. This process of cellular senescence could also act to prevent cancer initiation and progression by limiting growth of cells after DNA damage. Lowe and colleagues [87] found a distinct heterochromatic structure that accumulates in senescent human fibroblasts, designated senescence-associated heterochromatic foci (SAHF). They showed that SAHFs formed concomitantly with the Rb-E2F complex resulting in stable repression of *E2F* regulated genes. Importantly, the appearance of SAHFs and the simultaneous silencing of *E2F* target genes occurred only in cells undergoing cellular senescence. Cells treated with a reversible growth inhibitory reagent did not form SAHFs.

In summary, the Rb protein clearly impacts upon key cellular regulatory processes. Therefore, its inactivation during carcinogenesis results in pleiotrophic effects, many of which could promote tumor initiation and progression. Despite the large number of studies characterizing the normal function of this gene, many questions remain about the mechanisms involved in Rb loss and cancer etiology. Mouse models of Rb loss, gene expression profiling, and emerging structural biology studies may provide some insights into these unresolved issues.

6.6.2   *p53*

The *p53* gene was first identified due to its altered expression in human tumors [88]. Originally classified as an oncogene, subsequent studies have demonstrated that the wild-type protein acts

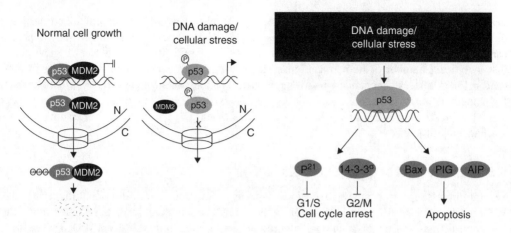

FIGURE 6.2   Regulation of the *p53* tumor suppressor gene pathway (Courtesy of Dr. Y. Xiong, University of North Carolina, Chapel Hill, NC).

as a tumor suppressor gene [89,90]. The *p53* gene shows low expression in virtually all normal tissues and participates in the development of a wide range of tumors from different developmental origins [91,92]. Reports have linked this protein to signal transduction, regulation of transcription, cell cycle control, and genomic instability [93–96]. Furthermore, the regulation of p53 protein expression also occurs at many levels including protein stability, posttranslational modifications, and protein–protein interactions [72,97,98].

The key role that the p53 protein plays in normal cells, that accounts for its tumor suppressor activity, lies in directing cellular responses to environmental stress and insults. Normally, p53 protein possesses a short half-life (<20 min) due to ubiquitination and degradation promoted by an association with the MDM2 protein (Figure 6.2) [99]. However, upon cellular damage, p53 protein levels rise due to increased protein stability. Several mechanisms account for this increase including induction of p14^{ARF} that inhibits MDM2–p53 interaction as well as phosphorylation and acetylation of p53 [97,99,100]. Once stabilized, the increased level of p53 protein increases its transcriptional activity and induces the transcription of the *CDKI* gene p21$^{WAF1/CIP1}$ [101]. The rise in p21 protein levels leads to a cell cycle arrest at both G1 and G2. In cases of severe DNA damage, where repair does not appear possible, p53 can induce expression of apoptotic-promoting genes such as *bax* and initiate programmed cell death [102]. Because of the critical role it plays in preventing the accumulation of mutations after DNA damage, Dr. Arnold Levine designated p53 as the cellular gatekeeper for growth and division [103].

The *p53* provides a good example of the many ways tumor suppressor genes can undergo inactivation during carcinogenesis. The most obvious event involves deletions or mutations of the *p53* gene. Because p53 binds to DNA as a tetramer to activate gene expression, partial or total deletions and nonsense or splice site mutations can prevent oligomerization, abrogating its activity. Point mutations in p53 also occur frequently in human tumors. Some of these mutations affect the ability of p53 to bind to DNA, to itself, or to other proteins. Therefore, many of these mutant proteins are considered "dominant-negative" in nature. In other words, the mutant protein can override the normal activity of a wild-type protein resulting in a p53-deficient phenotype. However, one unresolved issue concerns whether some of these mutations may actually confer new properties upon p53, that is, a "gain of function" mutation. Some reports have indicated that certain p53 mutant proteins have an altered affinity for DNA. In other words, they might activate expression of a different set of genes than the wild-type protein. Therefore, the most effective treatment for patients with tumors possessing these mutations may differ significantly from those with tumors that lack p53 entirely.

An alternative mechanism for p53 inactivation in cervical cancers involves the expression of the *E6* gene of human papillomavirus (HPV). The HPV E6 protein promotes rapid degradation through the ubiquitin pathway [104]. The HPV E7 protein also inactivates the Rb protein, albeit through a different mechanism [105]. In a similar fashion to HPV E6, over expression of the MDM2 protein can induce rapid degradation of the p53 protein [106]. Amplification of the *MDM2* gene occurs in a significant number of human sarcomas, accounting for the absence of p53 activity in these cancers [107]. The last mechanism involves epigenetic silencing of the p53 promoter, usually through DNA methylation. Although a relatively uncommon event for p53 inactivation in human tumors, loss of expression of other tumor suppressor genes, such as p16 and VHL, often involves this mechanism [108].

6.6.3 OTHER TYPES OF TUMOR SUPPRESSOR GENES

Several tumor suppressor genes, such as *p53*, appear to function as transcriptional regulators. The WT-1 protein, isolated by its association with Wilms' tumor, contains a zinc-finger motif that can control transcription of a number of growth factors including IGFII and PDGF [109,110]. Recently, loss of a subunit of the SWI/SNF chromatin remodeling complex, snf5/ini1/BAF47, was found in virtually all pediatric rhabdoid tumors [111]. This complex regulates transcription by altering the chromatin structure in the promoter region of genes [112,113]. How inactivation of a general transcription factor leads to this unusual childhood malignancy remains an open question. However, several reports indicate that induction of the p16 tumor suppressor gene might require the snf5 protein providing a possible mechanism for tumor development [114–116].

Another group of tumor suppressor genes may operate through signal transduction associated with normal development. The *DCC* gene was isolated because of its location on chromosome 18 which showed LOH in human colorectal carcinomas [117]. It shares sequence homology with the N-CAM protein suggesting an involvement in cell adhesion [117]. It now appears that this gene participates in normal neural development and may play a role in human tumor development [118]. Loss of the *APC* gene contributes to the onset of both sporadic colorectal cancers and in tumors found in family with adenomatous polyposis coli (see Section 6.5.2). This gene product forms part of the intercellular junctions found in most epithelial cells including E-cadherin and ß-catenin. Originally thought to represent a structural element, studies have shown that this protein participates in the wnt signaling pathway, important in normal tissue development [119]. Another tumor suppressor gene, *NF-2*, bears striking homology to the cytoskeletal proteins moesin, ezrin, and radixin [120]. The normal function of this gene may lie in the control of signal transduction pathways by physical manipulation of cell surface molecules such as growth factor receptors or through altered rac signaling [120]. Loss of the patched (*ptc*) tumor suppressor gene leads to basal cell nevus syndrome or Gorlin syndrome [121,122]. The *ptc* gene codes for a receptor for the sonic hedgehog pathway, which is another critical developmental signaling pathway [123,124].

Some tumor suppressor genes represent members of known signal transduction pathways. The *sMAD4* gene product helps transmit the growth inhibitory signal from the TGFß receptor to the nucleus after ligand binding [62]. The PTEN/MMAC1 tumor suppressor protein possesses phosphatase activity, exerting an antiproliferative effect on cells by inhibiting pAKT signaling [125–127]. Another tumor suppressor gene, *NF-1*, contributes to regulation of *ras* oncogene functions, another central proliferative and developmental signal transduction pathway [128,129].

A large number of tumor suppressor genes play key roles in DNA repair and replication. As discussed above, the tumor suppressor genes, *BRCA1* and *BRCA2*, play important roles in familial ovarian and breast cancer development. Initial studies suggested these proteins might possess transcriptional regulatory activities due to the presence of a ring-finger motif. However, recent studies have established a role for these proteins in DNA repair complexes [130–132]. Other tumor suppressor genes including the Fanconi's Anemia, ATM, MRE11, Bloom's Syndrome, Xeroderma Pigmentosum, and Werner's Syndrome code for proteins that also participate in other facets of

DNA repair and replication [20,133]. Thus, deficiencies in the ability of a cell to recognize and mend DNA damage generally increase the chances for malignant transformation.

Finally, several tumor suppressor genes fall into somewhat unique classes. Loss of the VHL tumor suppressor gene leads to the development of both sporadic and familial renal cell carcinomas [57]. Its gene product functions as a regulator of protein degradation of several important genes including vascular endothelial growth factor through its association with a ubiquitin complex [57]. Another type of familial colorectal cancer arises from mutations in a family of genes involved in DNA mismatch repair [134]. Loss of the *msh2* or *mlh1* tumor suppressor genes results in aberrations in dinucleotide repeat lengths [135–137]. Expansions of these tracts in genes such as TGFßRII and *bax* lead to their inactivation [138].

6.6.4 METASTASIS SUPPRESSOR GENES

Most of the tumor suppressor genes identified to date presumably act at the early stages of tumor development. Introduction of these genes into tumor cells causes inhibition of growth in culture and in animal models. However, in most solid tumors it is the metastatic disease rather than the primary tumor that eventually causes mortality. Several studies have shown that acquisition of metastatic potential results from accumulation of several molecular defects distinct from initial tumor formation [139,140]. Thus, characterization of the later molecular events in aggressive tumor cells could present several potential benefits including an understanding of the interplay between different tumor suppressor genes and cell-cycle genes like *p53*, the contributions of genetic susceptibility, and the impact of environmental factors on the development of metastasis. Therefore, knowledge of genetic loci whose loss or inactivation contributes to metastasis development could help with decisions of treatment and prognosis. However, identification of putative metastasis suppressor genes has proven quite difficult due to the high level of genetic instability found in metastatic tumor cells. Only a handful of genes have emerged whose loss appears to increase metastatic potential in human tumor cells including E-cadherin, NM23, Kai1, Kiss1, and Brms1 [141]. The mechanism of action of most of these genes remains unclear although altered cellular adhesion may play a role in certain cases.

6.6.5 TUMOR SUPPRESSOR PATHWAYS — THE FOCUS OF THE FUTURE

As we have determined the mechanisms by which tumor suppressor genes regulate the normal growth and development of cells, we have recognized that the initiation of human cancer may require the loss of function of a pathway rather than a particular tumor suppressor gene. For example, loss of Rb function may occur through at least four different mechanisms — inactivation of the *Rb* gene itself, over expression of the *CDK4* gene, loss of function of the $p16^{INK4a}$ gene or loss of SWI/SNF chromatin remodeling complex (Figure 6.1) [81,82,142]. One of these events occurs in almost 100% of human cancers indicating the critical importance of the loss of this cell-cycle regulatory pathway in the progression of human cancer. A similar scenario also exists for the *p53* tumor suppressor gene. As mentioned earlier, over expression of the MDM2 protein can significantly lower the level of p53 protein in tumor cells. In a similar fashion, inactivation of $p14^{ARF}$ will also increase MDM2 activity sufficiently to abrogate p53 activity. The BRCA1 tumor suppressor protein is also a member of the Fanconi's Anemia complex, another group of tumor suppressor proteins involved in DNA replication and repair [143]. The p53 and Rb proteins have been shown to interact with many other tumor suppressor proteins including BRCA1, BRCA2, ATM, and msh2. Therefore, cancer research will focus more on the common pathways involved in tumor development during the next decade in order to gain a clearer understanding of how tumor suppressor gene loss contributes to carcinogenesis.

6.7 CONCLUDING REMARKS

The availability of the near-completed human genomic sequence as well as the renewed interest in cancer genetics should lead to the identification of a large number of new tumor suppressor genes during the next decade. The isolation of the first human tumor suppressor gene in 1986 fueled an immediate and continuing interest in gene replacement therapy as a novel treatment modality for human cancers [48]. The functional groundwork for the efficacy of this avenue of approach came from studies on the genetics of cancer using somatic cell genetics [144,145]. Unfortunately, the promise of effective treatment using tumor suppressor replacement therapies has not been fulfilled. Oncologists still face problems with optimizing delivery of potent tumor suppressor genes into tumor cells *in vivo* in order to render them quiescent or prime them for destruction by other methods. However, the knowledge gained from the characterization of these genes has provided new avenues for other types of treatments such as small peptide inhibitors as well as improvements in diagnosis and prognosis. Studies into the normal functions of these genes have also given us novel insights into how cells respond to environmental stress and insult and how development proceeds in an orderly fashion. Thus, understanding the mechanisms of action of new tumor suppressor genes should provide more efficacious targets for intervention in the treatment of cancer and will continue to expand our understanding of the mechanisms of molecular carcinogenesis.

REFERENCES

1. Temin, H.M., The protovirus hypothesis: Speculations on the significance of RNA-directed DNA synthesis for normal development and for carcinogenesis, *J. Natl. Cancer Inst.*, 46, 3–7, 1971.
2. Huebner, R.J. and Todaro, G.J., Oncogenes of RNA tumor viruses as determinants of cancer, *Proc. Natl. Acad. Sci. USA*, 64, 1087–1094, 1969.
3. Todaro, G.J. and Huebner, R.J., N.A.S. symposium: New evidence as the basis for increased efforts in cancer research, *Proc. Natl. Acad. Sci. USA*, 69, 1009–1015, 1972.
4. Boveri, T., *Frage der Entstehung Maligner Tumoren*. Gustav Fischer, Jena, 1914.
5. Comings, D.E., A general theory of carcinogenesis, *Proc. Natl. Acad. Sci. USA*, 70, 3324–3328, 1973.
6. Aaronson, S.A., Bassin, R.H., and Weaver, C., Comparison of murine sarcoma viruses in nonproducer and S + L-transformed cells, *J. Virol.*, 9, 701–704, 1972.
7. Chattopadhyay, S., Rowe, W.P., and Levine, A.S., Quantitative studies of integration of murine leukemia virus after exogenous infection, *Proc. Natl. Acad. Sci. USA*, 73, 4095–4099, 1976.
8. Gallo, R.C., Viruses and the pathogenesis of human leukemia, *Schweiz Med Wochenschr*, 107, 1436–1440, 1977.
9. Rous, P., A transmissible avian neoplasm (Sarcoma of the common fowl.), *J. Exp. Med.*, 12, 696–705, 1910.
10. Ellerman, V. and Bang, O., Experimentelle Luekamie bei Huhnern, *Z. Hyg. Infektionskr.*, 63, 231–272, 1908.
11. Gallo, R.C., Human retroviruses after 20 years: A perspective from the past and prospects for their future control, *Immunol. Rev.*, 185, 236–265, 2002.
12. Klein, G., Lymphoma development in mice and humans: Diversity of initiation is followed by convergent cytogenetic evolution, *Proc. Natl. Acad. Sci. USA*, 76, 2442–2446, 1979.
13. Knudson, A.G., Jr., Mutation and cancer: Statistical study of retinoblastoma, *Proc. Natl. Acad. Sci. USA*, 68, 820–832, 1971.
14. Frankel, A.E., Gilbert, J.H., Porzig, K.J., Scolnick, E.M., and Aaronson, S.A., Nature and distribution of feline sarcoma virus nucleotide sequences, *J. Virol.*, 30, 821–827, 1979.
15. Scolnick, E.M. and Parks, W.P., Harvey sarcoma virus: A second murine type C sarcoma virus with rat genetic information, *J. Virol.*, 13, 1211–1219, 1974.
16. Stehelin, D., Guntaka, R.V., Varmus, H.E., and Bishop, J.M., Purification of DNA complementary to nucleotide sequences required for neoplastic transformation of fibroblasts by avian sarcoma viruses, *J. Mol. Biol.*, 101, 349–365, 1976.
17. Nowell, P.C., The clonal evolution of tumor cell populations, *Science*, 194, 23–28, 1976.

18. Fialkow, P.J., Clonal origin of human tumors, *Annu. Rev. Med.*, 30, 135–143, 1979.
19. Knudson, A.G., Jr., Genetics and etiology of human cancer, *Adv. Hum. Genet.*, 8, 1–66, 1977.
20. Friedberg, E.C., McDaniel, L.D., and Schultz, R.A., The role of endogenous and exogenous DNA damage and mutagenesis, *Curr. Opin. Genet. Dev.*, 14, 5–10, 2004.
21. Drets, M.E. and Shaw, M.W., Specific banding patterns of human chromosomes, *Proc. Natl. Acad. Sci. USA*, 68, 2073–2077, 1971.
22. Seabright, M., A rapid banding technique for human chromosomes, *Lancet*, 2, 971–972, 1971.
23. Sumner, A.T., Evans, H.J., and Buckland, R.A., New technique for distinguishing between human chromosomes, *Nat. New Biol.*, 232, 31–32, 1971.
24. Caspersson, T., Zech, L., and Johansson, C., Differential binding of alkylating fluorochromes in human chromosomes, *Exp. Cell Res.*, 60, 315–319, 1970.
25. Imreh, S., Klein, G., and Zabarovsky, E.R., Search for unknown tumor-antagonizing genes, *Genes Chromosomes Cancer*, 38, 307–321, 2003.
26. Khosla, S., Patel, V.M., Hay, I.D., Schaid, D.J., Grant, C.S., van Heerden, J.A., and Thibodeau, S.N., Loss of heterozygosity suggests multiple genetic alterations in pheochromocytomas and medullary thyroid carcinomas, *J. Clin. Invest.*, 87, 1691–1699, 1991.
27. Balaban, G.B., Herlyn, M., Clark, W.H., Jr., and Nowell, P.C., Karyotypic evolution in human malignant melanoma, *Cancer Genet. Cytogenet.*, 19, 113–122, 1986.
28. Hahn, S.A., Seymour, A.B., Hoque, A.T., Schutte, M., da Costa, L.T., Redston, M.S., Caldas, C., Weinstein, C.L., Fischer, A., Yeo, C.J. et al., Allelotype of pancreatic adenocarcinoma using xenograft enrichment, *Cancer Res.*, 55, 4670–4675, 1995.
29. Mora, J., Cheung, N.K., Kushner, B.H., LaQuaglia, M.P., Kramer, K., Fazzari, M., Heller, G., Chen, L., and Gerald, W. L., Clinical categories of neuroblastoma are associated with different patterns of loss of heterozygosity on chromosome arm 1p, *J. Mol. Diagn.*, 2, 37–46, 2000.
30. Riccardi, V.M., Sujansky, E., Smith, A.C., and Francke, U., Chromosomal imbalance in the Aniridia-Wilms' tumor association: 11p interstitial deletion, *Pediatrics*, 61, 604–610, 1978.
31. Yunis, J.J. and Ramsay, N., Retinoblastoma and subband deletion of chromosome 13, *Am. J. Dis. Child*, 132, 161–163, 1978.
32. Wang, N., Methodologies in cancer cytogenetics and molecular cytogenetics, *Am. J. Med. Genet.*, 115, 118–124, 2002.
33. Botstein, D., White, R.L., Skolnick, M., and Davis, R.W., Construction of a genetic linkage map in man using restriction fragment length polymorphisms, *Am. J. Hum. Genet.*, 32, 314–331, 1980.
34. Wyman, A.R. and White, R., A highly polymorphic locus in human DNA, *Proc. Natl. Acad. Sci. USA*, 77, 6754–6758, 1980.
35. Cavenee, W.K., Dryja, T.P., Phillips, R.A., Benedict, W.F., Godbout, R., Gallie, B.L., Murphree, A.L., Strong, L.C., and White, R.L., Expression of recessive alleles by chromosomal mechanisms in retinoblastoma, *Nature*, 305, 779–784, 1983.
36. Knudson, A.G., Jr. and Strong, L.C., Mutation and cancer: A model for Wilms' tumor of the kidney, *J. Natl. Cancer Inst.*, 48, 313–324, 1972.
37. Weissman, B.E., The genetic behavior of tumorigenicity in human cancer, *Cancer Surveys-Genetics and Cancer*, in Cavenee, W., Ponder, B., and Solomon, E., Eds., Oxford University Press, Oxford, 1990, pp. 475–485.
38. Stanbridge, E.J., Flandemeyer, R., Daniels, D., and Nelson-Rees, W., Specific chromosome loss associated with the expression of tumorigenicity in human cell hybrids, *Somat. Cell Genet.*, 7, 699–712, 1981.
39. Benedict, W.F., Weissman, B.E., Mark, C., and Stanbridge, E.J., Tumorigenicity of human HT1080 fibrosarcoma X normal fibroblast hybrids: Chromosome dosage dependency, *Cancer Res.*, 44, 3471–3479, 1984.
40. Pereira-Smith, O.M. and Smith, J.R., Evidence for the recessive nature of cellular immortality, *Science*, 221, 964–966, 1983.
41. Tlsty, T., White, A., and Sanchez, J., Suppression of gene amplification in human cell hybrids, *Science*, 255, 1425–1427, 1992.
42. Stallard, H.B., The conservative treatment of retinoblastoma, *Trans. Ophthalmol. Soc. UK*, 82, 473–534, 1962.
43. Francke, U., Retinoblastoma and chromosome 13, *Cytogenet. Cell Genet.*, 16, 131–134, 1976.

44. Sparkes, R.S., Muller, H., and Klisak, I., Retinoblastoma with 13q-chromosomal deletion associated with maternal paracentric inversion of 13q, *Science*, 203, 1027–1029, 1979.
45. Sparkes, R.S., Sparkes, M.C., Wilson, M.G., Towner, J.W., Benedict, W., Murphree, A.L., and Yunis, J.J., Regional assignment of genes for human esterase D and retinoblastoma to chromosome band 13q14, *Science*, 208, 1042–1044, 1980.
46. Benedict, W.F., Murphree, A.L., Banerjee, A., Spina, C.A., Sparkes, M.C., and Sparkes, R.S., Patient with 13 chromosome deletion: Evidence that the retinoblastoma gene is a recessive cancer gene, *Science*, 219, 973–975, 1983.
47. Cavenee, W.K., Hansen, M.F., and Nordenskjold, M., Genetic origins of mutations predisposing to retinoblastoma, *Science*, 228, 501–503, 1985.
48. Friend, S.H., Bernards, R., Rogelj, S., Weinberg, R.A., Rapaport, J.M., Albert, D.M., and Dryja, T.P., A human DNA segment with properties of the gene that predisposes to retinoblastoma and osteosarcoma, *Nature*, 323, 643–646, 1986.
49. Fung, Y.-K., Murphree, A.L., Tang, A., Qian, J., Hinrichs, S.H., and Benedict, W.F., Structural evidence for the authenticity of the human retinoblastoma gene, *Science*, 236, 1657–1661, 1987.
50. Lee, W.-H., Bookstein, R., Hong, F., Young, L.J., Shew, J.Y., and Lee, E.Y., Human retinoblastoma susceptibility gene: Cloning, identification, and sequence, *Science*, 235, 1394–1399, 1987.
51. Ottman, R., Pike, M.C., King, M.C., Casagrande, J.T., and Henderson, B.E., Familial breast cancer in a population-based series, *Am. J. Epidemiol.*, 123, 15–21, 1986.
52. Hall, J.M., Lee, M.K., Newman, B., Morrow, J.E., Anderson, L.A., Huey, B., and King, M.C., Linkage of early-onset familial breast cancer to chromosome 17q21, *Science*, 250, 1684–1689, 1990.
53. Miki, Y., Swensen, J., Shattuck-Eidens, D., Futreal, P.A., Harshman, K., Tavtigian, S., Liu, Q., Cochran, C., Bennett, L.M., Ding, W., Bell, R., Rosenthal, J., Hussey, C., Tran, T., McClure, M., Frye, C., Hattier, T., Phelps, R., Haugen-Strano, A., Katcher, H., Yakumo, K., Gholami, Z., Shaffer, D., Stone, S., Bayer, S., Wray, C., Bogden, R., Dayananth, P., Ward, J., Tonin, P., Narod, S., Bristow, P.K., Norris, F.H., Helvering, L., Morrison, P., Rosteck, P., Lai, M., Barrett, J.C., Lewis, C., Neuhausen, S., Cannon-Albright, L., Goldgar, D., Wiseman, R., Kamb, A., and Skolnick, M.H., A strong candidate for the breast and ovarian cancer susceptibility gene *BRCA1*, *Science*, 266, 66–71, 1994.
54. Futreal, P.A., Liu, Q., Shattuck-Eidens, D., Cochran, C., Harshman, K., Tavtigian, S., Bennett, L.M., Haugen-Strano, A., Swensen, J., Miki, Y., Eddington, K., McClure, M., Frye, C., Weaver-Feldhaus, J., Ding, W., Gholami, Z., Soderkvist, P., Terry, L., Jhanwar, S., Berchuck, A., Iglehart, J.D., Marks, J., Ballinger, D.G., Barrett, J.C., Skolnick, M.H., Kamb, A., and Wiseman, R., *BRCA1* mutations in primary breast and ovarian carcinomas, *Science*, 266, 120–122, 1994.
55. Wooster, R., Bignell, G., Lancaster, J., Swift, S., Seal, S., Manglon, J., Collins, N., Gregory, S., Gumbs, G.M.G., Barfoot, R., Hamoudi, R., Patel, S., Rice, C., Biggs, P., Hashim, Y., Smith, A., Connor, F., Arason, A., Gudmundsson, J., Ficenec, D., Kelsell, D., Ford, D., Tonin, P., Bishop, D.T., Spurr, N.K., Ponder, B.A.J., Eeles, R., Peto, J., Devilee, P., Cornelisse, C., Lynch, H., Narod, S., Lenoir, G., Egilsson, V., Barkadottir, R.B., Easton, D.F., Bentley, D.R., Futreal, P.A., Ashworth, A., and Stratton, M.R., Identification of the breast cancer susceptibility gene *BRCA2*, *Nature*, 378, 789–792, 1995.
56. Savitsky, K., Bar-Shira, A., Gilad, S., Rotman, G., Ziv, Y., Vanagaite, L., Tagle, D.A., Smith, S., Uziel, T., Sfez, S., et al., A single ataxia telangiectasia gene with a product similar to PI-3 kinase, *Science*, 268, 1749–1753, 1995.
57. Latif, F., Tory, K., Gnarra, J., Yao, M., Duh, F.M., Orcutt, M.L., Stackhouse, T., Kuzmin, I., Modi, W., Geil, L. et al., Identification of the von Hippel-Lindau disease tumor suppressor gene, *Science*, 260, 1320–1357, 1993.
58. Gessler, M., Poustka, A., Cavenee, W., Neve, R.L., Orkin, S.H., and Bruns, G.A., Homozygous deletion in Wilms' tumors of a zinc-finger gene identified by chromosome jumping, *Nature*, 343, 774–778, 1990.
59. Cawthon, R.M., Weiss, R., and Xu, G., A major segment of the neurofibromatosis type 1 gene: cDNA sequence, genomic structure, and point mutations, *Cell*, 62, 193–201, 1990.
60. Call, K.M., Glaser, T., Ito, C.Y., Buckler, A.J., Pelletier, J., Haber, D.A., Rose, E.A., Kral, A., Yeger, H., Lewis, W.H., Jones, C., and Housman, D.E., Isolation and characterization of a zinc finger polypeptide gene at the human chromosome 11 Wilms' tumor locus, *Cell*, 60, 509–520, 1990.
61. Seymour, A.B., Hruban, R.H., Redston, M., Caldas, C., Powell, S.M., Kinzler, K.W., Yeo, C.J., and Kern, S.E., Allelotype of pancreatic adenocarcinoma, *Cancer Res.*, 54, 2761–2764, 1994.

62. Hahn, S.A., Schutte, M., Hoque, A.T.M.S., Moskaluk, C.A., da Costa, L.T., Rozenblum, E., Weinstein, C.L., Fischer, A., Yeo, C.J., Hruban, R.H., and Kern, S.E., DPC4, a candidate tumor suppressor gene at human chromosome 18q21.1, *Science*, 271, 350–353, 1996.

63. Howe, J.R., Roth, S., Ringold, J.C., Summers, R.W., Jarvinen, H.J., Sistonen, P., Tomlinson, I.P.,Houlston, R.S., Bevan, S., Mitros, F.A., Stone, E.M., and Aaltonen, L.A., Mutations in the *SMAD4/DPC4* gene in juvenile polyposis, *Science*, 280, 1086–1088, 1998.

64. Bodmer, W.F., Bailey, C.J., Bodmer, J., Bussey, H.J., Ellis, A., Gorman, P., Lucibello, F.C., Murday, V.A., Rider, S.H., Scambler, P. et al., Localization of the gene for familial adenomatous polyposis on chromosome 5, *Nature*, 328, 614–616, 1987.

65. Herrera, L., Kakati, S., Gibas, L., Pietrzak, E., and Sandberg, A.A., Gardner syndrome in a man with an interstitial deletion of 5q, *Am. J. Med. Genet.*, 25, 473–476, 1986.

66. Leppert, M., Dobbs, M., Scambler, P., O'Connell, P., Nakamura, Y., Stauffer, D., Woodward, S., Burt, R., Hughes, J., Gardner, E. et al., The gene for familial polyposis coli maps to the long arm of chromosome 5, *Science*, 238, 1411–1413, 1987.

67. Solomon, E., Voss, R., Hall, V., Bodmer, W.F., Jass, J.R., Jeffreys, A.J., Lucibello, F.C., Patel, I., and Rider, S.H., Chromosome 5 allele loss in human colorectal carcinomas, *Nature*, 328, 616–619, 1987.

68. Kinzler, K.W., Nilbert, M.C., Vogelstein, B., Bryan, T.M., Levy, D.B., Smith, K.J., Preisinger, A.C., Hamilton, S.R., Hedge, P., Markham, A., Carlson, M., Joslyn, G., Groden, J., White, R., Miki, Y., Miyoshi, Y., Nishisho, I., and Nakamura, Y., Identification of a gene located at chromosome 5q21 that is mutated in colorectal cancers, *Science*, 251, 1366–1370, 1991.

69. Kinzler, K.W., Nilbert, M.C., Su, L.-K., Vogelstein, B., Bryan, T.M., Levy, D.B., Smith, K.J., Preisinger, A.C., Hedge, P., McKechnie, D., Finniear, R., Markham, A., Groffen, J., Boguski, M.S., Altshul, S.F., Horii, A., Ando, H., Miyoshi, Y., Miki, Y., Nishisho, I., and Nakamura, Y., Identification of FAP locus genes from chromosome 5q21, *Science*, 253, 661–669, 1991.

70. Groden, J., Thilveris, A., Samowitz, W., Carlson, M., Gelbert, L., Albertsen, H., Joslyn, G., Stevens, J., Spirio, L., Robertson, M., Sargeant, L., Krapcho, K., Wolff, E., Burt, R., Hughes, J.P., Warrington, J., McPherson, J., Wasmuth, J., Le Paslier, D., Abderrahim, H., Cohen, D., Leppert, M., and White, R., Identification and characterization of the familiar adenomatous polyposis coli gene, *Cell*, 66, 589–600, 1991.

71. Yarbrough, W.G. and Xiong, Y., P16 and ARF: Crossroads of tumorigenesis, *Encyclopedia of Cancer*, in Bertino, J., Ed., Academic Press, San Diego, CA, 2001.

72. Sherr, C.J. and Weber, J.D., The ARF/p53 pathway, *Curr. Opin. Genet. Dev.*, 10, 94–99, 2000.

73. Kamb, A., Gruis, N.A., Weaver-Feldhaus, J., Liu, Q., Harshman, K., Tavitigian, S.V., Stochert, E., Day, R.S., III, Johnson, B.E., and Skolnick, M.H., A cell cycle regulator potentially involved in genesis of many tumor types, *Science*, 264, 436–440, 1994.

74. Kamb, A., Shattuuck-Eidens, D., Eeles, R., Liu, Q., Gruis, N.A., Ding, W., Hussey, C., Tran, T., Miki, Y., Weaver-Feldhaus, J., McClure, M., Aitken, J.F., Anderson, D.E., Bergman, W., Frants, R., Goldgar, D.E., Green, A., MacLennan, R., Martin, N.G., Meyer, L.J., Youl, P., Zone, J.J., Skolnick, M.H., and Cannon-Albright, L.A., Analysis of the *p16* gene (CDKN2) as a candidate for the chromosome 9p melanoma susceptibility locus, *Nat. Genet.*, 8, 22–26, 1994.

75. Nevins, J.R., E2F: A link between the Rb tumor suppressor protein and viral oncoproteins, *Science*, 258, 424–429, 1992.

76. DeCaprio, J.A., Ludlow, J.W., Lynch, D., Furukawa, Y., Griffin, J., Piwnica-Worms, H., Huang, C.M., and Livingston, D.M., The product of the retinoblastoma susceptibility gene has properties of a cell cycle regulatory element, *Cell*, 58, 1085–1095, 1989.

77. Buchkovich, K., Duffy, L.A., and Harlow, E., The retinoblastoma protein is phosphorylated during specific phases of the cell cycle, *Cell*, 58, 1097–1105, 1989.

78. Chen, P.L., Scully, P., Shew, J.Y., Wang, J.Y., and Lee, W.H., Phosphorylation of the retinoblastoma gene product is modulated during the cell cycle and cellular differentiation, *Cell*, 58, 1193–1198, 1989.

79. Lees, J.A., Saito, M., Vidal, M., Valentine, M., Look, T., Harlow, E., Dyson, N., and Helin, K., The retinoblastoma protein binds to a family of E2F transcription factors, *Mol. Cell Biol.*, 13, 7813–7825, 1993.

80. Zhang, H.S., Postigo, A.A., and Dean, D.C., Active transcriptional repression by the Rb-E2F complex mediates G1 arrest triggered by p16^{INK4a}, TGFbeta, and contact inhibition, *Cell*, 97, 53–61, 1999.

81. Zhang, H.S., Gavin, M., Dahiya, A., Postigo, A.A., Ma, D., Luo, R.X., Harbour, J.W., and Dean, D.C., Exit from G1 and S phase of the cell cycle is regulated by repressor complexes containing HDAC-Rb-hSWI/SNF and Rb-hSWI/SNF, *Cell*, 101, 79–89, 2000.

82. Strobeck, M.W., Knudsen, K.E., Fribourg, A.F., DeCristofaro, M.F., Weissman, B.E., Imbalzano, A.N., and Knudsen, E.S., BRG-1 is required for RB-mediated cell cycle arrest, *Proc. Natl. Acad. Sci. USA*, 97, 7748–7753, 2000.

83. Dahiya, A., Wong, S., Gonzalo, S., Gavin, M., and Dean, D.C., Linking the Rb and polycomb pathways, *Mol. Cell*, 8, 557–569, 2001.

84. Hsieh, J.K., Chan, F.S., O'Connor, D.J., Mittnacht, S., Zhong, S., and Lu, X., RB regulates the stability and the apoptotic function of p53 via MDM2, *Mol. Cell*, 3, 181–193, 1999.

85. Pennaneach, V., Salles-Passador, I., Munshi, A., Brickner, H., Regazzoni, K., Dick, F., Dyson, N., Chen, T.T., Wang, J.Y., Fotedar, R., and Fotedar, A., The large subunit of replication factor C promotes cell survival after DNA damage in an LxCxE motif- and Rb-dependent manner, *Mol. Cell*, 7, 715–727, 2001.

86. Hayflick, L., Subculturing human diploid fibroblast cultures, *Tissue Culture Methods and Applications*, in Kruse, P.F. Ed., Academic Press, New York, 1973, pp. 220–223.

87. Narita, M., Nunez, S., Heard, E., Lin, A.W., Hearn, S.A., Spector, D.L., Hannon, G.J., and Lowe, S.W., Rb-mediated heterochromatin formation and silencing of E2F target genes during cellular senescence, *Cell*, 113, 703–716, 2003.

88. Crawford, L.V., The 53,000-Dalton cellular protein and its role in transformation, *Int. Rev. Exp. Pathol.*, 25, 1–50, 1983.

89. Eliyahu, D., Michalovitz, D., Eliyahu, S., Pinhasi-Kimhi, O., and Oren, M., Wild-type p53 can inhibit oncogene-mediated focus formation, *Proc. Natl. Acad. Sci. USA*, 86, 8763–8767, 1989.

90. Finlay, C.A., Hinds, P.W., and Levine, A.J., The *p53* proto-oncogene can act as a suppressor of transformation, *Cell*, 57, 1082–1093, 1989.

91. Nigro, J.M., Baker, S.J., Preisinger, A.C., Jessup, J.M., Hostetter, R., Cleary, K., Bigner, S.H., Davidson, N., Baylin, S., and Devilee, P., Mutations in the *p53* gene occur in diverse human tumor types, *Nature*, 342, 705–708, 1989.

92. Weinberg, R.A., Tumor suppressor genes, *Science*, 254, 1138–1146, 1991.

93. Hartwell, L., Defects in a cell cycle checkpoint may be responsible for the genomic instability of cancer cells, *Cell*, 71, 543–546, 1992.

94. Pietenpol, J.A., Stein, R.W., Moran, E., Yaciuk, P., Schlegel, R., Lyons, R.M., Pittelkow, M.R., Munger, K., Howley, P.M., and Moses, H.L., TGF-beta 1 inhibition of c-myc transcription and growth in keratinocytes is abrogated by viral transforming proteins with pRB binding domains, *Cell*, 61, 777–785, 1990.

95. Rotter, V., Foord, O., and Navot, N., In search of the functions of normal p53 protein, *Trends Cell Biol*, 3(2), 46–49, 1993.

96. Vogelstein, B. and Kinzler, K.W., p53 function and dysfunction, *Cell*, 70, 523–529, 1992.

97. Shieh, S.Y., Ikeda, M., Taya, Y., and Prives, C., DNA damage-induced phosphorylation of p53 alleviates inhibition by MDM2, *Cell*, 91, 325–334, 1997.

98. Fingerman, I.M. and Briggs, S.D., p53-mediated transcriptional activation: From test tube to cell, *Cell*, 117, 690–691, 2004.

99. Zhang, Y., Xiong, Y., and Yarbrough, W.G., ARF promotes MDM2 degradation and stabilizes p53: ARF-INK4a locus deletion impairs both the Rb and p53 tumor suppression pathways, *Cell*, 92, 725–734, 1998.

100. An, W., Kim, J., and Roeder, R.G., Ordered cooperative functions of PRMT1, p300, and CARM1 in transcriptional activation by p53, *Cell*, 117, 735–748, 2004.

101. el-Deiry, W.S., Tokino, T., Velculescu, V.E., Levy, D.B., Parsons, R., Trent, J.M., Lin, D., Mercer, W.E., Kinzler, K.W., and Vogelstein, B., WAF1, a potential mediator of p53 tumor suppression, *Cell*, 75, 817–825, 1993.

102. Miyashita, T. and Reed, J.C., Tumor suppressor p53 is a direct transcriptional activator of the human *bax* gene, *Cell*, 80, 293–299, 1995.

103. Levine, A.J., p53, the cellular gatekeeper for growth and division, *Cell*, 88, 323–331, 1997.

104. Scheffner, M., Werness, B.A., Huibregtse, J.M., Levine, A.J., and Howley, P.M., The E6 oncoprotein encoded by human papillomavirus types 16 and 18 promotes the degradation of p53, *Cell*, 63, 1129–1136, 1990.

105. Munger, K., Werness, B.A., Dyson, N., Phelps, W.C., Harlow, E., and Howley, P.M., Complex formation of human papillomavirus E7 proteins with the retinoblastoma tumor suppressor gene product, *EMBO J.*, 8, 4099–4105, 1989.

106. Momand, J., Zambetti, G.P., Olsonk, D.C., George, D., and Levine, A.J., The mdm-2 oncogene product forms a complex with the p53 protein and inhibits p53 mediated transactivation, *Cell*, 69, 1237–1245, 1992.

107. Oliner, J.D., Kinzler, K.W., Meltzer, P.S., George, D., and Vogelstein, B., Amplification of a gene encoding a p53-associated protein in human sarcomas, *Nature*, 358, 80–83, 1992.

108. Jones, P.A. and Baylin, S.B., The fundamental role of epigenetic events in cancer, *Nat. Rev. Genet.*, 3, 415–428, 2002.

109. Wang, Z., Madden, S.L., Deuel, T.F., and Rauscher, F.J., III, The Wilms' tumor gene product, WT1 represses transcription of the platelet-derived growth factor A-chain gene, *J. Biol. Chem.*, 267(31), 21999–22002, 1992.

110. Drummond, I.A., Madden, S.L., Rohwer-Nutter, P., Bell, G.I., Sukhatme, V.P., and Rauscher, F.J., III, Repression of the insulin-like growth factor II gene by the Wilms' tumor suppressor WT1, *Science*, 257, 674–678, 1992.

111. Versteege, I., Sevenet, N., Lange, J., Rousseau-Merck, M.F., Ambros, P., Handgretinger, R., Aurias, A., and Delattre, O., Truncating mutations of hSNF5/INI1 in aggressive paediatric cancer, *Nature*, 394, 203–206, 1998.

112. Laurent, B.C. and Carlson, M., Yeast SNF2/SWI2, SNF5, and SNF6 proteins function coordinately with the gene-specific transcriptional activators GAL4 and Biccoid, *Genes. Dev.*, 6, 1707–1715, 1992.

113. Peterson, C.L. and Herskowitz, I., Characterization of the yeast SWI1, SWI2, and SWI3 genes, which encode a global activator of transcription, *Cell*, 68, 573–583, 1992.

114. Reincke, B.S., Rosson, G.B., Oswald, B.W., and Wright, C.F., INI1 expression induces cell cycle arrest and markers of senescence in malignant rhabdoid tumor cells, *J. Cell Physiol.*, 194, 303–313, 2003.

115. Oruetxebarria, I., Venturini, F., Kekarainen, T., Houweling, A., Zuijderduijn, L.M., Mohd-Sarip, A., Vries, R.G., Hoeben, R.C., and Verrijzer, C.P., P16^{INK4a} is required for hSNF5 chromatin remodeler-induced cellular senescence in malignant rhabdoid tumor cells, *J. Biol. Chem.*, 279, 3807–3816, 2004.

116. Betz, B.L., Strobeck, M.W., Reisman, D.N., Knudsen, E.S., and Weissman, B.E., Re-expression of hSNF5/INI1/BAF47 in pediatric tumor cells leads to G1 arrest associated with induction of p16^{INK4a} and activation of RB, *Oncogene*, 21, 5193–5203, 2002.

117. Fearon, E.R., Cho, K.R., Nigro, J.M., Kern, S.E., Simons, J.W., Rupert, J.M., Hamilton, S.R., Presidinger, A.C., Thomas, G., and Kinzler, K.W., Identification of a chromosome 18q gene that is altered in colorectal cancers, *Science*, 247, 49–56, 1990.

118. Mehlen, P. and Fearon, E.R., Role of the dependence receptor DCC in colorectal cancer pathogenesis, *J. Clin. Oncol.*, 22, 3420–3428, 2004.

119. Kikuchi, A., Tumor formation by genetic mutations in the components of the Wnt signaling pathway, *Cancer Sci.*, 94, 225–229, 2003.

120. Trofatter, J.A., MacCollin, M.M., Rutter, J.L., Murrell, J.R., Duyao, M.P., Parry, D.M., Eldridge, R., Kley, N., Menon, A.N., Pulaski, K., Haase, V.H., Ambrose, C.M., Munroe, D., Bove, C., Haines, J.L., Martuza, R.L., Macdonald, M.E., Seizinger, B.R., Short, M.P., Buckler, A.J., and Gusella, J.F., A novel Moesin-Ezrin-Radixin-like gene is a candidate for the neurofibromatosis 2 tumor suppressor gene, *Cell*, 72, 791–800, 1993.

121. Hahn, H., Wicking, C., Zaphiropoulous, P.G., Gailani, M.R., Shanley, S., Chidambaram, A., Vorechovsky, I., Holmberg, E., Unden, A.B., Gillies, S., Negus, K., Smyth, I., Pressman, C., Leffell, D.J., Gerrard, B., Goldstein, A.M., Dean, M., Toftgard, R., Chenevix-Trench, G., Wainwright, B., and Bale, A.E., Mutations of the human homolog of Drosophila patched in the nevoid basal cell carcinoma syndrome, *Cell*, 85, 841–851, 1996.

122. Johnson, R.L., Rothman, A.L., Xie, J., Goodrich, L.V., Bare, J.W., Bonifas, J.M., Quinn, A.G., Myers, R.M., Cox, D.R., Epstein, E.H., Jr., and Scott, M.P., Human homolog of patched, a candidate gene for the basal cell nevus syndrome, *Science*, 272, 1668–1671, 1996.

123. Stone, D.M., Hynes, M., Armanini, M., Swanson, T.A., Gu, Q., Johnson, R.L., Scott, M.P., Pennica, D., Goddard, A., Phillips, H., Noll, M., Hooper, J.E., de Sauvage, F., and Rosenthal, A., The tumour-suppressor gene patched encodes a candidate receptor for Sonic hedgehog, *Nature*, 384, 129–134, 1996.

124. Marigo, V., Davey, R.A., Zuo, Y., Cunningham, J.M., and Tabin, C.J., Biochemical evidence that patched is the Hedgehog receptor, *Nature*, 384, 176–179, 1996.

125. Sansal, I. and Sellers, W.R., The biology and clinical relevance of the PTEN tumor suppressor pathway, *J. Clin. Oncol.*, 22, 2954–2963, 2004.

126. Steck, P.A., Pershouse, M.A., Jasser, S.A., Yung, W.K., Lin, H., Ligon, A.H., Langford, L.A., Baumgard, M.L., Hattier, T., Davis, T., Frye, C., Hu, R., Swedlund, B., Teng, D.H., and Tavtigian, S.V., Identification of a candidate tumour suppressor gene, MMAC1, at chromosome 10q23.3 that is mutated in multiple advanced cancers, *Nat. Genet.*, 15, 356–362, 1997.

127. Li, J., Yen, C., Liaw, D., Podsypanina, K., Bose, S., Wang, S.I., Puc, J., Miliaresis, C., Rodgers, L., McCombie, R., Bigner, S.H., Giovanella, B.C., Ittmann, M., Tycko, B., Hibshoosh, H., Wigler, M.H., and Parsons, R., PTEN, a putative protein tyrosine phosphatase gene mutated in human brain, breast, and prostate cancer, *Science*, 275, 1943–1947, 1997.

128. Wallace, M.R., Marchuk, D.A., Andersen, L.B., Letcher, R., Odeh, H.M., Saulino, A.M., Fountain, J.W., Brereton, A., Niocholson, J., and Mitchell, A.L., Type 1 neurofibromatosis gene: Identification of a large transcript disrupted in three NF1 patients, *Science*, 249, 181–186, 1990.

129. Viskochil, D., Buchberg, A.M., Xu, G., Cawthon, R.M., Stevens, J., Wolff, R.K., Culver, M., Carey, J.C., Copeland, N.G., and Jenkins, N.A., Deletions and a translocation interrupt a cloned at the neurofibromatosis type 1 locus, *Cell*, 62, 187–192, 1990.

130. Sharan, S.K., Morimatsu, M., Albrecht, U., Lim, D.S., Regel, E., Dinh, C., Sands, A., Eichele, G., Hasty, P., and Bradley, A., Embryonic lethality and radiation hypersensitivity mediated by Rad51 in mice lacking BRCA2, *Nature*, 386, 804–810, 1997.

131. Zhong, Q., Chen, C.F., Li, S., Chen, Y., Wang, C.C., Xiao, J., Chen, P.L., Sharp, Z.D., and Lee, W.H., Association of BRCA1 with the hRad50-hMre11-p95 complex and the DNA damage response, *Science*, 285, 747–750, 1999.

132. Lee, J.S., Collins, K.M., Brown, A.L., Lee, C.H., and Chung, J.H., hCds1-mediated phosphorylation of BRCA1 regulates the DNA damage response, *Nature*, 404, 201–204, 2000.

133. Motoyama, N. and Naka, K., DNA damage tumor suppressor genes and genomic instability, *Curr. Opin. Genet. Dev.*, 14, 11–16, 2004.

134. Kolodner, R., Biochemistry and genetics of eukaryotic mismatch repair, *Genes Dev.*, 10, 1433–1442, 1996.

135. Papadopoulos, N., Nicolaides, N.C., Wei, Y.F., Ruben, S.M., Carter, K.C., Rosen, C.A., Haseltine, W.A., Fleischmann, R.D., Fraser, C.M., Adams, M.D. et al., Mutation of a mutL homolog in hereditary colon cancer, *Science*, 263, 1625–1629, 1994.

136. Ionov, Y., Peinado, M.A., Malkhosyan, S., Shibata, D., and Perucho, M., Ubiquitous somatic mutations in simple repeated sequences reveal a new mechanism for colonic carcinogenesis, *Nature*, 363, 558–561, 1993.

137. Fishel, R., Lescoe, M.K., Rao, M.R., Copeland, N.G., Jenkins, N.A., Garber, J., Kane, M., and Kolodner, R., The human mutator gene homolog MSH2 and its association with hereditary nonpolyposis colon cancer, *Cell*, 75, 1027–1038, 1993.

138. Perucho, M., Tumors with microsatellite instability: Many mutations, targets, and paradoxes, *Oncogene*, 22, 2223–2225, 2003.

139. Bremner, R. and Balmain, A., Genetic changes in skin tumor progression: Correlation between presence of a mutant *ras* gene and loss of heterozygosity on mouse chromosome 7, *Cell*, 61, 407–417, 1990.

140. Bishop, J.M., The molecular genetics of cancer, *Science*, 235, 305–311, 1987.

141. Shevde, L.A. and Welch, D.R., Metastasis suppressor pathways — an evolving paradigm, *Cancer Lett.*, 198, 1–20, 2003.

142. Sherr, C.J., The Pezcoller lecture: Cancer cell cycles revisited, *Cancer Res.*, 60, 3689–3695, 2000.

143. D'Andrea, A.D. and Grompe, M., The Fanconi anaemia/BRCA pathway, *Nat. Rev. Cancer*, 3, 23–34, 2003.

144. Harris, H., Miller, O.J., Klein, G., Worst, P., and Tachibana, T., Suppression of malignancy by cell fusion, *Nature*, 223, 363–368, 1969.

145. Stanbridge, E.J., Functional evidence for human tumor suppressor genes: Chromosomal and molecular genetic studies, *Cancer Surv.*, 12, 5–24, 1992.

7 Hepatitis B, Hepatitis C, and Hepatocellular Carcinoma

Jing-hsiung Ou

CONTENTS

7.1 INTRODUCTION

Hepatitis A, B, C, D, and E viruses are the five major etiologic agents of viral hepatitis. These five different viruses infect primarily the liver and are commonly referred to as the hepatitis viruses. Several other viruses, including GB-virus type A (GBV-A), GB-virus type B (GBV-B), GB virus type C (GBV-C, also known as hepatitis G virus [HGV]), and TT-virus, have also been isolated from hepatitis patients. However, the role of these viruses in the development of hepatitis appears to be minor, if there is any.

Hepatitis A virus (HAV) and hepatitis E virus (HEV) are transmitted by the fecal–oral route. They cause only self-limited acute infection in patients. These two viruses have a linear RNA genome with a length of slightly longer than 7 kilobases (kb). Hepatitis D virus (HDV), also known as the delta virus, is an unusual virus as it is defective and requires the structural proteins provided by hepatitis B virus (HBV) to mature. Therefore, HDV can only be cotransmitted with HBV to patients or superinfect HBV patients. This virus contains a 1.7 kb circular RNA genome.

The two remaining hepatitis viruses, hepatitis B virus (HBV) and hepatitis C virus (HCV), can cause chronic infection in addition to acute infection. Chronic infection by these two viruses frequently leads to severe liver diseases including hepatocellular carcinoma. As these two viruses are widespread in the human population, they pose serious health problems. These two viruses and their relationship to hepatocellular carcinoma are the focus of this chapter.

7.2 HEPATITIS B VIRUS

7.2.1 THE DISCOVERY OF THE VIRUS

The existence of two forms of hepatitis, the epidemic form and the parenterally transmitted form, have been recognized for approximately a century [1,2]. In 1947, the terms "hepatitis A" and "hepatitis B" were introduced by MacCallum [3] to categorize these two forms of hepatitis. In 1963, B. Blumberg at the National Institutes of Health discovered that a serum sample from an Australian aborigine contained an antigen that reacted with antibodies in the blood of American hemophiliacs [4]. This antigen, which Blumberg named Australia antigen, was linked to hepatitis B in 1967. This finding allowed the blood banks to screen the blood supplies for the virus and reduced the incidence of posttransfusion hepatitis. Blumberg received a Nobel Prize in Medicine/Physiology in 1976 for his research accomplishments.

In 1970, Dane et al. [5] detected a 42 nm, virus-like particle in hepatitis B patients (Figure 7.1[a]). This is the mature HBV virion, which has since also been called the Dane particle. Epidemiology studies have indicated that there are approximately 350 million HBV carriers in the world, resulting in slightly more than 1 million deaths every year [6].

Since the early 1980s, the genomes of additional animal viruses related to HBV have been isolated. These viruses, which include woodchuck hepatitis B virus [7], ground squirrel hepatitis B

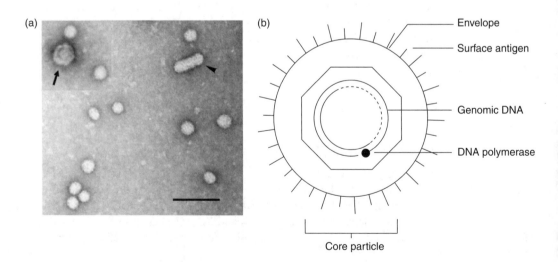

FIGURE 7.1   Structure of hepatitis B virus particles. (a) Electron microscopy of hepatitis B virus particles. The arrow denotes the mature HBV particle, and the arrowhead mark the location of a filamentous 22 nm particle. The rest of the particles seen in the picture are the 22 nm spherical particles. The scale bar represents 100 nm. (b) Schematic illustration of the structure of HBV.

virus [8], duck hepatitis B virus [9], heron hepatitis B virus [10], and wooly monkey hepatitis B virus [11], are genetically related, and all infect the hepatocytes of their respective hosts. Due to their relatedness, they have been grouped into one distinct viral family called *hepadnaviridae*. All the hepadnaviruses contain a partially double-stranded DNA genome with a size slightly larger than 3 Kb.

7.2.2 THE VIRAL STRUCTURE

HBV contains a lipid envelope (Figure 7.1[b]). There are three related envelope proteins named large (L), middle (M), and small (S) surface antigens (HBsAg). The HBsAg was the Australia antigen identified by B. Blumberg. The lipid envelope of HBV surrounds a core particle (Figure 7.1[b]). The major protein constituent of the core particle is the core protein (also known as the core antigen). This core particle also contains the genomic DNA and a DNA polymerase. A protein kinase is apparently also packaged in the core particle, as the core protein can be phosphorylated if the purified core particle is incubated with ^{32}P-labeled adenosine triphosphate [12]. The identity of this kinase activity has been controversial. It has been suggested to be protein kinase C [13], the kinase activity of glyceraldehyde phosphate dehydrogenase [14], a 46 kilodalton (kDa) protein associated with ribosomes [15], and SRPK1 and SRPK2 [16]. Clearly, more work is needed to identify this particular kinase and to understand its biological functions in the life cycle of HBV.

The HBV genomic DNA is a partially double-stranded molecule (Figure 7.1[b]). The long strand is approximately 3.2 Kb in length and is noncoding (the "minus-strand"). This strand is covalently linked to the viral DNA polymerase (Figure 7.1[b]) [17], which serves as a primer for viral DNA synthesis. The short strand is variable in length and is the coding strand (the "plus-strand"). Both the long strand and the short strand are held together by the complementary nucleotide sequences to form a circular structure.

7.2.3 THE VIRAL GENES

After entering cells, the HBV genomic DNA is transported to the nucleus where it is converted to a covalently closed circular DNA (cccDNA) by a not yet understood mechanism. The coding strand contains four open reading frames named S, C, P, and X. The coding sequences of these genes overlap extensively. Therefore, every nucleotide in the HBV genome participates in at least one coding function. These genes have different functions and are discussed separately.

7.2.3.1 The S Gene

The S gene coding sequence contains three in-frame ATG codons (Figure 7.2[a]). The translation initiating from these three ATG codons results in the production of L, M, and S HBsAg proteins. Hence, these three proteins share the same carboxy-terminal sequence but have different amino-terminal extensions. The viral envelope can also be secreted without packaging the core particle. These empty envelope particles are termed 22 nm particles due to their diameter (Figure 7.1[a]). The 22 nm particles are either spherical or filamentous in shape (Figure 7.1[a]). The spherical 22 nm particle contains few LHBsAg molecules, whereas between 10 and 20% of the HBsAg molecules in the filamentous 22 nm particle, and the mature HBV virion are LHBsAg [18].

All three HBsAg proteins are transmembrane proteins and glycosylated [19–22]. The LHBsAg protein is unusual, as its amino terminus is located in the cytosolic side of the endoplasmic reticulum (ER) membrane when it is newly synthesized. However, a fraction of the LHBsAg protein undergoes topological change when the virus matures, and its amino-terminus gets translocated across the membrane and becomes exposed on the viral surface [23–25]. The LHBsAg protein is myristylated [26].

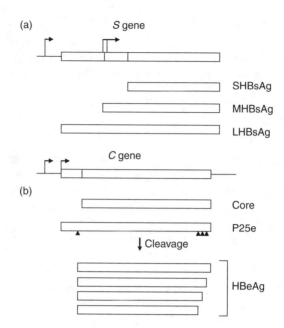

FIGURE 7.2   The HBV *S* and *C* genes. (a) The *S* gene. (b) The *C* gene. The vertical lines in the *S* gene and the *C* gene coding sequences represent the in-phase translation initiation codons. The rightward arrows mark the location of the transcription initiation sites of the *S* gene and the *C* gene RNA transcripts. The arrowheads in (b) mark the locations of proteolytic cleavage sites for the generation of HBeAg.

This myristylation is important for the viral infectivity, as mutations that affected myristylation did not affect the maturation of the virion but rendered the virus noninfectious [27–29].

7.2.3.2 The *C* Gene Products

Similar to the *S* gene, the *C* gene contains two in-phase ATG codons separated by 28 codons (Figure 7.2[b]). The translation of the 21 kDa core protein is initiated from the downstream ATG codon. Hence the sequence between these two in-phase ATG codons has been termed the "precore region." The translation from the upstream ATG codon generates a related 25 kDa protein, which has been called the precore protein. The amino-terminal 19 amino acids of the precore protein constitute a signal sequence for the translocation of the precore protein across the ER membrane for secretion [30–32]. During the translocation, the signal sequence is removed and the precore protein derivative is further cleaved by proteases at its carboxy-terminus to generate a heterogeneous population of proteins ranging from 17 to 22 kDa in size [33]. These secreted precore protein derivatives are known as the e-antigen (HBeAg). HBeAg is not required for the replication of HBV, as HBV mutants that cannot express the precore protein (e.g., mutants that contain nonsense mutations in the precore region) are frequently isolated from chronic HBV patients [34–36]. Clinical studies indicate that HBeAg may be important for establishing persistent infection following neonatal infection, as babies born to HBeAg-negative mothers often develop only self-limited acute infection, whereas those born to HBeAg-positive mothers frequently become chronic carriers [37]. The role of HBeAg in establishing persistent infection is supported by the experiments with the related woodchuck hepatitis B virus (WHV) [38]. It was found that while both HBeAg-positive and HBeAg-negative WHV could initiate infection in neonatal woodchucks, only the former could lead to chronic infection [38]. The ability of HBeAg to help establish chronic infection, which can increase the possibility of transmission of the virus between individuals, is probably the reason why HBeAg is conserved among different hepadnaviruses. How HBeAg helps HBV to establish chronic infection remains elusive (for a review, see Reference 38).

The core protein can form the core particle either in the presence or in absence of the viral pregenomic RNA. This protein contains a highly basic carboxy-terminal sequence for interaction with the viral genomic DNA. This sequence contains at least three phosphorylation sites [16,39]. Recent studies indicate that phosphorylation of the core protein may play important roles in regulating the packaging of the pregenomic RNA and the replication of the viral DNA [40,41].

7.2.3.3 The DNA Polymerase Gene

The HBV DNA polymerase gene encodes a protein over 90 kDa in size. This protein contains a reverse transcriptase activity and an RNaseH activity. In addition, this protein also serves as the primer for reverse transcription for the synthesis of the minus-strand DNA. The reverse transcription process is initiated by the binding of the polymerase to a conserved stem-loop structure located in the precore region near the 5′ end of the pregenomic RNA. This stem-loop structure, named ε, also serves as the encapsidation initiation signal [42–44]. The polymerase binds to the ε structure and serves as the protein primer for the initiation of the minus-stranded DNA synthesis [45,46]. The nucleotide sequence at the bulge of the ε structure was used as the template for the incorporation of the first four nucleotides AACA [46]. A phosphodiester bond is formed between the 5′-phosphate of the first nucleotide adenosine and the hydroxyl group of tyrosine located at amino acid 63 of the polymerase [46,47]. After the incorporation of these four nucleotides, the polymerase as well as the AACA sequence will be transferred to a complementary UUGU sequence located near the 3′ end of the pregenomic RNA to continue the synthesis of the minus-stranded DNA [48,49]. The pregenomic RNA is degraded by the RNaseH activity of the polymerase and its 5′ end is used as the primer for initiating the synthesis of the plus-stranded DNA [50].

7.2.3.4 The *X* Gene

The *X* gene encodes a protein approximately 17 kDa in size. This protein has plural regulatory functions, which can enhance the expression of polymerase I, polymerase II, and polymerase III-dependent genes [51–54]. The HBV X protein does not bind directly to DNA and is localized predominantly to the cytoplasm (for a review, see Reference 51, also unpublished observation). The X protein may activate gene expression through different pathways [55]. It has been shown to bind to different transcription factors including AP1, AP2, ATF-2, CREB, HNF-1, and TBP [56–59], and also to RBP5, a subunit of all three mammalian RNA polymerases [60]. It has also been shown to activate the ras-raf-MAPK signaling pathway, the calcium signaling pathway, and the NF-kB pathway [61–65], and, although controversial, may also be involved in the activation of the protein kinase C (PKC) pathway [66,67]. The X protein has also been shown to bind to proteasomes to indirectly affect gene expression in cells [68,69].

Experiments conducted using established human hepatoma cell lines indicated that the X protein was not required for the replication of HBV in cells [70,71]. The introduction of the wild-type WHV genomic DNA into the liver of woodchucks could result in the infection of these animals by WHV. The infection was unapparent if a missense mutation was introduced first to remove the X protein initiation codon [71,72]. Thus, the X protein apparently plays an important role for the replication of HBV in its natural host. In our recent transgenic mouse studies, we found that the mutations introduced to abolish the expression of the X protein did not prevent the replication of HBV in the mice carrying the mutated HBV genome [73]. However, if these mice were crossed to a mouse line carrying the HBV *X* gene, then the HBV replication rate was increased in the crossbred progeny mice. Further studies indicated that the *X* gene could enhance both HBV RNA transcription and DNA replication in transgenic mice [73]. These studies indicate that the *X* gene could enhance

HBV replication. Indeed, in a recent study using woodchucks, it was suggested that WHV incapable of expressing the *X* gene might only be attenuated [74].

7.2.4 THE REPLICATION CYCLE OF HBV

7.2.4.1 Adsorption and Penetration

Little is known about the early stage of the HBV infection cycle due to the lack of a cell culture system for infection studies. A number of candidate receptors for HBV have been proposed [75–86]. However, none of them has been conclusively demonstrated to be the true HBV receptors. The receptor of the related DHBV receptor appears to be carboxypeptidase D (gp180) [87–89], as this protein could bind strongly to the preS region of the DHBV HBsAg, and its truncated soluble form could suppress DHBV infection. The expression of carboxypeptidase D in heterologous hepatoma cells would allow the attachment and the internalization of DHBV but not the subsequent steps of viral replication [87]. These observations suggest the need of a cofactor for DHBV infection.

The infectivity of DHBV to primary duck hepatocytes is enhanced by the 22 nm HBsAg particles [90]. How this subviral particle enhances the viral infectivity is unclear. After binding to its receptor, the virus is internalized into cells. Many enveloped viruses use the acidic environment of lysosomes to expose fusogenic sequences of their surface proteins, which then trigger the fusion of the viral envelope with the lysosomal membranes to release the viral core particle into the cytosol [91]. This pathway is apparently not used by HBV as its internalization is pH-independent [92].

After infection, the HBV genomic DNA will be transported into the nucleus and converted to cccDNA form [93]. The molecular mechanisms for the nuclear import and the repair of the viral genomic DNA remain elusive. The carboxy-terminus of the HBV core protein contains a nuclear localization signal (NLS) and may be involved in the nuclear import of the genomic DNA [94,95]. It has been proposed that the phosphorylation of the core particle results in the exposure of the NLS, which then binds to the nuclear pore via the importin-mediated pathway [96]. This binding to the nuclear pore eventually leads to the release of viral genomic DNA into the nucleus [96].

7.2.4.2 Gene Expression

The HBV cccDNA is the template for the transcription of viral genes. This cccDNA contains two enhancers and four promoters (Figure 7.3) (for reviews, see References 97 and 98). The two enhancers, named ENI and ENII, are liver-specific due to the requirements of liver-enriched transcription factors for their activities [99–105]. These two enhancers regulate the activities of the four promoters [106,107]. The four promoters, SP1, SP2, XP, and CP control the transcription of HBV genes. The SP1 promoter directs the transcription of the LHBsAg mRNA, and the SP2 promoter directs the transcription of MHBsAg and SHBsAg mRNAs. The RNA transcripts derived from the SP2 promoter have multiple transcription initiation sites that flank the ATG codon of the MHBsAg protein (Figure 7.2). Hence, those initiating upstream of the ATG codon can encode the MHBsAg protein and those initiating downstream can only encode the SHBsAg protein. The XP directs the transcription of the X mRNA. The CP controls the transcription of precore protein and core protein mRNAs. Similar to SP2, the CP transcripts have multiple transcription initiation sites that flank the ATG codon of the precore protein (Figure 7.2). Thus, those initiating upstream of the ATG codon can encode the precore protein and those initiating downstream can only encode the core protein.

There is only one polyadenylation (polyA) site in the HBV genome (Figure 7.3). This site, which is used by all the HBV transcripts, is located in the core protein coding region. This poses an interesting dilemma, as the transcription of precore protein and core protein mRNAs must bypass this site once before it uses this site for polyadenylation. Studies with the transcription regulation suggest that the distance between the promoter and the polyA site as well as *cis*-acting sequence elements

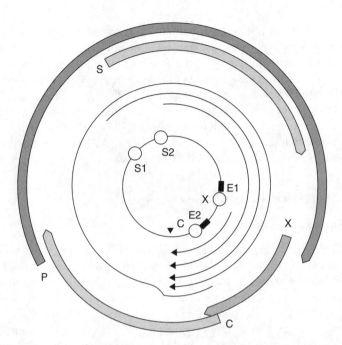

FIGURE 7.3　Transcriptional regulatory elements of the HBV genome. S1, S2, X, and C represent SP1, SP2, XP, and CP promoters, respectively. E1 and E2 are ENI and ENII enhancers. The unique polyadenylation site is denoted by an arrowhead. The arrows represent the HBV RNA transcripts, and the bars represent the HBV coding sequences.

located upstream of CP are important factors for regulating the polyadenylation of HBV transcripts [108–110]. Due to this polyadenylation process, both precore protein and core protein mRNAs contain the entire genomic information plus a terminal direct repeat of approximately 250 nucleotides.

After their transcription, the HBV RNAs are exported from the nucleus to the cytoplasm. The efficient nuclear export of HBV RNAs requires a sequence overlapping the X protein coding region [111–113]. This sequence, which is termed post-transcriptional regulatory element (PRE), apparently substitutes introns for the proper export of the intronless HBV RNAs. Indeed, the function of the PRE could be replaced by an intron of the β-globin gene for the nuclear export of HBV RNAs [113].

While the S gene, the X gene, and the C gene products are synthesized from their respective mRNAs, the HBV DNA polymerase does not have its specific mRNA (Figure 7.3). It is synthesized from the core protein mRNA by *de novo* initiation of translation. The initiation codon of the polymerase coding sequence is located near the 3′ end of the core protein coding sequence. Therefore, the translation of the polymerase from the core protein mRNA requires internal initiation. No internal ribosomal entry sequence (IRES) that may be used for the initiation of the polymerase translation has been identified. Thus, it has been proposed that the translation of the polymerase utilizes either a leaky ribosomal scanning or a ribosomal reinitiation mechanism [114–116]. Despite the fact that the precore protein mRNA is very similar in structure to the core protein mRNA, it is not used for polymerase synthesis [115,117]. The reason for this is unclear. It may be because the precore ATG codon is more efficient for translation initiation than the core protein ATG codon [117]. This higher translation initiation rate for the upstream precore protein sequence may have suppressed the translation of the downstream polymerase coding sequence from the precore protein mRNA [117].

7.2.4.3　Maturation and Release of Virions

After its synthesis the core protein encapsidates its own mRNA, which is also known as the pregenomic RNA. Interestingly, in spite of its close resemblance to the core protein mRNA, the precore protein mRNA is not encapsidated [118]. It has been suggested that the translation from the precore protein initiation codon might have overridden the ε signal located in the precore region and prevented it from being recognized by the core protein [119]. As noted above, the DNA polymerase, which also binds to the ε signal to initiate DNA synthesis, is also packaged in the core particle. The DNA polymerase converts the pregenomic RNA in the core particle to the partially double-stranded DNA genome in the core particle [120]. The core particle will then interact with LHBsAg that has been inserted into the ER membrane [121–124]. Electron microscopy studies indicate that the core particle buds through the membrane of the rough ER to acquire its envelope [125,126]. The virus is presumably released from the infected hepatocytes via the secretory pathway.

7.2.5 HBV-Induced Hepatocellular Oncogenesis

7.2.5.1 Epidemiology Studies

HBV was causally linked to hepatocellular carcinoma (HCC) in the late 1970s when it became apparent that those areas in the world with high HBV carrier populations also had high numbers of HCC cases [127]. In the early 1980s, Beasely et al. [128,129] reported the results of their seminal epidemiology study and conclusively demonstrated that HBV carriers had a more than 200-fold risk of developing HCC than the control population.

The causal role of HBV in the development of HCC was further demonstrated by a recent epidemiology study. In that study, it was found that the universal vaccination program for newborn infants initiated in the mid-1980s in Taiwan reduced the percentage of HBV carriage in 6-year old children from 10 to <1% [130,131]. This reduction of carrier population was coupled by a significant reduction of the incidence of HCC in the pediatric population in Taiwan. This study, which indicates that the prevention of HBV infection can reduce significantly the incidence of HCC, further establishes a role of HBV in the development of HCC.

Many studies have been conducted during the past two decades to study the molecular mechanism of HBV-induced hepatocellular oncogenesis. The studies indicate that HBV may directly or indirectly induce HCC. It is important to note that these two different pathways are not mutually exclusive, and it is highly likely that they are both responsible for the development of HCC in HBV patients.

7.2.5.2 The Direct Mechanism

7.2.5.2.1 Insertional Mutagenesis
The HBV DNA was found integrated in the cellular DNA of most of the HCC isolated from chronic HBV patients [132]. Unlike retroviruses, the integration of HBV DNA into the host chromosome is not an essential step for HBV replication and is likely that it involves illegitimate recombination between the viral DNA and the cellular DNA. The integration junctions of the host chromosomes and the HBV DNA contain limited sequence homology [133], and there are frequently sequence deletions, duplications, and rearrangements in both viral and host sequences at the integration junctions [134–136]. As the HBV DNA contains enhancers and promoters that can activate gene expression in *cis*, it may activate cellular oncogenes via DNA integration to initiate hepatocellular oncogenesis. Indeed, the studies of woodchuck hepatoma tissues indicated that up to 40% of the hepatoma tissues contained the activated *N-myc* oncogene due to the integration of WHV DNA within, or in the vicinity of, the *N-myc* gene [137,138]. The introduction of one such activated *N-myc* gene together with its adjacent viral DNA into the mouse embryos resulted in the development of HCC in virtually all the transgenic mice produced [139], indicating the importance of insertional mutagenesis by the viral DNA in the development of HCC in woodchucks. However, the studies of human HCC tissues have failed to reveal any cellular oncogenes that are commonly activated by the integrated HBV DNA [135,140,141]. These observations indicate that WHV and HBV initiate hepatocellular oncogenesis through different mechanisms. Nonetheless, isolated examples of HBV DNA integration

next to c-erbA and cyclin A have been reported [142,143], suggesting the possibility of HBV DNA integration to deregulate cellular gene expression for the initiation of oncogenesis.

7.2.5.2.2   The HBV Gene Products

Two HBV genes, the X gene and the truncated middle HBsAg ($MHBs^t$) gene, have been implicated in hepatocellular transformation. The X gene, whether it is in the HBV genome or expressed under the control of a heterologous promoter, can transform NIH3T3 cells that have been immortalized by the large T antigen of simian virus 40 (SV40) [144]. In addition, upon treatment with chemical carcinogens the transgenic mice carrying the HBV X gene could develop hepatocellular carcinoma either spontaneously or with an efficiency higher than the control [145,146]. The X gene has also been shown to potentiate c-myc-induced hepatocellular oncogenesis in transgenic mice [147]. These studies indicate the oncogenic potential of the X gene. The X protein can also interact with UV-DDB, a protein involved in DNA repair [148]. Hence, it has been postulated that this interaction might reduce the efficiency of cellular DNA repair and increase the possibility of gene mutations, which in turn leads to hepatocellular oncogenesis [148]. Experimental results indicated that the X protein could also functionally interact with the tumor suppressor p53, although the outcome of the interaction differed drastically between different experiments. In two different reports, it was indicated that the X protein sequestered p53 in the cytoplasm and suppressed p53-induced apoptosis [149,150]. However, in a third report it was shown that the X protein sensitized the cells to apoptosis in a p53-dependent manner after exposure to DNA damaging agents [151]. The proapoptotic effect of the X gene has also been demonstrated in another study [152]. Thus, the X protein apparently has both oncogenic and proapoptotic activities.

Besides the X protein gene, the truncated middle surface antigen, MHBs^t, has also been suggested to be an oncogenic factor for the development of hepatocellular carcinoma [153]. The $MHBsAg$ gene contains the entire SHBsAg sequence plus an amino-terminal extension of 55 amino acids (Figure 7.2). It has been shown that MHBsAg is not a transcriptional transactivator unless it is truncated at various locations in the SHBsAg sequence [154]. The truncated MHBs^t can bind to and activate protein kinase C (PKC) α and β, and, through this pathway, can also activate various transcription factors such as AP1, AP2, and NFkB [154]. It can also sequester p53 in the cytoplasm and suppresses its nuclear transport. Many hepatoma cells contain the disrupted $MHBsAg$ gene, and hence it was proposed that MHBs^t might be involved in hepatocellular oncogenesis through its transactivation activities. This postulation is supported by a transgenic mouse study. In that study, it was found that the expression of MHBs^t in the mouse liver using the albumin gene promoter could result in the constitutive activation of the Raf-1/Erk2 pathway, the increase of the proliferation rate of hepatocytes, and an increased incidence of liver tumors in MHBs^t transgenic mice older than 15 months [155].

7.2.5.3   The Indirect Mechanism

In this hypothesis, chronic liver injury caused by HBV infection was suggested to be the primary cause of hepatocellular carcinoma [156]. It has been postulated that the continuous liver injury caused by HBV infection and the immune response during chronic infection can result in the continuous liver regeneration, which creates a repeated mitogenic process of hepatocytes and facilitates the accumulation of random genetic mutations and chromosomal alterations, and eventually lead to the development of HCC. The genetic mutations may be facilitated by oxygen radicals produced by the immune system or by the environmental factors such as aflatoxin. This indirect mechanism is supported by the observation that chronic hepatitis induced by carbon tetrachloride or genetic disorders such as copper deficiency can lead to the formation of HCC in laboratory animals [157,158]. In addition, it is also supported by transgenic mouse studies, in which chronic immune response induced against the liver of transgenic mice carrying the HBV S gene was found to develop HCC at high frequencies [156].

7.3 HEPATITIS C VIRUS

7.3.1 THE DISCOVERY OF THE VIRUS

The development of the diagnostic blood test for HBV in the early 1970s reduced the incidence of posttransfusion viral hepatitis to 10% [159]. These remaining hepatitis cases were termed the "posttransfusion non-A, non-B hepatitis." In 1989, M. Houghton and his colleagues at Chiron Corporation used an expression vector to express cDNA prepared from the serum of a chimpanzee that had been infected with the non-A, non-B hepatitis. They subsequently identified a cDNA clone that reacted with the sera of non-A, non-B hepatitis patients [160]. By using this cDNA clone as the probe, they were able to isolate most of the genomic sequence of the non-A, non-B virus, which they renamed HCV [161]. HCV is believed to be an enveloped virus, as it could be inactivated by chloroform [162,163]. The genomic organization of HCV showed a striking homology to those of the members of the flavivirus family [161]. This virus was hence classified as a flavivirus. More than 10,000 HCV sequences have been reported since then. There are substantial genomic sequence variations among different HCV isolates. Based on the phylogenetic relationship of their genomic sequences, HCV has been grouped into six major genotypes and many more subtypes [164,165]. The cloning of the HCV genome represents a technology breakthrough, as it is the first time that scientists isolated the genome of a virus before the virus was even identified.

The isolation of the HCV cDNA clone led to the development of the diagnostic assays for screening the HCV-contaminated blood [166,167]. This has drastically reduced the incidence of posttransfusion hepatitis to one in 100,000 units of blood transfused [159].

The research with HCV has been hampered by the lack of a cell culture system for propagating the virus and by the lack of a convenient animal model for pathogenesis studies. Nevertheless, the recent development of a HCV RNA replicon system [168,169], which allowed subgenomic HCV RNA to replicate in Huh7 hepatoma cells, has offered hopes for studying the replication of HCV RNA and for the potential development of new anti-HCV drugs [168].

7.3.2 THE VIRAL GENOMIC ORGANIZATION

Like the other flaviviruses, HCV is a positive-stranded RNA virus with a genome size of nearly 10 kb. The HCV genomic RNA contains a long coding sequence of approximately 9 kb, flanked by a 5' noncoding region (NCR) of approximately 340 nucleotides and a 3' NCR with a variable length. The 5' NCR contains extensive stem-loop structures. These stem-loop structures as well as the first few codons of the coding sequence constitute an IRES for the cap-independent translation of the HCV RNA [170–174]. The 3' NCR is divided into three regions: a highly conserved and structured 98-nucleotide sequence at the extreme 3' end [175,176], a polyU sequence interspersed with C, and a region with a variable sequence and length located between the polyU sequence and the coding sequence. In contrast to the 98-nucleotide sequence and the polyU sequence, which are important for the infectivity of HCV in chimpanzees, the sequence of the 3' NCR variable region is not critical for HCV infectivity [177].

The coding region of the HCV RNA is slightly more than 9 kb in length. It encodes a polyprotein with a length of more than 3000 amino acids. This polyprotein is cleaved by cellular and viral proteases to generate at least ten final gene products. In addition, HCV also expresses another gene product by translational ribosomal frameshift. The genomic organization of HCV is shown in Figure 7.4.

7.3.3 THE HCV GENE PRODUCTS

The core protein, which packages the genomic RNA to form the viral nucleocapsid, is located at the amino-terminus of the polyprotein sequence. This protein is 191 amino acids in length. It is separated from the subsequent E1 envelope protein sequence by the signal peptidase located in

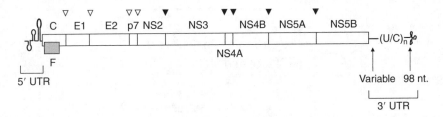

FIGURE 7.4   The HCV genomic organization. The 5′ UTR is highly structured and the 3′ UTR contains a variable region, a U/C-rich region, and a highly conserved 98-nucleotide sequence. The HCV genomic RNA encodes a polyprotein, which is cleaved by cellular proteases (denoted by empty arrowheads) or viral proteases (denoted by solid arrowheads), to generate ten different viral protein products. The F protein is encoded by an alternative reading frame.

the lumen of the ER [178]. The core protein expressed in *Escherichia coli* could form capsid-like particles [179,180]. The core protein can form a homodimer [181,182], which is presumably the building subunit of the viral capsid. The core protein can be further cleaved, apparently between amino acids 172 to 182, by the signal peptide peptidase [183]. The analysis of the core protein present in the HCV virion indicated that this truncated core protein is likely the core protein present in the HCV virion [184].

The core protein sequence is followed by E1 and E2 envelope protein sequences, which are separated from each other by signal peptidase [185,186]. Both E1 and E2 are extensively modified by N-linked glycosylation [187]. These two proteins can form a heterodimer [185,188–190]. E1 can also interact with the core protein [191]. These interactions are likely important for the formation of the mature virion. Recently, it was demonstrated that E2 could bind to CD81, a cell surface molecule [192]. However, the expression of human CD81 in transgenic mice did not render the mice susceptible to HCV infection, indicating that CD81 by itself is insufficient to initiate the infection cycle of HCV [193].

Following the E2 protein sequence is the p7 protein. This protein is a small 7 kDa protein, which can form an ion channel [194,195]. It may or may not be separated from the E2 sequence [196,197]. This protein contains a signal sequence for the subsequent NS2 protein, and is separated from the subsequent NS2 sequence by signal peptidase [198]. NS2 is an integral membrane protein [199,200], and is likely that it has at least four transmembrane domains [200]. Most of its sequence as well as part of its downstream NS3 sequence is required for the separation of NS2 and NS3 [201,202]. As the cleavage at the junction of NS2 and NS3 requires zinc, it has been suggested that the combined NS2-NS3 sequence contains a metalloproteinase activity [201]. The function of NS2 in the HCV life cycle is unclear; but it is known that it is not required for the HCV RNA replication [168].

The NS3 sequence follows that of NS2. The amino-terminal one-third of NS3 contains a serine protease activity [201,203–207], and the carboxy-terminal two-thirds contains NTPase and helicase activities [208–214]. The crystallographic structures of NS3 protease and helicase domains have been determined [215–218]. The NS3 protease domain binds a divalent zinc cation, which may be the reason why zinc is required for the protease activity of the combined NS2-NS3 sequence [219]. The serine protease activity of NS3 is distinct from the NS2-NS3 zinc protease activity [201,202,205]. This serine protease activity cleaves at the NS3-NS4A, NS4A-NS4B, NS4B-NS5A, and NS5A-NS5B junctions. These cleavage junctions are illustrated in Table 7.1. As shown in the table, there is a preference for cysteine at the P1 position, for serine or alanine at the P1′ position, and for an acidic amino acid at the P6 position. While mutagenesis experiments indicated that the P1 position at the NS3-NS4A *cis* junction was flexible, this position required cysteine for efficient cleavage at the NS4A-NS4B, NS4B-NS5A, and NS5A-NS5B *trans* junctions [220]. Similarly, the amino acids at P1′ and P6 positions have also been found to be critical for efficient NS3 cleavage [221–223].

The NS3 sequence is followed by the NS4A sequence, which is a small protein with a length of 54 amino acids. NS4A binds to NS3 and serves as a cofactor for the NS3 protease activity

TABLE 7.1

The Cleavage Junctions of the
HCV NS3 Serine Protease[a]

NS3/4A	DLEVVT/STW
NS4A/4B	DEMEEC/ASQ
NS4B/NS5A	DCSTPC/SQS
NS5A/NS5B	EDVVCC/SMS

[a] The sequences shown represent the most common sequences observed at the various cleavage junctions of the reported HCV sequences. Some variations of the sequences are possible.

[215,224–228]. The NS4A N-terminal region is highly hydrophobic and may contain a membrane-association domain to bring NS3 close to the membrane [215,227–229]. The central region of NS4A forms a beta strand and intercalates into the N-terminal domain of NS3 [215]. This interaction with NS3 apparently stabilizes the NS3 protease structure for its activity.

The function of NS4B has not been determined. However, recent studies have indicated that it is membrane associated and possibly has up to five transmembrane domains [230]. NS4B can form a complex with NS3 and NS5B and is likely that it plays an important role in the regulation of HCV RNA replication [231].

Following the NS4B sequence is the NS5A protein. NS5A is a phosphoprotein [232]. It can be induced to become hyperphosphorylated by the cleavage of its preceding NS2-3 sequence or by the presence of NS4A sequence [233,234]. The function of NS5A and its phosphorylation in the life cycle of HCV are unclear. However, NS5A can form a complex with NS5B and, thus, similar to NS4B, may be an essential component of the replication complex of the HCV RNA [235]. Indeed, in recent studies, the mutations in NS5A have been shown to enhance the replication of HCV RNA replicon in Huh7 human hepatoma cells [236,237]. Approximately 10 to 15% of HCV patients had sustained response to the interferon-α treatment [238,239]. Based on the sequence variations in NS5A, it has been proposed that the NS5A sequence located between amino acids 2209 and 2248 plays an important role in determining the responsiveness to interferon-α [240]. This sequence hence has been named the interferon sensitivity-determining regions (ISDR). The correlation between the NS5A ISDR and the interferon sensitivity appears to be restricted to the HCV 1b genotype in the Japanese patient population [240–245], as similar studies conducted with the European and American HCV patients have revealed no such correlation [246–252]. NS5A can bind to PKR, an interferon-inducible kinase, and suppress its kinase activity [253,254]. Thus, it has been postulated that the resistance to interferon is mediated by the binding of NS5A to PKR [253,254], and the previous controversy could be resolved by expanding the ISDR sequence to include the downstream NS5A sequence [254]. Further studies will be required to address this issue. In addition to its possible role in mediating interferon resistance, NS5A truncated at its amino-terminus has been shown to be a transcriptional transactivator [255,256]. As this truncated NS5A is not a natural HCV gene product, the significance of this transactivation function of NS5A is unclear.

Next to the NS5A sequence is the NS5B RNA polymerase sequence. NS5B expressed in *E. coli* or in insect cells using the baculovirus vector has been shown to possess an RNA-dependent RNA polymerase activity [257–260]. NS5B can bind to the 3′-coding region of the HCV RNA [261], and can also use the HCV genomic RNA as a template for RNA synthesis *in vitro* [257,260].

It is likely that the NS5B activity *in vivo* requires the interaction with cellular and other HCV proteins [231,235,262–265].

In addition to the polyprotein sequence, HCV also encodes another protein using an alternative reading frame that overlaps with the core protein coding sequence [266–268]. This protein is expressed by $-2/+1$ ribosomal frameshifting during the translation of the core protein sequence [267–269]. This protein has been named the F protein for "frameshifting" or the ARFP for "alternative reading frame protein" [266,267]. The length of the F protein or ARFP varies from 128 to 161 amino acids in length in a genotype-dependent manner [267]. For example, it is predominantly 162 amino acids in length for genotype 1a and 126 amino acids in length for gentoype 2a. The F protein is expressed during natural HCV infection [266–268]. It is a basic and labile protein associated with the ER [270]. Its function in the HCV life cycle remains unclear.

7.3.4 HCV-INDUCED HEPATOCELLULAR CARCINOMA

Approximately 85% of patients infected by HCV fail to clear the viral infection and become chronic carriers [159]. This is in contrast to HBV infection, which results in only 5% of chronic infection in adult patients. Why the immune system fails to clear the HCV infection in most of the patients is an interesting question that is not yet resolved. Recent studies with the HCV core protein indicate that it can modulate the immune system by interacting with members of the TNF receptor family [271–274]. HCV may establish chronic infection through this type of viral and cellular interactions.

Chronic HCV infection frequently leads to liver cirrhosis. It has been estimated that about 20% of HCV patients will require liver transplantation for survival within two decades from the onset of infection [159]. Chronic HCV patients also have an increased risk of developing hepatocellular carcinoma. The risk for the chronic HCV patients to develop HCC appears to be 1 to 5% after 20 years [159]. Depending on the geographical locations, between 6 to 75% of HCC patients are anti-HCV positive [275].

How HCV induces HCC remains unknown. It is possible that, similar to HBV, HCV induces HCC by the indirect mechanism of inducing chronic liver injury and regeneration. This is perhaps the reason why the onset of liver cirrhosis is correlated with an increased risk for the development of HCC [276]. It is also likely that some of the HCV gene products are oncogenic. For example, the HCV core protein has been shown to increase the oncogenic potential of primary cells [277,278], to enhance cell growth by activating the signaling pathway [279], and to induce hepatocellular carcinoma in one transgenic mouse study [280]. NS3 has also been shown to transform NIH3T3 cells [281]. Another HCV protein, the NS5A protein, which binds to PKR and suppresses its activity, is also antiapoptotic and can transform cells [282]. As mentioned above, the truncated NS5A protein possesses a transcriptional transactivator activity. This truncated protein has also been suggested to be a possible candidate responsible for the oncogenic process of HCC [255,256]. Note that while the HCV core protein has been shown to enhance cell growth, it has also been found to suppress cell growth in a p53-dependent manner and to sensitize cells to apoptosis [283,284]. These apparently opposite activities of the HCV core protein are reminiscent of the activities of the HBV X protein (see Section 7.2.3.3) and those of other oncogenes such as *myc* and *ras*. *Ras* can transform immortalized cells and can also activate p53 and $p21^{waf1/Cip1}$ expression to induce senescence in primary cells [285], and *myc* can promote cell growth and also sensitize cells to apoptosis [286].

It is conceivable that the observed increased risk of HCV patients to HCC is the combined effects of chronic liver injury and HCV gene functions.

7.4 CONCLUSION

HBV and HCV are two important human viral pathogens. These two viruses can both cause acute and chronic infections. The chronic infection by either HBV or HCV frequently leads to severe liver diseases including liver cirrhosis and HCC [287]. HBV and HCV may indirectly cause HCC by inducing chronic liver injury and regeneration, which has been shown to increase the risk for hepatocellular carcinogenesis. Alternatively, these two viruses may produce oncogenic viral gene products, which eventually cause hepatocellular transformation. While both HBV and HCV are the two major etiologic agents of HCC, it remains controversial regarding whether patients with dual HBV and HCV infection have a higher risk of developing HCC than patients with single HBV or HCV infection [288,289]. In addition to HBV and HCV, aflatoxin, cigarette smoking, heavy alcohol consumption, inorganic arsenic compounds, oral contraceptives, and a number of other factors are also considered HCC risk factors [290]. Epidemiology studies have indicated that HBV patients would further increase their risk for HCC if they are exposed to aflatoxin [291]. It is conceivable that all these environmental risk factors could interact with either HBV or HCV to facilitate the development of HCC in patients.

REFERENCES

1. Cockayne, E.A., Catarrhal jaundice, sporadic and epidemic, and its relation to acute yellow atrophy of the liver, *Q. J. Med.*, 6, 1, 1912.
2. Lurman, A., Eine icterusepidemie, *Berl. Klin. Wochenschr.*, 22, 20, 1885.
3. MacCallum, F.O., Homologous serum jaundice, *Lancet*, 2, 691, 1947.
4. Blumberg, B.S., Gerstley, B.J.S., Hungerford, D.A., London, W.T., and Sutnick, A.I., A serum antigen (Australia antigen) in Down's Syndrome, leukemia and hepatitis, *Ann. Intern. Med.*, 66, 924, 1967.
5. Dane, D.S., Cameron, C.H., and Briggs, M., Virus-like particles in serum of patients with Australia-antigen-associated hepatitis, *Lancet*, 1, 695, 1970.
6. Young, P., White House to expand response to infectious diseases, *ASM News*, 62, 450, 1996.
7. Galibert, F., Chen, T.-N., and Mandart, E., Nucleotide sequence of a cloned woodchuck hepatitis virus genome: Comparison with the hepatitis B virus sequence, *J. Virol.*, 41, 51, 1982.
8. Seeger, C., Ganem, D., and Varmus, H.E., Nucleotide sequence of an infectious molecularly cloned genome of ground squirrel hepatitis virus, *J. Virol.*, 51, 367, 1984.
9. Mason, W.S., Seal, G., and Summers, J., Virus of Pekin ducks with structural and biological relatedness to human hepatitis B virus, *J. Virol.*, 36, 829, 1980.
10. Sprengel, R., Kaleta, E.F., and Will, H. Isolation and characterization of a hepatitis B virus endemic in herons, *J. Virol.*, 62, 3832, 1988.
11. Lanford, R.E., Chavez, D., Brasky, K.M., Burns, R.B., 3rd, and Rico-Hesse, R., Isolation of a hepadnavirus from the woolly monkey, a New World primate, *Proc. Natl. Acad. Sci. USA*, 95, 5757, 1998.
12. Albin, C. and Robinson, W.S., Protein kinase activity in hepatitis B virus, *J. Virol.*, 34, 297, 1980.
13. Kann, M. and Gerlich, W.H., Effect of core protein phosphorylation by protein kinase C on encapsidation of RNA within core particles of hepatitis B virus, *J. Virol.*, 68, 7993, 1994.
14. Duclos-Vallee, J.C., Capel, F., Mabit, H., and Petit, M.A., Phosphorylation of the hepatitis B virus core protein by glyceraldehye-3-phosphate dehydrogenase protein kinase activity, *J. Gen. Virol.*, 79, 1665, 1998.
15. Kau, J.H. and Ting, L.P., Phosphorylation of the core protein of hepatitis B virus by a 46-kilodalton serine kinase, *J. Virol.*, 72, 3796, 1998.
16. Daub, H., Blencke, S., Habenberger, P., Kurtenbach, A., Dennenmoser, J., Wissing, J., Ullrich, A., and Cotton, M., Identification of SRPK1 and SRPK2 as the major cellular protein kinases phosphorylating hepatitis B virus core protein, *J. Virol.*, 76, 8124, 2002.
17. Bosch, V., Bartenschlager, R., Radziwill, G., and Schaller, H., The duck hepatitis B virus *P*-gene codes for protein strongly associated with the 5′-end of the viral DNA minus strand, *Virology*, 166, 475, 1988.
18. Heermann, K.H., Goldmann, U., Schwartz, W., Seyffarth, T., Baumgarten, H., and Gerlich, W.H., Large surface proteins of hepatitis B virus containing the Pre-S sequence, *J. Virol.*, 52, 396, 1984.
19. Eble, B.E., Lingappa, V.R., and Ganem, D., Hepatitis B surface antigen: An unusual secreted protein initially synthesized as a transmembrane polypeptide, *Mol. Cell. Biol.*, 6, 1454, 1986.

20. Eble, B.E., Macrae, D.R., Lingappa, V.R., and Ganem, D., Multiple topogenic sequences determine the transmembrane orientation of hepatitis B surface antigen, *Mol. Cell. Biol.*, 7, 3591, 1987.

21. Standring, D.N., Ou, J.H., and Rutter, W.J., Assembly of viral particles in Xenopus oocytes: Presurface-antigens regulate secretion of the hepatitis B viral surface envelope particle, *Proc. Natl. Acad. Sci. USA*, 83, 9339, 1986.

22. Ou, J.H. and Rutter, W.J. Regulation of secretion of the hepatitis B virus major surface antigen by the preS-1 protein, *J. Virol.*, 61, 782, 1987.

23. Bruss, V., Lu, X., Thomssen, R., and Gerlich, W.H., Post-translational alterations in transmembrane topology of the hepatitis B virus large envelope protein, *EMBO J.*, 13, 2273, 1994.

24. Ostapchuck, P., Hearing, P., and Ganem, D., A dramatic shift in the transmembrane topology of a viral envelope glycoprotein accompanies hepatitis B viral morphogenesis, *EMBO J.*, 13, 1048, 1994.

25. Prange, R. and Streeck, R.E., Novel transmembrane topology of the hepatitis B virus envelope proteins, *EMBO J.*, 14, 247, 1995.

26. Persing, D., Varmus, H., and Ganem, D., The preS1 protein of hepatitis B virus is acylated at its amino terminus with mysristic acid, *J. Virol.*, 61, 1672, 1987.

27. Bruss, V. and Ganem, D., The role of envelope proteins in hepatitis B virus assembly, *Proc. Natl. Acad. Sci. USA*, 88, 1059, 1991.

28. Macrae, D., Bruss, V., and Ganem, D., Myristylation of duck hepatitis B virus envelope protein is essential for infectivity but not for virus assembly, *Virology*, 181, 359, 1991.

29. Summers, J., Smith, P., Huang, M., and Yu, M., Morphogenetic and regulatory effects of mutations in the envelope proteins of an avian hepadnavirus, *J. Virol.*, 65, 1310, 1991.

30. Ou, J.H., Laub, O., and Rutter, W.J., Hepatitis B virus gene functions: The precore region targets the core antigen to cellullar membranes and causes the secretion of the e antigen., *Proc. Natl. Acad. Sci. USA*, 83, 1578, 1986.

31. Roossnick, M.J., Jameel, S., Loukin, S.H., and Siddiqui, A., Expression of hepatitis B viral core regions in mammalian cells, *Mol. Cell. Biol.*, 6, 1391, 1986.

32. Garcia, P.D., Ou, J.H., Rutter, W.J., and Walter, P., Targeting of the hepatitis B virus precore protein to the endoplasmic reticulum membrane: After signal peptide cleavage translocation can be aborted and the product released into the cytoplasm, *J. Cell Biol.*, 106, 1093, 1988.

33. Standring, D.N., Ou, J.H., Masiarz, F., and Rutter, W.J., A signal peptide encoded within the precore region of hepatitis B virus directs the secretion of a heterogeneous population of e antigen in Xenopus oocytes, *Proc. Natl. Acad. Sci. USA*, 85, 8405, 1988.

34. Carman, W., Thomas, H., and Domingo, E., Viral genetic variation: Hepatitis B virus as a clinical example, *Lancet*, 341, 349, 1993.

35. Ou, J.H., Molecular biology of hepatitis B virus e antigen, *J. Gastroenterol. Hepatol.*, 12 (Suppl.), S178, 1997.

36. Buckwold, V.E. and Ou, J.H., Hepatitis B virus C-gene expression and function: The lessons learned from viral mutants, *Curr. Top. Virol.*, 1, 71, 1999.

37. Okada, K., Kamiyama, I., Inomata, M., Miyakawa, Y., and Mayumi, M., e-Antigen and anti-e in the serum of asymptomatic carrier mothers as indicators of positive and negative transmission of hepatitis B virus to their infants, *N. Engl. J. Med.*, 294, 746, 1976.

38. Chen, H.-S., Kew, M.C., Hornbuckle, W.E., Tennant, B.C., Cote, P.J., Gerin, J.L., Purcell, R.H., and Miller, R.H., The precore gene of woodchuck hepatitis virus genome is not essential for viral replication in the natural host, *J. Virol.*, 66, 5682, 1992.

39. Liao, W.-Y. and Ou, J.H., Phosphorylation and nuclear localization of hepatitis B virus core protein: Significance of serine in the three repeated SPRRR motifs, *J. Virol.*, 69, 1025, 1995.

40. Lan, Y.T., Li, J., Liao, W.-Y., and Ou, J.H., Roles of the three major phosphorylation sites of hepatitis B virus core protein in viral replication, *Virology*, 259, 342, 1999.

41. Gazina, E.V., Fielding, J.E., Lin, B., and Anderson, D.A., Core protein phosphorylation modulates pregenomic RNA encapsidation to different extents in human and duck hepatitis B viruses, *J. Virol.*, 74, 4721, 2000.

42. Junker-Niepmann, M., Bartenschlager, R., and Schaller, H.A., A short cis-acting sequence is required for hepatitis B virus pregenome encapsidation and sufficient for packaging of foreign RNA, *EMBO J.*, 9, 3389, 1990.

43. Chiang, P.-W., Jeng, K.-S., Hu, C.-P., and Chang, C., Characterization of a cis element required for packaging and replication of the human hepatitis B virus, *Virology*, 186, 701, 1992.
44. Pollack, J. and Ganem, D., Site-specific RNA binding by a hepatitis B virus reverse transcriptase initiates two reactions: RNA packaging and DNA synthesis, *J. Virol.*, 68, 5579, 1993.
45. Bartenschlager, R. and Schaller, H., Hepadnaviral assembly is initiated by polymerase binding to the encapsidation signal in the viral RNA genome, *EMBO J.*, 11, 3413, 1992.
46. Wang, G.-H. and Seeger, C., The reverse transcriptase of hepatitis B virus acts as a protein primer for viral DNA synthesis, *Cell*, 71, 663, 1992.
47. Lanford, R.E., Notvall, L., Lee, H., and Beames, B., Transcomplementation of nucleotide priming and reverse transcription between independently expressed TP and RT domains of the hepatitis B virus reverse transcriptase, *J. Virol.*, 71, 2996, 1997.
48. Wang, G.H. and Seeger, C., Novel mechanism for reverse transcription in hepatitis B viruses, *J. Virol.*, 67, 6507, 1993.
49. Tavis, J.E., Perri, S., and Ganem, D., Hepadnavirus reverse transcription initiates within the stem-loop of the RNA packaging signal and employs a novel strand transfer, *J. Virol.*, 68, 3536, 1994.
50. Lien, J.M., Aldrich, C.E., and Mason, W.S., Evidence that a capped oligoribonucleotide is the primer for duck hepatitis B virus plus-strand DNA synthesis, *J. Virol.*, 57, 229, 1986.
51. Spandau, D.F. and Lee, C.H., Trans-activation of viral enhancers by the hepatitis B virus X protein, *J. Virol.*, 62, 427, 1988.
52. Seto, E., Yen, T.S.B., Peterlin, B.M., and Ou, J.H., Trans-activation of the human immunodeficiency virus long terminal repeat by the hepatitis B virus X protein, *Proc. Natl. Acad. Sci. USA*, 85, 8286, 1988.
53. Wang, H.D., Trivedi, A., and Johnson, D.L., Hepatitis B virus X protein induces RNA polymerase III-dependent gene transcription and increases cellular TATA-binding protein by activating the Ras signaling pathway, *Mol. Cell. Biol.*, 17, 6838, 1997.
54. Wang, H.D., Trivedi, A., and Johnson, D.L., Regulation of RNA polymerase I-dependent promoters by the hepatitis B virus X protein via activated Ras and TATA-binding protein, *Mol. Cell. Biol.*, 18, 7086, 1998.
55. Yen, T.S.B., Hepadnaviral X protein: Review of recent progress, *J. Biomed. Sci.*, 3, 20, 1996.
56. Seto, E., Mitchell, P.J., and Yen, T.S.B., Transactivation by the hepatitis B virus X protein depends on AP2 and other transcription factors, *Nature*, 344, 72, 1990.
57. Maguire, H.F., Hoeffler, J.P., and Siddiqui, A., HBV X protein alters the DNA binding specificity of CREB and ATF-2 by protein–protein interactions, *Science*, 252, 842, 1991.
58. Qadri, I., Maguire, H.F., and Siddiqui, A., Hepatitis B virus transactivator protein X interacts with the TATA-binding protein, *Proc. Natl. Acad. Sci. USA*, 92, 1003, 1995.
59. Li, J., Xu, Z., Zheng, Y., Johnson, D.L., and Ou, J.H., Regulation of hepatocyte nuclear factor-1 activity by the hepatitis B virus X protein and its frequent mutant, *J. Virol.*, 76, 5875, 2002.
60. Cheong, J.H., Yi, M., Lin, Y., and Murakami, S., Human RPB5, a subunit shared by eukaryotic nuclear RNA polymerases, binds human hepatitis B virus X protein and may play a role in X transactivation, *EMBO J.*, 14, 142, 1995.
61. Benn, J. and Schneider, R.J., Hepatitis B virus HBx protein activates Ras-GTP complex formation and establishes a Ras, Raf, MAP kinase signaling cascade, *Proc. Natl. Acad. Sci. USA*, 91, 10350, 1994.
62. Natoli, G., Avantaggiati, M.L., Chirillo, P. Puri, P.L., Iani, A., Balsano, C., and Levrero, M., Ras- and Raf-dependent activation of c-jun transcriptional activity by the hepatitis B virus transactivator pX, *Oncogene*, 9, 2837, 1994.
63. Twu, J.S., Rosen, C.A., Haseltine, W.A., and Robinson, W.S., Identification of a region within the human immunodeficiency virus type 1 long terminal repeat that is essential for transactivation by the hepatitis B virus gene X, *J. Virol.*, 63, 2857, 1989.
64. Lucito, R. and Schneider, R.J., Hepatitis B virus X protein activates transcription factor NF-kappa B without a requirement for protein kinase C, *J. Virol.*, 66, 983, 1992.
65. Bouchard, M.J., Wang, L.H., and Schneider, R.J., Calcium signaling by HBx protein in hepatitis B virus DNA replication, *Science*, 294, 2376, 2001.
66. Kekule, A.S., Lauer, U., Weiss, L., Luber, B., and Hofschneider, P.H., Hepatitis B virus transactivator HBx uses a tumour promoter signalling pathway, *Nature*, 361, 742, 1993.

67. Murakami, S., Cheong, J., Ohno, S., Matsushima, K., and Kaneko, S., Transactivation of human hepatitis B virus X protein, HBx, operates through a mechanism distinct from protein kinase C and okadaic acid activation pathways, *Virology*, 199, 243, 1994.
68. Fischer, M., Runkel, L., and Schaller, H., HBx protein of hepatitis B virus interacts with the C-terminal portion of a novel human proteasome alpha-subunit, *Virus Genes*, 10, 99, 1995.
69. Huang, J., Kwong, J., Sun, E.C., and Liang, T.J., Proteasome complex as a potential cellular target of hepatitis B virus X protein, *J. Virol.*, 70, 5582, 1996.
70. Blum, H.E., Zhang, Z.-S., Galun, E., von Weizsacker, F., Garner, B., Liang, T.J., and Wands, J.R., Hepatitis B virus X protein is not central to the viral life cycle in vitro, *J. Virol.*, 66, 1223, 1992.
71. Zoulim, F., Saputelli, J., and Seeger, C., Woodchuck hepatitis virus X protein is required for viral infection in vivo, *J. Virol.*, 68, 2806, 1993.
72. Chen, H.-S., Kaneko, S., Girones, R., Anderson, R.W., Hornbuckle, W.E., Tennant, B.C., Cote, P.J., Gerin, J.L., Purcell, R.H., and Miller, R.H., The woodchuck hepatitis virus X gene is important for establishment of virus infection in woodchucks, *J. Virol.*, 67, 1218, 1993.
73. Xu, Z., Yen, T.S.B., Madden, C.R., Wu, L., Tan, W., Slagle, B.L., and Ou, J.H., Enhancement of hepatitis B virus replication by its X protein in transgenic mice, *J. Virol.*, 76, 2579, 2002.
74. Zhang, Z., Torii, N., Hu, Z., Jacob, J., and Liang, T.J., X-deficient woodchuck hepatitis virus mutants behave like attenuated viruses and induce protective immunity in vivo, *J. Clin. Invest.*, 108, 1523, 2001.
75. Thung, S.N. and Gerber, M.A., Hepatitis B virus and polyalbumin binding sites, *Gastroenterology*, 85, 466, 1983.
76. Komai, K., Kaplan, M., and Peeles, M.E., The Vero cell receptor for the hepatitis B virus small S protein is a sialoglycoprotein, *Virology*, 163, 629, 1988.
77. Pontisso, P., Petit, M.A., Bankowski, M.J., and Peeples, M.E., Human liver plasma membranes contain receptors for the hepatitis B virus pre-S1 region and, via polymerized human serum albumin, for the pre-S2 region, *J. Virol.*, 63, 1981, 1989.
78. Franco, A., Paroli, M., Tresta, U., Benvenuto, R., Peschle, C., Balsano, F., and Barnaba, V., Transferrin receptor mediates uptake and presentation of hepatitis B envelope antigen by T lymphocytes, *J. Exp. Med.*, 175, 1095, 1992.
79. Neurath, A.R., Strick, N., and Li, Y., Cells transfected with human IL6 cDNA acquire binding sites for the hepatitis B virus envelope protein, *J. Exp. Med.*, 176, 1561, 1992.
80. Petit, M.A., Capel, F., Dubanchet, S., and Mabit, H., PreS1-specific binding proteins as potential receptors for hepatitis B virus in human hepatocytes, *Virology*, 187, 211, 1992.
81. Pontisso, P., Ruoletto, M., Tiribelli, C., Gerlich, W., Ruol, A., and Alberti, A., The preS1 domain of hepatitis B virus and IgA cross-react in their binding to the hepatocyte surface, *J. Gen. Virol.*, 73, 2041, 1992.
82. Budkowska, A., Quan, C., Groh, F., Bedossa, P., Dubreuil, P., Bouvet, J.P., and Pillot, J., Hepatitis B virus (HBV) binding factor in human serum: Candidate for a soluble form of hepatocyte HBV receptor, *J. Virol.*, 67, 4316, 1993.
83. Hertogs, K., Leenders, W.P., Depla, E., De Bruin, W.C., Meheus, L., Raymackers, J., Moshage, H., and Yap, S.H., Endonexin II, present on human liver plasma membranes, is a specific binding protein of small hepatitis B virus (HBV) envelope protein, *Virology*, 197, 549, 1993.
84. Treichel, U., Meyer zum Buschenfelde, K.H., Stockert, R.J., Poralla, T., and Gerken, G., The asialogly-coprotein receptor mediates hepatic binding and uptake of natural hepatitis B virus particles derived from viraemic carriers, *J. Gen. Virol.*, 75, 3021, 1994.
85. Mehdi, H., Yang, X., and Peeples, M.E., An altered form of apolipoprotein H binds hepatitis B virus surface antigen most efficiently, *Virology*, 217, 58, 1996.
86. Treichel, U., Meyer, zum Buschenfelde, K.H., Dienes, H.P., and Gerken, G., Receptor-mediated entry of hepatitis B virus particles into liver cells, *Arch. Virol.*, 142, 493, 1997.
87. Breiner, K.M., Urban, S., and Schaller, H., Carboxypeptidase D (gp180), a Golgi-resident protein, functions in the attachment and entry of avian hepatitis B viruses, *J. Virol.*, 72, 8098, 1998.
88. Eng, F.J., Novikova, E.G., Kuroki, K., Ganem, D., and Fricker, L.D., gp180, a protein that binds duck hepatitis B virus particles, has metallocarboxypeptidase D-like enzymatic activity, *J. Biol. Chem.*, 273, 8382, 1998.

89. Urban, S., Breiner, K.M., Fehler, F., Klingmuller, U., and Schaller, H., Avian hepatitis B virus infection is initiated by the interaction of a distinct pre-S subdomain with the cellular receptor gp180, *J. Virol.*, 72, 8089, 1998.

90. Bruns, M., Miska, S., Chassot, S., and Will, H., Enhancement of hepatitis B virus infection by noninfectious subviral particles, *J. Virol.*, 72, 1462, 1998.

91. White, J., Matlin, K., and Helenius, A., Cell fusion by Semliki Forest, influenza, vesicular stomatitis viruses, *J. Cell Biol.*, 89, 674, 1981.

92. Rigg, R.J. and Schaller, H., Duck hepatitis B virus infection of hepatocytes is not dependent on low pH, *J. Virol.*, 66, 2829, 1992.

93. Tuttleman, J., Pourcel, C., and Summers, J., Formation of the pool of covalently closed circular viral DNA in hepadnavirus-infected cells, *Cell*, 47, 451, 1986.

94. Yeh, C.-T., Liaw, Y.-F., and Ou, J.H., The arginine-rich domain of hepatitis B virus precore and core proteins contains a signal for nuclear transport, *J. Virol.*, 64, 6141, 1990.

95. Eckhardt, S.G., Milich, D.R., and McLachlan, A., Hepatitis B virus core antigen has two nuclear localization sequences in the arginine-rich caboxyl terminus, *J. Virol.*, 65, 575, 1991.

96. Kann, M., Sodeik, B., Vlachou, A., Gerlich, W.H., and Helenius, A., Phosphorylation-dependent binding of hepatitis B virus core particles to the nuclear pore complex, *J. Cell Biol.*, 145, 45, 1999.

97. Yen, T.S.B., Regulation of hepatitis B virus gene expression, *Semin. Virol.*, 4, 33, 1993.

98. Yen, T.S.B., Posttranscriptional regulation of gene expression in hepadnaviruses, *Semin. Virol.*, 8, 319, 1998.

99. Chen, M., Hieng, S., Qian, X., Costa, R., and Ou, J.H., Regulation of hepatitis B virus ENI enhancer activity by hepatocyte-enriched transcription factor HNF3, *Virology*, 205, 127, 1994.

100. Ori, A., and Shaul, Y., Hepatitis B virus enhancer binds and is activated by the hepatocyte nuclear factor 3, *Virology*, 207, 98, 1995.

101. Kosovsky, M.J., Huan, B., and Siddiqui, A., Purification and properties of rat liver nuclear proteins that interact with the hepatitis B virus enhancer 1, *J. Biol. Chem.*, 271, 21859, 1996.

102. Li, M., Xie, Y., Wu, X., Kong, Y., and Wang, Y., HNF3 binds and activates the second enhcner, ENII, of hepatitis B virus, *Virology*, 214, 371, 1995.

103. Guo, W., Chen, M., Yen, T.S.B., and Ou, J.H., Hepatocyte-specific expression of the hepatitis B virus core promoter depends on both positive and negative regulation, *Mol. Cell. Biol.*, 12, 443, 1993.

104. Lopez-Cabrera, M., Letovsky, J., Hu, K.Q., and Siddiqui, A., Transcriptional factor C/EBP binds to and transactivates the enhancer element II of the hepatitis B virus, *Virology*, 183, 825, 1991.

105. Yuh, C.H. and Ting, L.P., C/EBP-like proteins bindings to the functional box-alpha and box-beta of the second enhancer of hepatitis B virus, *Mol. Cell. Biol.*, 11, 5044, 1991.

106. Antonucci, T. and Rutter, W.J., Hepatitis B virus promoters are regulated by the HBV enhancer in a tissue-specific manner, *J. Virol.*, 63, 579, 1989.

107. Su, H. and Yee, J.K., Regulation of hepatitis B virus gene expression by its two enhancers, *Proc. Natl. Acad. Sci. USA*, 89, 2708, 1992.

108. Russnak, R. and Ganem, D., Sequences 5′ to the poly A signal mediate differential poly A site use in hepatitis B viruses, *Genes Dev.*, 4, 764, 1990.

109. Guo, W., Wang, J., Tam, G., Yen, T.S.B., and Ou, J.H., Leaky transcription termination produces larger and smaller than genome size hepatitis B virus X gene transcripts, *Virology*, 181, 630, 1991.

110. Cherrington, J., Russnak, R., and Ganem, D., Upstream sequences and cap proximity in the regulation of polyadenylation in ground squirrel hepatitis virus, *J. Virol.*, 66, 7589, 1992.

111. Huang, J. and Liang, T.J., A novel hepatitis B virus (HBV) genetic element with Rev response element-like properties that is essential for expression of HBV gene products, *Mol. Cell. Biol.*, 13, 7476, 1993.

112. Huang, Z.M. and Yen, T.S.B., Hepatitis B virus RNA element that facilitates accumulation of surface gene transcripts in the cytoplasm, *J. Virol.*, 68, 3193, 1994.

113. Huang, Z.M. and Yen, T.S.B., Role of the hepatitis B virus posttranscriptional regulatory element in export of intronless transcripts, *Mol. Cell. Biol.*, 15, 3864, 1995.

114. Fouillot, N., Tlouzeau, S., Rossignol, J.M., and Jean-Jean, O., Translation of the hepatitis B virus P gene by ribosomal scanning as an alternative to internal initiation, *J. Virol.*, 67, 4886, 1993.

115. Fouillot, N. and Rossignol, J.M., Translational stop codons in the precore sequence of hepatitis B virus pre-C RNA allow translation reinitiation at downstream AUGs, *J. Gen. Virol.*, 77, 1123, 1996.

116. Hwang, W.L. and Su, T.S., Translational regulation of hepatitis B virus polymerase gene by termination–reinitiation of an upstream minicistron in a length-dependent manner, *J. Gen. Virol.*, 79, 2181, 1998.

117. Ou, J.H., Bao, H., Shih, C., and Tahara, S.M., Preferred translation of hepatitis B virus polymerase from core protein but not precore protein-specific transcript, *J. Virol.*, 64, 4578, 1990.

118. Enders, G.H., Ganem, D., and Varmus, H.E., 5′-Terminal sequences influence the segregation of ground squirrel hepatitis virus RNAs into polyribosomes and viral core particles, *J. Virol.*, 61, 35, 1987.

119. Nassal, M., Junker-Niepmann, M., and Schaller, H., Translational inactivation of RNA function: Discrimination againt a subset of genomic transcripts during HBV nucleocapsid assembly, *Cell*, 63, 1357, 1990.

120. Summers, J. and Mason, W.S., Replication of the genome of a hepatitis B-like virus by reverse transcription of an RNA intermediate, *Cell*, 29, 403, 1982.

121. Bruss, V., A short linear sequence in the pre-S domain of the large hepatitis B virus envelope protein required for virion formation, *J. Virol.*, 71, 9350, 1997.

122. Poisson, F., Severac, A., Hourioux, C., Goudeau, A., and Roingeard, P., Both pre-S1 and S domains of hepatitis B virus envelope proteins interact with the core particle, *Virology*, 228, 115, 1997.

123. Le Seyec, J., Chouteau, P., Cannie, I., Guguen-Guillouzo, C., and Gripon, P., Role of the pre-S2 domain of the large envelope protein in hepatitis B virus assembly and infectivity, *J. Virol.*, 72, 5573, 1998.

124. Le Seyec, J., Chouteau, P., Canie, I., Guguen-Guillouzo, C., and Gripon, P., Infection process of the hepatitis B virus depends on the presence of a defined sequence in the pre-S1 domain, *J. Virol.*, 73, 2052, 1999.

125. Kamimura, T., Yoshikawa, A., Ichida, F., and Sasaki, H., Electron microscopic studies of Dane particles in hepatocytes with special reference to intracellular development of Dane particles and their relation with HBeAg in serum, *Hepatology*, 1, 392, 1981.

126. Yamada, G., Sakamoto, Y., Mizuno, M., Nishihara, T., Kobayashi, T., Takahashi, T., and Ngashima, H., Electron and immunoelectron microscopic study of Dane particle formation in chronic hepatitis B virus infection, *Gastroenterology*, 83, 348, 1982.

127. Szmuness, W., Hepatocellular carcinoma and the hepatitis B virus: Evidence for a causal association, *Prog. Med. Virol.*, 24, 40, 1978.

128. Beasley, R.P., Hwang, L.Y., Lin, C.C., and Chien, C.S., Hepatocellular carcinoma and hepatitis B virus, a prospective study of 22,707 men in Taiwan, *Lancet*, 2, 1129, 1981.

129. Beasley, R.P. and Hwang, L.-Y., Epidemiology of hepatocellular carcinoma, in *Viral Hepatitis and Liver Disease*, Vyas, G.N., Ed., Grune and Stratton, New York, 1984, p. 209.

130. Chang, M.H., Chen, C.J., Lai, M.S., Hsu, H.M., Wu, T.C., Kong, M.S., Liang, D.C., Shau, W.Y., and Chen, D.S., Universal hepatitis B vaccination in Taiwan and the incidence of hepatocellular carcinoma in children *New Eng. J. Med.*, 336, 1855, 1997.

131. Zuckerman, A.J., Prevention of primary liver cancer by immunization, *New Eng. J. Med.*, 336, 1906, 1997.

132. Pineau, P., Marchio, A., Mattei, M.G., Kim, W.H., Youn, J.K., Tiollais, P., and Dejean, A., Extensive analysis of duplicated-inverted hepatitis B virus integrations in human hepatocellular carcinoma, *J. Gen. Virol.*, 79, 591, 1998.

133. Shih, C., Burke, K., Chou, M.J., Zeldis, J.B., Yang, C.S., Lee, C.S., Isselbacher, K.J., Wands, J.R., and Goodman, H.M., Tight clustering of human hepatitis B virus integration sites in hepatomas near a triple-stranded region *J. Virol.*, 61, 3491, 1987.

134. Edman, J.C., Gray, P., Valenzuela, P., Rall, L.B., and Rutter, W.J., Integration of hepatitis B virus sequences and their expression in a human hepatoma cell, *Nature*, 286, 535, 1980.

135. Shaul, Y., Ziemer, M., Garcia, P.D., Crawford, R., Hsu, H., Valenzuela, P., and Rutter, W.J., Cloning and analysis of integrated hepatitis virus sequences from a human hepatoma cell line, *J. Virol.*, 51, 776, 1984.

136. Ziemer, M., Garcia, P., Shaul, Y., and Rutter, W.J., Sequence of hepatitis B virus DNA incorporated into the genome of a human hepatoma cell line, *J. Virol.*, 53, 885, 1985.

137. Fourel, G., Trepo, C., Bougueleret, L., Henglein, B., Ponzetto, A., Tiollais, P., and Buendia, M.A., Frequent activation of N-myc genes by hepadnavirus insertion in woodchuck liver tumours, *Nature*, 347, 294, 1990.

138. Hansen, L.J., Tennant, B.C., Seeger, C., and Ganem, D., Differential activation of myc gene family members in hepatic carcinogenesis by closely related hepatitis B viruses, *Mol. Cell. Biol.*, 13, 659, 1993.

139. Etiemble, J., Degott, C., Renard, C.A., Fourel, G., Shamoon, B., Vitvitski-Trepo, L., Hsu, T.Y., Tiollais, P., Babinet, C., and Buendia, M.A., Liver-specific expression and high oncogenic efficiency of a c-myc transgene activated by woodchuck hepatitis virus insertion, *Oncogene*, 9, 727, 1994.

140. Ou, J.H. and Rutter, W.J., Hybrid hepatitis B virus-host transcripts in a human hepatoma cell, *Proc. Natl. Acad. Sci. USA*, 82, 83, 1995.

141. Tokino, T., and Matsubara, K., Chromosomal sites for hepatitis B virus integration in human hepatocellular carcinoma, *J. Virol.*, 65, 6761, 1991.

142. Dejean, A., Bouguelleret, L., Grzeschick, K., and Tiollais, P., Hepatitis B virus DNA integration in a sequence homologous to v-erbA and steroid receptor genes in a hepatocellular carcinoma, *Nature*, 322, 70, 1986.

143. Wang, J., Chenivesse, X., Henglein, B., and Brechot, C., Hepatitis B virus integration in a cyclin A gene in a hepatocellular carcinoma, *Nature*, 343, 555, 1990.

144. Seifer, M., Hohne, M., Schaefer, S., and Gerlich, W.H., In vitro tumorigenicity of hepatitis B virus DNA and HBx protein, *J. Hepatol.*, 13 (Suppl. 4), S61, 1991.

145. Kim, C.-H., Koike, K., Saito, I., Miyamura, T., and Jay, G., HBx gene of hepatitis B virus induces liver cancer in transgenic mice, *Nature*, 353, 317, 1991.

146. Slagle, B.L., Lee, T.H., Medina, D., Finegold, M.J., and Butel, J.S., Increased sensitivity to the hepatocarcinogen diethylnitrosamine in transgenic mice carrying the hepatitis B virus X gene, *Mol. Carcinog.*, 15, 261, 1996.

147. Terradillos, O., Billet, O., Renard, C.A., Levy, R., Molina, T., Briand, P., and Buendia, M.A., The hepatitis B virus X gene potentiates c-myc-induced liver oncogenesis in transgenic mice, *Oncogene*, 14, 395, 1997.

148. Becker, S.A., Lee, T.H., Butel, J.S., and Slagle, B.L., Hepatitis B virus X protein interferes with cellular DNA repair, *J. Virol.*, 72, 266, 1998.

149. Takada, S., Kaneniwa, N., Tsuchida, N., and Koike, K., Cytoplasmic retention of the p53 tumor suppressor gene product is observed in the hepatitis B virus X gene-transfected cells, *Oncogene*, 15, 1895, 1997.

150. Elmore, L.W., Hancock, A.R., Chang, S.F., Wang, X.W., Chang, S., Callahan, C.P., Geller, D.A., Will, H., and Harris, C.C., Hepatitis B virus X protein and p53 tumor suppressor interactions in the modulation of apoptosis, *Proc. Natl. Acad. Sci. USA*, 94, 14707, 1997.

151. Chirillo, P., Pagano, S., Natoli, G., Puri, P.L., Burgio, V.L., Balsano, C., and Levrero, M., The hepatitis B virus X gene induces p53-mediated programmed cell death, *Proc. Natl. Acad. Sci. USA*, 94, 8162, 1997.

152. Pollicino, T., Terradillos, O., Lecoeur, H., Gougeon, M.L., and Buendia, M.A., Pro-apoptotic effect of the hepatitis B virus X gene, *Biomed. Pharmacother.*, 52, 363, 1998.

153. Kekule, A.S., Lauer, U., Meyer, M., Caselmann, W.H., Hofschneider, P.H., and Koshy, R., The preS2/S region of integrated hepatitis B virus DNA encodes a transcriptional transactivator, *Nature*, 343, 457, 1990.

154. Hildt, E. and Hofschneider, P.H., The PreS2 activators of the hepatitis B virus: Activators of tumour promoter pathways, *Recent Results Cancer Res.*, 154, 315, 1998.

155. Hildt, E., Munz, B., Saher, G., Reifenberg, K., and Hofschneider, P.H., The PreS2 activator MHBs(t) of hepatitis B virus activates c-raf-1/Erk2 signaling in transgenic mice, *EMBO J.*, 21, 525, 2002.

156. Nakamoto, Y., Guidotti, L.G., Kuhlen, C.V., Fowler, P., and Chisari, F.V., Immune pathogenesis of hepatocellular carcinoma, *J. Exp. Med.*, 188, 341, 1998.

157. Sawaki, M., Hattori, A., Tsuzuki, N., Sugawara, N., Enomoto, K., Sawada, N., and Mori, M., Chronic liver injury promotes hepatocarcinogenesis of the LEC rat, *Carcinogenesis*, 19, 331, 1998.

158. Edwards, J.E., Hepatomas in mice induced with carbon tetrachloride, *J. Natl. Cancer Inst.*, 2, 197, 1941.

159. Management of hepatitis C, *NIH consensus statement*, 2002.

160. Choo, Q.-L., Kuo, G., Weiner, A.J., Overby, L.R., Bradley, D.W., and Houghton, M., Isolation of a cDNA clone derived from a blood-borne non-A, non-B viral hepatitis genome, *Science*, 244, 359, 1989.

161. Choo, Q.L., Richman, K.H., Han, J.H., Berger, K., Lee, C., Dong, C. Gallegos, C., Coit, D., Medina-Selby, A., Barr, P.J., Weiner, A.J., Bradley, D.W., Kuo, G., and Houghton, M., Genetic organization and diversity of the hepatitis C virus, *Proc. Natl. Acad. Sci. USA*, 88, 2451, 1991.

162. Bradley, D.W., Maynard, J.E., Popper, H., Cook, E.H., Ebert, J.W., McCaustland, K.A., Schable, C.A., and Fields, H.A., Posttransfusion non-A, non-B hepatitis: Physicochemical properties of two distinct agents, *J. Infect. Dis.*, 148, 254, 1983.

163. Feinstone, S.M., Mihalik, K.B., Kamimura, T., Alter, H.J., London, W.T., and Purcell, R.H., Inactivation of hepatitis B virus and non-A, non-B hepatitis by chloroform, *Infect. Immun.*, 41, 816, 1983.

164. Simmonds, P., Holmes, E.C., Cha, T.A., Chan, S.W., McOmish, F., Irvine, B., Beall, E., Yap, P.L., Kolberg, J., and Urdea, M.S., Classification of hepatitis C virus into six major genotypes and a series of subtypes by phylogenetic analysis of the NS-5 region, *J. Gen. Virol.*, 74, 2391, 1993.

165. Smith, D.B. and Simmonds, P., Hepatitis C virus: Types, subtypes and beyond, *Methods Mol. Med.*, 19, 133, 1998.

166. Kuo, G., Choo, Q.L., Alter, H.J., Gitnick, G.L., Redeker, A.G., Purcell, R.H., Miyamura, T., Dienstag, J.L., Alter, M.J., Stevens, C.E., Tegtmeier, G.E., Bonino, F., Colombo, M., Lee, W.-S., Kuo, C., Berger, K., Shuster, J.R., Overby, L.R., Bradley, D.W., and Houghton, M., An assay for circulating antibodies to a major etiologic virus of human non-A, non-B hepatitis, *Science*, 244, 362, 1989.

167. McHutchison, J.G., Person, J.L., Govindarajan, S., Valinluck, B., Gore, T., Lee, S.R., Nelles, M., Polito, A., Chien, D., and DiNello, R. Improved detection of hepatitis C virus antibodies in high-risk populations, *Hepatology*, 15, 19, 1992.

168. Lohmann, V., Korner, F., Koch, J.-O., Herian, U., Theilmann, L., and Bartenschlager, R., Replication of subgenomic hepatitis C virus RNAs in a hepatoma cell line, *Science*, 285, 110, 1999.

169. Blight, K.J., Kolykhalov, A.A., and Rice, C.M., Efficient initiation of HCV RNA replication in cell culture, *Science*, 290, 1972, 2000.

170. Wang, C., Sarnow, P., and Siddiqui, A., A conserved helical element is essential for internal initiation of translation of hepatitis C virus RNA, *J. Virol.*, 68, 7301, 1994.

171. Honda, M., Ping, L.H., Rijnbrand, R.C., Amphlett, E., Clarke, B., Rowlands, D., and Lemon, S.M., Structural requirements for initiation of translation by internal ribosome entry within genome-length hepatitis C virus RNA, *Virology*, 222, 31, 1996.

172. Lu, H.H. and Wimmer, E., Poliovirus chimeras replicating under the translational control of genetic elements of hepatitis C virus reveal unusual properties of the internal ribosomal entry site of hepatitis C virus, *Proc. Natl. Acad. Sci. USA*, 93, 1412, 1996.

173. Hwang, L.H., Hsieh, C.L., Yen, A., Chung, Y.L., and Chen, D.S., Involvement of the 5′ proximal coding sequences of hepatitis C virus with internal initiation of viral translation, *Biochem. Biophys. Res. Commun.*, 252, 455, 1998.

174. Honda, M., Beard, M.R., Ping, L.H., and Lemon, S.M., A phylogenetically conserved stem-loop structure at the 5′ border of the internal ribosome entry site of hepatitis C virus is required for cap-independent viral translation, *J. Virol.*, 73, 1165, 1999.

175. Tanaka, T., Kato, N., Cho, M.-J., Sugiyama, K., and Shimotohno, K., Structure of the 3′-terminus of the hepatitis C virus genome, *J. Virol.*, 70, 3307, 1996.

176. Blight, K.J. and Rice, C.M., Secondary structure determination of the conserved 98-base sequence at the 3′ terminus of hepatitis C virus genome RNA, *J. Virol.*, 71, 7345, 1997.

177. Yanagi, M., St Claire, M., Emerson, S.U., Purcell, R.H., and Bukh, J., In vivo analysis of the 3′ untranslated region of the hepatitis C virus after *in vitro* mutagenesis of an infectious cDNA clone, *Proc. Natl. Acad. Sci., USA*, 96, 2291, 1999.

178. Liu, Q., Tqackney, C., Bhat, R.A., Prince, A.M., and Zhang, P., Regulated processing of hepatitis C virus core protein is linked to subcellular localization, *J. Virol.*, 71, 657, 1997.

179. Kunkel, M., Lorinczi, M., Rijnbrand, R., Lemon, S.M., and Watowich, S.J., Self-assembly of nucleocapsid-like particles from recombinant hepatitis C virus core protein, *J. Virol.*, 75, 2119, 2001.

180. Kunkel, M. and Watowich, S.J., Conformational changes accompanying self-assembly of the hepatitis C virus core protein, *Virology*, 294, 239, 2002.

181. Matsumoto, M., Hwang, S.B., Jeng, K.S., Zhu, N., and Lai, M.M., Homotypic interaction and multimerization of hepatitis C virus core protein, *Virology*, 218, 43, 1996.

182. Lo, S.-Y. and Ou, J.H., Expression and dimerization of hepatitis C virus core protein in *E. coli*, *Methods Mol. Med.*, 19, 325, 1998.

183. Choi, J., Lu, W., and Ou, J.H., Structure and functions of hepatitis C virus core protein, *Recent Res. Devel. Virol.*, 3, 105, 2001.

184. Yasui, K., Wakita, T., Tsukiyama-Kohara, K., Funahashi, S.-I., Ichikawa, M., Kajita, T., Moradpour, D., Wands, J.R., and Kohara, M. The native form and maturation process of hepatitis C virus core protein, *J. Virol.*, 72, 6048, 1998.

185. Grakoui, A., Wychowski, C., Lin, C., Feinstone, S.M., and Rice, C.M., Expression and identification of hepatitis C virus polyprotein cleavage products. *J. Virol.*, 67, 1385, 1993.

186. Hijikata, M., Kato, N., Ootsuyama, Y., Kakagawa, M., and Shimotohno, K., Gene mapping of the putative structural region of the hepatitis C virus genome by in vitro processing analysis, *Proc. Natl. Acad. Sci. USA*, 88, 5547, 1991.

187. Meunier, J.C., Fournillier, A., Choukhi, A., Cahour, A., Cocquerel, L., Dubuisson, J., and Wychowski, C., Analysis of the glycosylation sites of hepatitis C virus (HCV) glycoprotein E1 and the influence of E1 glycans on the formation of the HCV glycoprotein complex, *J. Gen. Virol.*, 80, 887, 1999.

188. Dubuisson, J. and Rice, C.M., Hepatitis C virus glycoprotein folding: Disulfide bond formation and association with calnexin, *J. Virol.*, 70, 778, 1996.

189. Matsuura, Y., Suzuki, T., Suzuki, R., Sato, M., Aizaki, H., Saito, I., and Miyamura, T., Processing of E1 and E2 glycoproteins of hepatitis C virus expressed in mammalian and insect cells, *Virology*, 205, 141, 1994.

190. Ralston, R., Thudium, K., Berger, K., Kuo, C., Bervase, B., Hall, J., Selby, M., Kuo, G., Houghton, M., and Choo, Q.-L., Characterization of hepatitis C virus envelope glycoprotein complexes expressed by recombinant vaccinia viruses, *J. Virol.*, 67, 6753, 1993.

191. Lo, S.-Y., Selby, M.J., and Ou, J.H., Interaction between hepatitis C virus core protein and E1 envelope protein, *J. Virol.*, 70, 5177, 1996.

192. Pileri, P., Uematsu, Y., Campagnoli, S., Galli, G., Falugi, F., Petracca, R., Weiner, A.J., Houghton, M., Rosa, D., Grandi, G., and Abrignani, S., Binding of hepatitis C virus to CD81, *Science*, 282, 938, 1998.

193. Masciopinto, F., Freer, G., Burgio, V.L., Levy, S., Galli-Stampino, L., Bendinelli, M., Houghton, M., Abrignani, S., and Uematsu, Y., Expression of human CD81 in transgenic mice does not confer susceptibility to hepatitis C virus infection, *Virology*, 304, 187, 2002.

194. Griffin, S.D., Beales, L.P., Clarke, D.S., Worsfold, O., Evans, S.D., Jaeger, J., Harris, M.P., and Rowlands, D.J., The p7 protein of hepatitis C virus forms an ion channel that is blocked by the antiviral drug, Amantadine, *FEBS Lett.*, 535, 34, 2003.

195. Pavlovic', D., Neville, D.C., Argaud, O., Blumberg, B., Dwek, R.A., Fischer, W.B., and Zitzmann, N., The hepatitis C virus p7 protein forms an ion channel that is inhibited by long-alkyl-chain iminosugar derivatives, *Proc. Natl. Acad. Sci. USA*, 100, 6104, 2003.

196. Lin, C., Lindenbach, B.D., Pragai, B.M., McCourt, D.W., and Rcie, C.M., Processing in the hepatitis C virus E2-NS2 region: Identification of p7 and two distinct E2-specific products with different termini, *J. Virol.*, 68, 5063, 1994.

197. Mizushima, H., Hijikata, M., Asabe, S.-I., Hirota, M., Kimura, K., and Shimotohno, K., Two hepatitis C virus glycoprotein E2 products with different termini, *J. Virol.*, 68, 6214, 1994.

198. Mizushima, H., Hijikata, M., Tanji, Y., Kimura, K., and Shimotohno, K., Analysis of N-terminal processing of hepatitis C virus nonstructural protein NS2, *J. Virol.*, 68, 2731, 1994.

199. Santolini, E., Pacini, L., Fipaldini, C., Migliccio, G., and La Monica, N., The NS2 protein of hepatitis C virus is a transmembrane polypeptide, *J. Virol.*, 69, 7461, 1995.

200. Yamaga, A.K. and Ou, J.H., Membrane topology of the hepatitis C virus NS2 protein, *J. Biol. Chem.*, 277, 33228, 2002.

201. Grakoui, A., McCourt, D.W., Wychowski, C., Feinstone, S.M., and Rice, C.M., A second hepatitis C virus-encoded protease, *Proc. Natl. Acad. Sci. USA*, 90, 10583, 1993.

202. Hijikata, M., Mizushima, H., Akagi, T., Mori, S., Kakiuchi, N., Kato, N., Tanaka, T., Kimura, K., and Shimotohno, K., Two distinct proteinase activities required for the processing of a putative nonstructural precursor protein of hepatitis C virus, *J. Virol.*, 67, 4665, 1993.

203. Bartenschlager, R., Ahlborn-Laaker, L., Mous, J., and Jacobsen, H., Nonstructural protein 3 of the hepatitis C virus encodes a serine type proteinase required for cleavage at the NS3/4 and NS4/5 junctions, *J. Virol.*, 67, 3835, 1993.

204. Eckart, M.R., Selby, M., Masiarz, F., Lee, C., Berger, K., Crawford, K., Kuo, G., Houghton, M., and Choo, Q.-L., The hepatitis C virus encodes a serine protease involved in processing of the putative nonstructural proteins from the viral polyprotein precursor, *Biochem. Biophys. Res. Commun.*, 192, 399, 1993.

205. Grakoui, A., McCourt, D.W., Wychowski, C., Feinstone, S.M., and Rice, C.M., Characterization of the hepatitis C virus-encoded serine proteinase: Determination of proteinase-dependent polyprotein cleavage sites, *J. Virol.*, 67, 2832, 1993.

206. Tomei, L., Failla, C., Santolini, E., De Francesco, R., and La Monica, N., NS3 is a serine protease required for processing of hepatitis C virus polyprotein, *J. Virol.*, 67, 4017, 1993.

207. Manabe, S., Fuke, I., Tanishita, O., Kaji, C., Gomi, Y., Yoshida, S., Mori, C., Takamizawa, A., Yosida, I., and Okayama, H., Production of nonstructural proteins of hepatitis C virus requires a putative viral protease encoded by NS3, *Virology*, 198, 636, 1994.

208. Suzich, J.A., Tamura, J.K., Palmer-Hill, F., Warrener, P., Grakoui, A., Rice, C.M., Feinstone, S.M., and Collett, M.S., Hepatitis C virus NS3 protein polynucleotide-stimulated nucleoside triphosphatase and comparison with the related pestivirus and flavivirus enzymes, *J. Virol.*, 67, 6152, 1993.

209. Kanai, A., Tanabe, K., and Kohara, M., Poly(U) binding activity of hepatitis C virus NS3 protein, a putative RNA helicase, *FEBS Lett.*, 376, 221, 1995.

210. Kim, D.W., Gwack, Y., Han, J.H., and Choe, J., C-terminal domain of the hepatitis C virus NS3 protein contains an RNA helicase activity, *Biochem. Biophys. Res. Commun.*, 215, 160, 1995.

211. Gwack, Y., Kim, D.W., Han, J.H., and Choe, J., The hepatitis C virus NS3 protein, *Biochem. Biophys. Res. Commun.*, 225, 654, 1996.

212. Preugschat, F., Averett, D.R., Clarke, B.E., and Porter, D.J.T., A steady-state and pre-steady-state kinetic analysis of the NTPase activity associated with the hepatitis C virus NS3 helicase domain, *J. Biol. Chem.*, 271, 24449, 1996.

213. Tai, C.L., Chi, W.K., Chen, D.S., and Hwang, L.H., The helicase activity associated with hepatitis C virus nonstructural protein 3 (NS3), *J. Virol.*, 70, 8477, 1996.

214. Kim, D.W., Kim, J., Gwack, Y., Han, J.H., and Choe, J., Mutational analysis of the hepatitis C virus RNA helicase, *J. Virol.*, 71, 9400, 1997.

215. Kim, J.L., Morgenstern, K.A., Lin, C., Fox, T., Dwyer, M.D., Landro, J.A., Chambers, S.P., Markland, W., Lepre, C.A., O'Malley, E.T., Harbeson, S.L., Rice, C.M., Murcko, M.A., Caron, P.R., and Thomson, J.A., Crystal structure of the hepatitis C virus NS3 protease domain complexed with a synthetic NS4A cofactor peptide, *Cell*, 87, 343, 1996.

216. Love, R.A., Parge, H.E., Wickersham, J.A., Hostomsky, Z., Habuka, N., Moomaw, E.W., Adachi, T., and Hostomska, Z., The crystal structure of hepatitis C virus NS3 proteinase reveals a trypsin-like fold and a structural zinc binding site, *Cell*, 87, 331, 1996.

217. Kim, J.L., Morgenstern, K.A., Griffith, J.P., Dwyer, M.D., Thomson, J.A., Murcko, M.A., Lin, C., and Caron, P.R., Hepatitis C virus NS3 RNA helicase domain with a bound oligonucleotide: The crystal structure provides insights into the mode of unwinding, *Structure*, 6, 89, 1998.

218. Kang, L.W., Cho, H.S., Cha, S.S., Chung, K.M., Back, S.H., Jang, S.K., and Oh, B.H., Crystallization and preliminary x-ray crystallographic analysis of the helicase domain of hepatitis C virus NS3 protein, *Acta Crystallogr. D Biol. Crystallogr.*, 54, 121, 1998.

219. Wu, Z., Yao, N., Le, H.V., and Weber, P.C., Mechanism of autoproteolysis at the NS2–NS3 junction of the hepatitis C virus polyprotein, *TIBS*, 23, 92, 1998.

220. Zhang, R., Durkin, J., Windsor, W.T., McNemar, C., Ramanathan, L., and Le, H.V., Probing the substrate specificity of hepatitis C virus NS3 serine protease by using synthetic peptides, *J. Virol.*, 71, 6208, 1997.

221. Kolykhalov, A.A., Agapov., E.V., and Rice, C.M., Specificity of the hepatitis C virus NS3 serine protease: Effects of substitutions at the 3/4A, 4A/4B, 4B/5A and 5A/5B cleavage sites on polyprotein processing, *J. Virol.*, 68, 7525, 1994.

222. Komoda, Y., Hijikata, M., Sato, S., Asabe, S.-I., Kimura, K., and Shimotohno, K., Substrate requirements of hepatitis C virus serine proteinase for intermolecular polypeptide cleavage in *E. coli*, *J. Virol.*, 68, 7351, 1994.

223. Steinkuhler, C., Urbani, A., Tomei, L., Biasiol, G., Sardana, M., Bianchi, E., Pessi, A., and De Francesco, R., Activity of purified hepatitis C virus protease NS3 on peptide substrates, *J. Virol.*, 70, 6694, 1996.

224. Bartenschlager, R., Ahlborn-Laake, L., Mous, J., and Jacobsen, H., Kinetic and structural analyses of hepatitis C virus polyprotein processing, *J. Virol.*, 68, 5045, 1994.

225. Lin, C., Pragai, B.M., Grakoui, A. Xu, J., and Rice, C.M., Hepatitis C virus NS3 proteinase: Trans-cleavage requirements and processing kinetics, *J. Virol.*, 68, 8147, 1994.

226. Lin, C., Thomson, J.A., and Rice, C.M., Hepatitis C virus NS4A protein allows formation of an active NS3-NS4A serine protease complex *in vivo* and *in vitro*, *J. Virol.*, 69, 4373, 1995.

227. Failla, C., Tomei, L., and De Francesco, R., Both NS3 and NS4A are required for proteolytic processing of hepatitis C virus nonstructural proteins, *J. Virol.*, 68, 3753, 1994.

228. Tanji, Y., Hijikata, M., Satoh, S., Kaneko, T., and Shimotohno, K., Hepatitis C virus-encoded nonstructural protein NS4A has versatile functions in viral protein processing, *J. Virol.*, 69, 1575, 1995.

229. Bartenschlager, R., Lohmann, V., Wilkinson, T., and Koch, J.O., Complex formation between the NS3 serine-type proteinase of the hepatitis C virus and NS4A and its importance for polyprotein maturation, *J. Virol.*, 69, 7519, 1995.

230. Lundin, M., Monne, M., Widell, A., Von Heijne, G., and Persson, M.A., Topology of the membrane-associated hepatitis C virus protein NS4B, *J. Virol.*, 77, 5428, 2003.

231. Piccininni, S., Varaklioti, A., Nardelli, M., Dave, B., Raney, K.D., and McCarthy, J.E., Modulation of the hepatitis C virus RNA-dependent RNA polymerase activity by the non-structural (NS) 3 helicase and the NS4B membrane protein, *J. Biol. Chem.*, 277, 45670, 2002.

232. Reed, K.E., Gorbalenya, A.E., and Rice, C.M., The NS5A/NS5 proteins of viruses from three genera of the family flaviviridae are phosphorylated by associated serine/threonine kinases, *J. Virol.*, 72, 6199, 1998.

233. Liu, Q., Bhat, R.A., Prince, A.M., and Zhang, P., The hepatitis C virus NS2 protein generated by NS2-3 autocleavage is required for NS5A phosphorylation, *Biochem. Biophys. Res. Commun.*, 254, 572, 1999.

234. Asabe, S.I., Tanji, Y., Satoh, S., Kaneko, T., Kimura, K., and Shimotohno, K., The N-terminal region of hepatitis C virus-encoded NS5A is important for NS4A-dependent phosphorylation, *J. Virol.*, 71, 790, 1997.

235. Tu, H., Gao, L., Shi, S.T., Taylor, D.R., Yang, T., Mircheff, A.K., Wen, Y., Gorbalenya, A.E., Hwang, S.B., and Lai, M.M., Hepatitis C virus RNA polymerase and NS5A complex with a SNARE-like protein, *Virology*, 263, 30, 1999.

236. Lohmann, V., Korner, F., Dobierzewska, A., and Bartenschlager, R., Mutations in hepatitis C virus RNAs conferring cell culture adaptation, *J. Virol.*, 75, 1437, 2001.

237. Gu, B., Gates, A.T., Isken, O., Behrens, S.E., and Sarisky, R.T., Replication studies using genotype 1a subgenomic hepatitis C virus replicons, *J. Virol.*, 77, 5352, 2003.

238. Liang, T.J., Combination therapy for hepatitis C infection, *New Eng. J. Med.*, 339, 1549, 1998.

239. Gish, R.G., Future directions in the treatment of patients with chronic hepatitis C virus infection, *Can. J. Gastroenterol.*, 13, 57, 1999.

240. Enomoto, N., Sakuma, I., Asahina, Y., Kurosaki, M., Murakami, T., Yamamoto, C., Izumi, N., Marumo, F., and Sato, C., Comparison of full-length sequences of interferon-sensitive and resistant hepatitis C virus 1b. Sensitivity to interferon is conferred by amino acid substitutions in the NS5A region, *J. Clin. Invest.*, 96, 224, 1995.

241. Enomoto, N., Sakuma, I., Asahina, Y., Kurosaki, M., Murakami, T., Yamamoto, C., Ogura, Y., Izumi, N., Marumo, F., and Sato, C., Mutations in the nonstructural protein 5A gene and response to interferon in patients with chronic hepatitis C virus 1b infection, *N. Engl. J. Med.*, 334, 77, 1996.

242. Murakami, T., Enomoto, N., Asahina, Y., Izumi, N., Marumo, F., and Sato, C., Mutations in NS5A region and response to interferon in HCV genotype 2 infection, *Hepatology*, 24, 158A, 1996.

243. Chayama, K., Tsubota, A., Kobayashi, M., Okamoto, K., Hashimoto, M., Miyano, Y., Koike, H., Kobayashi, M., Koida, I., Arase, Y., Saitoh, S., Suzuki, Y., Murashima, N., Ikeda, K., and Kumada, H., Pretreatment virus load and multiple amino acid substitutions in the interferon sensitivity-determining region predict the outcome of interferon treatment in patients with chronic genotype 1b hepatitis C virus infection, *Hepatology*, 25, 745, 1997.

244. Kurosaki, M., Enomoto, N., Murakai, T., Skuma, I., Asahina, Y., Yamomoto, C., Ikeda, T., Tozuka, S., Izumi, N., Marumo, F., and Sato, C., Analysis of genotypes and amino acid residues 2209 to 2248 of

the NS5A region of hepatitis C virus in relation to the response to interferon-β therapy, *Hepatology*, 25, 750, 1997.

245. Song, J., Fujii, M., Wang, F., Itoh, M., and Hotta, H., The NS5A protein of hepatitis C virus partially inhibits the antiviral activity of interferon, *J. Gen. Virol.*, 80, 879, 1999.
246. Halimi, G., Halfon, P., Gerolami, V., Castets, F., Khiri, H., Bourliere, M., Gauthier, A.P., and Cartouzou, G., Mutations in NS5A region and interferon response in hepatitis C-1b-infected patients, *Hepatology*, 24, 162A, 1996.
247. Hofgartner, W.T., Polyak, S.J., Sullivan, D., Carithers, R.L., Jr., and Gretch, D.R., Mutations in the NS5A gene of hepatitis C virus in North American patients infected with HCV genotype 1a or 1b, *J. Med. Virol.*, 53, 118, 1997.
248. Khorsi, H., Castelain, S., Wyseur, A., Izopet, J., Canva, V., Rombout, A., Capron, D., Capron, J.P., Lunel, F., Stuyver, L., and Duverlie, G., Mutations of hepatitis C virus 1b NS5A 2209–2248 amino acid sequences do not predict the response to recombinant interferon alpha therapy in French patients, *J. Hepatol.*, 27, 72, 1997.
249. Squadrito, G., Leone, F., Sartori, M., Nalpas, B., Berthelot, P., Raimondo, G., Pol, S., and Brechot, C., Mutations in the nonstructural 5A region of hepatitis C virus and response of chronic hepatitis C to interferon alpha, *Gastroenterology*, 113, 567, 1997.
250. Zeuzem, S., Lee, J.H., and Roth, W.K., Mutations in the nonstructural 5A gene of European hepatitis C virus isolates and response to interferon alpha, *Hepatology*, 25, 740, 1997.
251. Polyak, S.J., McCardle, S., Lui, S.-L., Sullivan, D., Chung, M., Hofgartner, W.T., Carithers, R.L., Jr., McMahon, B.M., Mullins, J.I., Corey, L., and Gretch, D.R., Evolution of hepatitis C virus quasispecies in hypervariable region 1 and the putative interferon sensitivity-determining region during interferon therapy and natural infection, *J. Virol.*, 72, 4288, 1998.
252. Duverlie, G., Khorsi, H., Castelain, S., Jaillon, O., Izopet, J., Lunel, F., Eb, F., Penin, F., and Wychowski, C., Sequence analysis of the NS5A protein of European hepatitis C virus 1b isolates and relation to interferon sensitivity, *J. Gen. Virol.*, 79, 1373, 1998.
253. Gale, M.J., Jr., Korth, M.J., Tang, N.M., Tan, S.L., Hopkins, D.A., Dever, T.E., Polyak, S.J., Gretch, D.R., and Katze, M.G., Evidence that hepatitis C virus resistance to interferon is mediated through repression of the PKR protein kinase by the nonstructural 5A protein, *Virology*, 230, 217, 1997.
254. Gale, M., Jr., Blakely, C.M., Kwieciszewski, B., Tan, S.L., Dossett, M., Tang, N.M., Korth, M.J., Polyak, S.J., Gretch, D.R., and Katze, M.G., Control of PKR protein kinase by hepatitis C virus nonstructural 5A protein: Molecular mechanisms of kinase regulation, *Mol. Cell. Biol.*, 18, 5208, 1998.
255. Kato, N., Lan, K.H., Ono-Nita, S.K., Shiratori, Y., and Omata, M., Hepatitis C virus nonstructural region 5A protein is a potent transcriptional activator, *J. Virol.*, 71, 8856, 1997.
256. Tanimoto, A., Ide, Y., Arima, N., Sasaguri, Y., and Padmanabhan, R., The amino terminal deletion mutants of hepatitis C virus nonstructural protein NS5A function as transcriptional activators in yeast, *Biochem. Biophys. Res. Commun.*, 236, 360, 1997.
257. Behrens, S.E., Tomei, L., and De Francesco, R., Identification and properties of the RNA-dependent RNA polymerase of hepatitis C virus, *EMBO J.*, 15, 12, 1996.
258. Yamashita, T., Kaneko, S., Shirota, Y., Qin W., Nomura, T., Kobayashi, K., and Murakami, S., RNA-dependent RNA polymerase activity of the soluble recombinant hepatitis C virus NS5B protein truncated at the C-terminal region, *J. Biol. Chem.*, 273, 15479, 1998.
259. Ferrari, E., Wright-Minogue, J., Fang, J.W., Baroudy, B.M., Lau, J.Y., and Hong, Z., Characterization of soluble hepatitis C virus RNA-dependent RNA polymerase expressed in *Escherichia coli*, *J. Virol.*, 73, 1649, 1999.
260. Lohmann, V., Overton, H., and Bartenschlager, R., Selective stimulation of hepatitis C virus and pestivirus NS5B RNA polymerase activity by GTP, *J. Biol. Chem.*, 274, 10807, 1999.
261. Cheng, J.C., Chang, M.F., and Chang, S.C., Specific interaction between the hepatitis C virus NS5B RNA polymerase and the 3′ end of the viral RNA, *J. Virol.*, 73, 7044, 1999.
262. Ito, T. and Lai, M.M.C., Determination of the secondary structure of and cellular protein binding to the 3′-untranslated region of the hepatitis C virus RNA genome, *J. Virol.*, 71, 8698, 1997.
263. Chung, R.T. and Kaplan, L.M., Heterogeneous nuclear ribonucleoprotein I (hnRNP-I/PTB) selectively binds the conserved 3′ terminus of hepatitis C viral RNA, *Biochem. Biophy. Res. Commun.*, 254, 351, 1999.

264. Luo, G., Cellular proteins bind to the poly(U) tract of the 3′ untranslated region of hepatitis C virus RNA genome, *Virology*, 256, 105, 1999.

265. Gontarek, R.R., Gutshall, L.L., Herold, K.M., Tsai, J., Sathe, G.M., Mao, J., Prescott, C., and Del Vecchio, A.M., hnRNP C and polypyrimidine tract-binding protein specifically interact with the pyrimidine-rich region within the 3′NTR of the HCV RNA genome, *Nucleic Acids Res.*, 27, 1457, 1999.

266. Walewski, J.L., Keller, T.R., Stump, D.D., and Branch, A.D. Evidence for a new hepatitis C virus antigen encoded in an overlapping reading frame, *RNA*, 7, 710, 2001.

267. Xu, Z., Choi, J., Yen, T.S.B., Lu, W., Strohecker, A., Govingdarajan, S., Selby, M., Chien, D., and Ou, J.H. Synthesis of a novel hepatitis C virus protein by ribosomal frameshift, *EMBO J.*, 20, 3840, 2001.

268. Varaklioti, A., Vassilaki, N., Georgopoulou, U., and Mavromara, P., Alternate translation occurs within the core coding region of the hepatitis C viral genome, *J. Biol. Chem.*, 277, 17713, 2002.

269. Choi, J., Xu, Z., and Ou, J.H., Triple decoding of hepatitis C virus RNA by programmed translational frameshifting. *Mol. Cell. Biol.*, 23, 1489, 2003.

270. Xu, Z., Choi, J., Lu, W., and Ou, J.H., Hepatitis C virus F protein is a short-lived protein associated with the endoplasmic reticulum, *J. Virol.*, 77, 1578, 2003.

271. Matsumoto, M., Hsieh, T.Y., Zhu, N., VanArsdale, T., Hwang, S.B., Jeng, K.S., Gorbalenya, A.E., Lo, S.Y., Ou, J.H., Ware, C.F., and Lai, M.M.C., Hepatitis C virus core protein interacts with the cytoplasmic tail of lymphotoxin-beta receptor, *J. Virol.*, 71, 1301, 1997.

272. Chen, C.M., You, L.R., Hwang, L.H., and Lee, Y.H.W., Direct interaction of hepatitis C virus core protein with the cellular lymphotoxin-beta receptor modulates the signal pathway of the lymphotoxin-beta receptor, *J. Virol.*, 71, 9417, 1997.

273. Zhu, N., Khoshnan, A., Schneider, R., Matsumoto, M., Dennert, G., Ware, C., and Lai, M.M.C., Hepatitis C virus core protein binds to the cytoplasmic domain of tumor necrosis factor (TNF) receptor 1 and enhances TNF-induced apoptosis, *J. Virol.*, 72, 3691, 1998.

274. You, L.R., Chen, C.M., and Lee, Y.H.W., Hepatitis C virus core protein enhances NF-kappaB signal pathway triggering by lymphotoxin-beta receptor ligand and tumor necrosis factor alpha, *J. Virol.*, 73, 1672, 1999.

275. Colombo, M., The role of hepatitis C virus in hepatocellular carcinoma, *Recent Results Cancer Res.*, 154, 337, 1998.

276. Ince, N. and Wands, J.R., The increasing incidence of hepatocellular carcinoma, *New Engl. J. Med.*, 340, 798, 1999.

277. Ray, R.B., Lagging, L.M., Meyer, K., and Ray, R., Hepatitis C virus core protein cooperates with ras and transforms primary rat embryo fibroblasts to tumorigenic phenotype, *J. Virol.*, 70, 4438, 1996.

278. Chang, J., Yang, S.-H., Cho, Y.-G., Hwang, S.B., Hahn, Y.S., and Sung, Y.C., Hepatitis C virus core from two different genotypes has an oncogenic potential but is not sufficient for transforming primary rat embryo fibroblasts in cooperation with the H-ras oncogene, *J. Virol.*, 72, 3060, 1998.

279. Tsuchihara, K., Hijikata, M., Fukuda, K., Kuroki, T., Yamamoto, N., and Shimotohno, K., Hepatitis C virus core protein regulates cell growth and signal transduction pathway transmitting growth stimuli, *Virology*, 258, 100, 1999.

280. Moriya, K., Fujie, H., Shintani, Y., Yotsuyanagi, H., Tsutsumi, T., Ishibashi, K., Matsuura, Y., Kimura, S., Miyamura, T., and Koike, K., The core protein of hepatitis C virus induces hepatocellular carcinoma in transgenic mice, *Nat. Med.*, 4, 1065, 1998.

281. Sakamuro, D., Furukawa, T., and Takegami, T., Hepatitis C virus nonstructural protein NS3 transforms NIH 3T3 cells, *J. Virol.*, 69, 3893, 1995.

282. Gale, M., Jr., Kwieciszewski, B., Dossett, M., Nakao, H., and Katze, M.G., Antiapoptotic and oncogenic potentials of hepatitis C virus are linked to interferon resistance by viral repression of the PKR protein kinase, *J. Virol.*, 73, 6506, 1999.

283. Lu, W., Lo, S.-Y., Wu, K., Fung, Y.-K., and Ou, J.H., Activation of p53 by hepatitis C virus core protein, *Virology*, 264, 134, 1999.

284. Ruggieri, A., Harada, T., Matsuura, Y., and Miyamura, T., Sensitization to Fas-mediated apoptosis by hepatitis C virus core protein, *Virology*, 229, 68, 1997.

285. Serrano, M., Lin, A.W., McCurrach, M.E., Beach, D., and Lowe, S.W., Oncogenic ras provokes premature senescence associated with accumulation of p53 and p16^{INK4a}, *Cell*, 88, 593, 1997.

286. Thompson, E.B., The many roles of c-myc in apoptosis, *Ann. Rev. Physiol.*, 60, 575, 1998.

287. Tabor, E., Viral hepatitis and liver cancer, in *Pathology of Viral Hepatitis*, Goldin, R.D., Thomas, H.C., and Gerber, M.A., Eds., Arnold, London, 1998, chap. 8.

288. Shiratori, Y., Shiina, S., Zhang, P.Y., Ohno, E., Okudaira, T., Payawal, D.A., Ono-Nita, S.K., Imamura, M., Kato, N., and Omata, M., Does dual infection by hepatitis B and C viruses play an important role in the pathogenesis of hepatocellular carcinoma in Japan, *Cancer*, 80, 2060, 1997.

289. Kew, M.C., Yu, M.C., Kedda, M.A., Coppin, A., Sarkin, A., and Hodkinson, J., The relative roles of hepatitis B and C viruses in the etiology of hepatocellular carcinoma in southern African blacks, *Gastroenterology*, 112, 184, 1997.

290. Chen, C.-J., Yu, M.-W., and Liaw, Y.-F. Epidemiological characteristics and risk factors of hepatocellular carcinoma, *J. Gastroenterol. Hepatol.*, 12 (Suppl.), S294, 1997.

291. Yeh, F.S., Yu, M.C., Mo, C.C., Luo, S., Tong, M.J., and Henderson, B.E., Hepatitis B virus, aflatoxins, and hepatocellular carcinoma in southern Guangxi, China, *Cancer Res.*, 49, 2506, 1989.

8 Oncogenesis in Transgenic and Knockout Mice

Laura R. Erker and Anthony Wynshaw-Boris

CONTENTS

SUMMARY

In order to treat and prevent cancer we must first understand how the activation of oncogenes and the loss of tumor suppressors play a role in tumorigenesis. Thus, the ability to accurately model cancer in a multicellular organism brings great insight to the questions surrounding tumor growth and treatment in humans. The ability to model cancer in multicellular organisms was a tremendous breakthrough in cancer research and manipulation of the mouse genome made a large impact in this field. It is now possible to model human cancer in the mouse by overexpressing oncogenes or inactivating tumor suppressor genes. New approaches in this field have improved the accuracy of modeling somatic cancer in the mouse and analyzing the genomic instability that occurs in murine tumors. Such models have provided much of our *in vivo* understanding of cancer. This chapter discusses how retroviral gene delivery, chromosome engineering, and inducible transgenes have been used to manipulate the mouse genome in a more precise spatial and temporal pattern. New tools for tumor analysis such as spectral karyotyping, fluorescence *in situ* hybridization and comparative genome hybridization have improved our ability to detect chromosomal rearrangements, while microarray analysis generates global expression patterns to decipher complex gene patterns that occur in cancers. We have attempted to highlight methodologies that have resulted in the improvement of genetic manipulation of the mouse with respect to tumorigenesis, as well as tools developed to demonstrated and analyze these tumors.

8.1 INTRODUCTION

Understanding how the loss of tumor suppressors and activation of oncogenes lead to cell transformation and eventually to cancer development has been the focus of cancer research for the past decade. Initial studies were done in cell culture and provided tight control of gene expression in ways that were not possible at the multicellular level. However, the ultimate goal of much cancer research remains to define new treatments and preventative strategies to combat cancer in humans. Therefore, researchers have investigated ways to study cancer development in intact model organisms such as the mouse. Mice proved to be the "workhorse" for the job, since they have a short gestation period, are small in size (for ease of housing), and have large litter sizes. In addition, the number of human and mouse genes is virtually identical and displays a high level of synteny with the human genome. Most importantly, it is possible to manipulate the genomes of mice in ways not possible in other mammals. As such, cancer research in the mouse has been a valuable approach in understanding oncogenesis and will aid in the development of cancer treatments in the future.

The basic experimental design differs when studying either oncogenes or tumor suppressor genes in the mouse. For the study of oncogenes, transgenic mice are produced that overexpress these genes in a tissue-specific manner by putting them under the control of tissue-specific promoters, based on the hypothesis that tissues with oncogene overexpression will develop tumors. Due to the powerful effects of promoter-based overexpression, tumors arise in tissues with uniformly high levels of oncogene expression. This differs from how tumor development occurs in sporadic cancers, in which each cell in a tumor can have different levels of onocogene expression. For the study of tumor suppressor genes, an alternative strategy of gene inactivation is used. This method has proved powerful, but it was soon discovered that there are several potential problems with gene inactivation. First, if there are additional mouse homologs for the gene that has been inactivated then there may not be a phenotype. Second, if the gene is necessary during embryonic development or for the viability of the organism then a special conditional knockout (KO) must be produced. Conditional KOs circumvent lethality by inactivating the gene of interest at defined times of development, usually after it is needed for viability, or by inactivating the gene in specific tissues that will not result in mouse morbidity.

This chapter discusses how such transgenic mice are engineered. Examples of application of these techniques for the study of oncogenes and tumor suppressors are presented. Finally, methods for analyzing these oncogenic mouse models are discussed.

8.2 MANIPULATION OF THE MOUSE GENOME

The ability to manipulate the mouse genome was pioneered by scientists who developed techniques to culture mouse embryos and embryonic stem cells (ES cells) [1]. Two techniques were developed for producing these transgenic mice, pronuclear injection, and gene targeting in ES cells.

8.2.1 PRODUCTION OF TRANSGENIC MICE BY PRONUCLEAR INJECTION

Pronuclear injection and gene targeting place a gene of interest into the mouse genome, allowing endogenous tissue-specific gene expression. Both have their specific uses for developing mouse

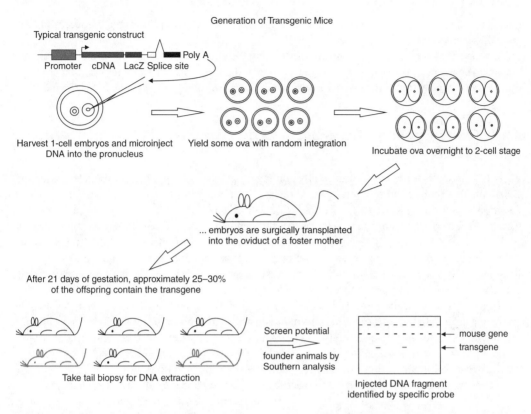

FIGURE 8.1   The structure of a typical transgenic construct (upper left). The transgene consists of a promoter, the cDNA to be expressed, a splice site, and a polyA addition site. Addition of a splice site (intron) greatly increases transgene expression. The flow of the experiment in generating transgenic mice is shown by the arrows.

models of cancer. Pronuclear injection involves the microinjection of DNA fragments into the pronucleus of a fertilized mouse egg (Figure 8.1). The DNA fragment does not need to be homologous to the host DNA and integrates randomly into the genome, usually in multiple copies oriented in a head-to-tail fashion. The DNA fragment is organized in the form of an expressed eukaryotic gene. This "transgene construct" is produced via standard molecular biological cloning techniques (Figure 8.1, top left). The engineered genes that are used for injection are called transgenes, and mice made from the insertion of transgenes are referred to as transgenic mice. In order to express these genes at a prescribed time and in a defined tissue in the mouse, promoter/enhancer elements are put upstream of these genes, which drive their expression in a temporal and tissue-specific manner. The expression pattern of promoter/enhancer cassettes can be monitored by placing a reporter gene within the DNA fragment, such as *LacZ*, which codes for a protein product that can be detected histochemically. At present there are many defined promoters that reproducibly give restricted expression patterns in transgenic mice and can be used to express oncogenic sequences.

Even with carefully chosen promoters, expression can be highly variable among founder lines, since the site of integration can greatly influence the level of expression. Enhancer elements at the site of integration can sometimes direct transgene expression to unexpected tissues in unpredictable spatial and temporal patterns. This problem can be circumvented by using larger DNA fragments for injection. Such fragments can be constructed in bacterial artificial chromosomes (BACs) or yeast artificial chromosomes (YACs). Such large fragments allow important regulatory regions to be included on the injected DNA, which can give a more predictable expression pattern. YACs can accommodate larger inserts of DNA (several hundred kilobases) compared with BACs (100 to 150 kb). However, compared with BACs, YACs are more unstable during manipulation and transgenesis, so BACs have become the vector of choice for producing transgenic mice from large DNA fragments.

Pronuclear injection of DNA is ideal for studying the effect of overexpression of endogenous or foreign genes. Since expression of oncogenes is one of the steps in cancer progression, scientists use this method to measure the effect of the overexpression of certain genes in cancer development.

8.2.2 GENE TARGETING VIA HOMOLOGOUS RECOMBINATION IN ES CELLS

In contrast to the random DNA insertion by nonhomologous recombination in pronuclear injection, DNA can also be inserted in a site-specific manner by homologous recombination. Such targeted insertion allows for the integration of DNA into a defined site of the mouse genome as a single copy that can make specific and precise changes in the genome. DNA is transferred into undifferentiated, pluripotent ES cells by electroporation (Figure 8.2). Insertion of the DNA fragment is based on homologous recombination between the targeting construct and matching genomic sequences. In order for this to occur successfully there must be several kilobases of homology between the exogenous and genomic DNA. ES cells with integration of the DNA via homologous recombination can be selected with positive selectable markers (e.g., antibiotic resistance genes) that are included on the electroporated construct (some common constructs are shown in Figure 8.3). In addition, it is prudent to include negative selectable markers to select against cells that have incorporated DNA by nonhomologous recombination. Using homologous recombination in ES cells, it is possible to make a variety of modifications to specific genomic loci (Figure 8.4).

Gene targeting is commonly used to inactivate a specific gene completely. A selectable marker (e.g., *neomycin phosphotransferase*, or *neo*, expressed from a constitutively active promoter) can be placed into a convenient restriction enzyme site of a gene and inserted directly into an exon. A neo cassette can also be used to completely replace a gene, if small, or to remove critical exons in larger

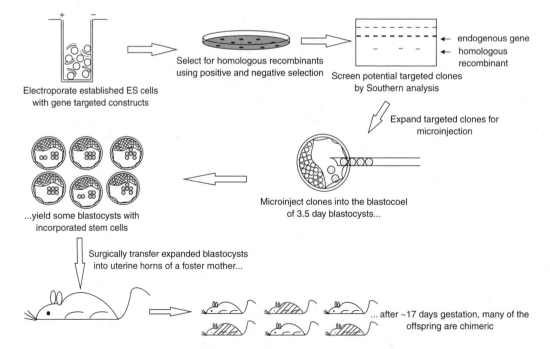

FIGURE 8.2   Construction of transgenic mice using ES cells.

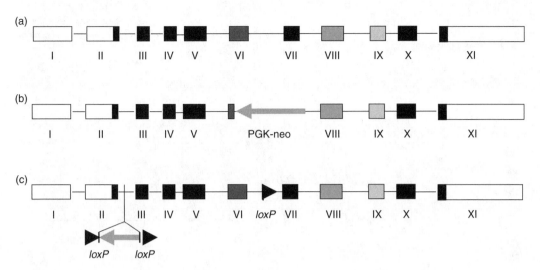

FIGURE 8.3 Common gene targeting constructs. (a) Wild-type allele. (b) A null KO allele is produced by inserting a neo cassette in reverse orientation. This prevents successful production of the protein that is encoded by this gene. (c) A conditional KO allele is produced by the addition of *loxP* sites (▶), which can be induced to recombine in the transgenic of Cre recombinase to produce a KO allele.

genes. Placement of the neo cassette within the targeting construct introduces stop codons into all three reading frames of the gene. This will result in complete loss of the gene product, as long as a stable truncated protein is not produced or splicing does not occur around the exon containing the inserted *neo* gene. If no gene product is produced then a KO mouse for this particular gene has been successfully produced (i.e., the gene of interest has essentially been "knocked out" of the entire organism). The goal of these gene disruptions is to create complete loss-of-function or null alleles. For the study of cancer, this type of gene targeting is generally used for studying how the loss of tumor suppressors effect cancer development in mice.

Problems can arise, however, when the genes that have been removed are essential for development or viability of the mouse. In these cases a gene that can be conditionally excised or expressed is constructed and used to target that gene locus in ES cells. Conditional alleles are created by placing recombination sites, either *loxP* or FRT sites, on either side of an exon or exons. When these sites are recognized by Cre or Flip recombinase enzymes respectively, recombination occurs between these two sites removing the exons (Figure 8.4 and Figure 8.5). This excision blocks successful transcription of the gene, creating a KO allele. This excision can be controlled by the expression of either Cre or Flip recombinase by temporal- and tissue-specific promoters. Currently, laboratories use Cre-*loxP*-mediated and Flip–FRT recombination of gene expression to produce subtle genetic changes. The tissue- and time-specific loss of genes in carcinogenesis can be addressed with these conditional gene-targeting methods.

A similar strategy can be used to create large deletions in the mouse germline using Cre-*loxP* technology. *LoxP* sites are introduced into the genome, at sites on the same chromosome. Recombination between these two sites will produce a large deletion in cell lines or in animals. Using these approaches, deletions of more than 1 Mb have been made *in vivo*. This strategy, pioneered by Allan Bradley and colleagues, allows for studying loss of heterozygosity (LOH) effects in both defined and undefined cancer-causing regions of the genome.

Besides knocking out an entire chromosomal region or gene, demonstrating the role of more subtle genetic changes is also possible using gene targeting. Critical protein domains can be removed by removing the exons that code for these domains. Even smaller changes, such as point mutations, can be introduced by gene targeting. Various strategies have been used to introduce point mutations at

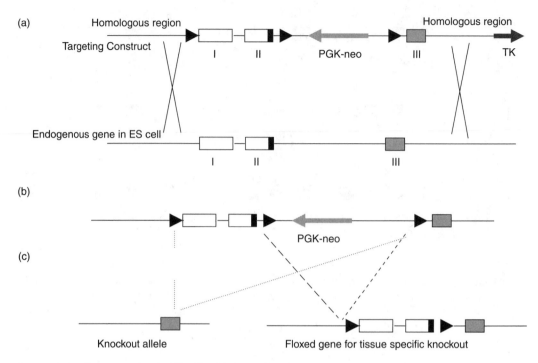

FIGURE 8.4 Gene targeting in ES cells. (a) Recombination occurs between the homologous regions of the target construct and the endogenous gene when the cells are electroporated. (b) Cells that have undergone homologous recombination are selected for with the neo cassette (PGK-neo). Those cells that have undergone nonhomologous recombination and kept the TK cassette are negatively selected for with Gancyclovir treatment. (c) *Cre* recombinase is then transfected into these cells to induce recombination between the *loxP* sites (▶) to produce a KO allele (left) or to create a conditional tissue-specific KO allele (right). The cells that have recombined to produce a KO or a conditional allele can be selected by Southern Blot or PCR. Cells that have not undergone recombination, and lost the TK cassette, are selected against with Gancyclovir.

specific residues of several genes. A point mutation can be introduced into the exon of a gene, closely linked to a removable selectable marker. Clones that have undergone homologous recombination can be screened for the presence of the point mutation by restriction mapping or polymerase chain reaction (PCR); the selectable marker can then be removed. These types of mutations can be used to study specific genetic diseases that are caused by critical cancer-causing genes.

8.3 MODELING ONCOGENESIS IN MICE: USING TRANSGENIC AND KO MICE TO STUDY THE FUNCTION OF CANCER-CAUSING GENES

This section describes different strategies that have been applied for modeling cancer in mice. Examples are presented that illustrate the value of genetic manipulation in the mouse to study the role of oncogenes and tumor suppressor genes.

8.3.1 STUDYING ONCOGENES *IN VIVO* IN TRANSGENIC MICE

In many cancerous tissues, it was discovered that some genes were overexpressed. Such misexpression can be modeled by producing transgenic mice via pronuclear injection, as discussed earlier. For example, a promoter from the mouse mammary tumor virus (MMTV) has been used

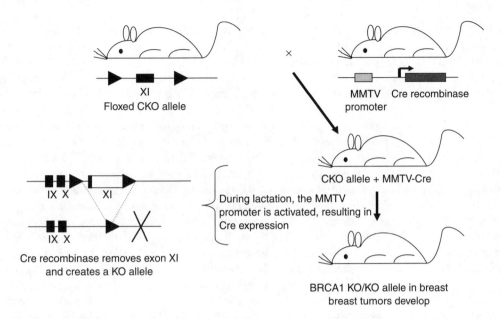

FIGURE 8.5 Cre-*loxP*-mediated recombination to produce a conditional KO of *BRCA1* in the breast epithelium of mice. The MMTV promoter activates expression of Cre recombinase during lactation. This Cre expression causes recombination between the *loxP* sites of the transgene, removing exon XI and producing *BRCA1*-deficient cells in a time- and tissue-specific manner.

to overexpress genes in the breast. MMTV is a retrovirus that, like other retroviruses, integrates randomly into DNA. This integration event induces mammary tumors in a high proportion of animals that have been infected with this virus. It was found that MMTV harbors regulatory sequences in the long terminal repeats (LTRs) that allow for high levels of expression in breast tissue. When this LTR randomly integrates near the site of an oncogene, it occasionally increases expression of the oncogene. If overexpression of this gene can promote breast cancer development then mammary tumorigenesis will occur. The LTR have been isolated from the provirus and characterized in transgenic mice. It is highly efficient in directing transgene expression to the mammary ductal epithelium.

Investigators have taken advantage of the properties of the LTR promoter from MMTV to make transgenic mice that overexpress oncogenes in breast tissue. The first of these experiments was performed in the laboratory of Philip Leder [2] at Harvard. Transgenic constructs were produced that fused the normal mouse *c-Myc* gene with various lengths of its own promoter to the MMTV promoter. *c-Myc* is a basic helix-loop-helix (bHLH) transcription factor that heterodimerizes with another bHLH transcription factor Max to activate transcription at E-box sites in promoters. It was one of the first identified oncogenes. In these first studies, two of the female founders of thirteen transgenic lines developed spontaneous mammary adenocarcinomas during an early pregnancy. The tumors and the breast tissue of these founder animals expressed RNA transcripts of the MMTV/*c-myc* fusion gene. In addition, all of the female progeny from these two founder lines that inherited the MMTV/*c-myc* transgene also developed mammary adenocarcinomas during their second or third pregnancies [2]. These mice were the first to demonstrate the utility of the MMTV promoter in directing expression of an oncogene in the breast of transgenic mice.

8.3.2 INDUCIBLE TRANSGENES (TETRACYCLINE ON/OFF)

As discussed in the previous section, transgenic mice that have global, high level expression of oncogenes are very powerful models for studying cancer development. However, this situation is rarely seen in spontaneous tumorigenesis in humans, in which a single cell can give rise to a tumor. To more accurately model this situation, inducible systems were developed to achieve tighter spatial and temporal control on the expression of transgenes.

The most widely used conditional expression systems use tetracycline (tet) inducible promoters. These systems were developed in the laboratory of Hermann Bujard in the early 1990s [3,4]. A tissue-specific promoter is used to express a fusion of the *Escherichia coli* Tn10Tc protein and HSV *VP16* transactivation domain, known as "tTA." This protein binds to the Tet operator (TetO) sequence and suppresses expression of the associated transgene in the presence of doxycycline. Suppression is released when doxycycline is removed from the system. Alternatively, a reverse tTA (rtTA) has also been constructed where doxycycline treatment releases suppression. The reverse method has an advantage in that one does not have to depend on drug clearance kinetics for gene activation.

These inducible systems have been used to study tumorigenesis in mice by controlling oncogene expression through administration of the drug in the animal's diet. Accordingly, this gene expression can be cycled between on and off states. The SV40 T antigen (T-Ag) was the first oncogene to be analyzed using inducible expression in the mammary gland of mice. At various times in the adult female, T-Ag was induced and then suppressed. Early in oncogene induction the breast tissue could reverse the hyperplasia. However, after prolonged induction T-Ag expression was no longer necessary for maintenance of epithelial hyperplasia [5]. Subsequently, several groups used the Tet-system to activate oncogenes in specific tissues. Induction of *c-Myc* in the epidermis resulted in papillomas, *Ha-Ras* and *Ink4a/p19Arf* induction led to melanoma development [6]. These studies suggested that the prolonged expression of the oncogene resulted in genetic changes that eventually become independent of the expression of the original oncogene [7].

Producing transgenic mice with reproducible inducible gene expression can be a great challenge. *Cis*-acting regulatory elements near the site of insertion of the transgene can greatly influence its expression. Regions of heterochromatin in the vicinity of the insertion can also hinder the precision of the system, affecting the "leakiness" of such aberrant expression. It is important to consider these factors for tight control of gene expression. In addition, turning off the expression of a gene is not always rapid. For example, full induction can take up to a week in the rtTA system due to doxycycline clearance kinetics. There are also indications that the repeated switching between a state of expression and repression can cause the system to become unresponsive to further stimulation. Undoubtedly, inducible systems are powerful tools to model cancer in mice and will continue to be exploited and refined in the future to improve our understanding of oncogenesis.

8.3.3 GENE INACTIVATION: THE KO MOUSE

Studying the role of tumor suppressor genes *in vivo* is possible through gene inactivation. As discussed earlier the gene product is completely removed from these mice via gene targeting. The role of the inactivated gene in cancer development can then be assessed directly in the transgenic mice that are produced. Gene inactivation is ideal for studying the effect of loss of tumor suppressor genes.

A prime example of how gene inactivation can be used to model cancer in mice was the p53 KO mouse. p53 is a tumor suppressor involved in the cellular response to DNA damage, arresting cell cycle progression so that DNA can be repaired and if damage is extensive, activating an apoptotic response. As such, p53 plays a crucial role in regulating the distinct pathways that control responses to DNA damage from diverse agents such as ultraviolet light, ionizing radiation (IR), and chemical carcinogens. p53 is maintained at a low level in normal, undamaged cells. When DNA damage occurs, a signal is transmitted to p53 via upstream molecules that sense and respond to DNA damage. The activation of p53 regulates distinct downstream cell cycle checkpoint and apoptotic pathways. The fact that p53 mutation is the most common genetic abnormality in human cancer highlights this protein as a key player in DNA damage response. Families with Li-Fraumeni syndrome carry heterozygous p53 mutations and are predisposed to a wide variety of tumors [8–10].

p53 is a transcription factor that mediates its effects by binding to DNA and regulating the transcription of several genes. DNA binding occurs as a tetramer at p53-specific promoter elements that are upstream of several genes involved in the DNA damage response. At least six of these genes have been shown to be direct transcriptional targets of p53: *p21/WAF1/Cip1*, *mdm-2*, *GADD45*, *cyclin G*, *Bax*, and *IGF-BP3*. Several of these genes have potential roles in cell cycle control and apoptosis, particularly *p21/WAF1/Cip1* (*p21*) and *Bax* [11]. Transcriptional activation of *p21* mediates, in part, a p53-dependent G1 arrest. This arrest occurs when *p21* binds to cyclin/CDK complexes. At high concentrations, *p21* inhibits their activity, arresting the cell cycle at G1. *Bax* forms a heterodimer with Bcl-2 and other family members, and these two proteins have opposing effects on apoptosis: *Bax* enhances apoptosis, while *Bcl-2* promotes cell survival.

p53 activation can be induced by increasing protein levels or by activating an inactive form of p53. DNA damage induces increased levels of p53 protein. This occurs post-transcriptionally from decreased protein turnover or increased translation. p53 can also be converted from inactive to active forms post-translationally *in vitro* by a variety of agents, including antibodies to specific regions of p53, redox conditions, single stranded DNA, and phosphorylation, suggesting that p53 activation may occur *in vivo*. Altered phosphorylation is an attractive mechanism for rapid regulation of p53 activity. Several phosphorylation sites at the amino- and carboxy-terminal regions of p53 have been identified. The existence of these regions of the protein suggests a potential role for phosphorylated forms of p53 in cell cycle responses.

Because of the central role of p53 in choreographing the DNA damage response, animals deficient for p53 were produced in two different laboratories. These animals were produced by gene targeting in the mouse (Figure 8.6) [12,13]. Both laboratories found that p53 homozygous null mice were viable and developed normally, but they displayed the spontaneous development of a variety of

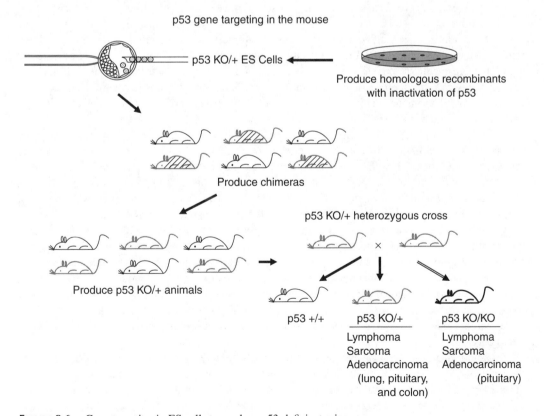

FIGURE 8.6   Gene targeting in ES cells to produce p53-deficient mice.

cancers by 4 to 6 months of age. Similar to humans with Li-Fraumeni syndrome, heterozygous mice also have an increased cancer risk, although with a longer latency to tumor formation than homozygous KOs. The distribution of tumor types in p53 heterozygous animals differs from that in homozygous mutants. In most cases, tumorigenesis in heterozygous animals is accompanied by loss of the wild-type *p53* allele.

8.3.4 CONDITIONAL GENE INACTIVATION

Creating null mice for essential genes results in embryonic lethality, making it impossible to study cancer development in a mouse that is not viable after birth. To circumvent this problem, investigators developed conditional KO strategies in which the gene of interest can be removed at a specific time or in certain tissues of the mouse [14].

This strategy was used to develop the *BRCA1* conditional KO mouse, generating a mouse model for human familial breast cancer. *BRCA1* is a commonly mutated gene in familial human breast cancers. Approximately 5% of women who develop breast cancer display an inherited form of breast cancer caused by loss-of-function mutations in *BRCA1*. In fact, 90% of women with familial breast and ovarian cancers and about 50% of women with familial breast cancer display heterozygous germline mutations in *BRCA1*. Consequently, many groups attempted to produce a mouse model for this important human disease. Unfortunately, complete loss of *BRCA1* causes early embryonic death in mice, and although the heterozygous mice survived they did not develop tumors [15].

To circumvent this early lethality, the *Brca1* gene was inactivated only in the mammary epithelial cells of female mice after embryogenesis, using the Cre-*loxP* recombination system (Figure 8.5). The mice were generated to carry one *Brca1* null allele and one conditional allele. The conditional allele was created so that exon 11 was flanked by *loxP* recombinase sites. These *loxP* sites recombine with each other in the presence of Cre recombinase, excising exon 11 to eliminate a large part of the protein coding region. Thus, crossing these mice containing the conditional *Brca1* allele with mice that are transgenic for Cre that is driven by a mammary epithelium-specific promoter, in this case whey acidic protein (WAP) or the MMTV-LTR promoter, will result in *Brca1*-deficient breast tissue. It was found that Cre-mediated excision of *Brca1* occurred in a majority of the breast tissue cells and was accompanied by a sharp reduction in *Brca1* transcripts. The reduction in *Brca1* caused a decrease in ductal outgrowth during pregnancy. This was frequently accompanied by apoptosis and the ducts did not fully penetrate the fat pad. These results demonstrate that *BRCA1* is involved in ductal elongation and branching morphogenesis during mammary gland development. Most importantly, mammary tumor formation did occur in these mice after a long (10 to 12 months) latency. These tumors showed genomic instability that was characterized by chromosomal rearrangements, aneuploidy, and alterations in p53 transcription. Mating of these mice with p53-deficient mice resulted in a marked acceleration in the formation of mammary tumors. This example demonstrates the utility of conditional KOs, especially when the gene is essential for normal development [16].

Cre-mediated excision of a number of genes has been used to avoid the embryonic lethal phenotype of null alleles of some tumor suppressor genes. For example, gene inactivation of *Nf2*, *Apc*, and *Rb* in mice displayed lethal phenotypes. When these genes were inactivated in specific targeted cell types and tissues, the cancer phenotypes that were produced in the mice modeled, in many ways, the tumors seen in human cancers with similar genetic lesions. These studies have shown the value of using Cre-mediated excision of tumor suppressors to model human tumorigenesis in the mouse. Such mouse models can now be used to understand cancer development and to screen anticancer treatments.

It should also be noted that further refinements to the Cre-*loxP* system are needed to reduce the leaky and varied expression of Cre, as well as the toxicity that is observed in some cases after elevated expression of Cre. Some of these problems can be overcome by using BACs to insulate the transgene from integration effects or by knocking in Cre to endogenous loci. Such modifications will allow more tightly controlled gene inactivation that will more faithfully model human cancer.

8.3.5 Strategies for Modeling Oncogenesis in "Real Time": TV-A Retroviral Delivery

A novel method for cell- and tissue-specific gene delivery has been developed in the laboratories of Harold Varmus and Stephen Hughes [17–20]. As noted previously, it is possible to use viruses to deliver genes to specific tissues. This method uses avian virus-mediated gene delivery to cells and tissues expressing the avian retroviral receptor. While mice do not express the tv-a receptor and are resistant to avian retroviral infection, they can be engineered to express the tv-a receptor in specific tissues. Transgenic mice are engineered to express tv-a under the control of cell- and tissue-specific promoters, rendering these tissues susceptible to avian retroviral infection. An avian pseudotyped virus that recognizes the tv-a receptor can then be used to deliver the gene(s) of interest to the tv-a expressing cells. The gene of interest is usually cloned into a Rous sarcoma-based avian proviral vector (RCAS). This application is limited to small cDNAs since the maximum size of the insert is 2.5 kb. This RCAS proviral vector is then transfected into avian virus producer cells *in vitro* to produce the high titer, replication competent, pseudotyped virus. These cells can be maintained in culture as they infect and reinfect the cells in culture. This virus contains the avian leucosis virus (ALV) coat protein, which recognizes the tv-a receptor, allowing viral entry to the cell. The virus is then delivered into the mice by injection of the virus producing cells or the virus itself. For further information on this method please see the review by Fisher et al. [19].

The feasibility of using this system was established in a study that was designed to determine which cells give rise to glial tumors, such as astrocytomas and oligodendrogliomas. Transgenic mice were produced with tv-a receptor expression under the control of the glial fibrillary protein (GFAP) promoter, so that tv-a would only be expressed in the astrocytes of these mice. When these mice were infected with the virus containing polyoma virus middle T antigen they developed oligodendrogliomas, astrocytomas, and mixed gliomas. This suggested that all of these tumors can form from astrocytes that express GFAP [20].

This system can be used to more closely mimic the genetic changes seen in cancer, where there is a heterogenous population of cells containing different genetic lesions within the tumor mass. Using the tv-a system, cells can be infected with multiple genes simultaneously since tv-a receptors are not blocked by viral infection. Such multiple infections will produce a heterogenous population of cells expressing different levels and combinations of the genes of interest. Further, the introduction of these cells somatically to a small group of cells at the location of infection also models the situation seen in tumors where a single cell can give rise to a tumor locally. The tv-a system may provide the flexibility needed to model mammalian cancer progression and tumorigenesis in somatic cells more accurately.

There are, however, some limitations to TV-A/RCAS retroviral gene delivery. First, injection of the virus producing cells or the virus itself can lead to an immune response. Second, the size of the gene that can be inserted into the proviral vector is limited to 2.5 kb. Finally, the virus can only infect actively replicating cells. These limitations may be addressed in the future by using the ALV virus with *gag* and *pol* genes from lentiviruses, which are capable of infecting nonreplicating cells [21].

This system will prove to be useful for studying tumor progression since new target genes can be produced readily and injected anywhere into transgenic mice expressing the tv-a receptor. Further contributing to the ease of use of this system, there are a number of vectors and tv-a expressing transgenic mice that have already been engineered.

8.3.6 Chromosome Engineering: Another Strategy for Modeling Somatic Mutations

When the tumor suppressor gene is known, germline or somatic mutations in this specific gene can be used to model cancer. However, in many cancers recurrent chromosomal aberrations and deletions are seen. These usually occur at the somatic level and are responsible for tumor progression. For example, a mutation often occurs somatically in one allele of a known tumor suppressor locus, when this is followed by LOH or mutation of the other allele, then tumor progression occurs. It is often difficult to identify the gene that was lost or mutated in such cases, and it has been difficult to recapitulate such events in a model system.

If mice that are hemizygous for a tumor suppressor gene are engineered, the probability of generating the critical somatic mutation event on the other allele is greatly increased. Allan Bradley and colleagues developed a method to produce regional haploidy within the mouse genome. The *loxP*/Cre recombinase technology was used to generate very large deletions (up to 4 Mb) and inversions (24 cM) in chromosomes, an approach they termed "chromosome engineering" [13,22]. The first *loxP* insert is tagged with the 5′ half of a hypoxanthine phosphoribosyl transferase (*hprt*) cDNA cassette and the other *loxP* insert is fused to the 3′ portion of the *hprt* cassette. These complimenting *loxP* sites are placed at either end of the region of DNA to be deleted. When Cre recombinase is activated in *hprt*-deficient ES cells in culture, a recombination event between the two sites results in a functional *hprt* allele, leading to hypoxanthine amniopterin thymidine (HAT)-resistant ES cells. This allows positive selection of cells that have the precise deletions of interest when such deletions are created in *hprt*-deficient ES cells [22].

Allan Bradley's group first tested chromosome engineering by making targeted deletions, duplications, and inversions within a small portion of Chromosome 11 that ranged in size from 1 Mb to 22 cM. The mice with a 1 Mb duplication in this region developed corneal hyperplasia and thymic tumors. This work demonstrated that mice could be produced with defined deletions and duplications, and one of the deletions resulted in cancer [13,23].

Creating large regions of chromosomal haploidy can also be combined with a wide-range of mutagenesis strategies. *N*-ethyl-*N*-nitrosourea (ENU) is a popular mutagen that induces intragenic mutations in spermatogonia at a frequency of 1.5 to 6×10^{-3} [24,25]. ENU mutagenized male mice are mated with wild-type females to generate males with scattered point mutations. These male mice are mated to females carrying large genomic deletions and the offspring are analyzed for phenotypic expression of cancer. This method can uncover recessive tumor suppressors within these missing regions.

To define the location and identify a tumor suppressor within such a large deleted region can be very challenging. To address this issue the Bradley group devised a strategy that creates variable length deletions in the same chromosome. The 5′ half of a *loxP/hprt* cassette is inserted at a defined location via homologous recombination. Downstream of this, the 3′ half of the *loxP/hprt* cassette is inserted randomly using a retroviral vector. In this way, variable nested deletions are produced within the same chromosome. This strategy takes advantage of the precision of Cre-*loxP*-mediated recombination and uses a retrovirus to randomly generate variable length deletions.

The ability to generate large deletions in the mouse genome is a powerful tool to define and locate recessive tumor suppressors. It should be noted that the laboratories of John Schimenti and Terry Magnuson have also developed other strategies to generate deletions in ES cells.

Besides the use of chromosome engineering to produce deletions in order to define tumor suppressors, other studies have attempted to produce defined translocations in somatic cells, since conserved sites of chromosomal rearrangement and fusion have been documented in a number of tumors. Such reciprocal chromosomal translocations have been documented between the *MLL* gene and the *ENL* gene in human myeloid and lymphoid leukemias [26]. The laboratory of Terence Rabbitts created a mouse model for these somatic translocations. They engineered transgenic constructs with *loxP* recombination sites within the *Mll* and *Enl* genes at the location corresponding to the human translocations [27]. These mice were crossed with a mouse line expressing Cre under the hematopoietic *Lmo2* gene [28]. Cre-*loxP* expression induced interchromosomal recombination between these two loci and resulted in rapid leukemogenesis in the mice [29]. This strategy produces *de novo* lesions that directly mimic those that occur spontaneously in human cancer-associated translocations.

8.4 ANALYZING MOUSE MODELS OF CANCER

After these mouse models of cancer are produced, careful analysis and observation are necessary to accurately study cancer development. Initial screening of these mice involves regular observation, checking for tumors, and performing histology on any tumors that form. However, understanding the mechanisms of genomic instability and the changes in gene expression that resulted in tumorigenesis is the main goal. Several methods to measure genomic instability have been developed and refined in the mouse. Both fluorescent *in situ* hybridization (FISH) and spectral karyotyping (SKY) can be used to examine chromosomal translocations. Global gene expression can be measured with oligonucleotide and cDNA microarrays. In order to determine areas of genomic duplication and deletion in tumors, comparative genomic hybridization (CGH) and array CGH are used. These methods for cancer analysis are discussed below.

8.4.1 INITIAL OBSERVATIONS

After a mouse model for cancer has been produced, it must be closely observed to determine the time at which tumor growth occurs. Close observation of the mice allows the investigators to determine the onset of tumor growth. Tumors that are located more externally in the epidermal layers, such as breast tumors, can be easily seen on the mouse. The size and growth of the tumor can be measured without sacrificing the animal. For tumors that are within the body of the animal, survival analysis can be used to determine tumor growth. In this case, the size of the tumor can be determined after the death of the animal. Survival analysis can be represented with a Kaplan–Meier plot. The age at which animals succumb to tumors is recorded and plotted over time. Using this plot, the age at which 50% of the animals survive is determined as a simple comparative measure of tumorigenesis between two populations. The histological type of cancer that develops is determined. Generally, the tumor is removed from the animal after death and histology is performed. Trained pathologists can then identify the type of cancer based on cellular morphology, and cell markers can be used if further diagnostic refinement is desired. These methods are used as primary screens for measuring tumorigenesis and identifying the cancer.

8.4.2 CYTOGENETIC APPROACHES: SKY AND FISH

Genomic instability is a hallmark of cancer, and it can be measured by several approaches [30]. The most common cytogenetic approaches are multicolor FISH, SKY, and CGH. SKY was developed in the laboratory of Thomas Ried [31,32], and has proven an invaluable tool in detecting gross chromosomal translocations. SKY is based on labeling chromosomes with chromosome-specific probes that are made with different combinations of fluorophores (i.e., Cy3, Cy5, Cy5.5, Texas red, and Spectrum green) either alone or in combination. These probes are hybridized to metaphase chromosomes simultaneously. The multicolored chromosome spreads are visualized by exciting the fluorophores and after excitation a unique and specific wavelength is emitted for each chromosome (Figure 8.7) (See Color Figure 8.7 following page 272). The spectral emission of a metaphase spread is determined pixel-by-pixel using a spectrophotometer attached to a charge-coupled device (CCD) camera. This spectral pattern is translated into chromosome-specific pseudocolors by Fourier transformation. The output represents each chromosome as a different color. Translocation events can be easily identified as a chromosome that consists of more than one color. This method greatly simplifies identifying translocations and common breakpoints. SKY is especially useful for identifying translocations in mice since their telocentric chromosomes are inherently difficult to identify by conventional banding methods. It should be noted however, that while SKY is very useful it does have a limited sensitivity for small rearrangements and cannot visualize translocations that are <1500 kb. Usually after SKY narrows the break point to a particular chromosomal band, a more sensitive technique like FISH can be used to pinpoint the site of a translocation breakpoint (Figure 8.7). The area of the translocation, as determined by SKY, usually suggests candidate genes that can be used to produce FISH probes

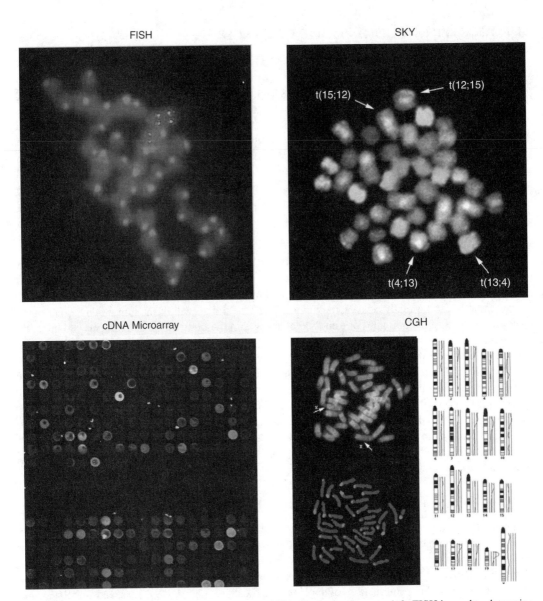

FIGURE 8.7 **(See Color Figure 8.7 following page 272**.) Clockwise from top left. FISH is used to determine amplification or loss of a gene. Here Granzyme C and TCRVα are found at a translocation breakpoint. SKY analysis detects chromosomal translocations, here four translocations: t(15;12), t(12;15), t(4;13), and t(13;4) are seen in a metaphase spread from a tumor. Global gene expression can be examined with cDNA microarrays. Different levels of hybridization are seen in this cDNA microarray. CGH is used to detect regions of chromosome gains and losses in the bottom panel.

and to determine what gene is affected. Once the genes that are affected by the translocation are identified, the breakpoint can be cloned using conventional molecular biological techniques. The availability of the mouse genome sequence greatly facilitates gene searches in such regions.

8.4.3 CGH AND ARRAY CGH

While SKY and FISH measure genomic instability via chromosomal rearrangement, CGH measures this instability by identifying regions of amplification and loss. This method compares DNA copy number variation across the genome. Tumor and nontumor control genomic DNA is labeled with different fluors and hybridized to the same normal metaphase chromosomes. The DNA copy number variation between the tumor and normal somatic cells is represented by the fluorescent ratio across the chromosomes (Figure 8.7). Important insights into regions of chromosome amplification and loss, in a variety of tumors, have been provided by this method [33].

One drawback to CGH is that the resolution is fairly low, because it is done on tightly packed metaphase chromosomes. To increase resolution in detection of copy number changes, array CGH was developed. Array CGH uses cDNA microarrays for the hybridization step, to pinpoint areas of copy number variation based on the known chromosomal location of a cDNA. An advantage to using cDNA microarrays is that it allows the direct identification of candidate genes that may be important for tumor development [34]. Array CGH has shown great potential in the analysis of complex gene expression patterns in cancerous cells [35,36]. This method will be valuable in the analysis of many of the mouse models of cancer that are currently being studied. It will also give greater insight into the similarities and differences in gene amplification and deletion among the many models that are being studied.

8.4.4 OLIGONUCLEOTIDE AND cDNA MICROARRAYS

Microarray technology has allowed scientists to catalog the expression differences responsible for tumorigenesis by analyzing global changes in gene expression. The goal is to develop arrays containing every gene in a genome to quantitatively assess mRNA expression of all genes simultaneously. Microarray analysis was given a boost when the complete genome sequence for the mouse was published and made available online by the end of 2002 [37]. Arrays are constructed by attaching large numbers of oligonucleotide sequences or cDNA clones to substrates such as glass slides. Several thousand of these oligonucleotide sequences or cDNA clones that represent a significant fraction of all genes can be "spotted" on a slide. As such, these arrays allow scientists to monitor changes in expression of thousands of genes simultaneously.

Gene expression is analyzed by comparing tumor tissue or cells with a wild-type control or comparing histologically different tumor types or subtypes. This is done with cDNA arrays by labeling RNA probes from each sample with different fluorescent dyes, such as Cy3 and Cy5 (visible as green and red, respectively). These differently labeled probes are then simultaneously hybridized to the constructed array(s) and compete for hybridization to the same spots. This competition allows for higher copy probes to hybridize more efficiently resulting in different combinations of red and green hybridizing to the spot on the slide. These color differences are analyzed with a confocal reader and the hybridization signal on each spot is measured. When the probes bind equally well to a spot, they appear as a yellow spot — a pseudocolor mixture of the green and red fluors generated by the confocal reader (Figure 8.7). Using this method it is possible to measure subtle differences in binding and to determine both single gene and complex genetic differences in expression between tumor and wild-type cells.

Affymetrix arrays, which are based on one chip and one sample, offer an even more quantitative method to analyze gene expression [38–40]. With these arrays, each gene is represented by spotted probe pairs, one that is a perfect match to the gene of interest and the other differing by a single base pair that is located in the center of the probe. Such match/mismatch probe pairing allows for the identification and subtraction of nonspecific background hybridization signals. The level of transcript can be more accurately quantified subtracting out this nonspecific background.

Much microarray analysis work is being done in the area of modeling cancer in the mouse. One example is the use of this technology to identify the mechanism of action of a fusion between PAX3 and FKHR that has transforming properties. This mechanism was discovered by comparing global gene expression between cells transfected with either the fusion PAX3/FKHR or PAX3 alone [33]. Expression analysis of various tumors will provide molecular fingerprints to diagnose cancer and detect similarities and differences in the origin of specific tumors. In the mouse, such diagnostic

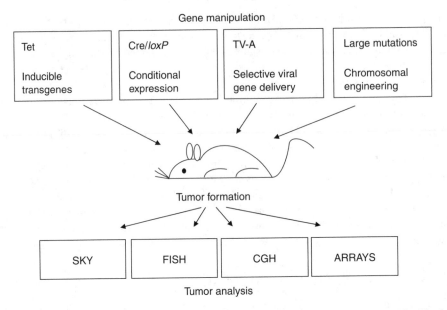

FIGURE 8.8   A schematic overview of the gene manipulations and tumor analysis discussed in this chapter.

fingerprints will be used to differentiate tumors that result from mutations in similar or dissimilar genetic pathways.

8.5 PERSPECTIVES

The development and use of transgenic and gene targeting techniques to create genetically modified mice has resulted in tremendous insights into the genetic processes that cause cancer. In this chapter, we have attempted to highlight recent methodologies that have resulted in the improvement of genetic manipulation of the mouse with respect to tumorigenesis as well as tools developed to demonstrate and analyze genomic instability in these tumors (Figure 8.8). These are powerful tools that will advance our ability to use the mouse to model human cancer more precisely. With the human and mouse genomes completely sequenced, scientists now have at their disposal a catalog of all the genes in these mammalian organisms. It is now possible to identify genes that are potentially involved in tumorigenesis. The more difficult task will be to prove that such genes are important to human cancer. Mouse models will be one very important tool that will be brought to bear on this question.

ACKNOWLEDGMENTS

The authors wish to thank Joy Greer, Jeff Long, T.J. Bowen, and Saugata Ray for their critical comments on the manuscript. Authors would especially like to thank Thomas Ried for the SKY and CGH photos, T.J. Bowen for the FISH photo, and Elaine Cheung for the cDNA microarray photo.

REFERENCES

1. Babinet, C. and Cohen-Tannoudji, M. (2001) Genome engineering via homologous recombination in mouse embryonic stem (ES) cells: an amazingly versatile tool for the study of mammalian biology. *An. Acad. Bras. Cienc.*, 73, 365–383.

2. Stewart, T.A., Pattengale, P.K., and Leder, P. (1984) Spontaneous mammary adenocarcinomas in transgenic mice that carry and express MTV/myc fusion genes. *Cell*, 38, 627–637.
3. Gossen, M. and Bujard, H. (1992) Tight control of gene expression in mammalian cells by tetracycline-responsive promoters. *Proc. Natl. Acad. Sci. USA*, 89, 5547–5551.
4. Kistner, A., Gossen, M., Zimmermann, F., Jerecic, J., Ullmer, C., Lubbert, H., and Bujard, H. (1996) Doxycycline-mediated quantitative and tissue-specific control of gene expression in transgenic mice. *Proc. Natl. Acad. Sci. USA*, 93, 10933–10938.
5. Ewald, D., Li, M., Efrat, S., Auer, G., Wall, R.J., Furth, P.A., and Henninghausen, L. (1996) Time-sensitive reversal of hyperplasia in transgenic mice expressing SV40 T antigen. *Science*, 273, 1384–1386.
6. Chin, L., Tam, A., Pomerantz, J., Wong, M., Holash, J., Bardeesy, N., Shen, Q., O'Hagan, R., Pantginis, J., Zhou, H. et al. (1999) Essential role for oncogenic Ras in tumour maintenance. *Nature*, 400, 468–472.
7. Zhu, Z., Zheng, T., Lee, C.G., Homer, R.J., and Elias, J.A. (2002) Tetracycline-controlled transcriptional regulation systems: advances and application in transgenic animal modeling. *Semin. Cell Dev. Biol.*, 13, 121–128.
8. Ko, L.J. and Prives, C. (1996) p53: puzzle and paradigm. *Genes Dev.*, 10, 1054–1072.
9. Donehower, L.A., Harvey, M., Slagle, B.L., McArthur, M.J., Montgomery, C.A., Jr., Butel, J.S., and Bradley, A. (1992) Mice deficient for p53 are developmentally normal but susceptible to spontaneous tumours. *Nature*, 356, 215–221.
10. Jacks, T., Remington, L., Williams, B.O., Schmitt, E.M., Halachmi, S., Bronson, R.T. and Weinberg, R.A. (1994) Tumor spectrum analysis in p53-mutant mice. *Curr. Biol.*, 4, 1–7.
11. Levine, A.J. (1997) p53, the cellular gatekeeper for growth and division. *Cell*, 88, 323–331.
12. Marino, S., Vooijs, M., Van der Gulden, H., Jonkers, J., and Berns, A. (2000) Induction of medulloblastomas in p53-null mutant mice by somatic inactivation of Rb in the external granular layer cells of the cerebellum. *Genes Dev.*, 22, 994–1004.
13. Ramirez-Solis, R., Liu, P., and Bradley, A. (1995) Chromosome engineering in mice. *Nature*, 378, 720–724.
14. Lewandoski, M. (2001) Conditional control of gene expression in the mouse. *Nat. Rev. Genet.*, 2, 743–755.
15. Xu, X., Wagner, K.U., Larson, D., Weaver, Z., Li, C., Ried, T., Hennighausen, L., Wynshaw-Boris, A., and Deng, C.X. (1999) Conditional mutation of Brca1 in mammary epithelial cells results in blunted ductal morphogenesis and tumour formation. *Nat. Genet.*, 22, 37–43.
16. Deng, C.X. (2002) Tumor formation in Brca1 conditional mutant mice. *Environ. Mol. Mutagen*, 39, 171–177.
17. Barsov, E. and Hughes, S. (1996) Gene transfer into mammalian cells by a Rous sarcoma virus-based retroviral vector with the host range of the amphotropic murin leukemia virus. *Proc. Natl. Acad. Sci. USA*, 70, 3922–3929.
18. Federspiel, M., Swing, D., Eagleson, B., Reid, S., and Hughes, S. (1996) Expression of transduced genes in mice generated by infecting blastocysts with Avian Leukosis virus-based retroviral vectors. *Proc. Natl. Acad. Sci. USA*, 93, 4931–4936.
19. Fisher, G.H., Orsulic, S., Holland, E., Hively, W.P., Li, Y., Lewis, B.C., Williams, B.O., and Varmus, H.E. (1999) Development of a flexible and specific gene delivery system for production of murine tumor models. *Oncogene*, 18, 5253–5260.
20. Holland, E. and Varmus, H.E. (1998) Basic fibroblast growth factor induces cell migration and proliferation after glia-specific gene transfer in mice. *Proc. Natl. Acad. Sci. USA*, 95, 1218–1223.
21. Holland, E., Li, Y., Celestino, J., Dai, C., Schaefer, L., Sawaya, R., and Fuller, G. (2000) Astrocytes give rise to oligodendrogliomas and astrocytomas after gene transfer of Polyoma Virus middle T antigen *in vivo*. *Am. J. Pathol.*, 157, 1031–1037.
22. Mills, A.A. and Bradley, A. (2001) From mouse to man: generating megabase chromosome rearrangements. *Trends Genet.*, 17, 331–339.
23. Liu, P., Zhang, H., McLellan, A., Vogel, H., and Bradley, A. (1998) Embryonic lethality and tumorigenesis caused by segmental aneuploidy on mouse chromosome 11. *Genetics*, 150, 1155–1168.

24. Justice, M.J., Zheng, B., Woychik, R.P., and Bradley, A. (1997) Using targeted large deletions and high-efficiency N-ethyl-N-nitrosourea mutagenesis for functional analyses of the mammalian genome. *Methods*, 13, 423–436.

25. Justice, M.J., Noveroske, J.K., Weber, J.S., Zheng, B., and Bradley, A. (1999) Mouse ENU mutagenesis. *Hum. Mol. Genet.*, 8, 1955–1963.

26. Ayton, P. and Cleary, M. (2001) Molecular mechanisms of leukemognesis mediated by MLL fusion proteins. *Oncogene*, 20, 5695–5707.

27. Collins, E., Pannell, R., Simpson, E., Forster, A., and Rabbitts, T. (2000) Inter-chromosomal recombination of *Mll* and *Af9* genes mediated by Cre-*loxP* in mouse development. *EMBO Rep.*, 1, 127–132.

28. Warren, A., Colledge, W., Carlton, M., Evans, M., Smith, A., and Rabbitts, T. (1994) The oncogenic cysteine-rich LIM domain protein rbtn2 is essential for erythroid development. *Cell*, 78, 45–58.

29. Forster, A., Pannell, R., Drynan, L., McCormack, M., Collins, E., Daser, A., and Rabbitts, T. (2003) Engineering de novo reciprocal chromosomal translocations associated with *Mll* to replicate primary events of human cancer. *Cancer Cell*, 3, 449–457.

30. Richards, R.I. (2001) Fragile and unstable chromosomes in cancer: causes and consequences. *Trends Genet.*, 17, 339–345.

31. Liyanage, M., Weaver, Z., Barlow, C., Coleman, A., Pankratz, D.G., Anderson, S., Wynshaw-Boris, A., and Ried, T. (2000) Abnormal rearrangement within the alpha/delta T-cell receptor locus in lymphomas from Atm-deficient mice. *Blood*, 96, 1940–1946.

32. Schrock, E., du Manoir, S., Veldman, T., Schoell, B., Wienberg, J., Ferguson-Smith, M.A., Ning, Y., Ledbetter, D.H., Bar-Am, I., Soenksen, D. et al. (1996) Multicolour spectral karyotyping of human chromosomes. *Science*, 273, 494–497.

33. Kallioniemi, A., Kallioniemi, O.P., Sudar, D., Rutovitz, D., Gray, J.W., Waldman, F., and Pinkel, D. (1992) Comparitive genomic hybridization for molecular cytogenetic analysis of solid tumors. *Science*, 258, 818–821.

34. Duggan, D.J., Bittner, M., Chen, Y., Meltzer, P., and Trent, J.M. (1999) Expression profiling using cDNA microarrays. *Nat. Genet.*, 21, 10–14.

35. Pinkel, D., Segraves, R., Sudar, D., Clark, S., Poole, I., Kowbel, D., Collins, C., Kuo, W.L., Chen, C., Zhai, Y. et al. (1998) High resolution analysis of DNA copy number variation using comparative genomic hybridization to microarrays. *Nat. Genet.*, 20, 207–211.

36. Pollack, J.R., Perou, C.M., Alizadeh, A.A., Eisen, M.B., Pergamenschikov, A., Williams, C.F., Jeffrey, S.S., Botstein, D., and Brown, P.O. (1999) Genome-wide analysis of DNA copy-number changes using cDNA microarrays. *Nat. Genet.*, 23, 41–46.

37. Waterston, E.A. (2002) Initial sequencing and comparative analysis of the mouse genome. *Nature*, 420, 520–562.

38. Chee, M., Yang, R., Hubbell, E., Berno, A., Huang, X.C., Stern, D., Winkler, J., Lockhart, D.J., Morris, M.S., and Fodor, S.P. (1996) Accessing genetic information with high-density DNA arrays. *Science*, 274, 610–614.

39. DeRisi, J., Penland, L., Brown, P.O., Bittner, M., Meltzer, P.S., Ray, M., Chen, Y., Su, Y.A., and Trent, J.M. (1996) Use of a cDNA microarray to analyse gene expression patterns in human cancer. *Nat. Genet.*, 14, 457–460.

40. Lockhart, D.J., Dong, H., Byrne, M.C., Follettie, M.T., Gallo, M.V., Chee, M.S., Mittmann, M., Wang, C., Kobayashi, M., Horton, H. et al. (1996) Expression monitoring by hybridization to high-density oligonucleotide arrays. *Nat. Biotechnol.*, 14, 1675–1680.

FURTHER READING

1. Hogan, B., Beddington, R., Costantini, F., and Lacy E. (1994) *Manipulating the Mouse Embryo: A Laboratory Manual*. Cold Spring Harbor Laboratory Press, Cold Spring Harbor.

2. Joyner, A.L. (1993) *Gene Targeting: A Practical Approach*. Oxford University Press, Oxford.

3. Wassarman, P.M. and DePamphilis, M.L. (1993) *Guide to Techniques in Mouse Development. Methods in Enzymology*, Vol. 225. Academic Press, San Diego.

WEB PAGES

Mouse genome browsers:

> www.ensembl.org
>
> www.ncbi.nih.gov/genome/guide/mouse
>
> www.ucsc.edu

The Jackson Laboratories: http://www.jax.org/
Induced Mutant Resource: http://www.jax.org/resources/documents/imr/
Oak Ridge National Laboratories Mutant Mouse Database: http://www.bio.ornl.gov/htmouse/
Trans-NIH Mouse Initiative: http://www.nih.gov/science/models/mouse/
The Whole Mouse Catalogue: http://www.rodentia.com/wmc/

GLOSSARY

Embryonic stem cells   Pluripotent cells derived from the inner cell mass of blastocysts capable of contributing to all mouse tissues including the germline. These are the cells used for gene targeting to produce KO mice.

Gene targeting   The introduction of specific mutations into the chromosomal locus of a gene by homologous recombination.

Conditional gene targeting   The introduction of recombinase sites into the chromosomal locus of a gene by homologous recombination, without removing any of the gene sequence. Often this will lead to a normally functioning gene, but it could also lead to a reduced function or hypomorphic allele. A hypomorphic allele is sometimes called a "knock-down."

Transgenic mice   Mice containing a gene made by recombinant DNA technology. Although it usually refers to mice made by pronuclear injection, it can also refer to mice made by gene targeting. The latter mice are often called "KO mice" if the mice have inactivated gene function.

Knockout mice   Transgenic mice made by gene targeting in embryonic stem cells, usually referring to complete gene inactivation (nulls). If the gene is conditionally inactivated, these mice are often called "conditional KOs."

Chromosome engineering   A term coined by the laboratory of Allan Bradley referring to the generation of embryonic stem cells and mice with gene deletions, inversions, and duplications.

9 Chemically Induced Morphological and Neoplastic Transformation in C3H/10T1/2 Cl 8 Mouse Embryo Cells

Joseph R. Landolph, Jr.

CONTENTS

9.1 INTRODUCTION

The topic of chemical transformation in cultured rodent cells has been reviewed previously [1–5]. Our purpose here is to update and review the utility of the currently used cell transformation assays in C3H/10T1/2 Cl 8 mouse embryo cells for detecting chemical carcinogens and for studying the molecular mechanisms of chemically induced morphological and neoplastic cell transformation to gain insight into the molecular mechanisms of chemical carcinogenesis.

9.2 CELL TYPES RELEVANT TO CELL TRANSFORMATION

There are three basic cell types relevant to cell transformation studies. First, there is the fibroblast. Fibroblasts are derived from the mesoderm of the embryo and are connective tissue cells (reviewed in References 1 to 6). They are not very fastidious in their nutritional requirements and they are also relatively resistant to high concentrations of calcium ion that usually cause epithelial cells to terminally differentiate. Hence, fibroblasts are very easily cultured. For this reason, most early cell culture studies were conducted on fibroblastic cell types [1–6]. In culture, fibroblasts are stellate or cigar-shaped at low cell density. At high cell densities, fibroblasts appear diamond-shaped [1–6]. They need to attach themselves to a substrate in culture, the plastic cell-culture dish, to synthesize DNA and divide. Hence, normal or nontransformed fibroblasts will not grow in three-dimensional suspension in soft agar [1–6]. However, some types of fibroblasts synthesize collagen in cell culture.

The second important cell type is the epithelial cell. Epithelial cells form the coverings of the skin, the lungs, and other organs. They are derived from the ectoderm of the embryo. They are very fastidious in their nutritional requirements and require very specialized nutritional components. At low cell density, epithelial cells grow as islands and have a cobblestone or pavement stone morphology. Epithelial cells are very difficult to culture. Many types of epithelial cells are killed by high concentrations of calcium ions or will undergo terminal differentiation when cultured in high concentrations of calcium ions. Normal epithelial cells also need to attach to a substratum to synthesize DNA and to divide. Hence, normal epithelial cells too will not grow in three-dimensional suspensions in soft agar.

A third general type of cells relevant to cell-culture and cell-transformation studies are cells of the lymphoid series. Cells of the lymphoid series are different from fibroblasts and epithelial cells, because lymphoid cells normally grow in suspension in the blood of mammals. Hence, as cells that normally grow in suspension in the blood, normal cells of the lymphoid series can also grow normally in three-dimensional suspension in soft agar.

9.3 BIOLOGICAL AND CELL CULTURE TERMS RELEVANT TO CHEMICAL TRANSFORMATION

It is important to define a number of cell culture terms relevant to chemical transformation. First, a primary cell culture is one that is derived by removing an embryo from a pregnant animal, and then mincing and trypsinizing the embryo to form a suspension of cells ([7,8]; reviewed in References 1 to 6). When this cell suspension is plated into cell culture dishes, the resultant culture is called a primary embryo cell culture. Examples of this are primary chicken embryo cell cultures, primary mouse embryo cell cultures [7], and hamster embryo cell cultures [9]. A secondary culture results when a primary culture has been trypsinized and passaged at a particular dilution into other tissue culture flasks. These secondary culture cells may then be passaged to form tertiary cultures, and so on.

A cell strain is a culture that has been passaged many times, but has not yet been proved to be an immortal cell line. After a mammalian cell culture has been passaged many times, and has gone through approximately 60 population doublings for murine cells, it passes through a stage called crisis or senescence. At this point the cells cease to replicate. Senescence is believed to be a process of programmed cell death in which the telomeres of the chromosomes have shortened to a length sufficient to cause the cells to cease replication.

Normal cells have a normal or euploid number of chromosomes. For the mouse this is $2n = 40$ and for humans $2n = 46$ chromosomes. An aneuploid number of chromosomes is an aberrant number of chromosomes, defined as other than the euploid number of chromosomes. In cultures of murine cells, when cells senesce, it is common for a small fraction (0.001) of the cells to become tetraploid (possessing 80 chromosomes). The cells double their chromosome number but do not divide. During passaging of mammalian cells in cell culture, it is difficult for tetraploid cells to partition their number of chromosomes equally during cell division. Hence, tetraploid cells often become aneuploid [7], due to centric fusion and loss or gain of chromosomes during cell division. It is common that murine cell lines passaged in culture for long periods of time often have sub-tetraploid numbers of chromosomes.

At some point during passaging, after the murine cells have become aneuploid, they often become immortalized at high frequencies. This leads to the evolution of a permanent or immortalized cell line. An example of this is the C3H/10T1/2 Cl 8 (10T1/2) mouse embryo cell line, derived by Reznikoff and colleagues in the laboratory of the late Professor Charles Heidelberger ([7,8]; reviewed in References 1 to 6).

9.4 RATIONALE FOR THE DEVELOPMENT OF CELL TRANSFORMATION SYSTEMS

In the late 1960s and early 1970s, scientists studying chemical carcinogenesis decided to investigate whether mammalian cells in culture could be converted into tumor cells, or transformed, after they were treated with oncogenic viruses or chemical carcinogens. The rationale for these experiments was the goal of the investigators to study the mechanisms of the conversion of normal cells into tumor cells, or cell transformation, in a relatively simple system such as cell culture. Such a simple cellular system would be independent of complicating host factors such as the immune response and the complex control mechanisms that govern homeostasis at the tissue and organ levels in the whole animal. It would be expected that conversion of normal cells into tumor cells in a cell-culture dish would be a relatively simple process to analyze, compared to the complex process of carcinogenesis in the animal. A number of such cell culture systems have since been developed, in which treatment of cultured mammalian cells by chemical carcinogens can convert normal or nontransformed cells into tumor cells. One of these, the C3H/10T1/2 Cl 8 (10T1/2) mouse embryo cell line, is discussed in Section 9.5.

9.5 CHEMICAL TRANSFORMATION OF C3H/10T1/2 CL 8 MOUSE EMBRYO CELLS

9.5.1 DERIVATION AND DESCRIPTION OF THE C3H/10T1/2 Cl 8 MOUSE EMBRYO CELL LINE

In 1973, Reznikoff et al. [7] derived a contact-inhibited, permanent mouse fibroblast cell line from C3H mouse embryo cells, and designated this cell line C3H/10T1/2 (10T1/2). These investigators established many clones of this cell line. One of these, the eighth clone, had a low saturation density and a low frequency of spontaneous morphological transformation, and was designated C3H/10T1/2 Cl 8 [7]. This cell line spontaneously became immortalized and aneuploid [7]. These cells are near tetraploid ($n = 81$) and have a low saturation density (approximately 0.8 million/25 cm^2 T-flask), do not grow in soft agar, and do not form tumors when injected into Balb/c nude mice or into immunosuppressed mice (x-irradiated Balb/c mice) [7]. The derivation of the name of this cell line is based on the nomenclature for Balb/c 3T3 cells. Balb/c 3T3 cells are derived from Balb/c mice, and the cells are passaged every 3 days at 3×10^5 cells per 25 cm^2 T-flask. C3H/10T1/2 cells are derived from C3H mice, and are passaged every ten days at $\frac{1}{2} \times 10^5$ cells per 25 cm^2 T-flask.

When 10T1/2 cells are treated with chemical carcinogens, such as 3-methylcholanthrene (MCA) or benzo[a]pyrene (BaP), transformed foci develop against the contact-inhibited monolayer at confluence [8]. These transformed foci can be cloned by the glass-cylinder technique and expanded into transformed cell lines [8]. When these morphologically transformed 10T1/2 cell lines are injected into nude or immunosuppressed Balb/c mice, they form fibroscarcomas. Hence, 10T1/2 cells are often thought to be fibroblasts [8]. However, when nontransformed 10T1/2 cells are treated with 5-azacytidine, which is not appreciably mutagenic with regard to these cells [10], the cells can differentiate into chrondrocytes, myocytes, and adipocytes, which is due to an inhibition of DNA methylation [11–14]. Hence, 10T1/2 cells are now thought to be an immortalized, primitive mesenchymal stem cell line.

9.5.2 Chemically Induced Morphological Transformation in C3H/10T1/2 Cl 8 Mouse Embryo Cells

Treatment of contact-inhibited, nontransformed 10T1/2 mouse embryo cells with the chemical carcinogens BaP, MCA, or 7,12-dimethylbenz[a]anthracene (DMBA) induces foci of morphologically transformed cells that stand out prominently against the faintly staining monolayer of nontransformed, contact-inhibited 10T1/2 cells after the cells are fixed with methanol and stained with Giemsa, a nuclear stain ([8,15,16]; reviewed in References 1 to 6). The chemically induced foci are derived from chemically induced, morphologically transformed cells which have lost the property of contact inhibition of cell division and grow above the monolayer to form a focus after the six weeks duration of the transformation assay [1–8].

Chemically induced transformed foci in 10T1/2 cells can be grouped into three major classes, type I foci, type II foci, and type III foci ([8]; reviewed in Reference 6). Type I foci represent a slight mottling and dark staining of the monolayer and are regarded as not having substantially altered biological properties from nontransformed C3H/10T1/2 cells. As such, type I foci are not usually scored in cell transformation assays. Type II foci represent foci of piled-up cells that stain a dark purple color with Giemsa stain as there are a large number of cells per unit area, but type II foci have relatively smooth edges (Figure 9.1). Type III foci are multi-layered or piled-up areas of darkly staining cells in which cells crisscross at the edges of the focus. Type III foci often stain a deep blue color with Giemsa stain, compared to the usual purple color of the Giemsa-stained Type II focus (Figure 9.2). A fraction of the Type II foci and Type III foci produce tumors when injected into immunosuppressed C3H mice (Figure 9.3, References 8 and 15, reviewed in Reference 1 to 6). The fraction of transformed cell lines derived from foci that induce tumors in nude mice usually varies depending on the transformation-inducing agent.

A wide variety of chemical carcinogens, including polycyclic aromatic hydrocarbons (PAH) [8,15,16], 5-azacytidine [17], N-acetoxy-acetylaminofluorene [15], aflatoxin B_1 [18,19], and cigarette smoke condensate [20], induce morphological transformation of asynchronous 10T1/2 cells. Our laboratory has also shown that metabolites of BaP [16], lead chromate [21], and specific insoluble nickel compounds, such as green (high temperature) nickel oxide, crystalline nickel sulfide, and nickel subsulfide [22], the anti-helminthic drug, flubendazole [23], and the cancer chemotherapeutic drug, harringtonine [23], induce morphological and neoplastic transformation of 10T1/2 cells.

A number of other carcinogens have relatively short half-lives, and it is necessary to first synchronize 10T1/2 cells before treating them with these agents in order to induce morphological transformation of 10T1/2 cells with these agents. N-methyl-N'-nitro-N-nitrosoguanidine (MNNG) [24–26], fluorodeoxyuridine (FudR) [27], and arabinofuranosylcytosine (Ara-C) [28] induce significant yields of morphologically transformed foci in 10T1/2 cells if the cells are first synchronized by various means and then treated with these carcinogens during the S-phase or G_1–S boundary of the cell cycle (reviewed in Reference 6). The laboratory of Nesnow has shown that treating 10T1/2 cells

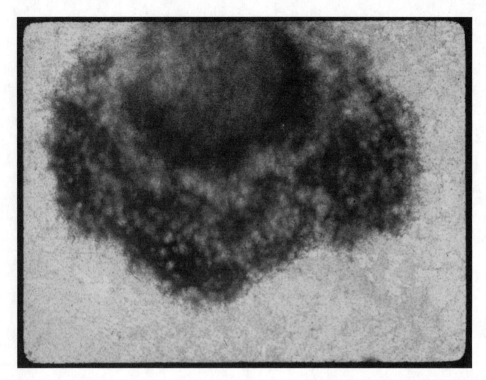

FIGURE 9.1 A photograph of a Type II focus of morphologically transformed 10T1/2 cells observed under the light microscope. This focus was induced by treating 10T1/2 cells with 1 μg/ml of 3-methylcholanthrene in the standard transformation assay, and then fixing the cells with methanol and staining the cells with Giemsa stain.

five days after seeding, rather than 24 h after seeding, can lead to transformation of 10T1/2 cells by MNNG, presumably because more of the cells are in the S-phase of the cell cycle at 5 days post-seeding compared to 1 day post-seeding [29].

In addition, it has also been shown that x-rays [30], x-rays plus 12-*O*-tetradecanoyl-phorbol-13-acetate [31], and ultraviolet light (UV) [32] induce morphological and neoplastic transformation of 10T1/2 cells. Some of the chemical compounds and radiations that induce morphological transformation in 10T1/2 cells are listed in Table 9.1.

9.5.3 ADVANTAGES OF THE ASSAY FOR CHEMICALLY INDUCED MORPHOLOGICAL TRANSFORMATION IN C3H/10T1/2 MOUSE EMBRYO CELLS

Because they are a permanent cell line, 10T1/2 cells can be easily manipulated with minimal time and effort. These cells are contact-inhibited and have a low saturation density and a very low frequency of spontaneous transformation at passage numbers less than 15. Therefore, 10T1/2 cells are used from passages 5 to 15 and then discarded. A new vial of 10T1/2 cells at as low a passage as possible, from 5 to 8, is then thawed and used. This serves to minimize the frequency of spontaneous transformation in the assay (reviewed in Reference 6).

This cell-transformation assay utilizes scoring of transformed foci against a contact-inhibited monolayer of nontransformed 10T1/2 cells. In general, this thin monolayer stains weakly with Giemsa, whereas the chemically transformed foci stain darkly. Hence, it is relatively easy to identify the foci in this system. The use of 1 μg/ml of the carcinogen, MCA, as a positive control in the transformation assay gives the investigator confidence that the transformation assay is

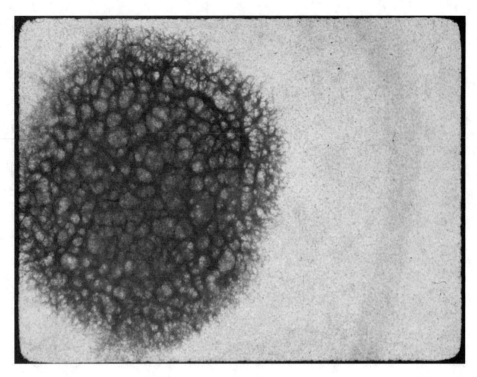

FIGURE 9.2   A photograph of a Type III focus of morphologically transformed 10T1/2 cells observed under the light microscope. This focus was induced by treating 10T1/2 cells with 1 μg/ml of 3-methylcholanthrene in the standard transformation assay, and then fixing the cells with methanol and staining them with Giemsa stain.

functioning correctly. Foci generated by treatment of cells with 1 μg/ml MCA are distinct and consist approximately 2/3 Type-II foci (Figure 9.1) and 1/3 Type-III foci (Figure 9.2).

These cells possess cytochrome P450, and hence can metabolically activate [33,34] and detect the mutagenicity [15] and morphological transformation [8,15,16] induced by PAH, BaP, and MCA (Figure 9.3). The induction of morphological cell transformation in these cells is generally dose-dependent upon the concentration of carcinogen, that is whether the concentration used is below the level at which endogenous cytochrome P450 activity is saturated or below the concentration at which the compound becomes insoluble [8,15,16]. In addition, the cells also detect the weak transformation caused by the mycotoxin aflatoxin B1 [18,19,35,36].

The delineation of foci in this cell transformation system may mark various stages in the progression of cells, from partial to full transformation. Hence, this cell transformation system has useful mechanistic applications in delineating various steps in neoplastic transformation caused by chemical carcinogens. Using temperature-sensitive mutants of chemically transformed 10T1/2 cells, Boreiko and Heidelberger [37] showed that the phenotypes of focus formation and anchorage independence could be dissociated.

9.5.4   PROCEDURES FOR CONDUCTING THE 10T1/2 CELL-TRANSFORMATION ASSAY TO DETECT CHEMICALLY INDUCED MORPHOLOGICAL AND NEOPLASTIC TRANSFORMATION

9.5.4.1   Practical Considerations

There are a number of practical considerations that investigators using the 10T1/2 cell transformation assay should be aware of. First of all, the standard transformation assay [8] has a duration of

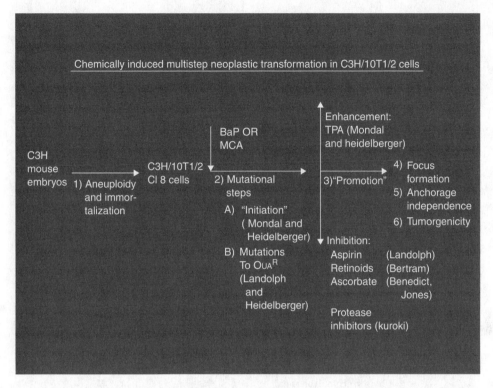

FIGURE 9.3   Schematic mechanistic diagram of the establishment of the C3H/10T1/2 Cl 8 mouse embryo cell line, and the steps this cell line undergoes after it is treated with BaP or MCA that lead to the development of morphological, anchorage-independent neoplastic cell transformation.

6 weeks. The investigator's facilities should be equipped with HEPA-filtered air and the investigator's personnel well trained in sterile cell-culture technique to minimize loss of assays due to bacterial or fungal contamination during this long assay time.

Further, 10T1/2 cells require specific batches of prescreened fetal calf serum to allow cells to attach properly to cell-culture dishes and grow optimally. The cells plate poorly and give poor plating efficiencies in some serum batches, and hence these sera cannot be used. Oshiro et al. [38] reported that batches of fetal calf sera occasionally give good plating efficiencies but poor yields of transformed foci with MCA. Hence, these sera cannot be used for transformation studies. This has also been our laboratory's experience. Most investigators routinely screen sera for plating efficiency, cell attachment, growth rate, and transformation with 1 μg/ml MCA. Fewer than half of the fetal calf serum batches tested are appropriate for use in the 10T1/2 cell transformation assay. Periodic shortages of fetal calf sera occasionally make it necessary to anticipate serum requirements in advance of the experiments (reviewed in Reference 6).

Water of high purity should be used to prepare tissue culture medium, as the plating efficiency of 10T1/2 cells decreases substantially in Basal Medium Eagle (BME) cell-culture medium prepared with water containing large amounts of calcium and magnesium ions. Hence, it is necessary to deionize and distill the water used to prepare BME medium for culturing 10T1/2 cells. The pH of the cultures is also important. If dishes remain outside of the incubator for too long and the medium becomes too basic, the plating efficiency decreases drastically and the transformation assay does not work well (reviewed in Reference 6).

One disadvantage of the transformation assay in 10T1/2 cells is that the plating efficiency is usually only 30%, although we have occasionally obtained a 35 to 40% plating efficiency in our laboratory. In the standard transformation assay testing toxic concentrations of carcinogens, the plating efficiency is reduced even further and the number of survivors can become too few for

TABLE 9.1

Some of the Chemical Carcinogens and Radiations that Induce Morphological and Neoplastic Transformation in C3H10T1/2 Mouse Embryo Cells

Compound or agent	Reference
a. In asynchronous cells	
3-Methylcholanthrene	8, 15, 16
Benzo[a]pyrene	8, 15, 16
Benzo[a]pyrene metabolites	15, 16
Aflatoxin B1	18, 19, 36, 60
5-Azacytidine	17
Lead chromate	21
Green (high temperature) nickel oxide	22
Crystalline nickel monosulfide	22
Nickel subsulfide	22
X-rays	30, 31
UV light	32
Cigarette smoke condensate	20
N-acetoxy-acetylaminofluorene	15
MNNG (modified assay of Nesnow et al.)	29
Flubendazole (antihelminthic drug)	23
Harringtonine (antineoplastic drug)	23
b. In synchronized cells	
MNNG (synchronized cells)	24, 25, 26
Fluorodeoxyuridine (Fudr) (synchronized cells)	27
Arabinofuranosylcytosine (synchronized cells)	28

a valid transformation assay. Incubators in which 10T1/2 cells are grown and transformation assays conducted should be cleaned with mild cleaning agents that do not leave toxic vapors or volatile residues. Such residues and vapors decrease the plating efficiency of the 10T1/2 cells. To clean incubators to reduce microbial contamination, we recommend a quaternary amine salt for initial sterilization followed by rinsing the inside of the incubator with sterile hot water. Toxic compounds such as formaldehyde and betadine should be avoided, as residues of these compounds can be toxic to 10T1/2 cells.

PAH can be tested for transformation because 10T1/2 cells have cytochrome P450 [33,34], which metabolically activates PAH to diol epoxides. For other carcinogens, such as aromatic amines and nitrosamines, exogenous metabolic activation systems such as S9 or rat liver hepatocytes must be provided.

With compounds that can be metabolized by 10T1/2 cells, such as PAH, or which are added with S9 to 10T1/2 cells, asynchronous 10T1/2 cells can be used to assay transformation [8,15]. However, with activated compounds that have short half-lives in cell-culture medium, such as MNNG [24] and certain nucleosides such as FudR [27] and cytosine arabinoside [28], marginal or no transformation is observed in asynchronous cells, and cells must be synchronized to obtain a significant number of foci. An alternative, for MNNG, is to use the modified protocol of Nesnow et al. [29].

One of the greatest practical concerns is the subjectivity in the scoring of foci. Reznikoff et al. [8] originally described three discrete types of foci — Type I, Type II, and Type III. Type-I foci are not very distinct and are not scored in this assay, while Type-II and Type-III foci are scored. Most definitive Type-II foci are easy to score (Figure 9.1), as are most definitive Type-III foci (Figure 9.3). However, there is actually a continuous distribution of transformed focus morphology, and cases arise in which it is difficult to decide whether a focus is Type II or Type III. Occasionally, foci arise that are intermediate between Type II and Type III, and it is difficult to classify them into either category. Further, we have occasionally observed pleiomorphic or heterogeneous foci composed partly of areas that are Type II-like and partly of areas that are Type III-like. This leads to some ambiguity in the scoring of foci. Both the variation of focus morphology and the pleiomorphic foci probably reflect the heterogeneity of transformation responses induced by chemical carcinogens. We have also noted, as have others, that occasionally there appear foci or focal-like areas not previously described. In particular, we have noted what we choose to call a "flat" Type-III focus, in which cells at the edges are crisscrossed but not as piled-up as in the usual Type-III focus. The biological significance of these types of foci is not clear, and they must be examined for anchorage independence and tumorigenicity (reviewed in Reference 6).

9.5.4.2 Theoretical Considerations

Some have questioned the validity of using permanent cell lines such as 10T1/2 in assay transformation because they are not normal cells and, in particular, because the cells are near-tetraploid. In this context, scientists from three laboratories have shown that transformation in primary mammalian cell cultures is a multistep process (Figure 9.4). One group of transfected oncogenes immortalizes primary mammalian cell cultures and converts them into permanent cell lines, while a second group causes focus formation and anchorage independence in permanent cell lines [9,39,40]. Transfection of DNA from chemically transformed 10T1/2 cells into NIH 3T3 cells induces foci in NIH 3T3 cells [41]. Hence, the chemical transformation events measured in 10T1/2 cells as focus formation are likely due to mutations in a second class of oncogenes, such as Ki-*ras*, and therefore there is a late step or delay in transformation.

Further, not all, but approximately 50% of Type-II foci and 80% of Type-III foci are tumorigenic. Hence, induction of morphological transformation in this transformation system is frequently but not always equivalent to induction of tumorigenicity [8].

The 10T1/2 cell transformation assay has very low frequencies of both spontaneous and induced transformation. Hence, when transformation is not observed with a particular carcinogen using 2,000 10T1/2 cells in the classical assay [8,15], investigators have resorted to treating asynchronous cells 5 days after seeding [29] or seeding higher cell numbers such as 10,000, to ensure that negative responses for chemicals studied in the 10T1/2 cell transformation assay are not due to insensitivity of the assay. In a further modification of the assay, called a "level-II assay," 10,000 cells are seeded and treated as in the classical or level-I assay [8,15], and then reseeded after four weeks when they have become confluent [42]. This procedure increases the yield of foci with MNNG,

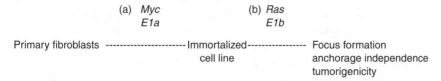

Transfection of oncogenes into primary mammalian fibroblasts and into immortalized
mammalian cell lines

(a) *Myc*   (b) *Ras*
     E1a   *E1b*

Primary fibroblasts ----------------------- Immortalized ----------------- Focus formation
                                  cell line   anchorage independence
                                                         tumorigenicity

FIGURE 9.4 Schematic diagram of the conversion of (a) primary fibroblasts into immortalized cell lines by transfecting them with the *myc* or adenovirus *E1a* genes, and (b) conversion of the immortalized cell lines into transformed cell lines that can form foci, colonies in soft agar, and tumors in immunosuppressed mice, following transfection of the immortalized cell lines with a mutated Ki-*ras* gene or an adenovirus *E1b* gene.

2-acetoxy-acetylaminofluorene, N-nitrosodimethylamine and dibenz[a,h]anthracene, which induce weak or no transformation in the standard transformation assay of Reznikoff et al. [8]. The transformation assay is thus evolving at this time, and the most sensitive transformation protocol may need to be defined for specific carcinogens, particularly those with low cell transforming or carcinogenic potential. At this point of time, we recommend utilizing the standard assay of Reznikoff et al. at first [8]. If a chemical does not yield morphological transformation in the standard Reznikoff et al. assay [8], modified to plate 2000 cells per 60-mm dish and to treat cells for 48 h with the candidate carcinogen [15], then the modified assay of Nesnow et al. [29] should be used when using this assay to screen chemicals to determine whether they are potential carcinogens (reviewed in Reference 6).

Finally, there is a strong dependence of transformation frequency on surviving cell number. Transformation frequency decreases monotonically with increasing surviving cell number [8,43]. The inverse dependence of transformation frequency on cell number is not yet understood theoretically, and makes it difficult to utilize a calculation of transformation frequency. At present, it is recommended that foci be scored, and the total numbers of foci, not transformation frequency, be reported. Attempts have been made to interpret the dependence of the frequency of morphological cell transformation on probabilistic [44] or epigenetic [45] bases, but no clear consensus has as yet emerged.

Currently, it is not recommended that transformation frequency be calculated in this assay, due to the strong dependence of transformation frequency on cell density [43]. Instead, the number of Type-II and Type-III foci per total number of dishes treated and the total number of dishes containing Type-II or Type-III foci out of the total number of dishes treated are tabulated and reported ([21,22]; reviewed in Reference 6). Investigators usually seed 1000 or 2000 cells (300 or 600 surviving cells) per 60-mm dish in the transformation assays, the cell density at which the assay is thought to be free from cell density effects [8,43]. This has not yet been rigorously proved. Moreover, this low number of cells at risk may reduce the sensitivity of the assay. If no morphological transformation is induced by treatment with a specific candidate carcinogen, the assay should be repeated utilizing the more recent protocol of Nesnow et al. [29].

9.5.5 EVOLUTION OF AND RECOMMENDATIONS FOR IMPROVEMENT IN THE TRANSFORMATION ASSAY PROTOCOL

From our own experience and that of our colleagues using the 10T1/2 cell transformation assay, the following procedures have evolved in the routine use of 10T1/2 cells to detect morphological transformation caused by suspected chemical carcinogens. The original assay procedure of Reznikoff et al. [8], which determines the number of foci present against a confluent monolayer of cells, is standardized and well accepted and should continue to be used initially. An interesting variation of this assay proposed by Nelson and Garry [45] utilizes the appearance of isolated transformed colonies. This assay requires further validation and standardization. This colony assay loses the advantage of scoring transformed foci against a background of normal cells. Scoring transformed foci against a monolayer aids in visualization of transformed cells. Since scoring isolated transformed colonies is subjective, it is more difficult to do that in our laboratory than scoring transformed foci against a monolayer as originally described in the transformation assay of Reznikoff et al. [8].

In the standard transformation assay of Reznikoff et al. [8] twenty 60-mm dishes, each seeded with 1000 cells per dish, are employed for testing each concentration of the compound. Cytotoxicity assays should be conducted simultaneously with transformation assays and the same concentrations of compound tested in both assays. In cytotoxicity determinations, five 60-mm dishes are each seeded with 200 cells for survival fractions from 1.0 to 0.1, and five with 2000 cells per dish for survival fractions below 0.1.

For routine transformation assays of suspected carcinogens, we have increased time of treatment with test compound from the original 24 h [8] to 48 h and the number of cells from 1000 to 2000 cells per dish to increase the sensitivity of the assay [15]. This allows 24 h for induction of cytochrome P450 and a further 24 h for P450 to metabolize carcinogens, giving cells more time to metabolize and process transforming metabolites. This yields a higher transformation response in 10T1/2 cells treated with with 0.1 μg/ml MCA compared to a 24 h treatment (Landolph, J.R., unpublished data). In another variation of the treatment regimen, Oshiro and Balwierz [36] used two 3-day exposures to carcinogens and obtained significant transformation with aflatoxin B1, benz[a]anthracene, and 4-nitroquinoline-1-oxide compounds that give marginal or weak transformation in the original assay of Reznikoff et al. [8].

Chemicals for which 10T1/2 cells lack appropriate activating enzymes should be tested for transformation following activation of the compound by added exogenous activating enzymes, such as S9 containing cytochrome P450 or rat liver hepatocytes [46,47]. Chemically reactive compounds with short half lives such as alkylating agents (e.g., MNNG) should be studied with the modified procedure of Nesnow et al. [29] or in synchronized 10T1/2 cells as per McCormick and Bertram [26].

At present, the transformation assay protocol is being improved so as to increase its sensitivity for detecting weak carcinogens. If no transformation is observed when 2000 cells are seeded per 60-mm dish and treated 1 day after seeding, then alternative, more sensitive strategies should be tried. To start with, the cells can be seeded and then treated 5 days later when there are more cells per dish [29]. This procedure is convenient and increases the sensitivity of the assay for certain carcinogens such as aflatoxin B1 and BaP [29]. This modified assay also detects transformation by MNNG and eliminates the tedious requirement to use synchronized cells for such activated compounds [29].

An alternative strategy is to seed 10,000 cells per 60-mm dish to increase the number of survivors and to treat cells one day after seeding, as in the standard Reznikoff et al. assay [46–48]. It is not yet clear whether the procedure of Nesnow et al. [29] or the latter procedure [46–48] is more sensitive in generating more number of total foci. Further work needs to be done to determine this. If these newer strategies are ineffective, the time of exposure to carcinogen can also be increased by employing two sequential 3-day exposures to candidate carcinogens [36].

If no transformation is obtained with these procedures, one can also employ the level-II assay [42,47]. In this assay, cells are seeded at 10,000 per 60-mm dish, treated as in the standard assay of Reznikoff et al. [8], and maintained until four weeks into the assay. At this point, the cells are split at a density of 1×10^5 into each of five new 60-mm dishes. These newly seeded dishes are maintained for a further period of 2 weeks, and then the foci are scored. Schechtman et al. [42] and Tu et al. [48] showed that for certain carcinogens, such as MNNG, lead acetate, dibenz[a]anthracene, and beryllium sulfate, this procedure is more sensitive than the original assay of Reznikoff et al. [8]. However, this level-II assay needs further validation, standardization, development into a quantitative assay, and theoretical justification as to why the number of foci increases, before it is widely accepted.

If a compound is such a weak carcinogen that its measured transformation activity in all these assays remains ambiguous, assays for base substitution mutation to Oua^r in 10T1/2 cells can be conducted [15]. Evidence of mutation could confirm genotoxic properties for mutagenic carcinogens that weakly transform 10T1/2 cells.

9.5.6 PROTOCOLS FOR SCORING TRANSFORMED FOCI

Investigators beginning to use the 10T1/2 cell transformation assay should consult the original paper of Reznikoff et al. [8], describing the morphology of Type-II and Type-III foci and should study the photomicrographs in that publication, both before and during the scoring of the assay to calibrate the scorer. All personnel scoring transformation assays should be trained by the principal investigator of the laboratory and whenever possible the principal investigator should score the transformation assay. Alternatively, two or more people should score the same assay and compare results.

In acceptable assays with properly functioning negative and positive controls, all areas of the dish should first be examined by eye, and foci or focus-like areas circled with a glass marking pencil. These areas should then be examined under a dissecting microscope and scored as Type II or Type III foci. When in doubt, these foci should be examined under higher magnification (20× or 100×), and particular attention should be paid to the morphology of the edge of the focus. Crisscrossing of cells at the focus edge serves to confirm the identity of Type-III foci. Occasionally, Type-III foci also stain a deep-blue color compared to the usual purple color of Giemsa-stained Type-II foci.

Mixed or pleiomorphic foci occasionally occur in the transformation assay. When mixed foci occur that are mixtures of Type-II and Type-III focal areas, they should be scored as Type-III foci. Similarly, when foci occur that are mixtures of Type-II and Type-I focal areas, they should be scored as Type-II foci.

Frequently, uniform foci occur that appear to be intermediate between the previously described Type-I, Type-II, and Type-III foci. When a focus is homogeneous and morphologically intermediate between a Type-III and a Type-II focus, we recommend scoring it conservatively as a Type-II focus. Similarly, when a focus is homogeneous and morphologically intermediate between a Type-I and a Type-II focus, we recommend scoring it conservatively as a Type-I focus and hence not score it in the transformation assay. Finally, we recommend that the investigator score a transformation assay, put it aside for a while, and then rescore it again at a later date to be confident of the validity of the scoring. Due to the complexity and the poorly understood nature of the cell density dependence of transformation frequency, we recommend reporting the number of Type-III foci/total dishes, the number of Type-II or Type-III foci/total dishes, the numbers of dishes with Type-II or Type-III foci out of the total number of dishes, and the number of dishes with Type-III foci out of the total number of dishes. We do not recommend calculating the transformation frequencies at this time.

When feasible, representative Type-II and Type-III foci induced in 10T1/2 cells by a new compound should be studied for their ability to grow in soft agarose and to cause tumors in nude mice when 1×10^6 cells are injected intrascapulary.

9.5.7 Recommended Future Research to Enhance the Utility of the 10T1/2 Cell Transformation Assay

In the short-term, a series of validation studies should be conducted in which sensitivity of the 10T1/2 cell transformation assay is determined using the standard (Reznikoff) assay [8] as modified by seeding and treating 2,000 cells per 60-mm dish for 48 h [15], the assay employing 10,000 cells per dish [42,46–48], the assay in which cells are treated on day five after seeding [29], the assay employing two sequential 3-day treatments [36], and the level-II assay [42,47]. A series of standard compounds should be constructed and tested in all four assays in different laboratories. After agreement among the laboratories, the most sensitive assay method should be adopted. Second, more research should be directed toward understanding theoretically the cell density dependence of transformation frequency.

Third, more effort should be devoted to develop inexpensive serum-free media in which to grow 10T1/2 cells. This should reduce the serum-dependent variability of carcinogen-induced focus formation and make the initiation–promotion assays more reproducible and more widely used. Finally, newer quantitative methods for scoring foci should be developed. In particular, immunological methods and techniques to analyze perturbations in oncogene expression should be explored to develop quantitative methods to determine the number of transformed cells in a population of 10T1/2 cells treated with a chemical carcinogen.

9.6 SIMULTANEOUS CHEMICAL INDUCTION OF MORPHOLOGICAL TRANSFORMATION, MUTATION, AND CHROMOSOMAL ABERRATIONS IN C3H/10T1/2 CELLS

We developed an assay for chemically induced mutation to ouabain resistance (Oua^r) in 10T1/2 cells in our laboratories many years ago [15]. We showed that chemically induced Oua^r mutants of

10T1/2 cells are stably resistant to ouabain [15,49], possess a Oua^r ^{86}Rubidium (Rb) transport (reflecting a Oua^r potassium transport) [49], and possess a Oua^r sodium, potassium adenosinetriphosphatase ([Na, K] ATPase) activity [50]. Further, these chemically induced Oua^r mutants are specifically resistant to ouabain but not to unrelated drugs [51]. We showed that Oua^r mutants are induced by base-substitution mutagens (MNNG) but not by frameshift mutagens (ICR 191 and ICR 170) [10]. We have mapped chemically induced Oua^r mutations to mouse chromosome 3 by microcell-mediated chromosome transfer [52]. Therefore, phenotypic, genetic, and biochemical evidences indicate that chemically induced Oua^r in 10T1/2 cells is due to specific types of base substitution mutations in the (Na, K) ATPase of 10T1/2 cells (reviewed in References 6, 49, 50, and 53).

By using this mutation assay, we studied chemically induced mutation, chemically induced chromosomal aberrations, and chemically induced morphological transformation simultaneously in 10T1/2 cells. The two chemical carcinogens BaP and *N*-acetoxyacetylaminofluorene induced concentration-dependent morphological transformation as well as mutation to Oua^r over the same concentration ranges (Figure 9.3 [15]; reviewed in Reference 6). This suggested that mutation is a likely step that 10T1/2 cells undergo during transformation by mutagenic organic chemical carcinogens ([15]; reviewed in References 6 and 53). The induction of cytotoxicity, mutation to Oua^r, morphological transformation, and chromosomal aberrations by BaP and its metabolites has been studied in 10T1/2 cells (Table 9.2). Furthermore, BaP and its metabolite, the ±(*trans*)-7,8-dihydrodiol of BaP, induce mutation to Oua^r, chromosomal aberrations, and morphological transformation in 10T1/2 cells [15,16]. Chan and Little [32] have also shown that UV light induced morphological transformation and mutation to Oua^r over the same concentration ranges as in 10T1/2 cells. Whether taken independently or together, the results of Landolph and Heidelberger [15] and Chan and Little [32] suggest that it is likely that mutation is a part of the molecular mechanism by which these mutagenic chemical carcinogens such as *N*-acetoxy-AAF and BaP [15] and

TABLE 9.2

Induction of Morphological Transformation, Mutation to Oua^r, and Chromosomal Aberrations in 10T1/2 Cells

Compound/ treatment	Induction of morphological transformation in 10T1/2 cells	Induction of mutation to ouabain resistance in 10T1/2 cells	Induction of chromosome aberrations in 10T1/2 cells	References
BaP	+	+	+	15, 16
N-acetoxy-AAF	+	+		15
BaP 7,8-dihydrodiol	+	+	+	16
UV light	+	+		32
MNNG	+	+		15, 24, 25, 26
Nickel subsulfide	+	−	+	22, 55
Green nickel oxide	+	−	+	22, unpublished
Crystalline nickel	+	−		22
Nickel refinery dust monosulfide	+		+	63
Lead chromate	+	−		21
5-azacytidine	+	−	+	10,17
Flubendazole	+		+	23
Harringtonine	+		+	23

UV light [32] induce morphological transformation in 10T1/2 cells [15]. The ability of BaP and its (trans)-7,8-dihydrodiol metabolite to additionally induce chromosomal aberrations [16] also suggests that mutation, which in proto-oncogenes is likely to activate them into oncogenes, or which in tumor suppressor genes thereby inactivates them, and chromosomal breakage, causing loss of parts of chromosomes bearing tumor suppressor genes, probably contributes to the mechanisms of chemically induced morphological and neoplastic cell transformation in 10T1/2 cells ([15,16,32]; reviewed in Reference 53).

This assay for chemically or radiation induced mutation to Oua^r can therefore be used to detect the mutagenicity of mutagenic carcinogens and to study mechanisms of morphological transformation in 10T1/2 cells by assaying mutation over the same concentration ranges as morphological transformation that is induced by various mutagenic carcinogens ([15]; reviewed in References 6 and 53).

For BaP and the ±(trans)-7,8-dihydrodiol of BaP, there was a correlation between their abilities to induce cytotoxicity, mutation to Oua^r, chromosome aberrations, and morphological cell transformation [15,16]. For the other BaP metabolites studied, except the ±(anti)-7,8-dihydrodiol-9,10 epoxide, there was no detectable mutation, transformation, or chromosomal aberration. The ± (anti)-7,8-dihydrodiol-9,10 epoxide of BaP caused strong mutation to Oua^r but only weak transformation and little or no chromosome damage in asynchronous 10T1/2 cells [15,16]. Presumably, this is in part because this anti diol epoxide may be strictly an initiator of neoplastic transformation [15,16]. In addition, due to its short half-life in cell culture medium, this compound needs to be tested in synchronized 10T1/2 cells or in the modified assay of Nesnow et al. [29] in order to accurately detect its ability to induce morphological cell transformation.

Interestingly, we found that the alkylating agent, MNNG, effectively mutates asynchronous 10T1/2 cells to Oua^r [15,54], but it does not induce significant amounts of morphological cell transformation in asynchronous 10T1/2 cells in the standard cell transformation assay [54]. MNNG will induce morphological transformation of asynchronous 10T1/2 cells in the modified transformation assay procedure of Nesnow et al. [29], wherein the cells are seeded, and treated with MNNG 5 days post-seeding. The greater efficiency of induction of morphological cell transformation by treating cells on day five post-seeding compared to the standard cell transformation assay, where cells are treated with MNNG one day post-seeding, is presumably due to the larger number of cells in S-phase than in the standard cell transformation assay. Consistent with this hypothesis and the observation of increased sensitivity of 10T1/2 to MNNG-induced morphological transformation in the modified transformation assay of Nesnow et al. [29]; McCormick and Bertram [26] demonstrated that MNNG efficiently induces morphological transformation and mutation to Oua^r in synchronized 10T1/2 cells.

The cancer chemotherapeutic agent, 5-azacytidine, induces morphological transformation of 10T1/2 cells [17]. However, 5-azacytidine does not induce detectable mutation of 10T1/2 cells to Oua^r nor does it significantly mutate V79 cells to either Oua^r or to resistance to 8-azaguanine (Azg^r) [10]. In addition, nickel subsulfide [22,55], green nickel oxide [22], and crystalline nickel monosulfide [22] induce morphological transformation in 10T1/2 cells, but these three compounds do not induce mutation to Oua^r in 10T1/2 cells over the same concentration ranges that they induce morphological transformation in 10T1/2 cells ([22]; reviewed in References 53 and 56). Similarly, our laboratory has shown that lead chromate induces morphological transformation in 10T1/2 cells, but did not induce mutation to Oua^r in these cells ([21]; reviewed in References 53 and 56).

Hence, certain carcinogens — BaP, activated BaP metabolites, and N-acetoxyacetylaminofluorene — simultaneously induce base substitution mutations, chromosomal breakage, and morphological and neoplastic transformation. Other carcinogens, such as nickel subsulfide, green (high temperature) nickel oxide, crystalline nickel monosulifde, and lead chromate, induce morphological transformation but no detectable base substitution mutation to Oua^r in 10T1/2 cells [22]. The insoluble nickel compounds also induce chromosomal aberrations in 10T1/2 cells (Verma, R., and Landolph, J.R., manuscript in preparation). Hence, it is likely that these insoluble nickel compounds induce morphological cell transformation partly through chromosomal aberrations, which probably cause inactivation of loss of tumor suppressor genes, and also through additional nonmutagenic mechanisms (reviewed in References 53 and 56).

The transformation assay in 10T1/2 cells has also been used to demonstrate that retinoids [57,58] and ascorbate [59] specifically inhibit chemically induced morphological transformation in 10T1/2 cells, probably at the promotional stages of chemically induced transformation (Figure 9.3).

An assay for initiation and promotion of transformation, analogous to that which occurs *in vivo*, was established [60] and used to demonstrate phorbol ester-induced promotion of PAH- [60], x-ray- [30], and aflatoxin B1- and MNNG-initiated [61] transformation of 10T1/2 cells (Figure 9.3). However, the variability of serum batches to support promotion of transformation [61] has limited the utility of this promotion assay.

9.7 MOLECULAR BIOLOGY OF CHEMICALLY INDUCED MORPHOLOGICAL AND NEOPLASTIC CELL TRANSFORMATION IN C3H/10T1/2 CELLS

Efforts in our laboratory have been directed toward understanding the molecular biology of chemically induced morphological and neoplastic transformation in 10T1/2 cells. We were able to show that there were increased steady-state levels of c-myc RNA in two bleomycin-induced, transformed 10T1/2 cell lines, in two UV-induced, transformed 10T1/2 cell lines, in two MCA-induced, transformed 10T1/2 cell lines, in one x-ray-induced, transformed 10T1/2 cell line, and in one neutron-induced, transformed 10T1/2 cell line [62]. The increases in the steady-state levels of c-myc RNA in the transformed 10T1/2 cell lines ranged from 2.5-fold to 7-fold compared to the levels in the nontransformed 10T1/2 cells. Interestingly, eight out of eight transformed cell lines had increased steady-state levels of c-myc RNA [62]. There was no amplification of the *c-myc* gene in the transformed cell lines, and no changes in methylation status of the *c-myc* gene in the transformed cell lines [62]. In a second paper from our laboratory, we were able to demonstrate that there were increased steady-state levels of c-myc mRNA in two methycholanthrene-induced, transformed 10T1/2 cell lines, in two bleomycin-induced, transformed 10T1/2 lines, in two UV light-induced transformed 10T1/2 cell lines, in three x-ray induced, transformed 10T1/2 cell lines, and in one neutron-induced, transformed 10T1/2 cell line [63]. Hence, in an additional 11 out of 12 transformed 10T1/2 cell lines, there were increased steady-state levels of c-myc mRNA. These increases in the steady-state levels of c-myc RNA in the transformed 10T1/2 cell lines ranged from 38-fold to 8-fold [63]. There was no amplification of, or changes in the methylation status of the *c-myc* gene in these transformed cell lines [63]. Hence, we proposed that increased steady-state levels of c-myc RNA were necessary for the induction and maintenance of the transformed state of these transformed, tumorigenic 10T1/2 cell lines [62,63].

Moreover, we have been studying the molecular mechanisms of morphological and neoplastic transformation induced by carcinogenic, insoluble nickel compounds in 10T1/2 cells. We have shown that nickel subsulfide, crystalline nickel monosulfide, and green (high temperature) nickel oxide induced morphological and anchorage-independent transformation in 10T1/2 cells without inducing mutation to Oua^r ([22]; reviewed in References 53 and 56). One transformed cell line derived from a focus induced by green nickel oxide formed tumors in nude mice [22].

Recently, we also studied nickel refinery samples from the International Nickel Company's nickel refinery in Clydach, Wales, in the United Kingdom. This refinery has been in operation since 1901. At this refinery, 365 respiratory cancers have occurred in refinery workers since the 1920s. From 1901 to 1923, incidences of these cancers were high. In 1923, the refining process was modified. This modification included removing arsenic-contaminated sulfuric acid from the refining process thereby eliminating a nickel arsenide, Ni_5As_2, called orcelite, from the refining process. Incidences of respiratory cancers then decreased significantly between 1925 to 1929 (reviewed in Reference 64). The orcelite content of the archived 1920 sample was 10%, and that of the 1929 sample was 1%.

We hypothesized that the 10% orcelite content of the 1920 sample was at least partially responsible for inducing nasal and respiratory cancers in the workers [64]. We found that the archived 1920 nickel refinery sample induced strong morphological transformation in 10T1/2 cells in a dose-dependent manner, while the 1929 sample did not induce any detectable morphological transformation in 10T1/2 cells [64]. These findings correlated with the high incidence of respiratory cancer at this nickel refinery before 1923, and the decreased cancer incidence after 1923. Hence, this data supports our hypothesis that the high content of orcelite in the 1920 sample contributed to its carcinogenicity, which affected the nickel refinery workers [63].

Next, we began to study the molecular biology of morphological and neoplastic cell transformation in 10T1/2 cells induced by specific pure insoluble, carcinogenic nickel compounds. In order to do this, we first derived four transformed 10T1/2 cell lines induced by nickel monosulfide (three cell lines) and one transformed cell line induced by green (high temperature) nickel oxide [22]. We utilized the technique of random arbitrarily primed (RAP) polymerase chain reaction (PCR), or the RAP-PCR technique [65,66]. With this technique, we demonstrated that nine genes were differentially expressed between the four nickel compound-induced, transformed 10T1/2 cell lines compared to nontransformed 10T1/2 cells in 6% of the total mRNA that we analyzed [65,66]. Therefore, in 100% of the mRNA, by extrapolation, 130 genes would be differentially expressed between the nickel compound-induced, transformed cell lines and nontransformed 10T1/2 cells [65,66]. This indicates that these nickel compound-induced, transformed 10T1/2 cell lines have global aberrations in gene expression, and that control of gene expression is substantially disrupted in these nickel compound transformed cell lines.

Second, we showed that a number of these genes were expressed at higher steady-state levels in the nickel compound-induced, transformed 10T1/2 cell lines. We were able to identify a number of these genes whose expression was altered by isolating them from the differential display gels, subcloning the gene fragments we isolated, sequencing these gene fragments, and entering the sequences of these gene fragments into the NCBI GenBank database. We then performed a Basic Local Alignment Search Tool (BLAST) search to identify any homologies between our gene fragments and those in the database. We identified the *calnexin* gene as one gene that was expressed at higher steady-state levels in the nickel compound transformed 10T1/2 cell lines [65,66]. *Calnexin* is a gene that encodes a 90 kDa Type-I integral membrane phosphoprotein that binds transiently to a number of soluble and membrane-bound nascent polypeptides. It functions as a molecular chaperone in transporting secretory glycoproteins from the endoplasmic reticulum (ER) to the outer cellular membrane [67–71]. It utilizes adenosine triphosphate (ATP) and calcium ion as two cofactors involved in substrate binding. It binds to a large number of membrane-bound and soluble proteins in the ER, including subunits of the T-cell receptor, the B-cell antigen receptor, the major histocompatibility complex (MHC) class I and II protein, viral glycoproteins, and integrins ([67–71]; reviewed in References 65 and 66). It also aids in proper oligomerization of a subset of integral membrane multisubunit proteins [67–71]. The exact role of calnexin in intracellular processes, however, is not clearly understood. We hypothesize that *calnexin* helps, along with other calcium binding proteins, to mediate a complex equilibrium of the distribution of intracellular calcium ions [65,66]. *Calnexin* is expressed at low levels in nontransformed 10T1/2 cells and at substantially increased steady-state levels in morphologically transformed 10T1/2 cell lines, induced by carcinogenic, insoluble nickel compounds ([65,66] and Ramnath, J.R., and Landolph, J.R., manuscript in preparation). It may also be functioning as a stress protein in the nickel transformed cell lines.

A second protein that we identified was expressed at higher steady-state levels in the nickel transformed 10T1/2 cell lines was the *ect-2* proto-oncogene ([65,66], and Clemens, F., et al., manuscript in press). The ect-2 protein product is a GDP-GTP exchange factor for rho, which aids in microtubule reorganization [72]. The *ect-2* gene is mutated in specific human tumors [72]. We have found that there are higher steady-state levels of ect-2 mRNA and protein in nickel transformed 10T1/2 cell lines, and that there is amplification of the *ect-2* gene in the transformed cell lines (Clemens et al., manuscript in press). Our current working hypothesis is that amplification of the *ect-2* proto-oncogene leads to higher steady state levels of ect-2 mRNA and protein, and this leads to higher steady-state levels of rho-GTP. Higher steady-state levels of rho-GTP would lead to microtubular disaggregation, forcing the transformed cells into S-phase and DNA synthesis. In addition, we also identified the stress-inducible gene, *Wdr1*, to be expressed at higher steady-state levels in the nickel compound-induced, transformed cell lines [65,66]. *Wdr1* was previously identified by Dr. Margaret Lomax's laboratory as being expressed at higher steady-state levels in the ears of chicks exposed to acoustic shock [73]. The function of this gene is not yet known. It may be functioning like a stress-response gene in nickel compound-induced, transformed 10T1/2 cell lines [65,66].

We also identified six genes that were not expressed at detectable levels or that were expressed at substantially lower steady-states levels in the nickel compound-induced, transformed 10T1/2 cell lines. These genes that were not expressed or expressed at substantially lower steady-state levels in the transformed cell lines include the vitamin D interacting protein/thyroid hormone activating protein (*DRIP/TRAP-80*) gene [74,75], the insulin-like growth factor receptor 1 (*IGFR1*) gene, the small nuclear activating protein (*SNAP C3*) gene, and three unknown genes. The DRIP/TRAP-80 protein is interesting, because it forms a part of a complex that binds to the vitamin D/vitamin D receptor complex, and makes this large complex function as a transcription factor [74,75]. Deletion of the DRIP/TRAP-80 protein in the transformed cells may cause deletion of the expression of a number of protein products mediating calcium metabolism in the cell, with consequent dysregulation of calcium metabolism and cell growth [64,65].

We therefore hypothesized that higher steady-state levels of expression of the *calnexin* gene, the *ect-2* proto-oncogene, and the stress-inducible gene, *Wdr1*, and concomitant under-expression or lack of expression of the *DRIP/TRAP-80* gene, the *IGFR1* gene, the *SNAP C3* gene and the three unknown genes, together contribute to the induction and maintenance of morphological, anchorage-independent, and neoplastic transformation induced by specific insoluble nickel compounds in 10T1/2 cells [65,66]. Our current working hypothesis is that the insoluble nickel compounds we studied (crystalline nickel monosulfide and green nickel oxide) caused amplification of the *ect-2* gene and expression of its mRNA and protein product at higher steady-state levels, mutational activation of additional proto-oncogenes, and mutational inactivation or deletion of a number of tumor suppressor genes. We believe that these nickel compounds have induced mutational activation of oncogenes and mutational inactivation of tumor suppressor genes, and that approximately eight of these events have occurred in total in the transformed 10T1/2 cell lines. We also postulate that each oncogene that was activated has caused 20 additional genes to become expressed at higher steady-state levels. We further postulate that each tumor suppressor gene that has become inactivated by these nickel compounds has caused an additional 20 genes to become transcriptionally silent. Hence, in total, we would expect $8 \times 20 = 160$ genes to become aberrantly expressed in the transformed cell lines, which is close to the 130 aberrations in gene expression that we observed in these nickel compound-induced, transformed 10T1/2 cell lines [65,66]. Studies to test these hypotheses critically are in progress in our laboratory and should lead to further insight into the molecular mechanisms of carcinogenesis induced by specific carcinogenic, insoluble nickel compounds.

REFERENCES

1. Heidelberger, C. Oncogenic transformation of rodent cell lines by chemical carcinogens. In *Origins of Human Cancer, Part C, Human Risk Assessment*, Hiatt, J.H., Watson, J.D., and Winsten, J.A., Eds., Cold Spring Harbor, NY, Cold Spring Harbor Laboratory, 1977, pp. 1513–1520.
2. Heidelberger, C. and Mondal, S. In vitro chemical carcinogenesis. In *Carcinogens Identification and Mechanisms of Action*, Griffin, A.C. and Shaw, C.R., Eds., New York, Raven Press, 1979, pp. 83–92.
3. Heidelberger, C. Initiation, promotion, transformation, and mutagenesis of mouse embryo cell lines. In *Proceedings of the 13th International Cancer Congress, Part B, Biology of Cancer*, Mirand, E.A., Hutchinson, W.B., and Mihich, E., Eds., New York, Alan R. Liss, 1983, pp. 83–89.

4. Heidelberger, C., Freeman, A.E., Pienta, R.J., Sivak, A., Bertram, J.S., Casto, B.C., Dunkel, V.C., Francis, M.W., Kakunaga, T., Little, J.B., and Schechtman, L.M. Cell transformation by chemical agents — review and analysis of the literature. A report of the U.S. Environmental Protection Agency Gene-Tox Program. *Mutat. Res.*, *114*: 283–385, 1983.

5. Heidelberger, C., Landolph, J.R., Fournier, R.E.K., Fernandez, A., and Peterson, A.R. Genetic and probability aspects of cell transformation by chemical carcinogens. *Prog. Nucleic Acid Res. Mol. Biol.*, *2*: 87–98, 1983.

6. Landolph, J.R. Chemical transformation in C3H/10T1/2 Cl 8 mouse embryo fibroblasts: Historical background, assessment of the transformation assay, and evolution and optimization of the transformation assay protocol. In *Transformation Assay of Established Cell Lines: Mechanisms and Applications*, Kakunaga, T. and Yamasaki, H., Eds., IARC Scientific Publications, No. 67, International Agency for Research on Cancer, Lyon, France, 1985, pp. 185–198.

7. Reznikoff, C.A., Brankow, D.W., and Heidelberger, C. Establishment and characterization of a cloned line of C3H mouse embryo cells sensitive to postconfluence inhibition of division. *Cancer Res.*, *33*: 3231–3238, 1973.

8. Reznikoff, C.A., Bertram, J.S., Brankow, D.W., and Heidelberger, C. Qualitative and quantitative studies of chemical transformation of cloned C3H mouse embryo cells sensitive to postconfluence inhibition of cell division. *Cancer Res.*, *33*: 3239–3249, 1973.

9. Ruley, H.E. Adenovirus early region 1A enables viral and cellular transforming genes to transform primary cells in culture. *Nature*, *304*: 601–607, 1983.

10. Landolph, J.R. and Jones, P.A. Mutagenicity of 5-azacytidine and related nucleosides in C3H/10T1/2 Clone 8 and V79 cells. *Cancer Res.*, *42*: 817–823, 1982.

11. Constantinides, P.G., Jones, P.A., and Gevers, W. Functional striated muscle cells form non-myoblast precursors following 5-azacytidine treatment. *Nature*, *267*: 364–366, 1977.

12. Constantinides, P.G., Taylor, S.M., and Jones, P.A. Phenotypic conversion of cultured mouse embryo cells by Aza pyrimidine nucleosides. *Dev. Biol.*, *66*: 57–71, 1978.

13. Taylor, S.M. and Jones, P.A. Multiple new phenotypes induced in 10T1/2 and 3T3 cells treated with 5-azacytidine. *Cell*, *17*: 771–779, 1990.

14. Jones, P.A. and Taylor, S.M. Cellular differentiation, cytidine analogs, and DNA methylation. *Cell*, *20*: 85–93, 1980.

15. Landolph, J.R. and Heidelberger, C. Chemical carcinogens produce mutations to ouabain resistance in transformable C3H/10T1/2 Cl 8 mouse fibroblasts. *Proc. Natl. Acad. Sci. USA*, *76*: 930–934, 1979.

16. Gehly, E.B., Landolph, J.R., Heidelberger, C., Nagasawa, H., and Little, J.B. Induction of cytotoxicity, mutation, cytogenetic changes, and neoplastic transformation by benzo(a)pyrene and derivatives in C3H/10T1/2 Clone 8 mouse fibroblasts. *Cancer Res.*, *42*: 1866–1875, 1982.

17. Benedict, W.F., Banerjee, A., Gardner, A., and Jones, P.A. Induction of morphological transformation in mouse C3H/10T1/2 Clone 8 cells and chromosomal damage in hamster A(T$_1$)Cl-3 cells by cancer chemotherapeutic agents. *Cancer Res.*, *37*: 2202–2208, 1977.

18. Billings, P.C., Uwaifo, A., and Heidelberger, C. Influence of benzoflavone on aflatoxin B1-induced cytotoxicity, mutation and transformation of C3H/10T1/2 cells. *Cancer Res.*, *43*: 2659–2663, 1983.

19. Billings, P.C, Heidelberger, C., and Landolph, J.R. S-9 metabolic activation enhances aflatoxin-mediated transformation of C3H/10T1/2 cells. *Toxicol. Appl. Pharmacol.*, *77*: 58–65, 1985.

20. Benedict, W.F., Rucker, N., Faust, J., and Kouri, R .E. Malignant transformation of mouse cells by cigarette smoke condensate. *Cancer Res.*, *35*: 857–860, 1975.

21. Patierno, S.R., Bahn, D., and Landolph, J.R. Transformation of C3H/10T1/2 mouse embryo cells to focus formation and anchorage independence by insoluble lead chromate but not by soluble calcium chromate: Relationship to mutagenesis and internalization of lead chromate particles. *Cancer Res.*, *48*: 5280–5288, 1988.

22. Miura, T., Patierno, S.R., Sakuramoto, T., and Landolph, J.R. Morphological and neoplastic transformation of C3H/10T1/2 Cl 8 mouse embryo cells by insoluble carcinogenic nickel compounds. *Environ. Mol. Mutagen.*, *14*: 65–78, 1989.

23. Nianjun, H., Cerapnalkoski, L., Nwankwo, J.O., Dews, M., and Landolph, J.R. Induction of chromosomal aberrations, cytotoxicity, and morphological transformation in mammalian cells by the antiparasitic drug, flubendazole, and the antineoplastic drug, harringtonine. *Fundam. Appl. Toxicol.*, *22*: 304–313, 1994.

24. Bertram, J.S. and Heidelberger, C. Cell cycle dependency of oncogenic transformation induced by N-methyl-N-nitro-N-nitrosoguanidine in culture. *Cancer Res.*, *34*: 526–537, 1974.

25. Grisham, J.W., Greenburg, D.S., Kaufman, D.G., and Smith, G. Cell cycle related toxicity and transformation in 10T1/2 cells treated with N-methyl-N'-nitro-N-nitrosoguanidine. *Proc. Natl. Acad. Sci. USA*, *77*: 4813–4817, 1980.

26. McCormick, P. and Bertram, J.S. Differential cell cycle phase specificity for neoplastic transformation and mutation to ouabain resistance induced by N-methyl-N-nitro-N-nitrosoguanidine. *Proc. Natl. Acad. Sci. USA*, *79*: 4342–4346, 1982.

27. Jones, P.A., Benedict, W.F., Baker, M.S., Mondal, S., Rapp, U., and Heidelberger, C. Oncogenic transformation of C3H/10T1/2 Clone 8 mouse embryo cells by halogenated pyrimidine nucleosides. *Cancer Res.*, *36*: 101–107, 1976.

28. Jones, P.A., Baker, M.S., Bertram, J.S., and Benedict, W.F. Cell-cycle specific oncogenic transformation of C3H/10T1/2 Clone 8 mouse embryo cells by 1-β-D-arabinofuranosylcytosine. *Cancer Res.*, *37*: 2214–2217, 1977.

29. Nesnow, S., Garland, H., and Curtis, G. Improved transformation of C3H/10T1/2 Cl cells by direct and indirect-acting carcinogens. *Carcinogenesis*, *3*: 377–380, 1982.

30. Kennedy, A.R., Mondal, S., Heidelberger, C., and Little, J.B. Enhancement of x-ray transformation by 12-O-tetradecanoyl-phorbol-13 acetate in a cloned line of C3H mouse embryo cells. *Cancer Res.*, *38*: 439–443, 1978.

31. Kennedy, A.R., Fox, M., Murphy, G., and Little, J.B. Relationship between x-ray exposure and malignant transformation in C3H/10T1/2 cells. *Proc. Natl. Acad. Sci. USA*, *77*: 7262–7266, 1980.

32. Chan, G.L. and Little, G.B. Induction of ouabain resistance mutations in C3H/10T1/2 mouse cells by ultraviolet light. *Proc. Natl. Acad. Sci. USA*, *75*: 3363–3366, 1978.

33. Gehly, E.B., Fahl, W.F., Jefcoate, C.R., and Heidelberger, C. The metabolism of benzo(a)pyrene by cytochrome P-450 in transformable and nontransformable C3H mouse fibroblasts. *J. Biol. Chem.*, *254*: 5041–5048, 1979.

34. Gehly, E.B. and Heidelberger, C. Metabolic activation of benzo(a)pyrene by transformable and nontransformable C3H mouse fibroblasts in culture. *Cancer Res.*, *42*: 2697–2704, 1982.

35. Wang, T.V. and Cerruti, P.A. Effect of formation and removal of aflatoxin B1: DNA adducts in 10T1/2 mouse embryo fibroblasts on cell viability. *Cancer Res.*, *40*: 2904–2909, 1980.

36. Oshiro, Y. and Balweirz, P.S. Morphological transformation of C3H/10T1/2 cells by procarcinogens. *Environ. Mutagen.*, *4*: 105–108, 1982.

37. Boreiko, C. and Heidelberger, C. Isolation of mutants temperature-sensitive for expression of the transformed state from chemically transformed C3H/10T1/2 cells. *Carcinogenesis*, *1*: 1059–1073, 1980.

38. Oshiro, Y., Balwierz, P.S., and Piper, C.E. Selection of fetal bovine serum for use in the C3H/10T1/2 Cl 8 cell transformation assay system. *Environ. Mutagen.*, *4*: 569–574, 1982.

39. Land, H., Parada, L.F., and Weinberg, R.A. Tumorigenic conversion of primary embryo fibroblasts requires at least the cooperating oncogenes. *Nature*, *304*: 596–602, 1983.

40. Newbold, R.F. and Overell, R.W. Fibroblast immortality is a prerequisite for transformation by EJ *c-Ha-ras* oncogene. *Nature*, *304*: 648–651, 1983.

41. Shih, C., Shilo, B., Goldfarb, M.P., Dannenberg, A., and Weinberg, R.A. Passage of the phenotypes of chemically transformed cells via transfection of DNA and chromatin. *Proc. Natl. Acad. Sci. USA*, *76*: 5714–5718, 1979.

42. Schechtman, L.M., Kiss, F., McCarvill, J., Gallagher, M., Kouri, R.E., and Lubet, R.A. A method for amplification of sensitivity of the C3H/10T1/2 cell transformation assay. *Proc. Am. Assoc. Cancer Res.*, *23*: 74, 1982.

43. Haber, D.A., Fox, D.A., Dynan, W.S., and Thilly, W.G. Cell density dependence of focus formation in the C3H/10T1/2 transformation. *Cancer Res.*, *37*: 1644–1648, 1977.

44. Fernandez, A., Mondal, S., and Heidelberger, C. Probabilistic view of the transformation of cultured C3H/10T1/2 mouse embryo fibroblasts by 3-methylcholanthrene. *Proc. Natl. Acad. Sci. USA*, *77*: 7272–7276, 1980.

45. Nelson, R.L. and Garry, V.F. Colony transformation in C3H/10T1/2 cells: Stepwise development of neoplastic change *in vitro*. *In Vitro*, *19*: 551–558, 1983.

46. Sivak, A. and Tu, A.S. Use of rodent hepatocytes for metabolic activation in transformation assays. In *Transformation Assay of Established Cell Lines: Mechanisms and Application*, Kakunaga, T. and Yamasaki, H., Eds., IARC Scientific Publications No. 67., International Agency for Research on Cancer, Lyon, France, 1985, pp. 121–136.

47. Schechtman, L.M. Metabolic activation of procarcinogens by subcellular enzyme fractions in the C3H/10T1/2 and Balb/c3T3 cell transformation systems. In *Transformation Assay of Established Cell Lines: Mechanisms and Application*, Kakunaga, T. and Yamasaki, H., Eds., IARC Scientific Publications No. 67, International Agency for Research on Cancer, Lyon, France, 1985, pp. 137–162.

48. Tu, A., Breen, P., Pallotta, S., and Sivak, A. Examination of modified procedures for the C3H/10T1/2 neoplastic transformation assay. *Proc. Am. Assoc. Cancer Res.*, *24*: 103, 1983.

49. Landolph, J.R., Telfer, N., and Heidelberger, C. Further evidence that ouabain-resistant variants induced by chemical carcinogens in transformable C3H/10T1/2 Cl 8 mouse fibroblasts are mutants. *Mutat. Res.*, *72*: 295–310, 1980.

50. Shibuya, M.L., Miura, T., Lillehaug, J.R., Farley, R.A., and Landolph, J.R. Ouabain-resistant (Na, K)-ATPase enzyme activity in chemically induced ouabain-resistant C3H/10T1/2 cells. *J. Mol. Toxicol.*, *2*: 75–98, 1989.

51. Landolph, J.R., Bhatt, R.S., Telfer, N., and Heidelberger, C. Adriamycin and ouabain induced cyto-toxicity and inhibition of [86]rubidium transport in wild-type and ouabain-resistant C3H/10T1/2 mouse fibroblasts. *Cancer Res.*, *40*: 4581–4588, 1980.

52. Landolph, J.R. and Fournier, R.E.K. Microcell-mediated transfer of carcinogen-induced ouabain resistance from C3H/10T1/2 Cl 8 mouse fibroblasts to human cells. *Mutat. Res.*, *107*: 447–463, 1983.

53. Landolph, J.R. Neoplastic transformation of mammalian cells by carcinogenic metal compounds: Cellular and molecular mechanisms. In *Biological Effects of Heavy Metals, Volume II: Metal Carcinogenesis*, Foulkes, E.C., Ed., CRC Press, Boca Raton, FL, 1990, pp. 2–18, chap. 1.

54. Peterson, A.R., Landolph, J.R., Peterson, H., Spears, C.P., and Heidelberger, C. Oncogenic transform-ation and mutation of C3H/10T1/2 Clone 8 mouse embryo fibroblasts by alkylating agents. *Cancer Res.*, *41*: 3095–3099, 1981.

55. Saxholm, H.J.K., Reith, A., and Brogger, A. Oncogenic transformation and cell lysis in C3H/10T1/2 cells and increased sister chromatid exchange in human lymphocytes by nickel subsulfide. *Cancer Res.*, *41*: 4136–4139, 1981.

56. Landolph, J.R. Molecular mechanisms of transformation of C3H/10T1/2 Cl 8 mouse embyro cells and diploid human fibroblasts by carcinogenic metal compounds. *Environ. Health Perspect.*, *102* (Suppl. 3): 119–125, 1994.

57. Merriman, R.L. and Bertram, J.S. Reversible inhibition by retinoids of 3-methylcholanthrene-induced neoplastic transformation in C3H/10T1/2 Cl 8 cells. *Cancer Res.*, *39*: 1661–1666, 1979.

58. Mordan, L.J., Bersin, L.M., Budnick, J.L., Meegan, R.R., and Bertram, J.S. Isolation of methylcholanthrene-'initiated' C3H/10T1/2 cells by inhibiting neoplastic progression with retinyl acetate. *Carcinogenesis*, *3*: 249–285, 1982.

59. Benedict, W.F., Wheatley, W.I., and Jones, P.A. Inhibition of chemically induced morphological trans-formation and reversion of the transformed phenotype by ascorbic acid in C3H/10T1/2 cells. *Cancer Res.*, *40*: 2796–2801, 1980.

60. Mondal, S., Brankow, D.M., and Heidelberger, C. Two-stage chemical carcinogenesis in cultures of C3H/10T1/2 cells. *Cancer Res.*, *36*: 2254–2260, 1976.

61. Boreiko, C.J., Rajan, D.L., Abernethy, D.J., and Frazelle, J.H. Initiation of C3H/10T1/2 cell trans-formation by *N*-methyl-*N'*-nitro-*N*-nitrosoguanidine and aflatoxin B1. *Carcinogenesis*, *3*: 391–395, 1982.

62. Shuin, T., Billings, P.C., Lillehaug, J.R., Patierno, S.R., Roy-Burman, P., and Landolph, J.R. Enhanced expression of *c-myc* and decreased expression of *c-fos* proto-oncogenes in chemically and radiation-transformed C3H/10T1/2 mouse embryo cell lines. *Cancer Res.*, *46*: 5302–5311, 1986.

63. Billings, P.C., Shuin, T., Lillehaug, J., Miura, T, Roy-Burman, P., and Landolph, J.R. Enhanced expression and state of the *c-myc* oncogene in chemically and x-ray-transformed C3H/10T1/2 Cl 8 mouse embryo fibroblasts. *Cancer Res.*, *47*: 3643–3649, 1987.

64. Clemens, F. and Landolph, J.R. Genotoxicity of samples of nickel refinery dust. *Toxicol. Sci.*, 73: 114–123, 2003.

65. Landolph, J.R., Verma, A., Ramnath, J., and Clemens, F. Molecular biology of deregulated gene expression in transformed C3H/10T1/2 mouse embryo cell lines induced by specific insoluble carcinogenic nickel compounds. *Environ. Health Perspect.*, *110* (Suppl. 5): 845–850, 2002.

66. Verma, R., Ramnath, J., Clemens, F., Kaspin, L.C., and Landolph, J.R. Molecular biology of nickel carcinogenesis: Identification of differentially expressed genes in morphologically transformed C3H/10T1/2 Cl 8 mouse embryo fibroblast cell lines induced by specific insoluble nickel compounds. *Mol. Cell. Biochem.*, *255*: 203–216, 2004.

67. Ora, A. and Henenius, A. Calnexin fails to associate with substrate proteins in gluconidase-deficient cell lines. *J. Biol. Chem.*, *270*: 26060–26062, 1995.

68. Galvin, K., Krishna, S., Pnochel, F., Frolich, M., Cummings, D.E., Carlson, R., Wands, J.R., Isselbacher, J.K., Pillai, S., and Ozturk, M. The major histocompatibility complex class I antigen-biding protein p88 is the product of the calnexin gene. *Proc. Natl. Acad. Sci.*, *89*: 8452–8456, 1992.

69. Tjoelker, L.W., Seyfried, C.E., Eddy, R.L., Jr., Byers, M.G., Shows, T.B., Calderon, J. Schreiber, R.B., and Gray, P.W. Human, mouse, and rat calnexin cDNA cloning: Identification of potential calcium binding motifs and gene localization to human chromosome 5. *Biochemistry*, *33*: 3229–3236, 1994.

70. Schreiber, K.L., Bell, M.P., Huntoon, C.J., Rajagopalan, S. Benner, M.B., and McKean, D.J. Class II histocompatibility molecules associate with calnexin during assembly in the endoplasmic reticulum. *Int. Immunol.*, *6*: 101–111, 1994.

71. Honore, B., Rassmussen, H.H., Celis, A, Leffers, H., Madsen, P., and Celis, J.E. The molecular chaperones HSP28, GRP78, endoplasmin, and calnexin exhibit strikingly different levels in quiescent keratinocytes as compared to their proliferating normal and transformed counterparts; cDNA cloning and expression of calnexin. *Electrophoresis*, *15*: 482–490, 1999.

72. Miki, T., Smith, C.L., Long, J.E., Eva, A., and Fleming, T.P. Oncogene *ect2* is related to regulators of small FTP-binding proteins. *Nature*, *362*: 462–465, 1993.

73. Adler, J.J., Winnicki, R.S., Gong, T.L., and Lomax, M.I. A gene up-regulated in the acoustically damaged chick basilar papilla encodes a novel WD40 repeat protein. *Genomics*, *56*: 56–59, 1999.

74. Rachez, C., Suldan, Z., Bromleigh, V., Gamble, M., and Naar, A.M., Erdjument-Bromage, H., Tempst, P., and Freedman, L.P. Ligand-dependent transcription activation by nuclear receptors requires the DRIP complex. *Nature*, *398*: 824–827, 1999.

75. Ito, M., Yuan, C., Malik, S., Gu, W., Fondell, J.D., Yamamura, S., Fu, Z., Zhang, X., Qin, J., and Roeder, R.G. Identity between TRAP and SMCC complexes indicates novel pathways for the function of nuclear receptors and diverse mammalian activators. *Mol. Cell*, *3*: 361–370, 1999.

10 Radiation Carcinogenesis

Colin K. Hill and Donna M. Williams-Hill

CONTENTS

SUMMARY

This chapter is concerned with the biology of radiation-induced carcinogenic events in animal and *in vitro* cellular studies. In particular, studies are presented to demonstrate the consensus view of the present state of science on the probability of a cancer event occurring when an organism is exposed to radiation and also that there is a dose–response function for such an exposure. The chapter also shows studies that demonstrate the importance of the time-line of exposure, the type of exposure, and the long-term effects of exposure to ionizing radiation. The risk or probability arising from an exposure rests on the biological knowledge that science has accumulated on radiation effects. Unlike the law profession where evidence is presented and a judge or jury makes a definite determination of truth or falsehood, science relies on an accumulation of data, which is often inconsistent when viewed in isolation, to build a consensus that can then be framed as a theory. Theories or hypotheses only hold as long as the data support them. Often new data do not support a hypothesis or theory and science designs newer ones. In biology there are few theories and many hypotheses.

    The human organism is very complex and we do not yet have a firm handle on how all the complexities interact and work together. It is a paradigm or tenet, "if you will" of biological research that there are uncertainties in any biological measurement. Radiation effects are not immune from this paradigm. Much of the research being performed in radiation biology today is aimed in one way or another at reducing uncertainties either by increasing sensitivity of the measurements being made or searching for mechanistic understanding of a phenomenon using new molecular tools which in itself will remove a source of uncertainty. In what follows, it will be clear that there is at present a consensus among researchers about the carcinogenic effects of radiation, but significant uncertainties remain. Due to the random nature of damage produced by radiations when impinging on cells and tissues, there is no defined molecular signature or pathway for radiation-induced carcinogenesis. Rather the many pathways and genes that are involved in the etiology of many cancers that are described in other chapters in this book are also possible targets for radiation-induced damage. Thus, many known tumor-promoter and tumor-suppressor genes have been implicated in cancers initiated by radiation exposure. Any decision about the likely effects on a person or persons who have a history of exposure should be made conservatively.

In brief, the present "Consensus State" of knowledge regarding exposure to radiation at low doses or protracted low doses tells the following about carcinogenic events:

1. These events are linear with dose (in most tissue types).
2. There is no threshold in dose for carcinogenic events.
3. There is a lifetime risk associated with any exposure (the size of the risk may change with time however).
4. Age at exposure is important. In most cases the risk of a cancer is higher if you are young when exposed, than if you are an adult.
5. There is a latent period between exposure and the appearance of the carcinogenic event.
6. Total body risk is not a good measure of overall risk, if exposure is for some reason concentrated in one or more organs.
7. Dose should not always be reported as total body exposure if one organ received most of the exposure; the organ dose should be used as the denominator for dose reporting.

10.1 INTRODUCTION

The possibility of radiation-induced carcinogenesis is still of major concern to society even 50 plus years after the Atomic bombs were dropped on Japan. Unfortunately contamination of our environment in many localized and not so localized instances gives rise to a significant number of people being exposed either occupationally or in the general public to doses above those considered part of normal background. This concern in people has boiled down to arguments over what should or should not be considered the dogma in deciding the extent of risk of an individual to radiation-induced disease from small exposures. For the purposes of this chapter by "disease" we mean any human condition not accepted as part of normal life. In this chapter we review the biology of radiation-induced damage and relate this to several aspects of carcinogenic risk estimation that are still ambiguous due to uncertainties. We have specifically included studies that might be able to shed light on such questions as:

1. Is any dose of radiation, no matter how small, carcinogenic?
2. Is there a clear dose response for carcinogenesis and if so, is it linear or not, and is there evidence for a threshold dose before any effect is seen? If not, is the "linear no threshold theory" still the strongest predictor of carcinogenic response to radiation?
3. Does the dose rate of radiation exposure alter the frequency of carcinogenic events and is the dose response still linear?
4. Do small doses of radiation shorten life span, and if so, what is the primary cause of the shortening?
5. Does radiation damage act independently of other causative agents, environment, or genetics?
6. And finally, how strongly can scientific data derived from nonhuman subjects be applied to the human situation?

These questions encompass the major aspects of radiation-induced carcinogenesis, which any student involved in radiation usage should be aware. We have used examples from the work of one of the Authors (Colin K. Hill) where prudent, and from the work of others, to illustrate the major discoveries and knowledge of radiation-induced carcinogenesis, as it exists at the beginning of the 21st century.

There are two aspects to cancer incidence that one must consider after radiation exposure. One is the likely frequency of a cancer-initiating event and the second is how long it will be before the cancer becomes apparent. Radiation and other environmental factors may interact to reduce the latency period between initial damage and the manifestation of the disease. Therefore, it is also important to examine life-span studies after irradiation to not only determine if the burden of disease increases, but also if that burden shortens the overall life span. For example, one could conceive of a situation where there was only a small increase in life threatening tumor burden for a particular tumor type, due to radiation exposure. However, the overall life span was decreased significantly because this small number of tumors appeared much earlier in the life span than in subjects who were not irradiated. Therefore one must consider the possibility that even if an exposure were not enough to cause a cancer in an average lifetime, it may still have significantly decreased the latency period of a possible cancer if a subsequent interacting damaging event occurred and thus shortened the disease-free life of the individual.

10.2 ANIMAL STUDIES

We will begin by reviewing some of the animal data because most of the animal studies were started before cellular assays were possible. There are two types of data from animals that are relevant to human population exposures. These two types are tumor-incidence studies and life-shortening studies. The former is relevant for obvious reasons; the latter is relevant because after exposure to radiation the vast majority of the life-span shortening is due to excess tumor burden [1].

Prior to the early 1970s, most radiation exposure studies were of the acute form. That is, acute toxic effects were measured. These types of studies, of which there were many, are not relevant to considerations of carcinogenesis at low doses because the radiation doses used were so large as to cause acute responses leading to death. After these early studies there were a number of studies looking at tumor induction for specific tumor endpoints, mostly in rats and mice. In most of these the exposure doses were fairly large due to the need to get good quantitative data without using too many animals. In these types of experiments it soon became obvious that the tumor incidence or percentage of animals with tumors increased with dose. However, at a certain exposure level the tumor incidence would begin to fall again. This was soon found to be due to the death of animals from acute radiation effects. These animals would have developed tumors if they had lived longer. In Figure 10.1 we show an example of such a type of response taken from an early report by Dr. Upton [2] on the incidence of myeloid leukemia in RF mice.

It is clear from this early study that the incidence of leukemia increases with increasing dose up to about 250 cGy. Above that the incidence of leukemia decreases because mice start dying from the acute effects (death by hemopoietic or gastrointestinal failure) of radiation before the leukemia can manifest itself. By the way, leukemia is one of the few neoplasms whose incidence curve is more curvilinear than linear at low doses as can be seen in the figure. At the time this data was collected there were no models for radiation-induced carcinogenesis, so Dr. Upton used interpolation to draw a curve through his data.

Some time after the studies by Upton, a very elegant series of studies on tumor incidence was published by Mole et al. [3]. The main figure from this study is reproduced in Figure 10.2, showing a very clear bell-shaped relationship between dose and tumor incidence for myeloid leukemia in male CBA/H mice exposed to 250 kVp x-rays, for which Mole suggested a quadratic relationship. Again let us make it clear that the descending portion of this curve is due to the mice dying of acute radiation effects before tumors could appear. We should also mention that leukemia was of particular interest because it is an early-appearing disease after radiation exposure in humans [4].

At about the same time as the tumor incidence studies were beginning to appear, the first life-span studies were published indicating that life can be shortened by exposure to radiation. In Figure 10.3 we reproduce one of the earliest studies by Rotblat and Lindop [5] in 1961 where they show a clear relationship between exposure dose and length of life shortening in mice exposed to radiation. Interestingly, for this endpoint, which at that time was called "non-specific life shortening," there is a good fit to a linear relationship between exposure dose and the length of the life span shortening. However, the smallest dose was 50 cGy and it was not known at the time of publication what

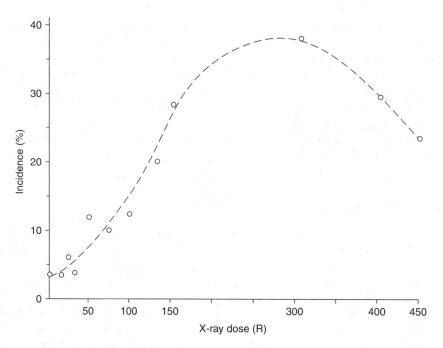

FIGURE 10.1   Incidence of myeloid leukemia in RF mice as a function of whole-body dose of radiation (With permission from Upton, A.C. [1961] *Cancer Res.*, 21, 717–729).

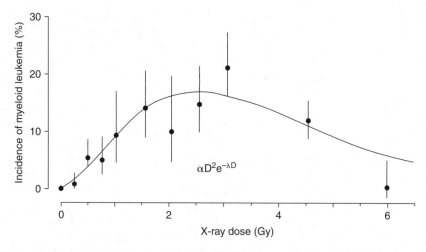

FIGURE 10.2   Incidence of myeloid leukemia in male CBA/H mice as a function of whole body radiation with 250 kVp x-rays (With permission from Mole, R.H., Papworth, D.G., and Corp, M.J. [1983] *Br. J. Cancer*, 47, 285–289).

exactly caused the life shortening. Thereafter followed a series of life-span studies that looked at life shortening due to radiation exposure. These studies were for the most part performed by science groups at several different National Laboratories run by the U.S. Department of Energy (DOE). These expensive and time-consuming studies examined not only single acute exposures but went on to look at exposures given weekly, daily, or continuously throughout an animal's life. It was in these studies that pathologists were brought in to examine the mice and other animals when they died to determine the cause of death.

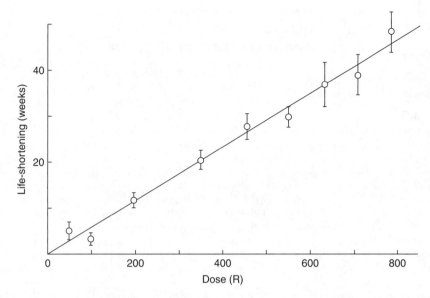

FIGURE 10.3   Life shortening plotted against dose of radiation in mice exposed early in life and monitored for their entire life span (With permission from Rotblat, J. and Lindop, P. [1961] *Proc. R. Soc. Lond.* [Biol.], 154, 350–368).

It would take too much space and be somewhat repetitive to mention all the studies. A thorough review of the animal data was published in Radiation Research in 1985 by Broerse et al. [6]. However, we will describe in some detail the series of experiments that were performed at Argonne National laboratory over the course of 15 years. This series of studies was one of the most comprehensive ones done with mice and came to be known as the "mega mouse" studies. They were so named because they involved several thousands of mice and later hundreds of beagle dogs in life-span studies. It soon became clear, in these and other studies, that life shortening was almost always due to excess induction of neoplasms [1]. This was particularly true for small to medium doses where other late effects such as fibrosis were not evident.

To illustrate the type of response these studies measured and the chronology or duration of the work done, we have provided some figures and tables from the papers published by the group performing the studies. In particular, we have reproduced a few of the results that clearly reinforce some of the points mentioned in the summary as the current consensus for low-dose carcinogenesis.

In Table 10.1 we have summarized the life shortening data from Table 10.2 in the paper by Thomson et al. [7] showing that there is a substantial decrease in life span even when a dose is fractionated over 24 weeks.

In this series the untreated B6CF1 mice lived an average of 848.4 days if they were male and 835.3 if they were female. In Figure 10.4 the results are plotted as excess mortality versus total dose. It is striking that the results fit a straight line indicating a linear response between exposure and effect. Because the total doses were in the medium to high range in this study, it was even more interesting to look at the second paper where studies were undertaken to look at longer-term fractionation with smaller doses as well as at continuous low-dose-rate exposure.

In Table 10.2, we have summarized some of the results from Reference 8, showing the life shortening from very protracted gamma-ray exposures.

Again, obviously, even at the much-reduced dose accumulation used in these experiments, there was clear evidence for a reduction in life span. When these experiments were first published, the enormous task of performing an autopsy on each mouse to determine cause of death and of classifying the tumors that were found had not been completed. It was some time before the same group published a series of papers, the main thrust of which was to show that in the low to moderate dose range almost

TABLE 10.1
Life-Shortening Data From B6CFI Mice

Total dose (rad)	Dose per fraction	Life shortening days	
		Males	Females
0	0	—	
855	35.6	154.6	155.6
206	8.6	73.9	—
417	17.4	125.7	107.1
1110	46.2	229.5	225.3

Source: Abstracted with permission from table II in the paper by Thomson, J.F., Williamson, F.S., Grahn, D., and Ainsworth, E.J. (1981) *Radiat. Res.*, 86, 559–572.

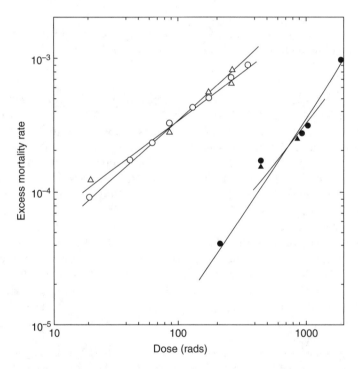

FIGURE 10.4 Excess mortality in B6CFI mice exposed to various fractionated radiation regimens (With permission from Kennedy, A.R., [1985] *Carcinog. Compr. Surv.*, 9, 355–364).

all the life shortening was due to excess tumor burden. All the data on the tumorigenic effect of gamma and neutron exposure was summarized and compared in a paper published in 1992 [9]. In this paper the authors summarized data from pathology reports on 32,000 animals and histopathology on 19,000 animals that were allowed to live a full natural lifetime and from the series of single, fractionated, and continuous dose rate exposures given to B6CF1 mice. They found that at least 85% of all the mice died with one or more tumors present. In their concluding remarks they reemphasized that the overwhelming cause of excess (or early) mortality was due to tumors appearing in the irradiated mice. They also stated very clearly that the data in the low to medium dose range were consistent with

---

TABLE 10.2

Life Shortening in B6CFI Mice Exposed to Various Fractioned Doses of Radiation Over Long Periods of Time and for the Duration of Life

Dose rate rad/weekly fraction	Number of fractions	Mean accumulation dose (rad)	Life shortening in days	
			Males	Females
7	60	417	36.5	65.8
7	DOL	813	80.0	32.4
17.4	DOL	1880	142.7	181.6

DOL = duration of life.

Source: With permission from Thomson, J.F., Williamson, F.S., Grahn, D., and Ainswoth, E.J. (1981) *Radiat. Res.*, 86, 573–579.

---

a simple linear response for carcinogenesis induced by low linear energy transfer (LET) radiation (such as gamma rays). They also noted that no unusual or unique neoplastic lesions were caused by radiation. The inference from such a study is that radiation introduces damage in a stochastic manner and thus has an equal chance of causing any type of tumor. This fact, of course, makes it unlikely that any specific molecular signature produced by radiation damage will be found. Grahn et al. [6] published 14 papers over a 15-year period on life-shortening and tumor induction by radiation and they also list 9 papers by other groups (principally by Ullrich and colleagues) on tumors induced by radiation.

At the same time as the mice were being studied in the 1980s, a major life-span study of chronic radiation exposure was also under way at Argonne National Laboratory using Beagle dogs. This study was primarily undertaken for two reasons. First, there was a question of the relevance of mouse studies to the human situation and second, there was a need to study the leukemia incidence in a large mammal subjected to chronic irradiation. This need was realized after the early onset and quite a large occurrence of leukemia in the atomic bomb survivors in Japan [4]. Although data on leukemia incidence and etiology in the chronically irradiated dogs was published some time ago (e.g., see Reference 10) the detailed life-span analysis has only recently been published because many of the dogs lived to a very old age. Some of them surpassed 17 years of age before dying. In fact some of the dogs outlived the experimental funding and were transferred to another DOE funded facility when DOE terminated the studies housed at Argonne National Laboratory.

Carnes and Fritz [11,12] have published two major papers on the overall results of the Beagle studies. Groups of dogs were exposed to gamma rays continuously at different dose rates to either a certain total dose or throughout their whole life span. The analysis is still ongoing. The main conclusions published so far are:

1. The early acute hematopoietic failures were dependent on dose rate.
2. Late occurring deaths (both solid tumor and nontumor) depended only on total dose received.
3. Both total dose and dose rate influenced the rate of leukemia induction [13].
4. As was observed in mice, when a dose or dose rate is low enough for the animal to survive the acute effects, the predominant cause of late occurring death is tumor induction.

The most important observation to come to light in these large animal studies is the clear evidence that late occurring death due to tumors in the very low-dose rate exposure groups is *only* dependent on the total dose received and is *not dependent* on dose rate. This observation also applies to mouse studies, but to date, has not been deduced from human studies. The only human study nearing completion is the analysis of tumor induction in the atomic bomb survivor cohort. However, all the subjects in that cohort received their dose as a single acute-dose-rate exposure. The reason for the apparent lack of dose-rate dependence for tumor induction in the beagle cohort has not been totally determined. However, Carnes and Fritz [11,12] point out that death occurs during a life span for different reasons at different times. They suggest that radiation rather than causing unique ways of killing an animal actually increases the risk of dying from an existing reason. The death occurs in one of the naturally occurring time "windows" during the life span, depending on the total dose received before the animal reaches that "window." Thus, for example, if the total dose is low enough not to cause any form of acute or chronic tissue failure, the animal dies of a tumor in the "window" of time leading up to the end of its natural life span. The effect of radiation exposure is to increase the frequency of these tumors, which in and of itself leads to an overall shortening of life span. So far these authors have not reported that tumors occur with any increased frequency earlier in the life of the irradiated dogs except in the last "window" of aging. Please remember that exposures in the beagles were of chronic whole-body radiation. Large doses deposited in a single organ like the thyroid may have other dependencies for tumor induction such as the age at exposure [14].

10.3 CELLULAR STUDIES

In the early 1970s, several new cell lines were discovered in which the cells behaved somewhat like normal tissue cells until they received an insult (injury). In response to the injury a small proportion of the cells change morphology and develop into a malignant tumor phenotype. A race ensued to turn such observations into a quantitative method for estimating the number of neoplastic transformations (conversion from a normal to a malignant cancerous cell type) that were occurring in an exposed population. Three types of cell lines were identified and widely used in the late 1970s and throughout the 1980s and 1990s. Eventually other cell types were discovered and human cell transformation became possible in a dish.

In this section we review some of the cell transformation data collected using these cell lines. The cell models, particularly, have proven very helpful in gathering large amounts of data about dose response curves for carcinogenic events for different types of radiation and for radiation exposure delivered at different dose rates. A large number of these studies were driven by and funded because of the perceived need to obtain better data on the shape of dose response curves and the frequency of events at lower doses. Studies have been performed primarily with three different cell preparations. The NIH/BALB/3T3 cell line [15] was extensively used early on but more recently has been used less for quantitative studies due to a high and variable background transformation frequency. The other two, the Syrian hamster embryo cell-culture system [16,17] and the mouse embryo C3H10T1/2 cell line [18] have been in continuous use for more than 30 years.

All the three-cell systems rely on morphological changes of the cells growing in a dish to identify the transformed cell, but the transformed cell in the dish can also be transferred to a syngeneic animal and will form tumors. In the case of the Syrian hamster assay, primary isolates from Syrian hamster embryos are used for the experiments. The cells grow into colonies in petri dishes after treatment and are scored for morphology. In parallel, the number of colonies that arise from a known number of cells can also be used to determine survival. This system is considered to be useful for looking at early events in the carcinogenic process. In most instances the morphologically transformed clones will not immediately form tumors when injected in syngeneic mice. They must be grown for many generations before they become malignant. However this assay has been widely used in studies of both chemical and radiation-induced transformation [17]. In one of their very earliest publications Borek and Hall [13] demonstrated the ability of such an assay to measure radiation-induced transformation after low doses of radiation in a quantitative manner. In Figure 10.5 we reproduce some of the data from that early paper showing how well such assays lend themselves to precise quantitative measurements.

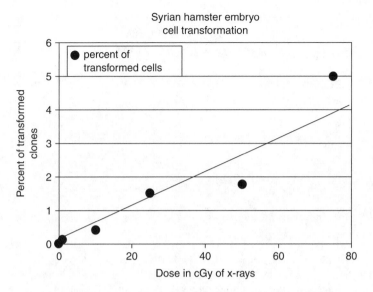

FIGURE 10.5 Percent of cells developing morphologically transformed clones versus the dose of 250 kVp x-rays (With permission from Borek, C. and Hall, E.J. [1973] *Nature*, 243, 450–453).

The data from Borek and Hall [13] have been replotted on a linear–linear scale to show that at low doses their data strongly indicated a linear response with increasing dose. The lowest dose in this study was 1 cGy.

In these early studies the background frequency or spontaneous transformation frequency had not been determined. Borek and Hall looked at more than 30,000 clones without seeing a spontaneous transformation. It was later determined that background frequencies are in the range of 1 in 100,000 cells in such cell lines. It is also interesting to note that (1) the dose to get a factor of two increase above the background level would be much less than 1 cGy in these studies and (2) the dose to increase this endpoint by a factor of two, going from two transformed clones to four transformed clones, for example, is about 7 to 10 cGy.

The paper by Borek and Hall essentially set the stage for a dramatic increase in studies of neoplastic transformation. The laboratory of Dr. Mortimer Elkind and collaborators, Drs. Antun Han and Colin Hill, began a series of careful studies of single, fractionated, and continuous low-dose-rate exposures to gamma rays and neutrons. Rather than belabor the points we are about to make with similar data from at least three labs, we show only data from studies performed in the Elkind laboratory and our own for this section and support it with references to the work of others. These studies used the C3H10T1/2 cell line that was first described by Reznikoff et al. [18]. This cell line has a high degree of contact inhibition when grown normally in the petri dish. The cells were originally isolated from an embryonic mouse and as such exhibit many of the properties of stem cells. That is, depending upon the stimuli, these cells can be induced to form several different cell types and are immortal in culture. The property of contact inhibition allows the identification of transformed cells as colonies, termed foci, growing on top of the contact inhibited sheet of normal cells following a damaging insult like an exposure to radiation. In the C3H10T1/2 cell line, cells of low passage are used for these assays and these grow as a single cell layer on the dish that is barely visible when stained. The transformed foci consist of piled up cells that crisscross and stain very darkly. Transformed foci are scored for aggressiveness according to their morphology. Foci in this assay classed as type II and type III clones form fibrosarcoma tumors when injected into the flanks of C3H mice. Type I clones are not so aggressive. The C3H10T1/2 assay, then, is a complete surrogate for measuring tumor induction in a whole animal. To get a quantitative measure of the number of foci induced by a particular insult, a known number of cells are seeded into many dishes. Then at

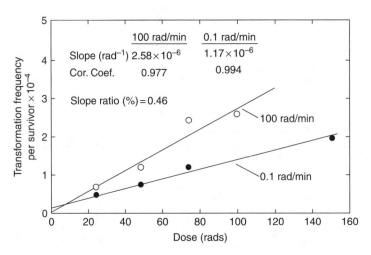

FIGURE 10.6   Transformation frequency of C3H/10T1/2 cells induced by ^{60}Co γ-rays at 100 (○) and at 0.1 (●) rad/min. The data lines were fitted by linear regression analysis to each set of data points ($p < .01$). CORR COEF is correlation coefficient (With permission from Han, A., Hill, C.K., and Elkind, M.M. [1980] *Cancer Res.*, 40, 3328–3332).

the end of the experiment the number of type II and III foci are divided by the total cells seeded to determine the transformation frequency. In parallel experiments, the surviving fraction of cells for that dose is also determined so that the transformation frequency can ultimately be reported as the number of transformed cells per surviving cell. This is a significant qualifying factor since, as stated earlier, a cancer cannot arise from a dead cell.

There are several studies of C3H10T1/2 cells pertinent to the problem posed by the low-dose exposure of a number of public and occupational groups of people. First is the question of the shape of the dose response curve. A study published in 1980 examined two different dose rates of gamma-ray exposures [19]. In Figure 10.6 the low-dose response data from that study is shown on linear coordinates. The data clearly indicates that the initial part of the dose response curve for neoplastic transformation is linear.

This conclusion strongly supports the "no threshold" theory for carcinogenic events. It should also be noted that although reducing the dose rate decreased the transformation frequency there was still a measurable response and the response remained linear. In this data set we were not able to clearly define the size of the errors on the individual points in the linear plot and they were thus left out of the low-dose plot (they were included in the log-linear plots; see Reference 19). Similar studies have been published by Hall's group at Columbia University [20,21] and by Dr. Jack Little and colleagues at the Harvard School of Public Health [22,23].

In a second follow up series of studies, also funded by the National Cancer Institute and DOE, the question of linearity in the low-dose region and for low-dose rate and fractionated exposure was examined more carefully. In this study we examined the effect of gamma rays given at a high-dose rate (100 cGy/min), at a low-dose rate (0.1 cGy/min), and in fractions (50 cGy/min given once per day for a total of five fractions). The study concentrated on doses in the low to medium range where little cell killing occurs [24,25]. In Figure 10.7 we show a reproduction of the published data. In each case the data was a compilation of several independent experiments. The work took about two years to complete with tens of thousands of dishes used to obtain sufficient data to reduce the error bars to the size shown. This set of data clearly supports the earlier one and the animal data. The lowest dose used was only 25 cGy at high-dose rate and in the fractionated regimen each dose was only 50 cGy. Clearly the linear fit to all of these three dose responses is very good. Moreover, though fractionation allows significant repair of radiation-induced damage to occur, there is still

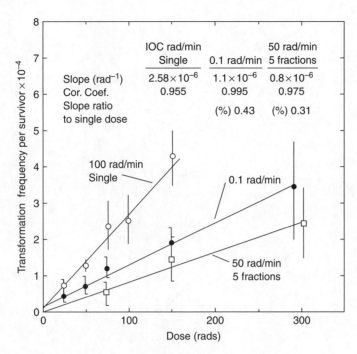

Slope (rad⁻¹)	IOC rad/min Single	0.1 rad/min	50 rad/min 5 fractions
Slope (rad^{-1})	2.58×10^{-6}	1.1×10^{-6}	0.8×10^{-6}
Cor. Coef.	0.955	0.995	0.975
Slope ratio to single dose		(%) 0.43	(%) 0.31

FIGURE 10.7   Transformation frequency of C3H10T1/2 cells induced by ^{60}Co γ-rays at 100 (o) and at 0.1 (●) rad/min and at 50 rad/min given in five fractions at 24 h intervals (□) Lines were fitted as in Figure 10.6 (With permission from Han, A., Hill, C.K., and Elkind, M.M. [1984] *Br. J. Cancer*, 49, 91–96 and Hill, C.K., Elkind, M.M., and Han, A. [1985] *Carcinog. Compr. Surv.*, 9, 379–397).

a measurable level of transformation even after the prolonged fractionation. In these experiments, the background or spontaneous neoplastic transformation frequency was approximately 1.05 per 100,000 cells. At 25 cGy the transformation frequency measured was 8.0 per 100,000 cells (eight times higher than background) for the high-dose rate. For fractionated exposures the transformation frequency at 50 cGy was 2.5 per 100,000 cells exposed.

How well does a neoplastic transformation event in an individual cell relate to tumor formation in an animal? Well, so far our understanding of a neoplastic event in a cell is that this is a required step for the cell to go on and form a tumor. At the genetic level it is now generally believed that to get a fully malignant tumor a cell must undergo several mutations [26,27]. These mutations can occur in any combination of a large number of oncogenic and tumor suppressor genes. These mutations may also be accumulated over a long period of time [26,27]. It is now thought that an end point such as morphological transformation in C3H10T1/2 cells can be reached by many different pathways involving several possible combinations of damaged genes. Therefore measuring transformation essentially gives one an estimate of the total number of tumor producing events that have occurred in a cell population. In an animal, a comparable measurement would be the tumor incidence per animal. However, in the animal we do not know how many target cells were killed by the radiation, so we cannot measure the number of tumors per surviving cell. We can only measure the number of tumors per animal. In cell culture assays we can plot our data in a similar fashion by calculating the transformation frequency per cell exposed.

In Figure 10.8 we show this for data from a series of experiments where we exposed cells to gamma rays at 100 cGy/min [24].

In Figure 10.8, it is very clear that the bell-shaped curve, so common in tumor incidence data (e.g., see Figure 10.1), can be demonstrated at the cell level. It is also clear that this bell-shaped curve is caused by the competing processes of tumor cell induction versus the probability of cell death.

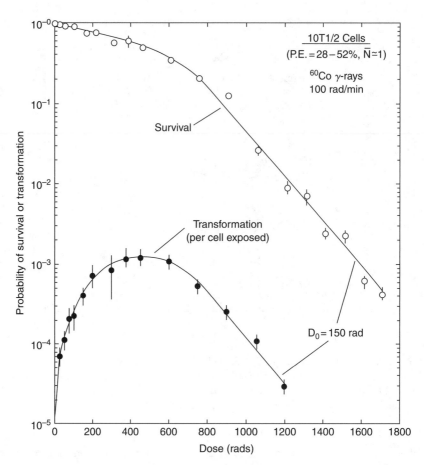

FIGURE 10.8   Transformation frequency (•) and surviving fraction (○) of C3H10T1/2 cells induced by ^{60}Co γ-rays at 100 rad/min.

In the realm of radiation-induced cancer risk estimates we are only concerned with the low-dose region of these curves. In essence, cell killing is not important at such doses. So originally in Elkind's group, and more recently in ours as well as in other groups, there have been efforts to define more clearly the shape of the initial response (hence the data we showed in Figure 10.6 and Figure 10.7). Now, having demonstrated that the dose–response for neoplastic transformation is most likely to be linear and that even at low-dose rates or for protracted doses there is a neoplastic response and it remained linear, we still wanted to answer two questions. First was to determine whether there was a dose rate that would produce such a low response that it was not measurable and second, can the expression of damage by radiation be modified by other environmental factors? There are two pieces of scientific evidence that led us to suspect that the number of cells damaged was larger than the number of transformed foci we can measure in the first instance. First, experiments by both Hall's group at Columbia and by Kennedy's group at Harvard showed that there is a high likelihood of most cells being initiated by exposure to even small doses of radiation [28,29]. The second is the often observed fact that radiation-induced damage can be promoted by agents such as 12-*O*-tetradecanoylphorbol-13-acetate, termed TPA [30–33]. These agents increase the number of expressed transformed foci produced by a damaging agent, but do not induce transformation on their own. A number of such agents are known to occur naturally and are called tumor promoters. They may play a role in determining the length of the latency period between initiation and growth of a tumorigenic cell. By this we are alluding to the well-known fact that in humans and many other

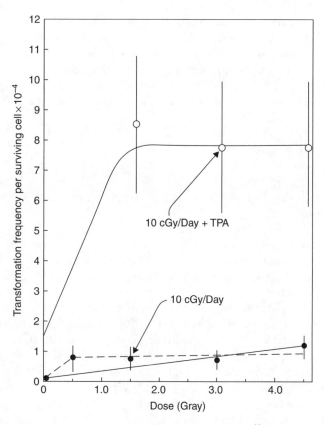

FIGURE 10.9   Transformation frequency of C3H10T1/2 cells induced by ^{60}Co γ-rays at 10 cGy/day ($\bullet$) and ^{60}Co γ-rays at 10 cGy/day plus the promoter tetra phorol myristate acetate (TPA) given 24 h after radiation exposure ends and once weekly in the medium thereafter (With permission from Hill, C.K., Han, A., and Elkind, M.M. [1989] *Radiat. Res.*, 119, 348–355).

animals the initiated cell may not express itself as a tumor for many years (we come back to this point in the summary of human data later).

To address the question of the effects of very low-dose rate exposures our laboratory staff performed a series of experiments where the C3H10T1/2 cells were exposed to gamma rays at only 10 cGy /day continuously for up to 45 days [34]. The transformation frequency was measured at different total doses during this exposure and in some instances the promoter TPA was added 24 h after the end of the exposure to see if there was initial unexpressed damage remaining. Figure 10.9 illustrates this point [34]. In this figure the dose–response is probably modified somewhat by the length of time it takes to give the radiation exposure. By that we mean that cell growth and divisions during the radiation may have had a diluting effect on our ability to measure the true frequency at higher accumulated doses. Even so, as can be seen in Figure 10.9 there is a significant increase in transformation frequency above the spontaneous rate at the lowest total dose measured (50 cGy).

More striking is the fact that even though the exposure was given at a very low-dose rate and took several generations of cells to deliver, it was still possible to greatly enhance the transformation frequency by adding the promoter TPA 24 h after the end of a particular exposure. In fact, on an average, the frequency was increased by a factor of ten. Our conclusion from this data is again that there is no evidence of a threshold and also that the data supports the idea that the number of initiated cells that can potentially form a neoplastically transformed cell is much larger than the initial measured transformation frequency would predict. This indicates that in a long-lived animal or in humans there is more risk of an eventual tumorigenic event than the initial dose response would suggest. This type of idea is now gaining credence as the solid tumors induced in the atomic bomb survivor cohorts are only now reaching a maximum incidence rate, some 50 years after the event [35].

So far the biology we have presented makes clear that carcinogenesis induced by ionizing radiation:

1. does not have a threshold
2. is dose dependent with a linear function, although the slope may vary with dose rate, or other modulating factors
3. that even small doses of radiation raise the frequency of neoplastic events above the spontaneous level.

Moreover, as was the case with the animal data, it is necessary to measure tumorigenic events over a lifetime as some expression may be delayed for a very long period of time. The tumor promoter data also indicate that an additional insult from the environment may trigger a radiation-damaged cell into producing a tumor.

Now, all that has been discussed so far relates only to low linear-energy ionizing radiation like the gamma rays. When highly ionizing radiation is the damaging agent the effects may be quite different. Although generally exposure to highly ionizing types of radiation is much lower, as we see later, this type of radiation can cause a disproportionate amount of damage. We illustrate the effects of highly ionizing radiation with one study on fission neutrons from the laboratory of Hill [36] and one on alpha particles from the work of Bettega et al. [37].

At first Hill and coworkers studied the transformation frequency induced by high and low-dose-rate fission spectrum neutrons, a highly ionizing radiation. This study was performed to further elucidate the observation first published by the Elkind group in *Nature* [38] that for certain high linear energy transfer radiations, reducing the dose rate actually increased the carcinogenic effect. We have reproduced the main figure from the low-dose neutron paper [36] in Figure 10.10.

In this study we used doses given at low-dose rate (0.1 cGy/min) of as small as 2.3 cGy. The data not only show that these neutrons are more effective at low-dose rate than at high-dose rate but also that both responses are best fitted by a linear function in the low-dose region. In Figure 10.10 the neutron data is compared to gamma-ray data to drive home the point about how much more effective high linear-transfer radiation can be in the low to moderate dose region of the response curve [39].

In Figure 10.10, if one compares the initial slopes of the low-dose rate gamma versus the low-dose rate neutrons you will see that the neutrons are about 70 times more effective in inducing neoplastic transformation. Alpha particles also have a very high linear energy transfer and in Figure 10.11 the data from the alpha particle study by Bettega et al. [37] is replotted with data from one of our gamma ray studies [24] to show the relative effectiveness of the alpha particles.

This comparison indicates that for C3H10T1/2 cells alpha particles are about ten times more effective than gamma rays. Plutonium and uranium isotopes released into the atmosphere by atomic and hydrogen bomb tests and found in uranium mining and processing facilitates decay by emitting alpha particles. Thus even though the amounts are small, the effective dose from highly ionizing radiation exposure from contaminants will continue to accumulate if the particles are lodged in the body due to long half-lives of some of the isotopes. The relative biological effectiveness of these highly ionizing exposures is possibly of an order of magnitude higher than gamma rays. What this means in real terms is that 1 cGy of alpha particle exposure might be as effective as 10 cGy of gamma rays. In the very low-dose region of exposure we really do not know how high the relative effect might be. However, for neutrons the effectiveness compared to gamma rays increased as the dose decreased [38].

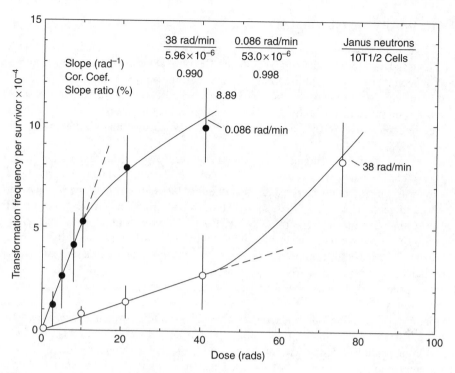

FIGURE 10.10   Transformation frequency of C3H10T1/2 cells induced by "JANUS" fission spectrum neutrons at 0.086 rad/min (•) and 38 rad/min (o). The initials slopes were fitted to the data points by linear regression analysis as in Figure 10.6 (With permission from Hill, C.K., Han, A., and Elkind, M.M. [1984] *Int. J. Radiat. Biol.*, 46, 11–15).

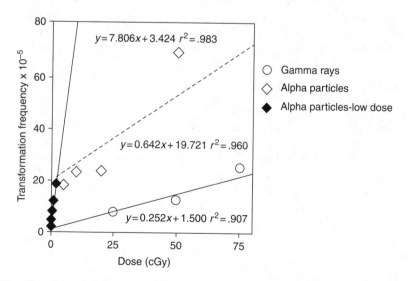

FIGURE 10.11   Thansformation frequency in C3H10T1/2 cells replotted from work by Dr. Bettega et al. [37], alpha particle data (closed diamonds (♦) and open diamonds (◊) [very low doses], compared to data from Dr. Hills laboratory [24], gamma radiation at 100 rad/min (○). The equations are the best linear-regression fits to each set of data points (With permission [24,37]).

10.4 SUMMARY OF HUMAN EFFECTS

Most of our understanding of the carcinogenic effects of exposure to radiation in humans comes from the longitudinal study of the survivors of the Hiroshima and Nagasaki atomic bomb attacks. There have been many papers published on the effects of the atomic bombs, too many to mention them all here. However, some of the more recent reviews can be found in References 4, 40 to 43. These data are supplemented with limited data from long-term follow up studies of nuclear workers from several countries and other groups who have received chronic low-level exposure to various radiation sources [35,44–46].

Most of the latest studies examine specific tumor sites, probably due to the type of the radioisotope causing the exposure. An extensive analysis of human thyroid cancer induction and risk by exposure to radiation was recently published as a result of the first International Meeting on Thyroid Cancer held in Cambridge in the summer of 1998 [14]. Moreover, a well-written review of most of the data prior to the Cambridge meeting was presented by Ron [35]. (Excess lifetime risk of thyroid cancer per unit absorbed dose for individuals exposed to iodine 131.) All the data we have presented for animals and in particular for cellular transformation also point strongly to a linear relationship for radiation-induced cancer particularly at low doses. According to the members of the Cambridge meeting, the excess risk per unit dose for thyroid cancer is somewhere around 10 to 20 per million people in a year or about 1000 per million in a lifetime. All the data presented at the Cambridge meeting indicated that a linear dose response for radiation-induced thyroid cancer is now accepted. Authors such as Ron [35], Shore [47], Williams and colleagues [14], and a host of studies from Belarus, Ukraine, and Russia support this conclusion (collected in Reference 14).

In the description of cellular data we included evidence that high LET radiation (which includes the radiation given off by plutonium) is much more effective than gamma rays in inducing carcinogenesis. There are few studies in humans of plutonium 239 induced carcinogenesis. Perhaps the most convincing evidence for an elevation of risk at very low doses is presented in a paper published by Whitehouse et al. [48] in 1998. In it the authors discuss the excess induction of chromosome aberrations by plutonium from an internal source and show strong evidence that there is an excess or increase in the number of symmetrical exchanges in peripheral lymphocytes [48].

There continues to be a steady accumulation of solid tumors arising in the surviving population that was exposed to atomic bomb radiation events. In a very recent paper from the Radiation Effects Research Foundation in Hiroshima, Pierce and Preston analyzed the radiation risk of cancer from low-dose exposure [40]. They showed quite clearly that the data for solid tumors does not depart from a linear interpretation. And in fact at low total dose there may be evidence for a steeper slope than for the overall dose response.

In Figure 10.12, their analysis of relative risk of a cancer event versus dose is reproduced [40]. In their discussion they say that the data are strong support for a linear response for tumor induction in the 0.0 to 10 cGy range of dose. Quoting from Pierce and Preston's discussion, "There is direct, statistically significant evidence of risk in the dose range of approximately 0.0 to 10 cGy." Even though Pierce and Preston admit that one cannot put great statistical weight on the data they go on to say, "In the presence of available data, it is neither sound statistical interpretation nor prudent risk evaluation to take the view that risk should be considered as zero in some low-dose range due to lack of statistical significance when restricting attention to that range. In particular the absence of any indication of departure from linearity should also be given substantial weight in the assessment."

Pierce and Preston have based their analysis on the latest solid tumor numbers for the Hiroshima and Nagasaki cohorts that received low doses at certain distances from the epicenters of the explosions. So far there are an estimated 724 excess cancers from a total of almost 7000 cancers in subjects who received a dose of 0.005 Sv (0.5 cGy) or above.

Since almost half of the exposed population are still alive even 55 years after the exposure occurred there will no doubt be many more cancers to analyze as time goes by. An exciting possibility is that

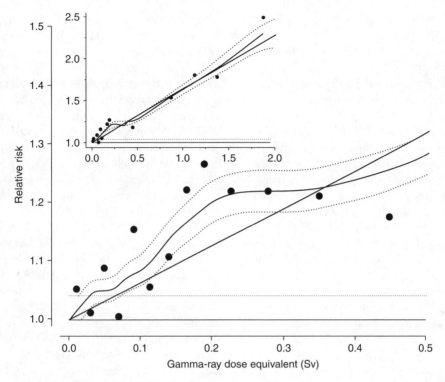

FIGURE 10.12   Estimated low-dose relative risks. Age specific cancer rates over the 1958–1994 follow-up period relative to those for an unexposed person averaged over the follow up and over sex, and for age at exposure. The dashed curves represent +1 standard error for the smoothed curve. The straight line is the linear risk estimate computed from the range 0–2.5 Sv. Because of an apparent distinction between distal and proximal zero dose cancer rates, the unity baseline corresponds to zero dose survivors within 3 km of the bombs. The horizontal dotted line represents the alternative baseline if the distal survivors were not omitted. The inset shows the same information for the fuller dose range (data and legend with permission [40]).

enough cancers may appear for the scientists to sector the cohorts by age and estimate the risk for age at exposure.

At the cellular level progress in understanding the molecular events caused by radiation exposure that lead to an observable cancer has been slow. In particular although normal human cells can be transformed by exposure to radiation, it has proved almost impossible to get a dose response for any of the experimental methods available. In most studies getting a human cell to transform is not only a very low-frequency event, but often requires more than one exposure and then the cell has to grow for a long period of time before the cancer phenotype shows itself. Nevertheless there have been some recent advances in the study of specific human cells and the discovery of certain processes that occur at low doses that may influence the outcome as to whether it is carcinogenic or not.

10.5   BYSTANDER AND ABSCOPAL EFFECTS MEASURED IN CELLULAR SYSTEMS

There has been a longstanding debate as to whether there is a threshold or not for cancer causing effects of radiation in living tissue, mostly fueled by the lack of data below ∼5 to 10 cGy for most assays. Critics of the current no-threshold theory (the BEIR V report of 1990, concluded that the no-threshold theory of radiation action was still the most reliable, Reference 43) point to the fact that in many cases effects at such low doses have been extrapolated from much higher doses by making best fits to the available data. Therefore, this area of low dose and low-dose rate exposure has been receiving much more attention recently and U.S. DOE has a program mandated by congress to explore the effects of doses below 10 cGy (see DOE web site, 2004). As part of this effort in the last 10 years there has developed a new area of intense research in cells and tissue that have received small doses of radiation. Drs. Little and Nagasawa (Harvard University, 1992; Reference 49) were the first to publish a report suggesting that cells that had not been hit by radiation but growing in the same dish could later show damage to their DNA. This became known as the bystander effect. In quite elegant experiments using alpha particles that can be controlled and individually delivered, the group at Columbia University showed that if a cell was hit with a single alpha particle, one could find more than one cell with damage. Dr. T. Hei and colleagues at Columbia have shown that this "bystander effect" can be shown for mutation endpoints, chromosome damage endpoints, and neoplastic transformation endpoints (cancer) (citations from *Proceedings of National Academy of Science* 1999, 2000, and 2001; References 50 to 52[†]). In the 1999 paper Hei and colleagues [52] showed that bystander effects can be seen in the DNA of nearby cells by irradiating only the cytoplasm of a cell, thus showing that radiation damage sensing and signaling does not have to begin with DNA damage. Others have shown that in some cases this bystander effect is mediated by direct communication between hit and not-hit cells via gap junctions but there is also strong evidence that in many cases cells must be secreting something into the medium as there is no cell touching the hit cell (References 53 and 54, respectively). One of the most attractive theories about how radiation-induced damage can eventually lead to the malignant cell phenotype is that of genomic instability. In the paper by Mothersill's group [54] they show that one route of the bystander effect is the initiation of apoptotic events in cells that were not irradiated. They conclude that genomic instability may be one result of the bystander activity. In recent years, others have shown similar "bystander" like effects in tissues. An example is a study by R.P. Hill from Canada using a lung model. These studies have shown that irradiating the lower lobe of the lung not only causes free radical damage in the radiation field but also causes free radical damage in the upper (unexposed) portion of the lung [55]. In further studies Dr. R.P. Hill also showed that waves of oxidative damage continue to occur in the irradiated and unexposed lung tissue long after the radiation exposure was given [56]. This work suggests that oxidative stress response genes (some of these are involved in inflammatory responses) continue to respond long after an initial insult.

Most of the recent work on bystander effects has been done using charged particles. One may wonder if the studies with alpha particles are relevant to radiations such as x-rays and gamma rays. Charged particles were used for practical reasons, as no machinery was available to irradiate single cells with x-rays. Recently this situation has changed and there are now several centers that have x-ray microbeams available. Dr. Prise and colleagues of the Gray Cancer Institute in England have just published a review in the *Journal Oncogene* of Low LET bystander studies [57]. It is clear that bystander effects do occur after exposure to low LET radiations [57]. To date the accumulated evidence suggests strongly that bystander effects are a general phenomenon induced by all types of radiation that can modify the effect on cells in the surrounding tissue that were not hit by the radiation. In particular the bystander effect appears to be an important determinant of the number of cells that are initiated by exposure to radiation and thus to the number of cells that eventually become malignant cancer. These types of responses were cited several times at a recent meeting sponsored by the American Association of Cancer Research (AACR) called Radiation Biology and Cancer (Laguna Cliffs Marriot Resort, Dana Point February 18–22, 2004, Reference 58). In fact several speakers noted that there is accumulating evidence that there may be an upward curvature to the cancer induction curve (above linear) in the very low-dose region for some types of tissue and

---

[†]It is notable that in the work of Hei et al., there is an emphasis on effects with very small doses of radiation. In many cases, only one particle traverses an individual cell.

radiation delivery. It remains to be seen if gaining a molecular understanding of such phenomenon can explain the tantalizing upward curvature shown in the analysis by Pierce and colleagues of the human data from the atomic bomb cohorts (see Figure 10.12; Reference 4).

10.6 CONCLUSIONS REGARDING CANCER RISK FROM RADIATION EXPOSURE

It is clear from the foregoing discussion, within the limits of experimental methods available to us, any exposure to radiation induces potentially cancer-causing damage in exposed cells. Further, the data from cells, animals, and from humans all lead us to conclude that there is a dose–response for carcinogenesis after exposure to radiation. That dose–response is linear with no threshold for most carcinogenic end points and most tissues that have been studied. The dose–response appears to be linear, particularly in the low-dose region of exposure. However, the initial slope can be modified by changing the dose rate, or the other factors that can alter radiation response (like the presence of chemical promoters or inhibitors), but still remains linear. Recent studies of bystander effects at very low doses (below 10 cGy) are beginning to suggest that at such low doses there may be instances when the dose response is an upward bending curve (i.e., greater than linear).

All the evidence collected so far supports the conclusion that carcinogenic initiation is a frequent event when cells are irradiated. However, all the evidence also show that expression of the malignant phenotype may be delayed for a long time. Thus in a complex organism such as a human, tumor incidence should be measured for the rest of a lifetime after an exposure. Expression of damage can occur at anytime after an exposure. We still do not know all the parameters that determine when that expression will occur. These last two facts offer, perhaps, a route into the future for molecular biology studies to find ways to prevent or stop the malignant expression of cells damaged by radiation.

The data presented in this report support the human data, but also provide substantial evidence at very low doses (1 to 10 cGy) for which not much human data exists. The animal and cellular data clearly show that at the lowest doses of radiation at which we can measure biological end points, there is still production of carcinogenic events above the background of spontaneous level. They also show that life shortening does occur after quite small chronically delivered doses and that it is all due to excess solid-tumor events.

In this time of "waiting" for the inevitable first dirty radiological bomb to be detonated by a terrorist group there remains very little scientific evaluation of the interaction of two or more mixed radio-nuclide-derived exposures. All the limited experimental data for mixed radiation show the effects are at least additive and time between doses does not remove the effect of an earlier dose completely and thus effects are cumulative. There is always some residual unexpressed damage after a radiation exposure.

The present state of radiobiological science strongly supports the no-threshold theory of radiation-induced cancer. There remains much to be learned about the molecular events that can be triggered by radiation damage, that lead to the development of a malignant tumor. However radiation-induced damage is randomly deposited at the outset, and later modified by cellular process that are determined by the individual genetic make-up of the exposed person. It remains to be seen if the individual genetic make-up alters the individual cancer risk significantly compared to the calculations of average risk now used to determine allowable exposure levels.

REFERENCES

1. Thomson, J.F. and Grahn, D. (1989) Life shortening in mice exposed to fission neutrons and γ-rays. *Radiat. Res.*, 118, 151–160.
2. Upton, A.C. (1961) The dose response relation in radiation induced cancer. *Cancer Res.*, 21, 717–729.

3. Mole, R.H., Papworth, D.G., and Corp, M.J. (1983) The dose response for x-ray induction of myeloid leukemia in male CBA/H mice. *Br. J. Cancer*, 47, 285–289.

4. Pierce, D.A., Shimizu, Y., Preston, D.L., Vaeth, M., and Mabuchi, K. (1996) Studies of the mortality of atomic bomb survivors. Report 12, Part I. Cancer: 1950–1990. *Radiat. Res.*, 146, 1–27.

5. Rotblat, J. and Lindop, P. (1961) Long-term effects of a single whole-body exposure of mice to ionizing radiations. II. Causes of death. *Proc. R. Soc. Lond.* (Biol)., 154, 350–368.

6. Broerse, J.J., Hennen, L.A., and van Zwieten, M.J. (1985) Radiation carcinogenesis in experimental animals and its implications for radiation protection. *Radiat. Res.*, 48, 167–187.

7. Thomson, J.F., Williamson, F.S., Grahn, D., and Ainsworth, E.J. (1981) Life shortening in mice exposed to fission neutrons and gamma rays: I. Single and short-term fractionated exposures. *Radiat. Res.*, 86, 559–572.

8. Thomson, J.F., Williamson, F.S., Grahn, D., and Ainswoth, E.J. (1981) Life shortening in mice exposed to fission neutrons and gamma gays: II. Duration-of-life and long-term fractionated exposures. *Radiat. Res.*, 86, 573–579.

9. Grahn, D., Lombard, L.S., and Carnes, B.A. (1992) The comparative tumorigenic effects of fission neutrons and Cobalt-60 gamma rays in the B6CF1 mouse. *Radiat. Res.*, 129, 19–36.

10. Seed, T.M., Kaspar, L.V., Fritz, T.E., and Tolle D.V. (1985) Cellular responses in chronic radiation leukemogenesis. In *Carcinogenesis*, Vol. 10, Huberman, E. and Barr, S.H., Eds., Raven Press, New York, pp. 363–379.

11. Carnes, B.A. and Fritz, T.E. (1991) Responses of the beagle to protracted irradiation: 1. Effect of total dose and dose rate. *Radiat. Res.*, 128, 125–132.

12. Carnes, B.A. and Fritz, T.E. (1993) Continuous irradiation of beagles with gamma rays. *Radiat. Res.*, 136, 103–110.

13. Borek, C. and Hall, E.J. (1973) Transformation of mammalian cells in vitro by low doses of x-rays. *Nature,* 243, 450–453.

14. Thomas, G., Karaoglou, A., and Williams, E.D. Eds. (1999) *Radiation and Thyroid Cancer*, World Scientific Publishing, London.

15. Little, J.B. (1979) Quantitative studies of radiation transformation with the A31-11 mouse BALB/3T3 cell line. *Cancer Res.*, 39, 1474–1480.

16. Berwald, Y. and Sachs, L. (1963) *Nature*, 200, 1182–1184.

17. Barrett, J.C., Hesterberg, T.W., Oshimura, M., and Tsutsui, T. (1985) Role of chemically induced mutagenic events in neoplastic transformation of Syrian hamster Embryo cells. *Carcinog. Compr. Surv.*, 9, 123–137.

18. Reznikoff, C.A., Bertram, J.S., Brankow, D.W., and Heidelberger, C. (1973) Quantitative and qualitative studies of chemical transformation of cloned C3H mouse embryo cells sensitive to post confluence inhibition of cell division. *Cancer Res.*, 33, 3239–3249.

19. Han, A., Hill, C.K., and Elkind. M.M. (1980) Repair of cell killing and neoplastic transformation at reduced dose rates of Cobalt 60 gamma rays. *Cancer Res.*, 40, 3328–3332.

20. Miller, R.J. and Hall, E.J. (1978) Effect of x-ray dose fractionation on the induction of oncogenic transformation *in vitro* in C3H10T1/2 mouse embryo cells. *Nature*, 72, 58–60.

21. Miller, R.J., Hall, E.J., and Rossi, H.H. (1979) Oncogenic transformation of mammalian cells *in vitro* with split doses of x-rays. *Proc. Natl. Acad. Sci. USA*, 76, 5755–5758.

22. Terzaghi, M. and Little, J.B. (1976) X-irradiation-induced transformation in a C3H mouse embryo-derived cell line. *Cancer Res.*, 36, 1367–1374.

23. Terzaghi, M. and Little, J.B. (1976) Oncogenic transformation after split-dose X-irradiation. *Int. J. Radiat. Biol. Relat. Stud. Phys. Chem.*, 29, 583–587.

24. Han, A., Hill, C.K., and Elkind, M.M. (1984) Repair of neoplastic transformation damage following protracted exposures to cobalt 60 gamma rays. *Br. J. Cancer*, 49, 91–96.

25. Hill, C.K., Elkind, M.M., and Han, A. (1985) Role of repair processes in neoplastic transformation induced by ionizing radiation in C3H10T1/2 cells. *Carcinog. Compr. Surv.*, 9, 379–397.

26. Vogelstein, B., Fearon, E.R., Hamilton, S.R., Kern, S.E., Preisinger, A.C., Leppert, M., Nakamura, Y., White, R., Smits, A.M., and Bos J.L. (1988) Genetic alterations during colorectal-tumor development. *N. Engl. J. Med.*, 319, 525–532.

27. Vogelstein, B. and Kinzler, K.W. (1993) The multi-step nature of cancer. *Trends Genet.*, 9, 138–141.

28. Rossi, H.H. and Hall, E.J. (1984) The multicellular nature of radiation carcinogenesis. In *Radiation Carcinogenesis: Epidemiology and Biological Significance*, Boice, J.D., Jr. and Fraumeni, J.F., Jr., Eds., Raven Press, New York, pp. 359–367.

29. Kennedy, A.R. (1985) Evidence that the first step leading to carcinogen-induced malignant transformation is a high frequency, common event. *Carcinog. Compr. Surv.*, 9, 355–364.

30. Kennedy, A.R. (1985) The conditions for the modification of radiation transformation *in vitro* by a tumor promoter and protease inhibitors. *Carcinogenesis*, 6, 1441–1445.

31. Han, A. and Elkind, M.M. (1982) Enhanced transformation of mouse 10T1/2 cells by 12-*O*-tetradecanoylphorbol-13-acetate following exposure to x-rays or fission-spectrum neutrons. *Cancer Res.*, 42, 477–483.

32. Hill, C.K., Han, A., and Elkind M.M. (1987) Promotion, dose rate, and repair processes in radiation-induced neoplastic transformation. *Radiat. Res.*, 109, 347–351.

33. Kennedy, A.R., Murphy, G., and Little, J.B. (1980) Effect of time and duration of exposure to 12-*O*-tetradecanoylphorbol-13-acetate on x-ray transformation of C3H10T1/2 cells. *Cancer Res.*, 40, 1915–1920.

34. Hill, C.K., Han, A., and Elkind, M.M. (1989) Promoter-enhanced neoplastic transformation after gamma ray exposure at 10 cGy/day. *Radiat. Res.*, 119, 348–355.

35. Ron, E. (1998) Ionizing radiation and cancer risk: Evidence from epidemiology. *Radiat. Res.*, 150, S30–S41.

36. Hill, C.K., Han, A., and Elkind, M.M. (1984). Fission-spectrum neutrons at a low dose rate enhance neoplastic transformation in the linear, low dose region (0–10 cGy). *Int. J. Radiat. Biol.*, 46, 11–15.

37. Bettega, D., Calzolari, P., Noris Chiorda, G., and Tallone Lombardi, L. (1992) Transformation of C3H10T1/2 cells with alpha particles at low doses: Effects of single and fractionated doses. *Radiat. Res.*, 131, 66–71.

38. Hill, C.K., Buonaguro, F.M., Myers, C.P., Han, A., and Elkind M.M. (1982) Fission-spectrum neutrons at reduced dose rates enhance neoplastic transformation. *Nature*, 298, 67–69.

39. Hill, C.K. and Elkind, M.M. (1987) Promotion and initiation are independent processes in radiation-induced neoplastic transformation. *Recent Advances in Leukemia and Lymphoma*. Alan R. Liss. Inc. New York, pp 399–406.

40. Pierce, D.A. and Preston, D.L. (2000) Radiation-related cancer risks at low doses among Atomic Bomb survivors. *Radiat. Res.*, 154, 178–186.

41. Goldman, M. (1996) Cancer risk of low-level exposure. *Science*, 271, 1821–1822.

42. United Nations Scientific Committee on the effects of Atomic Radiation (UNSCEAR), *Sources and Effects of Ionizing Radiation*. United Nations, New York, 1994.

43. National Research Council, Committee on the biological effects of ionizing radiation (1990) *Health Effects of Exposure to Low Levels of Ionizing Radiation* (BEIRV). National Academy Press, Washington DC.

44. Kreisheimer, M., Koshurnikova, N.A., Nekolla, E., Khokhryakov, V.F. Romanov, S.A., Sokolnikov, M.E., Shilnikova, N.S., Okatenko, P.V., and Kellerer, A.M. (2000) Lung cancer mortality among male nuclear workers of the Mayak facilities in the former Soviet Union. *Radiat. Res.*, 154, 3–11.

45. National Research Council, Committee on the biological effects of ionizing radiation (1988) *Health Risks of Radon and other Internally Deposited Alpha Emitters* (BEIRIV). National Academy Press, Washington DC.

46. United Nations Scientific Committee on the effects of Atomic Radiation (UNSCEAR) (1993) *Sources and Effects of Ionizing Radiation*. United Nations, New York.

47. Shore, R.E. (1992) Issues and epidemiological evidence regarding radiation-induced thyroid cancer. *Radiat. Res.*, 131, 98–111.

48. Whitehouse, C.A., Tawn, E.J., and Riddell, A.E. (1998) Chromosome aberrations in radiation workers with internal deposits of plutonium. *Radiat. Res.*, 150, 459–468.

49. Nagasawa, H. and Little, J.B. (1992) Induction of sister chromatid exchanges by extremely low doses of alpha particles. *Cancer Res.*, 52, 6394–6396.

50. Zhou, H., Randers-Pehrson, G., Waldren, C.A., Vannais, D., Hall, E.J., and Hei, T.K. (2000) Induction of a bystander mutagenic effect of alpha particles in mammalian cells. *Proc. Natl. Acad. Sci. USA*, 97, 2099–2104.

51. Zhou, H., Suzuki, M., Randers-Pehrson, G., Vannais, D., Chen, G., Trosko, J.E., Waldren, C.A., and Hei, T.K. (2001) Radiation risk to low fluences of alpha particles may be greater than we thought. *Proc. Natl. Acad. Sci. USA*, 98, 14410–14415.

52. Wu, L.J., Randers-Pehrson, G., Xu, A., Waldren, C.A., Geard, C.R., Yu, Z., and Hei, T.K. (1999) Targeted cytoplasmic irradiation with alpha particles induces mutations in mammalian cells. *Proc. Natl. Acad. Sci. USA*, 96, 4959–4964.

53. Little, J.B., Azzam, E.I., de Toledo, S.M., and Nagasawa, H. (2002) Bystander effects: Intercellular transmission of radiation damage signals. *Radiat. Prot. Dosimetry*, 99, 159–162.

54. Lyng, F.M., Seymour, C.B., and Mothersill, C. (2002) Initiation of apoptosis in cells exposed to medium from the progeny of irradiated cells: A possible mechanism for bystander-induced genomic instability? *Radiat. Res.*, 157, 365–370.

55. Moisenko, W., Battista, J.J., Hill, R.P., Travis, E.L., and Van Dyk, J. (2000) In field and out-of-field effects in partial volume lung irradiation in rodents: Possible correlation between early DNA damage and functional endpoints. *Int. J. Radiat. Oncol. Biol. Phys.*, 48, 1529–1548.

56. Khan, M.A., Van dyk, J., Yeung, I.W., and Hill, R.P. (2003) Partial volume rat lung irradiation: Assessment of early DNA damage in different lung regions and effect of radical scavengers. *Radiother. Oncol.*, 66, 95–102.

57. Prise, K.M., Folkard, M., and Michael, B.D. (2003) Bystander responses induced by low LET radiation. *Oncogene*, 22, 7043–7049.

58. AACR symposium on Radiation Biology and Cancer, from Molecular Responses to the Clinic, February 18–22, 2004, Laguna Cliffs Marriot Resort, Dana Point CA, USA.

11 DNA Microarrays and Computational Analysis of DNA Microarray Data in Cancer Research

Mario Medvedovic and Jonathan S. Wiest

CONTENTS

SUMMARY

The advent of DNA microarray technology has added a new dimension to the field of molecular carcinogenesis research. DNA microarrays have been used as a tool for identifying changes in gene expression and genomic alterations that are attributable to various stages of tumor development. Patterns defined by expression levels of multiple genes across different types of cancerous and normal tissue samples have been used to examine relationships between different genes, and as the tool for molecular classification of different types of tumor. The analysis of relatively large datasets generated in a typical microarray experiment generally requires at least some level of computer-aided automation. On the other hand, the large number of hypotheses that are implicitly tested during the data analysis, especially when identifying patterns of expression through supervised and unsupervised learning approaches, require careful assessment of statistical significance of obtained results. These basic requirements have brought to the forefront the need for developing statistical models and corresponding computational tools that are specifically tailored for the analysis of microarray data. Such models need to be able to differentiate between faint, yet statistically significant and biologically important signals, and patterns that are generated by random fluctuations in the data. In this endeavor, it is important to keep in mind the abundance of already existing statistical and machine-learning methodologies which can serve as the starting point for developing more specialized techniques. Here we describe different uses of DNA microarray technology in molecular carcinogenesis research and related methodological approaches for analyzing and interpreting DNA microarray data obtained in such experiments.

11.1 INTRODUCTION

Novel molecular biology technologies for performing large numbers of biological measurements in parallel provide an unprecedented opportunity for uncovering the molecular basis of cancer and mechanisms of cancer induction by carcinogens. The large volume of data generated by experiments utilizing such assays, as well as relatively high experimental noise often associated with them require a careful statistical/computational analysis. The most prominent of such novel technologies are DNA microarrays, which facilitate the assessment of a whole transcriptome of a cell population in single experiments.

DNA microarrays are glass slides on which a large number of DNA probes, each corresponding to a specific mRNA species, or a genomic DNA region, are placed at predefined positions. DNA probes are either synthesized *in situ* [1,2], or they are presynthesized and then spotted on the slide [3]. Two most commonly used technologies are Affymetrix *in situ* synthesized microarrays and the spotted microarray technology developed at Stanford University. In gene expression experiments RNA is extracted from the biologic sample, reverse transcribed into cDNA and fluorescently labeled. Such labeled cDNA representing the transcriptome of the biologic sample is then hybridized on the microarray. The amount of the labeled cDNA that hybridizes to each probe on the microarray is proportional to the relative abundance of the corresponding cDNA. The expression of all genes is then quantified by measuring the intensity of the dye used to label the RNA. The most common experimental protocol used with spotted microarrays consists of labeling two RNA extracts with different dyes and co-hybridizing the samples to the same microarray (two-channel microarrays). While this approach introduces some restrictions on the experimental design [4], the overall principles of the two major technologies are the same.

Quantification of individual gene expressions proceeds by various normalization procedures whose role is to remove systematic biases and to rescale measurements on different arrays to be directly comparable. The development of an appropriate normalization procedure is still an active research topic [5–7]. Normalized data is used to identify genes differentially expressed in different tissues, to identify groups of genes with similar pattern of expression across different biological states, and to construct rules for classifying different biological samples based on their expression profiles.

In general, computational analysis of microarray data can be separated into a single gene at a time analysis, in which the data for each gene is analyzed independently of the data for any other gene, and multiple gene at a time analyses in which the data for all or a subgroup of genes is jointly analyzed. In a single gene at a time analysis, the goal is generally to identify genes that are differentially expressed in different tissues. In a multiple gene at a time analysis, the information from multiple genes is combined to identify global patterns of expression that can offer additional insights not available by looking at genes separately.

11.2 APPLICATIONS OF MICROARRAYS

Although the most common application of microarrays is in monitoring gene expression, the other extremely relevant application in the context of cancer research is the microarray based Loss of Heterozygosity (LOH) analysis [8,9].

Gene expression data generated using microarrays is generally used to identify genes that are differentially expressed under different experimental conditions, identify groups of genes with similar expression profiles across different experimental conditions (co-expressed genes), and classify the biologic sample based on the pattern of expression of all or a subset of genes on the microarray. Differentially expressed genes as well as groups of co-expressed genes can be used to hypothesize which pathways were involved in a particular biologic process. Additionally, clusters of co-expressed genes can be used to hypothesize the functional relationship of a clustered gene, and as a starting point of dissecting regulatory mechanisms underlying the co-expression. In the context of tumor classification, gene expression profiles have been used as complex biomarkers defining the tumor as well as different subclasses of tumor.

Genomic instability is central to the development of cancer. Gene amplifications and deletions are a major factor in tumorigenesis. Copy number changes are important in the understanding of cancer biology, diagnosis, and progression. These genetic alterations can lead to expression changes in oncogenes or in tumor suppressor genes. Changes in gene expression as a result of these alterations are likely to be the driving force behind many of the amplifications and deletions that occur giving the transformed cell a growth advantage.

Comparative genomic hybridization (CGH) is a technique that analyzes the global genetic alterations in cells. The procedure detects both deletions and amplifications of the genome and allows for the global analysis of genetic alterations in tumors. The traditional CGH uses differentially labeled test and reference genomes that are co-hybridized to normal metaphase chromosome spreads. The fluorescent ratio of the labeled DNAs is then measured over the length of the chromosomes to determine regions of gain or loss. Analyzing the data indicates amplified and deleted regions of the genome based on the intensity of each fluorescent signal. Analyzing numerous samples demonstrates the frequency of the genomic aberrations. One disadvantage of the technique is that the sizes of the alterations need to be fairly large, for example on the order of 5 to 10 Mb, to be detected [10]. An additional problem is that the procedure is very labor intensive and not amenable to analyzing large numbers of samples. Other methods, including microsatellite marker analysis and fluorescent *in situ* hybridization provide a higher resolution map, but are also labor intensive and may not be applicable to whole genome analysis.

The advantages of newly developed microarray based CGH assays are numerous. This technology is theoretically capable of assessing relatively small genomic aberrations and is capable of the high-throughput analysis. A clear demonstration of the improved resolution of the microarray CGH over the traditional approach was offered in microarray-based CGH analysis of the SKBR3 breast cancer cell line [11]. In these experiments, microarray-based CGH analysis improved the resolution of amplicons in the 8q regions over the traditional analysis [12].

11.3 ANALYSIS OF GENE EXPRESSION MICROARRAYS

At this point we assume that the quantification of fluorescence intensities of individual spots on the microarray has been completed. The analysis of microarray data in most situations proceeds by normalizing data, identifying genes whose expression changes between different experimental conditions, and performing multivariate analyses, such as clustering and classifying.

11.3.1 DATA NORMALIZATION

The first step in the computational analysis of microarray data almost always consists of performing various transformations with the aim of reducing systematic variability. Although the optimal procedures are still being developed, a certain consensus is emerging on the appropriate ways to perform initial data normalizations in microarray experiments [13]. In the case of the spotted two-channel arrays, two major sources of the systematic variability are the spot-specific local-background fluorescence and the difference in the overall intensities of the two fluorescent dyes (Cy3 and Cy5).

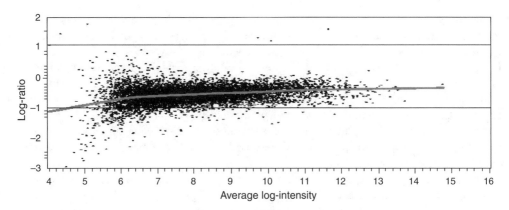

FIGURE 11.1   Scatter plot of log-ratios versus average log-intensity in a typical microarray experiment. The line represents the local regression line used for centering log-ratios.

The process of normalization generally proceeds by subtracting the local background and centering the log-ratios of two channel intensities around zero. In Figure 11.1, log-ratios of background-subtracted intensities in two channels are plotted against their average. The line describing the average behavior of data is the local regression (loess) curve [14].

Initially, the common practice was to center the log-ratios by subtracting the overall median value. However, it is fairly obvious from Figure 11.1 that such an adjustment is likely to "over-adjust" high-intensity spots and "under-adjust" the low-intensity spots. It turns out that the local regression based normalization, which subtracts the fitted loess curve value from the corresponding log-ratio, generally does a better job of reducing this channel bias [15,16] and is gaining wide acceptance. In the case of the Affymetrix data, similar strategy of scaling data on all microarrays in the experiment to a "control" chip so that all chips have equal median intensities has been commonly used. Recently, alternative approaches based on the intensity-specific normalizations have been introduced as well [17].

11.3.2 DETECTING DIFFERENTIALLY EXPRESSED GENES ACROSS DIFFERENT SAMPLES

The purpose of the statistical analysis in the process of identifying differentially expressed genes is to assess the reproducibility of observed changes in gene expression by assessing their statistical significance. This is done by comparing the magnitude of the observed changes in gene expression to the magnitude of random fluctuations in the data. For example, in the traditional t-test analysis, the average differential expression observed in replicated experiments is divided by its standard error and the obtained quantity (t-statistic) is compared to its theoretical distribution under the assumption that the observed average differential expression is a result of random fluctuations in the data. In the context of the cancer-related microarray data, paired t-test and the step-down Bonferonni adjustment was used to identify genes whose expression is affected in testis of mice that were exposed to diethylstilbestrol during gestation and lactation [18]. Furthermore, identifying genes that are differentially expressed between different classes of tumor tissues is often the first step in identifying relevant genes for the purpose of cluster analysis and tumor classification [19,20].

For any kind of analysis to be successful in assessing reproducibility of observed results, it is necessary to apply an appropriate experimental design in the process of gathering data. The key requirements for the appropriate experimental design are that it addresses all relevant sources of variability. Let us suppose that we want to identify genes that are differentially expressed between two different types of tumors using two-channel spotted microarrays. The logical requirement for the implicated genes is that they are, on an average, differentially expressed between the two tissue types. Several decisions that are made prior to performing experiments are going to significantly impact the reproducibility of the results of the experiment regardless of the subsequent statistical analysis. Assuming that we performed appropriate normalization of the data, two unavoidable types of variability will be present in the data. One is the technical variability that is introduced in the process of isolating and labeling RNA, fabrication of microarrays, scanning process, etc. The other is the biological variability between the different tissue samples of the same kind used in the analysis.

In most biological applications, the biologic variability dominates the technical variability. For example, the variability between different measurements of the same tissue sample will be much smaller than the variability between different tissue samples of the same kind (e.g., same tumor type from multiple individuals). To reduce the overall variability in our hypothetical experiment, one could be tempted to use only one tumor sample of each kind and perform several technical replicates. Due to the lower variability, than if different tumors are used in replicated experiments, such an approach is likely to result in more genes being pronounced differentially expressed. The obvious problem is, though, that such results cannot be generalized to the whole population of these two types of tumors and consequently will not be reproducible.

On the other hand, there are several sources of variability that can be efficiently removed from the estimates of differential expression using the factorial Analysis of Variance (ANOVA) approach. Two such sources that are more commonly addressed in the statistical analysis of microarray data are gene-specific dye effects and the array effect. These effects are manifested in the fact that fluorescence measurement of one dye are reproducibly higher than the other dye in gene-specific fashion, meaning that the effect varies from one gene to another. If this source of variability is not taken into account when the experiment is being designed, it could result in falsely implicating nondifferentially expressed genes as well as in missing truly differentially expressed genes. One way to deal with this problem is to perform "dye-flips," meaning that different RNA samples from the same tissue type are labeled with different dyes. If the number of replicates labeled with Cy3 is equal to the number of replicates labeled with Cy5, this will remove the systematic bias from the analysis. However, if this new source of variability is not extracted in the ANOVA analysis, it can seriously inflate the variability of the differential expression estimates.

In the factorial ANOVA one estimates contributions of different systematic sources of variability and extracts them from the estimates of the effect of interest. For example the simplest linear model that allows for the extraction of the gene-specific dye effect using ANOVA is

$$Y_{ijk} = \mu + T_i + D_j + A_k + \varepsilon_{ijk},$$

where Y_{ijk} is the expression measurement on the kth microarray of the tissue type i labeled by the jth dye ($j = 1$ for Cy3 and $j = 2$ for Cy5). μ is the overall expression level for this gene, ε_{ijk} is the random error in Y_{ijk} unexplained by factors in the model, and T_i is the effect of the ith tissue type on the expression level adjusted for the dye effect (D_j) and microarray (A_k) measuring the differential expression of the gene between different tissue types. By estimating differential expression after adjusting for the dye effect, one effectively removes the variability introduced by "flipping dyes" from the analysis. A more thorough review of experimental design issues in microarray experiments can be found elsewhere [21]. Issues relating to using ANOVA in analyzing microarray data are discussed in the context of the fixed-effect model [22], and in the context of the mixed-effect model [23].

Statistical methods for identifying differentially expressed genes have come a long way from initial heuristic attempts [24], through the realizations that rigorous statistical analysis of replicated data is needed [25], to sophisticated statistical modeling using frequentist and Bayesian approaches [26]. Generic statistical methods of the analysis of variance [22] and mixed models [23] are complemented with specialized maximum likelihood approaches [27], and Bayesian analysis flavored approaches [28–30]. While the consensus about the optimal method has still not been reached, the intense statistical research is a promising sign.

One of the most daunting issues in the process of identifying differentially expressed genes is the severe problem of multiple comparisons. Presently, expression level of up to more than 20,000 different genes can be assessed on a single microarray. Searching for genes whose expression change is statistically significant corresponds to testing 20,000 hypotheses simultaneously. If each of these tests is performed at the commonly used significance level of $\alpha = .05$, meaning that we expect for 5% of genes that are not differentially expressed to be falsely implicated, we expect on average 1,000 falsely implicated genes. The simplest way to deal with this multiple comparison problem is to divide the significance level by the number of hypotheses testing (Bonferonni adjustment). This will mean that in the case of 20,000 hypotheses, individual hypotheses will be tested at the significance level of $\alpha = .0000025$. Such a level is virtually unattainable in simple experiments with few experimental replicates. While the multiple comparison issue cannot be avoided, a better balance can be struck between the need to avoid false positives and false negatives. The False Discovery Rate (FDR) adjustment [31] keeps the balance between the specificity and the sensitivity of microarray data analysis [30]. In contrast with traditional adjustments that control the probability of a single false positive in the whole experiment, the FDR approach controls the proportion of false positives among the implicated genes. For example, if 20 genes are selected using FDR $= .05$, one of them will on average be a false positive regardless of the total number of genes. The traditional (e.g., Bonferonni) adjustment will limit the probability of a single false positive to .05, resulting in a possibly conservative testing procedure.

11.3.3 IDENTIFYING CLUSTERS OF CO-EXPRESSED GENES

11.3.3.1 Overview of Clustering Approaches

The high-dimensional nature of microarray data has prompted the widespread use of various multivariate analytical approaches aimed at identifying and modeling patterns of expression behavior. In cancer research, cluster analysis has been commonly utilized to identify groups of genes with a common pattern of expression across different tissues as well as to group tissues with similar genetic expression profiles. Results of the cluster analysis have been used to infer common biologic function and the co-regulation of co-expressed genes in response to mutagenic treatments [32] and p53-specific DNA damage response [33], to identify genes and groups of genes whose pattern of expression can serve as the marker of various stages in tumor progression [19], and to assess the possibility of classifying different kinds of tumors based on their gene expression profiles [34–37]. Relevance of different clusters obtained by hierarchically clustering breast carcinomas was confirmed by correlating them with the mutational status of the *P53* gene and the clinical outcome [20].

The power of the clustering approach in interpreting patterns of expression of groups of genes is demonstrated in Figure 11.2. The data in Figure 11.2 comes from the publicly available yeast cell-cycle dataset [38]. If we just observe expression patterns of two genes in Figure 11.2(a), about which we know very little, we would conclude that their expression profiles are highly correlated. This might lead us to conjecture that these two genes are participating in the same crucial point of the cell-cycle progression. On the other hand, if we knew that there are two different expression patterns that our two genes could be associated with, given in Figure 11.2(b), we would probably conclude that these two genes are actually representative of two different expression patterns. If we don't know of the existence of such two patterns but are given 74 genes that define these two patterns, a simple hierarchical clustering procedure will easily identify two clusters in Figure 11.2(c) and 11.2(d), defining two patterns in Figure 11.2(b) and associating our two genes to distinct clusters. Since the advent of the microarray technology, virtually all traditional clustering approaches have been applied in this context and numerous new clustering approaches have been developed.

Hierarchical clustering procedures were the first to be applied in the analysis of microarray data [39] and are still the most commonly used clustering procedure in this context. Such methods rely on the calculation of pairwise distances or similarities between the gene profiles. Various

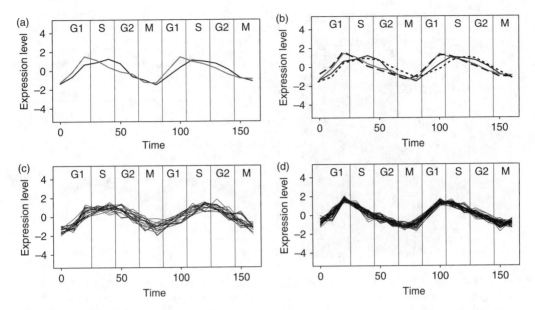

FIGURE 11.2 Clustering cell-cycle expression data: (a) two individual gene expression profile alone; (b) the same two profiles in the background of two major underlying expression patterns; (c,d) clusters defining underlying patterns of expression.

correlation coefficients are the most commonly used measures of similarity. Hierarchical agglomerative methods generally proceed by grouping genes and groups of genes based on such pairwise measures of similarity. In this process, the distance between two groups of genes is calculated as a function of individual pairwise distances of genes in two groups using different "linkage" functions. "Single-linkage" corresponds to the minimum pairwise distance between genes in two different groups, "complete-linkage" corresponds to the maximum distance, and "average-linkage" corresponds to the average distance [40]. Virtually every publication related to utilizing microarrays for gene expression profiling of tumor tissues and cell lines contains a figure with genes and tissues organized in this fashion.

Partitioning approaches, on the other hand, work by iteratively reassigning profiles in a pre-specified number of clusters with the goal of optimizing an overall measure of fit. Two of the most commonly used traditional approaches are k-means algorithm and self-organizing map (SOM) method, first applied in this context by Tavazoie et al. [41] and Tamayo et al. [42], respectively. One of the problems with clustering methods that are based on pairwise distances of expression profiles is that, at least in the initial steps, only data from two profiles is used at a time. That is, the information about relationships between the two profiles and the rest of the profiles is not taken into account although these relationships can be very informative about the association between the profiles. The major drawback of partitioning approaches is the need to specify the number of clusters. For example, given that we know that there are two clusters in the data in Figure 11.3, both k-means and SOM's will uncover the two clusters of interest. However, while in the hierarchical structure in Figure 11.3, it is obvious that there are two clusters of data, both k-means and SOM's require this to be known prior to the analysis.

In a model-based approach to clustering, the probability distribution of observed data is approximated by a statistical model. Parameters in such a model define clusters of similar observations and the cluster analysis is performed by estimating these parameters from the data. In a Gaussian mixture model approach [43], similar individual profiles are assumed to have been generated by the common underlying "pattern" represented by a multivariate Gaussian random variable [44,45]. In the situation where the number of clusters is not known, this approach relies on ones ability to identify the correct

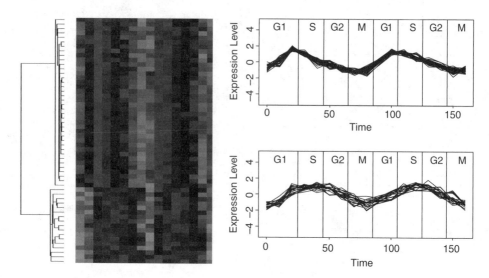

FIGURE 11.3   (**See Color Figure 11.3 following page 272.**) Hierarchical clustering of the cell-cycle genes from Figure 11.2. Each line in the color-coded display corresponds to the expression profile of one gene.

number of mixture components. A mixture based method for clustering expression profiles that produces clusters by integrating models with all possible number of clusters was developed [46]. In this approach, the joint distribution of the data is modeled by a specific hierarchical Bayesian model and the posterior distribution of clusterings is generated using a Gibbs sampler.

Model-based clustering procedures have been shown to have desirable properties in various comparative studies examining properties of different clustering procedures [46,47]. Recently, finite mixture models as implemented in the AutoClass software package [48] were used to refine the clustering of gene expression profiles of human lung carcinomas produced by the hierarchical procedures [49]. A similar method was applied to identify genes related to malignancy of colorectal carcinomas [50].

11.3.3.2   Assessing Statistical Significance of Observed Patterns

A reliable assessment of reproducibility of observed expression patterns and gene clusters is one of the burning issues in cluster analysis. Since cluster analysis has generally been used as an exploratory analysis tool, establishing statistical significance of observed results has not been a priority. However, just as in the case of establishing the statistical significance of differential expressions, an assessment of the reproducibility of observed patterns is necessary before one can take them seriously. Unfortunately, establishing the statistical significance of different features of observed clusters is a much more difficult problem than establishing differential expression of individual genes. Two exceptions are the significance of the existence of the overall clustering structure and the significance of pairwise associations between individual profiles. However, even the pairwise association between individual profiles is a difficult problem if one assumes possible but unknown clustering structures.

This can be illustrated in the analysis of the two genes of interest in Figure 11.2. Suppose we are asking the question whether or not these two genes are co-expressed. Or, in other words, do expression profiles of these two genes belong to the same underlying pattern of expression? If we use correlation as a measure of similarity of these two profiles, the Pearson's correlation turns out to be equal to 0.83. In the context of randomly chosen pairwise correlations for all genes in this dataset this turns out to be statistically significant. As a result, we could be tempted to say that these two genes are co-expressed. However, if we analyze the whole group of genes using hierarchical clustering

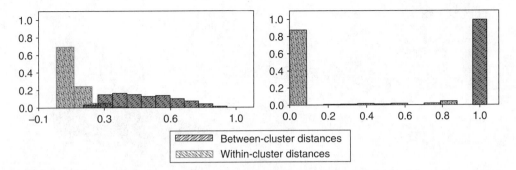

FIGURE 11.4   Right: Distribution of between- and within-clusters distances based on pairwise correlations. Left: Distribution of between- and within-clusters distances based on posterior probabilities of expression calculated from the Bayesian Infinite Mixture Model.

(Figure 11.2 and Figure 11.3) (See Color Figure 11.3 following page 272.), it seems that the two genes belong to two distinct patterns of expression. In this respect some of the newly developed statistical approaches offer a glimmer of hope. For example, in the Bayesian Infinite Mixture Model approach, the posterior probability of any particular clustering feature (say gene1 and gene2 are co-expressed) can be directly assessed from the output of the Gibbs sampler [46]. Such a model-based approach is capable of producing an objective measure of confidence in any such feature after incorporating sources of uncertainty in the process of clustering microarray data (i.e., experimental variability and unknown number of clusters).

When we apply the Bayesian Infinite Mixture (BIM) in the context of two genes in Figure 11.1, the result is rather unambiguous. First of all the posterior distribution of "distances," which are in this context defined as 1-Posterior Probability of Co-Expression, indicates strongly that there are actually two clusters in the data (Figure 11.4). Furthermore, the posterior probability of the feature of interest, which is that these two genes are co-expressed, after averaging over models with all possible number of clusters, is equal to 0 indicating that data actually offers strong evidence that these two genes are not co-expressed. The higher precision of the model based on posterior probabilities calculated from the BIM is illustrated by comparing distributions of between- and within-cluster distances for the two clusters in Figure 11.3 obtained by simple correlation and based on BIM model.

11.3.4   GENE EXPRESSION BASED TUMOR CLASSIFICATION

Classification of tumor samples based on gene transcription profiling has been one of the earliest and one of the most promising areas of microarray technology applications in cancer research. The concept of using the gene expression profiles as complex markers in classifying different types of cancers has been initially demonstrated by classifying different types of acute leukemias [37] and distinguishing between the tumor and normal colon tissues [36]. This approach has also been shown to have a great potential for clinical applications in the areas of tumor classification and toxicity screens of potential drug compounds [51–53].

In general, a "classifier" is a mathematical formula that uses as input the values of distinct features of an object and produces an output that can be used to predict to which of the predefined classes the object belongs. In terms of the gene expression data based tumor classification, the objects are tissue samples and the features are genes and their expression levels. The construction of a classifier generally proceeds by selecting an informative set of genes that can distinguish various classes, choosing an appropriate mathematical model for the classifier, and estimating parameters of the model based on the "training set" of tissues whose classification we know in advance. Finally, the specificity and the sensitivity of the classifier is tested on the data that was not used in the process of constructing the classifier.

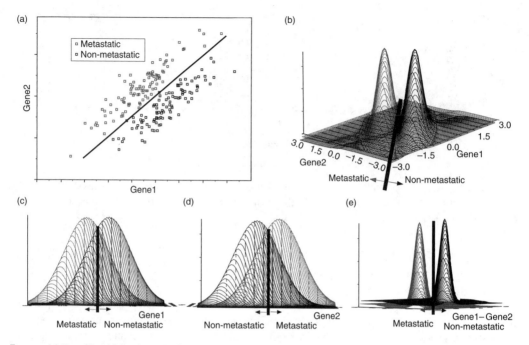

FIGURE 11.5 Classifying hypothetical tumor tissues based on the expression levels of two genes. (a) Scatter plot of data for 100 samples. (b) Underlying probability distribution of expression data for two classes. (c) Separating two classes based on Gene1 data only. (d) Separating two classes based on Gene2 data only. (e) Optimal linear classifier for the two classes based on the linear combination of Gene1 and Gene2 expression data.

The simplest classifier one can envision consists of a single gene and a cut-off value x_c such that a sample is classified in one class if the expression level of this gene is smaller than x_c and in the other class if it exceeds x_c. In the case when multiple features/genes are used, measurements from all of them are again summarized into a single number by using a variety of multivariate models. Such a summary value is then used in a similar fashion as one would use a single gene value. A hypothetical example of advantages of using expression levels of multiple genes for classifying "metastatic" and "nonmetastatic" tumors is depicted in Figure 11.5. While expression of any single of the two hypothetical genes in Figure 11.5 are not sufficient for reliably predicting whether the sample is "metastatic" or "nonmetastatic" (Figure 11.5(c) and 11.5(d)), their combination constructed by subtracting the expression of the Gene2 from the expression of Gene1 can separate the two classes almost perfectly (Figure 11.5(e)).

Various approaches to selecting informative genes can be coarsely grouped in methods that assess the classification abilities of a single gene at a time and methods that choose groups of genes based on their joint ability to distinguish between different tumor classes. The most common methods in practice to date have been based on choosing genes in one-gene at a time fashion based on the statistical significance or the magnitude of their differential expression between different classes of tumors [19,20]. The combinatorial explosion of possible number of different groups of genes generally makes the second approach of choosing groups of genes based on their joint discriminative capacity very difficult. Comparing all possible subgroups among 20,000 different genes is clearly impossible. Alternatives to the exhaustive comparison are heuristic optimization techniques such as Genetic Algorithm [54] or constructing groups of gene in a stepwise fashion. Both of these approaches will not necessarily identify the optimal group of genes but have been shown to often perform quite well in this context.

Mathematical models that have been used so far in constructing tumor classifiers can generally be divided in nonparametric methods such as k-nearest neighbor (KNN) [54], Fisher's linear discriminant analysis (FLDA) [55], and support vector machines (SVM) [56], and the methods based on the statistical model for data in individual classes such as various Gaussian model based classifiers, logistic regression [57], and artificial neural networks (ANN) [58]. Excellent descriptions and introductions into various classification approaches are given elsewhere [59,60].

In a KNN classifier for the two-classes situation, the distance of the sample expression profile of the sample to be classified from individual profiles of all training data is first calculated. The sample is then classified to the class having the most members within the k-closest neighbors of the sample. Fisher's linear discriminant function is based on identifying the direction in the k-dimensional space (where k is the number of genes used for the classification) that separates the two classes the best, in the sense that it maximizes the ratio of between-classes and within-classes variability. SVM classifiers are based on the idea of the "optimal separating hyper-plane" in the k-dimensional space. It chooses the hyper-plane so that the distance from the hyper-plane to the closest point in each class is maximized. For example, the linear (hyper-plane) SVM when $k = 2$ (two-genes classifier) will select the straight line such that the traditional Euclidian distance between the line and the closest points in two classes is maximized. In this sense the linear SVM classifiers are similar to the Fisher's linear discriminant function based classifier except that the two methods use different criteria to select the separating hyper-plane. ANN-based classifiers will generally fit a nonlinear hyper-plane to separate two classes of objects (tissues in our case). Probabilistic models based classifiers generally proceed by estimating the distribution of the features of the classes of objects to be classified. The classification is then based on the identification of the most likely of such distribution that have had generated the sample to be classified. In the hypothetical example in Figure 11.5, all above mentioned procedures will likely perform very well. Strictly speaking the linear discriminator depicted in Figure 11.5(a) to 11.5(c) corresponds to the Fisher's linear discriminant.

Validation of the predictive accuracy of any particular classifier is an essential step in the classification analysis. The optimal way of validating a classifier's performance is to test it on the samples that were not used in any way in the process of building the classifier. A commonly used strategy is to perform a "leave-one-out" analysis in which each of the samples is left out in the process of building the classifier and then used to test its predictive ability. Average predictive ability of the classification procedure can then be summarized as the proportion of the correctly classified samples in the "leave-one-out" analysis. Ideally, predictive ability is then compared to the predictive ability of the equivalent classifier on the randomized data. This is particularly important when we have different number of samples in different classes. For example, a trivial classification rule of always classifying samples into a single class will have 90% correct predictive rate if 90% of the samples that are being classified come from this class.

How to identify the best set of genes as well as questions related to the optimal mathematical model for constructing classifiers are two of the intense research areas of computational biology. Results of a comparison study of several traditional classification methods in relation to microarray data based tumor classification [61] suggest the need to base classifiers on statistical theory. For example, it was shown that the maximum likelihood-based classifier clearly outperforms a popular heuristic equivalent [37].

An alternative approach to generating optimal classification features is to use some of the dimension-reduction techniques. The most commonly used method is the Principal Component Analysis (PCA). In the PCA analysis, one seeks a small number of linear combinations of the initial features that in a sense condense the predictive information of the whole set of features. Linear combination of k values $(x_1, \ldots, x_k)$ is defined as $a_1 x_1 + a_2 x_2 + \cdots + a_k x_k$, where $(a_1, \ldots, a_k)$ are corresponding linear coefficients. PCA identifies linear combinations of features that maximize the variability between different objects. While this heuristic argument behind PCA works in many situations, in some situations it fails completely. For example, in our hypothetical example in Figure 11.5, the linear combination with the maximum variability is approximately equal to Gene1 + Gene2 and it actually results in worse separation than any of the original variables (Figure 11.6). Another related dimension-reduction technique is the Partial Least Squares (PLS) method [62], which extends

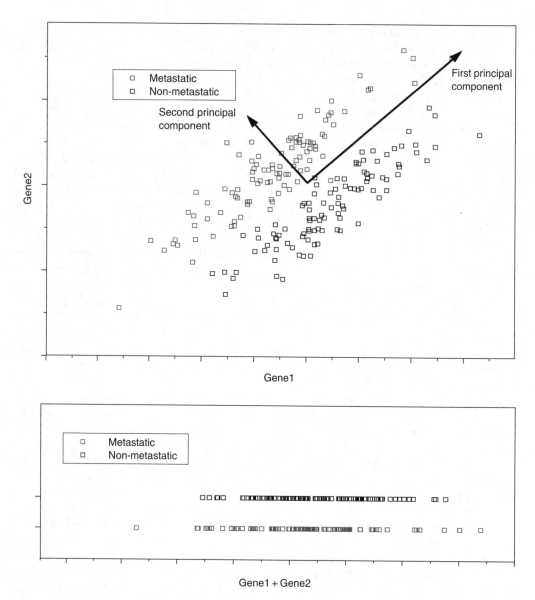

FIGURE 11.6   Principal component based classifier.

the PCA approach to incorporate the information about the correct classification in the process of identifying optimal linear combinations. Because the method chooses linear combinations that accentuate the relationship between the features and the classification of the training object, it generally results as a better classifier. Generally, use of such procedures in the tumor classification setting has an intuitive appeal that one can use information from a large number of genes without experiencing problems with the classification methods that perform best when the number of samples is significantly larger than the number of features used by the classifier.

11.4   ANALYSIS OF CGH MICROARRAY DATA

The CGH data is to a large extent similar to the analysis of gene expression arrays. The data still needs to be normalized, and the changes in copy numbers of different DNA regions represented on the microarray needs to be established by performing the statistical analysis. Similarly, as in the case of expression arrays, data can be clustered to identify patterns of common amplifications and deletions across all tissue samples. CGH microarray data can be also used to design classifiers in exactly the same way as described for the gene-expression data.

One specific feature that distinguishes this type of data is the correlation introduced by the linear organization of genome that can be utilized to improve the sensitivity of such analyses. The basic premise of such an analysis is that the closer two DNA regions are genomically, then if one of them is affected by a gross genomic aberration, it is highly likely that the other one will be affected as well. One way to explore such correlations is to use moving average estimates of fluorescence intensities of different DNA probes. Moving averages are calculated by averaging intensities of DNA probes corresponding to several neighboring DNA loci. The amount of "smoothing" induced by such a strategy is dependent on how many neighboring loci are averaged. Since such averages have a potential of completely concealing genomic aberrations covered by a single probe, one has to be careful about using it. Presumably, such an analysis can be used within a battery of different analytical approaches with performing experimental replicates still being the preferred approach to reducing variability in fluorescence measurements.

11.5 INTEGRATING CURRENT KNOWLEDGE AND VARIOUS TYPES OF EXPERIMENTAL DATA

Integration of the current knowledge with the new experimental data is done every time a biologist interprets results of a new experiment. Interpreting results of a microarray experiment that can yield hundreds of thousands of data points in the traditional informal way can be rather difficult. In this situation, one is forced to limit her/his attention to a subset of genes that were indicated in the initial statistical analysis. However, the sensitivity of the statistical analysis can be critically affected by the incorporation of the prior knowledge. For example, if one can make assumptions concerning the subset of genes most likely to be affected, this subset can be analyzed separately with higher statistical power due to fewer hypotheses that are being tested. Formal methods of integrating accumulated knowledge and information in the analysis are being developed. An example of such methods is the method for scoring likelihood of whole pathway involvement in the process under investigation based on integrating the analysis of expression levels of genes involved in the pathway and the existing pathway information [63]. Similarly, benefits of integrating genomic, functional genomic, and proteomic data have been demonstrated [64–67]. For example, a weak evidence of co-regulation implied by co-expression identified in a cluster analysis can be strengthened by the result from a two-hybrid assay, which indicated the two corresponding proteins interact, or by the shared regulatory elements in their promoter region. Statistical models capable of integrating such diverse data types have been proposed by several investigators [68–71], while the use of joint proteomic and functional genomic data after perturbing a biologic system to reverse engineer the underlying network of molecular interactions, in the context of the "systems biology" paradigm, has been demonstrated [72,73].

11.5.1 FROM CO-EXPRESSION TO CO-REGULATION

Transcriptional regulation is one of the crucial mechanisms used by a living system to regulate protein levels. It is estimated that 5 to 10% of the genes in eukaryotic genomes encode transcription factors that are dedicated to the complex task of deciding where, when, and which gene is to be expressed. Mechanisms applied by these factors range from the recruitment and the activation of the transcriptional preinitiation complex to necessary modulations of local chromatin structure. Two major determinants of gene expression specificity seem to be the composition of their *cis*-regulatory modules, and the presence/absence and phosphorylation status of *trans*-acting regulatory factors that interact with DNA regulatory modules and each other. However, the exact nature of the interactions between various components of the regulatory machinery is still largely unknown. Identification of co-expressed genes by a cluster analysis of gene expression profiles has often been utilized as a first step in identifying factors regulating expression of different genes. On the other hand, using information about presence of known regulatory elements can be applied to refine the cluster analysis of expression profiles and the simultaneous identification of known regulatory elements causing such co-regulation.

An indirect indication of co-regulation of co-expressed genes is the tendency of co-expressed genes to participate in the same biologic pathway as well as their tendency to code for proteins that interact with each other. In both of these situations, the mechanism of co-regulation might not be at the level of common *cis*-regulatory elements, yet the need for co-regulation and the actual presence of them is obvious. Actually, it has been shown that the particular expression regulatory mechanism can sometimes be a better determinant of the protein function than even its three-dimensional structure. All these suggest that analytical methods capable of integrating information about regulatory sequences, biologic pathways, protein–protein interactions, and expression data generated in microarrays experiments will be better able to create biologically meaningful clusters of genes than clustering expression data alone.

11.5.2 Integrating Microarray CGH and Expression Data

In terms of cancer research, tumors represent a naturally perturbed genomic system. Concurrent genomic and functional genomic investigations of tumors by the high-throughput microarray approaches can be used to dissect genetic networks involved in the process of tumorigenesis. In this context, microarrays can be used to both characterize the genomic aberrations and the gene expression in different tumors. Several models mentioned before are capable of integrating such information into a single powerful analysis.

The next logical step in high-throughput analysis is to combine the cDNA array analysis with CGH analysis of tumor samples. This has been accomplished in several recent reports. Fritz et al. [74] applied CGH and cDNA based arrays to liposarcomas and found that tumor subtypes were revealed more effectively by clustering genomic profiles than by clustering expression profiles. Weiss et al. [75] have shown that in gastric adenocarcinomas, microarray analysis of genomic copy number changes can predict the lymph node status and survival outcome in patient samples. In breast tumors, Pollack et al. [11] found that 62% of highly amplified genomic regions contain over expressed genes and in general that a 2-fold change in copy number corresponds to a 1.5-fold change in mRNA levels as detected on the cDNA arrays. Additionally, they report that 12% of all gene expression changes in breast tumors are directly attributed to changes in gene copy number.

In all these reports, the integration of CGH and gene expression data has been achieved by analysing them separately and correlating results of individual analysis. However, it is likely that unified analysis strategies, akin to already mentioned statistical models for joint analysis of expression and regulatory sequence data, will prove beneficial in this context as well.

11.5.3 Modeling Genetic Networks

The ultimate goal of integrating different types of experimental data and current biological knowledge into a mathematical framework is the construction of genetic network models that will help us understand and predict the global dynamics of complex biological processes that define a living cell. The traditional molecular biology approach to characterizing roles of different cellular components has been to collect information on a single gene, a single protein, or a single interaction at a time. However, some characteristics of behavior of the complex network of biochemical interactions defining the living system are unlikely to be recovered by such local approaches [72,76]. For example, the functional role of a gene whose expression is regulated by several transcription factors cannot be fully understood without simultaneously monitoring for the presence and activation status of all of them. The ability of DNA microarrays to generate at the same time measurements on a large number of molecules participating in such a network allows for assessing interactions of a substantial portion of the global network. The complete strategy for such analysis consists of a mathematical model describing interactions of various components of the network, experimental approaches to perturb the network, biologic assays for quantitating the effects of such perturbation, and the inference procedures for estimating parameters of the assumed model [77].

Mathematical models describing dynamics of biochemical networks include the deterministic ordinary differential equation (ODE) based models of kinetics of coupled chemical reactions [78,79], stochastic generalization of such models following the Gillespie's algorithm [80] for simulation approach to the chemical master equations [81–84], Boolean network models which reduce the information about the abundance of various interacting molecules to a binary variable representing on/off (0/1, present/absent) states [85], and Bayesian networks [86] and probabilistic graphical models in general [87]. All of these mathematical models have certain advantages and disadvantages depending on the goal of the analysis, available data, and the knowledge about interactions of various molecules in the network. A thorough overview of these models in the context of genetic regulatory networks can be found elsewhere [88].

The specification of an ODE model requires detailed knowledge of the modeled interactions and is intended for examination of the overall dynamics of the system when individual relationships between components of the network are more or less known. In this context, the data is primarily used for checking predictions based on such models and not necessarily for reconstructing the networks themselves. Similar conclusions can be made about various stochastic approaches to simulating behavior of such networks. However, such stochastic generalizations are likely to offer a more realistic result in biochemical networks involving molecules of very low abundance, as is the case in gene expression regulation. In contrast to these quantitative models, the Boolean network model offers a relatively straightforward approach to reconstructing the topology of the network based on discretized data (e.g., the expression data for each gene at each experiment is reduced to two states expressed/not expressed). However, it has been argued that the binary 0/1 representation of network components is inadequate in many situations.

Probabilistic graphical models, Bayesian networks in particular, seem to be capable of capturing the rich topological structure, integrating components operating on various scales and representing stochasticity both underlying biologic processes and the noise inherent in the data. In this statistical approach, the behavior of the network is expressed as the joint probability distribution of measurements that can be made on elements of the network. The structure of the network is described in terms of the Directed Acyclic Graph (DAG) [89]. Nodes in the network correspond to the elements of the network and directed edges specify the dependences between the components. DAG specifies the dependence structure of the network through the Markov assumption that the node is statistically independent of its nondescendants given its parents. Well-established inferential procedures allow for the data-driven reconstruction of the network topology and specific interactions along with the corresponding measures of confidence in the estimated structure and model parameters [90].

The hypothetical Bayesian network in Figure 11.7(a) describes the interaction of four different genes. Assuming that G1, G2, G3, and G4 are variables describing the level of expression of these genes, one of the probabilistic statements encoded by the topology of this network is that the expression level of Gene4 is independent of expression levels of Gene1 and Gene2 given the expression level of Gene3. In other words, while the expression levels of Gene1 and Gene2 do affect the expression level of Gene4, they do it only through their effect on Gene3, which in turn affects the expression level of Gene4. By just looking only at the expression of these four genes across different experiments, they would all appear to be correlated to various degrees. The goal of the analysis could then be to infer the most likely network topology explaining these correlations. For example, the network in Figure 11.7(b) would induce a similar pattern of correlations between these four genes. However, effects of Gene1 and Gene2 on Gene4 is direct and not through their regulation of Gene3.

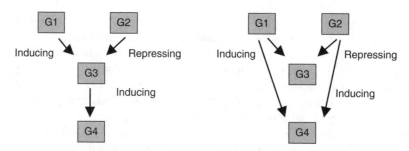

FIGURE 11.7   Two alternative Bayesian networks explaining the correlation structure between expression measurements of four hypothetical genes.

Given a sufficient amount of data, one can distinguish between these two structures and establish the most likely topology describing their interactions, which could then be tested experimentally.

The ability to incorporate various levels of prior knowledge through informative priors about the structure and the local probability distributions [91] allows Bayesian networks to potentially serve as the model of choice for encoding the current knowledge and the analysis of new data on the background of the current knowledge. Predictions about the future behavior of the network can take into account all sources of uncertainties — uncertainty about the estimated parameters of the networks, structure of the network, and the stochastic nature of the biologic system modeled by the network.

REFERENCES

1. Lockhart, D.J., Dong, H., Byrne, M.C., Follettie, M.T., Gallo, M.V., Chee, M.S., Mittmann, M., Wang, C., Kobayashi, M., Horton, H., and Brown, E.L. (1996) Expression monitoring by hybridization to high-density oligonucleotide arrays. *Nat. Biotechnol.*, 14, 1675–1680.
2. Hughes, T.R., Mao, M., Jones, A.R., Burchard, J., Marton, M.J., Shannon, K.W., Lefkowitz, S.M., Ziman, M., Schelter, J.M., Meyer, M.R., Kobayashi, S., Davis, C., Dai, H., He, Y.D., Stephaniants, S.B., Cavet, G., Walker, W.L., West, A., Coffey, E., Shoemaker, D.D., Stoughton, R., Blanchard, A.P., Friend, S.H., and Linsley, P.S. (2001) Expression profiling using microarrays fabricated by an ink-jet oligonucleotide synthesizer. *Nat. Biotechnol.*, 19, 342–347.
3. DeRisi, J.L., Iyer, V.R., and Brown, P.O. (1997) Exploring the metabolic and genetic control of gene expression on a genomic scale. *Science*, 278, 680–686.
4. Kerr, K.M. and Churchill, G.A. (2001) Experimental design for gene expression microarrays. *Biostatistics*, 2, 183–201.
5. Colantuoni, C., Henry, G., Zeger, S., and Pevsner, J. (2002) Local mean normalization of microarray element signal intensities across an array surface: Quality control and correction of spatially systematic artifacts. *Biotechniques*, 32, 1316–1320.
6. Hill, A.A., Brown, E.L., Whitley, M.Z., Tucker-Kellogg, G., Hunter, C.P., and Slonim, D.K. (2001) Evaluation of normalization procedures for oligonucleotide array data based on spiked cRNA controls. *Genome Biol.*, 2, RESEARCH0055.
7. Yang, Y., Dudoit, S., Luu, P., and Speed, T. (2000) Normalization for cDNA microarray data. SPIE BiOS 2001, San Jose, California.
8. Pinkel, D., Segraves, R., Sudar, D., Clark, S., Poole, I., Kowbel, D., Collins, C., Kuo, W.L., Chen, C., Zhai, Y., Dairkee, S.H., Ljung, B.M., Gray, J.W., and Albertson, D.G. (1998) High resolution analysis of DNA copy number variation using comparative genomic hybridization to microarrays. *Nat. Genet.*, 20, 207–211.
9. Pollack, J.R., Perou, C.M., Alizadeh, A.A., Eisen, M.B., Pergamenschikov, A., Williams, C.F., Jeffrey, S.S., Botstein, D., and Brown, P.O. (1999) Genome-wide analysis of DNA copy-number changes using cDNA microarrays. *Nat. Genet.*, 23, 41–46.
10. Forozan, F., Mahlamaki, E.H., Monni, O., Chen, Y., Veldman, R., Jiang, Y., Gooden, G.C., Ethier, S.P., Kallioniemi, A., and Kallioniemi, O.P. (2000) Comparative genomic hybridization analysis of 38

breast cancer cell lines: A basis for interpreting complementary DNA microarray data. *Cancer Res.*, 60, 4519–4525.

11. Pollack, J.R., Sorlie, T., Perou, C.M., Rees, C.A., Jeffrey, S.S., Lonning, P.E., Tibshirani, R., Botstein, D., Borresen-Dale, A.L., and Brown, P.O. (2002) Microarray analysis reveals a major direct role of DNA copy number alteration in the transcriptional program of human breast tumors. *Proc. Natl. Acad. Sci. USA*, 99, 12963–12968.

12. Fejzo, M.S., Godfrey, T., Chen, C., Waldman, F., and Gray, J.W. (1998) Molecular cytogenetic analysis of consistent abnormalities at 8q12–q22 in breast cancer. *Genes Chromosomes Cancer*, 22, 105–113.

13. Nadon, R. and Shoemaker, J. (2002) Statistical issues with microarrays: Processing and analysis. *Trends Genet.*, 18, 265–271.

14. Cleveland, W.S. and Devlin, S.J. (1988) Locally-weighted Regression: An approach to regression analysis by local fitting. *J. Am. Statist. Assoc.*, 83, 596–610.

15. Dudoit, S., Yang, Y., Callow, M.J., and Speed, T.P. (2002) Statistical methods for identifying differentially expressed genes in replicated cDNA microarray experiments. *Statist. Sinica*, 12, 111–139.

16. Yang, Y.H., Dudoit, S., Luu, P., Lin, D.M., Peng, V., Ngai, J., and Speed, T.P. (2002) Normalization for cDNA microarray data: A robust composite method addressing single and multiple slide systematic variation. *Nucleic Acids Res.*, 30, e15.

17. Bolstad, B.M., Irizarry, R.A., Astrand, M., and Speed, T.P. (2003) A comparison of normalization methods for high density oligonucleotide array data based on variance and bias. *Bioinformatics*, 19, 185–193.

18. Fielden, M.R., Halgren, R.G., Fong, C.J., Staub, C., Johnson, L., Chou, K., and Zacharewski, T.R. (2002) Gestational and lactational exposure of male mice to diethylstilbestrol causes long-term effects on the testis, sperm fertilizing ability in vitro, and testicular gene expression. *Endocrinology*, 143, 3044–3059.

19. Ma, X.J., Salunga, R., Tuggle, J.T., Gaudet, J., Enright, E., McQuary, P., Payette, T., Pistone, M., Stecker, K., Zhang, B.M., Zhou, Y.X., Varnholt, H., Smith, B., Gadd, M., Chatfield, E., Kessler, J., Baer, T.M., Erlander, M.G., and Sgroi, D.C. (2003) Gene expression profiles of human breast cancer progression. *Proc. Natl. Acad. Sci. USA*, 100, 5974–5979.

20. Sorlie, T., Perou, C.M., Tibshirani, R., Aas, T., Geisler, S., Johnsen, H., Hastie, T., Eisen, M.B., van de, R.M., Jeffrey, S.S., Thorsen, T., Quist, H., Matese, J.C., Brown, P.O., Botstein, D., Eystein, L.P., and Borresen-Dale, A.L. (2001) Gene expression patterns of breast carcinomas distinguish tumor subclasses with clinical implications. *Proc. Natl. Acad. Sci. USA*, 98, 10869–10874.

21. Churchill, G.A. (2002) Fundamentals of experimental design for cDNA microarrays. *Nat. Genet.*, 32 (Suppl.), 490–495.

22. Kerr, K.M., Martin, M., and Churchill, G.A. (2000) Analysis of variance for gene expression microarray data. *J. Comput. Biol.*, 7, 819–837.

23. Wolfinger, R.D., Gibson, G., Wolfinger, E.D., Bennett, L., Hamadeh, H., Bushel, P., Afshari, C., and Paules, R.S. (2001) Assessing gene significance from cDNA microarray expression data via mixed models. *J. Comput. Biol.*, 8, 625–637.

24. Schena, M., Shalon, D., Heller, R., Chai, A., Brown, P.O., and Davis, R.W. (1996) Parallel human genome analysis: Microarray-based expression monitoring of 1000 genes. *Proc. Natl. Acad. Sci. USA*, 93, 10614–10619.

25. Claverie, J.M. (1999) Computational methods for the identification of differential and coordinated gene expression. *Hum. Mol. Genet.*, 8, 1821–1832.

26. Cui, X. and Churchill, G.A. (2003) Statistical tests for differential expression in cDNA microarray experiments. *Genome Biol.*, 4, 210.

27. Ideker, T., Thorsson, V., Siegel, A.F., and Hood, L.E. (2000) Testing for differentially-expressed genes by maximum-likelihood analysis of microarray data. *J. Comput. Biol.*, 7, 805–817.

28. Efron, B., Tibshirani, R., Storey, J.D., and Tusher, V. (2001) Empirical Bayes analysis of a microarray experiment. *J. Am. Stat. Assoc.*, 96, 1151–1160.

29. Baldi, P. and Long, A.D. (2001) A Bayesian framework for the analysis of microarray expression data: Regularized t-test and statistical inferences of gene changes. *Bioinformatics*, 17, 509–519.

30. Tusher, V.G., Tibshirani, R., and Chu, G. (2001) Significance analysis of microarrays applied to the ionizing radiation response. *Proc. Natl. Acad. Sci. USA*, 98, 5116–5121.

31. Benjamini, Y. and Hochberg, Y. (1995) Controlling the false discovery rate: A practical and powerful approach to multiple testing. *J. R. Stat. Soc. B*, 57, 289–300.

32. Gasch, A.P., Spellman, P.T., Kao, C.M., Carmel-Harel, O., Eisen, M.B., Storz, G., Botstein, D., and Brown, P.O. (2000) Genomic expression programs in the response of yeast cells to environmental changes. *Mol. Biol. Cell*, 11, 4241–4257.

33. Zhao, R., Gish, K., Murphy, M., Yin, Y., Notterman, D., Hoffman, W.H., Tom, E., Mack, D.H., and Levine, A.J. (2000) Analysis of p53-regulated gene expression patterns using oligonucleotide arrays. *Genes Dev.*, 14, 981–993.

34. Perou, C.M., Jeffrey, S.S., van de, R.M., Rees, C.A., Eisen, M.B., Ross, D.T., Pergamenschikov, A., Williams, C.F., Zhu, S.X., Lee, J.C., Lashkari, D., Shalon, D., Brown, P.O., and Botstein, D. (1999) Distinctive gene expression patterns in human mammary epithelial cells and breast cancers. *Proc. Natl. Acad. Sci. USA*, 96, 9212–9217.

35. Alizadeh, A.A., Eisen, M.B., Davis, R.E., Ma, C., Lossos, I.S., Rosenwald, A., Boldrick, J.C., Sabet, H., Tran, T., Yu, X., Powell, J.I., Yang, L., Marti, G.E., Moore, T., Hudson, J., Jr., Lu, L., Lewis, D.B., Tibshirani, R., Sherlock, G., Chan, W.C., Greiner, T.C., Weisenburger, D.D., Armitage, J.O., Warnke, R., and Staudt, L.M. (2000) Distinct types of diffuse large B-cell lymphoma identified by gene expression profiling. *Nature*, 403, 503–511.

36. Alon, U., Barkai, N., Notterman, D.A., Gish, K., Ybarra, S., Mack, D., and Levine, A.J. (1999) Broad patterns of gene expression revealed by clustering analysis of tumor and normal colon tissues probed by oligonucleotide arrays. *Proc. Natl. Acad. Sci. USA*, 96, 6745–6750.

37. Golub, T.R., Slonim, D.K., Tamayo, P., Huard, C., Gaasenbeek, M., Mesirov, J.P., Coller, H., Loh, M.L., Downing, J.R., Caligiuri, M.A., Bloomfield, C.D., and Lander, E.S. (1999) Molecular classification of cancer: Class discovery and class prediction by gene expression monitoring. *Science*, 286, 531–537.

38. Cho, R.J., Campbell, M.J., Winzeler, E.A., Steinmetz, L., Conway, A., Wodicka, L., Wolfsberg, T.G., Gabrielian, A.E., Landsman, D., Lockhart, D.J., and Davis, R.W. (1998) A genome-wide transcriptional analysis of the mitotic cell cycle. *Mol. Cell*, 2, 65–73.

39. Eisen, M.B., Spellman, P.T., Brown, P.O., and Botstein, D. (1998) Cluster analysis and display of genome-wide expression patterns. *Proc. Natl. Acad. Sci. USA*, 95, 14863–14868.

40. Everitt, B.S. (1993) *Cluster Analysis*. Edward Arnold, London.

41. Tavazoie, S., Hughes, J.D., Campbell, M.J., Cho, R.J., and Church, G.M. (1999) Systematic determination of genetic network architecture. *Nat. Genet.*, 22, 281–285.

42. Tamayo, P., Slonim, D., Mesirov, J., Zhu, Q., Kitareewan, S., Dmitrovsky, E., Lander, E.S., and Golub, T.R. (1999) Interpreting patterns of gene expression with self-organizing maps: Methods and application to hematopoietic differentiation. *Proc. Natl. Acad. Sci. USA*, 96, 2907–2912.

43. McLachlan, J.G. and Basford, E.K. (1987) *Mixture Models: Inference and Applications to Clustering*. Marcel Dekker, New York.

44. Yeung, K.Y., Fraley, C., Murua, A., Raftery, A.E., and Ruzzo, W.L. (2001) Model-based clustering and data transformations for gene expression data. *Bioinformatics*, 17, 977–987.

45. McLachlan, G.J., Bean, R.W., and Peel, D. (2002) A mixture model-based approach to the clustering of microarray expression data. *Bioinformatics*, 18, 413–422.

46. Medvedovic, M. and Sivaganesan, S. (2002) Bayesian infinite mixture model based clustering of gene expression profiles. *Bioinformatics*, 18, 1194–1206.

47. Yeung, K.Y., Medvedovic, M., and Bumgarner, R.E. (2003) Clustering gene expression data with repeated measurements. *Genome Biol.*, 4, R34.

48. Cheeseman, P. and Stutz, J. (1996) Bayesian classification (AutoClass): Theory and results. In *Advances in Knowledge Discovery and Data Mining*, pp. 153–180.

49. Bhattacharjee, A., Richards, W.G., Staunton, J., Li, C., Monti, S., Vasa, P., Ladd, C., Beheshti, J., Bueno, R., Gillette, M., Loda, M., Weber, G., Mark, E.J., Lander, E.S., Wong, W., Johnson, B.E., Golub, T.R., Sugarbaker, D.J., and Meyerson, M. (2001) Classification of human lung carcinomas by mRNA expression profiling reveals distinct adenocarcinoma subclasses. *Proc. Natl. Acad. Sci. USA*, 98, 13790–13795.

50. Muro, S., Takemasa, I., Oba, S., Matoba, R., Ueno, N., Maruyama, C., Yamashita, R., Sekimoto, M., Yamamoto, H., Nakamori, S., Monden, M., Ishii, S., and Kato, K. (2003) Identification of expressed genes linked to malignancy of human colorectal carcinoma by parametric clustering of quantitative expression data. *Genome Biol.*, 4, R21.

51. Thomas, R.S., Rank, D.R., Penn, S.G., Zastrow, G.M., Hayes, K.R., Pande, K., Glover, E., Silander, T., Craven, M.W., Reddy, J.K., Jovanovich, S.B., and Bradfield, C.A. (2001) Identification of toxicologically predictive gene sets using cDNA microarrays. *Mol. Pharmacol.*, 60, 1189–1194.

52. Waring, J.F., Jolly, R.A., Ciurlionis, R., Lum, P.Y., Praestgaard, J.T., Morfitt, D.C., Buratto, B., Roberts, C., Schadt, E., and Ulrich, R.G. (2001) Clustering of hepatotoxins based on mechanism of toxicity using gene expression profiles. *Toxicol. Appl. Pharmacol.*, 175, 28–42.

53. Waring, J.F., Ciurlionis, R., Jolly, R.A., Heindel, M., and Ulrich, R.G. (2001) Microarray analysis of hepatotoxins in vitro reveals a correlation between gene expression profiles and mechanisms of toxicity. *Toxicol. Lett.*, 120, 359–368.

54. Li, L., Weinberg, C.R., Darden, T.A., and Pedersen, L.G. (2001) Gene selection for sample classification based on gene expression data: Study of sensitivity to choice of parameters of the GA/KNN method. *Bioinformatics*, 17, 1131–1142.

55. Cho, J.H., Lee, D., Park, J.H., Kim, K., and Lee, I.B. (2002) Optimal approach for classification of acute leukemia subtypes based on gene expression data. *Biotechnol. Prog.*, 18, 847–854.

56. Alizadeh, A.A., Ross, D.T., Perou, C.M., and van de, R.M. (2001) Towards a novel classification of human malignancies based on gene expression patterns. *J. Pathol.*, 195, 41–52.

57. West, M., Blanchette, C., Dressman, H., Huang, E., Ishida, S., Spang, R., Zuzan, H., Olson, J.A., Jr., Marks, J.R., and Nevins, J.R. (2001) Predicting the clinical status of human breast cancer by using gene expression profiles. *Proc. Natl. Acad. Sci. USA*, 98, 11462–11467.

58. Khan, J., Wei, J.S., Ringner, M., Saal, L.H., Ladanyi, M., Westermann, F., Berthold, F., Schwab, M., Antonescu, C.R., Peterson, C., and Meltzer, P.S. (2001) Classification and diagnostic prediction of cancers using gene expression profiling and artificial neural networks. *Nat. Med.*, 7, 673–679.

59. Hastie, T., Tibshirani, R. and Friedman, J. (2001) *The Elements of Statistical Learning: Data Mining, Inference, and Prediction*. Springer-Verlag, New York.

60. Webb, A. (1999) *Statistical Pattern Recognition*. Oxford University Press Inc., New York.

61. Dudoit, S., Fridlyand, J., and Speed, T.P. (2002) Comparison of Discrimination Methods for the Classification of Tumors Using Gene Expression Data. *J. Am. Stat. Assoc.*, 97, 77.

62. Nguyen, D.V. and Rocke, D.M. (2002) Tumor classification by partial least squares using microarray gene expression data. *Bioinformatics*, 18, 39–50.

63. Zien, A., Kuffner, R., Zimmer, R., and Lengauer, T. (2000) Analysis of gene expression data with pathway scores. *Proc. Int. Conf. Intell. Syst. Mol. Biol.*, 8, 407–417.

64. Boulton, S.J., Gartner, A., Reboul, J., Vaglio, P., Dyson, N., Hill, D.E., and Vidal, M. (2002) Combined functional genomic maps of the *C. elegans* DNA damage response. *Science*, 295, 127–131.

65. Ge, H., Liu, Z., Church, G.M., and Vidal, M. (2001) Correlation between transcriptome and interactome mapping data from *Saccharomyces cerevisiae. Nat. Genet.*, 29, 482–486.

66. Matthews, L.R., Vaglio, P., Reboul, J., Ge, H., Davis, B.P., Garrels, J., Vincent, S., and Vidal, M. (2001) Identification of potential interaction networks using sequence-based searches for conserved protein–protein interactions or "interologs." *Genome Res.*, 11, 2120–2126.

67. Vidal, M. (2001) A biological atlas of functional maps. *Cell*, 104, 333–339.

68. Holmes, I. and Bruno, W.J. (2000) Finding regulatory elements using joint likelihoods for sequence and expression profile data. *Proc. Int. Conf. Intell. Syst. Mol. Biol.*, 8, 202–210.

69. Barash, Y. and Friedman, N. (2002) Context-specific bayesian clustering for gene expression data. *J. Comput. Biol.*, 9, 169–191.

70. Segal, E., Yelensky, R., and Koller, D. (2003) Genome-wide discovery of transcriptional modules from DNA sequence and gene expression. *Bioinformatics*, 19 (Suppl. 1), I273–I282.

71. Segal, E., Wang, H., and Koller, D. (2003) Discovering molecular pathways from protein interaction and gene expression data. *Bioinformatics*, 19 (Suppl. 1), I264–I272.

72. Ideker, T., Galitski, T., and Hood, L. (2001) A new approach to decoding life: Systems biology. *Annu. Rev. Genomics Hum. Genet.*, 2, 343–372.

73. Ideker, T., Thorsson, V., Ranish, J.A., Christmas, R., Buhler, J., Eng, J.K., Bumgarner, R., Goodlett, D.R., Aebersold, R., and Hood, L. (2001) Integrated genomic and proteomic analyses of a systematically perturbed metabolic network. *Science*, 292, 929–934.

74. Fritz, B., Schubert, F., Wrobel, G., Schwaenen, C., Wessendorf, S., Nessling, M., Korz, C., Rieker, R.J., Montgomery, K., Kucherlapati, R., Mechtersheimer, G., Eils, R., Joos, S., and Lichter, P.

(2002) Microarray-based copy number and expression profiling in dedifferentiated and pleomorphic liposarcoma. *Cancer Res.*, 62, 2993–2998.

75. Weiss, M.M., Kuipers, E.J., Postma, C., Snijders, A.M., Siccama, I., Pinkel, D., Westerga, J., Meuwissen, S.G., Albertson, D.G., and Meijer, G.A. (2003) Genomic profiling of gastric cancer predicts lymph node status and survival. *Oncogene*, 22, 1872–1879.

76. Niehrs, C. and Meinhardt, H. (2002) Modular feedback. *Nature*, 417, 35–36.

77. Ideker, T.E., Thorsson, V., and Karp, R.M. (2000) Discovery of regulatory interactions through perturbation: Inference and experimental design. *Pac. Symp. Biocomput.*, 305–316.

78. Voit, E.O. (2002) Metabolic modeling: A tool of drug discovery in the post-genomic era. *Drug Discov. Today*, 7, 621–628.

79. Voit, E.O. and Radivoyevitch, T. (2000) Biochemical systems analysis of genome-wide expression data. *Bioinformatics*, 16, 1023–1037.

80. Gillespie, D.T. (1977) Exact stochastic simulation of cupled chemical reactions. *J. Phys. Chem.*, 81, 2340–2361.

81. Kierzek, A.M. (2002) STOCKS: STOChastic Kinetic Simulations of biochemical systems with Gillespie algorithm. *Bioinformatics*, 18, 470–481.

82. McAdams, H.H. and Arkin, A. (1999) It's a noisy business! Genetic regulation at the nanomolar scale. *Trends Genet.*, 15, 65–69.

83. McAdams, H.H. and Arkin, A. (1998) Simulation of prokaryotic genetic circuits. *Annu. Rev. Biophys. Biomol. Struct.*, 27, 199–224.

84. McAdams, H.H. and Arkin, A. (1997) Stochastic mechanisms in gene expression. *Proc. Natl. Acad. Sci. USA*, 94, 814–819.

85. D'haeseleer, P., Liang, S., and Somogyi, R. (2000) Genetic network inference: From co-expression clustering to reverse engineering. *Bioinformatics*, 16, 707–726.

86. Friedman, N., Linial, M., Nachman, I., and Pe'er, D. (2000) Using Bayesian networks to analyze expression data. *J. Comput. Biol.*, 7, 601–620.

87. Friedman, N. (2004) Inferring cellular networks using probabilistic graphical models. *Science*, 303, 799–805.

88. de Jong, H. (2002) Modeling and simulation of genetic regulatory systems: A literature review. *J. Comput. Biol.*, 9, 67–103.

89. Cowell, R.G., Dawid, P.A., Lauritzen, S.L., and Spiegelhalter, D.J. (1999) *Probabilistic Networks and Expert Systems*. Springer, New York.

90. Heckerman, D. (1998) A tutorial on learning with Bayesian networks. In *Learning in Graphical Models*, Jordan, M.I. (ed.), Kluwer Academic Publishers, Dordrecht, The Netherlands, pp. 301–354.

91. Heckerman, D., Geiger, D., and Chickering, D. (1995) Learning Bayesian networks: The combination of knowledge and statistical data. *Machine Learn.*, 20, 197–243.

12 Application of Proteomics in Basic Biological Sciences and Cancer

Christoph Borchers, Ting Chen, and Nouri Neamati

CONTENTS

12.1 INTRODUCTION

Proteomics (protein + omics) simply means many or groups of proteins. However, it refers to identification and quantification of complete sets of proteins expressed by the genome at any given time in a cell. The field has been divided into at least two disciplines: functional proteomics, which refers to the determination of the function of all the proteins encoded in a genome; and structural proteomics, which hopes to determine the structures of all these proteins [1,2]. Proteomics is therefore about understanding the structure and function of all proteins [3]. As a result, its application in biological sciences, especially cancer research and therapeutics, has been unprecedented [4–9]. The key reasons behind the progress and interest in proteomics are the facts that (1) protein expression levels are not always predictable from mRNA expression levels, (2) proteins undergo a variety of posttranslational modifications, and (3) proteins are dynamic and carry out a majority of the cellular functions [10]. Many cellular and environmental factors may alter gene and protein expression, and none of these factors can be identified solely from examining gene expression. On the other hand, proteins respond to altered conditions by changing their location within the cell, undergoing proteolysis, and adjusting their stability as well as changing their physical interactions with other macromolecules. The current cDNA microarray technology does not provide information regarding posttranscriptional control of gene expressions, changes in protein expression levels, or changes in protein synthesis and degradation rates. Therefore, a detailed understanding of the control of gene requires information about mRNA and protein expression levels.

Conventional approaches to protein studies relied on protein expression, protein chemistry, structure, and protein function. Such information generated from biochemical studies with protein or genetic studies were instrumental in elucidating biochemical pathways. However, at present, protein profiling in a genome necessitates high throughput techniques to systematically analyze and profile proteins in biological samples. The genome and the proteome of an organism can be influenced by cellular and environmental factors, such as stress, aging, disease, carcinogens, and drugs. Nowadays, alterations in gene expression caused by these factors are routinely assessed by high throughput microarray analysis. It is now well established that DNA microarray technology allows the simultaneous quantitation of the expression of thousands of genes. This methodology is now robust, reproducible, highly efficient, can evaluate defined cellular pathways, discover novel genes, and validate identified targets. This technology has been successfully used to identify gene expression changes under various experimental settings. The promise of proteomics is to perform just as microarrays and to identify all the proteins in human genome. Because of the inherent differences between protein and DNA, spotting thousands of proteins on glass slides similar to microarray technology has not been practical. Numerous technologies are under considerations to generate so-called protein chips using some of the technologies applied to DNA microarray methodologies [11–16]. Protein biochip technologies will assuredly use different ways to solve the inherent difficulties with proteins on solid support. However, current proteomic technologies rely on protein separation, identification, and characterization [17]. Because of alternative splicing of mRNA and posttranslational modification, one gene can yield multiple proteins. In bacteria, for example, a gene can produce one to two different proteins, in yeast up to three, and in humans up to six [18]. A gene on the other hand may be switched on, but its messenger RNA (mRNA) is not necessarily a translated protein. It has been estimated that more than 200 different types of modifications exist [19]. With all these possibilities, it has been estimated that the human proteome, at least, is in an order of magnitude more complex than the human genome. This multiplicity of proteins encoded by a single gene plus the varied and fragile nature of proteins, as well as the quantitative and qualitative changes of the proteome present formidable challenges in proteomics. Idealized proteomic technology will have to be high throughput with a high degree of sensitivity, ability to distinguish differentially modified proteins, and ability to quantitatively display and analyze all the proteins present in a sample. Knowing what proteins are actually present in a cell might not necessarily be the whole story — it is also useful to know which proteins interact with one another and how they do so. Such information is of paramount importance in identifying components of a particular enzymatic pathway and is also very useful in identifying new therapeutic targets.

12.1.1 Relevance of Proteomics to Cancer Research

Proteomic based approaches are continuing to play a major role in studying the natural history and treatment of cancer. For example, proteomics can facilitate discovery of new molecular targets for therapy, biomarkers for early detection, and new endpoints for therapeutic efficacy and toxicity [20].

Moreover, real-time information about the states of intracellular signaling pathways in the normal and in the diseased state before, during, and after therapy can be obtained to potentially guide individualized treatment. One such application that will be quite useful in future is in the area of clinical proteomics, which involves the application of proteomic technologies at the bedside; and cancer, in particular, is a model disease for studying such applications. The goals will be to detect cancer early by imaging biomarkers, to discover the next generation of targets and finally to tailor the therapy to the patient [21].

Proteomic technologies have been used for the study of cancer of various organs including lung, liver, prostate, breast, ovarian, bladder, esophagus, and stomach [22–25]. Various databases have been developed to facilitate classification schemes for cancer and the identification of novel markers for early diagnosis [25–28].

Identification of specific and sensitive markers of cancer is an important public health concern. Currently, there are no effective screening options available for many cancer patients [29]. The most attractive screening choices are the noninvasive means of proteomic analysis of bodily fluids. For example, proteomic analyses of nipple aspirates are being employed to identify differentially expressed proteins that could potentially predict the presence of breast cancer [30]. Breast cancer proteomics is breaking ground in identifying markers of potential clinical interest [31,32]. An understanding of the transition from normal epithelium to the first definable stage of cancer at the functional level of protein expression is hypothesized to contribute to improved detection, prevention, and treatment [33] (for a review see Reference 34). The 2-dimensional gel electrophoresis (2-DE) analysis of microdissected ovarian tumors has also revealed that certain proteins are uniquely over expressed in invasive human ovarian cancer [6,35]. Recently, it was shown by comparing protein maps of normal and malignant prostate that various proteins were identified in normal prostate that were lost in malignant transformation [5]. An area of intense study is the understanding of serum proteomic patterns for detection of prostate cancer [36] that could provide potential diagnostic/prognostic biomarkers for prostate and bladder cancers [4,37–39].

From these studies it is clear that cancer cells share discrete proteomic signatures that are more reflective of their biological phenotype than cellular heritage, highlighting that a common set of enzymes may support the progression of tumors from a variety of origins and thus represent attractive targets for the diagnosis and treatment of cancer [40].

12.2 PROTEOMICS TECHNOLOGIES

At present there are excellent reviews on this topic, and new journals and books are being published that are dedicated entirely to proteomics (see e.g., Reference 41). Therefore, no further attempts are made to discuss the historic progress and perspectives in this field. Since the technologies behind proteomics can be easily adapted to practically any discipline, we only describe what it takes to perform such experiments. Herein, we discuss three major technologies presently available in proteome analysis. First the *separation* of protein mixtures by 2-DE, which is the only technique currently available that can reveal hundreds of proteins at a time. Second, *characterization* of every single protein for which mass spectrometric (MS) technique is the methodology of choice in this regard. Third, *identification* of the proteins or polypeptides by MS followed by database searching using numerous computational algorithms is nowadays becoming routinely accessible.

12.2.1 PROTEIN SEPARATION

The most widely used technique for protein separation is 2-DE [42–44]. The first dimension separates proteins according to charge and the second according to mass (Figure 12.1). Traditional protein identification methods included immunoblotting, chemical sequencing of peptides, co-migration analysis of proteins, and over expression of genes of interest in some organisms. These methods are

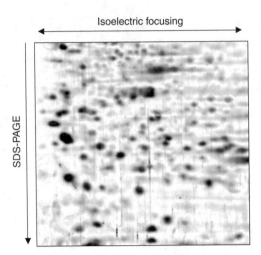

FIGURE 12.1 Example of a 2-DE. Protein fraction was extracted using the ReadyPrep Sequential extraction kit (Bio.Rad) and ran on a first dimension column containing a mixture of BioLyte ampholytes 3/10 and 5/7 (Pluymers and Neamati).

generally slow. In the last two decades, 2-DE has advanced to become a high-throughput method for the separation and purification of complex protein mixtures and the quantitation of proteins. However, its resolution can be lower than what is required for effective analysis, the reproducibility is sometimes questionable, and the analysis of 2D gel images is difficult. In addition, reliably measuring the quantitative changes in expression levels of low abundance proteins or hydrophobic membrane proteins is particularly difficult. This is why the integration of 2-DE and liquid chromatography combined with MS-based approaches is needed for proteomics to achieve its greatest power.

An alternative procedure for resolving complex mixtures of proteins takes advantage of the combination of free flow electrophoresis (FFE), a liquid-based isoelectric focusing (IEF) method, and sodium dodecyl sulphate-polyacrylamide gel electrophoresis (SDS-PAGE). It was recently shown that FFE is a powerful liquid-based IEF method for resolving proteins, because it is not limited by the amount of sample and is capable of fractionating intact protein complexes [45].

Since the dynamic range of protein abundance in a cell can span at least six orders of magnitude and most of the protein mass in a given cell comes from a small number of proteins it is necessary to fractionate the samples before running a 2-DE. A major limitation of high-throughput proteomics is to separate and identify all the low abundance proteins in a given mixture of sample. Another important aspect of proteomics is to elucidate protein–protein interaction. Because both dimensions in 2-DE are under denaturing conditions, all protein–protein interaction information is lost. It is important to bear in mind that supersensitive technologies are required to detect low abundant proteins, which in many cases have very important biological activities.

12.2.2 SAMPLE PREPARATION

The protocol for preparing sufficient and good quality sample has not changed much over the last 10 years ago. In brief, samples (cells or tissues) are disrupted by a variety of techniques such as sonication, homogenization, or high pressure and the proteins are solubilized with an appropriate lysis buffer containing detergent. Cell fractionation followed by centrifugation can be used to separate various proteins prior to gel electrophoresis. However, there is an increasing need to be able to extract all protein components, including hydrophobic membrane proteins, proteins with extreme isoelecric point pI (below pH 3 and above pH 10), and low copy number proteins in the presence of most abundant proteins [46–48]. Several groups are using optimized gel strips to separate such proteins (see e.g., References 49 to 53).

12.2.3 FIRST DIMENSION

Isoelectric focusing is a gentle, high-resolution technique capable of resolving proteins that differ in pI by fractions of a pH unit. Separation of proteins by IEF is based on the fact that all proteins have a pH-dependent net charge. The net charge is determined by the amino acid sequence of the protein and the pH of the environment. When a protein is electrophoresed through an established pH gradient, it will migrate until it reaches the pH where the net charge on the protein is zero; at that point it will stop migrating and is said to be focused at its pI. Ampholytes, which are low molecular weight, charged amphoteric molecules, are used to establish the pH gradients increasing in pH from anode to cathode. A mixture of ampholytes, each having a different pI is used. When voltage is applied to a system of ampholytes and proteins, all the components migrate to their respective pIs. Ampholytes rapidly establish the pH gradient and maintain it for long periods allowing the slower moving proteins to focus. Most common commercially available ampholytes are polycarboxylic acid polyamines ranging in molecular weight from 300 to 500. Their major disadvantage is the fact that they can bind tightly to many proteins. The classical 2-DE using tube gels with a pH gradient generated by carrier ampholytes were limited in their resolution and reproducibility. The gels are cast and run in glass tubes with an internal diameter of 1.5 mm to match the thickness of the second dimension gel. After the IEF gel has run, it is extruded from the tube and laid across the top of the second dimension gel.

The IEF gels can also be prepared as slabs, which allow an increased sample throughput. Slab gels can be cut into strips for loading onto second dimension gels. These slabs are also known as immobilized pH gradient (IPG) gel strips are also commercially available. The IPG strips are phasing out the tube gel technology because they are relatively simple to use, reproducible, and yield high-resolution gels. For high-throughput setting IEF is routinely performed in commercially available 3 mm wide and 13 to 20 cm long IPG gel strips and after equilibration with SDS buffer in the presence of urea, glycerol, and iodoacetamid, they are applied to a horizontal or vertical SDS-PAGE gel. For a complete protocol on preparing laboratory made IPG strips and running conditions see Reference 54.

12.2.4 SECOND DIMENSION

Electrophoretic separation of proteins based on their mass is routinely carried out using SDS-PAGE. The mobility of most polypeptide chains under these conditions is linearly proportional to the logarithm of their mass. Smaller proteins move rapidly through the gel, whereas large ones migrate slower. SDS-PAGE is rapid, sensitive, and capable of high degree of resolution. Masses determined by this technique are usually 90 to 95% accurate, the error range being within 5 to 10%.

The IEF strip containing proteins already separated on the basis of surface charge is placed adjacent to SDS-PAGE and proteins from the strip are electroeluted into the PAGE, in one continuous "well," and allowed to be resolved on the basis of differences in molecular weight. The 2D gel thus separates proteins, first on the basis of charge and then on the basis of molecular weight. The assumption, of course, is that most proteins that have a similar net charge will not have the same molecular weight and vice versa.

12.2.5 PROTEIN VISUALIZATION

To visualize the 2D gel, it needs to be properly stained. Staining is routinely performed using chemicals (Coomassie blue or fluorescent stains), metal-deposition stains (silver stain), or radiochemicals. When the bromophenol blue front completely migrates out of the SDS gel, the gel is normally fixed in ethanol/acetic acid solution and then stained. The Coomassie Blue G-250 can be used to detect proteins in 200 ng range within 1 to 2 h, whereas, Coommassie Blue R-250, which can take up to 16 h can detect 50 ng of protein. Silver staining, on the other hand is fast and more sensitive, and 1 ng or more protein can be visualized easily [55]. To remove most of the background staining, the gels are normally destained overnight.

Proteins can also be easily radiolabelled with ^{35}S methionine and visualized by phosphor imaging. After SDS-PAGE, the 2D gels are usually directly dried onto a sheet of filter paper under vacuum. The dried gels have to be exposed to phosphor imaging cassettes overnight. Nowadays with rapid progress in fluorescent dye and detection technology, more people are moving toward fluorescent-based techniques [56,57]. The three most popular fluorescent dyes are the SyproRuby™, Cy3, and Cy5. Fluorescent detection technologies have several advantages over the techniques mentioned earlier, such as sensitivity and dynamic range of fluorescent probes, possibility of multiple labeling and multicolor imaging (e.g., label one cell with Cy3 and other with Cy5, superimpose the gels and quantitate), stability of the fluorescently labeled molecules, low hazard, lower cost as compared with radioactivity, and commercial availability of a wide range of probes. The recent progress with the application of these dyes is the technique called differential gel electrophoresis (DIGE) [58], which is now commercially available from Amersham Biosciences. DIGE technology combined with DIGE data analysis circumvents the need to compare multiple gels. This technology was recently applied to esophageal cancer [59] and breast cancer [60] for the identification of protein biomarkers.

12.2.6 IMAGE ANALYSIS

Various commercially available software can be used to display images, adjust the contrast and brightness, annotate, and above all, quantitate all the spots. These image analysis tools allow numerous possibilities to compare each protein from different gels and are very useful for pattern analysis, fragment sizing, and generating analysis reports numerically and graphically. Some software packages also provide access to various databases or libraries of proteins or polypeptides for sample identification and query. If multiple fluorescent probes are used, image analysis becomes a very powerful tool to overlay images in multiple colors and to quantitate subtle differences. Image analysis software can also greatly facilitate protein identification by generating pick list files of the X-Y coordinates for all the spots of interest. All the annotated spots can be fed to a gel cutter and each spot is excised and transferred by a robot to microtiter plates for further in-gel digestion.

12.2.7 IN-GEL DIGESTION

The high demand for high-throughput capabilities and recent advances in robotics are the two main reasons for optimizing conditions for in-gel digestions. There are now automated systems to perform trypsin digests in microtiter plates and prepare the samples for MS analysis [61,62].

12.3 MASS SPECTROMETRY

12.3.1 INTRODUCTION

Less than 20 years ago, MS applications in biomedical research were limited by the lack of techniques for the transfer of higher molecular weight (MW) biomolecules into the gas phase without decomposition. The generation of gas-phase ions is a necessity for MS experiments, and had been the fundamental challenge. Since the development of two "soft ionization" techniques, matrix-assisted laser desorption/ionization (MALDI) [63–66] and electrospray ionization (ESI) [67,68] in the late 1980s, MS has become an important tool for the study of biomolecules on the molecular level, including RNA, DNA, and, especially, proteins and peptides.

Ongoing improvements in instrumentation over the last several years have led to a continuing enhancement of sensitivity and mass accuracy. Mass spectrometry now has the capability of determining the MW's of proteins and peptides with an accuracy of a few parts per million (ppm). In addition, novel MS methods and approaches have been developed. The most important of these are the techniques for MS sequencing of peptides. Based on these improvements and developments, together with the advances in computing power and knowledge of the genome, it is now possible to identify proteins at trace levels (down to the attomol range). Moreover, MS approaches combined with protein chemistry are very suitable for providing protein structural information. Mass spectrometry, therefore, has become an indispensable analytical technique for the study of proteins and peptides in living systems [69].

A mass spectrometer consists of three basic parts: the ion source, the mass analyzer, and the detector. In the ion source, the analyte is ionized and desorbed into the gas phase. The ions are then separated in the mass analyzer and recorded by the detector. The fundamentals of modern MS for the study of proteins and peptides, including an explanation of the ionization methods and mass analyzers, are described in Section 12.3.2.

12.3.2 BASIC MASS SPECTROMETRY

12.3.2.1 Ionization Techniques

12.3.2.1.1 Electrospray Ionization

Electrospray, a spray of small highly charged droplets created by means of a high voltage applied to a fine tip, is a well-known phenomenon discovered more than 100 years ago, and is widely used in the automobile industry for painting cars. In electrospray ionization mass spectrometry (ESI-MS), the analytes (e.g., peptides and proteins) are dissolved in an aqueous liquid that usually contains an organic solvent such as acetonitrile or methanol, and a low percentage of a volatile organic acid such as formic acid, acetic acid, or trifluoroacetic acid. The acids enhance the protonation of the analyte, which improves the sensitivity of MS analysis in the positive mode — that is, the detection of positive ions — which is the most common mode used in MS of proteins and peptides. The organic solvents assist in the formation of the electrospray.

In ESI-MS, the solvents flow through a conductive needle/capillary with a very fine tip to which high voltage (~900 to 2500 V) is applied. This results in the electrospray phenomenon, namely highly charged droplets with typical diameter of 10 μm (Figure 12.2[a]). These droplets are generated at atmospheric pressure and are introduced through a small hole (a few micrometer to a few millimeter in diameter) into the mass spectrometer, where increasing vacuum is applied. This results in desolvation of the droplets, whose volume decreases. As the solvent evaporates, the surface area of the droplet decreases to a point where there is an explosion, resulting in "naked," unsolvated charged ions.

Peptides are typically doubly or triply charged, making them easily distinguishable from signals resulting from chemical noise, which is commonly singly charged (Figure 12.2[b]). Under ESI conditions, proteins produce a distribution of highly charged ions (from 10 to over 100 charges) whose peaks are separated by exactly one charge (Figure 12.2[c]). A MW can be determined for each one of these charge states. Thus, a standard deviation can be calculated which reflects the mass accuracy of the MW determination. In addition, the shape of the envelope is related to the protein conformation, and can therefore provide structural information about the protein.

ESI-MS is divided into micro- and nano-ESI, according to the flow rate (l/min) at which the solution (in which the peptides and proteins are dissolved) passes through the electrospray needle/capillary. Since only a small portion of the electrospray actually enters the MS, due to the small dimension of the entry hole into the mass spectrometer, the approach providing the lowest flow rate and longest acquisition time is the most sensitive. Thus, nano-ESI-MS, where a typical flow rate of 20 nl/min and very fine needles (only a few micrometers ID at the tip) is used, achieves the highest sensitivity, which is one of the most important considerations in proteomics. The needles, however, are very fragile and require special care and handling. Furthermore, reproducibility of the commercially available needles is not always what one would desire. Another drawback of ESI-MS is that the electrospray and the ionization of peptides and proteins are very sensitive to salt and buffer concentrations. Detergents can also have a major influence and can suppress the detection of proteins and peptides. Consequently, sample preparation starting with the experimental design and including

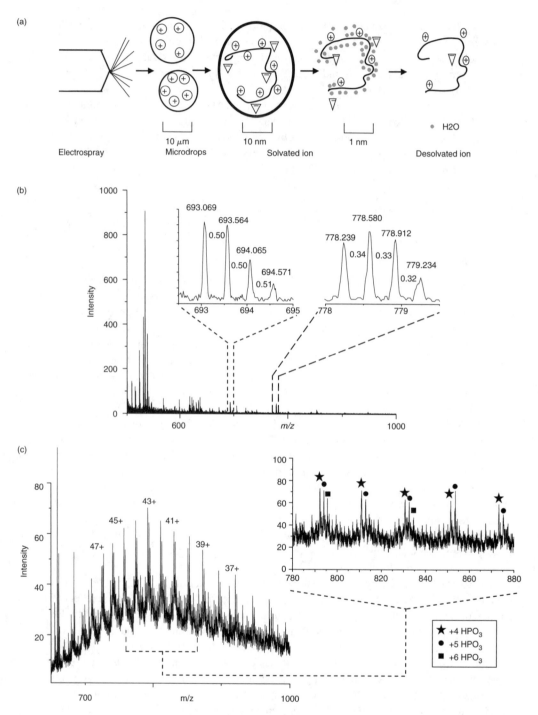

FIGURE 12.2 Mass spectrometric analysis using ESI. (a) Scheme of the mechanism in ESI-MS for producing desolvated ions by applying increasing vacuum to the highly charged droplets produced by the electrospray event. (b) ESI-MS analysis of a peptide mixture obtained from tryptic digestion of the Stem Loop Binding Protein (SLBP). The insets show a doubly charged ion at m/z 693 and a triply charged ion at m/z 778. The distance between the isotope signals of the doubly and triply charged ions correspond to 0.5 and 0.33 Da, respectively. (c) ESI-MS analysis of the Drosophila SLBP. The ion signals are annotated according to their charge states. The inset shows the ion signals from different SLBP isoforms possessing four, five, and six phosphoryl groups.

desalting steps is very important for ESI-MS. Desalting as well as concentration of the analyte can be achieved by direct on-line coupling of a liquid chromatography system to the mass spectrometer: a major advantage of the ESI-MS approach (see Section 12.3.2.1.2).

12.3.2.1.2 Matrix-Assisted Laser Desorption/Ionization

In contrast to ESI-MS, in MALDI-MS the peptides and proteins are introduced to the mass spectrometer as a solid-phase sample. A small drop of analyte solution (usually 0.5 to 1 μl) is mixed with a small drop of a solution containing a high molar excess ($\sim$5000:1) of a chemical compound, the matrix, and then dried on the MALDI-target, usually a stainless steel plate. Evaporation of the solvent leads to a spot of matrix crystals in which the analyte is embedded. For the analysis of proteins and peptides, the matrix is typically α-cyano-4-hydroxycinnamic acid (HCCA) or 2,5-dihydroxybenzoic acid (DHB) dissolved in solutions with a high percentage of organic solvents (commonly acetonitrile, methanol, ethanol, and acetone) and high proportion ($\sim$10%) of organic acids (Trifluoroacetic acid [TFA] or formic acid). These matrices have a UV-absorption maximum that is close to the wavelength of the N_2-laser ($\lambda = 337$ nm), which is the most commonly used laser in MALDI-mass spectrometry.

The target plate is introduced to the mass spectrometer and after achieving high vacuum the analysis is initiated by firing a laser pulse at the spot (Figure 12.3[a]). The matrix absorbs the energy of the laser and this leads to the ionization/desorption of proteins and peptides ultimately resulting in the transfer of intact ionized proteins and peptides into the gas phase. The MALDI-MS mechanism, in contrast to ESI-MS, leads to moderately charged ions. Peptides are typically singly charged (Figure 12.3[b]), while proteins can take up several charges, but still much fewer than in ESI-MS (Figure 12.3[c]). This simplifies the interpretation of MS-spectra, especially in the case of complex mixtures of proteins. Another difference from ESI-MS is that MALDI-MS is more tolerant of salt and detergents. It cannot, however, be coupled online to liquid-chromatography systems as easily as ESI-MS [70]. Thus, MALDI-MS and ESI-MS are complementary techniques, and the preference of their use mainly depends on the nature of the sample and the formulation of the analytical question.

12.3.2.2 Mass Analyzer

12.3.2.2.1 Time-of-Flight Mass Analyzers

The ions generated either by ESI or MALDI are accelerated by applying high voltage (20 to 30 kV) in such a way that all the ions (ideally) have the same kinetic energy. These ions are then extracted into a field-free tube, the time-of-flight (TOF) analyzer, where they are separated by mass linear mode. The velocity of the ions is inversely proportional to its mass to charge ratio (m/z); the greater the m/z, the slower the ions fly. In the linear mode, the detector at the end of the flight tube measures the "TOF" of the ions. The TOF is then converted into m/z and presented as intensity $= f(m/z)$ plot, the mass spectrum.

A very important criterion in MS is the resolution of the mass analyzer, which defines how well the instrument can separate ions that are very close in m/z values. The resolution of the TOF-analyzer (linear mode) is very poor, because ions at the same m/z value do not all have the same initial velocity due to the desorption and ionization process. This results in different flight times, which ultimately broaden the m/z signal. One way to compensate for the variations of initial velocity is by using a device called an ion mirror or reflectron at the end of the linear tube, which directs the ions into an electric field (Figure 12.5[b]). The higher the velocity, the deeper the ions fly into the reflectron. This results in a greater reduction in velocity and longer flight times. Ideally, the reflectron is adjusted so that the ions with same m/z value and different initial velocities exit the reflectron at the same time and fly to the detector with the same velocity, resulting in a very narrow m/z signal. Another instrumental development to improve the resolution is the delayed-extraction procedure, mainly used in MALDI-MS. In this procedure, the acceleration normally applied directly after the ions are transferred into the gas phase is slightly delayed, minimizing the variation in initial velocities.

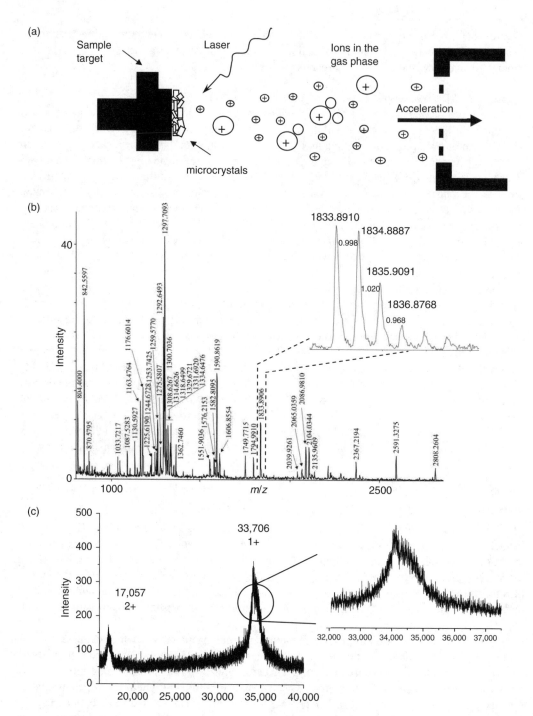

FIGURE 12.3   Mass spectrometric analysis using MALDI. (a) Schematic representation of the MALDI mechanism. A laser pulse strikes the peptides/proteins that are embedded in microcrystals of matrix, leading to the transfer of ions into the gas phase and then acceleration into the mass analyzer. (b) MALDI-MS analysis of a peptide mixture obtained by in-gel digestion of a protein separated by SDS-PAGE. The inset demonstrates that in MALDI-MS, singly charged peptides are predominantly formed, which have isotope signals separated by 1 Da. (c) MALDI-MS analysis of SLBP (see Fig. 12.2[c]). In contrast to the ESI-MS spectrum, singly and doubly charged ions are produced, and the resolution is not sufficient to distinguish between the different protein isoforms.

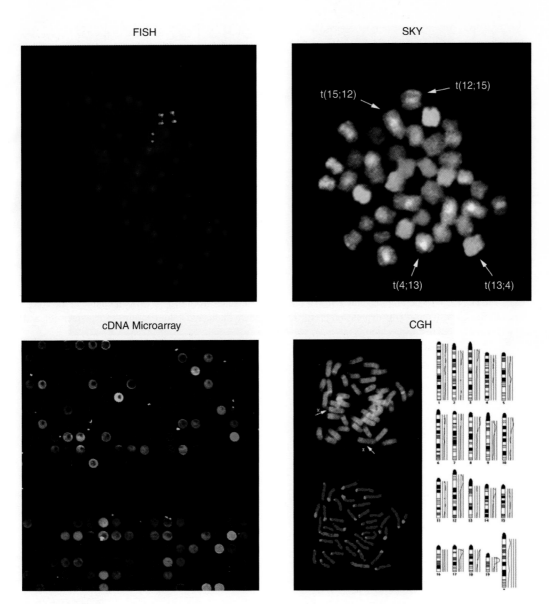

COLOR FIGURE 8.7 Clockwise from top left. FISH is used to determine amplification or loss of a gene. Here Granzyme C (green) and TCRVα (red) show an amplification of TCRVα. SKY analysis detects chromosomal translocations, here four translocations: t(15;12), t(12;15), t(4;13), and t(13;4) are seen in a metaphase spread from a tumor. Global gene expression can be examined with cDNA microarrays. Different levels of hybridization, ranging from green, yellow, to red are seen in this cDNA microarray. CGH is used to detect regions of chromosome gains and losses (seen in red) in the bottom panel.

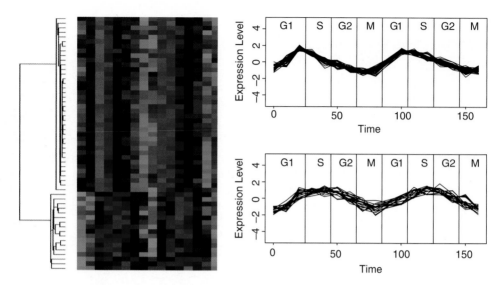

COLOR FIGURE 11.3   Hierarchical clustering of the cell-cycle genes from Figure 11.2. Each line in the color-coded display corresponds to the expression profile of one gene with the red color denoting high expression and green color denoting low expression.

The combination of delayed extraction and a reflectron provides a resolution $>10,000$, equivalent to the ability to resolve a peptide of m/z 10,000 from a peptide of m/z 10,001. In proteomics applications, this resolution is sufficient to achieve complete baseline resolution of the isotopic cluster of peptides. This means, the monoisotopic ion signal, which corresponds to the peptide ion possessing only C^{12}-, H^{1}-, O^{16}-atoms, etc., is clearly separated from other ion signals belonging to the same peptide, which are higher than the monoisotopic ion signal by multiples of 1 Da, since these ions contain C^{13}-atoms instead of C^{12}-atoms. The relative intensities of these ions compared to the monoisotopic signals reflect the natural abundance of the C^{13}-atom (0.1%). (See insets in Figure 12.2[b] and Figure 12.3[b].)

Better resolution coincides with better mass accuracy; therefore, current MALDI-TOF instruments with delayed extraction and a reflectron have the ability to determine monoisotopic peptide MWs within an error of 30 ppm ($\pm$0.03 Da for a 1000 Da peptide). Mass accuracy is one of the most important factors for identifying proteins using the combination of mass spectrometric molecular weight determination of peptides and the database searching algorithms described below in Section 12.4 [71]. In addition, MALDI-TOF MS instruments are very sensitive, and the detection of peptides present in low femtomoles amounts is routinely achieved. Moreover, MALDI-TOF MS instruments are robust and very user-friendly, thus making MALDI-TOF MS well suited to proteomic research.

12.3.2.2.2   Tandem Mass Analyzer

Besides MALDI-TOF MS, mass spectrometers with tandem mass analyzers are also frequently used in proteomic research, since those instruments have the capability of directly providing the sequence information of peptides. In general, tandem mass analyzers are able to perform two or more MS experiments (MS/MS or MS^n where n describes the number of the MS experiments) consecutively, in a single analysis. The first MS experiment is to isolate a specific ion, which then gains energy and dissociates into fragment ions via collision-induced dissociation (CID), the second MS experiment determines the MW of these fragment ions.

In most cases, low energy CID produces sequence-specific fragment ions by cleavage of the $-CO-NH-$ bond of the peptide backbone (Figure 12.4). Depending on the location of the positive charge in the fragment, fragment ions are formed, which contain the N- or C-terminus of the peptide. These ions, called b- or y-ions, are therefore complementary. The index i in "y_i" or "b_i" corresponds to the number of the peptide bond cleaved, starting from the terminus [72,73]. The differences in MW between two adjacent b- or y-ions (i.e., b_i and b_{i+1}, or y_i and y_{i+1}) corresponds to the MW of a specific amino acid residue. Thus, a series of b- and y-ions provide partial sequence information, which, together with the peptide MW, can be used for protein identification (see Section 12.4) and characterization, and for the identification of sites of modification, such as phosphorylation.

Another very significant type of fragment ion is the immonium ion, with a general formula of $H_2N^+=CHR$, where R is the side chain of a specific amino acid residue. It is called an immonium ion because it is similar in structure to the $H_2N^+=R'$ ion. Immonium ions provide information about the amino acid composition and the presence of specific amino acid modifications, such as phosphotyrosine and acetylated lysine residues, and can be used for specific detection of those modifications [74].

For proteomic research, three types of tandem mass analyzer performing low energy CID are now commonly used: the triple quadrupole, the quadrupole time-of-flight (Q-TOF), and the ion trap. Most of these tandem mass analyzers are connected to ESI-MS instruments. However, recently, a MALDI source has become commercially available for Q-TOF instruments.

1. *Triple quadrupole mass analyzer.* A quadrupole mass analyzer (Figure 12.5[a]) consists of four metal rods with two pairs of electronically connected rods, which face opposite direction from each other. By applying a combination of current and radio-frequency voltage to the two pairs of rods, a complex oscillating movement of the ions is produced that guides the ions through the mass

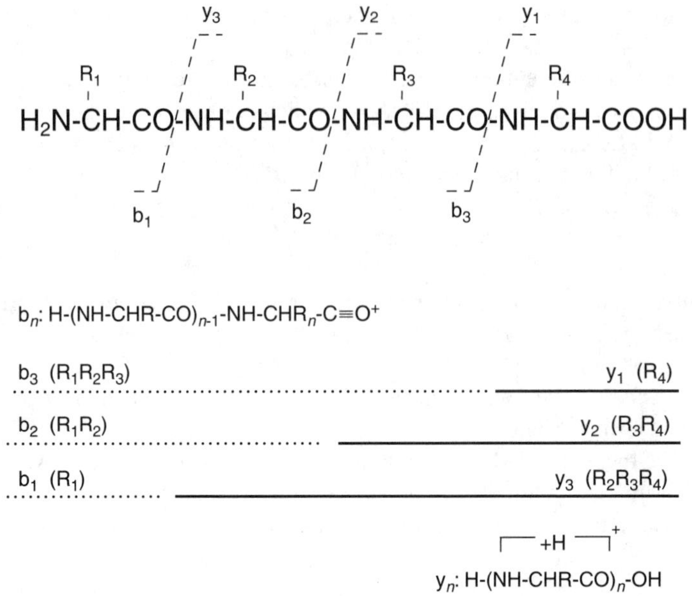

FIGURE 12.4 Nomenclature of the fragment ions formed by dissociation of the peptide bond in tandem MS (MS/MS) analysis of peptides. The localization of the positive charge determines how the fragment ions are named. Ions are called b_i ions if the positive charge is located at the N-terminus, and y_i ions if the positive charge is located at the C-terminus. The index i represent the number of the peptide bond dissociated. The b- and y-ions are complementary, and are used to determine the amino acid sequence of the peptide.

analyzer. For each m/z value a certain voltage at the rods is required for the ion to pass through, allowing the selection of specific ions. This mass analyzer, therefore, is also called the quadrupole mass filter.

For determination of the MWs of the peptides, the quadrupole mass filter scans from low m/z to high m/z at a certain rate to assure the detection of all the peptide ions within the mass range of the instrument. In a tandem mass spectrometric experiment using a triple quadrupole, the first quadrupole is locked at a certain m/z value, allowing the transmission of only those ions of that specific m/z value. The second quadrupole is filled with gas molecules (usually argon or nitrogen) with which the selected peptide ion collides. These collisions produce the fragmentation. Thus, the second quadrupole is also called a collision cell. The third quadrupole analyzes the fragment ions. This analysis can be performed in two different ways: (1) the third quadrupole can scan over a broad range of m/z values detecting all fragment ions produced (the product ion spectrum), or (2) the third quadrupole can be used to detect only specific fragment ions by selecting only certain m/z values. The second method increases the sensitivity of detection of those specific fragment ions, since the entire acquisition time is devoted specifically to these fragments. This procedure, for example, is used in neutral loss scanning experiments for highly sensitive detection of phosphorylated peptides. Here, the third quadrupole is scanned to pass ions 98 Da lower than those transmitted by the first quadrupole, in order to detect the loss of a phosphoryl group from the initial peptide — a loss that commonly occurs during collision-induced dissociation.

2. Q-TOF instruments. The main drawback of the triple quadrupole, however, is its moderate resolution and mass accuracy that hampers its usefulness in identifying proteins. This disadvantage is overcome by combining a quadrupole mass analyzer with a TOF analyzer (Q-TOF), which is arranged orthogonal to the quadrupole (Figure 12.5[b]). Analogous to the triple quadrupole, the first quadrupole acts as a mass filter to select a peptide of interest for sequencing. This selected ion is guided into the collision cell and the fragment ions are directed to the pusher, which accelerates them into the TOF. The TOF is typically equipped with a reflectron providing high mass accuracy and

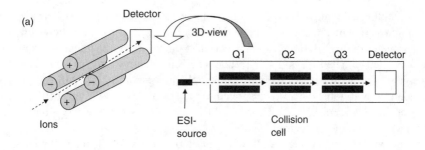

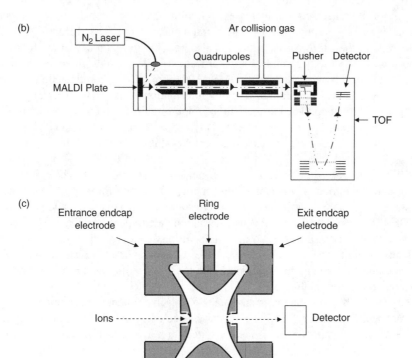

FIGURE 12.5   Schematic representation of a tandem mass analyzer. (a) Triple quadrupole analyzer. Three sequential mass filters (Q1, Q2, Q3), each consisting of four parallel rods driven by DC and radio-frequency fields and arranged so that the potential field on-axis is quadrupolar (saddle-shaped), and can be controlled by a computer to perform MS/MS experiments. Quadrupole Q1 selects an ion of a specific m/z value. Q2, which is filled with collision gas, is the site of the fragmentation, and Q3 analyzes the fragment ions. A triple-quadrupole mass analyzer is commonly equipped with an ESI source. (b) Q-TOF. As in a triple quadrupole, an ion of a specific m/z value is selected by a quadrupole and directed to the collision cell. After fragmentation the ions are pushed into the TOF analyzer, which is orthogonal to the quadrupole, and is equipped with a reflectron. A Q-TOF can be equipped with an ESI source or a MALDI source as shown in this figure. (c) Quadrupole ion trap mass analyzer. In the quadrupole ion trap, the ions do not pass through a mass analyzer. Here, the ions are collected, stored, and dissociated in the space within the electrodes and released for detection.

resolution; therefore, the Q-TOF is well suited for identifying proteins with high degree of certainty as well as, posttranslational modification sites.

3. *Ion traps.* In contrast to the triple quadrupole and Q-TOF, the quadrupole ion trap is not a combination of mass analyzers and does not select ions passing through the mass analyzer. The quadrupole

ion trap is three-dimensional, and consists of two endcap electrodes and a ring electrode surrounding the trapping space, in which the ions oscillate (Figure 12.5[c]). It collects and stores ions to perform the MS experiments through the application of current, voltage, and radio frequency to the electrodes. The ion trap is filled with helium gas, which reduces the ion velocity and helps control the distribution of ion energies. The helium also functions as the collision gas in the MS/MS experiments. After filling the ion trap with ions, the voltage applied to the electrodes is adjusted such that the ions are ejected out of the trap according to their m/z values. This allows the determination of the MWs of all the peptides present in the ion trap.

During the MS/MS experiment, only the ions of the m/z selected for sequencing are stored in the trap — all the other ions are ejected. By quickly increasing the voltages on the electrodes, the ion energy of the remaining ion increases and collisions with the helium gas molecules leads to the fragmentation of the peptide ions. These fragment ions are ejected from the trap, resulting in the acquisition of the MS/MS spectrum. It is also possible to retain one of the fragment ions in the trap and subject it to another fragmentation process. This leads to the dissociation of a fragment ion, which results in the acquisition of the MS/MS/MS or MS^3 spectrum. Higher-order tandem MS experiments can provide additional structural information. This can be especially useful for the analysis of glycosylated peptides, but are only of minor significance for proteomic research. Like the quadrupole mass analyzer, the ion trap has only moderate mass resolution and mass accuracy, but is distinguished by its high sensitivity, which makes this type of mass analyzer very useful for proteomics.

4. *TOF/TOF analyzer.* More recently, another tandem mass analyzer has been introduced to the proteomic research community, which uses high energy CID for performing MS/MS analyses. This tandem mass analyzer consists of two TOFs in series and a MALDI source, and is therefore called MALDI-TOF/TOF. Here, the first TOF is used as an ion gate to pass a selected m/z. These ions are then dissociated in the collision cell and the fragment ions are detected by the second TOF. Since the ion velocity is still high when the ions enter the collision cell, additional fragmentation, which requires higher energy, like the dissociation of side chains of amino acid residues, can occur. This is advantageous for differentiating between leucine and isoleucine, which cannot be distinguished by low energy CID. However, the major advantage of this instrumentation compared to the MALDI Q-TOF is the speed at which MS/MS spectra can be acquired, mainly due to the ten-times higher laser pulse rate. Thus, this instrument is ideally suited for performing high-throughput proteomics.

12.3.3 LIQUID CHROMATOGRAPHY COMBINED WITH MASS SPECTROMETRY

For MS proteomic analysis, separation is required to reduce the complexity of the proteome. This can be performed either on the protein level, for example, by 2D-PAGE [50] (see Section 12.1), or by separating the peptides obtained from a proteolytic digestion of the proteome. To separate proteins or peptides, high pressure liquid chromatography (HPLC) is advantageous because liquid handling is well-suited to automation, an important issue for high-throughput analysis [44]. In addition, HPLC can be coupled directly to the mass spectrometer using an ESI source.

As explained above, the sensitivity of ESI-MS is inversely proportional to the flow rate of the liquid in which the peptide is dissolved. To achieve a low flow rate and thereby the highest sensitivity, separation columns with inner diameters (ID's) as low as 75 μm are currently used and can be directly coupled to the ESI-source. These capillary columns typically use flow rates of 200 nl/min, but require the use of capillary LC pumps or liquid splitters to achieve this flow rate. The capillary columns are usually packed with C^{18} reversed-phase (RP) HPLC packing material (usually 3 or 5 μm diameter), because the solvents used for reversed-phase separations (acetonitrile/water/organic acid) are very compatible with ESI (see above).

The RP-HPLC can also be combined with another type of LC, for example, ion exchange chromatography (IEC). Here, the peptide or protein mixture is first applied to an IEC column, the peptides or proteins are then step eluted by solutions with increasing salt concentrations, and the eluate is loaded on capillary RP-column coupled to the ESI-source. After washing steps to remove the salt, the peptides/protein are eluted from the RP-HPLC column by applying a gradient with increasing amounts of the organic solvent. After completion of the first RP-HPLC run, the second step elution of the IEC column is performed, which initiates the second RP-HPLC run, and so on. Yates and coworkers [75] have demonstrated the performance of this 2-dimensional separation system on-line coupled to an ESI mass spectrometer by identifying more than 4000 proteins in yeast within one 24 h LC/LC-run of a complete proteome digestion. Unfortunately, the robustness of capillary LC systems equipped with a 75 μm capillary column is poor and the handling of these capillary systems require specific skills, so this system is not used on a routine basis.

Most modern ESI-mass spectrometers have the option of data-dependent switching between the MS and MS/MS analyses, allowing the sequencing of peptides as they elute from the LC column. As soon as the peptide ion signal reaches a certain threshold in the MS mode, the MS/MS experiment is initiated for a preselected time, and then the instrument is set back to the MS mode in preparation for a "full MS scan" of the next-eluting peptide (Figure 12.6). Currently available software systems allow the sequencing of co-eluting peptides by performing MS/MS analysis on several ions in succession, in order of the relative abundances of the ions. Even more sophisticated software programs provide features, which prevent sequencing the same peptide more than once. This allows more time for analyzing other peptides and thus increases the amount of information obtained from the analysis. Other novel software programs use fast switching between MS and MS/MS modes for improving the detection and analysis of specific peptides, like phosphopeptides. Here, a fast MS/MS analysis is performed on each peptide that elutes. If a preselected fragment ion is detected (e.g., the neutral loss of 98 Da for a phosphorylated peptide or the presence of 143 Da for the immonium ion of acetylated lysine), a comprehensive MS/MS spectrum of that peptide will be acquired. Another approach to increase the acquisition time for a specific peptide is to decrease the flow rate by effectively stopping the LC run. This method, commonly called "peak parking" can also be controlled in a data-dependent manner [76].

Despite these improvements, some peptides will not be sequenced during the LC-MS run, or if the acquisition time was not sufficient for obtaining an interpretable MS/MS spectrum. Reanalyzing the sample requires performing another LC-MS/MS experiment, which means spending more time and, even more importantly, consuming additional sample. For this reason, coupling HPLC offline with the new MALDI-TOF/TOF instruments, which have high-throughput capability and the ability to perform MS and MS/MS analysis on thousands of peptides from a given HPLC run has been shown to be a good alternative to ESI-LC-MS.

12.4 COMPUTATIONAL ASPECTS OF PROTEOMICS USING MASS SPECTROMETRY DATA

12.4.1 INTRODUCTION

In recent years, an increasing number of genomes of model organisms have been sequenced. Using these genomic sequences, researchers have been able to make tremendous progress in the study of genomes, such as numerous successes in the identification of genes, the detection of protein-binding DNA motifs, and the determination of gene regulations. Beyond these successes is the far more challenging and rewarding task of understanding proteomes by means of, for example, (1) discovering signal transduction pathways, (2) determining protein structures, (3) detecting protein–protein, protein–DNA, and protein–metabolite interactions, (4) detecting posttranslational modifications of proteins, and ultimately (5) elucidating the functions of genes and their protein products. Such challenges require increasing demand on modern computational algorithm to decipher information accurately and reliably. In the subsequent sections the state-of-the-art in understanding and interpreting MS data is described.

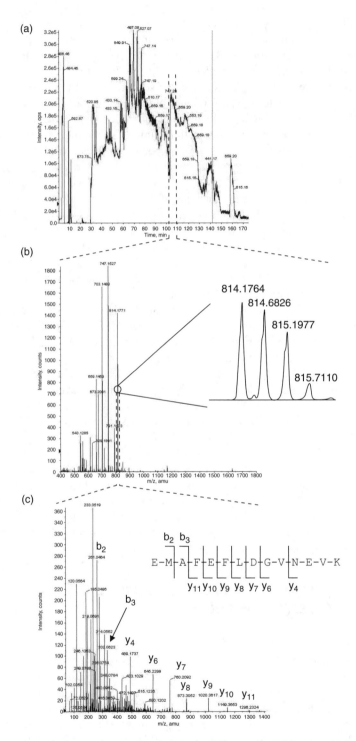

FIGURE 12.6 LC-MS/MS analysis of the proteolytic digest of Drosophila SLBP using a Q-TOF mass analyzer. (a) Total ion chromatogram (TIC). The TIC reflects the sum of the intensities of all ions at a given retention time. (b) MS analysis within the retention time window of 103 to 108 min. The inset shows that the ion at m/z 814 is doubly charged. (c) MS/MS analysis of the ion at m/z 814. The fragment ions corresponding to b- and y-ions have been assigned, demonstrating the complete sequencing of this peptide.

12.4.2 HIGH-THROUGHPUT PROTEIN IDENTIFICATION VIA TANDEM MASS SPECTROMETRY

There are two approaches to the identification of peptides for tandem mass spectra: the database search method and the *de novo* sequencing method. The first method requires a protein database or a genome database for the target organism, and the latter method makes no assumption of these. Both methods have to deal with the complexity of a tandem mass spectrum. In practice, many experimental factors can affect a tandem mass spectrum: (1) Some fragmented molecules may not be charged or some ions may not be obtained in the experiments. As a result, the corresponding mass peaks do not appear in the spectrum. (2) An ion may display two or three different mass peaks because of the distribution of isotopic carbons ^{12}C and ^{13}C in the molecules. (3) An ion may lose a water molecule or an ammonia molecule and display at a different mass peak from its normal one. (4) Random noise always appears in the spectrum. (5) Interpretation of abundance levels in the spectrum remains unsolved. (6) An amino acid at some unknown location of the peptide sequence is modified, which changes not only the peptide mass but also that of all the ions that contain this modification.

Computer programs for peptide identification such as SEQUEST [77], Protein Prospector [71], and PepFrag correlate peptide sequences in a protein database with the tandem mass spectrum, and then report all peptides with significant correlation scores. Some of them allow searching in a genomic database. The *de novo* peptide sequencing approach [78–80] extracts candidate peptide sequences directly from the spectral data without database search. MS/MS has also been applied to obtain accurate quantification of proteins by using isotope-coded affinity tags (ICATs) [81]; to analyze protein components in complex protein mixtures, to identify cross-linking sites [80,82], to identify protein–protein interactions, and to verify gene-finding programs [83].

The complication of tandem mass spectra makes it nontrivial to identify the correct peptide from databases for each spectrum. The *de novo* sequencing is even harder. Consequently, practically useful high-throughput identification of peptides requires advanced computer algorithms and mathematical models as well as their efficient implementations (Figure 12.7).

12.4.3 INTERPRETATION OF TANDEM MASS SPECTRA VIA DATABASE SEARCH

Mathematically, the problem of finding the optimal peptide from a database for a tandem mass spectrum can be formulated thus:

Given a database, a tandem mass spectrum S, and a scoring function $f()$, ask to find a peptide p from the database to maximize $f(S \mid p)$.

A typical approach [77] uses the predictability of peptide fragmentation patterns to identify amino acid sequences from tandem mass spectra. Amino acid sequences in the database are scanned to find the protease cleavage subsequences that are within some error tolerance of the peptide mass derived from MS/MS. When a subsequence within the mass tolerance is found, a hypothetical tandem mass spectrum is generated by calculating the masses of the predicted N- and C-terminal ions, and then assigning a relative abundance to each type of ions. Then, a cross-correlation function based on Fast Fourier Transform is used to compare the hypothetical spectrum with the experimental spectrum. Sequences with top scores are reported.

A heuristic approach [84] searches databases based on the information derived from partial interpretation of a tandem mass spectrum. It first derives some short strings of 2 to 4 amino acids in length from the distances between mass peaks in the spectrum. It then searches the database with these strings as well as other information such as the peptide's molecular weight and the protease cleavage specificity. Although none of the pieces of information is specific, their combination is

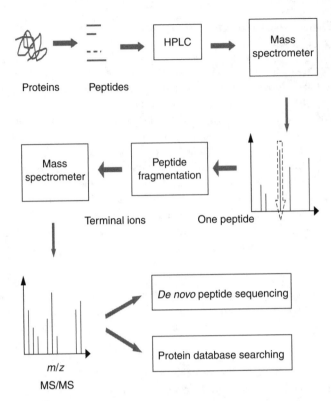

FIGURE 12.7 Protein analysis by HPLC-MS/MS: first, proteins are digested into short peptides. Then these peptides are separated by HPLC, ionized, and analyzed by a mass spectrometer. A peptide of specific mass/charge ratio is chosen for fragmentation, and the resulting fragmentation ions are further analyzed by the second mass spectrometer, which produces a tandem mass spectrum. The tandem mass spectrum can be used to identify the original peptide sequence by a database search program or a *de novo* sequencing program.

specific enough that only a small set of candidates satisfies the combined constraints. If a search does not identify a sequence, additional searches are carried out to consider the presence of an unknown amino acid modification or a sequence error in the database.

Interpretation of tandem mass spectra generally requires a protein database that is assumed to contain target proteins in an experiment. Most sequences in a protein database are derived from genomic sequences through gene-finding programs, and thus sequencing errors cannot be avoided. In spite of this, database search is still the most robust and accurate among all known methods. An alternative choice is to use a cDNA or genomic database instead of a protein database to overcome the low gene-finding accuracy in eukaryotic genomes. The protein database PDB now consists of about half a million proteins and hundreds of millions of amino acids. The human genome, representing a typical genomic database, has 3 billion nucleotide base pairs, corresponding to 6 billion amino acids when 2 strands and 6 frames of the genome sequences are translated. Searching databases of similar sizes is nontrivial.

A good scoring function is crucial to the identification of the right peptide from databases for each spectrum. It is difficult to determine whether a spectrum is generated from a given peptide because the underlying chemical and physical processes of the peptide fragmentation are not completely understood. A simple correlation method [77] has been used to score a spectrum against a peptide. More sophisticated probabilistic models [78,85,86] have been proposed to compute the statistics of the match between a spectrum and a peptide. Still, there always exist a large number of tandem mass spectra that cannot be linked to peptides.

One difficulty in the interpretation of tandem mass spectra is amino acid modification, because mass spectrometers are very sensitive to mass changes. However, finding the modified amino acids is of great interest to biologists because modifications are usually associated with protein functions. Almost every protein is posttranslationally modified, and more than 200 different types of modifications have been found. Proteins in complex biological systems such as signaling pathways and metabolic pathways are subject to enormous numbers of biological modifications such as phosphorylation and glycosylation. It is estimated that at least 1000 kinases exist in the human genome [87]. They bind to other proteins and phosphorylate one or multiple amino acids such as tyrosines, threonines, and serines. Phosphorylation is common in signal transduction and in regulation of enzyme activities. Some enzymes become active when phosphorylated and inactive when dephosphorylated or vice versa. A challenging problem in proteomics is to find out where those modifications are present in each protein as well as in which tissue, developmental stage, and environment they are present.

12.4.4 USING PROTEIN DATABASES VERSUS USING GENOME DATABASES

As more and more large eukaryotic genomes are being sequenced, the task of locating genes on genomes accurately has seriously fallen behind. Current gene-finding methods have been successful in bacterium and lower eukaryotic genome projects, but it is still very difficult to use them to make a complete and accurate prediction of novel genes in higher eukaryotic genomes because the underlying biological processes are not completely understood. These genomes are 10 to 100 times larger than lower eukaryotic genomes, and thus predicting signals with a low false-positive ratio is much more difficult. Also, a gene may have many relatively short exons separated by long introns. Most prediction methods are not sensitive enough to find these exons or to assemble these exons into a correct gene structure. A recent study [88] shows that the current best exon prediction in a well-studied 3 Mbpsec Drosophila genome sequence has 0.79 sensitivity and 0.68 specificity scores, and that the best gene structure prediction is even lower with 0.44 sensitivity and 0.30 specificity scores.

To overcome these limitations, interpretation of tandem mass spectra using genomic sequences instead of protein sequences is of great interest. First, it can provide a shortcut for gene prediction by giving the locations of genes in a genome. Second, it can lead to a potentially better interpretation than a method that uses a protein database with protein sequences derived from genomes through gene finding programs. An obvious shortcoming is that searching genome databases increases the number of false-positives because large areas of noncoding regions are also being compared.

Chen [83] designed an algorithm to compare tandem mass spectral data with a given genome sequence to locate DNA coding sequences such that their translated amino acid sequences are optimally correlated with these tandem mass spectra. The algorithm also searches for two DNA coding sequences separated by one gap, which corresponds to an intron, such that the concatenation of these two sequences generates one coding sequence. The algorithm has proven successful for simulated hypothetical tandem mass spectra, but it has not been tested on experimental data.

12.4.5 DE NOVO PEPTIDE SEQUENCING

Mathematically, the problem of *de novo* peptide sequencing can be formulated thus:

Given a tandem mass spectrum S and a scoring function $f()$, ask to find a peptide p to maximize $f(S \mid p)$.

De novo peptide sequencing extracts candidate sequences from tandem mass spectral data before validating them in a database or adding them into a database. If correct sequences are among the candidates, this approach can greatly improve the speed of peptide identification, because searching sequences in a database is much faster than searching a tandem mass spectrum in a database for an optimal match. If the target sequences are not in the protein database, a database search program will fail, but a program for *de novo* sequencing could find the novel proteins.

There are many further reasons why it is important to improve computer programs for *de novo* peptide sequencing. First, the protein sequences in a database are mostly derived from genomic sequences through gene prediction programs, which are not accurate [88]. *De novo* peptide sequencing does not rely on a database and thus could overcome the inaccuracy of gene-finding. Second, even if the gene prediction is perfect, transcription-level modifications (such as RNA editing, translation-level modifications, and posttranslation-level modifications) can change a protein sequence. Just one such mutation may completely alter a tandem mass spectrum. Most existing computer programs have difficulty handling such a change of mass.

Dancik et al. [78] handle *de novo* peptide sequencing as follows. First, the spectral data is transformed to a directed acyclic graph, called a *spectrum graph*, where (1) a node corresponds to a mass peak; (2) an edge, labeled by some amino acids, connects two nodes differing by the total mass of the amino acids in the label; (3) a mass peak is transformed into several nodes in the graph, and each node represents a possible prefix subsequence (i.e., ion) for the peak. Then, an algorithm is called to find the longest or highest-scoring path in the graph. The concatenation of edge labels in the path gives one or multiple candidate-peptide sequences. However, all known algorithms [89] for finding the longest paths tend to include multiple nodes associated with the same mass peak. This redundancy interprets a mass peak with multiple ions of a peptide sequence, which is rare in nature.

Another approach [80] provides efficient sequencing algorithms for a general interpretation of tandem mass data by restricting a path in the spectrum graph to contain at most one node for each mass peak. Therefore, each mass peak corresponds to either one terminal ion (b or y) but not both. First a spectrum graph is built for a given tandem mass spectrum, where a mass peak is transformed into two complementary nodes corresponding two mutually exclusive assumptions of this mass peak: either it is an N-terminal ion or it is an C-terminal ion. An edge between two nodes is derived and labeled with some amino acids if the distance between two nodes equals the mass sum of these amino acids. In this graph, a dynamic programming algorithm is employed to find a path that contains at most one node from every complementary pair. This algorithm solves the *de novo* peptide sequencing problem efficiently even for noisy spectra, and can be extended to find an amino acid at an unknown position and modified to an unknown mass.

12.4.6 ICAT Approach for Protein Quantitation

Protein quantitation using MS is difficult due to the incomplete proteolytic digestion of proteins, the differences in peptide ionization efficiencies, and the unequal loss of peptides MS runs. Even a subtle change in the experimental conditions may produce very different results. To overcome these obstacles, Aebersold and coworkers [81] designed a very clever experiment that mixed differentially labeled proteins from two different cell states and ran them through the same digestion-HPLC-tandem mass spectrometry process. Then the relative quantitation of the labeled proteins in the two states could be measured because all the above problems were evened out. This technique is called the isotope-coded affinity tags (ICAT).

The ICAT reagent has three components: a biotin, a linker, and a thiol-specific reactive group. The biotin is used for isolation of proteins and peptides; the thiol-specific reactive group covalently binds to cysteines in proteins; and the linker has eight hydrogens that can be replaced by eight heavier deuteriums. The light reagent containing hydrogens is called d0-ICAT and the heavy reagent containing eight deuteriums is called d8-ICAT. The chemical properties of a peptide labeled with either d0 or d8 are almost identical, however they differ in mass by 8.05 Da.

The strategy for quantifying differential protein expression is as follows: first, the cell state 1 is labeled by d0-ICAT and the cell state 2 is labeled by d8-ICAT. Then these two cell state samples are combined and digested by enzymes. Finally, the ICAT labeled peptides are isolated and analyzed by HPLC-MS/MS. Peptides can be quantified by mathematically computing the quantity from the abundance levels in MS. The ratio of the quantities of the two differentially labeled peptides gives the accurate relative quantitation. This technique has been widely used in many proteomic projects.

12.4.7 DETECTING PROTEIN CROSS-LINKER VIA MASS SPECTROMETRY

Cross-linking technology can connect two amino acids to a linker by covalent bonds. These two amino acids may come from one protein or from two interacting proteins. Combined with MS/MS, cross-linking technology becomes a powerful method that provides a rapid solution to the discovery of protein–protein interactions and protein structures. Traditionally, the three-dimensional structure of a protein is solved by x-ray crystallography or NMR. However, generating an accurate structure that satisfies all the constraints of experimental data can be extremely difficult. There has been some success with other computational methods that predict structures from energy functions, multiple alignments, and threading. Thus, accuracy and general applicability of these methods lags far behind the rate at which new protein sequences are being identified. Unfortunately, computational methods of studying protein–protein interactions often scale up poorly, make assumptions about the number of interacting proteins or their modification state, or lose details about environmental effects on complex formation.

Cross-linking technology combined with MS adds reliable distance constraints to protein structures. A carefully designed cross-linking technique has been able to produce low resolution interatomic distance constraints, which in conjunction with a protein threading algorithm, have led to the determination of the three-dimensional structure of a model protein by Young et al. [82]. They successfully identified many decorated and cross-linked peptides from mass spectra obtained from protein cross-linking experiments. The following figure shows four common cross-links: (a) two cross-linked peptides, (b) a decorated peptide, (c) a self-cross-linked peptide, and (d) two doubled cross-linked peptides. The structure of this protein was estimated using the threading algorithm and assuming that the detected labeled peptides were on the surface of the protein. A technique called isotopic labeling [90] has been proposed to separate and identify peptides and cross-links in MS. Through a proper design of labeling, a cross-link can be easily identified by its unique mass in the spectrum. Similar techniques can be applied to identify the "docking" structure of two interacting proteins.

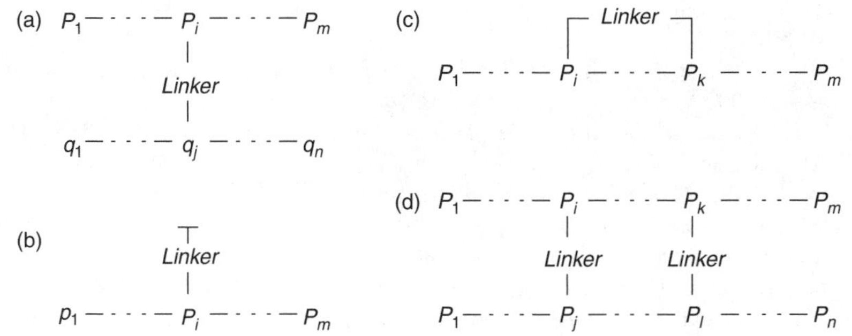

Another study [80] detects cross-linked peptides and cross-linked amino acids from tandem mass spectra. They performed an experiment that first chemically cross-linked human red blood hemoglobin proteins, which consist of 2 and 2 polypeptides, and then digested them with trypsin. The resulting peptides and cross-links were separated and identified via HPLC-MS/MS. Algorithms were designed to identify cross-linked peptides from the tandem mass spectral data. In this experiment, a cross-link between lysine 82 of both the subunits of hemoglobin was detected. Structurally, this would correspond to bridging the central channel of the Hb protein. The inter-residue distance between the two lysines was measured to be 8.18 Å (derived from the PDB file 1A3N), which is comparable with the 7.7 Å spacer length of the cross-linking agent, DSG.

Although MS/MS could potentially be a high-throughput solution for the identification of cross-linked peptides, the tandem mass spectra for cross-linked peptides are much more complicated that those of single peptides. Many mass peaks are unknown, so verifying these cross-links to discover the exact location on the proteins remains difficult. On the other hand, the general protein folding problem on 3D lattice is NP-hard. Adding constraints on surface amino acids does not change the nature of this problem. However, these constraints could greatly reduce the search space of many heuristic folding algorithms such as threading, simulated annealing, and genetic algorithms, among others.

12.4.8 PROTEIN–PROTEIN INTERACTIONS

Affinity chromatography, the two-hybrid approach, co-purification, immunoprecipitation, and cross-linking are some of the tools used to verify proteins that physically associate. Of these, the two-hybrid assay has been widely used to analyze protein–protein interactions [91–93]. The protein interaction profiles have made it possible to examine complexes comprising large number of proteins and to also functionally classify proteins of unknown function.

Various computational methods have also been developed to predict protein–protein interactions. Those approaches include the Rosetta stone/gene fusion method [94] and the phylogenetic profile method [95]. These two methods are based on protein sequence homologues to predict protein interactions and functional correlations. A hybrid method for computing interactions has been proposed to combine the results of these two methods and other sources such as protein pairs with correlated mRNA expression levels and experimentally determined protein–protein interactions. Other computational methods to predict protein–protein interaction have been presented based on different principles, including the interaction-domain pair profile method and the support vector machine-learning method [96].

REFERENCES

1. Norin, M. and Sundstrom, M., Structural proteomics: Developments in structure-to-function predictions, *Trends Biotechnol.*, 20, 79–84, 2002.
2. Jhoti, H., High-throughput structural proteomics using x-rays, *Trends Biotechnol.*, 19 (10 Suppl.), S67–S71, 2001.
3. Rappsilber, J. and Mann, M., What does it mean to identify a protein in proteomics? *Trends Biochem. Sci.*, 27, 74–78, 2002.
4. Ahram, M., Best, C.J., Flaig, M.J., Gillespie, J.W., Leiva, I.M., Chuaqui, R.F., Zhou, G., Shu, H., Duray, P.H., Linehan, W.M., Raffeld, M., Ornstein, D.K., Zhao, Y., Petricoin, E.F., 3rd, and Emmert-Buck, M.R., Proteomic analysis of human prostate cancer, *Mol. Carcinog.*, 33, 9–15, 2002.
5. Meehan, K.L., Holland, J.W., and Dawkins, H.J., Proteomic analysis of normal and malignant prostate tissue to identify novel proteins lost in cancer, *Prostate*, 50, 54–63, 2002.
6. Jones, M.B., Krutzsch, H., Shu, H., Zhao, Y., Liotta, L.A., Kohn, E.C., and Petricoin, E.F., 3rd, Proteomic analysis and identification of new biomarkers and therapeutic targets for invasive ovarian cancer, *Proteomics*, 2, 76–84, 2002.
7. Simpson, R.J. and Dorow, D.S., Cancer proteomics: From signaling networks to tumor markers, *Trends Biotechnol.*, 19 (10 Suppl.), S40–S48, 2001.
8. Jain, K.K., Proteomics: Delivering new routes to drug discovery — Part 1, *Drug Discov. Today*, 6, 772–774, 2001.
9. Jain, K.K., Proteomics: Delivering new routes to drug discovery — Part 2, *Drug Discov. Today*, 6, 829–832, 2001.
10. Graves, P.R. and Haystead, T.A., Molecular biologist's guide to proteomics, *Microbiol. Mol. Biol. Rev.*, 66, 39–63, 2002.

11. Stoll, D., Templin, M.F., Schrenk, M., Traub, P.C., Vohringer, C.F., and Joos, T.O., Protein microarray technology, *Front Biosci.*, 7, c13–c32, 2002.

12. Mirzabekov, A. and Kolchinsky, A., Emerging array-based technologies in proteomics, *Curr. Opin. Chem. Biol.*, 6, 70–75, 2002.

13. Mouradian, S., Lab-on-a-chip: Applications in proteomics, *Curr. Opin. Chem. Biol.*, 6, 51–56, 2002.

14. Schweitzer, B. and Kingsmore, S.F., Measuring proteins on microarrays, *Curr. Opin. Biotechnol.*, 13, 14–19, 2002.

15. Zhou, H., Roy, S., Schulman, H., and Natan, M.J., Solution and chip arrays in protein profiling, *Trends Biotechnol.*, 19 (10 Suppl.), S34–S39, 2001.

16. Figeys, D. and Pinto, D., Proteomics on a chip: Promising developments, *Electrophoresis*, 22, 208–216, 2001.

17. Hebestreit, H.F., Proteomics: An holistic analysis of nature's proteins, *Curr. Opin. Pharmacol.*, 1, 513–520, 2001.

18. Wilkins, M.R., Sanchez, J.C., Williams, K.L., and Hochstrasser, D.F., Current challenges and future applications for protein maps and post-translational vector maps in proteome projects, *Electrophoresis*, 17, 830–838, 1996.

19. Krishna, R.G. and Wold, F., Post-translational modification of proteins, *Adv. Enzymol. Relat. Areas Mol. Biol.*, 67, 265–298, 1993.

20. Bichsel, V.E., Liotta, L.A., and Petricoin, E.F., 3rd, Cancer proteomics: From biomarker discovery to signal pathway profiling, *Cancer J.*, 7, 69–78, 2001.

21. Petricoin, E.F., Zoon, K.C., Kohn, E.C., Barrett, J.C., and Liotta, L.A., Clinical proteomics: Translating benchside promise into bedside reality, *Nat. Rev. Drug Discov.*, 1, 683–695, 2002.

22. Ha, G.H., Lee, S.U., Kang, D.G., Ha, N.Y., Kim, S.H., Kim, J., Bae, J.M., Kim, J.W., and Lee, C.W., Proteome analysis of human stomach tissue: Separation of soluble proteins by two-dimensional polyacrylamide gel electrophoresis and identification by mass spectrometry, *Electrophoresis*, 23, 2513–2524, 2002.

23. Jain, K.K., Applications of proteomics in oncology, *Pharmacogenomics*, 1, 385–393, 2000.

24. Jain, K.K., Recent advances in oncoproteomics, *Curr. Opin. Mol. Ther.*, 4, 203–209, 2002.

25. Oh, J.M., Brichory, F., Puravs, E., Kuick, R., Wood, C., Rouillard, J.M., Tra, J., Kardia, S., Beer, D., and Hanash, S., A database of protein expression in lung cancer, *Proteomics*, 1, 1303–1319, 2001.

26. Chen, G., Gharib, T.G., Huang, C.C., Thomas, D.G., Shedden, K.A., Taylor, J.M., Kardia, S.L., Misek, D.E., Giordano, T.J., Iannettoni, M.D., Orringer, M.B., Hanash, S.M., and Beer, D.G., Proteomic analysis of lung adenocarcinoma: Identification of a highly expressed set of proteins in tumors, *Clin. Cancer Res.*, 8, 2298–2305, 2002.

27. Gharib, T.G., Chen, G., Wang, H., Huang, C.C., Prescott, M.S., Shedden, K., Misek, D.E., Thomas, D.G., Giordano, T.J., Taylor, J.M., Kardia, S., Yee, J., Orringer, M.B., Hanash, S., and Beer, D.G., Proteomic analysis of cytokeratin isoforms uncovers association with survival in lung adenocarcinoma, *Neoplasia*, 4, 440–448, 2002.

28. Hanash, S.M., Bobek, M.P., Rickman, D.S., Williams, T., Rouillard, J.M., Kuick, R., and Puravs, E., Integrating cancer genomics and proteomics in the post-genome era, *Proteomics*, 2, 69–75, 2002.

29. Ardekani, A.M., Liotta, L.A., and Petricoin, E.F., 3rd, Clinical potential of proteomics in the diagnosis of ovarian cancer, *Expert Rev. Mol. Diagn.*, 2, 312–320, 2002.

30. Sauter, E.R., Zhu, W., Fan, X.J., Wassell, R.P., Chervoneva, I., and Du Bois, G.C., Proteomic analysis of nipple aspirate fluid to detect biologic markers of breast cancer, *Br. J. Cancer*, 86, 1440–1443, 2002.

31. Hondermarck, H., Vercoutter-Edouart, A.S., Revillion, F., Lemoine, J., el-Yazidi-Belkoura, I., Nurcombe, V., and Peyrat, J.P., Proteomics of breast cancer for marker discovery and signal pathway profiling, *Proteomics*, 1, 1216–1232, 2001.

32. Pucci-Minafra, I., Fontana, S., Cancemi, P., Basirico, L., Caricato, S., and Minafra, S., A contribution to breast cancer cell proteomics: Detection of new sequences, *Proteomics*, 2, 919–927, 2002.

33. Wulfkuhle, J.D., Sgroi, D.C., Krutzsch, H., McLean, K., McGarvey, K., Knowlton, M., Chen, S., Shu, H., Sahin, A., Kurek, R., Wallwiener, D., Merino, M.J., Petricoin, E.F., 3rd, Zhao, Y., and Steeg, P.S., Proteomics of human breast ductal carcinoma *in situ*, *Cancer Res.*, 62, 6740–6749, 2002.

34. Wulfkuhle, J.D., McLean, K.C., Paweletz, C.P., Sgroi, D.C., Trock, B.J., Steeg, P.S., and Petricoin, E.F., 3rd, New approaches to proteomic analysis of breast cancer, *Proteomics*, 1, 1205–1215, 2001.

35. Petricoin, E.F., Ardekani, A.M., Hitt, B.A., Levine, P.J., Fusaro, V.A., Steinberg, S.M., Mills, G.B., Simone, C., Fishman, D.A., Kohn, E.C., and Liotta, L.A., Use of proteomic patterns in serum to identify ovarian cancer, *Lancet*, 359, 572–577, 2002.

36. Petricoin, E.F., 3rd, Ornstein, D.K., Paweletz, C.P., Ardekani, A., Hackett, P.S., Hitt, B.A., Velassco, A., Trucco, C., Wiegand, L., Wood, K., Simone, C.B., Levine, P.J., Linehan, W.M., Emmert-Buck, M.R., Steinberg, S.M., Kohn, E.C., and Liotta, L.A., Serum proteomic patterns for detection of prostate cancer, *J. Natl. Cancer Inst.*, 94, 1576–1578, 2002.

37. Adam, B.L., Vlahou, A., Semmes, O.J., and Wright, G.L., Jr., Proteomic approaches to biomarker discovery in prostate and bladder cancers, *Proteomics*, 1, 1264–1270, 2001.

38. Adam, B.L., Qu, Y., Davis, J.W., Ward, M.D., Clements, M.A., Cazares, L.H., Semmes, O.J., Schellhammer, P.F., Yasui, Y., Feng, Z., and Wright, G.L., Jr., Serum protein fingerprinting coupled with a pattern-matching algorithm distinguishes prostate cancer from benign prostate hyperplasia and healthy men, *Cancer Res.*, 62, 3609–3614, 2002.

39. Celis, J.E. and Gromov, P., Proteomics in translational cancer research: Toward an integrated approach, *Cancer Cell*, 3, 9–15, 2003.

40. Jessani, N., Liu, Y., Humphrey, M., and Cravatt, B.F., Enzyme activity profiles of the secreted and membrane proteome that depict cancer cell invasiveness, *Proc. Natl. Acad. Sci. USA*, 99, 10335–10340, 2002.

41. Tyers, M. and Mann, M., From genomics to proteomics, *Nature*, 422, 193–197, 2003.

42. Rabilloud, T., Two-dimensional gel electrophoresis in proteomics: Old, old fashioned, but it still climbs up the mountains, *Proteomics*, 2, 3–10, 2002.

43. Herbert, B.R., Harry, J.L., Packer, N.H., Gooley, A.A., Pedersen, S.K., and Williams, K.L., What place for polyacrylamide in proteomics?, *Trends Biotechnol.*, 19 (10 Suppl.), S3–S9, 2001.

44. Issaq, H.J., The role of separation science in proteomics research, *Electrophoresis*, 22, 3629–3638, 2001.

45. Knezevic, V., Leethanakul, C., Bichsel, V.E., Worth, J.M., Prabhu, V.V., Gutkind, J.S., Liotta, L.A., Munson, P.J., Petricoin, E.F., 3rd, and Krizman, D.B., Proteomic profiling of the cancer microenvironment by antibody arrays, *Proteomics*, 1, 1271–1278, 2001.

46. Santoni, V., Molloy, M., and Rabilloud, T., Membrane proteins and proteomics: Un amour impossible?, *Electrophoresis*, 21, 1054–1070 [pii], 2000.

47. Corthals, G.L., Wasinger, V.C., Hochstrasser, D.F., and Sanchez, J.C., The dynamic range of protein expression: A challenge for proteomic research, *Electrophoresis*, 21, 1104–1115, 2000.

48. Molloy, M.P., Two-dimensional electrophoresis of membrane proteins using immobilized pH gradients, *Anal. Biochem.*, 280, 1–10, 2000.

49. Wildgruber, R., Harder, A., Obermaier, C., Boguth, G., Weiss, W., Fey, S.J., Larsen, P.M., and Gorg, A., Towards higher resolution: Two-dimensional electrophoresis of Saccharomyces cerevisiae proteins using overlapping narrow immobilized pH gradients, *Electrophoresis*, 21, 2610–2616, 2000.

50. Lilley, K.S., Razzaq, A., and Dupree, P., Two-dimensional gel electrophoresis: Recent advances in sample preparation, detection and quantitation, *Curr. Opin. Chem. Biol.*, 6, 46, 2002.

51. Zuo, X. and Speicher, D.W., Comprehensive analysis of complex proteomes using microscale solution isoelectrofocusing prior to narrow pH range two-dimensional electrophoresis, *Proteomics*, 2, 58–68, 2002.

52. Hoving, S., Gerrits, B., Voshol, H., Muller, D., Roberts, R.C., and van Oostrum, J., Preparative two-dimensional gel electrophoresis at alkaline pH using narrow range immobilized pH gradients, *Proteomics*, 2, 127–134, 2002.

53. Hoving, S., Voshol, H., and van Oostrum, J., Towards high performance two-dimensional gel electrophoresis using ultrazoom gels, *Electrophoresis*, 21, 2617–2621, 2000.

54. Gorg, A., Obermaier, C., Boguth, G., Harder, A., Scheibe, B., Wildgruber, R., and Weiss, W., The current state of two-dimensional electrophoresis with immobilized pH gradients, *Electrophoresis*, 21, 1037–1053, 2000.

55. Sinha, P., Poland, J., Schnolzer, M., and Rabilloud, T., A new silver staining apparatus and procedure for matrix-assisted laser desorption/ionization-time of flight analysis of proteins after two-dimensional electrophoresis, *Proteomics*, 1, 835–840, 2001.

56. Patton, W.F. and Beechem, J.M., Rainbow's end: The quest for multiplexed fluorescence quantitative analysis in proteomics, *Curr. Opin. Chem. Biol.*, 6, 63–69, 2002.

57. Tonge, R., Shaw, J., Middleton, B., Rowlinson, R., Rayner, S., Young, J., Pognan, F., Hawkins, E., Currie, I., and Davison, M., Validation and development of fluorescence two-dimensional differential gel electrophoresis proteomics technology, *Proteomics*, 1, 377–396, 2001.

58. Unlu, M., Morgan, M.E., and Minden, J.S., Difference gel electrophoresis: A single gel method for detecting changes in protein extracts, *Electrophoresis*, 18, 2071–2077, 1997.

59. Zhou, G., Li, H., DeCamp, D., Chen, S., Shu, H., Gong, Y., Flaig, M., Gillespie, J.W., Hu, N., Taylor, P.R., Emmert-Buck, M.R., Liotta, L.A., Petricoin, E.F., 3rd, and Zhao, Y., 2D differential in-gel electrophoresis for the identification of esophageal scans cell cancer-specific protein markers, *Mol. Cell Proteomics*, 1, 117–124, 2002.

60. Gharbi, S., Gaffney, P., Yang, A., Zvelebil, M.J., Cramer, R., Waterfield, M.D., and Timms, J.F., Evaluation of two-dimensional differential gel electrophoresis for proteomic expression analysis of a model breast cancer cell system, *Mol. Cell Proteomics*, 1, 91–98, 2002.

61. Pluskal, M.G., Bogdanova, A., Lopez, M., Gutierrez, S., and Pitt, A.M., Multiwell in-gel protein digestion and microscale sample preparation for protein identification by mass spectrometry, *Proteomics*, 2, 145–150, 2002.

62. Katayama, H., Nagasu, T., and Oda, Y., Improvement of in-gel digestion protocol for peptide mass fingerprinting by matrix-assisted laser desorption/ionization time-of-flight mass spectrometry, *Rapid Commun. Mass Spectrom.*, 15, 1416–1421, 2001.

63. Karas, M., Bachmann, D., Bahr, U., and Hillenkamp, F., Matrix-assisted ultraviolet laser desorption of non-volatile compounds, *Int. J. Mass Spectrom Ion Processes*, 78, 53, 1987.

64. Karas, M., Bachmann, D., and Hillenkamp, F., Influence of the wavelength in high-irradiance ultraviolet laser desorption mass spectrometry of organic molecules, *Anal. Chem.*, 57, 2935, 1985.

65. Karas, M., Bahr, U., and Hillenkamp, F., UV laser matrix desorption/ionization mass spectrometry of proteins in the 100,000 Dalton range, *Int. J. Mass Spectrom Ion Processes*, 92, 231, 1989.

66. Karas, M. and Hillenkamp, F., Laser desorption ionization of proteins with molecular masses exceeding 10,000 Daltons, *Anal. Chem.*, 60, 2299–2301, 1988.

67. Yamashita, M. and Fenn, J.B., Electrospray ion source. Another variation of the free-jet theme, *J. Phys. Chem.*, 88, 4451, 1984.

68. Whitehouse, C.M., Dreyer, R.N., Yamashita, M., and Fenn, J.B., Electrospray interface for liquid chromatographs and mass spectrometers, *Anal. Chem.*, 57, 675, 1985.

69. Mann, M., Hendrickson, R.C., and Pandey, A., Analysis of proteins and proteomes by mass spectrometry, *Annu. Rev. Biochem.*, 70, 437–473, 2001.

70. Foret, F. and Preisler, P., Liquid phase interfacing and miniaturization in matrix-assisted laser desorption/ionization mass spectrometry, *Proteomics*, 2, 360, 2002.

71. Clauser, K.R., Baker, P., and Burlingame, A.L., Role of accurate mass measurement ($\pm$10 ppm) in protein identification strategies employing MS or MS/MS and database searching, *Anal. Chem.*, 71, 2871–2882, 1999.

72. Roepstorff, P. and Fohlman, J., Proposal for a common nomenclature for sequence ions in mass spectra of peptides, *Biomed. Mass Spectrom*, 11, 601, 1984.

73. Biemann, K., Contributions of mass spectrometry to peptide and protein structure, *Biomed. Environ. Mass Spectrom.*, 16, 99, 1988.

74. Borchers, C., Parker, C.E., Deterding, L.J., and Tomer, K.B., A preliminary comparison of precursor scans and LC/MS/MS on a hybrid quadrupole time-of-flight mass spectrometer, *J. Chromatogr.*, 854, 119, 1999.

75. Washburn, M.P., Wolters, D., and Yates, J.R., 3rd, Large-scale analysis of the yeast proteome by multidimensional protein identification technology, *Nat. Biotechnol.*, 19, 242–247, 2001.

76. Davis, M.T., Stahl, D.C., Hefta, S.A., and Lee, T.D., A microscale electrospray interface for on-line, capillary liquid chromatography/tandem mass spectrometry of complex peptide mixtures, *Anal. Chem.*, 67, 4549–4556, 1995.

77. Eng, J.K., McCormack, A.L., and Yates, J.R., An approach to correlate tandem mass spectral data of peptides with amino acid sequences in a protein database, *J. Am. Soc. Mass Spect.*, 5, 976–989, 1994.

78. Dancik, V., Addona, T.A., Clauser, K.R., Vath, J.E., and Pevzner, P.A., De novo peptide sequencing via tandem mass spectrometry, *J. Comput. Biol.*, 6, 327–342, 1999.

79. Taylor, J.A. and Johnson, R.S., Sequence database searches via de novo peptide sequencing by tandem mass spectrometry, *Rapid Commun. Mass Spectrom.*, 11, 1067–1075, 1997.

80. Chen, T., Kao, M.Y., Tepel, M., Rush, J., and Church, G.M., A dynamic programming approach to de novo peptide sequencing via tandem mass spectrometry, *J. Comput. Biol.*, 8, 325–337, 2001.

81. Gygi, S.P., Rist, B., Gerber, S.A., Turecek, F., Gelb, M.H., and Aebersold, R., Quantitative analysis of complex protein mixtures using isotope-coded affinity tags, *Nat. Biotechnol.*, 17, 994–999. java/Propub/biotech/nbt1099_994.fulltext java/Propub/biotech/nbt1099_994.abstract, 1999.

82. Young, M.M., Tang, N., Hempel, J.C., Oshiro, C.M., Taylor, E.W., Kuntz, I.D., Gibson, B.W., and Dollinger, G., High throughput protein fold identification by using experimental constraints derived from intramolecular cross-links and mass spectrometry, *Proc. Natl. Acad. Sci. USA*, 97, 5802–5806, 2000.

83. Chen, T., GeneFinding via tandem mass spectrometry, in *Proceedings of the ACM-SIGACT Fifth Annual International Conference on Computational Molecular Biology (RECOMB01)*, 2001, pp. 85–92.

84. Mann, M. and Wilm, M., Error-tolerant identification of peptides in sequence databases by peptide sequence tags, *Anal. Chem.*, 66, 4390–4399, 1994.

85. Bafna, V. and Edwards, N., SCOPE: A probabilistic model for scoring tandem mass spectra against a peptide database, *Bioinformatics*, 17 (Suppl. 1), S13–S21, 2001.

86. Perkins, D.N., Pappin, D.J., Creasy, D.M., and Cottrell, J.S., Probability-based protein identification by searching sequence databases using mass spectrometry data, *Electrophoresis*, 20, 3551–3567, 1999.

87. Pevzner, P.A., Dancik, V., and Tang, C.L., Mutation-tolerant protein identification by mass spectrometry, *J. Comput. Biol.*, 7, 777–787, 2000.

88. Reese, M.G., Hartzell, G., Harris, N.L., Ohler, U., Abril, J.F., and Lewis, S.E., Genome annotation assessment in *Drosophila melanogaster*, *Genome Res.*, 10, 483–501, 2000.

89. Comen, T.H., Leiserson, C.E., and Rivest, R.L., *Introduction to Algorithms*, The MIT Press, 1990.

90. Bailey-Kellogg, C., Kelley, J.J., Stein, C., and Donald, B.R., Reducing mass degeneracy in SAR by MS by stable isotopic labeling, *J. Comp. Biol.*, 8, 19–36, 2001.

91. Uetz, P., Giot, L., Cagney, G., Mansfield, T.A., Judson, R.S., Knight, J.R., Lockshon, D., Narayan, V., Srinivasan, M., Pochart, P., Qureshi-Emili, A., Li, Y., Godwin, B., Conover, D., Kalbfleisch, T., Vijayadamodar, G., Yang, M., Johnston, M., Fields, S., and Rothberg, J.M., A comprehensive analysis of protein–protein interactions in *Saccharomyces cerevisiae*, *Nature*, 403, 623–627, 2000.

92. Ito, T., Chiba, T., Ozawa, R., Yoshida, M., Hattori, M., and Sakaki, Y., A comprehensive two-hybrid analysis to explore the yeast protein interactome, *Proc. Natl. Acad. Sci. USA*, 98, 4569–4574, 2001.

93. Ito, T., Tashiro, K., Muta, S., Ozawa, R., Chiba, T., Nishizawa, M., Yamamoto, K., Kuhara, S., and Sakaki, Y., Toward a protein–protein interaction map of the budding yeast: A comprehensive system to examine two-hybrid interactions in all possible combinations between the yeast proteins, *Proc. Natl. Acad. Sci. USA*, 97, 1143–1147, 2000.

94. Enright, A.J., Iliopoulos, I., Kyrpides, N.C., and Ouzounis, C.A., Protein interaction maps for complete genomes based on gene fusion events, *Nature*, 402, 86–90, 1999.

95. Marcotte, E.M., Pellegrini, M., Ng, H.L., Rice, D.W., Yeates, T.O., and Eisenberg, D., Detecting protein function and protein–protein interactions from genome sequences, *Science*, 285, 751–753, 1999.

96. Ho, Y., Gruhler, A., Heilbut, A., Bader, G.D., Moore, L., Adams, S.L., Millar, A., Taylor, P., Bennett, K., Boutilier, K., Yang, L., Wolting, C., Donaldson, I., Schandorff, S., Shewnarane, J., Vo, M., Taggart, J., Goudreault, M., Muskat, B., Alfarano, C., Dewar, D., Lin, Z., Michalickova, K., Willems, A.R., Sassi, H., Nielsen, P.A., Rasmussen, K.J., Andersen, J.R., Johansen, L.E., Hansen, L.H., Jespersen, H., Podtelejnikov, A., Nielsen, E., Crawford, J., Poulsen, V., Sorensen, B.D., Matthiesen, J., Hendrickson, R.C., Gleeson, F., Pawson, T., Moran, M.F., Durocher, D., Mann, M., Hogue, C.W., Figeys, D., and Tyers, M., Systematic identification of protein complexes in *Saccharomyces cerevisiae* by mass spectrometry, *Nature*, 415, 180–183, 2002.

13 Overview of Human Cancer Induction and Human Exposure to Carcinogens

David Warshawsky and Joseph R. Landolph, Jr.

CONTENTS

13.1 INTRODUCTION

The purpose of this chapter is to give the reader an overview of human cancer induction and the proportional causes of human cancer. Cancer is a group of diseases characterized by uncontrolled growth and spread of abnormal cells, and can result in death [1–4]. It is estimated that approximately 30% of human cancer is caused by exposure to tobacco smoking and tobacco products (Figure 13.1; reviewed in Reference 3). The cancers induced by tobacco smoking include lung cancer, esophageal cancer, laryngeal cancer, pancreatic cancer, bladder cancer, and cancer of the oral cavity (reviewed in References 3 and 5).

Approximately another 15% of human cancer is thought to be due to various factors in human diets (Figure 13.1; reviewed in References 3, 5 to 9). These cancers include cancers of the gastrointestinal tract — stomach cancer, colon cancer, and rectal cancer — and nasopharyngeal cancer [3]. The various factors include consumption of an excess of animal fat (reviewed in Reference 3), excess of caloric intake [3,5–9], consumption of too few fruits and green-leafy and yellow vegetables and too little dietary fiber (reviewed in References 3, 5 to 8), and the presence of carcinogens in certain diets, such as mutagens in certain cooked foods [6–8]. A number of cooked-food mutagens are pyrolytic decomposition products of the amino acid, tryptophan [9,10]. Colon cancer in particular is believed to be related to an excess of animal fat and a deficiency of fiber in the diet (reviewed in Reference 3). An excess of calories in the diet is related to endometrial cancer (reviewed in Reference 3). Excessive alcohol intake correlates with increased risk of pharyngeal cancer, laryngeal cancer, liver cancer, esophageal cancer, and cancer of the oral cavity, colorectal cancer, and breast cancer (reviewed in Reference 3).

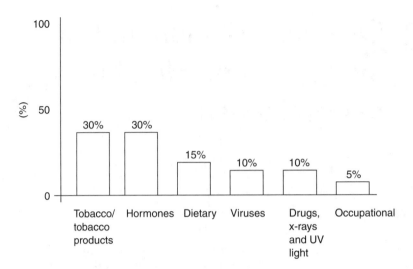

FIGURE 13.1   Causes of human cancer.

The influence of hormones is thought to be associated with approximately 30% of new cancer cases (Figure 13.1). These include endometrial cancer, which is caused by estrogens without progestogens (reviewed in Reference 3) and breast cancer, which is caused by estrogen (progesterone can increase this effect) (reviewed in Reference 3). Ovarian cancer can be induced by ovulation and its accompanying hormonal changes (reviewed in Reference 3). Prostate cancer arises due to the effects of testosterone or its metabolite, dihydrotestosterone, as mitogens for prostate tissue [3].

A certain fraction of human cancer is thought to be caused by oncogenic viruses, such as hepatitis B virus causing and contributing to liver cancer, an effect that can be synergistically enhanced by aflatoxin B1, a metabolite of the mold, *Aspergillus flavus* [3,11]. In addition, specific papilloma viruses cause cervical cancer [3,11]. Epstein Barr virus contributes to the induction of Burkitt's lymphoma (reviewed in References 3 and 11). Another example of viral carcinogenesis in humans is the linkage of infection with human T cell lymphotropic virus type 1 and specific T cell lymphomas [3]. The contribution of oncogenic viruses to human cancer is estimated to be approximately 10% (Figure 13.1; Reference 3).

An additional 10% of human cancer is estimated to be due to medical drugs and exposure to radiations, such as medical and dental x-rays, and UV light (Figure 13.1; reviewed in Reference 3). These include cancers induced by medical x-rays, secondary tumors induced by specific cancer chemotherapeutic agents, skin tumors induced by UV light, and tumors induced by specific drugs. Phenacetin was previously used as an analgesic and is carcinogenic to the lower urinary tract in humans (reviewed in Reference 3). Analgesics, including aspirin, phenacetin, and acetaminophen, have often been used in combinations in humans, and such usage correlates with an increased risk of cancer of the renal pelvis and ureter [12]. Acetaminophen is the major metabolite of phenacetin in mammals, and both phenacetin and acetaminophen can induce morphological transformation of cultured C3H/10T1/2 Cl 8 mouse embryo fibroblasts [13].

A final 5% of human cancer is estimated to be due to exposure to carcinogens in the workplace (occupationally induced cancer) (Figure 13.1; reviewed in Reference 3). A few of the numerous examples of occupational carcinogens include asbestos, which induces lung cancer and mesothelioma [3], benzene, which induces acute myelogenous leukemia, and hexavalent chromium, which can induce nasal and respiratory cancer in the chromate manufacturing and chrome plating industries (reviewed in References 14 to 18). Environmental carcinogens, such as hexavalent chromium ion, trichloroethylene, and perchloroethylene, also contribute to human cancer. However, no solid estimates of the fraction of human cancer that environmental carcinogens cause have yet been made.

Therefore, at least 45% of all human cancers are caused at least in part by chemical and radiation carcinogens (30% due to tobacco and tobacco products plus 10% due to drugs, x-rays, and sunlight plus 5% occupationally induced cancer) [3]. In addition, chemical mutagens arising in the diet formed from the cooking of foods also contribute to the fraction of human cancers caused by chemical carcinogens. Therefore, chemical carcinogens and radiation contribute to greater than 45% of the human cancer burden. It is stunning that current estimates are that 30% of all human cancers can be eliminated by elimination of exposure to tobacco and tobacco products. Lastly, several major cancers are associated with lifestyles that include obesity, inappropriate diets, and physical inactivity [1,3]. Hence, at least 60% of all human cancers are in theory preventable. This could be accomplished by removing exposure to tobacco and tobacco products (30%), by controlling caloric intake, adding the proper amount of dietary fiber to the diet, minimizing intake of animal fat and maximizing micronutrient and vegetable intake (15%), by controlling occupational exposure to carcinogens (5%), and by avoiding excess exposure to specific drugs, x-rays, and UV light (10%) (reviewed in References 3, 6 to 10).

13.2 CHARACTERISTICS OF HUMAN CANCER: UNCONTROLLED CELL GROWTH, BENIGN AND MALIGNANT TUMORS, INVASION, AND METASTASIS

Cells are the structural units of mammals, including humans. Each of us has trillions of cells. Cells cooperate structurally, metabolically, and functionally to form tissue networks and organs. In addition, the most fundamental characteristic of cells is their ability to reproduce and divide, making two daughter cells (Figure 13.2). The division of normal, healthy cells occurs in a regulated and systematic fashion. Normal cells continually divide to form new daughter cells to supply the material for growth and to replace injured, mutated, or dead cells (Figure 13.2; References 14 and 18 to 21).

However, all of these useful functions can only be conducted by normal cells that cooperate metabolically and structurally with their neighboring cells to form tissues and organs. Some cells stop cooperating with their neighboring cells, and become autonomous in their growth or are transformed (Figure 13.2). Some of the transformed cells eventually become tumor cells, and some of these tumor cells can later become cancer cells (Figure 13.2; References 11, 14, 15, 18 to 20, and 22 to 24).

There are two generic classes of tumors. The first type is the benign tumor. Benign tumors are frequently enclosed or encapsulated in a protective capsule of tissue [2,19,20,22,25]. A benign tumor remains encapsulated and localized to its origin. It does not spread to other parts of the body [25]. Benign tumors in general are not lethal. However, under certain conditions, benign tumors can occur in areas controlling important biological functions. An example would be a benign tumor of the brain, which could exert pressure against critical brain tissue controlling important brain functions, leading to death.

Malignant tumors can destroy the part of the body in which they originate. They can also invade neighboring tissue, and can later spread or metastasize to other parts of the human body [26], where they can initiate the growth of new secondary tumors, cause angiogenesis to occur to establish the blood supply of the new tumor [22], and induce further destruction of additional tissues and organs [25]. This characteristic distinguishes cancer from benign growths. Benign tumors may grow quite large and press on neighboring structures, but they do not spread to other parts of the body [25].

Malignant tumor cells, or cancer cells, can destroy their tissue of origin, and invade the surrounding tissues and through them into neighboring tissues [25]. In addition, in contrast to normal cells, cancer cells divide in a haphazard manner and typically pile up into a nonstructured mass or tumor [14,18–20,23,24,27]. Cancer cells can eventually spread or metastasize toward distant tissue,

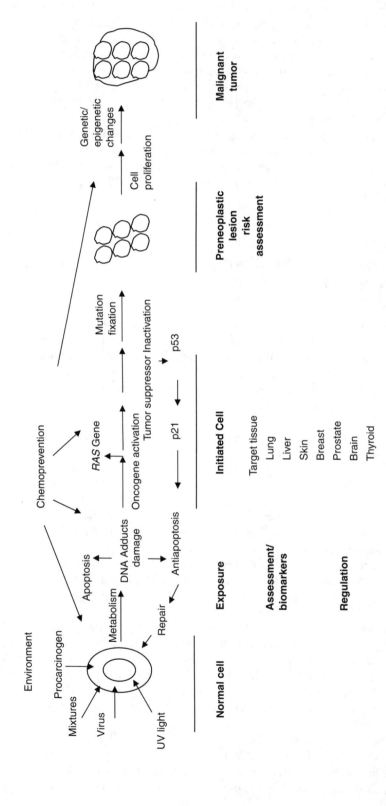

FIGURE 13.2 Mechanisms of chemical carcinogenesis.

and can destroy that distant tissue [25]. During metastasis, cancer cells escape from the primary tumor, enter the bloodstream or lymphatic system, and spread to nearby parts of the body and eventually to sites far away from the original tumor [25].

There are a number of different types of malignant tumors. Carcinomas are malignant tumors that develop in the tissue that lines the surfaces of certain organs such as the lung, liver, skin, and breast. Carcinomas constitute 92% of the human tumor burden world-wide. Sarcomas are malignant tumors that arise from mesenchymal cells, including cells in the bone, cartilage, fat, connective tissues, and muscle. Leukemias are cancers of the white blood cells. Together, leukemias and sarcomas comprise approximately 8% of the total human tumor burden. Lymphomas are cancers that develop in the lymphatic system.

Melanoma is a cancer of melanocytes, which make the pigment melanin. When melanocytes are transformed into cancer cells, they can form tumors in the skin, eyes, and other tissues. If a melanoma is identified early and surgically excised completely, it then presents no further clinical problems. However, if during surgical excision, some of this tumor is missed, the unexcised tumor will then metastasize to distal sites, and will frequently kill the patient.

13.3 THE HUMAN CANCER BURDEN

Heart disease is currently the leading cause of death in humans in the United States, followed closely by cancer deaths (reviewed in References 3 and 28). The rate of deaths due to heart disease in the United States is actually decreasing with time, because the U.S. population is exercising more, controlling their diets both as to total calories and the components of the diets, are also working to lower their cholesterol levels by dietary and pharmacological interventions, and are also controlling hypertension better (reviewed in References 3 and 28).

However, the cancer incidence in the United States has been increasing over the last 14 years due to the increase in the incidence of lung cancer in women. The American Cancer Society (ACS) has reported the Age-Adjusted Cancer Death Rates, Male and Female, at various organ sites in the United States from the period 1930 to 2000. Some of that data is reported in Table 13.1 and in Figure 13.3 [4]. The annual order of occurrence of the various cancers in humans is as follows: prostate cancer in males, with an incidence of 230,110 > breast cancer in females, with an incidence of 215,990 > cancer of the lungs and bronchus in males and females combined (173,770) > cancer in the colon and rectum in males and females combined (145,940) > lymphomas (62,250) > cancer of the urinary bladder in male and females combined (60,240) > melanoma (55,100) > leukemias (33,440) > cancer of the oral cavity and pharynx (38,260) > cancer of the ovary (25,580) > uterine cervix cancer (10,520) (Table 13.1) [3,4].

Some types of cancer are more treatable and curable than others. Hence, the order of incidence of various cancers does not precisely reflect the mortality due to cancers at the same sites. The annual order of mortalities of the various cancers by sites hence is lung and bronchial cancer (160,449) > cancer of the colon and rectum (56,730) > urinary bladder cancer (53,320) > breast cancer (40,110) > prostate cancer (29,500) > leukemia (32,200) > lymphomas (20,730) > ovarian cancer (16,090) > melanoma (7,910) > cancer of the oral cavity/pharynx (7,230) > uterine cancer (3,900) [4].

Lung and bronchus sites for both male and female have the highest death rates from cancer of all major sites (Figure 13.3; References 3 and 4): For males, the incidence of lung cancer has reached a plateau, and is thought to be in the declining phase as males have begun to stop smoking. Males began to smoke in the work force in the United States in large numbers much earlier than females. For females, who entered the work force later in large numbers in the United States around World War II, the lung cancer incidence curve rose later in time than that for males. The incidence curve for lung cancer in females is still in their ascending phase, due to smoking in females in the United States (Figure 13.3; References 3 and 4). The annual male lung-cancer death rate was approximately

TABLE 13.1

Occurrence, Estimated Deaths, Symptoms, Risk Factors, Early Detection, and Treatment of Selected Human Cancers

Selected cancer	Female estimated new cases	Female estimated deaths	Male estimated new cases	Male estimated deaths	Symptoms	Risk factors	Early detection	Treatment
Breast	215,990	40,110	—	—	Abnormality in mammary glands	Age, genetics	Mammography	Lumpectomy or mastectomy and removal of the lymph nodes under the arm
Lung and bronchus	80,660	68,510	93,110	91,930	Persistent coughs, sputum, chest pain	Cigarette smoking	Has not yet been demonstrated	Surgery, radiation therapy, and chemotherapy
Prostate	—	—	230,110	29,500	No symptoms	Age, ethnicity, and genetics	PSA blood test	Hormonal therapy, chemotherapy, and radiation
Colon and rectum	73,320	28,410	73,620	28,320	Change in bowl habits	90% over age 50	Should take blood test at age 50	Surgery
Ovary	25,580	16,090	—	—	Enlargement of abdomen	Body weight and peaks at age 70	No accurate screening test available	Surgery, radiation therapy, and chemotherapy
Urinary bladder	15,600	3,930	44,640	8,780	Blood and urine, increased frequency of urination	Smoking	Examination of bladder with a cystoscope	Surgery

(continued)

				Symptoms	Risk factors	Detection	Treatment
Leukemia	14,420	10,310	19,020	12,990	Fatigue, paleness, and weight loss. Bruises easily and nose bleeds	Myeloid leukemia is associated with benzene, cigarette smoke, and exposure to ionizing radiation	Difficult to diagnose	Chemotherapy
Lymphoma	29,070	9,640	33,180	11,090	Enlarged lymph nodes, fever, sweats, and weight loss	Factors largely unknown		Chemotherapy
Oral cavity and pharynx	9,710	2,400	18,550	4,830	Sores that bleed and do not heal easily	Cigarette and cigar or pipe smoking, smokeless tobacco		Radio therapy and surgery
Melanoma-skin	25,200	2,860	29,900	5,050	Any change in the skin in size, shape, or color	Genetics	Recognizing changes in skin growths or appearance of new growths	Excision of tissues, lymph nodes
Uterine cervix	10,520	3,900			Symptoms do not appear until cervical cells become cancerous. Abnormal vaginal bleeding	Linked to sexually transmitted infections	PAP test	Surgery or radiation
Uterine corpus	40,320	7,980			Abnormal urine bleeding	High cumulative exposure to estrogen	Post menopausal bleeding	Surgery, radiation, hormones, and chemotherapy

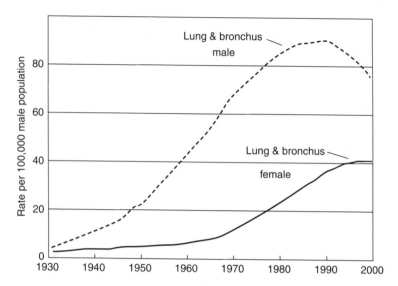

FIGURE 13.3   Age-adjusted cancer death rates, males and females, by site, United States, 1930–2000.

80 per 100,000 male population, whereas the female lung-cancer death rate was approximately one half that of the male, or 40 per 100,000 population, in the year 2000 [4].

For the most recent data in the year 2004, for cancer of the lung and bronchus, it is estimated there are to be 93,110 new cases of cancer and 91,930 deaths in males and 80,660 new cases of cancer and 68,510 new deaths in females. In males, it is estimated there will be 230,110 new cases of prostate cancer and 29,500 estimated deaths from prostate cancer in 2004. For females, it is estimated there will be 215,990 new cases of breast cancer and 40,110 deaths due to breast cancer in 2004. For cancers of the colon and rectum, it is estimated there will be 73,620 new cases of colon and rectal cancer and 28,410 deaths. In males, it is estimated to be 73,260 new cases of colon and rectal cancer and 28,320 deaths in 2004. The probable reason for the large number of new cases and estimated deaths for cancers of the breast, prostate, lung and bronchus, and colon and rectum is due to early detection. Additional information on this topic is found in Table 13.1 [3,4]. According to the American Association for Cancer Research, survival rates for cancer have increased from 40% in 1971 to 65% today [4]. However, there clearly is still a long way to go to eradicate cancer as a substantial cause of death in society. An additional approach is to identify and to eliminate exposure of humans to chemical, viral, and radiation carcinogens to the greatest extent practicable [3,4].

13.4   FACTORS CAUSING CANCER

13.4.1   FACTORS OUTSIDE THE BODY — EXOGENOUS CARCINOGENS

What causes cancer in humans, and what percentage does each human carcinogen identified contribute to the total human cancer burden? Factors both within (endogenous factors, such as hormones) and outside (exogenous factors, such as occupational carcinogens) the human body contribute to the development of human cancer. There are differences in cancer rates around the world and change in the cancer rates when people move from country to country, and this gives us clues to the origins of human cancer. For example, Asians, including Japanese, have low incidences of prostate and breast cancer and high rates of stomach cancer in their country of origin [3]. When Asian peoples immigrate to the United States, their prostate and breast cancer incidences increase over time until they are nearly the same rates as that found in the United States. It is thought that dietary components may account in part for these differences [3]. By the same token, the incidence of stomach cancer in the Asian immigrants to the United States tends to decrease to the same level of incidences present in the population of the United States [3]. Lack of physical exercise, poor diet (excessive red meat, and lack of green-leafy and yellow vegetables), and being overweight tend to play a role in breast and prostate cancer [3]. Infection with the *Helicobacter pylori* bacterium, a diet high in salt, and probable dietary carcinogens are causative factors in the induction of stomach cancer in humans (reviewed in Reference 3). Interestingly, the recent increase in the incidence of colorectal cancer in China and Japan suggests an environmental component associated with this type of cancer.

Specific types of human cancer are associated with exposures to specific carcinogens, as determined in epidemiological studies. For example, aromatic amines are associated with bladder cancer, while cigarette smoking is linked to pancreas, liver, cervix, throat, larynx, lip, stomach, kidney, colon, lung, and bladder cancers (reviewed in Reference 3). People exposed to carcinogenic mixtures such as cigarette smoking do not always contract lung cancer, and not all women who are infected with human papilloma virus develop cervical cancer [1,11]. This is because induction of human cancer is thought to be a probabilistic and stochastic process.

Quantitatively, what are thought to be the major causes of human cancer? Epidemiological studies give us information on this topic. A major cause of human cancer is thought to be cigarette smoking and exposure to tobacco and tobacco products [3]. This is thought to account for approximately 30% of all human cancer (Figure 3.1). Human cancers thought to be induced by exposure to tobacco and tobacco products include lung cancer, bladder cancer, pancreatic cancer, liver cancer, cervical cancer, laryngeal cancer, cancer of the lip, stomach cancer, kidney cancer, and colon cancer (reviewed in Reference 3).

A second major contributor to induction of human cancer is the category referred to as dietary factors (reviewed in References 3, 5 to 10). This category embodies many different factors. Excess caloric content and a diet high in red meats are believed to contribute to cancer induction. A paucity of green-leafy and yellow vegetables is correlated with a higher cancer incidence, and diet higher in green-leafy and yellow vegetables protects against various types of cancers (reviewed in References 3, 5 to 10, and 29). Dietary factors are thought to be responsible for approximately 15% of the human cancer burden (Figure 13.1; reviewed in References 3, 5 to 10).

There are also animal models that confirm many of the findings of the involvement of dietary factors in cancer causation. Kroes et al. [30] showed that hamsters on a high-fat diet had a higher yield of BaP-induced respiratory tumors than animals on a low-fat diet, in terms of the number of tumor-bearing animals and the multiplicity of respiratory tract tumors being increased in the high-fat-diet groups. This same group also investigated the effects of various fat-containing and fiber-containing diets on N-methyl-N'-nitro-N-nitroso-guanidine (MNNG) and dimethylhydrazine (DMH)-induced colon tumors in rats. They found that high-fat, low-fiber diets caused the greatest incidence of MNNG-induced colon tumors in Wistar rats. High-fat, medium-fiber diets also caused a high incidence of MNNG-induced colon tumors. Similarly, they found that a high-fat, low-fiber diet and a high-fat, medium-fiber diets also caused the largest number of colon tumors in DMH-treated Wistar rats [29]. These authors also noted that dietary factors also influence levels of circulating hormones in animals, indicating that dietary factors and hormonal factors are linked in carcinogenesis.

Infection with oncogenic viruses is also thought to contribute to the human cancer incidence [3,11]. This includes infections with hepatitis B virus, with or without aflatoxin B1 exposure, which causes hepatocellular carcinoma in The People's Republic of China, Africa, and Mozambique [3,11]. Infection with specific papilloma viruses leads to cervical carcinoma [1,3,11]. Infection with Epstein Barr virus leads to Burkitt's lymphoma, and also to nasopharyngeal carcinoma in The People's Republic of China, when certain dietary factors are present, such as consumption of Cantonese style salted fish [3,11]. Infection with Human T Cell Leukemia Virus (HTLV) causes leukemia in the Caribbean and in Japan [11]. We also note at this point that oncogenic viruses and chemical carcinogens can act together synergistically to induce specific human cancers [3,11]. The contribution of oncogenic viruses to human cancer is estimated at approximately 10% (Figure 13.1; derived from data in Reference 3).

Iatrogenic or medical causes of cancer, such as diagnostic x-rays and specific drugs, plus UV light from the sun, are also thought to contribute to human cancer. These factors together are thought to be responsible for approximately 10% of all human cancers (see chapter 10; reviewed in Reference 3). A number of cancer chemotherapeutic agents may eradicate a primary tumor, but can cause secondary tumors. Bleomycin, adriamycin, cyclophosphamide [3], and arsenic compounds [14,15,18] are all cancer chemotherapeutic agents that are also carcinogenic. However, the contribution of iatrogenic factors (drugs and x-rays) and UV light to human cancer induction is thought to be small, approximately 10%. It is likely to decrease with time due to use of newer and more efficacious cancer chemotherapeutic agents and drugs with fewer side effects and lesser or no carcinogenicity, better shielding of x-ray machines and lesser x-ray doses delivered with today's efficient x-ray machines, which deliver less doses of x-rays to patients than in earlier times.

Finally, occupational causes of cancer are thought to contribute approximately 5% of the total human cancer incidence (Figure 13.2). There is controversy over this estimate of 5%. The range of contribution that has been discussed has been from 2 to 32%. Since 2% appears low to us and 32% appears overly high, and many scientists agree that 5% is a reasonable figure for the contribution of occupational factors to human cancer incidence. Among the occupational carcinogens, benzene exposure causes acute myelogenous leukemia (AML) in humans. Exposure to vinyl chloride in the workplace has caused angiosarcoma, a rare tumor of the liver, among the workers. Occupational exposure to compounds containing hexavalent chromium has led to nasal and respiratory cancer (reviewed in References 15 to 18). Occupational exposure to mixtures of soluble and insoluble nickel compounds in nickel refinery workers who smoked cigarettes, has led to nasal and lung tumors (reviewed in References 14 to 16, 18, 23, and 26). In copper smelting operations, exposure of the workers to arsenic trioxide, formed when arsenic-containing copper ores are smelted, has led to lung cancer in the workers (reviewed in References 14, 15, and 18).

It is believed that environmental pollution with chemical carcinogens also contributes to the human cancer burden. However, no firm quantitative estimates as to how much this contributes to human cancer have been made. Examples of occupational chemical carcinogens that have also contaminated the environment, from industrial processes, include asbestos, arsenic compounds [14–18], nickel compounds [14–16,18,24,26,27], compounds containing hexavalent chromium [14–18], trichloroethylene (previously used as a degreaser of metal tools in industrial operations), and perchloroethylene (used in the dry-cleaning industry). In the State of California, serpentine is the state mineral, and is found in large deposits in the state. It can be aerosolized into the air when hikers walk on this material, and also when buildings containing asbestos are improperly remediated or destroyed. Benzene can arise from a number of industrial processes and also from gasoline, and is a toxic and carcinogenic air contaminant. Other chemical carcinogens, including natural products, are present in food [1,9,10].

13.4.2 FACTORS INSIDE THE BODY (ENDOGENOUS FACTORS) THAT CAUSES HUMAN CANCER

Certain factors inside the body, or endogenous factors, also contribute to human cancer development. Estrogens are a cause of human breast cancer, and testosterone and its metabolites are thought to contribute to human prostate cancer (reviewed in Reference 3). In addition, some people either inherit altered genes in the body's cells, such as the mutated Retinoblastoma (Rb) gene, which predisposes to induction of retinoblastoma, a tumor of the eye, in affected individuals (See Chapter 6). Some individuals have the disease Xeroderma Pigmentosum (XP), which involves mutated or missing genes in the DNA excision pathways and predisposes these individuals to sunlight-induced skin cancer. Others have a mutated ATM gene, which fails to cause mutated cells to pause in the cell cycle so they can be repaired properly. Certain individuals can have an intrinsically weakened immune system, as in the case of Severe Combined Immune Deficiency (SCID), or in the case of infection by the Human Immunodeficiency Virus (HIV). Each of these endogenous factors may make an individual more susceptible to cancer [1,19,20].

13.5 INTERACTION OF ENVIRONMENTAL FACTORS AND GENES

The chance that an individual will develop cancer in response to exposure to a particular environmental agent depends on several interacting factors — how long and how often a person is exposed to a particular carcinogen, his/her exposure to other modifying agents, such as tumor promoters, cocarcinogens, or inhibitors of carcinogenesis, genetic factors, diet, lifestyle, health, age, and gender (reviewed in Reference 3). There are specific combinations of genetic alterations and environmental carcinogen exposures that make people either more susceptible or more resistant to cancer. Consumption of factors such as fruits and green-leafy and yellow vegetables and vitamins in the diet help protect against specific cancers (reviewed in Reference 3). One of the challenging areas of research today is trying to identify the unique combinations of these factors that explain why one person will develop cancer and another will not [1,3,19,20]. The new science of molecular epidemiology both helps to predict why specific groups of people are sensitive to induction of specific cancers. Molecular epidemiology can also help to predict what specific carcinogen has contributed to induction of a specific tumor by sequencing specific genes in that tumor and attempting to relate the resultant mutations to the putative carcinogenic agent by virtue of the types of mutations this agent is known to cause.

13.6 THE CARCINOGENIC PROCESS

Normal cells are being exposed on a regular basis to chemical carcinogens, mixtures of chemical carcinogens, viruses, and UV light (for skin cells) (Reference 3, Figure 13.2). Such frequent exposure to carcinogens results in frequent damage to the DNA of the cells [5,19,20]. The damaged cell can undergo DNA repair and revert back to a normal cell [31]. Alternatively, the damaged cell can also undergo apoptosis, which is programmed cell death [21]. Based on a single strand break, the p53-pathway is activated, leading to apoptosis or antiapoptosis [21]. Apoptosis involves removal of irreparable damage by death of the cell in a programmed fashion [21]. On the other hand, a subsequent increase in p21 can allow the cell to survive the DNA repair process [31]. When there is too little apoptosis, the consequences can result in cancer or autoimmune diseases. When there is too much apoptosis, stroke or neuro-degenerative diseases, such as Alzheimer's disease, can result (reviewed in Reference 21).

Normal cells with DNA damage can escape apoptosis and can then be converted into initiated cells (Figure 13.2). Division of initiated cells can then result in the formation of preneoplastic lesions, which can eventually be promoted to malignant tumors by treating them with additional carcinogen or tumor promoter (Figure 13.2). To begin to understand the relationship between cancer and the environment, it is important to recognize the relationship of exposure assessment (biomarkers) to risk assessment so that regulations can be in place in order to control carcinogen emission into the environment and to minimize the exposure of humans to environmental carcinogens.

13.7 STRATEGIES FOR THE PREVENTION OF HUMAN CANCER

As mentioned previously, in humans, cancer is the second leading cause of death, exceeded only by heart disease [3]. The lifetime cancer risk for women is slightly greater than 1 in 3 or 33%, in the United States. For men, the lifetime cancer risk is slightly less than 1 in 2, or 50% [4]. Approximately 1,368,303 new cases of cancer will be diagnosed in 2004 [4].

Cancers that can be detected earlier by screening, such as breast cancer, prostate cancer, and colorectal cancer, account for approximately half of all new cases of human cancer [1,3,4]. The 5-year survival rate for these cancers is approximately 84% [4]. Hence, many cancers, particularly colorectal cancer, can be prevented by early and routine screening and early detection. Routine use of colonoscopy can detect colorectal cancer early, as early as the polyp stage, and allow its removal before it progresses on to metastatic adenocarcinoma of the colon, for which the prognosis is poor. The relative risk is a measure of the strength of the relationship between risk factors for cancer and occurrence of a particular cancer. For example, cigarette smoking causes approximately a 10-fold relative risk of incurring lung cancer in heavy smokers compared to nonsmokers [3]. As we indicated previously, cancer is causally associated with both endogenous and exogenous factors. More than 10 years can pass between carcinogen exposure and mutation induction and the induction and detection of cancer. All cancers caused by cigarette smoking and exposure to tobacco products, which is approximately 30% of human cancers, and those cancers caused by heavy drinking of alcohol, can be prevented. The ACS estimates that in 2004, more than 180,000 deaths can be expected to be associated to tobacco use [4]. In addition, approximately one third of the 563,700 deaths due to cancer in 2004 will be associated with lifestyle factors, including poor nutrition, obesity, and physical inactivity [4]. Some cancers, such as cervical cancer, liver cancer, and certain types of leukemia will be related to exposures to infectious oncogenic viruses that could be prevented by vaccines in the case of hepatitis B virus and liver cancer, or by avoidance of sexual promiscuity and by use of barrier protections to inhibit infections with these oncogenic viruses in the case of cervical adenocarcinoma and specific human papilloma viruses (reviewed in References 3 and 18). Other cancers, such as stomach cancer, which is caused by the bacterium *Helicobacter pylori*, a diet high in salt, and dietary carcinogens, could be prevented by eradication of the Helicobacter infection by antibiotic treatment [3]. In addition, more than one million cases of skin cancer can be prevented by taking precautions to avoid overexposure to sunlight's rays, such as covering exposed areas with clothing, avoidance of exposure to sunlight at the peak hours of 12:00 to 2:00 P.M., and use of suntan lotions with appropriate protective factors [1,3]. Hence, modification of human behavior can prevent a substantial fraction of human cancers, perhaps as much as 70% [32].

Harris reviewed progress in understanding the molecular mechanisms of chemical and physical carcinogenesis and advances and perspectives for the 1990s in *Cancer Research* [19]. In particular, he discussed advances in our understanding of cancer etiology, multistage carcinogenesis, and genetic and epigenetic mechanisms of carcinogenesis, as well as carcinogen metabolism and DNA damage and repair [19]. By the end of the 20th century, molecular epidemiology is beginning to have an impact on human cancer risk and in the development of cancer prevention strategies and chemoprevention therapy. In the year 2000, the scientific journal, *Carcinogenesis*, identified the major themes of cancer research and gathered eminent scientists to report on the field of cancer research, past, present, and future [11,19,20,22]. One particular area in which chemical and viral carcinogenesis converged was the discovery that many viral oncogenes were altered forms of normal cellular genes, called proto-oncogenes, that controlled growth, differentiation, or gene expression (reviewed in References 11, 19, and 20). These cellular genes are the targets for carcinogen-induced mutations during the process of carcinogenesis. Other areas of research that are making significant contributions toward elucidation of new knowledge that can be used in the prevention of carcinogenesis and cancer treatment include an understanding of the mechanisms of tumor angiogenesis [22], tumor progression and metastasis [25], DNA methylation [33], viral carcinogenesis [11], and a better knowledge of the proportional causes of human cancer (reviewed in Reference 3).

Lastly, it has been reported that preferential formation of adducts of benzo[a]pyrene occurs at lung cancer mutational hotspots in the *p53* gene [30]. These results provide a direct etiological link between a defined metabolite of a known carcinogen benzo[a]pyrene, which is a component of cigarette smoke, and human cancer [34]. It is hoped in the future that modifications of human behavior, reduction in exposure to carcinogenic mixtures such as tobacco and tobacco products, minimization of exposure to sunlight, better engineering controls in the workplace to minimize exposure of workers to carcinogens, and stronger regulations to prevent contamination of air, drinking water, and food with carcinogens, as well as improved treatment modalities, will together lead to significant decrease in the human cancer burden.

REFERENCES

1. Cancer and the Environment, What you need to know, what you can do. U.S. Department of Health and Human Services, NIH Publication No. 03-2039, Printed September, 2004.
2. http://training.seer.cancer.gov/module_cancer_disease/unit2whatscancer1_definition.html, 10/15/2004.
3. Henderson, B.E., Ross, R.K., and Pike, M.C. Toward the primary prevention of cancer. *Science*, 254: 1131–1138, 1991.
4. Cancer Facts and Figures 2004. American Cancer Society, 1–56, 2004.
5. Heber, D. Diet and cancer prevention. *UCLA Cancer Center Bull.*, 15: 3–5, 1987.
6. Lew, E.A. and Garfinkel, L. Variations in mortality by weight among 750,000 men and women. *J. Chronic Dis.*, 32: 563–5766, 1974.
7. Baxhitt, D.P. Epidemiology of cancer of the colon and rectum. *Cancer*, 28: 3–13, 1971.
8. Goldin, B.R., Adlercrentz, H., and Borbach, S. Estrogen excretion patterns and plasma levels in vegetarian and omnivorous women. *N. Engl. J. Med.*, 397: 1542–1547, 1982.
9. Hatch, F.T., MacGregor, J.T, and Zeiger, E. Review: Putative mutagens and carcinogens in foods. VII. Genetic toxicology of the diet. *Environ. Mutagen.* 8: 467–484, 1986.
10. Ames, B.N. Dietary carcinogens and anti-carcinogens. *Science*, 221: 1256–1263, 1983.
11. Butel, J.S., Viral carcinogenesis: Revelation of molecular mechanisms and etiology of human disease. *Carcinogenesis*, 21: 405–426, 2000.
12. Ross, R.K., Paganini-Hill, A., Landolph, J., Gerkins, V., and Henderson, B.E. Analgesics, cigarette smoking, and other risk factors for cancer of the renal pelvis and ureter. *Cancer Res.*, 49: 1045–1048, 1989.
13. Patierno, S.R., Lehman, N.L., Henderson, B.E., and Landolph, J.R. Study of the ability of phenacetin, acetaminophen, and aspirin to induce cytotoxicity, mutation, and morphological transformation in C3H/10T1/2 Clone 8 Mouse Embryo Cells. *Cancer Res.*, 49: 1038–1044, 1989.
14. Landolph, J.R. Neoplastic transformation of mammalian cells by carcinogenic metal compounds: Cellular and molecular mechanisms. In *Biological Effects of Heavy Metals. Vol. II. Metal Carcinogenesis*, Foulkes, E.C., Ed., CRC Press, Boca Raton, FL, 1990, pp. 1–18, Chapter 1.
15. Landolph, J.R., Dews, M., Ozbun, L., and Evans, D.P. Metal-induced gene expression and neoplastic transformation. In *Toxicology of Metals*, Chang, L.W., Ed., CRC Lewis Publishers, Boca Raton, FL 1996, pp. 321–329, Chapter 19.
16. Biedermann, K.A. and Landolph, J.R. Induction of anchorage independence in human diploid foreskin fibroblasts by carcinogenic metal salts. *Cancer Res.*, 47: 3815–3823, 1987.
17. Biedermann, K.A. and Landolph, J.R. Role of valence state and solubility of chromium compounds on induction of cytotoxicity, mutagenesis, and anchorage independence in diploid human fibroblasts. *Cancer Res.*, 50: 7835–7842, 1990.
18. Landolph, J.R. Molecular mechanisms of transformation of C3H/10T1/2 Cl 8 mouse embryo cells and diploid human fibroblasts by carcinogenic metal compounds. *Environ. Health Perspect.*, 102 (Suppl. 3): 119–125, 1994.
19. Harris, C.C. Chemical and physical carcinogenesis: Advances and perspectives for the 1990s. *Cancer Res.*, 51: 5023–5044, 1991.
20. Yuspa, S.H. Commentary overview of carcinogenesis: Past, present, and future. *Carcinogenesis*, 21: 341–344, 2000.
21. Barinaga, M. Apoptosis. *Science*, 281: 1303–1325, 1998.
22. Kerbel, R.S. Tumor angiogenesis: Past, present, and the near future. *Carcinogenesis*, 21: 505–515, 2000.

23. Landolph, J.R., Verma, A., Ramnath, J., and Clemens, F. Molecular biology of deregulated gene expression in transformed C3H/10T1/2 mouse embryo cell lines induced by specific insoluble carcinogenic nickel compounds. *Environ. Health Perspect.*, 110 (Suppl. 5): 845–850, 2002.

24. Verma, R., Ramnath, J., Clemens, F., Kaspin, L.C., and Landolph, J.R. Molecular biology of nickel carcinogenesis: Identification of differentially expressed genes in morphologically transformed C3H/10T1/2 Cl 8 mouse embryo fibroblast cell lines induced by specific insoluble nickel compounds. *Mol. Cell. Biochem.*, 255: 203–216, 2004.

25. Yokota, J. Tumor progression and metastasis. *Carcinogenesis*, 21: 497–503, 2000.

26. Clemens, F. and Landolph, J.R. Genotoxicity of samples of nickel refinery dust. *Toxicolog. Sci.*, 73: 114–123, 2003.

27. Miura, T., Patierno, S.R., Sakuramoto, T., and Landolph, J.R. Morphological and neoplastic transformation of C3H/10T1/2 Cl 8 mouse embryo cells by insoluble carcinogenic nickel compounds. *Environ. Mol. Mutagen.*, 14: 65–78, 1989.

28. Glass, R.I. New prospects for epidemiologic investigation. *Science*, 21: 951–955, 1986.

29. Cohen, L.A. Diet and cancer. *Scientific Am.*, 257: 42–48, 1987.

30. Kroes, R., Beems, R.B., Bosland, M.C., Bunnik, G.S.J, and Sinkeldam, E.J. Nutritional factors in lung, colon, and prostate carcinogenesis in animal models. *Federation Proc.*, 45: 136–141, 1986.

31. Weiss, R.H. p21[Waf1/Cip1] as a therapeutic target in breast and other cancers. *Cancer Cell*, 4: 425–429, 2003.

32. Willett, W.C. Balancing life-style and genomics research for disease prevention. *Science*, 296: 695–698, 2002.

33. Robertson, K.D. and Jones, P.A. DNA methylation: Past, present, and future directions. *Carcinogenesis*, 21: 461–467, 2000.

34. Denissenko, M.F., Pao, A., Tang, M., and Pfeifer, G.P. Preferential formation of benzo(a)pyrene adducts at lung cancer mutational hotspots in p53. *Science*, 274: 430–432, 1996.

14 Complex Mixtures of Chemical Carcinogens: Principles of Action and Human Cancer[1]

Stephen Nesnow

CONTENTS

[1] This chapter has been reviewed by the National Health and Environmental Effects Research Laboratory at the U.S. Environmental Protection Agency and approved for publication. The views expressed in this chapter are those of the authors and do not necessarily represent the views or policy of the U.S. Environmental Protection Agency. Mention of trade names or commercial products does not constitute endorsement or recommendation for use.

14.1 INTRODUCTION

There is strong epidemiological evidence supported by experimental animal data that complex environmental mixtures pose a risk to human health producing increases in cancer incidence. Understanding the chemical and biological properties of these mixtures leads to a clearer comprehension of the principles inherent in their carcinogenic activities. Rarely are humans exposed to single toxic agents in the environment. They are exposed to complex mixtures containing multiple carcinogens, and these exposures can come from multiple sources (e.g., vehicle exhaust fumes and water disinfection by-products) and by multiple routes (inhalation, oral, dermal). Many of the agents known to be carcinogenic are found in the air, water, and diet. Pharmaceutical exposures combined with these chemicals found in the environment may also have adverse health effects. Within each exposure source, there may be many agents that either directly or indirectly affect the induction, promotion, and progression of cancer. Interactions of noncarcinogens with carcinogens may enhance or inhibit the ultimate tumor response. This chapter focusses on and summarizes the major findings on the carcinogenic effects of complex environmental mixtures and discusses mixture interactions as they pertain to chemical carcinogenesis.

In general, most of the complex mixtures whose emissions are associated with human cancer are derived from the combustion of or high temperature processing of fossil fuels and vegetative matter. These emissions are comprised of three phases — volatile, semivolatile, and particulate — and humans are exposed to all three. Although each of the complex mixtures discussed below is unique in its composition of components in each phase, due to the differences in source materials and processing technology, there are several generalities that can be made. The volatile phase can consist of gases such as carbon dioxide, carbon monoxide, hydrogen cyanide, and sulfur and nitrogen oxides. The semivolatile phase may consist of hundreds to thousands of low boiling organic materials containing carbon, hydrogen, oxygen, nitrogen, halogens, and sulfur. Some of the most common components of the semivolatile phases are those with low molecular weight. Many well-known experimental carcinogens in this class are benzene, naphthalene, butadiene, acetaldehyde, malonaldehyde, and formaldehyde. The particulate phase of these mixtures usually contains a carbonaceous core coated with condensed higher boiling point organic materials containing carbon, hydrogen, oxygen, nitrogen, halogens, and sulfur. Common to most complex environmental mixture emissions from the combustion of fossil or vegetative fuels are the polycyclic aromatic hydrocarbons (PAHs), and members of this ubiquitous chemical class are potent experimental chemical carcinogens.

14.2 DISCOVERY OF COMPLEX ENVIRONMENTAL
MIXTURES AS HUMAN CARCINOGENS

The first reported example of a carcinogenic complex environmental mixture human carcinogen is ascribed to Sir John Percivall Pott. Pott was born in 1714 and became a respected surgeon who practiced at St. Bartholomew's Hospital, London, Great Britain. In his practice, Pott observed "sores" on the scrotums of chimney sweeps in London. While other surgeons presumed these were the results of venereal disease, Pott realized that they were some kind of skin cancer. He surmised that cause of the cancers were "a lodgement of soot in the ruggae of the scrotum." In 1775 he reported these findings in "Chirugical Observations Relative to the Cataract, the Polypus of the Nose, the Cancer of the Scrotum, the Different Kinds of Ruptures, and the Mortification of Toes and Feet." This publication was the first in epidemiology that related external exposures of coal tar/soot to a human cancer [1]. Originally termed Pott's cancer, it is more commonly referred to as chimney-sweeper's cancer.

One hundred and forty years later, the first experimental evidence that coal tar was carcinogenic came from two Japanese pathologists, Katsusaburo Yamagiwa and Koichi Ichikawa at the University of Tokyo in 1915 [2]. This was also the first example of experimental chemical carcinogenesis. Professors Yamagiwa and Ichikawa were testing the irritation theory of the Danish pathologist, Johannes Andreas Grib Fibiger, who achieved the first controlled induction of cancer in laboratory animals in 1913 by feeding mice with cockroaches infected with the worm *Gongylonema neoplasticum* [3]. Larvae of this worm induced a chronic inflammation of stomach tissue, eventually inducing gastric tumors. Fibiger received the Nobel Prize for this research in 1926. Yamagiwa and Ichikawa repeatedly painted coal tar on the ears of rabbits and succeeded in producing multiple squamous cell carcinomas, a range of benign and malignant hyperplastic lesions, and inflammatory changes in the painted areas. They were also the first to describe the complexity and progressive nature of the carcinogenesis process as they identified the conversion of less malignant to more malignant tumor cells as well as the regression of benign tumors [2].

Years later, the search for the active carcinogenic components in coal tar began in the laboratory of Ian Heiger at the Institute for Cancer Research in Great Britain, who isolated carcinogenic PAHs from coal tar [4]. Using two tons of coal tar pitch, they isolated several components, of which one was highly carcinogenic on mouse skin and was identified as benzo[a]pyrene (B[a]P). B[a]P was then synthesized and the synthetic material was also found to be highly tumorigenic on mouse skin [5].

Since then, B[a]P has remained the archetypical PAH used for the study of the mechanisms of chemical carcinogenesis and is the most widely and thoroughly studied PAH in this chemical class. As B[a]P is a product of the incomplete combustion of fossil fuels, it is pervasive in the environment. Structurally, B[a]P is a fused pentacyclic PAH and has been found to be tumorigenic in almost every species tested by many different routes of exposure: mice (dermal, subcutaneous, intraperitoneal, feed), rats (subcutaneous, inhalation, intratracheal), hamsters (intratracheal), rabbit (dermal), fish (water), and dogs (endobronchial). The target organs for B[a]P-induced neoplasia include skin, forestomach, lung, liver, mammary gland, esophagus, and tongue (see References 6 and 7 for details).

14.3 CARCINOGENIC HUMAN COMPLEX ENVIRONMENTAL MIXTURES

14.3.1 Diesel Emissions

14.3.1.1 Chemistry of Diesel Combustion Emissions

Diesel fuels are complex mixtures of compounds obtained from the refining of petroleum. They contain alkanes, cycloalkanes, and aromatic hydrocarbons with carbon numbers in the range of $C_9–C_{28}$ and with a boiling-range of 150 to 390 °C. Kerosene-type diesel fuel (diesel fuel No. 1) is manufactured from straight-run petroleum distillates, whereas automotive and railroad diesel fuel (diesel fuel No. 2) contains straight-run middle distillate with other added distillate fractions [8]. If diesel combustion technologies were 100% efficient, then the combustion by-products would be water, carbon dioxide, and nitrogen oxides. However, these combustion processes are less efficient and produce a myriad of incomplete combustion by-products and particles. A number of these materials are shown to be carcinogenic to humans and experimental animals including benzene, and carcinogenic to animals including formaldehyde, butadiene, and a number of PAHs and nitro-PAHs [9–12]. Gasoline and diesel engine emissions represent the largest source of human exposure to PAHs from combustion emissions using biomass fuels.

14.3.1.2 Cancer Epidemiology of Diesel Emission Exposures

The most cited studies of diesel-exposed individuals are by Garshick et al. [13] based on an epidemiology case control study of male railroad workers exposed to diesel exhaust. Railroad workers had significant exposure to heavy-duty diesel exhaust due to the use of diesel locomotives. The cohort assessed consisted of 55,407 white male railroad workers in the age group of 40–64 years in 1959 who had been in railroad service for 10–20 years. The cohort was followed until the end of 1980, and death certificates were obtained for 88% of 19,396 deaths resulting in 1,694 lung cancer cases. Although diesel exposures were not directly measured, they were estimated from an analysis of railroad job activities. A relative risk of 1.45 (95% confidence intervals [CI] = 1.11, 1.89) for lung cancer was obtained in the group of workers with the longest duration of diesel exposure. For a comprehensive review of studies on the carcinogenic effects of diesel emissions see *International Agency for Research on Cancer (IARC) Monograph*, Volume 46 (1989) [8].

14.3.2 COKE OVEN EMISSIONS

14.3.2.1 Chemistry of Coke Oven Emissions

Coke, a coal by-product, is used in the manufacture of steel. Carbon from coke is used in the blast furnace to generate some of the energy and reducing gases needed to reduce iron oxides to iron. The coke-making process involves carbonization of coal to high temperatures (1100 °C) in an oxygen-deficient atmosphere in order to concentrate the carbon. During this process at about 475 to 600 °C, there is an evolution of tar containing PAHs and other materials. This tar material is highly carcinogenic in experimental animals and is the substance to which coke oven workers are exposed.

14.3.2.2 Cancer Epidemiology of Coke Oven Emission Exposures

The key epidemiological studies that clearly related exposure to emissions from coking operations and human cancer (lung and kidney) are those of Redmond et al. [14]. Redmond studied a population of 4661 coke oven workers. This population exhibited a significant excess relative risk of lung cancer of 2.85 where initial exposure was estimated from work history. The exposure estimates were also refined by collecting particles at the coke-oven sites and these particles were extracted with benzene to remove the organic materials. The organic materials thus extracted (benzene soluble organics) were then used as a more accurate measure of exposure. Using this dose metric, Redmond found a dose response in exposure and excess relative lung cancer risk. For a comprehensive review of studies on the carcinogenic effects of coke oven emissions, see *IARC Monograph*, Volume 34 (1984) [15].

14.3.3 COAL TAR EMISSIONS

14.3.3.1 Chemistry of Coal Tar Emissions

Coal tar is another coal by-product produced by the coal gasification process — a process for converting coal partially or completely to combustible gases, mainly methane. The coal gasification plants operated between the mid-1800s through to the 1950s. During this time, almost all fuel gas distributed for residential or commercial use in the United States was produced by the gasification of coal or coke. In the 1940s, the growing availability of low-cost natural gas led to its substitution for gases derived from coal. Coal gasification plants were usually situated near cities, and they processed coal to produce coal gas for heat and light. The coal gasification process involves the destructive distillation of coal by air and steam that react with the coal to convert it to gas. By-products of the coal gasification process are volatile emissions of coal tar. Although coal tar emissions can arise from several sources, the coal gasification process gives the highest exposures.

14.3.3.2 Cancer Epidemiology of Coal Tar Emission Exposures

In a series of classic studies by Sir Richard Doll, British gas workers were evaluated for excess cancer risks. The health status of over 12,000 men were studied over the course of 20 years. Workers were categorized by their jobs in the various parts of the plants. Doll reported increased risks of skin and lung cancer and possibly bladder cancer for workers exposed to coal tar [16]. Another population exposed to coal tar as the higher boiling coal tar pitch are roofers and asphalt workers. Coal tar pitch comes from the processing of crude tars obtained from coke ovens or by other methods using the destructive distillation of coal. Hammond et al. [17] studied 5,939 members of a roofer's union for increased cancer risk related to coal tar exposure. He found increased risk for lung cancers and a suggested risk for laryngeal and oral cavity cancers. Partanen and Boffetta [18] reviewed and performed a meta analysis on a number of epidemiological studies related to cancer risk in asphalt workers and roofers. They found an increased risk in roofers for cancers of the lung (relative risk 1.8, 95% confidence interval 1.5–2.1), stomach (1.7, 1.1–2.5), skin (nonmelanoma) (4.0, 0.8–12), and leukemia (1.7, 0.9–2.9) and suggested that some of the excess risks may have been attributable to PAH from coal tar products. These findings have been confirmed by recent proportionate mortality study of 11,144 unionized roofers and waterproofers [19]. For a comprehensive review of studies on the carcinogenic effects of coal tar emissions, see *IARC Monographs* Volumes 34 and 35 (1984, 1985) [15,20].

14.3.4 SHALE OIL

14.3.4.1 Chemistry of Shale Oil

Shale oil is derived from the thermal processing (pyrolysis) of oil shale, a major sedimentary rock containing minerals and kerogen. Kerogen is a naturally occurring, solid, insoluble organic material consisting of organic constituents from algae and woody plant material. Shale oil is a complex mixture of aliphatic olefinic and aromatic hydrocarbons, and polar compounds containing carbon, hydrogen, nitrogen, oxygen, and sulfur. Few components have been identified, however a number of carcinogenic PAHs have been measured in shale oils (e.g., naphthalene and B[a]P) [20].

14.3.4.2 Epidemiology of Shale Oil Exposure

In a monograph review, IARC [20] found an association between shale oils and skin cancer, including scrotal cancer in humans. However, more recent epidemiological studies of some of the same populations, Scottish shale oil workers, and U.S. shale oil workers do not support that conclusion [21,22].

14.3.5 TOBACCO SMOKE

14.3.5.1 Chemistry of Tobacco Smoke

Of all environmental exposures to complex mixtures, tobacco smoke exposure represents the greatest human health hazard based on exposure levels and numbers of people exposed. Mainstream cigarette smoke contains thousands of chemicals representing almost every class of chemicals

TABLE 14.1

Complex Mixtures Associated with Human Cancer

Complex mixture	Study population	Tumor sites	Reference
Diesel emissions	Male railroad workers	Lung	IARC [8]
Coke oven emissions	Coke oven workers	Lung	IARC [15]
Coal tar emissions	British gas workers	Lung, skin, stomach	IARC [15,20]
	Roofers and sphalt workers		
Tobacco smoke	Many population studies	Upper respiratory tract, upper digestive tract, bladder, kidney, oral, liver	IARC [24]

known to induce cancer in experimental animals: aldehydes (e.g., formaldehyde), aromatics (e.g., benzene), aromatic amines (e.g., 2-naphthylamine), hydrazines (e.g., hydrazine), lactones (e.g., γ-butyrolactone), metals (e.g., chromium), nitroaliphatics (e.g., 2-nitropropane), nitrosamines (e.g., 4-(methylnitrosamino)-1-(3-pyridyl)-1-butanone (NNK)), acrylonitrile, urethane, vinyl chloride, and PAHs. Many carcinogenic PAHs have been identified including naphthalene, B[a]P, benz[a]anthracene, dibenzo[a, l]pyrene, and dibenz[a, h]anthracene to name a few. For a detailed review on the chemistry of tobacco smoke carcinogens, see Wynder and Hoffmann [23] and *IARC Monograph*, Volume 38 (1986) [24].

14.3.5.2 Cancer Epidemiology of Tobacco Smoke Exposure

Probably the most thoroughly and extensively studied human exposure to complex mixtures is to tobacco smoke, either by direct exposure or by second-hand smoke. Lung and bronchial cancer is the most important cause of cancer death accounting for just over a quarter of all cancer-related deaths. Approximately 90% of these lung cancer deaths are attributed to smoking. IARC has prepared an excellent monograph that summarizes and evaluates the major epidemiological studies on tobacco smoke [24]. They concluded that tobacco smoking was strongly associated with cancer of the upper respiratory tract, upper digestive tract, bladder, and kidney (IARC, 1986). Other tobacco smoking-related tumor sites were oral, liver, and possibly pancreas. These findings had been previously reported by the U.S. Surgeon General in a series of reports, the latest being "Women and Smoking, A Report of the Surgeon General-2001" [25]. A summary of the complex mixtures associated with human cancer is presented in Table 14.1.

14.4 EVIDENCE FROM EXPERIMENTAL ANIMAL STUDIES FOR THE CARCINOGENIC EFFECTS OF COMPLEX ENVIRONMENTAL MIXTURES

Experimental studies have provided important and convincing evidence that some complex mixtures associated with human cancer are potent animal carcinogens. Experimental animals using both inbred and outbred strains provide controlled exposures and observation periods whereby the cancer process can be monitored and measured. Moreover, the major tumor target sites for these mixtures are concordant in humans and in animals. Experimental animals also provide sources of tissues for ancillary studies on the mechanism(s) of action of complex mixtures. There are three predominant types of studies that have been used to characterize the tumorigenic effects of complex mixtures. These studies use different routes of administration: inhalation, lung implant, and dermal.

TABLE 14.2
Examples of Experimental Cancer Inhalation Studies of Complex Mixtures

Complex mixture	Strain/species	Tumor sites	Lung tumor types
Coal tar [27]	Wistar rat	Lung	Squamous cell tumor Adenocarcinoma, adenoma
PAH-rich exhausts [26]	NMRI/BR newborn mouse	Lung	Adenoma, adenocarcinoma, squamous cell carcinoma
Coke oven PAH-rich mixture [35]	Wistar rat NMRI mouse	Lung	Adenocarcinoma, adenoma
Whole diesel exhaust particles [36,37]	F344 rat	Lung	Adenocarcinoma, adenoma
Whole diesel exhaust [32]	CD-1 mouse	Lung	No tumors
Tobacco smoke [38]	F344 rat	Lung	Carcinoma, adenoma
Tobacco smoke [39]	Syrian golden hamster	Larynx	Carcinoma

14.4.1 INHALATION STUDIES

A limited number of inhalation studies with complex mixtures have been reported using experimental animals. PAH-enriched exhausts were studied in a 10-month inhalation study using newborn female mice to determine its pulmonary tumorigenicity [26]. At two exhaust concentrations, significant increases in bronchiolo-alveolar adenomas and malignant bronchiolo-alveolar adenocarcinomas were observed. Female Wistar rats were exposed to coal tar/pitch condensation aerosol. A 97% lung tumor rate was recorded at the highest concentration [27]. Both complex mixtures contained PAHs.

In a unique rodent inhalation study, mice and rats were exposed in the field to smoky coal and wood. Xuan Wei County of Yunnan Province in China has a high mortality rate of lung cancer in humans [28]. This cancer mortality is associated with the heating and cooking in unvented indoor environments with smoky coal or wood. In the study, mice and rats were placed in these environments in which they inhaled air containing smoky coal and wood for 15 to 19 months [29]. The incidences of lung cancer in mice in the control group, wood group, and coal group were 17.0% (29/171), 45.8% (81/177), and 89.5% (188/210) respectively. In rats the incidences were 0.9% (1/110), 0% (0/110), and 67.2% (84/125) respectively. Organic extracts of the smoky coal particulates were also found to be carcinogenic on mouse skin [30].

The carcinogenic effects of inhaled diesel exhaust have been reported in a number of studies by the Mauderly group at the Inhalation Toxicology Research Institute. Male and female F344 rats were exposed 7 h/day, 5 day/week for up to 30 months to automotive diesel engine exhaust. Focal fibrotic and proliferative lung disease accompanied a progressive accumulation of soot in the lung. There was a significant increase in the number of rats bearing lung tumors in the treated group over the control group. The tumor types observed were adenoma, adenocarcinoma, and squamous cell carcinoma [31]. Unlike rats, CD-1 mice exposed to whole diesel exhaust 7 h/day, 5 days/week for 24 months did not exhibit an increase in the incidence of lung neoplasms [32]. It is believed that the pulmonary carcinogenic effects of diesel exhaust are due to several components including PAHs adsorbed to the diesel particles. Organic extracts of these particles are tumorigenic on mouse skin [33]. Based on inhalation studies it is found that diesel particles without the coated organic materials are also carcinogenic in rat lung owing to particle overload. It has been proposed that reactive oxygen species (ROS) derived from the surface properties of particles or formed secondarily from lung inflammatory processes could be involved in the cancer process [34]. A summary of selected inhalation studies of complex mixtures is found in Table 14.2.

14.4.2 LUNG IMPLANT STUDIES

Lung implantation of carcinogens in rats has been used by the Grimmer group from the Biological Institute for Chemical Carcinogens in Germany. Osborne–Mendel rats were implanted with flue gas condensate from coal-fired residential furnaces in a trioctanoin/beeswax mixture. The implanted materials induced lung carcinomas and sarcomas in the rats [40]. A similar study was performed with diesel exhaust fractions [41]. They concluded that most of the carcinogenicity of diesel exhaust originated from the PAH consisting of four or more rings [41]. Grimmer also used this method to identify the contribution of PAHs in side-stream cigarette smoke. He found that most of the lung carcinogenic activity side-stream smoke was caused by the fraction containing PAHs of four or more rings [42].

14.4.3 MOUSE SKIN STUDIES

Mouse skin has been a favored system for studying the biological activities of complex mixtures as well as their modes of action. Specific mouse strains have been found to be quite sensitive to the tumorigenic effects of complex mixtures, individual PAHs, and simple mixtures of PAHs, which are frequent constituents of complex environmental mixtures from combustion sources. These systems are efficient with regard to sample quantity and assay time. In general two protocols are used, tumor initiation/promotion, and complete carcinogenesis. The tumor initiation/promotion protocol features a single treatment with the complex mixture onto the shaved backs of mice followed by twice-weekly treatments of the potent tumor promoter, 12-O-tetradecanoylphorbol-13-acetate (TPA). The incidence of papillomas rises generally from 5 to 10 weeks after initiation and plateaus after 20–26 weeks. An example of data from this protocol is found in Figure 14.1, in which an organic extract from smoky coal particles was administered to Sencar mice. Both the time course and dose response of papilloma formation were noted. In the complete carcinogenesis protocol, the backs of shaved mice are treated twice weekly for 1–2 years. Carcinomas usually appear between 1 and 2 years of

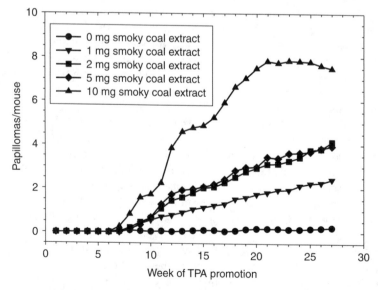

FIGURE 14.1   Time and dose response of papilloma formation in Sencar mice treated with an organic extract of smoky coal particles. The particles were collected in homes in Xuan Wei County, Yunnan Province, China which as an unusually high lung cancer mortality rate. In the tumor initiation protocol, mice were treated once with the extract and one week later treated twice weekly with TPA (2 μg/mouse) [30].

TABLE 14.3
Examples of Experimental Cancer Studies of Complex Mixtures Using Dermal
Applications to Mouse Skin

Complex mixture	Strain	Protocol	Dermal tumor types
Diesel emissions [43]	Sencar	Tumor initiation	Papilloma
Coke oven emissions [33]	Sencar	Tumor initiation	Papilloma
Roofing tar emissions [33]	Sencar	Tumor initiation	Papilloma
Gasoline emissions [33]	Sencar	Tumor initiation	Papilloma
Coal tar extract [44]	Sencar	Tumor initiation	Papilloma
Smoky coal extract [30]	Sencar	Tumor initiation	Papilloma
		Complete carcinogenesis	Carcinoma
Emission condensate from coal-fired residential furnaces [45]	CLFP	Complete carcinogenesis	Carcinoma
Emission condensate from brown coal-fired residential furnaces [46]	CLFP	Complete carcinogenesis	Carcinoma
Coke oven emissions [33]	Sencar	Complete carcinogenesis	Carcinoma
Roofing tar emissions [33]	Sencar	Complete carcinogenesis	Carcinoma
Condensed asphalt roofing fumes [47]	C3H/HeJ	Complete carcinogenesis	Papilloma
	Sencar		Carcinoma
Gasoline emissions [33]	Sencar	Complete carcinogenesis	Carcinoma
Cigarette smoke [48]	Swiss	Complete carcinogenesis	Carcinoma
Cigarette smoke [48,49]	Swiss	Complete carcinogenesis	Papilloma, carcinoma
Gasoline emissions [50]	Swiss	Complete carcinogenesis	Papilloma, carcinoma
Shale oil [51]	Swiss	Complete carcinogenesis	Papilloma, carcinoma
Automobile exhaust condensate [52]	CLFP	Complete carcinogenesis	Papilloma, carcinoma

treatment. A summary of the tumorigenic effects of complex mixtures as determined from mouse
skin studies is found in Table 14.3.

14.5 MIXTURE INTERACTIONS AND CANCER

14.5.1 INTRODUCTION TO COMPLEX MIXTURE INTERACTIONS

There are many reported examples of interactions between two chemical carcinogens (binary mixtures) that result in greater than or less than the sum of the biological activities of the two individual
components. This has been demonstrated for genotoxic agents using multiple bioassay systems.
However, with complex mixtures there is little evidence that there are large interactions between
multiple chemicals creating greater than expected tumorigenic effects. One of the first studies on
interactions within complex mixtures was reported by Kier et al. [53], using cigarette smoke condensate as the complex mixture and bacterial mutation in *Salmonella typhimurium* as the bioassay
endpoint. They fractionated whole cigarette smoke condensate according to an acid–base–neutral
scheme and tested the mutagenic activities of each fraction. They also reconstituted all of the fractions into one mixture and compared its mutagenic activity with that of the whole condensate mixture
(Table 14.4). They found that the reconstituted mixture could account for 89% of the total mutagenic
activity of the whole cigarette smoke condensate, suggesting that the fractionation procedure lost
only 11% of the mutagenic activity. More importantly, they found that the sum of all of the fractions

TABLE 14.4

Mutagenic Activity of Whole and Fractionated Cigarette Smoke Condensate

Sample	Mg/cigarette	Revertants/plate[a] (250 μg sample)	Calculated revertants/ cigarette	Activity of whole condensate, %
Whole condensate	23.5	310	29,100	100
Reconstituted condensate	23.0	280	25,800	89
Sum of individual fractions	23.0		18,200	62 (70)[b]

[a]Samples were bioassayed in *S. typhimurium* strain TA 1538 with an S9 mix.
[b]Based on recovery.

Source: Kier, L.D., Yamasaki, E., and Ames, B.N., *Proc. Natl. Acad. Sci. USA*, 71, 4159, 1974. With permission.

was only 62% of the mutagenic activity of the whole cigarette smoke condensate (or 70% adjusted for recovery). This suggested that small interactions that enhanced the mutagenic activity were operative.

More extensive studies on fractionation and bioassay of complex mixtures obtained from organic extracts of particles from diesel exhausts, coke oven emissions, roofing tar emissions, and cigarette smoke condensate have been reported [54]. In these studies each of these extracts was fractionated and each of the fractions was bioassayed for mutagenic activity using *Salmonella typhimurium*. For the diesel sample, the sum of the individual mutagenic activities of each of the fractions exceed that of the unfractionated sample by about twofold after adjusting for recovery. This suggested that some mixture components in one fraction were reducing the mutagenic activities of components in another fraction or an inhibitory response. Similarly, for the coke oven sample the sum of mutagenic activities of the fractions exceeded that of the unfractionated sample by 1.6-fold, again indicative of inhibitory interactions. For the roofing tar sample, the sum of mutagenic activities of the fractions equaled the mutagenic activity of the unfractionated sample, suggesting no interactions. For the cigarette smoke condensate, the individual fractions accounted for 57% of the total activity, suggesting that some components in one fraction enhanced the mutagenic activities of another fraction. These results demonstrate that mixture interactions can lead to greater, equal, or lesser biological activity than would be expected from the sum of the individual components. However, the extent of these interactions is not very large.

The approaches most commonly taken to identify and quantitate interactions within mixtures are to create chemically defined mixtures of relatively simple complexity. By carefully designing the experimental protocol, interactions between individual components can be obtained. Another less frequently used approach is to add known agents to complex mixtures and to evaluate the effects of these agents on the biological activities of the mixtures [55].

14.5.2 THEORIES OF MIXTURE INTERACTIONS

When exposure consists of two or more chemicals, their resultant toxic effects may vary depending on the nature and strength of the individual interactions. In general, there are three types of overall responses: additive; greater than additive; and less than additive.

Additive responses are the result of the sum of the toxic responses of each individual component of the mixture. There are two classes of additivity: dose addition and response addition. Dose addition assumes that each chemical is a clone (in terms of toxicity) of the other. If each chemical is equipotent, then various proportionate mixtures of the two will give the same toxic outcomes.

If the two chemicals are non-equipotent, then admixtures of both will give varying toxic responses depending on the each chemical's contribution to the mixture. Response addition is simply the sum of the toxic responses (e.g., tumors, mutants) of each individual component.

Greater than additive responses can be due to effects such as synergism and potentiation. Synergistic effects are those in which each chemical induces toxic effects by different and interacting mechanisms. Examples of this outcome are studies in which a DNA-reactive chemical is administered to rodents simultaneously with a chemical that induces mitogenesis. The DNA-reactive chemical alone will induce tumors, as will the mitogenic chemical but by different mechanisms, and the outcome of these interacting mechanisms are increased tumor yields. Potentiative effects are those in which one chemical's effects are enhanced by a second chemical (the second chemical has no direct toxic effect itself). An example of potentiation is the treatment of animals with two chemicals. One chemical requires enzymatic metabolic activation to exert its toxic effects, and the second is a nontoxic chemical that modifies the enzymatic complex to enhance the amounts of DNA-reactive metabolic forms.

Less than additive responses can be due to effects such as antagonism and inhibition. Antagonism is where two active components interact in such a manner as to produce reduced responses for each component. An example of this effect is where two carcinogens compete for the same receptor. If the less potent carcinogen is a more strongly bound to the receptor than the other, the result will be fewer tumors. Inhibition is where the effects of one active component are lessened by the effects of a second nontoxic chemical. A simple example of this effect is a nontoxic enzyme inhibitor that reduces the extent of metabolic activation of a toxicant that requires metabolic activation to exert its toxic effects.

14.5.3 TYPES OF MIXTURE ANALYSES

There are a number of different types of methods used to analyze mixture interactions: isobolographic analyses [56], interaction indices, [57] and response surface approaches [58]. The last two methods have found use in identifying and quantitating chemical and drug interactions. Isoboles are combinations of two component or binary mixtures that give equivalent responses. Isobolographic analyses involve the plotting of these dose addition responses in a specific manner. The resultant shapes of the response curves can indicate additivity, supra-additivity, or antagonism. Interaction indices are mathematical methods by which additivity or deviations from additivity are calculated based on the number of interacting components and their fractional contributions to the aggregate toxic response. Response surface analyses feature mathematical and statistical analyses of complex interactions. Response surface graphs are graphic depictions of dose-continuous interaction indices between chemicals plotted as multidimensional surfaces. Inherent in response surface methodology are the use of statistical tools used for experimental design and analyses. Using factorial designs, experimental studies can be devised to identify the individual interaction parameters among individual components in a mixture. Software tools are available to calculate and visualize these interactions (see CombiTool, a Microsoft Windows program designed for the analysis of combination experiments with biologically active agents) [59].

14.5.4 EXAMPLES OF EXPERIMENTAL APPROACHES TO STUDY MIXTURE INTERACTIONS

14.5.4.1 An *In Vivo* Study Using a Defined Multicomponent Chemical Mixture of PAHs

Since PAHs dominate so many complex environmental mixtures, a defined chemical mixtures study was undertaken to identify and quantitate interactions. Five-component mixtures of environmental

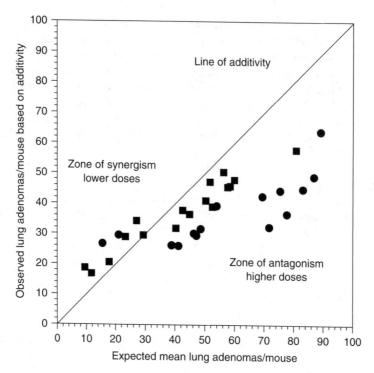

FIGURE 14.2   The correlation between the observed lung adenomas per mouse after treatment of male strain A/J mice with 32 mixtures (each containing five PAHs with different doses) and the expected lung adenomas per mouse based on response additivity. The data points represent the mean responses for each of the 32 mixtures of the five PAHs. The calculation of the expected additive responses for each PAH mixture was performed by summing the individual responses for each PAH within each mixture group based on the individual PAH dose response studies. The line (——) represents the expected relationship if the data fit additivity. ● represents a statistically significant ($p < .05$) difference between the observed and expected additive responses. ■ represents no statistically significant difference between the observed and expected additive responses [60].

PAHs were administered to strain A/J mice using a 2^5 factorial experimental design (5 PAHs at 2 doses each). Each mouse received a single dose intraperitoneal injection of the five-component mixture and lung tumors were counted at the end of the study. The PAHs used were B[a]P, benzo[b]fluoranthene, dibenz[a,h]anthracene, 5-methylchrysene, and cyclopenta[c,d]pyrene (CPP). All of these individual PAHs were tumorigenic in strain A/J mice producing lung tumors. To first analyze the data for overall interactions, the total number of tumors found in each of the 32 treatment groups were plotted against the number of tumors expected in that group. The expected numbers of tumors in each group was the sum of the numbers of tumors that were induced by each of the individual PAHs in the mixtures (response additivity). The results showed greater than additive (synergistic), additive, and less than additive (antagonistic) interactions (Figure 14.2). When total doses of the five PAHs in each mixture group were overlayed on the plot, antagonism was found to occur at higher doses and synergism at lower doses. Using response surface statistical analyses, interactions between the PAHs were identified. This analysis produced statistically significant values for 16 interactions. The binary interactions were dominated for the most part by dibenz[a, h]anthracene and were inhibitory. The response surface mathematical model predicted, to a significant degree, the observed lung tumorigenic responses of the five-component mixtures. This data suggests that although interactions between PAHs do occur, they are limited in extent [60].

14.5.4.2 An *In Vitro* Study Using a Defined Binary Chemical Mixture of PAHs

To identify PAH–PAH interactions, binary mixtures of B[*a*]P and dibenz[*a,h*]anthracene (DBA) were administered to transformable C3H10T1/2Cl8 (C3H10T1/2) mouse embryo fibroblast cells in culture. When these cells are treated with carcinogens they produce altered clones of cells (foci) that are heritably changed (known as morphological cell transformation or carcinogenesis *in vitro*). Both DBA and B[*a*]P each induced morphologically altered foci in these cells. However, when C3H10T1/2 cells were treated with binary mixtures of B[*a*]P and DBA fewer foci were observed than the sum of the foci produced by cells treated individually with B[*a*]P and DBA. These results were consistent with those reported in mice and rats on the antagonistic effects of B[*a*]P and DBA on tumorigenesis. ^{32}P-Postlabeling DNA adduct studies revealed that DBA reduced B[*a*]P–DNA adduct levels by 47% while B[*a*]P had no effect on DBA–DNA adduct levels. This suggests that the mechanism for the inhibition of morphological cell transformation of binary mixtures of B[*a*]P and DBA was antagonism due to alterations in the metabolic activation of B[*a*]P [61].

REFERENCES

1. Pott, P., *Chirugical Observations Relative to the Cataract, the Polypus of the Nose, the Cancer of the Scrotum, the Different Kinds of Ruptures, and the Mortification of Toes and Feet*, Hawkes L., Clarke W., and Collins R., London, 1775.
2. Yamagiwa, K. and Ichikawa, K., Uber die kunstliche von erzeugung von papillom. *V. Jap. Path. Ges.*, 5, 142, 1915.
3. Fibiger, J.A.G., Untersuchungen über eine Nematode *Spiroptera* sp. n.) und deren Fähigkeit papillomatöse und carcinomatöse Geschwulstbildungen im Magen der Ratte hervorzurufen, *Zeitschrift für Krebsforschung*, 13, 217, 1913.
4. Cook, J.W., Hewitt, C.L., and Heiger, I., The isolation of a cancer producing hydrocarbon from coal tar, *J. Chem. Soc.*, 395, 1933.
5. Kennaway, E., Identification of carcinogenic compound in coal tar, *Br. Med. J.*, 2, 749, 1955.
6. Osborne, M.R. and Crosby, N.T., *Benzopyrenes*, Cambridge University Press, Cambridge, England, 1987.
7. NTP, NTP Chemical Repository Benzo[*a*]pyrene, http://ntp-server.niehs.nih.gov/htdocs/CHEM_H&S/NTP_Chem5/Radian50-32-8.html, 2002.
8. IARC, *IARC Monographs on the Evaluation of the Carcinogenic Risk of Chemicals to Humans Diesel and Gasoline Exhausts and Some Nitroarenes*, International Agency for Research on Cancer, Lyon, France, 1989.
9. IARC, *IARC Monographs on the Evaluation of the Carcinogenic Risk of Chemicals to Humans Overall Evaluations of Carcinogenicity: An Updating of IARC Monographs Volumes 1 to 42*, International Agency for Research on Cancer, Lyon, France, 1987.
10. IARC, *IARC Monographs on the Evaluation of the Carcinogenic Risk of Chemicals to Humans Wood Dust and Formaldehyde*, International Agency for Research on Cancer, Lyon, France, 1995.
11. Miller, R.A., Melnick, R.L., and Boorman, G.A., Neoplastic lesions induced by 1,3-butadiene in B6C3F1 mice, *Exp. Pathol.*, 37, 136, 1989.
12. IARC, *IARC Monographs on the Evaluation of the Carcinogenic Risk of Chemicals to Humans Polynuclear Aromatic Compounds, Part 1, Chemical, Environmental, and Experimental Data*, International Agency for Research on Cancer, Lyon, France, 1983.
13. Garshick, E., Schenker, M.B., Munoz, A., Segal, M., Smith, T.J., Woskie, S.R., Hammond, S.K., and Speizer, F.E., A retrospective cohort study of lung cancer and diesel exhaust exposure in railroad workers, *Am. Rev. Respir. Dis.*, 137, 820, 1988.
14. Redmond, C.K., Strobino, B.R., and Cypess, R.H., Cancer experience among coke by-product workers, *Ann. N.Y. Acad. Sci.*, 271, 102, 1976.
15. IARC, *IARC Monographs on the Evaluation of the Carcinogenic Risk of Chemicals to Humans Polynuclear Aromatic Compounds, Part 3, Industrial Exposures in Aluminum Production, Coal Gasification,*

Coke Production, and Iron and Steel Founding, International Agency for Research on Cancer, Lyon, France, 1984.

16. Doll, R., Vessey, M.P., Beasley, R.W.R., Buckley, A.R., Fear, E.C., Fisher, R.E.W., Gammon, E.J., Gun, W., Hughes, G.O., Lee, K., and Norman-Smith, B., Mortality of gasworkers-final report of a prospective study, *Br. J. Ind. Med.*, 29, 394, 1972.

17. Hammond, E.C., Selikoff, I.J., Lawther, P.L., and Seidman, H., Inhalation of benzpyrene and cancer in man, *Ann. N.Y. Acad. Sci.*, 271, 116, 1976.

18. Partanen, T. and Boffetta, P., Cancer risk in asphalt workers and roofers: Review and meta-analysis of epidemiologic studies, *Am. J. Ind. Med.*, 26, 721, 1994.

19. Stern, F.B., Ruder, A.M., and Chen, G., Proportionate mortality among unionized roofers and waterproofers, *Am. J. Ind. Med.*, 37, 478, 2000.

20. IARC, *IARC Monographs on the Evaluation of the Carcinogenic Risk of Chemicals to Humans Polynuclear Aromatic Compounds, Part 4, Bitumens, Coal-Tars, and Derived Products, Shale Oils and Soots*, International Agency for Research on Cancer, Lyon, France, 1985.

21. Seaton, A., Louw, S.J., and Cowie, H.A., Epidemiologic studies of Scottish oil shale workers: I. Prevalence of skin disease and pneumoconiosis, *Am. J. Ind. Med.*, 9, 409, 1986.

22. Rom, W.N., Krueger, G., Zone, J., Attfield, M.D., Costello, J., Burkart, J., and Turner, E.R., Morbidity survey of U.S. oil shale workers employed during 1948–1969, *Arch. Environ. Health*, 40, 58, 1985.

23. Wynder, E.L. and Hoffmann, D., *Tobacco and Tobacco Smoke*, Academic Press, New York, 1967.

24. IARC, *IARC Monographs on the Evaluation of the Carcinogenic Risk of Chemicals to Humans Tobacco Smoking*, International Agency for Research on Cancer, Lyon, France, 1986.

25. Surgeon General, U.S., Women and Smoking. A Report of the Surgeon General-2001, http://www.cdc.gov/tobacco/sgr_forwomen.htm, 2001.

26. Schulte, A., Ernst, H., Peters, L., and Heinrich, U., Induction of squamous cell carcinomas in the mouse lung after long-term inhalation of polycyclic aromatic hydrocarbon-rich exhausts, *Exp. Toxicol. Pathol.*, 45, 415, 1994.

27. Heinrich, U., Roller, M., and Pott, F., Estimation of a lifetime unit lung cancer risk for benzo(a)pyrene based on tumour rates in rats exposed to coal tar/pitch condensation aerosol, *Toxicol. Lett.*, 72, 155, 1994.

28. Mumford, J.L., He, X.Z., Chapman, R.S., Cao, S.R., Harris, D.B., Li, X.M., Xian, Y.L., Jiang, W.Z., Xu, C.W., Chuang, J.C. et al., Lung cancer and indoor air pollution in Xuan Wei, China, *Science*, 235, 217, 1987.

29. Liang, C.K., Quan, N.Y., Cao, S.R., He, X.Z., and Ma, F., Natural inhalation exposure to coal smoke and wood smoke induces lung cancer in mice and rats, *Biomed. Environ. Sci.*, 1, 42, 1988.

30. Mumford, J.L., Helmes, C.T., Lee, X.M., Seidenberg, J., and Nesnow, S., Mouse skin tumorigenicity studies of indoor coal and wood combustion emissions from homes of residents in Xuan Wei, China with high lung cancer mortality, *Carcinogenesis*, 11, 397, 1990.

31. Mauderly, J.L., Jones, R.K., Griffith, W.C., Henderson, R.F., and McClellan, R.O., Diesel exhaust is a pulmonary carcinogen in rats exposed chronically by inhalation, *Fundam. Appl. Toxicol.*, 9, 208, 1987.

32. Mauderly, J.L., Banas, D.A., Griffith, W.C., Hahn, F.F., Henderson, R.F., and McClellan, R.O., Diesel exhaust is not a pulmonary carcinogen in CD-1 mice exposed under conditions carcinogenic to F344 rats, *Fundam. Appl. Toxicol.*, 30, 233, 1996.

33. Nesnow, S., Triplett, L.L., and Slaga, T.J., Mouse skin tumor initiation–promotion and complete carcinogenesis bioassays: Mechanisms and biological activities of emission samples, *Environ. Health Perspect.*, 47, 255, 1983.

34. Schins, R.P., Mechanisms of genotoxicity of particles and fibers, *Inhal. Toxicol.*, 14, 57, 2002.

35. Heinrich, U., Pott, F., Mohr, U., Fuhst, R., and Konig, J., Lung tumours in rats and mice after inhalation of PAH-rich emissions, *Exp. Pathol.*, 29, 29, 1986.

36. Iwai, K., Higuchi, K., Udagawa, T., Ohtomo, K., and Kawabata, Y., Lung tumor induced by long-term inhalation or intratracheal instillation of diesel exhaust particles, *Exp. Toxicol. Pathol.*, 49, 393, 1997.

37. Nikula, K.J., Snipes, M.B., Barr, E.B., Griffith, W.C., Henderson, R.F., and Mauderly, J.L., Comparative pulmonary toxicities and carcinogenicities of chronically inhaled diesel exhaust and carbon black in F344 rats, *Fundam. Appl. Toxicol.*, 25, 80, 1995.

38. Dalbey, W.E., Nettesheim, P., Griesemer, R., Caton, J.E., and Guerin, M.R., Chronic inhalation of cigarette smoke by F344 rats, *J. Natl. Cancer Inst.*, 64, 383, 1980.

39. Bernfeld, P., Homburger, F., Soto, E., and Pai, K.J., Cigarette smoke inhalation studies in inbred Syrian golden hamsters, *J. Natl. Cancer Inst.*, 63, 675, 1979.

40. Grimmer, G., Brune, H., Deutsch-Wenzel, R., Dettbarn, G., and Misfeld, J., Contribution of polycyclic aromatic hydrocarbons and polar polycyclic aromatic compounds to the carcinogenic impact of flue gas condensate from coal-fired residential furnaces evaluated by implantation into the rat lung, *J. Natl. Cancer Inst.*, 78, 935, 1987.

41. Grimmer, G., Brune, H., Deutsch-Wenzel, R., Dettbarn, G., Jacob, J., Naujack, K.W., Mohr, U., and Ernst, H., Contribution of polycyclic aromatic hydrocarbons and nitro-derivatives to the carcinogenic impact of diesel engine exhaust condensate evaluated by implantation into the lungs of rats, *Cancer Lett.*, 37, 173, 1987.

42. Grimmer, G., Brune, H., Dettbarn, G., Naujack, K.W., Mohr, U., and Wenzel-Hartung, R., Contribution of polycyclic aromatic compounds to the carcinogenicity of sidestream smoke of cigarettes evaluated by implantation into the lungs of rats, *Cancer Lett.*, 43, 173, 1988.

43. Nesnow, S., Triplett, L.L., and Slaga, T.J., Comparative tumor-initiating activity of complex mixtures from environmental particulate emissions on SENCAR mouse skin, *J. Natl. Cancer Inst.*, 68, 829, 1982.

44. Marston, C.P., Pereira, C., Ferguson, J., Fischer, K., Hedstrom, O., Dashwood, W.M., and Baird, W.M., Effect of a complex environmental mixture from coal tar containing polycyclic aromatic hydrocarbons (PAH) on the tumor initiation, PAH-DNA binding and metabolic activation of carcinogenic PAH in mouse epidermis, *Carcinogenesis*, 22, 1077, 2001.

45. Grimmer, G., Brune, H., Deutsch-Wenzel, R., Dettbarn, G., Misfeld, J., Abel, U., and Timm, J., The contribution of polycyclic aromatic hydrocarbons to the carcinogenic impact of emission condensate from coal-fired residential furnaces evaluated by topical application to the skin of mice, *Cancer Lett.*, 23, 167, 1984.

46. Deutsch-Wenzel, R.P., Brune, H., Grimmer, G., Dettbarn, G., Misfeld, J., and Timm, J., Investigation on the carcinogenicity of emission condensate from brown coal-fired residential furnaces applied to mouse skin, *Cancer Lett.*, 25, 103, 1984.

47. Sivak, A., Niemeier, R., Lynch, D., Beltis, K., Simon, S., Salomon, R., Latta, R., Belinky, B., Menzies, K., Lunsford, A., Cooper, C., Ross, A., and Bruner, R., Skin carcinogenicity of condensed asphalt roofing fumes and their fractions following dermal application to mice, *Cancer Lett.*, 117, 113, 1997.

48. Dontenwill, W., Chevalier, H.J., Harke, H.P., Klimisch, H.J., Reckzeh, G., Fleischmann, B., and Keller, W. [Experimental investigations on the tumorigenic activity of cigarette smoke condensate on mouse skin. VII. Comparative studies of condensates from different modified cigarettes (author's transl)], *Z. Krebsforsch. Klin. Onkol. Cancer Res. Clin. Oncol.*, 89, 145, 1977.

49. Hoffmann, D. and Wynder, E.L., A study of tobacco carcinogenesis. XI. Tumor initiators, tumor accelerators, and tumor promoting activity of condensate fractions, *Cancer*, 27, 848, 1971.

50. Wynder, E.L. and Hoffmann, D., A study of air pollution carcinogenesis. III. Carcinogenic activity of gasoline engine exhaust condensate, *Cancer*, 15, 103, 1962.

51. Rowland, J., Shubik, P., Wallcave, L., and Sellakumar, A., Carcinogenic bioassay of oil shale: Long-term percutaneous application in mice and intratracheal instillation in hamsters, *Toxicol. Appl. Pharmacol.*, 55, 522, 1980.

52. Grimmer, G., Brune, H., Deutsch-Wenzel, R., Naujack, K.W., Misfeld, J., and Timm, J., On the contribution of polycyclic aromatic hydrocarbons to the carcinogenic impact of automobile exhaust condensate evaluated by local application onto mouse skin, *Cancer Lett.*, 21, 105, 1983.

53. Kier, L.D., Yamasaki, E., and Ames, B.N., Detection of mutagenic activity in cigarette smoke condensates, *Proc. Natl. Acad. Sci. USA*, 71, 4159, 1974.

54. Austin, A.C., Claxton, L.D., and Lewtas, J., Mutagenicity of the fractionated organic emissions from diesel, cigarette smoke condensate, coke oven, and roofing tar in the Ames assay, *Environ. Mutagen.*, 7, 471, 1985.

55. Nesnow, S., Triplett, L.L., and Slaga, T.J., Studies on mouse skin tumor initiating, tumor promoting and tumor co-initiating properties of respiratory carcinogenesis, in *Cancer of the Respiratory*

Tract: Predisposing Factors, Mass, M., Kaufman, D., Siegfried, J., Steele, V., and Nesnow, S., Eds., Raven Press, New York, 1985, p. 257.

56. Gessner, P.K., Isobolographic analysis of interactions: An update on applications and utility, *Toxicology*, 105, 161, 1995.

57. Berenbaum, M.C., The expected effect of a combination of agents: The general solution, *J. Theor. Biol.*, 114, 413, 1985.

58. Carter, W.H., Jr. and Carchman, R.A., Mathematical and biostatistical methods for designing and analyzing complex chemical interactions, *Fundam. Appl. Toxicol.*, 10, 590, 1988.

59. Dressler, V., Müller, G., and Sühnel, G., CombiTool, http://www.imb-jena.de/www_bioc/CombiTool/, 1997.

60. Nesnow, S., Mass, M.J., Ross, J.A., Galati, A.J., Lambert, G.R., Gennings, C., Carter, W.H., Jr., and Stoner, G.D., Lung tumorigenic interactions in strain A/J mice of five environmental polycyclic aromatic hydrocarbons, *Environ. Health Perspect.*, 106 (Suppl. 6), 1337, 1998.

61. Nesnow, S., Davis, C., Pimentel, M., Mass, M.J., Nelson, G.B., and Ross, J.A., Interaction analyses of binary mixtures of carcinogenic PAHs using morphological cell transformation of C3H10T$^1/_2$Cl8 mouse embryo fibroblasts in culture, *Polycyclic Aromatic Compd.*, 21, 31, 2000.

15 Lung Cancer

Jay W. Tichelaar and George D. Leikauf

CONTENTS

SUMMARY

Lung cancer is the leading cause of cancer death in the United States and is increasing in incidence worldwide. The vast majority of lung cancer cases can be attributed to cigarette use with several other environmental agents accounting for the remainder of the cases. Human lung cancer consists of four main subtypes, small-cell lung cancer, squamous-cell lung cancer, lung adenocarcinoma, and large-cell lung cancer. Specific genetic and molecular attributes distinguish each type of lung cancer and indicate that treatment of each lung cancer type will require distinct therapies. In this chapter we briefly describe the major subtypes of lung cancer, consider the environmental agents that induce this neoplasm, and describe the genetic changes that have been linked with lung cancer susceptibility. The underlying molecular alterations involved in control of the multistep process of lung tumorigenesis are considered. The continued elucidation of the molecular and genetic changes observed in human lung cancer is critical to increase our understanding of this disease and to offer new avenues for improved treatment of lung cancer patients.

15.1 INTRODUCTION

Lung cancer is the leading cause of cancer mortality among both men and women in the United States, accounting for an estimated 160,000 deaths in 2004 [1], while worldwide it is estimated that over 1 million people die from lung cancer each year [2]. Cigarette smoking accounts for 85–90% of the risk for developing lung cancer [3,4], thus smoking prevention and cessation programs are extremely important in lowering the incidence of this disease. However, 46 million people in the United States and approximately 1 billion worldwide [5,6] use this addictive product, ensuring that cigarette smoking, and its health consequences, will be widespread for the foreseeable future. In addition, approximately half of new lung cancer cases in the United States now occur in former smokers [7], and improving the treatment options for these patients who have followed their physicians advice to give up smoking is a pressing need. While cigarette smoke and tobacco use account for the majority of lung cancer cases, other environmental toxicants, such as radon and asbestos, have also been implicated in the etiology of lung cancer. In addition, only 10–15% of lifetime smokers will develop lung cancer [8,9], indicating that genetic and environmental factors affect lung tumor formation. In this chapter we review the pathological classification and progression of cancers of the lung and bronchus, emphasizing the molecular and genetic changes that have been implicated in lung tumorigenesis.

15.2 CLASSIFICATION OF LUNG NEOPLASMS

Lung cancers fall mainly into two groups: small cell lung carcinoma (SCLC) that accounts for approximately 20% of lung cancers and non-small cell lung carcinoma (NSCLC) that accounts for the majority of the remaining cases. A small percentage of cases include carcinoids, undifferentiated carcinoma, and mixed tumor types. NSCLC is further divided into adenocarcinoma (including bronchioloalveolar carcinoma), squamous-cell carcinoma, and large-cell carcinoma (Table 15.1). While these classifications are historically based on pathological findings, recent results indicate that there are also distinct linkages (chromosomal loci) as well as genetic and gene expression differences between these tumor types [11–14]. SCLCs typically express markers of neuroendocrine differentiation and this may indicate that this tumor type derives from neuroepithelial bodies within the lung [15]. The subtypes of NSCLC are believed to arise from lung epithelial cells, although the exact precursor cells have not been identified. It is likely that different epithelial cell types can serve as precursors depending on the location of the initiating event within the lung. For example, squamous cell carcinoma is typically found in the conducting airways and likely arises from airway epithelial cells such as Clara cells or basal cells. In contrast, adenocarcinoma is typically localized in the lung parenchyma and likely arises from alveolar type II cells or Clara cells, a concept supported by the observation that differentiation markers from these cell types are frequently expressed in lung tumors [16,17].

TABLE 15.1
Frequency and 5-Year Survival Rates for the Four Major Histologic Types of Lung Cancer

Histologic type	Frequency (%)	5-year survival (%) (all stages)
Adenocarcinoma	32	17
Squamous cell carcinoma	29	15
Large cell carcinoma	9	11
Small cell carcinoma	18	5
Other and unspecified	11	—

Source: Data adapted from Minna, J. In *Harrison's Principles of Internal Medicine*, Braunwald, E., Fauci, A., Isselbacher, K., Kasper, D., Hauser, S., Longo, D., and Jameson, J., Eds., McGraw-Hill, New York, 2001, chap. 88.

Adenocarcinoma is now the most common type of lung cancer, replacing squamous-cell carcinoma over the last 20 to 30 years [18]. While the reasons for this change are not fully understood, it has been proposed that the introduction of reduced "tar" and nicotine cigarettes has played a role in this transition [19]. Though promoted by the tobacco companies as a "safer" cigarette, low tar/nicotine cigarettes can still be viewed as a nicotine delivery device. Altering nicotine levels in cigarettes may have resulted in a change in smoking behavior to maintain the same blood levels of nicotine in smokers. Smokers of low-nicotine cigarettes have been reported to smoke more cigarettes [20] and take larger and deeper puffs [21], delivering tobacco carcinogens to the distal lung where adenocarcinomas arise. The difference may also partly be gender based as adenocarcinoma accounts for a larger percentage of lung-cancer cases in women than in men and overall lung-cancer rates in women continued to rise over the last 20 years even after the rates in men had begun to decline. It is unclear whether this gender difference represents a biological difference, changes in the male:female ratio of the smoking population, or simply differences in smoking habits (i.e., low tar versus high tar cigarettes).

15.3 CAUSES OF LUNG CANCER

The dominant cause of lung cancer is tobacco smoke, accounting for approximately 90% of the attributable risk [3,4]. Cigarette smoke is a complex mixture of approximately 4000 chemicals, more than 60 carcinogens, and 10 known strong carcinogens, such as beta-naphthylamine, benzene, polonium 210, dibenzo[a]pyrene, benzo[a]pyrene and dibenz[c,g]carbazole [9,22,23]. Additionally many tumor promoters and cocarcinogens are present in cigarette smoke. During burning of the cigarette it is also thought that free radicals are formed by pyrolysis reactions.

Two classes of compounds have been suggested to be the primary initiators of lung tumorigenesis in tobacco smoke [9]; polycyclic aromatic hydrocarbons such as benzo[a]pyrene (B[a]P) and tobacco specific nitrosamines such as 4-(methylnitrosamino)-1-(3-pyridyl)-1-butanone (also known as nicotine-derived nitrosaminoketone or NNK). Metabolism of carcinogenic components of tobacco smoke results not only in elimination of the compounds but also bioactivation. Several groups of enzymes including cytochrome p450 (CYP), glutathione-S-transferase (GST), microsomal epoxide hydrolase (mEH), and NAD(P)H quinone oxidoreductase (NQO1) are involved in metabolism of cigarette smoke components to more carcinogenic forms. B[a]P is initially metabolized by CYP enzymes to benzo[a]pyrene diol epoxide (BPDE), markedly increasing its mutagenic potential [23]. The tobacco-specific nitrosamine, NNK, induces lung tumors in experimental animals regardless of the site of administration and is present in significant amounts in cigarette smoke [24]. In addition to cigarette smoking several other environmental agents contribute lesser amounts to lung cancer risk including radon [25], asbestos [26], metals [27], particulate matter [28,29], and air pollution [30].

15.4 GENETIC LINKAGES

Despite the clear evidence that cigarette smoking is a causative agent in lung cancer, only 10 to 15% of lifelong smokers develop the disease [8,9], suggesting that there is a genetic component to lung cancer susceptibility. Additional evidence for a genetic component in lung cancer susceptibility, or that combined environmental and genetic factors influence lung cancer susceptibility, can be inferred from the observation that smoking affects different populations in different ways [31]. Many studies have examined the relationship between genetic polymorphisms and the risk of lung cancer. Variable results have been observed linking polymorphisms in CYP, GST, and other metabolic genes to increased risk of lung cancer (for reviews see References 8 and 32). Individuals with multiple alterations in the pathways involved in activating or detoxifying tobacco smoke constituents may be at an elevated risk compared to individuals with a single alteration [32]. In addition to polymorphisms in metabolic genes, the involvement of DNA repair gene polymorphisms in lung cancer risk has been evaluated. Reduced nucleotide excision repair gene expression [33] and low DNA repair capacity [34,35] were both associated with increased lung cancer risk; but most polymorphisms examined to date have not been associated with lung cancer [8]. Therefore, caution is needed in assigning a causal relationship between lowered DNA repair capacity and lung cancer risk as this relationship may be a result of increased tumor burden.

Germ-line mutations of the tumor suppressor gene *p53* are found in patients with Li-Fraumeni syndrome, a disease associated with early onset of a variety of tumor types [36] although lung cancer accounts for only about 4% of tumors in these patients [37]. However, it has been reported that carriers of *p53* mutations have a significantly younger age of cancer diagnosis than the general population [38] and carriers of germ-line p53 mutations have an increased lung cancer risk compared to noncarriers [39]. A polymorphism at codon 72 of the *p53* gene has been reported to be variably associated with lung cancer risk. Meta-analysis demonstrated a small and marginally significant risk for all types of lung cancer, although the relative risk was higher for adenocarcinoma and small cell carcinoma, while the polymorphism was not associated with an increased risk of squamous cell carcinoma [8].

To identify chromosomal locations associated with lung cancer, a genome-wide linkage analysis was performed in families with multiple members suffering from lung, laryngeal, oropharyngeal, or hypopharyngeal cancer [40]. This population had a large number of early onset lung cancer patients. Proposed susceptibility loci were found on several chromosomes (including chromosomes 1, 4, 6, 9, 12, 20, and 21). Of these, the strongest association mapped to markers on the long arm of chromosome 6 (6q23–25), with suggestive linkages on chromosome 12q, 14, and 20. Numerous other tumor types (including mesothelioma and squamous cell carcinoma of the oral cavity) have been linked to allelic loss in the regions of 6q that overlap this interval. Further studies are needed to fine map this region and identify the genes that cause the predisposition to lung cancer.

15.5 MOLECULAR MECHANISMS

Exposure to cigarette smoke or other environmental carcinogens, coupled with genetic risk factors, contribute to the initiation of lung carcinogenesis. This sets in motion a progression from normal lung tissue to invasive carcinoma. Lung tumors contain many alterations and it has been proposed that about 20 to 30 events (mutation, chromosomal loss, gene amplification, promoter hypermethylation) will occur leading to the development of clinically apparent lung cancer [13,14]. Hanahan and Weinberg [41] have proposed a model in which cancer cells acquire six traits: (1) Self-sufficiency in growth signals, (2) insensitivity to antigrowth signals, (3) evasion of apoptosis, (4) limitless replicative potential, (5) sustained angiogenesis, and (6) tissue invasion and metastasis. The many genetic and molecular changes observed in human lung tumors cause dysregulation of most, if not all, of these characteristics, giving rise to clinical disease.

15.5.1 *RAS*

Mutations in *RAS* family genes (*HRAS*, *KRAS2*, and *NRAS*) are the most frequent mutation found in human cancers [42]. The large majority (90%) of *RAS* mutations in human lung cancer occur in the *KRAS2* gene with 85% of *KRAS* mutations in lung adenocarcinoma occurring specifically at codon 12 [43,44]. *KRAS2* is mutated in 30 to 50% of lung adenocarcinomas [44–49] and is associated with poor prognosis [43,50]. In contrast, *KRAS2* is infrequently mutated in other lung cancer types [46]. *In vitro* studies with human bronchial epithelial cells have shown that polycyclic aromatic hydrocarbons such as B[a]P that are found in cigarette smoke preferentially form DNA adducts at codon 12 and that damage at codon 12 in *KRAS2* is repaired at a lower efficiency rate than the other sites of damage [51,52]. Supporting the concept that cigarette smoke induces *KRAS2* mutations in lung cancer is the observation that *KRAS2* mutations are not observed in lung adenocarcinoma of nonsmokers [53].

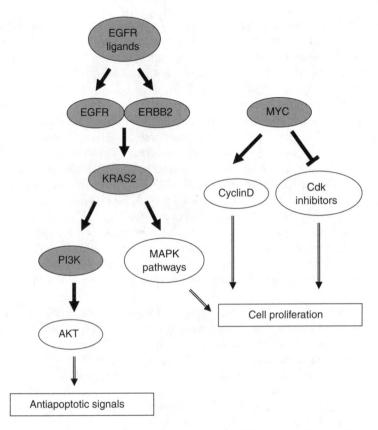

FIGURE 15.1 Growth promoting pathways in lung cancer. A simplified representation of oncogenes involved in stimulating cell proliferation during lung tumorigenesis. Shaded genes are activated in lung tumors, either by mutation (KRAS2), gene amplification (PI3K), increased expression (EGF ligands, EGFR, ERBB2, MYC), or increased activation (AKT). KRAS2 activates both mitogen activated protein kinase (MAPK) pathways, leading to cell proliferation, and the phoshphatidylinositol 3-kinase (PI3K) pathway leading to activation of AKT kinase, a central regulator of antiapoptotic signals. MYC activates genes involved in promoting cell cycle (CyclinD1) and also inhibits genes involved in blocking cell-cycle progression (Cyclin dependent kinase (Cdk) inhibitors such as p21 and p15^{INK4B}).

In unstimulated, nontransformed cells RAS proteins exist in the cytoplasm in an inactive, GDP-bound state. Growth signals transmitted through membrane-bound receptor tyrosine kinases recruit RAS, and through interactions with guanine nucleotide exchange factors (GEFs), convert RAS to a GTP bound form that is capable of transmitting the extracellular stimulus into the cell via multiple effector pathways [54]. The GTPase activity of RAS hydrolyzes GTP to GDP and the RAS molecule returns to a resting state. Activating mutations of RAS, such as the codon 12 mutation frequently seen in lung adenocarcinoma, inhibit the GTPase activity resulting in a continuously activated form of the protein. Activation of RAS leads to induction of multiple signaling pathways (e.g., Raf/mitogen activated protein kinase [Raf/MAPK], phosphoinositide-3-kinase [PI3K], phospholipase C [PLC], RalGEFs, etc.) involved in cell proliferation and cell survival with the specific pathways induced likely to depend on cell-specific factors as well as inherent differences in the RAS proteins (Figure 15.1) [55–57].

Activation of the PI3K pathway by KRAS2 can lead to the phosphorylation and activation of AKT, a serine/threonine kinase that is a key regulator of cell survival and antiapoptotic pathways in cells [58]. Gene amplification of PI3K has been reported in all major lung cancer types with the highest percentage found in squamous cell carcinomas and small cell lung cancer and this *PI3K* gene amplification correlated with increased AKT phosphorylation [59]. Increased phosphorylation of AKT has also been reported in preneoplastic bronchial metaplasia/dysplasia [60,61], suggesting that activation of this pathway is a relatively early event in the progression of lung tumorigenesis.

Mutations in the serine/threonine kinase BRAF, a RAS activated protein, are infrequent in NSCLC as they are detected in only 1 to 3% of tumors [62–64]. Interestingly, the BRAF mutational spectrum found in lung tumors is distinct from that seen in malignant melanoma occurring in distinct protein domains [63], suggesting that BRAF mutations in lung tumors are qualitatively different from those in melanoma and could require different therapeutic strategies directed toward RAF inhibition.

Studies in mice have also demonstrated an important role for RAS in lung tumorigenesis. Expression of oncogenic *Kras2* in the lungs of transgenic mice is sufficient to induce pulmonary adenocarcinoma and is also required to maintain the tumor as repression of transgene expression resulted in tumor resolution [65]. Gene-targeted mouse models have also been developed that allow for induction of tumors with a defined genetic alteration in the *KRAS2* gene. One such model allows for activation of a latent oncogenic *Kras2* allele through spontaneous recombination and leads to the formation of lung adenocarcinoma, thymic lymphoma, and skin papillomas [66]. Lung tumors are by far the most common tumor type occurring in 100% of the animals. Histopathological and immuno-histochemical analyses indicated that the lung tumors were adenocarcinomas with an alveolar type II cell origin.

Several other models have used Cre recombinase to activate an oncogenic allele of *Kras2*. Aden-oviral delivery of Cre to the respiratory tree resulted in progressive lung tumor formation [67,68]. Surprisingly, transgenic expression of Cre from a ubiquitously expressed cytomegalovirus (CMV) promoter resulted in lung tumor formation in 100% of mice, while most other tissues either did not develop tumors or developed tumors at low frequency although recombination was demonstrated in all the tissues examined [69]. Although not all cells expressing oncogenic *Kras2* in the lung develop into lung tumors, these results suggest that lung tissue is particularly susceptible to the oncogenic effects of *Kras2*.

While mutated *Kras2* is an established oncogene, wild-type Kras2 may function as a tumor suppressor. Complete ablation of Kras2 in mice was embryonic lethal [70,71] but mice lacking a single copy of the *Kras2* gene had increased susceptibility to chemical induction of lung tumorigenesis independent of mouse strain [72]. In addition, wild-type Kras2 inhibited colony formation and tumor development by transformed NIH/3T3 cells and a mouse lung-tumor cell line containing activated Kras2. In mouse tumors, loss of the wild-type allele was found in 67 to 100% of chemically induced lung adenocarcinomas that contained a mutant *Kras2* allele. Similarly, one study reported that human lung adenocarcinomas harboring a mutant *KRAS2* allele have lost the remaining wild-type allele [73], although this was not observed in a separate study [74].

15.5.2 *MYC*

The *MYC* proto-oncogene family consists of *MYC*, *MYCN*, and *MYCL1*. These gene family members are members of the basic-helix-loop-helix (bHLH) superfamily of nuclear transcription factors. Evidence suggests that *RAS* and *MYC* can collaborate in inducing cell transformation [75], potentially through *RAS* stabilization of MYC protein [76]. MYC proteins heterodimerize with the protein MAX to regulate transcription by binding specific hexameric DNA sequences known as E boxes. Conversely, heterodimers of MAX and Max dimerization (MAD) proteins repress transcription from the same promoter elements [77]. In addition, it is now clear that MYC/MAX heterodimers recruit histone acetylases that play an important role in MYC transcriptional activation and conversely MAD/MAX dimers recruit histone deacetylases that act as transcriptional corepressors [78]. Functions of MYC that could effect lung tumor cell proliferation include increased expression of CyclinD1 [79] or repression of cyclin dependent kinase inhibitors p15^{INK4B} [80,81] and p21 [82] (Figure 15.1). Increased expression of MYC has been reported in both SCLC and NSCLC [83,84].

Transgenic expression of Myc in the lung under control of a Clara cell secretory protein promoter fragment resulted in Clara cell hyperplasia [85] while expression of Myc under control of a surfactant protein C promoter fragment caused hyperplasia and bronchioloalveolar adenomas and adenocarcinomas. Combination of Myc with transgenic over expression of soluble epidermal growth factor (EGF) in the lung caused earlier onset and a higher percentage of animals with adenocarcinomas [86] and these mice were more susceptible to the carcinogenic effects of NNK [87].

15.5.3 EGFR and ERBB2

Epidermal growth factor receptor (EGFR, also known as ErbB1) is a receptor tyrosine kinase that is stimulated by a family of related ligands including EGF, transforming growth factor-α (TGF-α), amphiregulin, epiregulin, betacellulin, and heparin-binding EGF (HB-EGF). Ligand binding to these receptor tyrosine kinases activates RAS and subsequent downstream signaling cascades (Figure 15.1). The ERBB2 protein (also known as HER2/Neu) is a related transmembrane tyrosine kinase but by itself has no ability to bind ligand and functions by heterodimerizing with the closely related proteins ERBB3 or ERBB4 and binding neuregulins [88]. Alterations in EGFR and ERBB2 in lung cancer typically involve over expression of these genes without gene amplification [89–93]. Elevated expression of ERBB3 has also been observed in 20% of NSCLC and is associated with poor prognosis [94]. In addition to increased expression of the receptors, elevated expression of EGFR ligands has been reported in lung cancer including TGF-α [89,95] and amphiregulin [96].

Cancer therapies aimed at blocking signaling of the EGFR and ERBB2/ERBB3 signaling pathways have recently received considerable attention. These therapeutics can be broadly divided into two categories, monoclonal antibodies or small molecule inhibitors of tyrosine kinase activity [97], and several compounds have been approved for human use including the small molecule inhibitor gefitinib in NSCLC. Several phase II clinical trials have examined the efficacy of gefitinib in NSCLC with response rates in the range of 10 to 18% [98–100]. Until recently, the reason for differences in response between patients was not known, but it now appears that patients who respond to gefitinib possess specific mutations in the tyrosine kinase domain of EGFR that render them sensitive to the anticancer effect of this drug [101,102]. In a related study immunohistochemistry was used to detect the phosphorylated (active) form of the kinases, mitogen activated protein kinase (MAPK) and AKT, in resected lung tumors prior to initiation of treatment. Positive staining for phosphorylated AKT, but not phosphorylated MAPK, was associated with an improved response to gefitinib [103]. To date, phase II studies using the monoclonal antibody trastumazab directed against ERBB2 have shown a response in only a small percentage of patients expressing high levels of ERBB2 in their tumors, while most patients have shown no difference in response relative to standard treatment [104]. These results point out that treatment of lung cancer, or any other neoplasm, is likely to be specific for the mutational spectrum and cellular environment that is present within a particular tumor and "designer drugs" may be required for effective treatment of individual patients.

15.5.4 TP53

The tumor suppressor p53 (TP53) functions in regulation of the cell cycle, control of apoptosis, and cellular response to DNA damage and so plays a critical role in several aspects of tumorigenesis. The p53 protein is a transcription factor that can induce transcription of multiple gene targets and inactivating mutations occur most frequently in the DNA binding domain, eliminating the ability of the protein to bind DNA and activate transcription. p53 regulates multiple pathways that normally function coordinately to protect the cell from DNA damage and thus has been referred to as the guardian of the genome (Figure 15.2). Activation of p53 can occur by multiple signals including DNA damage [105], hyperploidy [106,107], and hypoxia [108]. In response to DNA damage, such as those that occur from tobacco carcinogens, p53 is activated directly by ataxia telangiectasia mutated (ATM) kinase [109] and indirectly by ataxia telangiectasia and Rad3-related (ATR) kinase [110],

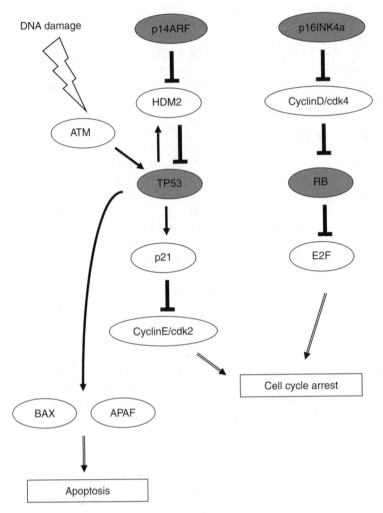

FIGURE 15.2 Inhibition of the cell cycle and apoptosis in lung cancer. Pathways involved in control of cell cycle and apoptosis during lung tumorigenesis. Both the p14^{ARF}/TP53 pathway and the p16^{INK4A}/Rb pathway can inhibit activity of cyclin/cyclin dependent kinase (CDK) complexes involved in cell-cycle control. Reduced expression or mutations in these proteins that negatively regulate cell-cycle progression results in unregulated cell proliferation. TP53 is also a key mediator of apoptosis in cells in response to DNA damage, thus mutations in TP53 allow damaged cells to escape apoptosis in response to DNA damage. Shaded genes are frequently inactivated during lung tumorigenesis.

key enzymes in the regulation of cellular response to DNA damage. Phosphorylation of p53 by ATM and ATR kinase, as well as additional kinases [111], at specific serine residues induces transcriptional activity of p53. Activation of p53 induces transcription of genes that negatively regulate cell-cycle progression such as the cyclin dependent kinase inhibitor p21 and 14-3-3σ [112,113] and genes that induce apoptosis such as BAX and APAF-1 [114–116]. Thus, when a cell senses DNA damage p53 plays a critical role in blocking cell division and allowing the DNA repair processes to occur. If repair is unsuccessful p53 is involved in initiating an apoptotic cascade that will remove the cell containing the mutation.

Inactivation of p53 occurs in approximately 90% of SCLC and 40 to 70% of NSCLC [117] by point mutations that alter conserved regions of the protein. The spectrum of p53 mutation hotspots found in lung cancer is distinct from other human cancers and the differences are likely to be induced by tobacco smoke carcinogens [118]. For example codon 157 is a mutational hotspot in lung cancer from smokers but not from nonsmokers [117]. Mutational hotspots in lung cancer contain a methylated CpG dinucleotide that serves as a target for DNA adduct formation caused by the mutagenic B[a]P metabolite BPDE that preferentially forms adducts at these sites [119–121]. The observation that BPDE induces the specific pattern of p53 mutation observed in human lung cancer lends strong evidence for a causative role of tobacco-smoke carcinogens in the formation of the p53 mutational spectra observed in human lung cancer.

The protein HDM2 (Mdm2 in mice), an ubiquitin E3 ligase, is a key negative regulator of p53 activation (Figure 15.2). Phosphorylation of the N-terminus of p53 disrupts the interaction of p53 and HDM2 allowing p53 to activate its downstream target genes. HDM2 itself is inhibited by the tumor suppressor gene $p14^{ARF}$, an alternate protein generated from the same locus as the $p16^{INK4A}$ tumor suppressor. HDM2 is over expressed in 25 to 50% of NSCLC [122,123] and has been reported to be associated with a favorable prognosis [123], a result that was unexpected given that this protein inhibits the activity of the p53 tumor suppressor gene. Expression of the $p14^{ARF}$ tumor suppressor is lost in approximately 60% of human lung cancer [124]. It was originally hypothesized that since $p14^{ARF}$ functions to inactivate p53, the presence of these mutations would be inversely correlated. However, such an inverse correlation has not been observed in NSCLC [125,126].

Studies in mice have confirmed an important role for p53 during lung tumorigenesis. Mice with targeted ablation of p53 are viable but develop multiple types of spontaneous tumors at an early age, especially lymphomas and sarcomas [127]. Mice lacking a single copy of the $p53$ gene have more and larger lung tumors following administration of the carcinogen B[a]P and this effect is amplified if the mice also lack a single copy of the $p16^{INK4A}/p14^{ARF}$ locus [128]. Mice expressing a mutant form of p53 from a lung-specific promoter were more likely to develop spontaneous lung adenocarcinomas than nontransgenic littermates [129]. Recently it has been demonstrated that while lung-specific ablation of p53 alone resulted in the formation of spontaneous adenocarcinomas, lung-specific ablation of both p53 and Rb in mice caused lesions consistent with SCLC including metastasis to distant organ sites [130]. It is clear from these studies that p53 is important in lung cancer development in these models and cooperation between p53 and other tumor suppressor pathways can influence the progression and histological subtype of lung cancer.

15.5.5 Rb-p16^{INK4A} Pathway

Mutations in tumor suppressor genes that make up the retinoblastoma (Rb) and p16^{INK4A} pathway are also frequently found in human lung tumors. Mutations in the Rb gene are especially common (>90%) in SCLC but are rare in NSCLC [131–133]. In contrast, decreased levels of cyclin-dependent kinase inhibitor p16^{INK4A} are frequent in NSCLC but are rarely observed in SCLC. The decreased levels of p16^{INK4A} observed in NSCLC are controlled by the epigenetic mechanism of promoter hypermethylation [134–136]. The inverse correlation of Rb and p16^{INK4A} mutations in lung cancer has led to the hypothesis that this pathway is mutated in most, if not all, human lung tumors [137].

Rb associates with several other proteins in a phosphorylation-dependent manner resulting in the regulation of the cell cycle (Figure 15.2). Rb protein is hypophosphorylated in resting cells, or cells arrested at the G1/S transition of the cell cycle, but is hyperphosphorylated by cyclin-dependent kinases upon the entry into the cell cycle [138–140]. The activating E2F transcription factors are the best characterized binding partners for Rb and are key regulators of the cell cycle [141,142]. Hypophosphorylated Rb binds to and inactivates E2F family members thus inhibiting cell cycle progression while phosphorylation of Rb by cyclin/cyclin-dependent kinase complexes releases E2F from the complex. p16^{INK4A} controls the activity of Rb by inactivating the cyclinD/cyclin-dependent kinase 4 complex and preventing phosphorylation of Rb [143].

15.5.6 TGF-β SIGNALING

Transforming growth factor-β (TGF-β) has well characterized growth inhibitory properties on epithelial cells including lung epithelial cells [144,145]. Signaling through the TGF-β receptor, TGFβRII, results in the phosphorylation of a family of proteins named the Smads that can positively or negatively regulate target gene transcription [146]. Expression of TGFβRII is reduced in NSCLC compared to normal lung [147–149], indicating the loss of a growth inhibitory mechanism in lung tumors. Mutations in *TGFβRII* and *SMAD* genes are detected infrequently in human lung tumors [150,151]. Expression of TGFβRII can be lost in lung cancer cells due to an epigenetic mechanisms involving histone deacetylation and altered chromatin structure [152] or promoter hypermethylation [147] suggesting alternate mechanisms for the loss of TGFβRII expression observed in lung tumors.

15.5.7 EVASION OF APOPTOSIS

One of the hallmarks of cancer is the ability to evade apoptosis [41]. As described above, p53 is a key regulator of apoptosis in response to DNA damage and other injuries and is lost or mutated in a large percentage of both SCLC and NSCLC. One mechanism used by p53 to initiate apoptosis in normal cells is by activation of proapoptotic members and repression of antiapoptotic members of the *Bcl*-2 family of genes. The antiapoptotic Bcl-2 protein is expressed in both SCLC and NSCLC [153,154] with some studies demonstrating a survival benefit for patients with Bcl-2 positive tumors [155] while other studies did not [156]. Interestingly, it has been reported that nicotine can induce phosphorylation of Bcl-2 in a SCLC cell line with the result being enhanced survival and resistance to apoptotic drugs [157]. While nicotine also induces phosphorylation of the proapoptotic molecule Bad, in this case the result is inhibition of its proapoptotic function [158].

15.5.8 ANGIOGENESIS

Lung tumors, as all solid tumors, require a developing blood supply to sustain continued growth. Microvessel density in early operable NSCLCs is higher in the invading front of the tumors compared to inner tumor areas or surrounding normal tissue [159]. Microvessel density count is a highly significant adverse prognostic factor in patients with NSCLC [160]. The angiogenic growth factor vascular endothelial growth factor (VEGF) is frequently expressed in lung tumors [161] and is associated with new vessel formation and an unfavorable prognosis in NSCLC [162]. Basic fibroblast growth factor (bFGF or FGF2) is also a strong angiogenic factor but the relationship between lung tumor expression of bFGF and prognosis is unclear with some studies indicating it is a poor prognostic factor [163] and others finding no association [164]. It has been reported that when VEGF, bFGF, and platelet derived endothelial cell growth factor are measured simultaneously there is a significant trend between increased numbers of angiogenic factors expressed in lung tumors and lymph node involvement [165]. It has also been reported that serum bFGF levels are increased in SCLC and NSCLC [166] and that high pretreatment levels of serum bFGF is a predictor of poor prognosis in SCLC [167].

15.5.9 TELOMERASE ACTIVITY

The ends of chromosomes are capped with a series of hexameric repeats called telomeres. During normal somatic cell division an absence of telomerase activity results in progressive telomere shortening and this process contributes to cell senescence and death. Telomerase is normally expressed in germ cells and some stem cells and is widely reexpressed in tumor cells [168,169]. The majority of SCLC and NSCLC have increased telomerase activity [170,171]. The reexpression of this activity in tumor cells is a major factor in the acquisition of limitless replicative potential, one of the hallmarks of cancer cells [41]. Because reactivation of telomerase is common in a wide variety of tumors, antitelomerase drugs are in development as new treatments for a variety of cancers [172,173].

15.6 METASTATIC DISEASE AND THE NEED FOR EARLY DETECTION

While survival rates for lung cancer have remained low, prognosis improves dramatically if the disease is detected before regional or distant metastasis. Unfortunately, with no accepted method for early screening only 16% of lung tumors are detected while they are still localized [1]. Surgical resection, often with radiation or chemotherapy, remains the primary treatment for lung tumors and if metastasis has not occurred 5-year survival rates are approximately 50%, a dramatic improvement from the overall 5-year survival rate of approximately 15% for lung cancer. Five-year survival rates in the United States for the years 1992 to 1999 were reported to be 16% for lung cancer with regional metastasis and 2% for lung cancer with distant metastasis [1].

The presence of a number of proteins has been associated with lung cancer metastasis. Matrix metalloproteinases (MMPs) have been reported to be associated with metastasis in several tumor types and are expressed in lung tumors [174]. While elevated expression of MMP2 and MMP9 in lung tumors correlated with poor prognosis [175,176], the MMP tissue inhibitor of metalloproteinase 1 (TIMP1) also has been shown to correlate with poor prognosis when expressed in lung tumors [177,178]. Genes that are upregulated in metastatic lung cancer compared to localized lung cancer include metastasis associated gene 1 (*MTA1*) [179], cyclin E [180], the calcium-binding proteins S100A2 and S100P, and a trypsinogen protease [181]. In lung adenocarcinoma, but not squamous cell carcinoma, co-expression of fibroblast growth factor-7 (FGF7) and its receptor correlate with lymph node metastasis and shortened survival suggesting that establishment of this autocrine signaling loop may play a role in disease progression [182].

The discrepancy in 5-year survival rates for patients with localized disease compared to metastatic disease points out the need for early detection. Low-dose spiral computerized tomograph (CT) has been studied as an early detection tool for lung cancer screening and has improved sensitivity compared to standard chest x-ray [183,184]. A limitation of this screening method is that noncalcified nodules are detected in a majority of patients and require periodic follow up by CT, with only 1 to 2% of small nodules proving to be cancerous subsequently [185,186]. Autofluorescence bronchoscopy can be used to detect preneoplastic lesions in upper airways and is more sensitive than white-light bronchoscopy [187]. A combination of these approaches may be complementary and the best current method of screening for early lung-cancer lesions [186].

The use of RNA microarray and proteomic technology to search for a molecular lung cancer "signature," particularly in readily available biological samples such as serum, holds promise for improved early detection of the disease. Several studies have demonstrated unique gene-expression profiles between normal lung tissue and lung tumor tissue and also between early- and late-stage lung tumors in both humans and experimental animals [149,188–191]. In a number of cases gene-expression profiles that correlate with patient prognosis [192–194] or metastasis [181,195] have been described. The use of mass spectrometry proteomic techniques can also detect differences in protein profiles between tumor types and has been shown to accurately classify tumors based on nodal involvement and survival [196]. These studies are important in defining the molecular mechanisms that occur during lung tumorigenesis and may also prove useful in determining the treatment to be provided following surgical resection. However, they rely on tumor tissue obtained during surgery and thus are not an effective screening procedure. To overcome this limitation investigators have undertaken studies examining the gene expression or protein profile in more readily available biological samples such as serum. Recent studies have used Surface Enhanced Laser Desorption/Ionization (SELDI) technology to develop a model that tested serum samples from lung-cancer patients and normal controls and yielded a sensitivity of 93% and a specificity of 97% [197,198]. This technique has also been used with samples obtained by laser capture microdissection to demonstrate protein pattern differences in malignant versus premalignant lung neoplasms [199]. These studies demonstrate the utility of this system for identifying protein expression patterns even in small samples and the possibilities for its use in early cancer screening.

15.7 CONCLUSION

Lung cancer is the leading cause of cancer mortality in the United States and will remain so for the foreseeable future. Thus, control of lung cancer is necessary to continue to reduce overall cancer mortality rates. Public policy and education strategies designed to prevent smoking initiation and encourage smoking cessation will continue to be critical in reducing the impact of this deadly disease. Given the current prevalence of this tumor type and the continued large number of smokers worldwide, a better understanding of the molecular and genetic abnormalities underlying this disease is necessary to improve treatments for lung cancer and to design successful treatments for prevention of the disease in high risk populations. Lung cancer is a diverse group of neoplasms with distinct molecular and genetic signatures and successful treatment of this disease will probably require interventions that target the specific mutations and abnormalities found in a given tumor type.

REFERENCES

1. Jemal, A., Tiwari, R.C., Murray, T., Ghafoor, A., Samuels, A., Ward, E., Feuer, E.J., and Thun, M.J. Cancer statistics. *CA Cancer J. Clin.*, *54*: 8–29, 2004.
2. Parkin, D.M., Bray, F., Ferlay, J., and Pisani, P. Estimating the world cancer burden: Globocan 2000. *Int. J. Cancer*, *94*: 153–156, 2001.
3. Alberg, A.J. and Samet, J.M. Epidemiology of lung cancer. *Chest*, *123*: 21S–49S, 2003.
4. Beckett, W.S. Epidemiology and etiology of lung cancer. *Clin. Chest. Med.*, *14*: 1–15, 1993.
5. CDC Cigarette smoking among adults — United States, 2002. *Morbidity and Mortality Weekly Report*, *53*: 427–431, 2004.
6. Proctor, R.N. Tobacco and the global lung cancer epidemic. *Nat. Rev. Cancer*, *1*: 82–86, 2001.
7. Tong, L., Spitz, M.R., Fueger, J.J., and Amos, C.A. Lung carcinoma in former smokers. *Cancer*, *78*: 1004–1010, 1996.
8. Kiyohara, C., Otsu, A., Shirakawa, T., Fukuda, S., and Hopkin, J.M. Genetic polymorphisms and lung cancer susceptibility: A review. *Lung Cancer*, *37*: 241–256, 2002.
9. Hecht, S.S. Cigarette smoking and lung cancer: Chemical mechanisms and approaches to prevention. *Lancet Oncol.*, *3*: 461–469, 2002.
10. Minna, J. Neoplasms of the lung. In *Harrison's Principles of Internal Medicine*, Braunwald, E., Fauci, A., Isselbacher, K., Kasper, D., Hauser, S., Longo, D., and Jameson, J., Eds., McGraw-Hill, New York, 2001, chapter 88.
11. Mitsuuchi, Y. and Testa, J.R. Cytogenetics and molecular genetics of lung cancer. *Am. J. Med. Genet.*, *115*: 183–188, 2002.
12. Massion, P.P. and Carbone, D.P. The molecular basis of lung cancer: Molecular abnormalities and therapeutic implications. *Respir. Res.*, *4*: 12, 2003.
13. Zochbauer-Muller, S., Gazdar, A.F., and Minna, J.D. Molecular pathogenesis of lung cancer. *Ann. Rev. Physiol.*, *64*: 681–708, 2002.
14. Girard, L., Zochbauer-Muller, S., Virmani, A.K., Gazdar, A.F., and Minna, J.D. Genome-wide allelotyping of lung cancer identifies new regions of allelic loss, differences between small cell lung cancer and non-small cell lung cancer, and loci clustering. *Cancer Res.*, *60*: 4894–4906, 2000.
15. Brambilla, E., Lantuejoul, S., and Sturm, N. Divergent differentiation in neuroendocrine lung tumors. *Semin. Diagn. Pathol.*, *17*: 138–148, 2000.
16. Linnoila, R.I., Jensen, S.M., Steinberg, S.M., Mulshine, J.L., Eggleston, J.C., and Gazdar, A.F. Peripheral airway cell marker expression in non-small cell lung carcinoma. Association with distinct clinicopathologic features. *Am. J. Clin. Pathol.*, *97*: 233–243, 1992.

17. Bejarano, P.A., Baughman, R.P., Biddinger, P.W., Miller, M.A., Fenoglio-Preiser, C., al-Kafaji, B., Di Lauro, R., and Whitsett, J.A. Surfactant proteins and thyroid transcription factor-1 in pulmonary and breast carcinomas. *Mod. Pathol.*, *9*: 445–452, 1996.

18. Wynder, E.L. and Muscat, J.E. The changing epidemiology of smoking and lung cancer histology. *Environ. Health Perspect.*, *103 (Suppl.)*: 143–148, 1995.

19. Thun, M.J. and Burns, D.M. Health impact of "reduced yield" cigarettes: A critical assessment of the epidemiological evidence. *Tob Control*, *10 (Suppl)*: 14–11, 2001.

20. Risks associated with smoking cigarettes with low machine-measured yields of tar and nicotine. US Department of Health and Human Services, National Cancer Institute, National Institutes of Health, Bethesda, MD, 2001.

21. Herning, R.I., Jones, R.T., Bachman, J., and Mines, A.H. Puff volume increases when low-nicotine cigarettes are smoked. *Br. Med. J. (Clin. Res. Ed.)*, *283*: 187–189, 1981.

22. Hoffmann, D., Hoffmann, I., and El-Bayoumy, K. The less harmful cigarette: A controversial issue. A tribute to Ernst L. Wynder. *Chem. Res. Toxicol.*, *14*: 767–790, 2001.

23. Hecht, S.S. Tobacco smoke carcinogens and lung cancer. *J. Natl. Cancer Inst.*, *91*: 1194–1210, 1999.

24. Hecht, S.S. Biochemistry, biology, and carcinogenicity of tobacco-specific *N*-nitrosamines. *Chem. Res. Toxicol.*, *11*: 559–603, 1998.

25. Pawel, D.J. and Puskin, J.S. The U.S. Environmental Protection Agency's assessment of risks from indoor radon. *Health Phys.*, *87*: 68–74, 2004.

26. LaDou, J. The asbestos cancer epidemic. *Environ. Health Perspect.*, *112*: 285–290, 2004.

27. Kasprzak, K.S., Sunderman, F.W., Jr., and Salnikow, K. Nickel carcinogenesis. *Mutat. Res.*, *533*: 67–97, 2003.

28. Knaapen, A.M., Borm, P.J., Albrecht, C., and Schins, R.P. Inhaled particles and lung cancer. Part A: Mechanisms. *Int. J. Cancer*, *109*: 799–809, 2004.

29. Borm, P.J., Schins, R.P., and Albrecht, C. Inhaled particles and lung cancer, part B: Paradigms and risk assessment. *Int. J. Cancer*, *110*: 3–14, 2004.

30. Vineis, P., Forastiere, F., Hoek, G., and Lipsett, M. Outdoor air pollution and lung cancer: Recent epidemiologic evidence. *Int. J. Cancer*, *111*: 647–652, 2004.

31. Peto, J. Cancer epidemiology in the last century and the next decade. *Nature*, *411*: 390–395, 2001.

32. Bartsch, H., Nair, U., Risch, A., Rojas, M., Wikman, H., and Alexandrov, K. Genetic polymorphism of CYP genes, alone or in combination, as a risk modifier of tobacco-related cancers. *Cancer Epidemiol. Biomarkers Prev.*, *9*: 3–28, 2000.

33. Cheng, L., Spitz, M.R., Hong, W.K., and Wei, Q. Reduced expression levels of nucleotide excision repair genes in lung cancer: A case–control analysis. *Carcinogenesis*, *21*: 1527–1530, 2000.

34. Wei, Q., Cheng, L., Hong, W.K., and Spitz, M.R. Reduced DNA repair capacity in lung cancer patients. *Cancer Res.*, *56*: 4103–4107, 1996.

35. Wei, Q., Cheng, L., Amos, C.I., Wang, L.E., Guo, Z., Hong, W.K., and Spitz, M.R. Repair of tobacco carcinogen-induced DNA adducts and lung cancer risk: A molecular epidemiologic study. *J. Natl. Cancer Inst.*, *92*: 1764–1772, 2000.

36. Strong, L.C., Williams, W.R., and Tainsky, M.A. The Li-Fraumeni syndrome: From clinical epidemiology to molecular genetics. *Am. J. Epidemiol.*, *135*: 190–199, 1992.

37. Kleihues, P., Schauble, B., zur Hausen, A., Esteve, J., and Ohgaki, H. Tumors associated with p53 germline mutations: A synopsis of 91 families. *Am. J. Pathol.*, *150*: 1–13, 1997.

38. Nichols, K.E., Malkin, D., Garber, J.E., Fraumeni, J.F., Jr., and Li, F.P. Germ-line p53 mutations predispose to a wide spectrum of early-onset cancers. *Cancer Epidemiol. Biomarkers Prev.*, *10*: 83–87, 2001.

39. Hwang, S.J., Cheng, L.S., Lozano, G., Amos, C.I., Gu, X., and Strong, L.C. Lung cancer risk in germline p53 mutation carriers: Association between an inherited cancer predisposition, cigarette smoking, and cancer risk. *Hum. Genet.*, *113*: 238–243, 2003.

40. Bailey-Wilson, J.E., Amos, C.I., Pinney, S.M., Petersen, G.M., De Andrade, M., Wiest, J.S., Fain, P., Schwartz, A.G., You, M., Franklin, W., Klein, C., Gazdar, A., Rothschild, H., Mandal, D., Coons, T., Slusser, J., Lee, J., Gaba, C., Kupert, E., Perez, A., Zhou, X., Zeng, D., Liu, Q., Zhang, Q., Seminara, D., Minna, J., and Anderson, M.W. A major lung cancer susceptibility locus maps to chromosome 6q23–25. *Am. J. Hum. Genet.*, *75*: 460–474, 2004.

41. Hanahan, D. and Weinberg, R.A. The hallmarks of cancer. *Cell*, *100*: 57–70, 2000.

42. Anderson, M.W., Reynolds, S.H., You, M., and Maronpot, R.M. Role of proto-oncogene activation in carcinogenesis. *Environ. Health Perspect.*, *98*: 13–24, 1992.

43. Slebos, R.J., Kibbelaar, R.E., Dalesio, O., Kooistra, A., Stam, J., Meijer, C.J., Wagenaar, S.S., Vanderschueren, R.G., van Zandwijk, N., Mooi, W.J. et al. K-ras oncogene activation as a prognostic marker in adenocarcinoma of the lung. *N. Engl. J. Med.*, *323*: 561–565, 1990.

44. Mills, N.E., Fishman, C.L., Rom, W.N., Dubin, N., and Jacobson, D.R. Increased prevalence of K-ras oncogene mutations in lung adenocarcinoma. *Cancer Res.*, *55*: 1444–1447, 1995.

45. Rodenhuis, S., Slebos, R.J., Boot, A.J., Evers, S.G., Mooi, W.J., Wagenaar, S.S., van Bodegom, P.C., and Bos, J.L. Incidence and possible clinical significance of K-ras oncogene activation in adenocarcinoma of the human lung. *Cancer Res.*, *48*: 5738–5741, 1988.

46. Rodenhuis, S. and Slebos, R.J. Clinical significance of ras oncogene activation in human lung cancer. *Cancer Res.*, *52*: 2665s–2669s, 1992.

47. Reynolds, S.H., Anna, C.K., Brown, K.C., Wiest, J.S., Beattie, E.J., Pero, R.W., Iglehart, J.D., and Anderson, M.W. Activated protooncogenes in human lung tumors from smokers. *Proc. Natl. Acad. Sci. USA*, *88*: 1085–1089, 1991.

48. Suzuki, Y., Orita, M., Shiraishi, M., Hayashi, K., and Sekiya, T. Detection of ras gene mutations in human lung cancers by single-strand conformation polymorphism analysis of polymerase chain reaction products. *Oncogene*, *5*: 1037–1043, 1990.

49. Li, S., Rosell, R., Urban, A., Font, A., Ariza, A., Armengol, P., Abad, A., Navas, J.J., and Monzo, M. K-ras gene point mutation: A stable tumor marker in non-small cell lung carcinoma. *Lung Cancer*, *11*: 19–27, 1994.

50. Kern, J.A., Slebos, R.J., Top, B., Rodenhuis, S., Lager, D., Robinson, R.A., Weiner, D., and Schwartz, D.A. C-erbB-2 expression and codon 12 K-ras mutations both predict shortened survival for patients with pulmonary adenocarcinomas. *J. Clin. Invest.*, *93*: 516–520, 1994.

51. Feng, Z., Hu, W., Chen, J.X., Pao, A., Li, H., Rom, W., Hung, M.C., and Tang, M.S. Preferential DNA damage and poor repair determine ras gene mutational hotspot in human cancer. *J. Natl. Cancer Inst.*, *94*: 1527–1536, 2002.

52. Hu, W., Feng, Z., and Tang, M.S. Preferential carcinogen-DNA adduct formation at codons 12 and 14 in the human K-ras gene and their possible mechanisms. *Biochemistry*, *42*: 10012–10023, 2003.

53. Ahrendt, S.A., Decker, P.A., Alawi, E.A., Zhu Yr, Y.R., Sanchez-Cespedes, M., Yang, S.C., Haasler, G.B., Kajdacsy-Balla, A., Demeure, M.J., and Sidransky, D. Cigarette smoking is strongly associated with mutation of the K-ras gene in patients with primary adenocarcinoma of the lung. *Cancer*, *92*: 1525–1530, 2001.

54. Malumbres, M. and Barbacid, M. RAS oncogenes: The first 30 years. *Nat. Rev. Cancer*, *3*: 459–465, 2003.

55. Hancock, J.F. Ras proteins: Different signals from different locations. *Nat. Rev. Mol. Cell Biol.*, *4*: 373–384, 2003.

56. Hingorani, S.R. and Tuveson, D.A. Ras redux: Rethinking how and where Ras acts. *Curr. Opin. Genet. Dev.*, *13*: 6–13, 2003.

57. Olson, M.F. and Marais, R. Ras protein signalling. *Semin. Immunol.*, *12*: 63–73, 2000.

58. Nicholson, K.M. and Anderson, N.G. The protein kinase B/Akt signalling pathway in human malignancy. *Cell Signal.*, *14*: 381–395, 2002.

59. Massion, P.P., Taflan, P.M., Shyr, Y., Rahman, S.M., Yildiz, P., Shakthour, B., Edgerton, M.E., Ninan, M., Andersen, J.J., and Gonzalez, A.L. Early involvement of the phosphatidylinositol 3-kinase/Akt pathway in lung cancer progression. *Am. J. Respir. Crit. Care Med.*, *170*: 1088–1094, 2004.

60. Tsao, A.S., McDonnell, T., Lam, S., Putnam, J.B., Bekele, N., Hong, W.K., and Kurie, J.M. Increased phospho-AKT (Ser(473)) expression in bronchial dysplasia: Implications for lung cancer prevention studies. *Cancer Epidemiol. Biomarkers Prev.*, *12*: 660–664, 2003.

61. Balsara, B.R., Pei, J., Mitsuuchi, Y., Page, R., Klein-Szanto, A., Wang, H., Unger, M., and Testa, J.R. Frequent activation of AKT in non-small cell lung carcinomas and preneoplastic bronchial lesions. *Carcinogenesis*, 2004.

62. Cohen, Y., Xing, M., Mambo, E., Guo, Z., Wu, G., Trink, B., Beller, U., Westra, W.H., Ladenson, P.W., and Sidransky, D. BRAF mutation in papillary thyroid carcinoma. *J. Natl. Cancer Inst.*, *95*: 625–627, 2003.

63. Naoki, K., Chen, T.H., Richards, W.G., Sugarbaker, D.J., and Meyerson, M. Missense mutations of the BRAF gene in human lung adenocarcinoma. *Cancer Res.*, *62*: 7001–7003, 2002.

64. Brose, M.S., Volpe, P., Feldman, M., Kumar, M., Rishi, I., Gerrero, R., Einhorn, E., Herlyn, M., Minna, J., Nicholson, A., Roth, J.A., Albelda, S.M., Davies, H., Cox, C., Brignell, G., Stephens, P., Futreal, P.A., Wooster, R., Stratton, M.R., and Weber, B.L. BRAF and RAS mutations in human lung cancer and melanoma. *Cancer Res.*, *62*: 6997–7000, 2002.

65. Fisher, G.H., Wellen, S.L., Klimstra, D., Lenczowski, J.M., Tichelaar, J.W., Lizak, M.J., Whitsett, J.A., Koretsky, A., and Varmus, H.E. Induction and apoptotic regression of lung adenocarcinomas by regulation of a K-ras transgene in the presence and absence of tumor suppressor genes. *Genes Dev.*, *15*: 3249–3262, 2001.

66. Johnson, L., Mercer, K., Greenbaum, D., Bronson, R.T., Crowley, D., Tuveson, D.A., and Jacks, T. Somatic activation of the K-ras oncogene causes early onset lung cancer in mice. *Nature*, *410*: 1111–1116, 2001.

67. Jackson, E.L., Willis, N., Mercer, K., Bronson, R.T., Crowley, D., Montoya, R., Jacks, T., and Tuveson, D.A. Analysis of lung tumor initiation and progression using conditional expression of oncogenic K-ras. *Genes Dev.*, *15*: 3243–3248, 2001.

68. Meuwissen, R., Linn, S.C., van der Valk, M., Mooi, W.J., and Berns, A. Mouse model for lung tumorigenesis through Cre/lox controlled sporadic activation of the K-Ras oncogene. *Oncogene*, *20*: 6551–6558, 2001.

69. Guerra, C., Mijimolle, N., Dhawahir, A., Dubus, P., Barradas, M., Serrano, M., Campuzano, V., and Barbacid, M. Tumor induction by an endogenous K-ras oncogene is highly dependent on cellular context. *Cancer Cell.*, *4*: 111–120, 2003.

70. Johnson, L., Greenbaum, D., Cichowski, K., Mercer, K., Murphy, E., Schmitt, E., Bronson, R.T., Umanoff, H., Edelmann, W., Kucherlapati, R., and Jacks, T. K-ras is an essential gene in the mouse with partial functional overlap with N-ras. *Genes Dev.*, *11*: 2468–2481, 1997.

71. Koera, K., Nakamura, K., Nakao, K., Miyoshi, J., Toyoshima, K., Hatta, T., Otani, H., Aiba, A., and Katsuki, M. K-ras is essential for the development of the mouse embryo. *Oncogene*, *15*: 1151–1159, 1997.

72. Zhang, Z., Wang, Y., Vikis, H.G., Johnson, L., Liu, G., Li, J., Anderson, M.W., Sills, R.C., Hong, H.L., Devereux, T.R., Jacks, T., Guan, K.L., and You, M. Wildtype Kras2 can inhibit lung carcinogenesis in mice. *Nat. Genet.*, *29*: 25–33, 2001.

73. Li, J., Zhang, Z., Dai, Z., Plass, C., Morrison, C., Wang, Y., Wiest, J.S., Anderson, M.W., and You, M. LOH of chromosome 12p correlates with Kras2 mutation in non-small cell lung cancer. *Oncogene*, *22*: 1243–1246, 2003.

74. Uchiyama, M., Usami, N., Kondo, M., Mori, S., Ito, M., Ito, G., Yoshioka, H., Imaizumi, M., Ueda, Y., Takahashi, M., Minna, J.D., Shimokata, K., and Sekido, Y. Loss of heterozygosity of chromosome 12p does not correlate with KRAS mutation in non-small cell lung cancer. *Int. J. Cancer*, *107*: 962–969, 2003.

75. Leone, G., DeGregori, J., Sears, R., Jakoi, L., and Nevins, J.R. Myc and Ras collaborate in inducing accumulation of active cyclin E/Cdk2 and E2F. *Nature*, *387*: 422–426, 1997.

76. Sears, R., Leone, G., DeGregori, J., and Nevins, J.R. Ras enhances Myc protein stability. *Mol. Cell*, *3*: 169–179, 1999.

77. Grandori, C., Cowley, S.M., James, L.P., and Eisenman, R.N. The Myc/Max/Mad network and the transcriptional control of cell behavior. *Ann. Rev. Cell Dev. Biol.*, *16*: 653–699, 2000.

78. Eisenman, R.N. Deconstructing myc. *Genes Dev.*, *15*: 2023–2030, 2001.

79. Perez-Roger, I., Kim, S.H., Griffiths, B., Sewing, A., and Land, H. Cyclins D1 and D2 mediate myc-induced proliferation via sequestration of p27(Kip1) and p21(Cip1). *Embo J.*, *18*: 5310–5320, 1999.

80. Staller, P., Peukert, K., Kiermaier, A., Seoane, J., Lukas, J., Karsunky, H., Moroy, T., Bartek, J., Massague, J., Hanel, F., and Eilers, M. Repression of p15INK4b expression by Myc through association with Miz-1. *Nat. Cell Biol.*, *3*: 392–399, 2001.

81. Seoane, J., Pouponnot, C., Staller, P., Schader, M., Eilers, M., and Massague, J. TGFbeta influences Myc, Miz-1 and Smad to control the CDK inhibitor p15INK4b. *Nat. Cell Biol.*, *3*: 400–408, 2001.

82. Gartel, A.L., Ye, X., Goufman, E., Shianov, P., Hay, N., Najmabadi, F., and Tyner, A.L. Myc represses the p21(WAF1/CIP1) promoter and interacts with Sp1/Sp3. *Proc. Natl. Acad. Sci. USA*, 98: 4510–4515, 2001.

83. Gazzeri, S., Brambilla, E., Caron de Fromentel, C., Gouyer, V., Moro, D., Perron, P., Berger, F., and Brambilla, C. p53 genetic abnormalities and myc activation in human lung carcinoma. *Int. J. Cancer*, 58: 24–32, 1994.

84. Broers, J.L., Viallet, J., Jensen, S.M., Pass, H., Travis, W.D., Minna, J.D., and Linnoila, R.I. Expression of c-myc in progenitor cells of the bronchopulmonary epithelium and in a large number of non-small cell lung cancers. *Am. J. Resp. Cell Mol. Biol.*, 9: 33–43, 1993.

85. Geick, A., Redecker, P., Ehrhardt, A., Klocke, R., Paul, D., and Halter, R. Uteroglobin promoter-targeted c-MYC expression in transgenic mice cause hyperplasia of Clara cells and malignant transformation of T-lymphoblasts and tubular epithelial cells. *Transgenic Res.*, 10: 501–511, 2001.

86. Ehrhardt, A., Bartels, T., Geick, A., Klocke, R., Paul, D., and Halter, R. Development of pulmonary bronchiolo-alveolar adenocarcinomas in transgenic mice overexpressing murine c-myc and epidermal growth factor in alveolar type II pneumocytes. *Br. J. Cancer*, 84: 813–818, 2001.

87. Ehrhardt, A., Bartels, T., Klocke, R., Paul, D., and Halter, R. Increased susceptibility to the tobacco carcinogen 4-(methylnitrosamino)-1-(3-pyridyl)-1-butanone in transgenic mice overexpressing c-myc and epidermal growth factor in alveolar type II cells. *J. Cancer Res. Clin. Oncol.*, 129: 71–75, 2003.

88. Citri, A., Skaria, K.B., and Yarden, Y. The deaf and the dumb: The biology of ErbB-2 and ErbB-3. *Exp. Cell Res.*, 284: 54–65, 2003.

89. Rusch, V., Baselga, J., Cordon-Cardo, C., Orazem, J., Zaman, M., Hoda, S., McIntosh, J., Kurie, J., and Dmitrovsky, E. Differential expression of the epidermal growth factor receptor and its ligands in primary non-small cell lung cancers and adjacent benign lung. *Cancer Res.*, 53: 2379–2385, 1993.

90. Rachwal, W.J., Bongiorno, P.F., Orringer, M.B., Whyte, R.I., Ethier, S.P., and Beer, D.G. Expression and activation of erbB-2 and epidermal growth factor receptor in lung adenocarcinomas. *Br. J. Cancer*, 72: 56–64, 1995.

91. Kern, J.A., Schwartz, D.A., Nordberg, J.E., Weiner, D.B., Greene, M.I., Torney, L., and Robinson, R.A. p185neu expression in human lung adenocarcinomas predicts shortened survival. *Cancer Res.*, 50: 5184–5187, 1990.

92. Weiner, D.B., Nordberg, J., Robinson, R., Nowell, P.C., Gazdar, A., Greene, M.I., Williams, W.V., Cohen, J.A., and Kern, J.A. Expression of the neu gene-encoded protein (P185neu) in human non-small cell carcinomas of the lung. *Cancer Res.*, 50: 421–425, 1990.

93. Schneider, P.M., Hung, M.C., Chiocca, S.M., Manning, J., Zhao, X.Y., Fang, K., and Roth, J.A. Differential expression of the c-erbB-2 gene in human small cell and non-small cell lung cancer. *Cancer Res.*, 49: 4968–4971, 1989.

94. Yi, E.S., Harclerode, D., Gondo, M., Stephenson, M., Brown, R.W., Younes, M., and Cagle, P.T. High c-erbB-3 protein expression is associated with shorter survival in advanced non-small cell lung carcinomas. *Mod. Pathol.*, 10: 142–148, 1997.

95. Rusch, V., Klimstra, D., Venkatraman, E., Pisters, P.W., Langenfeld, J., and Dmitrovsky, E. Over-expression of the epidermal growth factor receptor and its ligand transforming growth factor alpha is frequent in resectable non-small cell lung cancer but does not predict tumor progression. *Clin. Cancer Res.*, 3: 515–522, 1997.

96. Fontanini, G., De Laurentiis, M., Vignati, S., Chine, S., Lucchi, M., Silvestri, V., Mussi, A., De Placido, S., Tortora, G., Bianco, A.R., Gullick, W., Angeletti, C.A., Bevilacqua, G., and Ciardiello, F. Evaluation of epidermal growth factor-related growth factors and receptors and of neoangiogenesis in completely resected stage I–IIIA non-small-cell lung cancer: Amphiregulin and microvessel count are independent prognostic indicators of survival. *Clin. Cancer Res.*, 4: 241–249, 1998.

97. Gschwind, A., Fischer, O.M., and Ullrich, A. The discovery of receptor tyrosine kinases: Targets for cancer therapy. *Nat. Rev. Cancer*, 4: 361–370, 2004.

98. Cohen, M.H., Williams, G.A., Sridhara, R., Chen, G., and Pazdur, R. FDA drug approval summary: Gefitinib (ZD1839) (Iressa) tablets. *Oncologist*, 8: 303–306, 2003.

99. Kris, M.G., Natale, R.B., Herbst, R.S., Lynch, T.J., Jr., Prager, D., Belani, C.P., Schiller, J.H., Kelly, K., Spiridonidis, H., Sandler, A., Albain, K.S., Cella, D., Wolf, M.K., Averbuch, S.D., Ochs, J.J., and

Kay, A.C. Efficacy of gefitinib, an inhibitor of the epidermal growth factor receptor tyrosine kinase, in symptomatic patients with non-small cell lung cancer: A randomized trial. *JAMA, 290*: 2149–2158, 2003.

100. Fukuoka, M., Yano, S., Giaccone, G., Tamura, T., Nakagawa, K., Douillard, J.Y., Nishiwaki, Y., Vansteenkiste, J., Kudoh, S., Rischin, D., Eek, R., Horai, T., Noda, K., Takata, I., Smit, E., Averbuch, S., Macleod, A., Feyereislova, A., Dong, R.P., and Baselga, J. Multi-institutional randomized phase II trial of gefitinib for previously treated patients with advanced non-small-cell lung cancer. *J. Clin. Oncol., 21*: 2237–2246, 2003.

101. Lynch, T.J., Bell, D.W., Sordella, R., Gurubhagavatula, S., Okimoto, R.A., Brannigan, B.W., Harris, P.L., Haserlat, S.M., Supko, J.G., Haluska, F.G., Louis, D.N., Christiani, D.C., Settleman, J., and Haber, D.A. Activating mutations in the epidermal growth factor receptor underlying responsiveness of non-small-cell lung cancer to gefitinib. *N. Engl. J. Med., 350*: 2129–2139, 2004.

102. Paez, J.G., Janne, P.A., Lee, J.C., Tracy, S., Greulich, H., Gabriel, S., Herman, P., Kaye, F.J., Lindeman, N., Boggon, T.J., Naoki, K., Sasaki, H., Fujii, Y., Eck, M.J., Sellers, W.R., Johnson, B.E., and Meyerson, M. EGFR mutations in lung cancer: Correlation with clinical response to gefitinib therapy. *Science, 305*: 1497–1500, 2004.

103. Cappuzzo, F., Magrini, E., Ceresoli, G.L., Bartolini, S., Rossi, E., Ludovini, V., Gregorc, V., Ligorio, C., Cancellieri, A., Damiani, S., Spreafico, A., Paties, C.T., Lombardo, L., Calandri, C., Bellezza, G., Tonato, M., and Crino, L. Akt phosphorylation and gefitinib efficacy in patients with advanced non-small-cell lung cancer. *J. Natl. Cancer Inst., 96*: 1133–1141, 2004.

104. Langer, C.J., Stephenson, P., Thor, A., Vangel, M., and Johnson, D.H. Trastuzumab in the treatment of advanced non-small-cell lung cancer: Is there a role? Focus on Eastern Cooperative Oncology Group study 2598. *J. Clin. Oncol., 22*: 1180–1187, 2004.

105. Kastan, M.B., Onyekwere, O., Sidransky, D., Vogelstein, B., and Craig, R.W. Participation of p53 protein in the cellular response to DNA damage. *Cancer Res., 51*: 6304–6311, 1991.

106. Cross, S.M., Sanchez, C.A., Morgan, C.A., Schimke, M.K., Ramel, S., Idzerda, R.L., Raskind, W.H., and Reid, B.J. A p53-dependent mouse spindle checkpoint. *Science, 267*: 1353–1356, 1995.

107. Di Leonardo, A., Khan, S.H., Linke, S.P., Greco, V., Seidita, G., and Wahl, G.M. DNA rereplication in the presence of mitotic spindle inhibitors in human and mouse fibroblasts lacking either p53 or pRb function. *Cancer Res., 57*: 1013–1019, 1997.

108. Graeber, T.G., Peterson, J.F., Tsai, M., Monica, K., Fornace, A.J., Jr., and Giaccia, A.J. Hypoxia induces accumulation of p53 protein, but activation of a G1-phase checkpoint by low-oxygen conditions is independent of p53 status. *Mol. Cell Biol., 14*: 6264–6277, 1994.

109. Saito, S., Goodarzi, A.A., Higashimoto, Y., Noda, Y., Lees-Miller, S.P., Appella, E., and Anderson, C.W. ATM mediates phosphorylation at multiple p53 sites, including Ser(46), in response to ionizing radiation. *J. Biol. Chem., 277*: 12491–12494, 2002.

110. Tibbetts, R.S., Brumbaugh, K.M., Williams, J.M., Sarkaria, J.N., Cliby, W.A., Shieh, S.Y., Taya, Y., Prives, C., and Abraham, R.T. A role for ATR in the DNA damage-induced phosphorylation of p53. *Genes Dev., 13*: 152–157, 1999.

111. Appella, E. and Anderson, C.W. Post-translational modifications and activation of p53 by genotoxic stresses. *Eur. J. Biochem., 268*: 2764–2772, 2001.

112. Hermeking, H., Lengauer, C., Polyak, K., He, T.C., Zhang, L., Thiagalingam, S., Kinzler, K.W., and Vogelstein, B. 14-3-3 sigma is a p53-regulated inhibitor of G2/M progression. *Mol. Cell, 1*: 3–11, 1997.

113. el-Deiry, W.S., Tokino, T., Velculescu, V.E., Levy, D.B., Parsons, R., Trent, J.M., Lin, D., Mercer, W.E., Kinzler, K.W., and Vogelstein, B. WAF1, a potential mediator of p53 tumor suppression. *Cell, 75*: 817–825, 1993.

114. Miyashita, T. and Reed, J.C. Tumor suppressor p53 is a direct transcriptional activator of the human bax gene. *Cell, 80*: 293–299, 1995.

115. Moroni, M.C., Hickman, E.S., Denchi, E.L., Caprara, G., Colli, E., Cecconi, F., Muller, H., and Helin, K. Apaf-1 is a transcriptional target for E2F and p53. *Nat. Cell Biol., 3*: 552–558, 2001.

116. Robles, A.I., Bemmels, N.A., Foraker, A.B., and Harris, C.C. APAF-1 is a transcriptional target of p53 in DNA damage-induced apoptosis. *Cancer Res., 61*: 6660–6664, 2001.

117. Robles, A.I., Linke, S.P., and Harris, C.C. The p53 network in lung carcinogenesis. *Oncogene, 21*: 6898–6907, 2002.

118. Hainaut, P. and Pfeifer, G.P. Patterns of p53 G –> T transversions in lung cancers reflect the primary mutagenic signature of DNA-damage by tobacco smoke. *Carcinogenesis, 22*: 367–374, 2001.

119. Harris, C.C. Structure and function of the p53 tumor suppressor gene: Clues for rational cancer therapeutic strategies. *J. Natl. Cancer Inst., 88*: 1442–1455, 1996.

120. Denissenko, M.F., Pao, A., Tang, M., and Pfeifer, G.P. Preferential formation of benzo[a]pyrene adducts at lung cancer mutational hotspots in P53. *Science, 274*: 430–432, 1996.

121. Denissenko, M.F., Chen, J.X., Tang, M.S., and Pfeifer, G.P. Cytosine methylation determines hot spots of DNA damage in the human P53 gene. *Proc. Natl. Acad. Sci. USA, 94*: 3893–3898, 1997.

122. Higashiyama, M., Doi, O., Kodama, K., Yokouchi, H., Kasugai, T., Ishiguro, S., Takami, K., Nakayama, T., and Nishisho, I. MDM2 gene amplification and expression in non-small-cell lung cancer: Immunohistochemical expression of its protein is a favourable prognostic marker in patients without p53 protein accumulation. *Br. J. Cancer, 75*: 1302–1308, 1997.

123. Ko, J.L., Cheng, Y.W., Chang, S.L., Su, J.M., Chen, C.Y., and Lee, H. MDM2 mRNA expression is a favorable prognostic factor in non-small-cell lung cancer. *Int. J. Cancer, 89*: 265–270, 2000.

124. Sanchez-Cespedes, M., Reed, A.L., Buta, M., Wu, L., Westra, W.H., Herman, J.G., Yang, S.C., Jen, J., and Sidransky, D. Inactivation of the INK4A/ARF locus frequently coexists with TP53 mutations in non-small cell lung cancer. *Oncogene, 18*: 5843–5849, 1999.

125. Nicholson, S.A., Okby, N.T., Khan, M.A., Welsh, J.A., McMenamin, M.G., Travis, W.D., Jett, J.R., Tazelaar, H.D., Trastek, V., Pairolero, P.C., Corn, P.G., Herman, J.G., Liotta, L.A., Caporaso, N.E., and Harris, C.C. Alterations of p14ARF, p53, and p73 genes involved in the E2F-1-mediated apoptotic pathways in non-small cell lung carcinoma. *Cancer Res., 61*: 5636–5643, 2001.

126. Park, M.J., Shimizu, K., Nakano, T., Park, Y.B., Kohno, T., Tani, M., and Yokota, J. Pathogenetic and biologic significance of TP14ARF alterations in nonsmall cell lung carcinoma. *Cancer Genet. Cytogenet., 141*: 5–13, 2003.

127. Donehower, L.A., Harvey, M., Slagle, B.L., McArthur, M.J., Montgomery, C.A., Jr., Butel, J.S., and Bradley, A. Mice deficient for p53 are developmentally normal but susceptible to spontaneous tumours. *Nature, 356*: 215–221, 1992.

128. Wang, Y., Zhang, Z., Kastens, E., Lubet, R.A., and You, M. Mice with alterations in both p53 and Ink4a/Arf display a striking increase in lung tumor multiplicity and progression: Differential chemopreventive effect of budesonide in wild-type and mutant A/J mice. *Cancer Res., 63*: 4389–4395, 2003.

129. Duan, W., Ding, H., Subler, M.A., Zhu, W.G., Zhang, H., Stoner, G.D., Windle, J.J., Otterson, G.A., and Villalona-Calero, M.A. Lung-specific expression of human mutant p53–273H is associated with a high frequency of lung adenocarcinoma in transgenic mice. *Oncogene, 21*: 7831–7838, 2002.

130. Meuwissen, R., Linn, S.C., Linnoila, R.I., Zevenhoven, J., Mooi, W.J., and Berns, A. Induction of small cell lung cancer by somatic inactivation of both Trp53 and Rb1 in a conditional mouse model. *Cancer Cell, 4*: 181–189, 2003.

131. Kelley, M.J., Nakagawa, K., Steinberg, S.M., Mulshine, J.L., Kamb, A., and Johnson, B.E. Differential inactivation of CDKN2 and Rb protein in non-small-cell and small-cell lung cancer cell lines. *J. Natl. Cancer Inst., 87*: 756–761, 1995.

132. Otterson, G.A., Kratzke, R.A., Coxon, A., Kim, Y.W., and Kaye, F.J. Absence of p16INK4 protein is restricted to the subset of lung cancer lines that retains wildtype RB. *Oncogene, 9*: 3375–3378, 1994.

133. Shapiro, G.I., Edwards, C.D., Kobzik, L., Godleski, J., Richards, W., Sugarbaker, D.J., and Rollins, B.J. Reciprocal Rb inactivation and p16INK4 expression in primary lung cancers and cell lines. *Cancer Res., 55*: 505–509, 1995.

134. Herman, J.G., Merlo, A., Mao, L., Lapidus, R.G., Issa, J.P., Davidson, N.E., Sidransky, D., and Baylin, S.B. Inactivation of the CDKN2/p16/MTS1 gene is frequently associated with aberrant DNA methylation in all common human cancers. *Cancer Res., 55*: 4525–4530, 1995.

135. Merlo, A., Herman, J.G., Mao, L., Lee, D.J., Gabrielson, E., Burger, P.C., Baylin, S.B., and Sidransky, D. 5′ CpG island methylation is associated with transcriptional silencing of the tumour suppressor p16/CDKN2/MTS1 in human cancers. *Nat. Med., 1*: 686–692, 1995.

136. Otterson, G.A., Khleif, S.N., Chen, W., Coxon, A.B., and Kaye, F.J. CDKN2 gene silencing in lung cancer by DNA hypermethylation and kinetics of p16INK4 protein induction by 5-aza 2′deoxycytidine. *Oncogene, 11*: 1211–1216, 1995.

137. Kaye, F.J. RB and cyclin dependent kinase pathways: Defining a distinction between RB and p16 loss in lung cancer. *Oncogene*, *21*: 6908–6914, 2002.

138. Chen, P.L., Scully, P., Shew, J.Y., Wang, J.Y., and Lee, W.H. Phosphorylation of the retinoblastoma gene product is modulated during the cell cycle and cellular differentiation. *Cell*, *58*: 1193–1198, 1989.

139. DeCaprio, J.A., Ludlow, J.W., Lynch, D., Furukawa, Y., Griffin, J., Piwnica-Worms, H., Huang, C.M., and Livingston, D.M. The product of the retinoblastoma susceptibility gene has properties of a cell cycle regulatory element. *Cell*, *58*: 1085–1095, 1989.

140. Mihara, K., Cao, X.R., Yen, A., Chandler, S., Driscoll, B., Murphree, A.L., T'Ang, A., and Fung, Y.K. Cell cycle-dependent regulation of phosphorylation of the human retinoblastoma gene product. *Science*, *246*: 1300–1303, 1989.

141. Harbour, J.W. and Dean, D.C. The Rb/E2F pathway: Expanding roles and emerging paradigms. *Genes Dev.*, *14*: 2393–2409, 2000.

142. Mundle, S.D. and Saberwal, G. Evolving intricacies and implications of E2F1 regulation. *Faseb J.*, *17*: 569–574, 2003.

143. Kamb, A., Gruis, N.A., Weaver-Feldhaus, J., Liu, Q., Harshman, K., Tavtigian, S.V., Stockert, E., Day, R.S., III, Johnson, B.E., and Skolnick, M.H. A cell cycle regulator potentially involved in genesis of many tumor types. *Science*, *264*: 436–440, 1994.

144. Massague, J., Blain, S.W., and Lo, R.S. TGFbeta signaling in growth control, cancer, and heritable disorders. *Cell*, *103*: 295–309, 2000.

145. Bartram, U. and Speer, C.P. The role of transforming growth factor beta in lung development and disease. *Chest*, *125*: 754–765, 2004.

146. Derynck, R. and Zhang, Y.E. Smad-dependent and Smad-independent pathways in TGF-beta family signalling. *Nature*, *425*: 577–584, 2003.

147. Zhang, H.T., Chen, X.F., Wang, M.H., Wang, J.C., Qi, Q.Y., Zhang, R.M., Xu, W.Q., Fei, Q.Y., Wang, F., Cheng, Q.Q., Chen, F., Zhu, C.S., Tao, S.H., and Luo, Z. Defective expression of transforming growth factor beta receptor type II is associated with CpG methylated promoter in primary non-small cell lung cancer. *Clin. Cancer Res.*, *10*: 2359–2367, 2004.

148. Kang, Y., Prentice, M.A., Mariano, J.M., Davarya, S., Linnoila, R.I., Moody, T.W., Wakefield, L.M., and Jakowlew, S.B. Transforming growth factor-beta 1 and its receptors in human lung cancer and mouse lung carcinogenesis. *Exp. Lung Res.*, *26*: 685–707, 2000.

149. Bhattacharjee, A., Richards, W.G., Staunton, J., Li, C., Monti, S., Vasa, P., Ladd, C., Beheshti, J., Bueno, R., Gillette, M., Loda, M., Weber, G., Mark, E.J., Lander, E.S., Wong, W., Johnson, B.E., Golub, T.R., Sugarbaker, D.J., and Meyerson, M. Classification of human lung carcinomas by mRNA expression profiling reveals distinct adenocarcinoma subclasses. *Proc. Natl. Acad. Sci. USA*, *98*: 13790–13795, 2001.

150. Uchida, K., Nagatake, M., Osada, H., Yatabe, Y., Kondo, M., Mitsudomi, T., Masuda, A., and Takahashi, T. Somatic in vivo alterations of the JV18-1 gene at 18q21 in human lung cancers. *Cancer Res.*, *56*: 5583–5585, 1996.

151. Nagatake, M., Takagi, Y., Osada, H., Uchida, K., Mitsudomi, T., Saji, S., Shimokata, K., and Takahashi, T. Somatic in vivo alterations of the DPC4 gene at 18q21 in human lung cancers. *Cancer Res.*, *56*: 2718–2720, 1996.

152. Osada, H., Tatematsu, Y., Masuda, A., Saito, T., Sugiyama, M., Yanagisawa, K., and Takahashi, T. Heterogeneous transforming growth factor (TGF)-beta unresponsiveness and loss of TGF-beta receptor type II expression caused by histone deacetylation in lung cancer cell lines. *Cancer Res.*, *61*: 8331–8339, 2001.

153. Pezzella, F., Turley, H., Kuzu, I., Tungekar, M.F., Dunnill, M.S., Pierce, C.B., Harris, A., Gatter, K.C., and Mason, D.Y. bcl-2 protein in non-small-cell lung carcinoma. *N. Engl. J. Med.*, *329*: 690–694, 1993.

154. Kaiser, U., Schilli, M., Haag, U., Neumann, K., Kreipe, H., Kogan, E., and Havemann, K. Expression of bcl-2 — protein in small cell lung cancer. *Lung Cancer*, *15*: 31–40, 1996.

155. Higashiyama, M., Doi, O., Kodama, K., Yokouchi, H., Nakamori, S., and Tateishi, R. bcl-2 oncoprotein in surgically resected non-small cell lung cancer: Possibly favorable prognostic factor in association with low incidence of distant metastasis. *J. Surg. Oncol.*, *64*: 48–54, 1997.

156. Anton, R.C., Brown, R.W., Younes, M., Gondo, M.M., Stephenson, M.A., and Cagle, P.T. Absence of prognostic significance of bcl-2 immunopositivity in non-small cell lung cancer: Analysis of 427 cases. *Hum. Pathol.*, *28*: 1079–1082, 1997.

157. Mai, H., May, W.S., Gao, F., Jin, Z., and Deng, X. A functional role for nicotine in Bcl2 phosphorylation and suppression of apoptosis. *J. Biol. Chem.*, *278*: 1886–1891, 2003.

158. Jin, Z., Gao, F., Flagg, T., and Deng, X. Nicotine induces multi-site phosphorylation of Bad in association with suppression of apoptosis. *J. Biol. Chem.*, *279*: 23837–23844, 2004.

159. Koukourakis, M.I., Giatromanolaki, A., Thorpe, P.E., Brekken, R.A., Sivridis, E., Kakolyris, S., Georgoulias, V., Gatter, K.C., and Harris, A.L. Vascular endothelial growth factor/KDR activated microvessel density versus CD31 standard microvessel density in non-small cell lung cancer. *Cancer Res.*, *60*: 3088–3095, 2000.

160. Fontanini, G., Lucchi, M., Vignati, S., Mussi, A., Ciardiello, F., De Laurentiis, M., De Placido, S., Basolo, F., Angeletti, C.A., and Bevilacqua, G. Angiogenesis as a prognostic indicator of survival in non-small-cell lung carcinoma: A prospective study. *J. Natl. Cancer Inst.*, *89*: 881–886, 1997.

161. O'Byrne, K.J., Koukourakis, M.I., Giatromanolaki, A., Cox, G., Turley, H., Steward, W.P., Gatter, K., and Harris, A.L. Vascular endothelial growth factor, platelet-derived endothelial cell growth factor and angiogenesis in non-small-cell lung cancer. *Br. J. Cancer*, *82*: 1427–1432, 2000.

162. Fontanini, G., Vignati, S., Boldrini, L., Chine, S., Silvestri, V., Lucchi, M., Mussi, A., Angeletti, C.A., and Bevilacqua, G. Vascular endothelial growth factor is associated with neovascularization and influences progression of non-small cell lung carcinoma. *Clin. Cancer Res.*, *3*: 861–865, 1997.

163. Takanami, I., Imamura, T., Hashizume, T., Kikuchi, K., Yamamoto, Y., Yamamoto, T., and Kodaira, S. Immunohistochemical detection of basic fibroblast growth factor as a prognostic indicator in pulmonary adenocarcinoma. *Jpn. J. Clin. Oncol.*, *26*: 293–297, 1996.

164. Volm, M., Koomagi, R., Mattern, J., and Stammler, G. Prognostic value of basic fibroblast growth factor and its receptor (FGFR-1) in patients with non-small cell lung carcinomas. *Eur. J. Cancer*, *33*: 691–693, 1997.

165. Volm, M., Koomagi, R., and Mattern, J. PD-ECGF, bFGF, and VEGF expression in non-small cell lung carcinomas and their association with lymph node metastasis. *Anticancer Res.*, *19*: 651–655, 1999.

166. Ueno, K., Inoue, Y., Kawaguchi, T., Hosoe, S., and Kawahara, M. Increased serum levels of basic fibroblast growth factor in lung cancer patients: Relevance to response of therapy and prognosis. *Lung Cancer*, *31*: 213–219, 2001.

167. Ruotsalainen, T., Joensuu, H., Mattson, K., and Salven, P. High pretreatment serum concentration of basic fibroblast growth factor is a predictor of poor prognosis in small cell lung cancer. *Cancer Epidemiol. Biomarkers Prev.*, *11*: 1492–1495, 2002.

168. Sharpless, N.E. and DePinho, R.A. Telomeres, stem cells, senescence, and cancer. *J. Clin. Invest.*, *113*: 160–168, 2004.

169. Hahn, W.C. Role of telomeres and telomerase in the pathogenesis of human cancer. *J. Clin. Oncol.*, *21*: 2034–2043, 2003.

170. Albanell, J., Lonardo, F., Rusch, V., Engelhardt, M., Langenfeld, J., Han, W., Klimstra, D., Venkatraman, E., Moore, M.A., and Dmitrovsky, E. High telomerase activity in primary lung cancers: Association with increased cell proliferation rates and advanced pathologic stage. *J. Natl. Cancer Inst.*, *89*: 1609–1615, 1997.

171. Hiyama, K., Hiyama, E., Ishioka, S., Yamakido, M., Inai, K., Gazdar, A.F., Piatyszek, M.A., and Shay, J.W. Telomerase activity in small-cell and non-small-cell lung cancers. *J. Natl. Cancer Inst.*, *87*: 895–902, 1995.

172. Incles, C.M., Schultes, C.M., and Neidle, S. Telomerase inhibitors in cancer therapy: Current status and future directions. *Curr. Opin. Investig. Drugs*, *4*: 675–685, 2003.

173. Saretzki, G. Telomerase inhibition as cancer therapy. *Cancer Lett.*, *194*: 209–219, 2003.

174. Pritchard, S.C., Nicolson, M.C., Lloret, C., McKay, J.A., Ross, V.G., Kerr, K.M., Murray, G.I., and McLeod, H.L. Expression of matrix metalloproteinases 1, 2, 9 and their tissue inhibitors in stage II non-small cell lung cancer: Implications for MMP inhibition therapy. *Oncol. Rep.*, *8*: 421–424, 2001.

175. Passlick, B., Sienel, W., Seen-Hibler, R., Wockel, W., Thetter, O., Mutschler, W., and Pantel, K. Overexpression of matrix metalloproteinase 2 predicts unfavorable outcome in early-stage non-small cell lung cancer. *Clin. Cancer Res.*, *6*: 3944–3948, 2000.

176. Sienel, W., Hellers, J., Morresi-Hauf, A., Lichtinghagen, R., Mutschler, W., Jochum, M., Klein, C., Passlick, B., and Pantel, K. Prognostic impact of matrix metalloproteinase-9 in operable non-small cell lung cancer. *Int. J. Cancer, 103*: 647–651, 2003.

177. Ylisirnio, S., Hoyhtya, M., and Turpeenniemi-Hujanen, T. Serum matrix metalloproteinases-2, -9 and tissue inhibitors of metalloproteinases-1, -2 in lung cancer — TIMP-1 as a prognostic marker. *Anticancer Res., 20*: 1311–1316, 2000.

178. Aljada, I.S., Ramnath, N., Donohue, K., Harvey, S., Brooks, J.J., Wiseman, S.M., Khoury, T., Loewen, G., Slocum, H.K., Anderson, T.M., Bepler, G., and Tan, D. Upregulation of the tissue inhibitor of metalloproteinase-1 protein is associated with progression of human non-small-cell lung cancer. *J. Clin. Oncol., 22*: 3218–3229, 2004.

179. Sasaki, H., Moriyama, S., Nakashima, Y., Kobayashi, Y., Yukiue, H., Kaji, M., Fukai, I., Kiriyama, M., Yamakawa, Y., and Fujii, Y. Expression of the MTA1 mRNA in advanced lung cancer. *Lung Cancer, 35*: 149–154, 2002.

180. Muller-Tidow, C., Metzger, R., Kugler, K., Diederichs, S., Idos, G., Thomas, M., Dockhorn-Dworniczak, B., Schneider, P.M., Koeffler, H.P., Berdel, W.E., and Serve, H. Cyclin E is the only cyclin-dependent kinase 2-associated cyclin that predicts metastasis and survival in early stage non-small cell lung cancer. *Cancer Res., 61*: 647–653, 2001.

181. Diederichs, S., Bulk, E., Steffen, B., Ji, P., Tickenbrock, L., Lang, K., Zanker, K.S., Metzger, R., Schneider, P.M., Gerke, V., Thomas, M., Berdel, W.E., Serve, H., and Muller-Tidow, C. S100 family members and trypsinogens are predictors of distant metastasis and survival in early-stage non-small cell lung cancer. *Cancer Res., 64*: 5564–5569, 2004.

182. Yamayoshi, T., Nagayasu, T., Matsumoto, K., Abo, T., Hishikawa, Y., and Koji, T. Expression of keratinocyte growth factor/fibroblast growth factor-7 and its receptor in human lung cancer: Correlation with tumour proliferative activity and patient prognosis. *J. Pathol., 204*: 110–118, 2004.

183. Kaneko, M., Eguchi, K., Ohmatsu, H., Kakinuma, R., Naruke, T., Suemasu, K., and Moriyama, N. Peripheral lung cancer: Screening and detection with low-dose spiral CT versus radiography. *Radiology, 201*: 798–802, 1996.

184. Henschke, C.I., McCauley, D.I., Yankelevitz, D.F., Naidich, D.P., McGuinness, G., Miettinen, O.S., Libby, D.M., Pasmantier, M.W., Koizumi, J., Altorki, N.K., and Smith, J.P. Early Lung Cancer Action Project: Overall design and findings from baseline screening. *Lancet, 354*: 99–105, 1999.

185. Swensen, S.J., Jett, J.R., Hartman, T.E., Midthun, D.E., Sloan, J.A., Sykes, A.M., Aughenbaugh, G.L., and Clemens, M.A. Lung cancer screening with CT: Mayo clinic experience. *Radiology, 226*: 756–761, 2003.

186. Jett, J.R. and Midthun, D.E. Screening for lung cancer: Current status and future directions: Thomas A. Neff lecture. *Chest, 125*: 158S–162S, 2004.

187. Lam, S., Kennedy, T., Unger, M., Miller, Y.E., Gelmont, D., Rusch, V., Gipe, B., Howard, D., LeRiche, J.C., Coldman, A., and Gazdar, A.F. Localization of bronchial intraepithelial neoplastic lesions by fluorescence bronchoscopy. *Chest, 113*: 696–702, 1998.

188. Heighway, J., Knapp, T., Boyce, L., Brennand, S., Field, J.K., Betticher, D.C., Ratschiller, D., Gugger, M., Donovan, M., Lasek, A., and Rickert, P. Expression profiling of primary non-small cell lung cancer for target identification. *Oncogene, 21*: 7749–7763, 2002.

189. Yao, R., Wang, Y., Lubet, R.A., and You, M. Differentially expressed genes associated with mouse lung tumor progression. *Oncogene, 21*: 5814–5821, 2002.

190. Bonner, A.E., Lemon, W.J., Devereux, T.R., Lubet, R.A., and You, M. Molecular profiling of mouse lung tumors: Association with tumor progression, lung development, and human lung adenocarcinomas. *Oncogene, 23*: 1166–1176, 2004.

191. Virtanen, C., Ishikawa, Y., Honjoh, D., Kimura, M., Shimane, M., Miyoshi, T., Nomura, H., and Jones, M.H. Integrated classification of lung tumors and cell lines by expression profiling. *Proc. Natl. Acad. Sci. USA, 99*: 12357–12362, 2002.

192. Beer, D.G., Kardia, S.L., Huang, C.C., Giordano, T.J., Levin, A.M., Misek, D.E., Lin, L., Chen, G., Gharib, T.G., Thomas, D.G., Lizyness, M.L., Kuick, R., Hayasaka, S., Taylor, J.M., Iannettoni, M.D., Orringer, M.B., and Hanash, S. Gene-expression profiles predict survival of patients with lung adenocarcinoma. *Nat. Med., 8*: 816–824, 2002.

193. Borczuk, A.C., Shah, L., Pearson, G.D., Walter, K.L., Wang, L., Austin, J.H., Friedman, R.A., and Powell, C.A. Molecular signatures in biopsy specimens of lung cancer. *Am. J. Respir. Crit. Care Med.*, *170*: 167–174, 2004.

194. Jones, M.H., Virtanen, C., Honjoh, D., Miyoshi, T., Satoh, Y., Okumura, S., Nakagawa, K., Nomura, H., and Ishikawa, Y. Two prognostically significant subtypes of high-grade lung neuroendocrine tumours independent of small-cell and large-cell neuroendocrine carcinomas identified by gene expression profiles. *Lancet*, *363*: 775–781, 2004.

195. Ramaswamy, S., Ross, K.N., Lander, E.S., and Golub, T.R. A molecular signature of metastasis in primary solid tumors. *Nat. Genet.*, *33*: 49–54, 2003.

196. Yanagisawa, K., Shyr, Y., Xu, B.J., Massion, P.P., Larsen, P.H., White, B.C., Roberts, J.R., Edgerton, M., Gonzalez, A., Nadaf, S., Moore, J.H., Caprioli, R.M., and Carbone, D.P. Proteomic patterns of tumour subsets in non-small-cell lung cancer. *Lancet*, *362*: 433–439, 2003.

197. Xiao, X., Liu, D., Tang, Y., Guo, F., Xia, L., Liu, J., and He, D. Development of proteomic patterns for detecting lung cancer. *Dis. Markers*, *19*: 33–39, 2003.

198. Xiao, X.Y., Tang, Y., Wei, X.P., and He, D.C. A preliminary analysis of non-small cell lung cancer biomarkers in serum. *Biomed. Environ. Sci.*, *16*: 140–148, 2003.

199. Zhukov, T.A., Johanson, R.A., Cantor, A.B., Clark, R.A., and Tockman, M.S. Discovery of distinct protein profiles specific for lung tumors and pre-malignant lung lesions by SELDI mass spectrometry. *Lung Cancer*, *40*: 267–279, 2003.

16 Breast Cancer

Sue C. Heffelfinger

CONTENTS

16.1 BREAST HISTOLOGY AND DEVELOPMENT — EFFECT OF HORMONES

As a prelude to discussing mechanisms of carcinogenesis, it is important to understand breast development and differentiation. The anatomical features of the mammary gland are illustrated in Figure 16.1. The human mammary gland is comprised of 12 to 20 segments defined as the branching ductular and lobular components emanating from a single lactiferous collecting duct. The human mammary gland undergoes unique phases of growth and differentiation in a particular age and in hormonally regulated fashion. At birth the breast is a palpable structure in both sexes, comprised of a nipple and collecting ducts. In addition there is a large interindividual variation that ranges from no additional branching to a well-developed dichotomous branching system with the formation of rudimentary lobules [1]. During the first two years of life in both sexes, the mammary gland undergoes a functional differentiation including production and secretion of casein, cystic dilatation with apocrine metaplasia, and involution with accumulation of a dense matrix surrounding atrophic ducts or lobules [1]. There is no apparent correlation between structural morphology and epithelial differentiation at this stage.

At puberty the female breast undergoes extensive elongation and branching of the ductular system. Lobular development begins at the periphery and extends centrally with the degree of lobular development being variable among individuals [2]. In nulliparous women three levels of lobular development are described, Types 1 to 3 [3]. Type 1 lobules have a rudimentary branching of intralobular ductules and are found in the first two years of life in some children. Lobule types 2 and 3 have progressively more levels of branching, making for increasing numbers of terminal alveoli. This developmental process occurs over years of menstrual cycling, although there is no correlation between age and degree of gland development [3].

Pregnancy is required for the development of type 4 or lactating lobules. During this stage of development there is a marked increase in branching, lobular epithelial area, and differentiation with respect to the synthesis of milk products. Upon weaning, the mammary gland undergoes massive cellular apoptosis with regression to lobule types 2 and 3. However, type 1 lobules may also be found

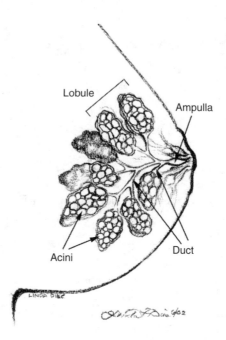

FIGURE 16.1   Each mammary segment is comprised of a collecting duct system with its complex branching of ductules and terminal ductolobular units. Each lobule has numerous acini, which increase in number with gland differentiation. Toward the nipple, the lactiferous ducts dilate into an ampulla for milk storage (Drawing by Linda Diec).

in cycling postparous women, suggesting that there is asynchronous breast development, that is, possibly not all lobular units differentiate with each pregnancy [4]. Then again, at menopause, there is further involution to rudimentary lobular structures. With involution there are also changes in stroma, which have been morphologically but not biochemically defined [5].

During development estrogen acts upon stromal cells, inducing ductal morphogenesis in a paracrine fashion [6,7]. Lobuloalveolar development is dependent upon epithelial expression of the progesterone receptor [8,9]. Estrogen and progesterone are critical determinants of breast development and proliferation; however, they do not act alone. *In vivo* pituitary hormones are absolutely required for ovarian hormone stimulation of proliferation or differentiation of breast epithelial cells [10].

16.2   CARCINOGENIC SUSCEPTIBILITY: MATURATION OR HORMONAL ENVIRONMENT?

One of the most important questions in mammary carcinogenesis is what cell types are capable of parenting a tumor? Extensive subgross and microscopic analysis of breast tumors showed that the site of origin is the terminal ductolobular unit [11]. Numerous chemical carcinogenesis models in rodents indicate that mammary gland immaturity, particularly the presence of the highly proliferative terminal end bud (rodents) or type 1 lobule (humans), is the site of tumor initiation [3,12]. In rodents the maximal level of chemically induced tumors occur when the carcinogen is given during the time of terminal end bud (TEB) morphogenesis into alveolar buds (AB) and terminal ductules [12]. In the well characterized 7,12-dimethylbenz[a]anthracene (DMBA) model system, exposure of TEB to the carcinogen results in progressive epithelial dysplasia and carcinomas morphologically similar to ductal carcinomas in humans. Exposure of AB to the carcinogen results in lobular tumors. Other carcinogens, such as *N*-nitroso-*N*-methylurea (NMU), produce similar results [13]. The cells in the TEB have the highest proliferative rate of any mammary structure, potentially explaining the greater effect of carcinogens [14]. In addition, the metabolic processing of DMBA differs in these cells relative to lobular cells, producing more potential DNA damaging epoxides and fewer phenolic metabolites [15,16].

The human equivalent of the TEB is the lobule type 1. When comparing parous women with breast cancer versus parous women without breast cancer, Russo et al. found that parous women with breast cancer had a considerably greater fraction of normal epithelium as lobule type 1 versus lobule types 2 and 3 seen in the women who were cancer free. These authors suggest that increased cancer susceptibility is correlated with the increased ratio of type 1 to type 3 lobules. It is not known whether in these parous women the type 1 lobules did not undergo maturation with pregnancy or whether they underwent a preferential dedifferentiation/atrophy after delivery. Interestingly, treatment of nulliparous rodents with placental hormones, such as human chorionic gonadotropin, is associated with mammary gland maturation and a decreased sensitivity to chemical carcinogenesis [17].

The concept that structurally immature breast tissue is most susceptible to carcinogenesis is consistent with numerous epidemiological observations. Infants who were irradiated in the first year of life for thymic enlargement showed a subsequent increased risk of breast cancer [18]. Mantle irradiation for Hodgkin's disease and fluoroscopy for tuberculosis also increased breast cancer risk particularly when performed in adolescence [19,20]. In addition, studies of the survivors from the Japanese atomic bomb blasts showed that breast cancer was highest in the group of women who were younger than ten at the time of exposure [21]. Furthermore morphological differentiation, as is seen postpartum, is a protective factor. First pregnancy occurring at less than 20 years of age decreases risk by 50% compared to nulliparous women [22]. Lactation duration is also inversely correlated with cancer risk [23].

However, there are other observations that indicate that gland maturation may not correlate with carcinogen susceptibility. In a series of studies by different groups, mice underwent pituitary iso-grafts and were then analyzed for mammary gland maturation and sensitivity to NMU carcinogenesis [24–26]. As expected with the early phase of lobuloalveolar development, tumor latency decreased and tumor frequency increased. However, even after maximal lobular development (determined by casein synthesis), the mice continued to be extremely sensitive to NMU carcinogenesis. Pituitary isografts were not required for tumor growth but were required at the time of carcinogen exposure. Interestingly, tumor incidence did not parallel epithelial proliferation kinetics, an observation seen by others [27]. Medina's group has also shown a disconnect between morphologic differentiation and resistance to carcinogenesis [28,29]. In these studies high-dose estrogen plus progesterone together were able to confer partial protection from NMU-induced tumors in rats when given for 10 days; whereas for lower doses 21 days of treatment were required. In the latter case, the extent of morpho-logic differentiation was the same at 10 and 21 days. Finally, nonhormonal induction of morphologic differentiation is unable to elicit this inhibition of carcinogenesis [30]. Therefore, something about the hormonal environment at the time of carcinogen exposure regulates the tumorigenic potential of the carcinogen. This effect is also seen *in vitro*, arguing that the hormonal effect is on the epithelium, not on the surrounding stroma. For instance, mitogenic stimulation of cultured mammary epithelial cells in an undifferentiated fashion using epidermal growth factor followed by NMU exposure results in fewer tumors when transplanted *in vivo* than when stimulated by progesterone and prolactin [31,32]. In the latter case the cells *in vitro* undergo differentiation and casein synthesis and yet are highly sensitive to NMU initiation.

Characterization of epithelial phenotypes during development led to the concept that there exist totipotent mammary stem cells in end buds and in the basal layer of ducts, as well as pluripotent stem cells, which may give rise to alveolar development [33]. Anderson et al. [34] have proposed that the estrogen receptor (ER) negative cells involved in the cyclical proliferation of the breast epithelium during the menstrual cycle are, in fact, stem cells. Epithelial transplant studies in mice have shown that the fully developed mammary gland contains stem cells with the capacity to uniquely reconstitute an entire mammary gland [35], and that these stem cells exist even postpartum [36]. Individual stem cells are calculated to have the potential for evolving 10^{12} to 10^{13} multipotent epithelial cells [35]. There are also distinct progenitors for ductal and lobular differentiation [37]. Using rat mammary cell transplants these progenitors were estimated to occur at frequencies of 1 lobular clonogen per 935 epithelial cells versus 1 ductal clonogen per 1572 epithelial cells [38]. In mice it has been estimated that a totipotent mammary stem cell comprises about 1 in 2500 epithelial cells [35]. Indeed, extensive studies of human breast epithelium have shown that entire ducto-lobular structures are derived from a single stem cell [39], and multilineage stem cells have been isolated from human breast tissue and characterized *in vitro* [40].

Several lines of evidence indicate that these stem cells are the cells of origin for carcinomas, including those in the mammary gland [41]. *In vitro* models of mammary epithelial cell transformation show that for a carcinogen to lead to full neoplastic transformation, the cells must be immortalized prior to exposure to the carcinogen [42]. Isolated human breast epithelial stem cells are highly susceptible to telomerase activation and immortalization after transfection with SV40 large *T* antigen [43]. Quantitative transplantation studies of rat mammary clonogens show that radiogenic mammary carcinogenesis is derived from clonogenic populations of the epithelium [44]. Furthermore, the differentiation of rat mammary clonogens into alveolar or ductal colonies is hormonally regulated, as is the susceptibility to radiation-induced carcinogenesis [45]. Finally, induction of premature stem cell senescence by transforming growth factor β transfection correlates with a decreased incidence of mouse mammary tumor virus-induced malignant transformation [46].

Most breast tumors are morphologically and biochemically related to luminal epithelial cells [47,48]. However, breast cancers that occur in pre versus postmenopausal women tend to be morphologically and biochemically distinct [49–51]. Generally, breast cancers in premenopausal patients are less well differentiated and do not express the estrogen and progesterone receptors [52]. In extremely young women (<35 years), there are some data to suggest that alterations in p53 function may be more frequent [49]. In contrast, breast cancers in postmenopausal patients are often well differentiated, expressing both estrogen and progesterone receptors. There are also data suggesting that in this population the incidence of invasive lobular carcinoma is rising [53]. These authors speculate the increase in this subset of cancers is being driven by the marked increase in estrogen plus progesterone hormone replacement therapy over the past two decades. Regardless of etiology, these functional differences in tumor behavior and biochemistry impact both prognosis and the treatment choices for these patients. One explanation for this difference in tumor type by age is the hypothesis that tumor genotype is dependent upon the hormonal milieu at the time of initiation [54]. In this study, cultured mammary epithelial cells were treated with agents that were equivalent mitogens but that otherwise had quite different mechanisms of action (summarized in Reference 54). NMU treatment resulted in mitogen-specific morphological variants and unique genotypes. These data suggest that not only is the sensitivity of mammary epithelial cells to a particular carcinogenic insult hormone dependent, but that the resulting tumor genotype is also dependent upon the hormonal milieu at the time of carcinogen exposure.

16.3 HYPOTHESES OF BREAST CARCINOGENESIS

Studies examining the mechanism of breast carcinogenesis can be framed by interrelated hypotheses that impute the importance of three factors: hormonal stimulation of the mammary tissue, exogenous carcinogen exposure, and hereditary susceptibility. Each of these influences is supported by epidemiological studies.

By far the strongest epidemiological correlates are with hormonal regulation and breast cancer. Breast cancer incidence rates increase linearly until about the age of 45. During the perimenopausal decade the incidence rates level off, after which, during the postmenopausal years the rate increases sharply, albeit with a slope less than during the premenopausal years [55]. The risk factors for breast cancer show a strong dependence upon female hormones and include being female (male breast cancers occur at 1% the rate of female breast cancers [56]), early menarche, late menopause, postmenopausal obesity, and hormone replacement therapy [57]. Protective factors are associated with decreased lifetime exposure to estrogen (such as prolonged lactation, surgical menopause, and exercise) or to a permanent alteration in mammary gland susceptibility to carcinogenesis, as observed in young age at first full-term pregnancy. However, the attributable risk to these hormonal risk factors accounts for only a quarter of breast cancer cases, and therefore other factors are thought to be involved [58].

Data correlating specific environmental exposures to chemical carcinogens are highly contradictory. During the era in which exogenous chemicals in the environment have been the greatest, as with many other types of cancers, the incidence in breast cancer has risen. Indeed the incidence has risen steadily since data have been well kept with a rate during 1987–1991 of about 113 per 100,000 women in the U.S. Some optimism is felt as it has been noted that during the 1990s this rate has remained flat and may be decreasing slightly [60]. Most investigators believe that the rise in incidence is largely explained by increased utilization of breast cancer screening [61,62]. In addition, there is a shift in population demographics with a large cohort of women surviving into the decades of maximum risk. However, additional factors are probable and environmental exposure to genotoxic and hormone-modifying agents has gained considerable attention [63].

Finally, about 10% of breast cancer cases are thought to be directly attributable to an inherited susceptibility gene, such as *BRCA1* or *BRCA2* [64]. Prevalence of these mutations is particularly high in specific populations [65]. In addition, there is growing evidence that inherited susceptibility may involve polygenic traits interacting with lifestyle choices and environmental insults. These interactions are only now beginning to be defined.

In this chapter, we look briefly at the potential role of endogenous hormones in carcinogenesis, the effects of genotoxic agents and environmental agents with hormonal effects, and finally inherited susceptibility. We are ignoring the role of diet in breast carcinogenesis, particularly the large number of estrogen-like compounds prevalent in soy products. Unfortunately, we are also unable to make a link with potential cause and mutation pathways taken by epithelial cells on their way to a fully malignant phenotype. These pathways are largely unknown for any breast tumor regardless of environmental determinants.

16.4 ESTROGENS AS CARCINOGENS

The data pointing to a hormonal influence in breast carcinogenesis are extensive. Among the established risk factors for breast cancer are, being female, early menarch and late menopause (factors that increase the duration of reproductive cycling), and hormone replacement therapy. Even more interesting are the protective factors such as young age at first-full term pregnancy, prolonged lactation, and exercise. Among the hormones that make theoretical sense for explaining these data, estrogen has been the most studied, although the role of progesterone, prolactin, and others are under intense investigation. Biochemical data also support the role of estrogen in breast carcinogenesis. Many studies have found a positive correlation between serum or urinary estrogens and risk of breast cancer, either across national boundaries in comparative population studies or prospectively within specific population groups [66–70]. For example, one large meta-analysis in postmenopausal women showed that individuals who developed breast cancer had serum estrogen levels 15% higher than unaffected women [69]. In addition, exogenous estrogens, particularly as hormone replacement in women with absent or reduced ovarian function, increase risk of breast cancer with duration of use [71]. In the presence of high endogenous estrogens, the addition of estrogen in the combination oral contraceptive preparations relates to breast cancer risk in a less obvious fashion. A recent meta-analysis indicates that while there may be slight increased risk (relative risk of 1.24), risk was greatest in the youngest users (<20 years old at initiation of use) [72]. In the case of hormonal replacement therapy the increased risk ended 5 years after cessation of therapy. In the case of combination oral contraceptives, it is not clear whether the risk continued even 10 years after cessation. Together these data give insights into the complexity with which estrogens may affect breast carcinogenesis.

Given the epidemiological data regarding estrogen associations with breast cancer, as well as endometrial cancer, and large numbers of rodent studies implicating estrogen in tumor induction, the International Agency for Research on Cancer concluded that estrogens, including estradiol, are carcinogens [73,74]. Numerous conflicting proposals have been proffered to explain how estrogen impacts breast cancer risk. The simplest explanation is that estrogen is a breast mitogen and therefore is merely promoting replication of cells that are already initiated. Indeed, frankly tumorigenic ER positive cell lines, such as MCF7, are stimulated to proliferate by exogenous estrogen [75]. However, normal mammary epithelium when proliferating does not express the ER [76,77], implying that normal cell proliferation is not directly ER-driven. More importantly, even highly proliferative terminal-end-bud mammary epithelium in rats has an intracellular dissociation between ER expression and thymidine incorporation [77]. Finally, breast proliferative rates were found to be higher in tissues removed from individuals during the luteal phase of the menstrual cycle than in the follicular phase [78,79], and mapping of cell proliferation throughout the menstrual cycle shows a peak at day 25 due to gradual increases in the number of proliferating lobules with progression during the luteal phase [80]. On the other hand, normal breast tissue explants in athymic nude mice are stimulated to proliferate in a dose-responsive fashion to physiological concentrations of estrogen [81]. However, the ER expression is very low in these tissues, and the proliferative response lags three days after treatment initiation, reaching a peak only after 6 to 7 days. These data suggest that estrogen is mitogenic but by some indirect mechanism, potentially via the induction of other growth factors [82]. In support of this paracrine action of estrogen, Cunha et al. [7] showed that $ER^{-/-}$ stroma could not support ductal morphogenesis of $ER^{+/+}$ epithelium. Conversely, $ER^{-/-}$ epithelium underwent ductal morphogenesis when placed with $ER^{+/+}$ stroma. Although these analyses pertain to normal epithelium, what is not known is the role of estrogen-induced proliferation in cells that have undergone some initiating event.

A related hypothesis is that elevated estrogen levels drive cell proliferation, increasing the opportunity for random genetic errors [57]. While estrogen may be an indirect epithelial mitogen, there is no direct evidence that fixation of random mutations via this mechanism is sufficient to elicit cell transformation. On the other hand, there is now abundant evidence that estrogen-metabolic products are directly genotoxic. Despite the abundant evidence that estrogen is carcinogenic in rodents and humans [73,74], neither estradiol nor any of its metabolites have been shown to produce point mutations in bacterial or mammalian assay systems [83–87]. However, there is now *in vivo* and *in vitro* evidence that estrogen and its metabolites cause other types of genetic damage. Figure 16.2. illustrates the chemical structure of estrogen. A number of studies have shown direct DNA adduction with various estrogens *in vivo*, particularly estrogen-3,4-quinones and 4-hydroxyestrogens [88,89]. There are also individual reports for estrogen-induced chromosomal abnormalities such as gene amplification, translocation, aneuploidy, and microsatellite instability (reviewed in Reference 90). Finally, there are a number of studies that support estrogen-induced oxidative DNA damage. Redox metabolism of estrogen produces oxygen radicals that have been documented to generate single strand breaks, 8-hydroxylation of guanine bases, and malondialdehyde-DNA adducts *in vivo* [91–94]. Estradiol, estrone 3,4-quinone, 2- and 4-hydroxyestradiol, 4-hydroxyestrone, and equilenin-3,4-quinone have all been implicated.

Having demonstrated the possibility of estrogen and its metabolites as being directly genotoxic, the next questions are which metabolites are elevated in breast tissue, which epidemiologically correlate with breast cancer risk, and what is the evidence that the reputed chemicals correlate with the genetic damage found early in tumorigenesis? While the data are far from complete, some suggest that estrogen metabolism is a key player in human disease. Estrogens are metabolized by phase 1 or P450 enzymes [95], as well as by phase 2 or conjugating enzymes, which may influence both excretion and synthesis of bioactive compounds [96]. The first step in the metabolism of estradiol is the oxidation of the 17β-hydroxyl group to produce estrone. Studies comparing normal breast

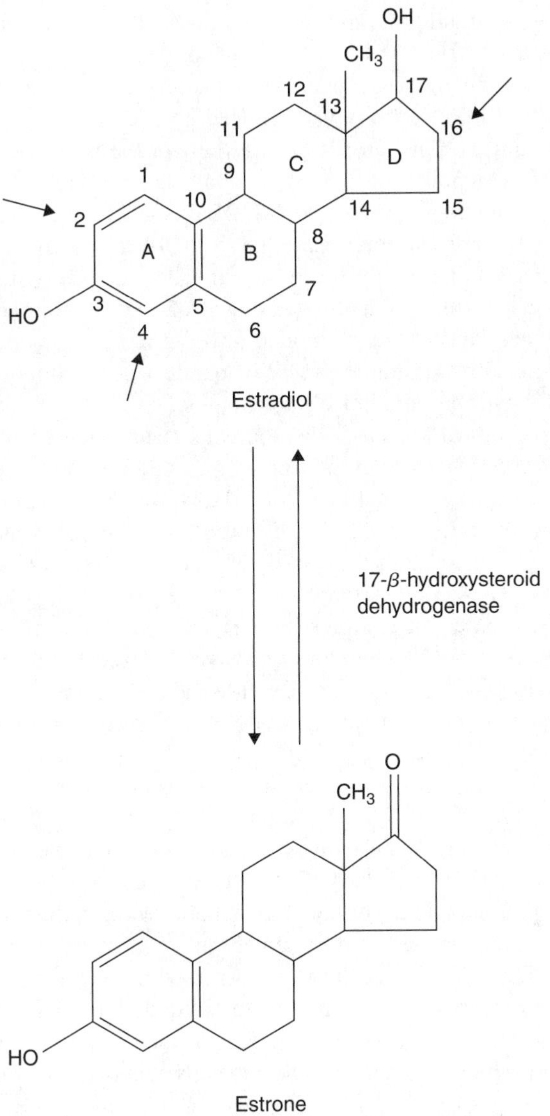

FIGURE 16.2 Arrows depict major sites of hydroxylation by P450 enzymes, which can use either estradiol or estrone as substrates. Hydroxylation at the 2 or 4 positions leads to catecholestrogens, the substrates for catechol-*O*-methyltransferase. Estradiol, estrone, and their many hydroxylated metabolites are all substrates for class 2 enzymes.

tissue to breast tumors indicate that the activity of 17β-hydroxysteroid dehydrogenase is increased in tumors but may actually have a propensity for a reverse equilibrium favoring the synthesis of the more potent estrogen, estradiol [97,98]. Following this initial hydroxylation, there are three major sites for hydroxylation (reviewed in Reference 99) (see Figure 18.2). In the liver, NADPH-dependent cytochrome P450 enzymes function primarily at C-2 and C-16. However, in extrahepatic tissues, C-4 is a common target. Hydroxylaton at C-2 or C-4 results in catecholestrogens [95], but these reactions are catalyzed by separate enzymes [100]. Multiple P450 have been identified, many of which are regulated during development, by endocrine levels, and by exogenous compounds [101].

When comparing women with breast cancer and their controls in the metabolism of tritiated estradiol, levels of 2-hydroxylated estrogens were found to be low in both groups [102]. Despite extensive

2-hydroxylation, these compounds are rapidly conjugated and excreted in the urine explaining their low levels in serological and tissue based assays, being more rapidly cleared than 4-hydroxyestrogens [103–105]. Locally formed 2-hydroxyestrogens, however, have significant effects. They bind to the classical ER, albeit with a lower affinity and reduced potency in the rat [106], but with similar affinity relative to estradiol with human ER [107]. Many of the well-studied effects are in neuro-endocrine regulation of dopamine receptor interactions and in secretion of prolactin, luteinizing hormone, and follicle stimulating hormone [108–112]. However, the evidence that they play a role in stimulating carcinogenesis is minimal. Like the 4-hydroxyestrogens, 2-hydroxyestrogens also undergo metabolic redox cycling to form free radicals and chemically reactive semiquinone and quinone intermediates [113]. Unlike 4-hydroxyestrogens, 2-hydroxyestrogens have no tumorigenic activity in the male Syrian hamster kidney model, a well-established model system for measuring estrogen-induced tumorigenesis, despite their ability to induce cell proliferation [85,114]. In fact the methylation product of 2-hydroxyestradiol is a potent inhibitor of cell proliferation and angiogenesis and shows promise in preclinical trials as a chemotherapeutic agent [115–118].

The role of 4-hydroxyestrogens in breast cancer is quite different. In human liver and placenta the 2-hydroxylation of estrogens predominates [100]. However, in organs with a propensity for estrogen-induced tumor formation, such as uterus, breast, and hamster kidney, the 4-hydroxylation predominates, primarily due to the action of P450 IB1 [100,119–121]. Not only is the ratio of 4- to 2-hydroxylation elevated in normal breast tissue relative to liver or placenta, but both fibroadenomas and breast carcinoma tissues have significantly elevated 4-hydroxyestrogens relative to normal breast tissue [122]. Unlike 2-hydroxyestrogens, 4-hydroxyestradiol is similar to estradiol in its ability to bind and activate the classical ER; however, the dissociation rate is reduced making 4-hydroxyestradiol potentially a longer-acting estrogen than estradiol [106,107,123]. In vivo 4-hydroxyestrogens are cleared via conjugation and excretion, but 4-hydroxyestradiol actually inhibits catechol O-methyltransferase activity on catecholamines [124]. On the other hand, 2-hydroxyestradiol will inhibit the O-methylation of 4-hydroxyestradiol, but the reverse is not true [105]. Unlike 2-hydroxyestradiol, 4-hydroxyestradiol is a potent carcinogen in the Syrian hamster kidney model [85,114]. It is not clear whether this is due to its direct stimulation of kidney proximal tubule cell proliferation [125] or due to metabolic redox cycling. In one study the reactive intermediate estrone-3,4-quinone produced hepatomas in young mice, whereas in another study this quinone was unable to induce breast carcinoma when directly injected into rat mammary glands despite the fact that others have shown the production of depurinating DNA adducts under similar conditions [88,126,127]. These data plus the observations that 4-hydroxyestradiol is the most common estrogen metabolite in human breast cancer specimens, makes this compound a likely candidate for breast carcinogenicity [128].

The other class of estrogens for which there is some evidence for a role in breast cancer is the class of 16α-hydroxyestrogens. These compounds retain potent ER-stimulating activity and in fact, can form covalent binding with ER [129,130]. In population-based studies using labeled estrogens, 16α-hydroxylation was found to be elevated in postmenopausal breast cancer patients and in women from high-risk families relative to healthy control subjects [131,132]. Furthermore, 16α-hydroxylation was elevated in isolated terminal ductolobular units isolated from mastectomy specimens compared to reduction mammoplasties [133]. This is in contrast to data from others who did not find any specific association between risk and this metabolite by comparing plasma or urinary estrogens between two populations with markedly different risks for breast cancer [70,134]. 16α-hydroxyestrone is quite carcinogenic in the Syrian hamster kidney model system, its carcinogenic effect correlating with its capacity to induce cellular proliferation [125]. In addition, the in vitro data are highly suggestive of some role for these compounds in breast cancer. First, two different types of mammary carcinogens, MMTV and DMBA, result in increased 16α-hydroxyestrogens in terminal ductolobular units [132,135,136]. In addition, treatment of mammary epithelial cells with 16α-hydroxyestrone results in genotoxic damage and anchorage-independent growth [137].

TABLE 16.1
Agents Implicated in the
Pathogenesis of Breast
Cancer

Estrogen and metabolites
DDT and metabolites
HCH
PAH
PCB
Endocrine disrupters

16.5 ENVIRONMENTAL AGENTS AS BREAST CARCINOGENS

Despite the direct evidence that radiation can induce breast cancer, there has not been irrefutable evidence for exogenous chemical carcinogenesis. Various lipophilic compounds that accumulate over a lifetime have been frequent targets for investigation, particularly since the mammary gland is largely composed of adipose tissue. These types of compounds include p,p′-dichloro(diphenyl)trichloroethane (DDT), p,p′-dichloro(diphenyl)dichloroethane (DDD), p,p′-dichloro(diphenyl)dichloroethene (DDE), hexachlorobenzine (HCB), α, β, and γ-hexachlorocyclohexane (HCH), polyaromatic hydrocarbons (PAH), and a variety of polychlorinated biphenyls (PCB). A summary of some of the agents implicated in breast carcinogenesis is presented in Table 16.1.

Numerous investigators have looked at human breast disease as it relates to serum, breast adipose tissue, or tumor levels and correlated these data with breast cancer incidence, either prospectively or retrospectively. In sum, the data are extraordinarily heterogeneous, making firm conclusions difficult. The reasons are as follows: Populations differ significantly in the basal level of exposure to these compounds such that, ignoring methodological differences, the levels of specific compounds may differ manyfold. For instance, HCB levels in breast adipose tissue from breast cancer patients in Toronto, Canada, and Central Hesse, Germany, were 32 μg/kg and 343 μg/kg, respectively [138,139]. Since many of these chemicals are now banned, the era in which the study is performed strongly influences the results. For instance, in one study comparing PCBs from serum in 1974 and 1989 found mean lipid-adjusted levels for PCBs of 663.6 ng/gm versus 332.9 ng/gm in control subjects, a drop of 50% over a period of 15 years. All of these chemicals accumulate over time and therefore data must be normalized to lipid content and age of the donor. Romieu et al. [140] illustrate the marked differences in lipid-adjusted serum levels of DDT and DDE in individuals aged between 20 and 83. Although there are statistically significant correlates between serum and adipose levels and among adipose tissues from different body sites, these correlations are crude, at best. In one recent study the correlation of DDE in serum and adipose tissue showed an r-value of only 0.364 [141]. Furthermore, absolute levels in serum, being much lower than in adipose tissue, call for technological heroics for measurement to any degree of confidence. Because adipose tissue is a major reservoir, many compounds can be identified in adipose biopsies that are below the limit of detection in serum [142] or measured values show no relation to each other [143,144]. Finally, there are no human data that examine adipose tissue metabolism relative to body site for xenobiotic accumulation.

Dewailly et al. [145] performed an interesting study in Greenlanders comparing organochlorine levels in brain, liver, and adipose tissue from a subcutaneous site versus omentum. Brain lipids were generally lower than other sources for all compounds tested. However, the accumulation relative to liver versus adipose tissue was compound specific. In general, omental and subcutaneous adipose tissues gave similar results. However, in this study there was no direct comparison with mammary fat. Despite this heterogeneity, when confounding variables are matched and samples sizes are adequate, fairly convincing data has been developed that shows positive correlations between hexachlorocyclohexane and DDE with risk of breast cancer [146,147].

Lipophilic compounds are known to be secreted in the milk. Since the ban on DTT in the early 1970s, milk surveys have shown a decreasing trend in milk secretion [148]. However, it is unclear whether the clearing of these compounds by lactation is correlated with a general redistribution throughout all adipose stores. A number of studies have demonstrated a decrease risk of breast cancer with increased duration of lactation [140,149,150]. Even more interesting is the study by Ing et al. [151] comparing women who experienced unilateral versus bilateral breastfeeding. Women with unilateral breast-feeding showed increased risk in the unsuckled breast. However, whether this risk relates to accumulated xenobiotics is, of course, unknown.

PAH are widely found in cooked foods, cigarette smoke, air pollution, and combustion exhaust particulates (reviewed in El-Bayoumy et al. [152]). Human mammary epithelial cells readily form DNA adducts upon exposure to PAH [153], and analyses of normal mammary tissue from breast cancer patients confirmed a higher level of PAH-DNA adducts relative to tissues from cancer-free patients [154]. However, direct epidemiological studies implicating PAH in breast carcinogenesis are lacking.

Among the most compelling data for a role of PCBs in breast cancer are studies in populations with occupational and extensive environmental exposure [155–159]. Furthermore, decreases in environmental exposure may also correlate with decreased breast cancer mortality [160]. However, as a group, these and other epidemiological studies have failed to provide convincing data linking PCBs as a genre of compounds to breast cancer. The caveat to this conclusion is that PCBs are a highly heterogeneous group of compounds with complex pharmacology, as discussed in Section 16.6.

16.6 XENOBIOTIC/HORMONE INTERACTIONS

Although it is not clear what mechanisms may be important in the carcinogenicity of xenobiotics or hormones, what is clear is that biochemically they may act in antagonistic fashions relative to each other. Because of the effect of these compounds on hormone synthesis and metabolism, they are often referred to as endocrine disrupters. Such compounds include specific herbicides, pesticides, petroleum products, and plasticizers (see Reference 161 for a review). DDT and its stable metabolite, DDE, as well as β-hexachlorocyclohexane have estrogenic activity [162–163]. PCB congeners may be classified by structure–activity relationships into three groups [164]. Briefly, group 1 are estrogenic. Group 2 are antiestrogenic and immunotoxic. Group 3 do not modulate estrogen activity but induce P450 enzymes. Given this breadth of known activities, it seems that studies examining PCBs in vivo are strengthened if subset analyses show relationships relative to specific groups. For instance, Aronson et al. [138] showed that the PCBs associated with breast cancer risk (PCBs 105, 118, and 156) are mono-ortho substituted dioxin-like compounds (Group 2).

Since steroid hormones are metabolized by the same monooxygenases that metabolize potential environmental carcinogens [165], it is not surprising that many xenobiotics and drugs have been found to influence estrogen metabolism. For instance, phenobarbital is a classical drug with profound effects on the function of many steroid hormones through upregulation of specific P450 activities [166]. Female smokers have lower serum and urinary estrogens with increased metabolism by 2-hydroxylation [167]. The formation of 4-hydroxyestrogens is stimulated due to the induction of P450 1B1 by 2,3,7,8-tetrachlorodibenzo-p-dioxin (TCDD) [120]. TCDD is an aryl hydrocarbon receptor agonist. Such agonists induce P450 IA1 and IA2 [101]. TCDD has an antiestrogen activity by stimulating metabolism of estradiol at the C-2, C-4, C-15α, and C-6α positions [168] and by inducing the proteosome-dependent degradation of ERα [169]. Given the potential chemopreventive activity of aryl hydrocarbon receptor stimulation, chronic administration of indole-3-carbinol was found to stimulate 2-hydroxylation of estradiol and to inhibit mammary preneoplasia in mice [170].

However, chronic administration of TCDD to female rats inhibited the formation of all spontaneous tumors in hormone sensitive organs (uterus, mammary gland, and pituitary), while increasing the incidence of liver tumors in an estrogen-dependent fashion [171], which may be due to the increase in 4-hydroxyestrogens as well as 2-hydroxyestrogens. Bradlow et al. [172] have shown an increased rate of 16α-hydroxylation of estrone relative to 2-hydroxylation in MCF-7 following incubation with numerous compounds, including DDT, atrazine, and coplanar PCBs. Therefore, correlating, not to mention predicting, the affect of a particular component on mammary neoplasia is extraordinarily complicated because stage of development and total animal physiology must be considered.

Given this panoply of activities, it is clear that no one mechanism will explain associated affects for these environmental compounds. Some compounds may act directly as endogenous hormones, their activity being modulated by site-specific concentrations and timing of exposure during the life-long changes in mammary gland development and differentiation. Some may block hormonal activity and feedback loops resulting in aberrant hormonal synthesis, metabolism, or activity. Others may have significant effects on endogenous hormone metabolism leading to increases in genotoxic estrogen metabolites. Any one or more of these activities could result in direct genotoxic damage or alter the hormonal milieu to the extent that the sensitivity of the epithelium to carcinogenic insults is altered. At this point all of this is speculative.

16.7 GENETIC PREDISPOSITION, HORMONES, AND XENOBIOTICS

Because endogenous metabolism of both xenobiotics and hormones may be important for the production of carcinogenic metabolites, there are hints that specific subsets of the population are more susceptible to breast cancer due to heritable polymorphisms in enzyme activity. P450 enzymes have been particularly highlighted because they metabolize both xenobiotics and endogenous hormones. On the other hand both xenobiotics and hormones regulate P450 activity. Li et al. [173] present a theoretical argument that specific polymorphisms in CYP1B1 may lead to increased breast cancer risk through increased synthesis of 4-hydroxyestradiol. Interestingly, a particular CYP1B1 polymorphism with low estradiol 4-hydroxylase activity is inversely associated with breast cancer incidence in a Japanese population study [174]. Other data regarding specific polymorphisms are contradictory [174,175]. Aromatase (CYP19) activity is higher in the breast adipose tissue of women with breast cancer than in women with benign breast disease [176]. Interestingly, this increased aromatase is higher in the quadrant in which the invasive cancer occurs, leading to the question of whether this change in enzyme activity is the cause or the effect. Baxter et al. [177] have shown an association with specific CYP19 polymorphisms in breast cancer patients relative to controls, although these data suggest linkage disequilibrium rather than functional changes in the enzyme. Others have confirmed the association between breast cancer risk and CYP19 polymorphisms, noting that exon switching may be involved in enzyme disregulation [178]. This association was not confirmed by others [179]. One study showed a weak association with specific genotypes for the aryl hydrocarbon hydroxylase (CYP1A1) and breast cancer risk that was specific to young women who smoke [180]. This is reminiscent of two earlier studies showing that breast cancer risk is higher in women with an extensive smoking history and a polymorphism in CYP1A1 [181] or in postmenopausal women who smoke and have the slow acetylator genotype. This risk is further increased in women who began smoking under the age of 16 [182].

Others have focused not on P450 hydroxylases but upon the estrogen metabolic pathway. For instance, catechol-O-methyltransferase (COMT) inactivates catechol estrogens. Specific polymorphisms in the population correlate with postmenopausal breast cancer, this risk being significantly increased when found in individuals who also have one or more polymorphisms/mutations in glutathione S-transferase [183]. Polymorphisms in glutathione S-transferase, another detoxifying enzyme, have been correlated with postmenopausal breast cancer in some studies [184], but not others [185].

The fact that genetic polymorphisms in enzyme pathways for endogenous hormone metabolites correlate with breast cancer risk is significant at several levels. First, not just one but many different enzymes may be involved, supporting the concept that rate of synthesis or degradation of specific estrogen metabolites plays a key role in breast tumorigenesis. In many of these studies, the correlations were to specific subpopulations, such as postmenopausal women or young women who smoke. These types of data may ultimately help us understand the biochemical events making pre- and postmenopausal carcinomas so different. Finally, whenever there is an inherited predisposition, the potential exists for early intervention, once risk factors are clearly defined.

16.8 EPIGENETIC EVENTS IN TUMORIGENESIS

Most of the data examining hormonal or xenobiotic events in breast tumorigenesis have focused on genotoxic damage. However, a number of studies have implicated epigenetic events in the earliest stages of breast carcinogenesis. An epigenetic event permanently or reversibly alters gene expression, by transcriptional, translational, or posttranslational mechanisms [186]. Biochemical mechanisms include the methylation and acetylation pathways as well as numerous other types of cellular regulation [187]. Data supporting epigenetic initiation are as follows: generally, transformation events are 1–3 orders of magnitude more frequent than mutational events at specific loci [188]. Estimates for mutation rates per base pair per cell generation are on the order of 10^{-10} [189,190] or a mutation rate of 10^{-5}–10^{-6} per gene per cell generation [191]. However, the rate of epigenetic changes or chromosomal mutations is much higher [192]. In addition, tumor mutation frequency events are time but not proliferation dependent [193].

Several authors have emphasized the role of epigenetic events in carcinogenesis [192,194,195]. Eldridge and Gould [196] compared the mutation frequencies at one locus and found that the mutation rate was the same in normal breast epithelial cells and breast tumors cells but tenfold lower than the mutation rate in immortalized human breast epithelial cells. In the 12-year studies examining spontaneous transformation events in the human breast epithelial cell line, HMT-3522, removal of epidermal growth factor (EGF) was a major contributor to tumorigenicity [197]. Indeed, the relationship of EGF/EGFR signaling and β1-integrin expression have been shown to be interrelated and profoundly alter the tumorigenic phenotype of mammary epithelial cells [198]. Using models of irradiated or NMU-treated mammary epithelial clonogens, Kamiya et al. [199] hypothesized that highly frequent epigenetic events are tumor initiators and that these events are suppressed by adjacent cells. In a series of studies Kennedy et al. [200,201] showed that exposure of rodent cells to ionizing radiation increase the probability that a cell will later undergo a mutation-driven transforming event. Given that stem cells are likely to be the source of tumor progenitor cells and that hormonal environment, with or without morphological differentiation, profoundly alters their susceptibility to carcinogenic insults, anything that alters this hormonal milieu, including endocrine disrupters, may impact tumorigenesis via this change in susceptibility, which is potentially an epigenetic event.

The fact that hormonally regulated epigenetic events may impact tumorigenesis is most poignantly brought up by the field of hormonal imprinting. To date, data are few and correlative relative to *in utero* effects of hormones on breast cancer risk. Human data examining maternal chemical exposure and fetal risk are entirely lacking. However, this is a field ripe for investigation. Epidemiological risk factors in the mothers of women with breast cancer suggest that in a subset of women early genetic/epigenetic events occur *in utero* or in the perinatal period. In all cases these correlations associate exposure to high estrogen levels and increased risk. For instance, several studies correlate increased birth weight with elevated risk of subsequent breast cancer, particularly in premenopausal women [202,203]. Birth weight has been directly correlated with levels of estrogen in pregnancy, and thus prenatal exposure to elevated estrogens may impact future susceptibility to breast cancer [204]. Severe prematurity and neonatal jaundice, two conditions associated with elevated estrogen in the newborn are also associated with an increased risk of subsequent breast cancer [205,206]. On the other hand, conditions with low levels of estrogen in pregnancy, such as preeclampsia or eclampsia, actually decrease subsequent risk for breast cancer in the child [205,207]. In support of the link between risk and prenatal environment, Sandson et al. have shown a correlation between specific patterns of cerebral asymmetry, developmentally related to intrauterine hormone levels, and breast cancer risk [208,209].

There are several hypotheses that may explain the affect of neonatal estrogen exposure. Anbazhagan et al. [1,210] have shown that not only is there marked heterogeneity in the degree to which neonatal mammary glands may be developed relative to the presence of terminal ductolob-ular units, but indeed this degree of development correlates with birth weight and length, measures of intrauterine estrogen exposure. One could speculate that hormone exposure *in utero* may impact the state of mammary gland differentiation during the critical period of childhood exposure to envir-onmental mutagens. Alternatively, the number and distribution of susceptible stem cells could be differentially regulated. It is known that the prenatal hormonal environment regulates neural-network organization [211], so one could suggest that hypothalamic–pituitary–gonadal pathways and thus the proliferation/differentiation of the mammary epithelium in the adult may be permanently altered. To date no experiments have been reported testing any of these hypotheses. However, there are data suggesting that these hypotheses are not unreasonable. For example, rodent-model studies have shown that maternal estrogen levels regulate bone turnover in adult offspring [212]. Similarly, pren-atal treatment with allylestrenol resulted in permanent changes in hormone levels in offspring of both sexes [213]. Maternal high-fat diet, associated with elevated estrogen in pregnancy, was cor-related with increased sensitivity to DMBA-induced mammary tumors in offspring [214]. Again the mechanisms through which these effects are mediated are completely unknown.

As expected there is now growing evidence that the interrelationship between hormones and xenobiotics may determine tumor risk *in utero*. For instance, prenatal and lactational exposure to TCDD in male rats is associated with significant alterations in androgen responsiveness in the prostate [215]. Similarly, neonatal Aroclor 1254 resulted in permanent changes in adult levels of various testosterone hydroxylase activities [216]. These studies suggest the possibility that hormonal imprinting *in utero* may be one mechanism through which xenobiotic exposure could alter breast cancer risk.

16.9 SUMMARY

Breast cancer risk is clearly influenced by factors that define hormonal control of mammary gland differentiation and development. Unfortunately, most of the known risk factors are beyond an indi-vidual's control. How exposure to exogenous chemicals, whether one-time exposures or lifetime accumulations affect breast cancer risk for a woman or her offspring is likely to be exquisitely compound-dependent. Defining the interrelationships among estrogen metabolic pathways, hor-monal regulation of susceptibility, xenobiotics, and inherited polymorphisms is a major task for the future.

Unanswered questions raised in this discussion include the following: What cell types parent a carcinoma and when are they present in a susceptible state in the human breast? Are the differences in pre versus postmenopausal breast cancer due to hormonal environments at initiation, developmental state of the breast tissue during certain stages of life, type of initiating event, or related to epigenetic events? Are there any important age-related windows for breast carcinogenesis in the human popula-tion? Given the complex interactions between many environmental chemicals and hormonal systems, how do we design epidemiological studies to optimize the identification of harmful environmental insults?

REFERENCES

1. Anbazhagan, R. et al., Growth and development of the human infant breast, *Am. J. Anatomy*, 192, 407, 1991.

2. Monaghan, P. et al., Peripubertal human breast development, *Anatom. Record.*, 226, 501, 1990.
3. Russo, J. et al., Biology of disease: Comparative study of human and rat mammary tumorigenesis, *Lab. Invest.*, 62, 244, 1990.
4. Levin, M. et al., Lactation and menstrual function as related to cancer of the breast, *Am. J. Public Health*, 54, 580, 1964.
5. Kenney, N. et al., The aged mammary gland, in *Pathobiology of the Aging Mouse*, Vol. 2, Mohs, U., Dungworth, D., Capen, C., Cailton, W., and Sundberg, J., Eds., ILSI Press, Washington, DC, 1996.
6. Daniel, C., Silberstein, G., and Strickland, P., Direct action of 17 beta-estradiol on mouse mammary ducts analyzed by sustained release implants and steroid autoradiography, *Cancer Res.*, 47, 6052, 1987.
7. Cunha, G. et al., Elucidation of a role for stromal steroid hormone receptors in mammary gland growth and development using tissue recombination experiments, *J. Mam. Gland Biol. Bioplasia*, 2, 393, 1997.
8. Lydon, J. et al., Mice lacking progesterone receptor exhibit pleiotropic reproductive abnormalities, *Genes Dev.*, 9, 2266, 1995.
9. Humpherys, R. et al., Mammary gland development is mediated by both stromal and epithelial progesterone receptors, *Mol. Endocrinol.*, 11, 801, 1997.
10. Nandi, S., Endocrine control of mammary gland development and function in the C3H/HeCrg1 mouse, *J. Natl. Cancer Inst.*, 21, 1039, 1958.
11. Wellings, S., Jensen, H., and Marcum, R., An atlas of subgross pathology of the human breast with special reference to possible precancerous lesions, *J. Natl. Cancer Inst.*, 55, 231, 1975.
12. Russo, J. and Russo, I., Biology of disease: Biological and molecular bases of mammary carcinogenesis, *Lab. Invest.*, 57, 112, 1987.
13. Lu, J. et al., Pathogenic characterization of 1-methyl-1-nitrosourea-induced mammary carcinomas in the rat, *Carcinogenesis*, 19, 223, 1998.
14. Russo, J. and Russo, I., Influence of differentiation and cell kinetics on the susceptibility of the rat mammary gland to carcinogenesis, *Cancer Res.*, 40, 2677, 1980.
15. Tay, L. and Russo, J., 7,12-dimethylbenz(a)anthracene-induced DNA binding and repair synthesis in susceptible and non-susceptible mammary epithelial cells in culture, *J. Natl. Cancer Inst.*, 67, 155, 1981.
16. Tay, L. and Russo, J., Formation and removal of 7,12-dimethylbenz(a)-anthracene-nucleic acid adducts in rat mammary epithelial cells with different susceptibility to carcinogenesis, *Carcinogenesis*, 2, 1327, 1981.
17. Russo, I., Koszalka, M., and Russo, J., Comparative study of the influence of pregnancy and hormonal treatment on mammary carcinogenesis, *Br. J. Cancer*, 64, 481, 1991.
18. Hildreth, N., Shore, R., and Dvoretsky, P., The risk of breast cancer after irradiation of the thymus in infancy, *N. Engl. J. Med.*, 321, 1281, 1989.
19. Boice, J.D.J. and Monson, R.R., Breast cancer in women after repeated fluoroscopic examinations of the chest, *J. Natl. Cancer Inst.*, 59, 823, 1977.
20. Miller, A. et al., Mortality from breast cancer after irradiation during fluoroscopic examinations in patients being treated for tuberculosis, *N. Engl. J. Med.*, 321, 1285, 1989.
21. Tokunga, M. et al., Incidence of female breast cancer among atomic bomb survivors, 1950–1985, *Radiat. Res.*, 138, 209, 1994.
22. McMahon, B. et al., Age at first birth and breast cancer risk, *Bull. WHO*, 43, 209, 1970.
23. Newcomb, P. et al., Lactation and a reduced risk of premenopausal breast cancer, *New. Engl. J. Med.*, 330, 81, 1994.
24. Haran-Ghera, N., The role of hormones in mammary tumor development, *Acta Unio. Int. Contra. Cancrum*, 18, 207, 1962.
25. Medina, D., Mammary tumorigenesis in chemical carcinogen-treated mice. II. Dependence on hormone stimulation for tumorigenesis, *J. Natl. Cancer Inst.*, 53, 223, 1974.
26. Swanson, S. et al., Pituitary-isografted mice are highly susceptible to MNU-induced mammary carcinogenesis irrespective of the level of alveolar differentiation, *Carcinogenesis*, 15, 1341, 1994.
27. Sinha, D., Pazik, J., and Dao, T., Progression of rat mammary development with age and its relationship to carcinogenesis by a chemical carcinogen, *Int. J. Cancer*, 31, 321, 1983.

28. Sivaraman, L. et al., Hormone-induced refractoriness to mammary carcinogenesis in Wistar-Furth rats, *Carcinogenesis*, 19, 1573, 1998.

29. Medina, D. et al., Short-term exposure to estrogen and progesterone induces partial protection against *N*-nitroso-*N*-methylurea-induced mammary tumorigenesis in Wistar-Furth rats, *Cancer Lett.*, 169, 1, 2001.

30. Guzman, R. et al., Hormonal prevention of breast cancer: Mimicking the protective effect of pregnancy, *Proc. Natl. Acad. Sci. USA*, 96, 2520, 1999.

31. Miyamoto, S. et al., Neoplastic transformation of mouse mammary epithelial cells by in vitro exposure to *N*-methyl-*N*-nitrosourea, *Proc. Natl. Acad. Sci. USA*, 85, 477, 1988.

32. Nandi, S., Guzman, E., and Miyamoto, S., Hormones, cell proliferation and mammary carcinogenesis, in *Hormonal Carcinogenesis*, Li, J., Nandi, S., and Li, S., Eds., Springer-Verlag, New York, 1992, p. 73.

33. Dulbecco, R. et al., Marker evolution during the development of the rat mammary gland: Stem cells identified by markers and the role of the myoepithelial cells, *Cancer Res.*, 46, 2449, 1986.

34. Anderson, E., Clarke, R., and Howell, A., Estrogen responsiveness and control of normal human breast proliferation, *J. Mammary Gland Biol. Neoplasia*, 3, 23, 1998.

35. Kordon, E. and Smith, G., An entire functional mammary gland may comprise the progeny from a single cell, *Development*, 125, 1921, 1998.

36. Young, L. et al., The influence of host and tissue age on life span and growth rate of serially transplanted mouse mammary gland, *Exp. Geront.*, 6, 49, 1971.

37. Smith, G., Experimental mammary epithelial morphogenesis in an in vivo model: Evidence for distinct cellular progenitors of the ductal and lobular phenotype, *Breast Cancer Res. Treat.*, 39, 21, 1996.

38. Kamiya, K., Gould, M., and Clifton, K., Quantitative studies of ductal versus alveolar differentiation from rat mammary clonogens, *Proc. Soc. Exp. Biol. Med.*, 219, 217, 1998.

39. Tsai, Y. et al., Contiguous patches of normal human mammary epithelium derived from a single stem cell: Implications for breast carcinogenesis, *Cancer Res.*, 56, 402, 1996.

40. Stingl, J. et al., Phenotypic and functional characterization in vitro of a multipotent epithelial cell present in the normal adult human breast, *Differentiation*, 63, 201, 1998.

41. Sell, S. and Pierce, G., Maturation arrest of stem cell differentiation is a common path for the cellular origin of teratocarcinomas and epithelial cancer, *Lab. Invest.*, 70, 6, 1994.

42. Russo, J., Calaf, G., and Russo, I., A critical approach to the malignant transformation of human breast epithelial cells with chemical carcinogens, *Crit. Rev. Oncogen.*, 44, 401, 1993.

43. Chang, C.-C. et al., A human breast epithelial cell type with stem cell characteristics as target cells for carcinogenesis, *Rad. Res.*, 155, 201, 2001.

44. Clifton, K., Tanner, M., and Gould, M., Assessment of radiogenic cancer initiation frequency per clonogenic rat mammary cell in vivo, *Cancer Res.*, 46, 2390, 1986.

45. Kamiya, K. et al., Control of ductal vs. alveolar differentiation of mammary clonogens and susceptibility to radiation-induced mammary cancer, *J. Rad. Res*, (Suppl.), Vol 32, 181, 1991.

46. Boulanger, C. and Smith, G., Reducing mammary cancer risk through premature stem cell senescence, *Oncogene*, 20, 2264, 2001.

47. Taylor-Papadimitriou, J. et al., Keratin expression in human mammary epithelial cells cultured from normal and malignant tissue: Relation to in vivo phenotypes and influence of medium, *J. Cell. Sci.*, 94, 403, 1987.

48. Taylor-Papadimitriou, J. et al., Patterns of reaction of monoclonal antibodies HMGF-1 and 2 with benign breast tissues and breast carcinomas, *J. Exp. Pathol.*, 2, 247, 1986.

49. Walker, R. et al., Breast carcinomas occurring in young women (<35 years) are different, *Br. J. Cancer*, 74, 1796, 1996.

50. Clark, G., The biology of breast cancer in older women, *J. Gerontol.*, 47, 19, 1992.

51. Diab, S., Elledge, R., and Clark, G., Tumor characteristics and clinical outcome of elderly women with breast cancer, *J. Natl. Cancer Inst.*, 92, 550, 2000.

52. Struse, K. et al., The estrogen receptor paradox in breast cancer: Association of high receptor concentrations with reduced overall survival, *Breast J.*, 6, 115, 2000.

53. Li, C. et al., Changing incidence rate of invasive lobular breast carcinoma among older women, *Cancer*, 88, 2561, 2000.

54. Nandi, S., Guzman, R., and Yang, J., Hormones and mammary carcinogenesis in mice, rats, and humans: A unifying hypothesis, *Proc. Natl. Acad. Sci. USA*, 92, 3650, 1995.

55. Clemmesen, J., Carcinoma of the breast: Results from statistical research, *Br. J. Radiol.*, 21, 583, 1948.

56. Ewertz, M. et al., Risk factors for male breast cancer — a case–control study from Scandinavia, *Acta Oncol.*, 40, 467, 2001.

57. Henderson, B. and Feigelson, H., Hormonal carcinogenesis, *Carcinogenesis*, 21, 427, 2000.

58. Seidman, H., Stellman, S., and Mushincki, M., A different perspective on breast cancer risk factors: Some implications of the nonattributable risk, *CA Cancer J. Clin.*, 32, 301, 1982.

59. Devesa, S. et al., Recent cancer trends in the United States, *J. Natl. Cancer. Inst.*, 87, 175, 1995.

60. Wingo, P. et al., Cancer incidence and mortality, 1973–1995, *Cancer*, 82, 1197, 1998.

61. Miller, B., Feuer, E., and Hankey, B., Recent incidence trends for breast cancer in women and the relevance of early detection: An update, *CA Cancer J. Clin.*, 43, 27, 1993.

62. Wun, L., Feuer, E., and Miller, B., Are increases in mammographic screening still a valid explanation for trends in breast cancer incidence in the United States?, *Cancer Causes Control*, 6, 135, 1995.

63. Krieger, N., Exposure, susceptibility, and breast cancer risk, *Breast Cancer Res. Treat.*, 13, 205, 1989.

64. Carter, R., BRCA1, BRCA2 and breast cancer: A concise clinical review, *Clin. Invest. Med.*, 24, 147, 2001.

65. Bahar, A. et al., The frequency of founder mutations in the BRCA1, BRCA2, and APC genes in Australian Ashkenazi Jews: Implication for the generality of U.S. population data, *Cancer*, 92, 440, 2001.

66. MacMahon, B. et al., Urine oestrogen profiles of Asian and North American women, *Int. J. Cancer*, 14, 161, 1974.

67. Bernstein, L. et al., Serum hormone levels in pre-menopausal Chinese women in Shanghai and white women in Los Angeles: Results from two breast cancer case–control studies, *Cancer Causes Control*, 1, 51, 1990.

68. Shimizu, H. et al., Serum oestrogen levels in postmenopausal women: Comparison of American whites and Japanese in Japan, *Br. J. Cancer*, 62, 451, 1990.

69. Thomas, H., Reeves, G., and Key, T., Endogenous estrogen and postmenopausal breast cancer: A quantitative review, *Cancer Causes Control*, 8, 922, 1997.

70. Adlercreutz, H. et al., Estrogen metabolism and excretion in Oriental and Caucasian women, *J. Natl. Cancer Inst.*, 86, 1076, 1994.

71. Collaborative Group on Hormonal Factors in Breast Cancer, Breast cancer and hormone replacement therapy: Collaborative reanalysis of data from 51 epidemiological studies of 52,705 women with breast cancer and 108,411 women without breast cancer, *Lancet*, 350, 1047, 1997.

72. Collaborative Group on Hormonal Factors in Breast Cancer, Breast cancer and hormonal contraceptives: Collaborative reanalysis of individual data on 53,297 women with breast cancer and 100,239 women without breast cancer from 54 epidemiological studies, *Lancet*, 347, 1713, 1996.

73. IARC, *IARC Monographs on the Evaluation of Carcinogenic Risks to Humans*, IARC, Lyon, France, 1987.

74. IARC, *IARC Monographs on the Evaluation of Carcinogenic Risks to Humans: Hormonal Contraception and Postmenopausal Hormone Therapy*, IARC, Lyon, France, 1999.

75. Kodama, M. and Kodama, T., A new trend of breast cancer research in the genome era, *Int. J. Mol. Med.*, 8, 291, 2001.

76. Clark, R., et al., Dissociation between steroid receptor expression and cell proliferation in human breast, *Cancer Res.*, 57, 4987, 1997.

77. Russo, J. et al., Pattern of distribution of cells positive for estrogen receptor alpha and progesterone receptor in relation to proliferative cells in the mammary gland, *Breast Cancer Res. Treat.*, 53, 217, 1999.

78. Olsson, H. et al., Proliferation of the breast epithelium in relation to menstrual cycle phase, hormonal use, and reproductive factors, *Breast Cancer Res. Treat.*, 40, 187, 1996.

79. Soderqvist, G. et al., Proliferation of breast epithelial cells in healthy women during the menstrual cycle, *Am. J. Obstet. Gynecol.*, 176, 123, 1997.

80. Ferguson, D. and Anderson, T., Morphological evaluation of cell turnover in relation to the menstrual cycle in the "resting" human breast, *Br. J. Cancer*, 44, 177, 1981.

81. Laidlaw, I. et al., The proliferation of normal human breast tissue implanted into athymic nude mice is stimulated by estrogen but not progesterone, *Endocrinology*, 136, 164, 1995.

82. Lu, R. and Serrero, G., Mediation of estrogen mitogenic effect in human breast cancer MCF-7 cells by PC-cell-derived growth factor (PCDGF/granulin precursor), *Proc. Natl. Acad. Sci. USA*, 98, 142, 2001.

83. Lang, R. and Redmann, U., Non-mutagenicity of some sex hormones in the Salmonella/microsome mutagenicity test, *Mutat. Res.*, 67, 361, 1979.

84. Drevon, C., Piccoli, C., and Montesano, R., Mutagenicity assays of estrogenic hormones in mammalian cells, *Mutat. Res.*, 89, 83, 1981.

85. Liehr, J. et al., Carcinogenicity of catechol estrogens in Syrian hamsters, *J. Steroid Biochem.*, 24, 353, 1986.

86. Lang, R. and Reiman, R., Studies for a genotoxic potential of some endogenous and exogenous sex steroids. I. Communication: Examination for the induction of gene mutations using the Ames Salmonella/microsome test and the HGPRT test in V79 cells, *Environ. Mol. Mutagen*, 21, 272, 1993.

87. Rajah, T. and Pento, J., The mutagenic potential of antiestrogens at the HPRT locus in V79 cells, *Res. Commun. Mol. Pathol. Pharmacol.*, 89, 85, 1995.

88. Cavalieri, E. et al., Molecular origin of cancer: Catechol estrogen-3,4-quinones as endogenous tumor initiators, *Proc. Natl. Acad. Sci. USA*, 94, 10937, 1997.

89. Devanesan, P. et al., Catechol estrogen conjugates and DNA adducts in the kidney of male Syrian Golden hamsters treated with 4-hydroxyestradiol: Potential biomarkers for estrogen-initiated cancer, *Carcinogenesis*, 22, 489, 2001.

90. Liehr, J., Role of DNA adducts in hormonal carcinogenesis, *Reg. Toxicol. Pharmacol.*, 32, 276, 2000.

91. Han, X. and Liehr, J., 8-Hydroxylation of guanine bases in kidney and liver DNA of hamsters treated with estradiol: Role of free radicals in estrogen-induced carcinogenesis, *Cancer Res.*, 54, 5515, 1994.

92. Han, X. and Liehr, J., DNA single strand breaks in kidneys of Syrian hamsters treated with steroidal estrogens. Hormone-induced free radical damage preceding renal malignancy, *Carcinogenesis*, 15, 977, 1994.

93. Ho, S.-M. and Roy, D., Sex hormone-induced nuclear DNA damage and lipid peroxidation in the dorsolateral prostates of Noble rats, *Cancer Lett.*, 84, 155, 1994.

94. Wang, M. and Liehr, J., Induction by estrogens of lipid peroxidation and lipid peroxide-derived malonaldehyde-DNA adducts in male Syrian hamsters: Role of lipid peroxidation in estrogen-induced kidney carcinogenesis, *Carcinogenesis*, 16, 1941, 1995.

95. Martucci, C. and Fishman, J., P450 enzymes of estrogen metabolism, *Pharmacol. Ther.*, 57, 237, 1993.

96. Raftogianis, R. et al., Estrogen metabolism by conjugation, *J. Natl. Cancer Inst. Monogr.*, 27, 113, 2000.

97. Bonney, R. et al., The relationship between 17 beta-hydroxysteroid dehydrogenase activity and oestrogen concentrations in human breast tumours and in normal breast tissue, *Clin. Endocrinol.*, 19, 727, 1983.

98. McNeill, J. et al., A comparison of the in vivo uptake and metabolism of ^{3}H-oestrone and ^{3}H-oestradiol by normal breast and breast tumour tissues in post-menopausal women, *Int. J. Cancer*, 38, 193, 1986.

99. Zhu, B. and Conney, A., Functional role of estrogen metabolism in target cells: Review and perspectives, *Carcinogenesis*, 19, 1, 1998.

100. Liehr, J. et al., 4-Hydroxylation of estradiol by human uterine myometrium and myoma microsomes: Implications for the mechanism of uterine tumorigenesis, *Proc. Natl. Acad. Sci. USA*, 92, 9220, 1995.

101. Nebert, D. and Gonzalez, F., P450 genes: Structure, evolution, and regulation, *Annu. Rev. Biochem.*, 56, 945, 1987.

102. Schneider, J. et al., Abnormal oxidative metabolism of estradiol in women with breast cancer, *Proc. Natl. Acad. Sci. USA*, 79, 3027, 1982.

103. Zhu, B. and Liehr, J., Inhibition of catechol-*O*-methyltransferase-catalyzed *O*-methylation of 2- and 4-hydroxyestradiol by quercetin, *J. Biol. Chem.*, 271, 1357, 1996.

104. Li, S., Purdy, R., and Li, J., Variations in catechol-*O*-methyltransferase activity in rodent tissues: Possible role in estrogen carcinogenicity, *Carcinogenesis*, 10, 63, 1989.

105. Roy, D., Weisz, J., and Liehr, J., The *O*-methylation of 4-hydroxyestradiol is inhibited by 2-hydroxyestradiol: Implications for estrogen-induced carcinogenesis, *Carcinogenesis*, 11, 459, 1990.

106. Ball, P. and Knuppen, R., Catecholoestrogens (2- and 4-hydroxyoestrogens): Chemistry, biogenesis, metabolism, occurrence and physiological significance, *Acta Endocrinol.*, 232, 1, 1980.

107. van Aswegen, C., Purdy, R., and Wittliff, J., Binding of 2-hydroxyestradiol and 4-hydroxestradiol to estrogen receptor human breast cancers, *J. Steroid Biochem.*, 32, 485, 1989.

108. Paden, C. et al., Comparison by estrogens for catecholamine receptor binding in vitro, *J. Neurochem.*, 39, 512, 1982.

109. Schaeffer, J. and Hsueh, A., 2-Hydroxyestradiol interactions with dopamine receptor binding in rat anterior pituitary, *J. Biol. Chem.*, 254, 5606, 1979.

110. Fishman, J. and Tulchinsky, D., Suppression of prolactin secretion in normal young women by 2-hydroxyestrone, *Science*, 210, 73, 1980.

111. Linton, E. et al., Hydroxyestradiol inhibits prolactin release from the superfused rat pituitary gland, *J. Endocrinol.*, 90, 315, 1981.

112. Martucci, C. and Fishman, J., Impact of continuously administered catechol estrogens on uterine growth and LH secretion, *J. Biol. Chem.*, 271, 1357, 1979.

113. Liehr, J. and Roy, D., Free radical generation by redox cycling of estrogens, *Free Radical Biol. Med.*, 8, 415, 1990.

114. Li, J. and Li, S., Estrogen carcinogenesis in Syrian hamster tissues: Role of metabolism, *Fed. Proc.*, 46, 1858, 1987.

115. Seegers, J. et al., The cytotoxic effects of estradiol-17beta, catecholestradiols and methoxyestradiols on dividing MCF-7 and Hela cells, *J. Steroid Biochem.*, 32, 797, 1989.

116. Fotsis, T. et al., The endogenous oestrogen metabolite 2-methoxyestradiol inhibits angiogenesis and suppresses tumor growth, *Nature*, 368, 237, 1994.

117. Schumacher, G. and Neuhaus, P., The physiological estrogen metabolite 2-methoxyestradiol reduces tumor growth and induces apoptosis in human solid tumors, *J. Cancer Res. Clin. Oncol.*, 127, 405, 2001.

118. Huober, J. et al., Oral administration of an estrogen metabolite-induced potentiation of radiation anti-tumor effects in presence of wild-type p53 in non-small-cell lung cancer, *Int. J. Radiat. Oncol. Biol. Phys.*, 48, 1127, 2000.

119. Newbold, R., Bullock, B., and MacLachlan, J., Uterine adenocarcinoma in mice following developmental treatment with estrogens: A model for hormonal carcinogenesis, *Cancer Res.*, 50, 7677, 1990.

120. Spink, D. et al., The effects of 2,3,7,8-tetrachlorodibenzo-p-dioxin on estrogen metabolism in MCF-7 breast cancer cells: Evidence for induction of a novel 17 beta-estradiol 4-hydroxylase, *J. Steroid Biochem. Mol. Biol.*, 51, 251, 1994.

121. Sutter, T. et al., Complete cDNA sequence of a human dioxin-inducible mRNA identifies a new gene subfamily of cytochrome P450 that maps to chromosome 2, *J. Biol. Chem.*, 269, 13092, 1994.

122. Liehr, J. and Ricci, M., 4-Hydroxylation of estrogens as markers of human mammary tumors, *Proc. Natl. Acad. Sci. USA*, 93, 3294, 1996.

123. Barnea, E., MacLusky, N., and Naftolin, F., Kinetics of catechol estrogen–estrogen receptor dissociation: A possible factor underlying differences in catechol estrogen biological activity, *Steroids*, 41, 643, 1983.

124. Ball, P. et al., Interactions between estrogens and catechol amines. III. Studies on the methylation of catechol estrogens, catechol amines and other catechols by the catechol *O*-methyltransferase of human liver, *J. Clin. Endocrinol. Metab.*, 34, 736, 1972.

125. Li, J. et al., Carcinogenic activites of various steroidal and nonsteroidal estrogens in the hamster kidney: Relation to hormonal activity and cell proliferation, *Cancer Res.*, 55, 4347, 1995.

126. Cavalieri, E., Minisymposium on endogenous carcinogens: The catechol estrogen pathway, *Polycyclic Aromatic Comp.*, 6, 223, 1994.

127. El-Bayoumy, K. et al., Lack of tumorigenicity of cholesterol epoxides and estrone-3,4-quinone in the rat mammary gland, *Cancer Res.*, 56, 1970, 1996.

128. Castagnetta, L. et al., Gas chromatography/mass spectrometry of catechol estrogens, *Steroids*, 57, 437, 1992.

129. Fishman, J. and Martucci, C., Biological properties of 16 alpha-hydroxyestrone: Implications in estrogen physiology and pathophysiology, *J. Clin. Endocrinol. Metab.*, 51, 611, 1980.

130. Swaneck, G. and Fishman, J., Covalent binding of the endogenous estrogen 16 alpha-hydroxyestrone to estradiol receptor in human breast cancer cells: Characterization and intranuclear localization, *Proc. Natl. Acad. Sci. USA*, 85, 7831, 1988.

131. Fishman, J. et al., Increased estrogen 16 alpha-hydroxylase activity in women with breast and endometrial cancer, *J. Steroid Biochem. Mol. Biol.*, 20, 1007, 1984.

132. Bradlow, H. et al., 16 alpha-hydroxylation of estradiol: A possible risk marker for breast cancer., *Ann. N.Y. Acad. Sci.*, 464, 138, 1986.

133. Osborne, M. et al., Upregulation of estradiol C16alpha-hydroxylation in human breast tissue: A potential biomarker of breast cancer risk, *J. Natl. Cancer Inst.*, 85, 1917, 1993.

134. Lemon, H., Heidel, J., and Rodriguez-Sierra, J., Increased catechol estrogen metabolism as a risk factor for nonfamilial breast cancer, *Cancer*, 69, 457, 1992.

135. Fishman, J., Osborne, M., and Telang, N., The role of estrogen in mammary carcinogenesis, *Ann. N.Y. Acad. Sci.*, 768, 91, 1995.

136. Telang, N. et al., Parallel enhancement of ras protooncogene expression and of estradiol-16 alpha-hydroxylation in human mammary terminal duct-lobular units (TDLU) by a carcinogen, *Breast Cancer Res. Treat.*, 12, 138, 1988.

137. Telang, N. et al., Induction by estrogen metabolite 16 alpha-hydroxyestrone of genotoxic damage and aberrant proliferation in mouse mammary epithelial cells, *J. Natl. Cancer Inst.*, 84, 634, 1992.

138. Aronson, K. et al., Breast adipose tissue concentration of polychlorinated biphenyls and other organochlorines and breast cancer risk, *Cancer, Epidemiol. Biomarkers Prevent.*, 9, 55, 2000.

139. Guttes, S. et al., Chlororganic pesticides and polychlorinated biphenyls in breast tissue of women with benign and malignant breast disease, *Arch. Environ. Contam. Toxicol.*, 35, 140, 1998.

140. Romieu, I. et al., Breast cancer, lactation history, and serum organochlorines, *Am. J. Epidemiol.*, 152, 363, 2000.

141. Lopez-Carrillo, L. et al., The adipose tissue to serum dichlorodiphenyldichloroethane (DDE) ratio: Some methodological considerations, *Environ. Res. A*, 81, 142, 1999.

142. Pauwels, A. et al., Comparison of persistent organic pollutant residues in serum and adipose tissue in a female population in Belgium, 1996–1998, *Arch. Environ. Contam. Toxicol.*, 39, 265, 2000.

143. Archibeque-Engle, S. et al., Comparison of organochlorine pesticide and polychlorinated biphenyl residues in human breast adipose tissue and serum, *J. Toxicol. Environ. Health*, 52, 285, 1997.

144. Needham, L. et al., Adipose tissue/serum partitioning of chlorinated hydrocarbon pesticides in humans, *Chemosphere*, 20, 975, 1990.

145. Dewailly, E. et al., Concentration of organochlorines in human brain, liver, and adipose tissue autopsy samples from Greenland, *Environ. Health Perspect.*, 107, 823, 1999.

146. Mussalo-Rauhamaa, H. et al., Occurrence of beta-hexachlorocyclohexane in breast cancer patients, *Cancer*, 66, 2124, 1990.

147. Wolff, M. et al., Blood levels of organochlorine residues and risk of breast cancer, *J. Natl. Cancer Inst.*, 85, 648, 1993.

148. Jensen, A., Levels and trends of environmental chemicals in human milk, in *Chemical Contaminants in Human Milk*, Jensen, A. and Slorach, S., Eds., CRC Press, Boca Raton, FL, 1991, p. 45.

149. Romieu, I. et al., Breast cancer and lactation history in Mexican women, *Am. J. Epidemiol.*, 143, 543, 1996.

150. Newcomb, P. et al., Cancer of the breast in relation to lactation history, *N. Engl. J. Med.*, 330, 81, 1994.

151. Ing, R., Petrakis, N., and Ho, J., Unilateral breast-feeding and breast cancer, *Lancet*, 2, 124, 1977.

152. El-Bayoumy, K. Environmental carcinogens that may be involved in human breast cancer etiology, *Chem. Res. Toxicol.*, 5, 585, 1992.

153. Seidman, L., Moore, C., and Gould, M. 32P-postlabeling analysis of DNA adducts in human and rat mammary epithelial cells, *Carcinogenesis*, 9, 1071, 1988.

154. Li, D., et al. DNA adducts in normal tissue adjacent to breast cancer: A review. *Cancer Detect. Prev.*, 23, 454, 1999.

155. Manz, A. et al., Cancer mortality among workers in a chemical plant contaminated with dioxin, *Lancet*, 338, 959, 1991.

156. Kuratsune, M. et al., A cohort study on mortality of "Yusho" patients: A preliminary report, *Int. Symp. Princess Takamatsu Cancer Res. Fund*, 18, 61, 1987.

157. Dean, A. et al., Adjusting morbidity ratios in two communities using risk factor prevalence in cases, *Am. J. Epidemiol.*, 127, 654, 1988.

158. Hall, N. and Rosenman, K., Cancer by industry: Analysis of a population-based cancer registry with an emphasis on blue collar workers, *Am. J. Ind. Med.*, 19, 145, 1991.

159. Griffith, J. et al., Cancer mortality in US counties with hazardous waste sites and groundwater pollution, *Arch. Environ. Health*, 44, 69, 1989.

160. Westin, J. and Richter, E., The Israeli breast cancer anomaly, *Ann. N.Y. Acad. Sci.*, 609, 269, 1990.

161. Davis, D. and Bradlow, H., Can environmental estrogens cause breast cancer?, *Sci. Am.*, 273, 166, 1995.

162. Smith, A., Chlorinated hydrocarbon insecticides, in *Handbook of Pesticide Toxicology*, Hayes, W. and Laws, Eds., Academic Press, San Diego, CA, 1991.

163. Mattison, D., Wohlieb, J., and To, T., Pesticide concentrations in Arkansas breast milk, *J. Ark. Med. Soc.*, 88, 553, 1992.

164. Wolff, M. et al., Proposed PCB congener groupings for epidemiological studies, *Environ. Health Perspect.*, 103, 141, 1995.

165. Kuntzman, R. et al., Similarities between oxidative drug-metabolizing enzymes and steroid hydroxylases in liver microsomes, *J. Pharmacol. Exp. Ther.*, 146, 280, 1964.

166. Anzenbacher, P. and Anzenbacherova, E., Cytochromes P450 and metabolism of xenobiotics, *Cell Mol. Life Sci.*, 58, 737, 2001.

167. Michnovicz, J. et al., Increased 2-hydroxylation of estradiol as a possible mechanism for the anti-estrogenic effect of cigarette smoking, *N. Engl. J. Med.*, 315, 1305, 1986.

168. Spink, D. et al., 2,3,7,8-Tetrachlorodibenzo-p-dioxin causes an extensive alteration of 17beta-estradiol metabolism in MCF-7 breast tumor cells, *Proc. Natl. Acad. Sci. USA*, 87, 6917, 1990.

169. Krishnan, V. et al., Molecular mechanism of inhibition of estrogen-induced cathepsin D gene expression by 2,3,7,8-tetrachlorodibenzo-p-dioxin (TCDD) in MCF-7 cells, *Mol. Cell Biol.*, 15, 6710, 1995.

170. Bradlow, H. et al., Effects of dietary indole-3-carbinol on estradiol metabolism and spontaneous mammary tumors in mice, *Carcinogenesis*, 12, 1571, 1991.

171. Kociba, R. et al., Results of a two-year chronic toxicity and oncogenecity study of 2,3,7,8-tetrachlorodibenzo-p-dioxin in rats, *Toxicol. Appl. Pharmacol.*, 46, 279, 1978.

172. Bradlow, H. et al., Effects of pesticides on the ratio of 16alpha/2-hydroxyestrone: A biological marker of breast cancer risk, *Environ. Health Perspect.*, 103, 147, 1995.

173. Li, D. et al., Polymorphisms in P450 CYP1B1 affect the conversion of estradiol to the potentially carcinogenic metabolite 4-hydroxyestradiol, *Pharmacogenetics*, 10, 343, 2000.

174. Watanabe, J. et al., Association of CYP1B1 genetic polymorphism with incidence to breast and lung cancer, *Pharmacogenetics*, 10, 25, 2000.

175. Zheng, W. et al., Genetic polymorphism of cytochrome P450-1B1 and risk of breast cancer, *Cancer Epidemiol. Biomarkers Prevent.*, 9, 147, 2000.

176. Miller, W. and Mullen, P., Factors influencing aromatase activity in the breast, *J. Steroid Biochem. Mol. Biol.*, 44, 597, 1993.

177. Baxter, S. et al., Polymorphic variation in CYP19 and the risk of breast cancer, *Carcinogenesis*, 22, 347, 2001.

178. Kristensen, V. et al., Genetic variants of CYP19 (aromatase) and breast cancer risk, *Oncogene*, 19, 1329, 2000.

179. Healey, C. et al., Polymorphisms in the human aromatase cytochrome P450 gene (CYP19) and breast cancer risk, *Carcinogenesis*, 21, 189, 2000.

180. Ishibe, N. et al., Cigarette smoking, cytochrome P450 1A1 polymorphisms, and breast cancer risk in the Nurses' Health Study, *Cancer Res.*, 58, 667, 1998.

181. Ambrosone, C. et al., Cytochrome P4501A1 and glutathione S-transferase (M1) genetic polymorphisms and postmenopausal breast cancer risk, *Cancer Res.*, 55, 3483, 1995.

182. Ambrosone, C. et al., Cigarette smoking, *N*-acetyltransferase 2 genetic polymorphism and breast cancer risk, *JAMA*, 276, 1494, 1996.

183. Lavigne, J. et al., An association between the allele coding for a low activity variant of catechol-*O*-methyltransferase and the risk for breast cancer, *Cancer Res.*, 57, 5493, 1997.

184. Helzlsouer, K. et al., Association between glutathione S transferase M1, P1, and T1 genetic polymorphisms and development of breast cancer, *J. Natl. Cancer Inst.*, 90, 512, 1998.

185. Zhong, S. et al., Relationship between the GSTM1 genetic polymorphism and susceptibility to bladder, breast and colon cancer, *Carcinogenesis*, 14, 1821, 1993.

186. Trosko, J., Hierarchical and cybernetic nature of biologic systems and their relevance to homeostatic adaptation to low-level exposures to oxidative stress-inducing agents, *Environ. Health. Perspect.*, 106, 331, 1998.

187. Hofmanova, J., Machala, M., and Kozubik, A., Epigenetic mechanisms of the carcinogenic effects of xenobiotics and in vitro methods of their detection, *Folia Biologica (Praha)*, 46, 165, 2000.

188. Parodi, S. and Brambilla, G., Relationship between mutation and transformation frequencies in mammalian cells treated "in vitro" with chemical carcinogens, *Mutat. Res.*, 47, 53, 1977.

189. Loeb, L., Mutator phenotype may be required for multistage carcinogenesis, *Cancer Res.*, 51, 3075, 1991.

190. Landolph, J. and Heidelberger, C. Chemical carcinogens produce mutations to ouabain resistance in transformable C3H/10T1/2 Cl8 mouse fibroblasts, *Proc. Natl. Acad. Sci. USA*, 76, 900, 1979.

191. Stein, W., Analysis of cancer incidence data on the basis of multistage and clonal growth models, *Adv. Cancer Res.*, 56, 161, 1991.

192. Holliday, R., Mutations and epimutations in mammalian cells, *Mutat. Res.*, 250, 351, 1991.

193. Strauss, B., The origin of point mutations in human tumor cells, *Cancer Res.*, 52, 249, 1992.

194. Rubin, H., Cellular epigenetics: Control of the size, shape, and spatial distribution of transformed foci by interactions between the transformed and nontransformed cells, *Proc. Natl. Acad. Sci. USA*, 91, 1039, 1994.

195. Kennedy, A., *Environ. Health Perspect.*, 93, 199, 1991.

196. Eldridge, S. and Gould, M., Comparison of spontaneous mutagenesis in early-passage human mammary cells from normal and malignant tissues, *Int. J. Cancer*, 50, 321, 1992.

197. Briand, P. and Lykkesfeldt, A., An in vitro model of human breast carcinogenesis: Epigenetic aspects, *Breast Cancer Res. Treat.*, 65, 179, 2001.

198. Wang, F. et al., Reciprocal interactions between beta1-integrin and epidermal growth factor receptor in three dimensional basement membrane breast cultures: A different perspective in epithelial biology, *Proc. Natl. Acad. Sci. USA*, 95, 14821, 1998.

199. Kamiya, K. et al., Evidence that carcinogenesis involves an imbalance between epigenetic high-frequency initiation and suppression of promotion, *Proc. Natl. Acad. Sci. USA*, 92, 1332, 1995.

200. Kennedy, A. et al., Relationship between x-ray exposure and malignant transformation in C3H10T1/2 cells, *Proc. Natl. Acad. Sci. USA*, 77, 7262, 1980.

201. Kennedy, A. and Little, J., Evidence that a second event in x-ray-induced oncogenic transformation in vitro occurs during cellular proliferation, *Radiat. Res.*, 99, 228, 1984.

202. Sanderson, M. et al., Perinatal factors and risk of breast cancer, *Epidemiology*, 7, 34, 1996.

203. Michels, K. et al., Birthweight as a risk factor for breast cancer, *Lancet*, 348, 1542, 1996.

204. Petridou, E. et al., Tobacco smoking, pregnancy estrogens and birth weight, *Epidemiology*, 1, 247, 1990.

205. Ekbom, A. et al., Intrauterine environment and breast cancer risk in women: A population-based study, *J. Natl. Cancer Inst.*, 88, 71, 1997.

206. Sedin, G., Bergquist, C., and Lindgren, P., Ovarian hyperstimulation in preterm infants, *Pediatr. Res.*, 19, 548, 1985.

207. Ekbom, A. et al., Evidence of prenatal influences on breast cancer risk, *Lancet*, 340, 1015, 1992.

208. Geschwind, N. and Galaburda, A., Cerebral lateralization, biological mechanisms, associations, and pathology, I: A hypothesis and a program for research, *Arch. Neurol.*, 42, 428, 1985.

209. Sandson, T., Wen, P., and Lemay, M., Reversed cerebral asymmetry in women with breast cancer, *Lancet*, 339, 523, 1992.

210. Anbazhagan, R., Nathan, B., and Gusterson, B., Prenatal influences and breast cancer, *Lancet*, 340, 1477, 1992.

211. McEven, B., Interactions between hormones and nerve tissue, *Sci. Am.*, 235, 48, 1976.

212. Migliaccio, S. et al., Alterations of maternal estrogen levels during gestation affect the skeleton of female offspring, *Endocrinology*, 137, 2118, 1996.

213. Pap, E. and Csaba, G., Effect of prenatal allylestrenol treatment (hormonal imprinting) on the serum testosterone and progesterone level in adult rats, *Gen. Pharmacol.*, 26, 365, 1995.

214. Hilakivi-Clarke, L., Clarke, R., and Lippman, M., Perinatal factors increase breast cancer risk, *Breast Cancer Res. Treat.*, 31, 273, 1994.

215. Bjerke, D. et al., Effects of in utero and lactational 2,3,7,8-tetrachlorodibenzo-p-dioxin exposure on responsiveness of the male rat reproductive system to testosterone stimulation in adulthood, *Toxicol. Appl. Pharmacol.*, 127, 250, 1994.

216. Haake-McMillan, J. and Safe, S. Neonatal exposure to Aroclor 1254: Effects on adult hepatic testosterone hydroxylase activities, *Xenobiotica*, 21, 481, 1991.

17 Cancer of the Prostate — Mechanisms of Molecular Carcinogenesis

Daniel Djakiew

CONTENTS

17.1 STRUCTURE OF THE PROSTATE

17.1.1 ANATOMY

The human prostate gland has been described to consist of three anatomically distinct zones [1,2], the transitional zone, the central zone, and the peripheral zone (Figure 17.1). From the bladder, the prostatic urethra courses through the transitional zone of the prostate forming the verumontanum (seminal colliculus) and receives the paired ejaculatory ducts. Adjacent to the point of entry of the ejaculatory ducts into the verumontanum is present the prostatic utricle, which represents an embryological remnant of Mullerian duct. The urethra then exits the prostate coursing through the penis. The transitional zone of the prostate comprises as little as 5% of the total glandular tissue, although it is the location for the development of virtually all benign prostatic hyperplasia (BPH) pathologies and up to 20% of prostate cancers [1,2]. The somewhat larger central zone of the prostate, consisting of approximately 20% of total glandular tissue, lies adjacent to the urethra and transitional zone, and contains the paired ejaculatory ducts as they course from the ductus (vas) deferens. Less than 10% of adenocarcinomas may arise within the central zone of the prostate. The peripheral zone accounts for approximately 75% of the total glandular mass of the prostate. It surrounds both the transitional and central zones, and is bounded on one surface by the anterior fibromuscular stroma. The peripheral zone of the prostate is the location for the development of the majority

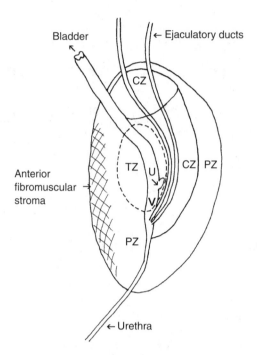

FIGURE 17.1   Gross anatomy of the human prostate. The prostate organ consists of three anatomical zones: the peripheral zone (PZ), the central zone (CZ), and the transitional zone (TZ). The ejaculatory ducts merge with the urethra in the region of the verumontanum (V), which lies adjacent to the prostatic utricle (U).

of adenocarcinomas. The posterior peripheral zone is accessible by digital rectal examination for palpation of nodular masses that may arise in the prostate.

A rich nerve supply for the prostate arises from both sympathetic and parasympathetic divisions of the autonomic nervous system [3] at the thoracic (T), lumbar (L), and sacral (S) levels of the vertebrae. Autonomic innervation to the human prostate is supplied by the pelvic plexus [4,5]. Parasympathetic visceral efferent preganglionic nerve fibers arise from S2 to S4 and enter the plexus via the pelvic splanchnic nerve. The sympathetic component emanates from T11 to L2 entering the plexus via the hypogastric nerve, with branches from the sacral sympathetic chain (S4 to S5) and branches that originate from the inferior mesenteric plexus that accompany the superior hemorrhoidal artery. Innervation from the neurovascular bundle coursing adjacent to the prostate form a superior pedicle and an inferior pedicle of nerves attached to the capsule of the prostate [6]. Nerves within the pedicles are surrounded with fibrofatty tissue forming localized thickenings that penetrate the capsule where they branch into within the parenchyma of the prostate [6]. Both cholinergic parasympathetic and noradrenergic nerves innervate smooth muscle around the ducts and acini of the prostate but do not appear to directly innervate the acinar epithelium [7]. Cholinergic parasympathetic stimulation increases secretory activity of the prostate, whereas noradrenergic sympathetic stimulation increases smooth muscle tone resulting in the expulsion of prostate secretions into the urethra [8]. The nerves that branch throughout the prostate form a route for perineural invasion by tumor cells, with subsequent capsular penetration along the nerve pedicles considered the dominant mechanism for metastasis of prostate tumor cells [6].

17.1.2   HISTOLOGY

The human prostate is a branched tubuloalveolar gland. Between 30 and 50 distinct glandular structures empty into the prostatic urethra that in turn is lined by a transitional epithelium. The human

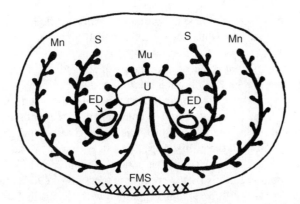

FIGURE 17.2   Glandular organization of the human prostate. The main (Mn), submucosal (S) and mucosal (Mu) glands, along with the ejaculatory ducts (ED) all empty into the urethra (U). A fibromuscular stroma (FMS) is prominent on one surface of the prostate gland.

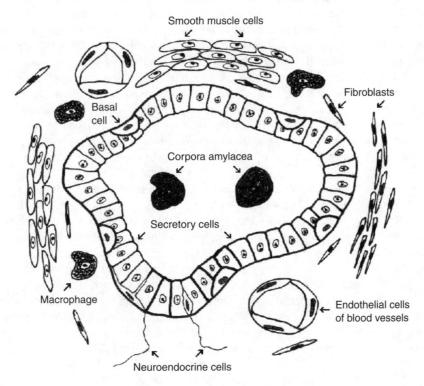

FIGURE 17.3   Histology of the normal human prostate. A cross section through a glandular acinus contains columnar secretory cells interspersed with basal cells and the occasional neuroendocrine cell that may extend into the surrounding stroma. Corpora amylacea may be observed within the lumen of an acinus. The stromal compartment contains an abundance of smooth muscle cells and fibroblasts, with occasional macrophages and endothelial cells lining the blood vessels.

prostate gland has been described as consisting of three concentric tubuloalveolar glands [9], the mucosal glands, the submucosal glands, and the main glands (Figure 17.2). Within the lumen of the glands can be occasionally observed spherical prostatic concretions (corpora amylacea). Corpora amylacea (Figure 17.3) form as concretions of glycoprotein secretions within the glandular acini. These corpora amylacea appear to be unique to the human prostate and accumulate with age. Both the normal and diseased human prostate have in common a major epithelial cell component, a major stromal cell component, and less abundant specialized cells interspersed amongst the epithelia and stroma.

The histology of the normal human prostate consists of an epithelial component, predominantly glandular, surrounded by an abundant fibromuscular component (Figure 17.3). The epithelial cell components of the human prostate are predominantly composed of the columnar secretory cells, either simple or psuedostratified, and the basal cells. The columnar secretory cells are highly differentiated and function in the synthesis and secretion of prostatic fluids. These columnar secretory cells have been distinguished from other cell types by the specific expression of cytokeratins 8 and 18 [10,11]. In addition, basal cells occur at the base of the acinar epithelium. These basal cells have been distinguished from the columnar cells by the specific expression of cytokeratins 5 and 14 [10,11], as well as several others. Basal cells may be intermingled with stem cells that co-localize to a stem cell compartment [12]. These stem cells are hypothesized to be capable of self-renewal and to give rise to the basal cells [13]. However, the presence of stem cells distinct from basal cells has not been universally accepted. In any event, the basal cells appear less differentiated, have a large cell body, and exhibit processes that intercalate between surrounding columnar cells [14]. An intermediate cell type (amplifying cell) that arises from the basal cells and differentiates into columnar secretory cells has also been postulated [12]. These intermediate (amplifying) cells can be distinguished by an absence of cytokeratin 14 (present in basal/stem cells) and immunoreactivity with a basal cell specific (K_{basal}) cytokeratin antibody (absent in luminal cells) [15]. In BPH the epithelial cell compartment contains both basal cells and columnar secretory cells, as determined by cytokeratin expression patterns [16]. In contrast, the malignant epithelial cells of both primary hormone responsive tumors and recurrent hormone refractory tumors across all pathologic phenotypes of glandular dedifferentiation, as defined by Gleason stages, express cytokeratin 18 in common with normal columnar secretory cells [15]. Hence, based upon singular or combinations of cytokeratin expression patterns, it is possible to distinguish between the various types of epithelial cells in the normal, hyperplastic, and cancerous human prostate.

Neuroendocrine cells are a less abundant specialized cell type interspersed amongst the acinar epithelia and extending into the stroma of the human prostate. Neuroendocrine cells are part of the widely dispersed diffuse neuroendocrine regulatory system, also known as endocrine–paracrine cells. In the human prostate, subpopulations of neuroendocrine cells have been identified based upon morphology and the secretory products (reviewed in Reference 17). Most neuroendocrine cells of the prostate contain the proteins serotonin, synaptophysin [18], and chromogranin A [19]. The absence of chromogranin A immunoreactive neuroendocrine cells in the rat, guinea pig, cat, and dog prostates challenge the validity of these animal models for physiological studies of neuroendocrine cells in the prostate gland [20]. The significance of neuroendocrine differentiation in human prostate cancer is uncertain. Proponents for a role of neuroendocrine differentiation in prostate cancer have shown that the neuroendocrine cell population enlarges with cancer progression [21], appears to correlate with Gleason sum and cancer stage [22], may be a prognostic factor for progression [22–24], and elevated serum levels of chromogranin A may detect prostate cancer in patients whose prostate specific antigen (PSA) is not elevated [19,25]. Conversely, the detractors for a role of neuroendocrine differentiation in prostate cancer have shown that the neuroendocrine cell population does not exhibit a relationship with co-localized MIB-1 estimates of cell proliferation [26], is not correlated with pathological stage of Gleason sum [26,27] or other measures of progression [27] such as increasing serum PSA [28], and is not correlated with cancer specific survival [29] and therefore has no independent prognostic significance for prostate cancer [27,29,30]. Irrespective of their prognostic and diagnostic significance, neuroendocrine cells are expressed in the normal and diseased human prostate.

The stromal cell components of the prostate include smooth muscle cells, fibroblasts, tissue macrophages [31], and endothelial cells (Figure 17.3). All of these stromal components of connective tissue differentiate from embryonic mesoderm. The majority of the stromal cells (fibroblasts and smooth muscle cells) are distinguished from the epithelia by their filiform morphological appearance typical of mesenchymal cells. In addition, these stromal cells typically exhibit an intense immuno-reactivity for vimentin intermediate filaments. The smooth muscle cell component of the stroma is further distinguished by an intense immunoreactivity for α-actin. Within the stromal tissue, macro-phages represent a transient cell population that occur as single cells dispersed throughout the stroma and occasionally migrate into the epithelial cell compartment. The dispersed macrophages are iden-tified immunohistochemically by the expression of CD11c. Endothelial cells derived from blood vessels within the stroma are typically identified by immunoreactivity for von Willebrand's factor (factor VIII) and pecam/CD31. The stromal cell components of the human prostate appear to be particularly relevant to the etiology of BPH where the overgrowth of stroma appears to be a major contributing factor for the formation of BPH nodules.

17.1.3 PATHOLOGY

The human prostate is relatively unique among mammals in that it is the site of origin of both cancer and BPH. Virtually no other mammals examined to date, with the qualified exception of dogs [32,33] normally develop prostate cancer and BPH with age [34]. Although rodents represent the animal models most widely used to study prostate pathologies *in vivo*, they normally do not develop cancer and BPH with age to any degree approaching that observed in man, if at all. Transgenically modified strains of mice [35], and inbred strains of the Brown Norway rat used as an aging model [36] develop epithelial prostate pathologies. However, the mechanisms activated to develop these pathologies in rodents may be unrelated to that documented to occur in the human prostate, so that the results derived from rodent models should be considered within the context of the caveats specific to species and method of transgenic induction.

Cancer of the prostate is thought to originate from a precancerous pathology of epithelial cells within the ducts and acini of the glands (Figure 17.4). Two major forms of precancerous lesion have been described. In proliferative inflammatory atrophy (PIA) the glandular acini contain two layers of attenuated epithelial cells (Figure 17.4(b)) composed of basal and secretory cells [37] that have a very basophilic appearance (hyperchromatic) due to the scant cytoplasm [38]. Many of the PIA epithelial cells exhibit an enhanced proliferative capacity [37] and are typically associated with inflammatory cells such as macrophages and lymphocytes [37]. The PIA epithelial cells may be exposed to inflammatory oxidants that may damage the genome [39] and promote transformation to prostatic intraepithelial neoplasia (PIN) or directly to cancer [37]. PIN lesions (dysplasia) exhibits cytologic features of nuclear and nucleolar enlargement (Figure 17.4(c)), aneuploidy, and increased proliferation index similar to cancer [40], but differs from cancer in that PIN retains an intact or fragmented basal cell layer [41]. PIN lesions remain within the acini where the dysplastic epithelial cells proliferate to form tufts and mounds of cells projecting into the lumen (Figure 17.4(c)). The incidence and extent of PIN increase with age predating the onset of cancer by more than five years [42]. Progression of PIN to cancer is characterized by progressive loss of basal cells, increased genetic instability, and increased rate of cell proliferation [43].

Transformation from PIN to a cancerous phenotype is strongly correlated with extensive and diverse morphologic changes in appearance of the glandular acini and epithelial cells. Within the organ confined prostate, progressive loss of normal glandular morphology accompanied by growth of irregular shaped masses of tumor cells has been described in the context of the Gleason grading system [44]. The classification system describes the sequential change in the pattern and architecture of the glandular acini that accompanies the loss of differentiated phenotype. Five Gleason grades with subtypes are used to describe the histologic patterns of the gland (Figure 17.5). Gleason grade 1 contains a homogeneous array of single, round, or oval glands that are separate and closely packed together. Gleason grade 2 is characterized by some loss of uniformity in glandular shape with some transformation from round to oval shape and an increase in the distance between the acini. Gleason

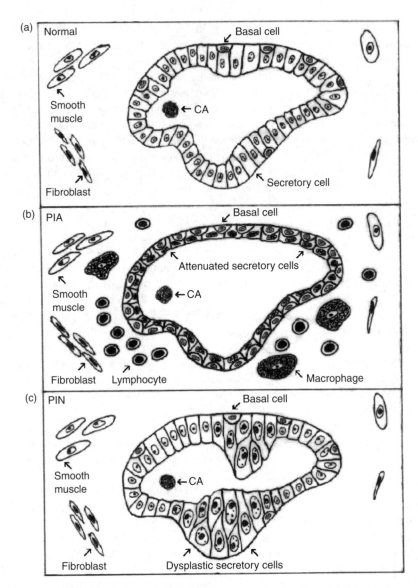

FIGURE 17.4   Premalignant pathologies of the human prostate. (a) The glandular acini of the normal prostate may contain corpora amylacea (CA) and is lined by basal cells and secretory columnar cells. The stroma surrounding the acini normally contains both smooth muscle and fibroblasts. (b) Glandular acini exhibiting proliferative inflammatory atrophy (PIA) may contain CA and are lined by basal cells and attenuated secretory epithelial cells that appear hyperchromatic. Interspersed between the smooth muscle cells and fibroblasts of the stroma occur increased numbers of inflammatory cells such as lymphocytes and macrophages. (c) Glandular acini exhibiting prostatic intraepithelial neoplasia (PIN) may contain CA and are lined by basal cells and dysplastic secretory cells that form tufts projecting into the lumen. The dysplastic secretory cells exhibit enlarged nuclei and nucleoli. The stroma surrounding PIN acini normally contains both smooth muscle and fibroblasts.

grade 3 contains three subpatterns with some degree of stromal invasion. The first subpattern (A) con-sists of glandular masses that form small to very small groups or chords. The second subpattern (B) is characterized by large, single but irregular shaped acini, where the masses are separated from one another by at least one diameter. The third subpattern (C) is characterized by rounded masses of papillary and cribiform epithelium within acini of irregular shape. Gleason grade 4 contains two

FIGURE 17.5 The Gleason grading system. Grade 1 contains homogeneous, single, rounded glands. Grade 2 is characterized by some loss of uniformity in glandular shape and an increase in distance between the acini. Grade 3 contains three subpatterns with some degree of stromal invasion. Grade 3A consists of glandular masses that form small to very small groups or cords. Grade 3B contains large, single but irregular shaped acini well separated apart. Grade 3C is characterized by masses of papillary and cribiform epithelium within acini of irregular shape. Gleason grade 4 contains two subpatterns with prominent stromal invasion. Grade 4A contains ragged, fused, or branched glandular masses. Grade 4B is similar to 4A with the addition of many large clear cells intermingled within the tumor masses. Gleason grade 5 contains two subpatterns. Grade 5A contains very large rounded groups of papillary and cribiform epithelium with some central necrosis. Grade 5B contains irregular masses of highly dedifferentiated tumor cells largely devoid of glandular architecture containing infiltrating masses of epithelium that has invaded and co-mingled with the surrounding stroma.

subpatterns. The first subpattern (A) is characterized by ragged glandular masses that appear to have either fused together or branched away from the mass. The second subpattern (B) is similar to the first with the addition of many large clear cells intermingled within the tumor masses. Stromal invasion is prominent contributing to loss of acinar integrity. Gleason grade 5 contains two subpatterns. The first subpattern (A) is characterized by very large rounded groups of papillary and cribiform epithelium that form mostly solid masses with some central necrosis. The second subpattern (B) contains irregular masses of highly dedifferentiated tumor cells largely devoid of glandular architecture containing infiltrating masses of epithelium that has invaded and comingled with the surrounding stroma. The typical grading strategy takes two Gleason grades from different areas of a tumor, and the sum of the two grades provides an estimate of the degree of malignancy out of a possible Gleason sum of 10. A concurrent pathologic staging system can also be used that takes account of the degree to which a tumor mass is palpable and whether it has disseminated outside of the prostate organ [45]. In this system, stage A represents a nonpalpable tumor detected histologically from biopsy, or following measurement of elevated PSA. Stage B represents a palpable, organ confined tumor mass. Stage C involves extracapsular or seminal vesicle invasion that is locally confined. Stage D represents a metastatic tumor, stage D1 with metastases to the lymph nodes, stage D2 with bony metastases to the either of the pelvis, ribs, spine, or metastases to the liver and lungs, and stage D3 representing hormone refractory metastases. Metastatic prostate cancers usually contain multiple histologic patterns consisting of cribiform glands, undifferentiated histologic patterns, and masses of cells with no glandular organization [45].

17.2   BENIGN PROSTATIC HYPERPLASIA

The term benign prostatic hyperplasia (BPH) has often been used synonymously with the term benign prostatic hypertrophy to describe the same clinical condition. Since BPH involves an increase in cell numbers (hyperplasia) and not an increase in cell size (hypertrophy) the accurate terminology for this disease is benign prostatic hyperplasia. BPH is a disease of the adult prostate in which considerable morbidity may result from obstruction of the urinary passage. However, it is distinct from prostate cancer in that BPH does not result from carcinogenesis of the prostate. The pathology of BPH is relatively complex with several histological forms of the disease being described under the aegis of BPH. In general, BPH has been described as a nodular growth containing a nonhomogeneous mixture of glandular and stromal tissue. The five types of histological nodules that have been described are (1) stromal, (2) fibromuscular, (3) muscular, (4) fibroadenoma, and (5) fibromyoadenoma [46]. In all of these pathological subtypes of BPH, the stroma appears to contribute a major component of the tissue. Indeed, several studies have shown that BPH tissues contain a significantly greater percentage of stroma and a significantly lower percentage of epithelium compared to normal tissue [47,48]. It has been suggested that BPH initially arises as a fibrostromal nodule that can have an inductive effect on the surrounding epithelial cells [49]. Hence, depending on the extent of induction, all of the five subtypes of BPH could potentially arise from a progenitor fibrostromal nodule. This further supports the concept that varying combinations of fibroblasts and smooth muscle cells constitute the major components of the stromal compartment in the human prostate.

17.3   RISK FACTORS

17.3.1   GENDER, THE PRESENCE OF GONADS AND STEROIDS

Since the prostate gland is a male reproductive organ, it is self-evident that diseases of the prostate are restricted to the male gender. *In utero*, embryological development of the indifferent gonads into either testis or ovary is determined by the presence (male) or absence (female) of a testis-determining factor (TDF) expressed exclusively on the Y chromosome. When the TDF is expressed, the primary sex cords condense to form seminiferous cords that subsequently develop into seminiferous tubules surrounded by Leydig cells within the developing testis. In the absence of TDF, sex cords initially develop, but then break up into clusters of primordial follicles within the developing ovary. The production of testosterone by Leydig cells drives the development of the prostate gland as well as other male reproductive glands. Removal of the testes protects against the development of prostate diseases, with the earlier age of testes removal providing the greater protective effect. This information is known from individuals who have been gonadectomized as a result of traumatic accidents and eunuchs who were intentionally castrated for cultural reasons. The Skoptzys, a Russian sect in which the males were ritually castrated at the age of 35 years, did not suffer from prostate diseases later in life [50].

The role of testosterone produced predominantly by the Leydig cells of the testis (95%), and to a much lesser extent by the adrenal gland (5%), appears fundamentally linked to the growth and development of both cancer and BPH of the prostate [51]. Within the prostate, testosterone is converted by 5α-reductase to 5α-dihydrotestosterone, the latter of which is in an order of magnitude more potent for stimulating prostate growth [51]. Removal of testosterone by gonadectomy or chemical castration using luteinizing hormone releasing hormone agonists causes involution of the prostate [52]. Since gonadectomy in an earlier age can prevent the development of prostate cancer and BPH, it has been widely suggested that testosterone is involved in the eventual development of these diseases, although the exact mechanism remains elusive. Interestingly, as men age coinciding with the increased occurrence of cancer and BPH, the levels of testosterone decline and the levels of estrogen rise [53]. A large portion (>75%) of estrogens in the aging male is derived by peripheral conversion from testosterone [51]. This age-associated decrease in the ratio of testosterone to estrogen has also been linked to early neoplastic transformation or initiation of aberrant prostate growth [54]. Hence, it can be concluded that the male gender with the associated presence of functional testes producing testosterone, are, at a minimum risk for the development of prostate cancer.

17.3.2 Age

The prevalence of clinically diagnosed prostate cancer and BPH increase significantly with age, and age is considered one of the most important risk factors for these diseases [55]. A major distinction can be made between microscopic (latent) disease that is diagnosed by viewing samples of pathological tissue obtained at autopsy and macroscopic (clinical) disease that is diagnosed upon presentation of clinical symptoms. Microscopic disease tends to predate macroscopic disease by a decade or more. However, the presence of microscopic disease has a poor correlation with age-related progression to macroscopic disease. Conversely, the incidence of clinically diagnosed disease increases significantly after the age of approximately 50 years and continues to increase with each succeeding decade of life. Suggestions for the mechanistic relationship between age and the etiology of prostate cancer have included changes in the steroidal milieu, the lifetime dependent accumulation of genotoxic insults to the prostate resulting from ingestion of potential chemical carcinogens, and oxidative stress resulting in free radical based DNA damage and adduct formation [54]. Hence, with improvements in a healthy lifestyle, disease prevention, and medical therapies resulting in extended life spans, a much larger population of aging men are projected to be at risk for the development of clinical prostate cancer and BPH.

17.3.3 Race

Differences in the incidence of prostate cancer are well-documented to occur between certain groups of men with Asian, Caucasian, and African-American background. Specifically, African-American men exhibit one of the highest occurrences of prostate cancer compared to nearly all other groupings. They tend to develop prostate cancer at an earlier age and have more aggressive tumors with associated higher rates of metastases and reduced survival rate [56]. This high incidence of prostate cancer in African-American men does not appear to be related to genetic factors since the incidence of prostate cancer in African men is quite low [57]. Hence, factors such as diet, environment, lifestyle, access and utilization of medical services have been suggested to compound the high incidence of prostate cancer in African-American men.

Japanese men have been reported to have a much lower incidence of prostate cancer compared to most other groupings [58]. American-Indians and Eskimos, both of whom are thought to have links to an Asian origin, also exhibit significantly lower incidence rates of prostate cancer [59]. Interestingly, men of Japanese origin who are first generation within the United States exhibited a higher rate of death to prostate cancer than Japanese men in Japan, but a lower death rate than Caucasian men within the United States [60]. Similar observations have been used to suggest that race may be intertwined with other factors such as diet, environment, lifestyle, access, and utilization of medical services to determine the overall incidence of prostate cancer. Another confounding variable to the determination of risk factors for prostate cancer is the possibility that some degree of under-reporting of prostate cancer in Asian societies may have occurred as a result of cultural impediments to the acknowledgement and acceptance of prostate cancer. Hence, the risk factors that contribute to racial differences between the incidence of prostate cancer may be cumulative and complex.

17.3.4 FAMILIAL GERM LINE MUTATION HYPOTHESIS

A family history of developing prostate cancer [59], defined as men whose father or brother developed prostate cancer, has formed the basis of numerous studies to identify mutated genes that are transmitted through heredity in the germ line. Linkage analysis of families with a high incidence of prostate cancer has identified several chromosomal loci thought to harbor prostate cancer susceptibility genes [61]. However, the relatively modest associations used for the statistical identification of potential hereditary genes have been suggestive rather than confirmatory of a familial germ-line mutation that is causal of prostate cancer. It cannot be excluded that given a large enough population of men, several families may develop a high incidence of sporadic prostate cancer, either by bad luck or common environmental exposure to carcinogens, that subsequently may have been interpreted as a family history of prostate cancer. The functions of several prostate cancer susceptibility genes that have been cloned suggest an indirect role in the development of prostate cancer. HPC1 identified as RNASE L [62] is a component of the interferon-inducible viral response program that degrades viral and cellular RNA. Another cloned prostate cancer susceptibility gene, the macrophage scavenger receptor 1 [63] is expressed in macrophages and may affect inflammation at sites of infection. Both of these prostate cancer susceptibility genes exhibit subtle phenotypes that require further clarification as to their role in the development of prostate cancer. Whether these rare, low penetrance genes represent localized clustering of sporadic prostate cancer genes or familial linkage of germ-line mutations may require more powerful statistical analysis of much larger populations of men who exhibit a family history of developing prostate cancer.

17.3.5 EPIGENETICS (DNA METHYLATION)

In prostate cancer cells, increased DNA methylation can occur in cytosine guanine dinucleotide (CpG) islands near the transcriptional regulatory regions of many genes [64]. The transcriptional silencing of tumor suppressor gene expression associated with such CpG island DNA hypermethylation represents an epigenetic lesion, rather than a mutational genetic alteration in the genome [64]. Tumor suppressor genes, when expressed in normal prostate cells, tend to retard growth by regulating cell-cycle effectors and regulating the level of apoptosis. Hence, the epigenetic silencing of tumor suppressor genes could contribute to the malignant transformation of prostate cells by promoting their proliferation and survival. Methylation of regulatory sequences near GSTP1, which encodes the pi class glutathione S-transferase, is one of the most common epigenetic alteration associated with prostate cancer [65]. More than 90% of prostate cancers exhibit extensive deoxycytidine methylation of the GST-Pi promotor, leading to transcriptional inactivation of the gene [66]. The methylation-induced inactivation of GST-Pi was restricted to cancer and not found in BPH. Since GSTP1 plays a critical role in reactive oxygen species detoxification, the epigenetic silencing of this activity is thought to increase the vulnerability of prostate epithelial cells to oxidative damage that in turn could promote the pathogenesis of the prostate. Significantly, it is generally accepted that an epigenetic alteration of the genome by DNA methylation progressively accumulates with age. Hence, DNA methylation may explain, in part, age as a risk factor in the development of prostate cancer.

17.3.6 DIET AND ENVIRONMENTAL AGENTS

Dietary factors may contribute to prostate cancer development according to descriptive epidemiologic studies of migrants, such as Japanese coming to the United States, and studies of geographic variation between populations [67]. A high positive correlation exists between prostate cancer incidence and the corresponding rates of several other cancers probably related to diet, including breast and colon cancer [68]. There is a strong association between the incidence of prostate cancer mortality and fat consumption [69]. One study measured fat intake adjusted for energy intake, and observed a positive association between increased fat intake and risk of advanced prostate cancer [70]. It is unclear how dietary fat may increase the risk of prostate cancer, but the mechanisms that have been proposed include dietary fat-induced alterations in hormonal profiles, the effect of fat metabolites as protein or DNA-reactive intermediates, dietary fat-induced elevation of oxidative stress, and as reservoirs of fat-soluble carcinogens.

Cadmium exposure is associated with a modest risk of developing prostate cancer [71]. Cadmium is a significant environmental contaminant that has many industrial uses. It is elevated as a consequence of sewage-sludge disposal and the combustion of municipal waste and fossil fuels [72]. It is used in pigments, batteries, stabilizers in plastics, metallurgy, the semiconductor industry, and as catalysts. In addition to occupational exposure in industrial areas, many individuals may be exposed to low doses of cadmium through consumption of contaminated fish, drinking water, contaminated air, and cigarette smoking.

Several studies have suggested that endocrine disruptors present in the environment in the form of estrogenic mimics may mediate carcinogenesis [73]. There are numerous man-made environmental compounds that have estrogenic activity, including industrial and agricultural chemicals, plastics, detergents, and certain dyes used in the food industry such as Red Dye No. 3. Use of some of these chemicals has been restricted for many years, but such compounds accumulate in the food chain and are highly lipophilic.

The heterocyclic amine, 2-amino-1-methyl-6-phenylimidazo(4,5-b) pyridine (PhiP), found in cooked fish and meat, has been suggested to be a carcinogen due to its induction of carcinoma in the ventral prostate of rodents [74]. This compound is also a carcinogen in the mammary gland of female F344 rats and in the colon of male rats [75]. Hence, small amounts of PhiP in a diet of cooked fish and meat may cause mutations in prostate tissue over a lifetime. Consequently, consumption of PhiP in cooked fish and meat might contribute to the high incidence of prostate cancer in Western men [74]. Since PhiP is mutagenic, and is a carcinogen in the rat prostate, this could provide evidence for a genotoxic mechanism of human prostate carcinogenesis.

17.4 CONCLUSIONS

We can definitively conclude that gender, the presence of testes that produce testosterone, age, and race are risk factors for the development of prostate cancer. It is also clear that testosterone is a permissive risk factor that allows the growth and development of the adult prostate. However, the mechanisms of carcinogenesis that are causal of prostate cancer during development and growth are even less well defined. During the growth of the prostate over a long life span, the presence of low penetrance germ-line mutations in conjunction with exposure to a variety of agents may promote the accumulation of genetic alterations that eventually manifest as prostate cancer. Current evidence suggests that a very modest aggregate effect of exposure to reactive oxygen species that form DNA adducts, epigenetic alterations resulting from DNA methylation, a high-fat diet

TABLE 17.1
Potential Mechanisms of
Prostate Carcinogenesis

Germ-line mutations
DNA methylation
Reactive-oxygen species
Endocrine disruptors/high-fat diet
PhiP
Cadmium
Other carcinogens/high-fat diet

that acts as a reservoir of endocrine disruptors and carcinogens, high exposure to cadmium, PhiP, and several other unidentified carcinogens may contribute to the development of prostate cancer (Table 17.1).

REFERENCES

1. McNeal, J.E. et al., Zonal distribution of prostatic adenocarcinoma, *Am. J. Surg. Pathol.*, 12, 897, 1988.
2. Stamey, T.A. and McNeal, J.E., Adenocarcomina of the prostate, in *Campbell's Urology*, 6th ed., Walsh, P.C., Retik, A., Stamey, T., and Vaughn, E., Eds., W.B. Saunders, Philadelphia, PA, 1992, p. 1159.
3. Baumgarten, H.G. et al., Adrenergic innervation of the human testis, epididymis, ductus deferens and prostate: A fluorescence microscopic and fluorimetric study, *Z. Zellforsch.*, 90, 81, 1968.
4. Walsh, P.C. and Donker, P.J., Impotence following radical prostatectomy: Insights into etiology and prevention, *J. Urol.*, 128, 497, 1982.
5. Lepor, H. et al., Precise localization of the autonomic nerves from the pelvic plexus to the corpora cavernosa: A detailed anatomical study of the adult male pelvis, *J. Urol.*, 133, 207, 1985.
6. Villers, A. et al., The role of perineural space invasion in the local spread of prostatic adenocarcinoma, *J. Urol.*, 142, 763, 1989.
7. Vaalasti, A. and Hervonen, A., Autonomic innervation of the human prostate, *Invest. Urol.*, 17, 293, 1980.
8. Bruschini, H. et al., Neurologic control of prostatic secretion in the dog, *Invest. Urol.*, 15, 288, 1978.
9. Franks, L.M., Benign nodular hyperplasia of the prostate, a revire, *Ann. R. Coll. Surg. Engl.*, 14, 92, 1954.
10. Verhagen, A.P.M. et al., Differential expression of keratins in the basal and luminal cell types cytokeratins compartments of rat prostatic epithelium during degeneration and regeneration, *Prostate*, 13, 25, 1988.
11. Sherwood, E.R. et al., Differential cytokeratin expression in normal, hyperplastic and malignant epithelial cells from the human prostate, *J. Urol.*, 143, 167, 1990.
12. Isaacs, J. and Coffee, D., Etiology and disease process of benign prostatic hyperplasia, *Prostate*, 2 (Suppl.), 33, 1989.
13. Foster, C.S. and Key, Y., Stem cells in prostatic epithelia, *Int. J. Exp. Pathol.*, 789, 311, 1997.
14. Hayward, S.W., Brody, J.R., and Cunha, G.R., An edgewise look at basal epithelial cells: Three dimensional views of the rat prostate, mammary gland and salivary gland, *Differentiation*, 60, 219, 1996.
15. Verhagen, A.P.M. et al., Colocalization of basal and luminal cell-type cytokeratins in human prostate cancer, *Cancer Res.*, 15, 6182, 1992.
16. Bonkhoff, H., Stein, U., and Remberger, K. The proliferative function of basal cells in the normal and hyperlpastic human prostate, *Prostate*, 24, 114, 1994.
17. Djakiew, D, Role of nerve growth factor-like protein in the paracrine regulation of prostate growth, *J. Androl.*, 13, 476, 1992.

18. Di Sant'Agnese, P.A. et al., Human prostatic endocrine-paracrine (APUD) cells. *Arch. Pathol. Lab. Med.*, 109, 607, 1985.
19. Kimura, N. et al., Plasma chromogranin A in prostatic carcinoma and neuroendocrine cancers. *J. Urol.*, 157, 565, 1997.
20. Angelsen, A. et al., Neuroendocrine cells in the prostate of the rat, guinea pig, cat, and dog, *Prostate*, 33, 8, 1997.
21. Van de Voorde, W.M. et al., Morphologic and neuroendocrine features of adenocarcinoma arising in the transition zone and in the peripheral zone of the prostate, *Modern Pathol.*, 8, 591, 1995.
22. Weistein, M.H. et al., Neuroendocrine differentiation in prostate cancer: Enhanced prediction of progression after radical prostatectomy, *Human Pathol.*, 27, 683, 1996.
23. Abrahamsson, P.A. et al., The course of neuroendocrine differentiation in prostatic carcinomas. An immunohistochemical study testing chromogranin A as an "endocrine marker," *Pathol. Res. Pract.*, 185, 373, 1989.
24. Di Sant'Agnese, P.A., Neuroendocrine differentiation in carcinoma of the prostate. Diagnosis, prognostic, and therapeutic implications, *Cancer*, 70, 254, 1992.
25. Deftos, L.J., Immunoassay and immunohistology studies of chromogranin A as a neuroendocrine marker in patients with carcinoma of the prostate, *Urology*, 48, 58, 1996.
26. Bubendoef, L. et al., Ki67 labeling index: An independent predictor of progression in prostate cancer treated by radical prostatectomy, *J. Pathol.*, 178, 437, 1996.
27. Noordzij, M.A. et al., The prognostic influence of neuroendocrine cells in prostate cancer: Results of a long-term follow-up study with patients treated by radical prostatectomy, *Int. J. Cancer*, 62, 252, 1995.
28. Cohen, M.K. et al., Neuroendocrine differentiation in prostatic adenocarcinoma and its relationship to tumor progression, *Cancer*, 74, 1899, 1994.
29. Aprikian, A.G. et al., Neuroendocrine differentiation in metastatic prostatic adenocarcinoma. *J. Urol.*, 151, 914, 1994.
30. Speights, V.O. et al., Neuroendocrine stains and proliferative indices of prostatic adenocarcinomas in transurethral resection samples, *Br. J. Urol.*, 80, 281, 1997.
31. Prins, G.S., Birch, L., and Greene, G.L., Androgen receptor localization in different cell types of the adult rat prostate, *Endocrinology*, 129, 3187, 1991.
32. Coffey, D.S., Similarities of prostate and breast cancer: Evolution, diet, and estrogens, *Urology*, 57, 31, 2001.
33. Teske, E. et al., Canine prostate carcinoma: Epidemiological evidence of an increased risk in castrated dogs, *Mol. Cell. Endocrinol.*, 197, 251, 2002.
34. Djakiew, D., Dysregulated expression of growth factors and their receptors in the development of prostate cancer, *Prostate*, 42, 150, 2000.
35. Greenberg, N.M. et al., Prostate cancer in a transgenic mouse, *Proc. Natl. Acad. Sci. USA*, 92, 3439, 1995.
36. Banerjee, P.P. et al., Age-dependent and lobe-specific spontaneous hyperplasia in the Brown Norway rat prostate, *Biol. Reprod.*, 59, 1163, 1998.
37. De Marzo, A.M. et al., Proliferative inflammatory atrophy of the prostate: Implications for prostatic carcinogenesis, *Am. J. Pathol.*, 155, 1985, 1999.
38. Ruska, K.M., Sauvageot, J., and Epstein, J., Histology and cellular kinetics of prostatic atrophy, *Am. J. Surg. Pathol.*, 22, 1073, 1998.
39. Nelson, W.G. et al., Preneoplastic prostate lesions: An opportunity for prostate cancer prevention, *Ann. N.Y. Acad. Sci.*, 952, 135, 2001.
40. McNeal, J.E. and Bostwick, D.G., Intraductal dysplasia: A premalignant lesion of the prostate, *Hum. Pathol.*, 17, 64, 1986.
41. Bostwick, D.G., High grade prostatic intraepithelial neoplasia: The most likely precursor of prostate cancer, *Cancer*, 75, 1823, 1995.
42. Billis, A., Age and race distribution of high-grade prostatic intraepithelial neoplasia. An autopsy study in Brazil (South America), *J. Urol. Pathol.*, 5, 1, 1996.
43. Bostwick, D.G., Pacelli, A., and Lopez-Beltran, A., Molecular biology of prostatic intraepithelial neoplasia, *Prostate*, 29, 117, 1996.

44. Gleason, D.F., Histologic grading of prostatic cancer: A perspective, *Hum. Pathol.*, 23, 273, 1992.

45. Brawn, P., Histologic features of metastatic prostate cancer, *Hum. Pathol.*, 23, 267–272, 1992.

46. Franks, L.M., Atrophy and hyperplasia in the prostate proper, *J. Pathol. Bacteriol.*, 68, 617, 1954.

47. Bartsch, G. et al., Light microscopic stereological analysis of the normal human prostate and of benign hyperplasia, *J. Urol.*, 122, 487, 1979.

48. Shapiro, E., The relative proportion of stromal and epithelial hyperplasia is related to the development of symptomatic benign prostate hyperplasia, *J. Urol.*, 147, 1293, 1992.

49. Lawsen, R.K., The natural history of benign prostatic hyperplasia, in *AUA Update Series*, 5(19), 1986, p. 1.

50. Geller, J., Pathogenesis and medical treatment of benign prostatic hyperplasia, *Prostate*, 2(Suppl.), 95, 1989.

51. Coffee, D.S. and Isaacs, J.T., Control of prostate growth, *Urology*, 17(Suppl.), 17 1981.

52. Peters, C.A. and Walsh, P.C., The effect of nafarelin acetate, a LHRH agonist, on BPH, *New Engl. J. Med.*, 317, 599, 1987.

53. Kreig, M., Nass, R., and Tunn, S., Effect of aging on endogenous level of 5-dihydrotestosterone, testosterone, estradiol and estrone in epithelium and stroma of normal and hyperplasic prostate, *J. Clin. Endocrinol. Metab.*, 77, 375, 1993.

54. Ho, S.-M., Lee, K.F., and Lane, K., Neoplastic transformation of the prostate, in *Prostate: Basic and Clinical Aspects*, Naz, R., Ed., CRC Press, Boca Raton, FL, 1997.

55. Ahmed, M.M., Lee, C.T., and Oesterling, J.E., Current trends in the epidemiology of prostatic diseases: Benign hyperplasia and adenocarcinoma, in *Prostate: Basic and Clinical Aspects*, Naz, R., Ed., CRC Press, Boca Raton, FL, 1997.

56. Greenwald, P., Prostate, in *Cancer Epidemiology and Prevention*, Schottenfeld, D. and Fraumeni J.F., Eds., WB Saunders, Philadelphia, PA, 1982, p. 938.

57. Herring, B.D., Cancer of the prostate in blacks, *J. Nat. Med. Assoc.*, 69, 165, 1977.

58. Scardino, P.T., Weaver, R., and Hudson, M.A., Early detection of prostate cancer, *Hum. Pathol.*, 23, 211, 1992.

59. Mandel, J.S. and Schumen, L.M., Epidemiology of cancer of the prostate, in *Reviews in Cancer Epidemiology*, Lilienfeld, A.M., Ed., Elsevier, NY, 1980, p. 1.

60. Haenszel, W. and Kurihara, I., Studies of Japanese migrants. I. Mortality from cancer and other diseases among Japanese in the United States, *J. Natl. Cancer Inst.*, 40, 43, 1968.

61. Verhage, B.A.S. et al., Allelic imbalance in hereditary and sporadic prostate cancer, *Prostate*, 54, 50, 2003.

62. Rokman, A. et al., Germline alterations of the RNASE L gene, a candidate HPC1 gene at 1q25, in patients and families with prostate cancer, *Am. J. Hum. Genet.*, 70, 1299, 2002.

63. Xu, J. et al., Germline mutations and sequence variants of the macrophage scavenger receptor 1 gene are associated with prostate cancer risk, *Nat. Genet.*, 32, 321, 2002.

64. Lin, X. et al., Reversal of GSTP1 CpG island hypermethylation and reactivation of pi-class glutathione S-transferase (GSTP1) expression in human prostate cancer cells by treatment with procainamide, *Cancer Res.*, 61, 8611, 2001.

65. Jeronimo, C. et al., Quantitation of GSTP1 methylation in non-neoplastic prostatic tissue and organ-confined prostate adenocarcinoma, *J. Natl. Cancer Inst.*, 93, 1747, 2001.

66. Lee, W.H. et al., Cystidine methylation of regulatory sequences near the -class glutathione S-transferase gene accompanies human prostate carcinogenesis, *Proc. Natl. Acad. Sci. USA*, 91, 11733, 1994.

67. Kolonel, L.N., Nutrition and prostate cancer, *Cancer Causes Control*, 7, 83, 1996.

68. Berg, J.W., Can nutrition explain the pattern of international epidemiology of hormone-dependent cancers? *Cancer Res.*, 35, 3345, 1975.

69. Blair, A. and Fraumeni, J.F., Geographic patterns of prostate cancer in the United States, *J. Natl. Cancer Inst.*, 61, 379, 1978.

70. Giovannucci, E. et al., A prospective study of dietary fat and risk of prostate cancer, *J. Natl. Cancer Inst.*, 85, 1571, 1993.

71. Pienta, K.J. and Esper, P.S., Risk factors for prostate cancer, *Ann. Int. Med.*, 118, 793, 1993.

72. Lloyd, O.L., Respiratory-cancer clustering associated with localised industrial air pollution, *Lancet*, 1, 318, 1978.

73. Watson, C.S., Pappas, T.C., and Gametchu, B., The other estrogen receptor in the plasma membrane: Implications for the actions of environmental estrogens, *Environ. Health Perpect.*, 103, 41, 1995.
74. Shirai, T. et al., The prostate: A target for carcinogenicity of 2-amino-1-methyl-6-phenylimidazo(4,5-b) (PhIP) derived from cooked foods, *Cancer Res.*, 57, 195, 1997.
75. Ito, N. et al., A new colon and mammary carcinogen in cooked food, 2-amino-1-methyl-6- phenylimid-azo-(4,5-b)pyridine (PhIP), *Carcinogenesis*, 14, 2553, 1991.

18 Skin Cancer: Epidemiology, Clinical Manifestations, and Genetic and Molecular Aspects

Zalfa A. Abdel-Malek, Ana Luisa Kadekaro,
Michelle A. Pipitone, and Diya F. Mutasim

CONTENTS

18.1   SKIN CANCER, MAIN TYPES, AND STATISTICS

The term skin cancer generally refers to melanoma and nonmelanoma skin cancer. The latter is further categorized as basal or squamous cell carcinoma (BCC and SCC, respectively). Melanoma arises due to the malignant transformation of melanocytes, the pigment cells in the cutaneous epithelium or epidermis. Melanoma is the most fatal form of skin cancer due to its metastatic potential and resistance to most forms of conventional chemotherapy or radiation. In contrast, BCC and SCC are malignancies of the keratinocytes, the structural cells of the epidermis. The latter two types of cancer are highly curable, but if left untreated, BCC can cause deformities, and SCC can become invasive and metastatic.

In the United States and Australia, skin cancer is the most common type of cancer [1]. Skin cancer is a major public health issue in the United States, particularly in the Southwest such as Arizona and California, because it accounts for almost half of the total cases of cancer. According to the statistics of the American Cancer Society, it was expected that more than 1 million new cases of skin cancer would be diagnosed in the United States in the year 2003, with men having twice the risk to develop these cancers than women (American Cancer Society, 2003, Facts and Figures). Of those new cases, 91,900 would be melanoma, a disease that has been increasing by about 5% every year. At this rate, one in every 39 Americans would be expected to develop melanoma sometime in his or her life. More than 77% of skin cancer deaths are from melanoma. In 2003, melanoma was expected to result in 7,600 deaths, mostly of older Caucasian males. Although death from nonmelanoma skin cancer is uncommon, still 2,200 people die annually from SCC and BCC. Melanoma is the most common cancer among young adults 25–29 years of age. In the United States, melanoma is the third most common cancer in individuals 15–39 years of age. Since the 1970s, the increase in the incidence of melanoma has been the highest compared to other types of cancer (Cancer Research U.K. website).

18.2 SUN EXPOSURE AND SKIN CANCER

The main etiological factor for all forms of skin cancer is sun exposure. SCC is strongly linked to chronic sun exposure, and the risk increases with increased cumulative exposure to solar ultraviolet (UV) radiation. On the other hand, the risk for BCC and melanoma is thought to correlate directly with acute sun exposure and the number of severe sunburns during childhood. Owing to the significance of sun exposure in skin cancer formation, these neoplasms may be preventable by sun-protection practices, such as wearing protective clothing, applying sunscreens, and avoiding sun exposure from midday until 4:00 P.M.

The most mutagenic spectrum of solar UV is the UVB spectrum (wavelength of 290 to 320 nm). Exposure to UVB results in direct damage to genomic DNA, mainly in the form of DNA photoproducts, primarily cyclobutane pyrimidine dimers, and to a lesser extent, pyrimidine 6,4-pyrimidone photoproducts [2]. There is overwhelming evidence that UVB is involved in the carcinogenic process that leads to SCC and BCC [3,4]. Evidence for UV-induced mutagenicity is provided by the presence of UVB signature mutations in SCC and BCC tumors, as well as in actinic keratosis, some of which are precursor lesions for SCC. Further evidence is provided by the disease xeroderma pigmentosum, which is characterized by defects in various steps of nucleotide excision repair that result in impairment of the repair of UVB-induced DNA photoproducts, and by an increased incidence and early onset of skin cancer tumors [5,6].

18.3 IMMEDIATE RESPONSE OF THE SKIN TO
UV EXPOSURE

Most sporadic skin cancers are the latent effects of excessive sun exposure. The immediate response of the skin to UV exposure includes a series of events that represent a stress-signaling pathway (Figure 18.1). Activation of this pathway allows the skin to cope with the damaging effects of UV by repairing DNA damage in order to prevent mutagenesis. Exposure of cultured human epidermal keratinocytes and melanocytes to UV initiates a stress-signaling pathway that includes the activation of the MAP kinases p38 and JNK/SAPK, and the transcription factor CREB [7,8]. This pathway also involves accumulation of p53 and increased expression of the cyclin/cdk inhibitor p21, resulting in cell-cycle arrest to allow for DNA repair, or in apoptosis of cells with DNA damage that surpasses the DNA repair capacity. Exposure of the skin to UV results in increased synthesis of a wide array of paracrine factors that participate in the cellular response to UV [9–11]. Some of these factors determine whether cells survive or undergo apoptosis. Activation of survival pathways in the skin, for example, the Akt/PKB pathway, limits the extent of apoptosis [12]. Also, increased melanin

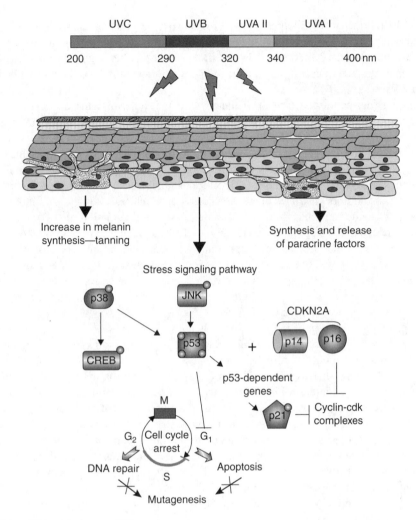

FIGURE 18.1 Direct response to UV.

synthesis, that is, tanning, is expected to reduce the extent of UV-induced DNA damage. Disruption of critical steps in the stress-signaling pathway in response to UV is expected to compromise the genomic stability of keratinocytes and melanocytes, and to increase the risk for mutagenesis and ultimately malignant transformation.

18.4 CUTANEOUS PIGMENTATION AND SKIN CANCER RISK

Important risk factors for skin cancer are light skin color, red hair, freckles, a large number of nevi, particularly atypical nevi, and family history or personal history of skin cancer [13–15]. Additional factors related to the response of the skin to sun exposure include susceptibility to burn, inability to tan, and frequent severe burns during childhood [16,17]. Cutaneous pigmentation conferred by melanin is considered the main photoprotective mechanism in the skin against UV-induced damage. The protective role of melanin against photocarcinogenesis is supported by numerous clinical observations. The incidence of skin cancer is markedly higher in individuals with fair skin that do not tan, that is, skin type I or II, than in those with dark skin that tan readily when exposed to the sun, that is, skin types III to VI (reviewed by Gilchrest et al. [18]; Sober et al., [19]; Pathak [20]). In the United States, Caucasians have 50-fold higher incidence of BCC and SCC, and 10-fold higher incidence of melanoma than African Americans [21–24]. In Australia that has high UV intensity, the Celtic population characterized by fair skin, red hair, and poor tanning ability, has the highest incidence of melanoma worldwide [25].

Melanin is the pigment synthesized by melanocytes in the epidermis and hair follicles. In the skin, melanin filters and scatters UVA and UVB rays, thus limits their penetration into the inner epidermal layers, and reduces the extent of UV-induced DNA damage [26,27]. In addition, melanin serves as a scavenger for reactive oxygen species that can result in oxidative DNA damage, which is potentially mutagenic. In the skin, melanin is deposited onto specialized organelles within melanocytes, termed melanosomes. When fully pigmented, mature melanosomes are transferred via the dendrites of melanocytes to keratinocytes. This interaction between the melanocytes and keratinocytes has led to the concept of epidermal melanin unit, which describes the association of one melanocyte with 32 to 36 keratinocytes in the epidermis, leading to uniform pigmentation of the skin. Melanosomes form supranuclear caps that protect the nuclei of melanocytes and keratinocytes from the impinging UV rays [28,29]. This protective effect is most evident in dark skin, where melanosomes remain intact throughout the epidermal layers. In lightly pigmented skin, however, photoprotection is compromised as melanosomes are degraded in the suprabasal layers and only melanin dust is detected in the upper layers of the epidermis [28–30].

There are two main forms of melanin, the brown–black eumelanin, and the red–yellow pheomelanin. In every individual, epidermal melanocytes synthesize both types of melanin [31,32]. However, the relative amounts of eumelanin and pheomelanin differ among individuals depending on their ethnic and genetic background. Constitutive pigmentation is mainly determined by three factors: (1) total melanin content, which is governed by the rate of melanin synthesis by melanocytes, (2) the ratio of eumelanin to pheomelanin, and (3) the rate of transfer of melanosomes from melanocytes to surrounding keratinocytes [33] (reviewed by Kadekaro et al. [34]). Accordingly, dark skin and hair have higher total melanin content and eumelanin to pheomelanin ratio than light-colored skin and hair. The relative amounts of eumelanin and pheomelanin determine the extent of photoprotection of the skin. Eumelanin is superior to pheomelanin in its photoprotective properties due to its resistance to photodegradation, and to its ability to scavenge reactive oxygen radicals that are generated in the skin after UV exposure [35,36]. Pheomelanin, on the other hand, is photolabile, and contributes to oxidative stress by the generation of superoxide anions and hydroxyl radicals [37–39].

18.5 ARSENIC EXPOSURE AND NONMELANOMA SKIN CANCER

Beside sun exposure, an environmental factor that has been linked to nonmelanoma skin cancer is arsenic. The association between arsenic exposure and skin cancer was made more than a century ago (reviewed by Leonard and Lauwery [40,41]). Arsenic is a metalloid that is prevalent in nature, in rocks, soil, metal ores, and artisan wells [41–43]. Human exposure to arsenic occurs through consumption of water and crops that contain high levels of arsenic, and from industrial sources, such as smelters, mines, or from the use of arsenic in pesticides and herbicides [40]. Until the 1960s, potassium arsenite, at a concentration of 7600 mg/L was used medicinally as Fowler's solution for various diseases, including leukemia, asthma, and psoriasis [44]. Trivalent arsenite is the most toxic form of arsenic. Inorganic arsenicals are the most prevalent forms in nature; however, in humans, these forms can be reduced to arsenite by arsenate reductase in the liver.

Chronic exposure to arsenic in several endemic areas, such as Taiwan and Mexico, presents itself in distinct cutaneous manifestations, characterized by hyperkeratosis, hyperpigmentation, and skin cancer lesions that include BCC, SCC, and Bowen's disease, which is a distinct form of SCC *in situ* [41,45–47]. These cutaneous changes have a strong positive correlation with the dose of, and duration of exposure to, arsenic [43,45,48]. Usually multiple skin cancer tumors arise due to chronic arsenic exposure, and can be a mixture of BCC and SCC. Histologically these tumors do not differ significantly from those induced by UV [46]. However, in general, arsenic-induced skin cancer tumors are more aggressive and metastatic than their UV-induced counterparts. The most distinguishing features of arsenic related skin cancer is the anatomical location of the tumors, which arise on sun exposed, as well as unexposed skin sites, such as the trunk, the palms, and the soles.

It is established that arsenic is a human carcinogen, yet by itself, it is not directly mutagenic. The exact mechanism by which arsenic exposure causes skin cancer is not known. Mutations in the tumor suppressor gene *p53* have been reported in arsenic-related skin cancer tumors and Bowen's disease in patients from endemic areas in Taiwan [49]. In addition, arsenic is known to induce the generation of reactive oxygen species and oxidative DNA damage has been detected in arsenic-exposed skin [50]. Arsenic is thought to function as a co-mutagen, which inhibits the repair of DNA damage induced by other mutagens, such as UV, as shown to be the case in a mouse model for arsenic carcinogenesis [51]. Arsenic has been shown to inhibit nucleotide excision repair after UV-irradiation [52]. Recently, these findings were further corroborated by the observation that arsenic inhibits the expression of critical genes involved in nucleotide excision repair, mainly *ERCC1*, *XPF*, and *XPB* [53].

18.6 IMMUNOSUPPRESSION AND SKIN CANCER

18.6.1 UV-INDUCED IMMUNOSUPPRESSION

Exposure to UV results in systemic immunosuppression, which contributes to skin carcinogenesis. The immunosuppressive effect of UV was first described more than 25 years ago [54], and was first observed in studies that revealed that UV-irradiated mice failed to reject highly antigenic transplantable UV-induced tumors, while unirradiated syngeneic mice rejected these tumors. These seminal studies suggested that immunosuppression by UV (UVB as well as UVA) allows for the growth of skin cancer tumors by inducing tumor tolerance.

It has been proposed that specific chromophores exist in the skin and mediate the systemic immunosuppressive effect of UV by functioning as immunomodulators. The best candidates for such a role are DNA damage, cell membrane peroxidation, and epidermal *cis*-urocanic acid [55,56]. The main target for UV in the skin is the epidermal Langerhans cells, which are inhibited by UV from presenting antigens or stimulating allogeneic type T1 cells. Additionally, UV induces the synthesis and release of an array of soluble immunosuppressive mediators by keratinocytes, dermal macrophages, and neutrophils [57]. There is convincing evidence that apoptosis of epidermal Langerhans cells and reactive T cells through the Fas/Fas ligand system contributes to the UV-induced immunosuppression.

18.6.2 SKIN CANCER IN IMMUNOCOMPROMISED PATIENTS

The role of immunosuppression in skin cancer is best evidenced in solid organ transplant recipients. In fact, nonmelanoma skin cancer is the most common malignancy following transplantation, and is thought to be due to infection with human papilloma virus (HPV). HPV DNA was detected more frequently in SCC of transplant recipients than in SCC in nonimmunosuppressed patients [58]. The prevalence of HPV types 5 and 8 in SCC than in precancerous or benign lesions in transplant recipients suggests that infection with these two types of HPV increases the risk for SCC.

18.7 BASAL CELL CARCINOMA: CLINICAL PRESENTATION AND HISTOLOGY

Basal cell carcinoma is the most common cancer in the Caucasian population [23]. BCC has varying clinical and histological morphologies. Typically, BCC is localized in sun-exposed areas of

TABLE 18.1
Risk Factors for Skin Cancer

Factor	Basal cell carcinoma	Squamous cell carcinoma	Melanoma
Pigmentary phenotype and inability to tan	+	+	+
Sun exposure	+	+	+
Arsenic	+	+	−
Immunosuppression	+	+	+
High number (>50) and presence of atypical nevi	+	+	+
Family and personal history of skin cancer	+	+	+

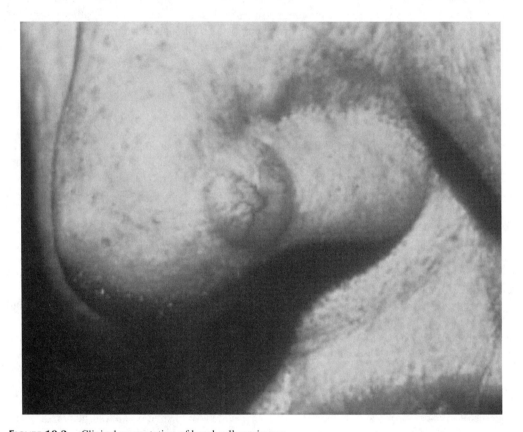

FIGURE 18.2   Clinical presentation of basal cell carcinoma.

the head and neck in older patients, but can present earlier if the patient has significant risk factors (summarized in Table 18.1). There are several subtypes of BCC, the most common of which is nodular BCC [59]; patients may present with a bleeding, nonhealing papule. Nodular BCC presents as a well-demarcated, pearly or translucent, telangiectatic papule, occasionally with a central depressed erosion or ulcer, and hemorrhagic crust (Figure 18.2). Occasionally, patients may have pigmented lesions, clinically presenting as brown or blue–black papules that mimic malignant melanoma. Micronodular BCC is clinically similar to nodular BCC and is distinguished histologically by smaller nodular aggregates of tumor cells. Superficial BCC is the second most common subtype, comprising 15% of all BCC, which presents as a well-demarcated, erythematous, peripherally-raised and pearly-bordered plaque on the trunk or extremities of middle-aged adults, most commonly females. This subtype differs from nodular BCC in that it is common in transplant recipients, appears on body locations of intermittent sun exposure, and occurs in younger patients [59,60].

A minor subtype that comprises about 5% of all BCC is infiltrative BCC, which is predominant on the head and neck of older adults [59]. This subtype does not have a specific clinical morphology, but rather, presents as either an ill-defined flat papule or plaque, similar to morpheaform BCC. Sclerosing or morpheaform BCC accounts for 3% of all BCC. The head and neck are the most common locations, and the morphology is characterized by indurated, flat or slightly raised, fibrotic, scar-like plaque with ill-defined margins.

On histological examination, BCC has aggregates of basophilic, cuboidal cells arranged in islands and cords, with the periphery of the island comprising cuboidal cells arranged in a parallel manner, also called peripheral palisade. The tumor cells are homogeneous, with scant cytoplasm, and the number of mitoses depends on the differentiation level of the tumor. There are characteristic clefts around tumor lobules, with surrounding fibrosis, myxoid stroma, and variable lymphocytic inflammatory infiltrate.

18.8 SQUAMOUS CELL CARCINOMA: CLINICAL PRESENTATION AND HISTOLOGY

Squamous cell carcinoma *in situ* generally presents as a well-defined erythematous scaly papule or plaque. Histologically, SCC *in situ* is a full thickness, intraepidermal carcinoma that has specific names depending on the clinical presentation and histological characteristics. Overall, SCC *in situ* follows an indolent course, but progression to invasive carcinoma occurs in up to 26%, with metastasis in 16% [61]. Lesions in sun-damaged locations historically have lower rates of invasion and metastasis.

Bowen's disease is a specific type of SCC *in situ*, characterized as a well-defined, brightly erythematous minimally scaly plaque, most commonly in sun-exposed areas, similar in appearance to psoriasis or dermatitis. Bowen's disease is histologically characterized by the bizarre, large atypical keratinocytes in all layers of an acanthotic epidermis, in the absence of dermal invasion. According to various reports, the incidence of invasive SCC arising from lesions of Bowen's disease varies from 1% to 11% [62].

Squamous cell carcinoma is most commonly located on sun-exposed skin, specifically the head, neck, and arms of adults. Ten percent of these tumors arise from actinic keratosis, which are scaly erythematous papules [63]. This type of SCC has the best prognosis, in contrast to SCC found in chronic, nonhealing wounds or areas of chronic inflammation. The occurrence of SCC on mucosal sites is associated with a poor prognosis due to the propensity of the tumor to metastasize.

Clinically, SCC presents as erythematous papules, plaques, and nodules with frequent erosions, ulcerations, crusting, and bleeding (Figure 18.3). Chronic, nonhealing ulcers should be monitored and biopsied for the development of carcinoma. Occasionally, SCC can appear as a verrucous, polypoid papule or plaque, termed a verrucous carcinoma, and has specific designations depending on its location.

The histology of SCC features anastomosing cords and nests of polygonal cells with eosinophilic cytoplasm and enlarged nuclei. Dyskeratosis is common, and mitoses are variable. SCC is classified according to the extent of nuclear atypia and keratinization, which determine the extent of differentiation of tumor cells. The poorly differentiated tumors tend to be the most aggressive. However, the well-differentiated tumors might also metastasize and cause death.

18.9 ETIOLOGY OF MELANOMA

Melanoma is thought to be associated with acute intermittent sun exposure that results in severe sunburn, rather than the cumulative result of chronic sun exposure [64–66]. Melanoma is mainly

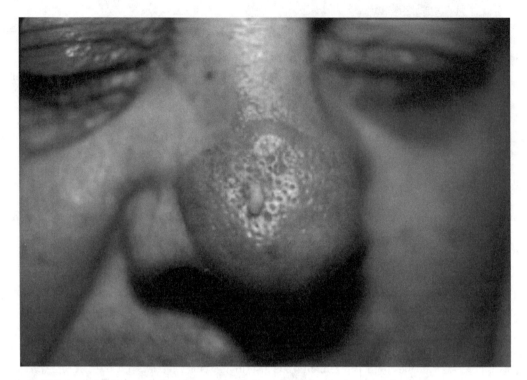

FIGURE 18.3   Clinical presentation of squamous cell carcinoma.

a disease of indoor workers, and is common in young and middle-aged professionals. The incidence of melanoma correlates directly with the number of severe sunburns during childhood, and the risk for melanoma is thought to be already determined around the second decade of life. A history of five or more severe sunburns during adolescence is estimated to double the risk for melanoma [67].

A role for UVA (320 to 400 nm wavelength) in the transformation of normal melanocytes to melanoma has been proposed [68,69]. UVA has longer wavelengths than UVB, thus can penetrate deeper through the epidermal layers to reach the basal layer where melanocytes reside (reviewed by Dillman [27,70,71]). UVA is far less energetic and mutagenic than UVB. However, the extent of UVA rays that reach the earth's surface is at least tenfold greater than that of UVB. UVA affects target cells mainly through the generation of reactive oxygen species that can cause oxidative DNA damage [72,73]. Compared to keratinocytes, melanocytes seem to be more sensitive to oxidative stress due to their reduced antioxidant defenses and therefore may be more vulnerable to oxidative DNA damage that can be mutagenic [74]. UVA has been shown to induce melanoma tumor formation in the Xiphophorus fish and in the South American opossum *Monodelphis domestica* [75,76]. Unlike BCC and SCC where UVB-signature mutations are common, melanoma tumors rarely express such mutations [77,78], suggesting a different mutagen, possibly UVA. So far, the contribution of UVA to human melanoma formation and progression is unclear, and needs to be further delineated.

18.10   MELANOMA: CLINICAL PRESENTATION AND HISTOLOGY

Melanoma is a malignant tumor of melanocytes and occasionally nevus cells in which one fifth of patients develop metastatic disease. Its incidence and mortality have been increasing, and it is one of the most common types of cancer in young adults [79]. Thin melanomas (<1 mm in depth) can be cured with surgical excision in over 90% of patients, but thicker melanomas portend a poorer diagnosis.

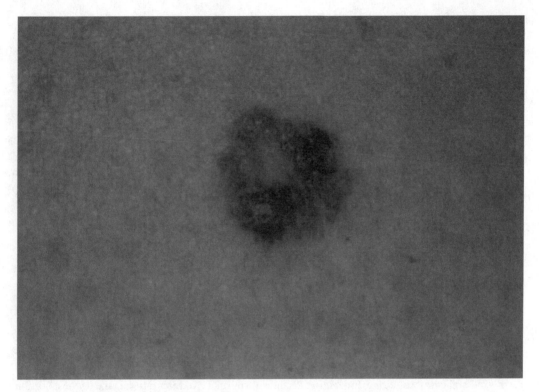

FIGURE 18.4   Clinical presentation of malignant melanoma.

Risk factors, which have been reviewed extensively and summarized in Table 18.1, include genetic mutations, UV radiation, large atypical melanocytic nevi, 50 or greater melanocytic nevi, large congenital nevi greater than 20 cm in diameter, family history of melanoma, personal history of melanoma, type I and II skin, and immunosuppression [80,81].

There are various clinical presentations of melanoma. Superficial spreading melanoma is the most common type and is seen in young and middle-aged adults. It is located most commonly on the legs of women and the trunks of men. It is a brown or black macule or patch, with irregular borders, asymmetry, and variable pigmentation, including hypo- and hyperpigmentation.

Nodular melanoma is the second most common type of melanoma and is diagnosed most commonly during the sixth decade of life (Figure 18.4). It is most frequently located on the trunk, head, and neck, and presents as a black, red, or skin-colored, rapidly growing papule or nodule.

Lentigo maligna (melanoma *in situ*) and lentigo maligna melanoma are diagnosed in the late adult years, on chronically sun-damaged skin, most commonly on the head and neck. Clinically, a lentigo maligna is a brown or black, irregularly pigmented and bordered patch, which typically has an indolent course. The incidence of progression of lentigo maligna to lentigo maligna melanoma is unknown [82]. Lentigo maligna melanoma is a brown–black, occasionally blue–red, variably pigmented, large patch with a focal papule(s) or nodule(s), indicating the deeper invasive portion. However, the clinical presentation may underestimate the depth or degree of invasion, especially in the early stages of the neoplasm.

Acral lentiginous melanoma (ALM) is a relatively uncommon type of melanoma, and is diagnosed in the late adult years. Brown–black asymmetric macules or patches develop on acral sites. It is believed that ALM may have a more aggressive biological behavior, but this finding is confounded by the fact that ALM occurs in areas that are difficult to observe, such as the soles of the feet, which result in a thicker, more advanced neoplasm upon presentation. Subungual melanoma is a variant of ALM (Table 18.2).

TABLE 18.2
Types of Skin Cancer

Basal cell carcinoma
Nodular
Superficial
Infiltrative
Sclerosing or morpheaform
Squamous cell carcinoma
In situ
   Bowen's disease
   Verrucous
Melanoma
Superficial spreading
Nodular
Lentigo maligna
Acral lentiginous

The histological criteria required in the diagnosis of melanoma include architectural and cyto-logical abnormalities. The neoplasm is asymmetric, with extension of the epidermal/horizontal component laterally beyond the vertical depth of the neoplasm. Typically, there is invasion of the epidermis by atypical melanocytes, confluence of noncohesive nests, sheets of melanocytes in the dermis, melanocytes extending down into the adnexa, and a lack of maturation in the deepest portion of the tumor, unlike benign melanocytic nevi. Cytologic abnormalities include atypical melanocytes with pleomorphic nuclei, presence of prominent nucleoli, mitotic figures, and necrotic melano-cytes. Other histologic findings include signs of regression, angiolymphatic invasion, and atypical distribution of melanin.

Certain histologic findings have been found to correlate with prognosis, and as such, have been incorporated in the staging of melanoma. These include the thickness of the lesion, angiolymphatic invasion, tumor infiltrating lymphocytes, and ulceration. In the sixth edition of the American Joint Committee on Cancer (AJCC), the tumor node metastasis staging has been revised to incorporate these histologic findings, in addition to the clinical features that affect prognosis. The major revisions emphasize the significance of ulceration, rather than the level of invasion, the number of metastatic lymph nodes, rather than their dimensions, and macroscopic or microscopic invasion of nodes. Other revisions include the site of metastasis, lactate dehydrogenase level, separating clinical and histologic staging to account for information obtained from the sentinel node biopsy procedure, and considering in-transit and satellite metastasis as Stage III.

18.11   MULTISTAGE MODEL OF CARCINOGENESIS

The multistage nature of cancer pathogenesis was established over 50 years ago [83]. Ever since, the experimental skin carcinogenesis model has provided remarkable insights into the biology, bio-chemistry, pharmacology, and genetics of carcinogenesis. The accessibility of the skin as an organ has made it an attractive model to study the process of carcinogenesis. Experimental carcinogenesis aimed at inducing tumors in the skin of mice following sequential topical applications of chemical agents on mouse skin, starting with a single subthreshold dose of a carcinogen, such as 7,12-dimethylbenz[a]anthracene (DMBA), followed by the repetitive application of a tumor promoter, such as 12-*O*-tetradecanoylphorbol-13-acetate (TPA) [84]. The results of these experiments led to the concept that carcinogenesis is a multistage process, which includes initiation, promotion, and finally progression (Fig. 18.5).

The observation that mutagens appear to be effective initiators implicates mutagenesis as the mechanism underlying the initiation stage. Benign squamous papillomas generally develop within 10 weeks and virtually all of these papillomas contain Harvey-ras (Ha-ras) mutations [85]. DMBA has been shown to induce point mutation (A-T) at the second position of codon 61 of the Ha-*ras* gene [85–87]. In fact, Ha-ras mutations that are possibly induced by UV have been found in nonmelanoma skin cancers and in benign skin neoplasms [88,89]. Mutagenesis is at least one mechanism of initiation; other mechanisms include stable epigenetic repression or gene activation.

It is assumed that cancer arises from genetic alterations initiated in one cell, as suggested by Knudson [90] and Nowell [91]. The second step in carcinogenesis is the promotion stage, which appears to involve the clonal expansion of initiated cells. Application of tumor promoters to initiated epidermis causes the selective clonal growth of cells to produce multiple benign squamous cell papillomas, each representing an expanded clone of nonterminally differentiated cells [92,93]. The most potent exogenous skin tumor promoters are the phorbol esters, which activate protein kinase C (PKC) [94]. Indeed, most cancer studies have been consistent with this clonal theory. Obviously clonal expansion requires stimulation of cell division that is usually triggered by growth factors. The loss of tumor suppressor genes by mutation, deletion, or silencing of their activity, allows cells to enter the progression phase. If one of these initiated and promoted cells, acquires additional genetic alterations or "hits" (e.g., other mutations or stable epigenetic changes) that allow the cell to become promoter-independent, then the third step of carcinogenesis, the malignant conversion proceeds (see Figure 18.5).

A small percentage of papillomas eventually progress to malignant SCC. The persistent exogenous exposure of initiated skin to carcinogens or clonal expansion of the initiated population can increase the probability for additional genetic changes that are required for malignant progression. In fact, premalignant progression and malignant conversion can be enhanced and accelerated by exposing animals bearing papillomas to mutagens [95]. The progression of papillomas to SCC has been characterized phenotypically by inappropriate expression or loss of expression of membrane receptors [96], growth factors [97,98], adhesion molecules [99], keratins [100,101], and cyclins or cyclin-dependent kinases [102].

Carcinogenesis involves the accumulation of multiple genetic lesions or hits in a single cell. Epidemiological studies on age-dependent cancer incidence have indicated that four to six genetic events are necessary for tumor development [103]. Numerous studies have indicated that a mutated phenotype or genetic instability is necessary for a single cell to accumulate these multiple discrete alterations [104]. The emerging concept is that genomes of cancer cells are unstable, and this

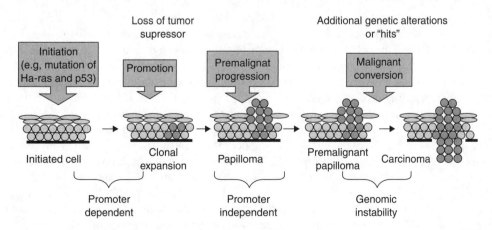

FIGURE 18.5 Multistage model of skin carcinogenesis.

instability results in a cascade of mutations, some of which enable cells to bypass the host regulatory processes that control cellular adhesion, motility, invasion, division, and survival.

In normal cells, proto-oncogenes and tumor suppressor genes are two classes of genes that guard against carcinogenesis [105]. In general, proto-oncogenes code for growth factors, growth factor receptors, transmembrane mitogenic signaling elements, and nuclear transcription factors, which are involved in regulating the proliferation and differentiation of cells. Uncontrolled activation of these proto-oncogenes by amplification or mutations converts them to oncogenes that transform a normal cell into a premalignant or malignant cell. Malignant conversion also occurs by deactivation of tumor suppressor genes that code for proteins that totally negate or inhibit the function of oncogenes. Given the oncogene/tumor suppressor gene paradigm, a balance between proto-oncogenes and tumor suppressor genes is needed for the control of cellular proliferation and differentiation.

18.11.1 Role of UV as Initiator and Promoter

Ultraviolet radiation is an "activator" of certain oncogenes and a "deactivator" of tumor suppressor genes. UV radiation acts as an initiator, since it induces specific mutations, coined UV signature mutations. Exposure to UVB typically results in the dimerization of adjacent pyrimidine residues in DNA, resulting in the formation of cyclobutane pyrimidine dimers or pyrimidine [6–4] pyrimidone photoproducts. Misrepair or failure to repair DNA photoproducts through nucleotide excision repair system results in mutations that frequently appear at dipyrimidine sites in DNA, resulting in the conversion of TT to CC. Mutations characteristic of UVB exposure have been detected in the *p53* gene in human SCC tumors [4] and in mouse skin tumors induced by UV exposure [106,107]. Furthermore, UV has been shown to cause immunosuppression and to induce mutations in certain genes, such as *ras*, through DNA damage at dipyrimidinic sites [108,109]. Additionally, UVA can act as an initiator, enhancing melanoma tumor formation in mice injected with melanoma tumor cells, and can also function as a tumor promoter, resulting in melanoma formation in mice initiated with DMBA [20,110].

18.12 GENETIC ALTERATIONS IN SKIN CANCER TUMORS

18.12.1 UV Signature Mutations in p53 in SCC and BCC

A gene that is commonly mutated in response to UVB exposure is the tumor suppressor p53. UVB signature mutations in hot spots of the *p53* gene are commonly found in BCC and SCC, and are considered early markers of these tumors [3,109]. Such mutations, however, are uncommon in melanoma tumors, suggesting a different or an additional etiological factor for melanomagenesis [77,78]. In accordance with the multistage model of tumorigenesis, UVB acts as an initiator as well as a promoter of SCC. Clones of p53-mutated keratinocytes have been detected in normal human skin, especially in sun-exposed anatomical sites [111]. It is postulated that UVB results in p53 mutations in a stem cell, which gives rise to a clone of p53-mutated keratinocytes. The clonal expansion of a mutated stem cell is favored by sun exposure, and in this case, UVB functions as a promoter, allowing for the expansion of a population of cells that is predisposed to cancer.

Wild-type p53 functions as a tumor suppressor, a cell cycle checkpoint and a sensor of DNA damage [112,113]. Exposure of cells to UV or other DNA damaging agents results in the accumulation of p53 by prolonging its half-life and inhibiting its degradation by HDM2, the product of the human double minute 2 gene [114–117]. In response to DNA damage, such as DNA photoproducts induced by UV, p53 undergoes a complex process of posttranslational modification that includes phosphorylation and acetylation of specific residues [118–120]. These modifications lead to the stabilization, accumulation, and transcriptional activation of p53. Accumulation of p53 arrests cells in G1 phase via the expression of p53-dependent genes, such as the cyclin-cdk inhibitor p21, in order to allow time for DNA repair prior to cellular replication [112,113,121,122]. Repair of UVB-induced DNA photoproducts by nuclear excision repair is critical for prevention of all forms of skin cancer. A strong evidence for the significance of UVB-induced DNA damage in skin cancer formation comes from patients with xeroderma pigmentosum, who have defects in nucleotide excision repair, and a strikingly higher incidence and earlier onset of all forms of skin cancer, compared to the general population [5,6]. P53 is also considered the guardian of the tissue, since high levels of p53 induce apoptosis of cells with extensive DNA damage that exceeds the cellular repair capacity [123]. Eliminating cells with extensive DNA damage is a mechanism to eliminate mutated cells and to preserve genomic stability in tissues.

18.12.2 Mutations in Genes in the Sonic Hedgehog Pathway in BCC and SCC

Mutations in critical genes in the sonic hedgehog (SHH) pathway are found in BCC and are critical for tumor formation. Germ-line mutations in the Patched (*PTCH*) gene were first identified in nevoid BCC syndrome, also called Gorlin syndrome [124,125]. Somatic mutations in the *PTCH* gene that inactivate the Patched protein have been identified in sporadic BCC, and in SCC from individuals with a history of multiple BCC tumors [126–129]. Some of those mutations are typical UVB-signature mutations, such as C → T substitutions at dipyrimidinic sites [126]. Patched acts as a tumor suppressor that inhibits the SHH pathway via its interaction with smoothened (Smo), a G-protein coupled receptor with seven transmembrane domains [130] (reviewed by Bale and Yu [131]). The repressive effect of Patched is relieved by its binding to SHH. This binding frees and activates Smo and increases the transcription of SHH target genes via Gli1 family of transcription factors. Mutations that inactivate PTCH result in unrestricted signaling of Smo and activation of the SHH pathway. Activation mutations in the genes for the oncogenic SHH and Smo have also been identified in sporadic BCC and result in the same outcome as PTCH inactivating mutations.

The role of the SHH pathway in the carcinogenic events that lead to BCC has been demonstrated experimentally. Over expression of SHH in engineered human epidermis recapitulated features of BCC *in vivo*, and resulted in hyperproliferation, lack of differentiation, and G0/G1 arrest of keratinocytes in response to calcium [132,133].

18.12.3 Mutations in the CDKN2A Locus and Melanoma

The tumor suppressor that is often mutated in melanoma is INK4A, also known as p16, which is encoded for by the CDKN2A locus (9p21 locus) (reviewed by Piepkorn [134]). Missense or deletion mutations of CDKN2A are found in the germ line of 40% of melanoma kindreds [135]. Deletion mutations in CDKN2A, which are typical UVB-signature mutations, are also prevalent in sporadic melanoma tumors [136,137]. CDKN2A codes for two different proteins, the tumor suppressor INK4A and p14 alternative reading frame (ARF), through the utilization of two separate reading frames [138,139]. Mutations in p16 result in loss of its function as an inhibitor of cyclin D1/cdk4 complexes, thus hindering G1 arrest that is needed for DNA repair [140,141]. The less common mutations in p14 ARF also disrupt the G1 cell cycle checkpoint, but via a mechanism different than that of mutated p16. Normally, p14 ARF activates p53 by binding and inactivating HDM2, which targets p53 to degradation by the ubiquitin pathway [142,143]. Mutations in p14 ARF prevent the accumulation of p53, thus disrupt the G1 restriction point and compromise DNA repair. Therefore, deletions at the CDKN2A locus disable two separate pathways that control cell growth, leading to unregulated proliferation and mutagenesis.

Germ line and sporadic mutations in CDK4 have been found in some melanoma tumors. These mutations disrupt the physical binding of INK4A to CDK4 and have the same outcome as mutations in CDKN2A, thus further confirming the significance of cell cycle regulation by CDKN2A in inhibiting the malignant transformation of melanocytes.

18.12.4 THE MELANOCORTIN 1 RECEPTOR GENE AND MELANOMA SUSCEPTIBILITY

The melanocortin 1 receptor (*MC1R*) gene is considered a skin cancer, particularly a melanoma, susceptibility gene. This gene has received tremendous interest during the past decade due to its significance in the diversity of human pigmentation [144,145]. The *MC1R* gene codes for the melanocortin 1 receptor, the receptor for melanocortins that is expressed on human melanocytes [146,147]. Activation of the gene by ligand binding results in stimulation of eumelanin synthesis [148]. The *MC1R* gene is highly polymorphic, having at least 35 different allelic variants. The wild-type gene is predominant in African populations, while the diverse alleles are mainly expressed in populations with light skin, such as Northern Europeans and Australians [144,145,149]. Some of the alleles are highly associated with red hair, fair skin, and poor tanning ability, and their expression is necessary but insufficient for red hair phenotype. Recently, those allelic variants have been associated with increased risk for melanoma and nonmelanoma skin cancer [150,151]. The allelic variants that are highly associated with red-hair phenotype and increased risk for skin cancers, namely Arg151Cys, Arg160Trp, and Asp294His, represent loss of function mutations in the MC1R [152]. Experiments on cultured human melanocytes revealed that natural expression of these variants in the homozygous or compound heterozygous state renders melanocytes refractory to melanocortins and increases their susceptibility to UV-induced cytotoxicity, presumably by compromising their DNA repair capacity. The *MC1R* genotype modifies the risk for melanoma in families with CDKN2A mutations [153,154]. Expression of *Arg151Cys*, *Arg160Trp*, or *Asp294His MC1R* alleles increases the penetrance of and results in the early onset of melanoma in carriers of CDKN2A mutations.

18.12.5 MUTATIONS IN THE RAS/MAP KINASE PATHWAY AND MELANOMA

Ten to fifteen percent of all melanoma tumors have activation mutations in Ras. These mutations mainly correlate with nodular lesions and sun exposure [155,156]. The presence of Ras mutations in some melanoma tumors suggests the involvement of oncogene activation in the malignant transformation of melanocytes. The Ras–Raf–MEK–MAP kinase pathway is activated by many melanocyte-specific mitogens [8,157–159]. Recently, mutations in BRAF, one of the three members of the RAF family of genes that code for Ras-regulated cytoplasmic serine threonine kinases, were found in 66% of melanoma tumors tested [160]. One mutation, namely V599E, seems to be specific to melanoma, and is a somatic mutation that results in 11-fold increase in BRAF kinase activity [161]. This same mutation is also expressed in all forms of nevi [162], suggesting that it might be an early event in melanocyte transformation.

Genomic analysis of various types of melanoma tumors revealed that BRAF mutations were significantly more common in melanomas occurring on skin that is intermittently sun-exposed, and not on chronically sun-damaged skin, or completely unexposed skin, such as the palms and soles. These findings suggest distinct transformation pathways for different types of melanomas [163] (Table 18.3).

18.13 ANIMAL MODELS FOR SKIN CANCER

Animal models for skin cancer are a valuable tool that allows for the identification of skin cancer genes, their role in initiation, promotion, and progression of skin cancer tumors, gene–gene and gene–environment (particularly gene–UV) interactions. A mouse model for BCC is transgenic mice that over express SHH in the skin and develop features of basal cell nevus syndrome [164]. Furthermore, mice heterozygous for PTCH inactivating mutations develop BCC-like tumors when exposed to UV or ionizing radiation [165]. This mouse model demonstrates that SHH expression is sufficient for induction of BCC in mice.

TABLE 18.3

Genetic Alterations Commonly Associated with Various Forms of Skin Cancer

Gene	Basal cell carcinoma	Squamous cell carcinoma	Melanoma
p53	+	+	−
PTCH[a]	+	+	−
CDKN2A[a]	−	−	+
CDK4[a]	−	−	+
MC1R	+	+	+
Ras mutations/BRAF	−	−	+

[a] Germ-line mutations in these genes are associated with familial skin cancer. Except for MC1R, somatic mutations in the above genes have been identified in sporadic cancer tumors.

Several mouse models of melanoma have been described (reviewed in Chin [166]). Ink4a and Arf knockout mice that express activating Ha-ras mutation in melanocytes develop melanoma tumors with a short latency [167,168]. The role of p53 in melanoma tumors has been controversial due to the rare occurrence of UVB-signature mutations in the *p53* gene in these tumors. However, transgenic mice over expressing SV40T antigen were shown to develop ocular and skin melanoma tumors [169]. Additionally, p53 knockout mice expressing activating Ras mutation in their melanocytes develop melanoma tumors despite retaining functional Ink4a [170]. These mouse models underscore the significance of p16 and p53 in melanoma suppression.

Hepatocyte growth factor/scatter factor (HGF/SC) transgenic mice are an interesting and relatively new model for the study of the carcinogenic effects of UV exposure. HGF/SF stimulates the proliferation, motility, and invasiveness of many cell types, including melanocytes. Normally HGF acts as a paracrine factor, however, autocrine production of HGF and activation of its tyrosine kinase receptor c-Met have been shown in some melanomas, and are associated with metastasis [171–173]. HGF transgenic mice develop ectopic intrafollicular melanocytes, and melanoma tumors arise as the mice age [174]. Chronic exposure of these mice to suberythemal doses of UV results in SCC [175]. However, exposure of neonatal transgenic mice to a sunburning dose of UV results in melanoma tumors as well as SCC [174]. The development and progression of melanoma tumors in the neonatal HGF transgenic mice recapitulates to a large extent these processes in human melanoma, in that these tumors arise from pigmented lesions that are induced as an early event by UV, then progress into epidermal tumors with junctional activity [174]. This animal model emphasizes the significance of acute sun exposure during childhood in melanoma formation.

18.14 SUMMARY

Skin cancer represents a multiplicity of neoplasms that arise from the malignant transformation of keratinocytes — in the case of BCC and SCC, or epidermal melanocytes — in the case of melanoma. The prevalence of skin cancer has made it a major public health issue due to the life-threatening impact of melanoma and the high cost of treatment. Etiological factors for sporadic skin cancer include exposure to environmental solar UV radiation and arsenic, and immunosuppression. There are many risk factors for skin cancer, which include cutaneous pigmentation that affects the response of an individual to UV. For the most part, skin cancer is preventable, and tremendous emphasis has been placed on prevention and early detection. Skin cancer research has led to remarkable progress in elucidating the complex mechanism(s) of carcinogenesis. The accessibility of the skin and its direct interphase with the environment have made it an attractive model to investigate gene–gene and gene–environment interactions that impact genomic stability and lead to cancer formation. Many challenges remain to be overcome in skin cancer research, particularly in understanding the biology of melanoma. The quest for specific genetic markers for melanoma should lead to its prevention and to the development of more effective therapies to conquer this devastating cancer.

REFERENCES

1. Miller, D.L. and Weinstock, M.A. Non-melanoma skin cancer in the United States: Incidence. *J. Am. Acad. Dermatol.*, *30*, 774–778, 1994.
2. Mitchell, D.L., Jen, J., and Cleaver, J.E. Relative induction of cyclobutane dimers and cytosine photo-hydrates in DNA irradiated in vitro and in vivo with ultraviolet-C and ultraviolet-B light. *Photochem. Photobiol.*, *54*, 741–746, 1991.
3. Brash, D.E., Rudolph, J.A., Simon, J.A., Lin, A., McKenna, G.J., Baden, H.P., Halperin, A.J., and Pontén, J. A role for sunlight in skin cancer: UV-induced p53 mutations in squamous cell carcinoma. *Proc. Natl. Acad. Sci. USA*, *88*, 10124–10128, 1991.
4. Ziegler, A., Leffell, D.J., Kunala, S., Sharma, H.W., Gailani, M., Simon, J.A., Halperin, A.J., Baden, H.P., Shapiro, P.E., and Bale, A.E. Mutation hotspots due to sunlight in the p53 gene of nonmelanoma skin cancers. *Proc. Natl. Acad. Sci. USA*, *90*, 4216–4220, 1993.
5. Cleaver, J.E. Defective repair replication of DNA in xeroderma pigmentosum. *Nature*, *218*, 652–656, 1968.
6. Kraemer, K.H., Lee, M.M., and Scotto, J. Xeroderma pigmentosum: cutaneous, ocular and neurologic abnormalities in 830 published cases. *Arch. Dermatol.*, *123*, 241–250, 1987.
7. Nomura, M., Kaji, A., Ma, W.-Y., Zhong, S., Liu, G., Bowden, T., Miyamoto, K.-I., and Dong, Z. Mitogen- and stress-activated protein kinase-1 mediates activation of Akt by ultraviolet B irradiation. *J. Biol. Chem.*, *276*, 25558–25567, 2001.
8. Tada, A., Pereira, E., Beitner Johnson, D., Kavanagh, R., and Abdel-Malek, Z.A. Mitogen and ultraviolet-B-induced signaling pathways in normal human melanocytes. *J. Invest. Dermatol.*, *118*, 316–322, 2002.
9. Halaban, R., Langdon, R., Birchall, N., Cuono, C., Baird, A., Scott, G., Moellmann, G., and McGuire, J. Basic fibroblast growth factor from human keratinocytes is a natural mitogen for melanocytes. *J. Cell Biol.*, *107*, 1611–1619, 1988.
10. Imokawa, G., Yada, Y., and Miyagishi, M. Endothelins secreted from human keratinocytes are intrinsic mitogens for human melanocytes. *J. Biol. Chem.*, *267*, 24675–24680, 1992.
11. Wakamatsu, K., Graham, A., Cook, D., and Thody, A.J. Characterization of ACTH peptides in human skin and their activation of the melanocortin-1 receptor. *Pigment Cell Res.*, *10*, 288–297, 1997.
12. Wan, Y.S., Wang, Z.Q., Shao, Y., Voorhees, J.J., and Fisher, G.J. Ultraviolet irradiation activates PI 3-kinase/AKT survival pathway via EGF receptors in human skin in vivo. *Int. J. Oncol.*, *18*, 461–466, 2001.
13. Epstein, J.H. Photocarcinogenesis, skin cancer and aging. *J. Am. Acad. Dermatol.*, *9*, 487–502, 1983.
14. Albert, L.S., Rhodes, A.R., and Sober, A.J. Dysplastic melanocytic nevi and cutaneous melanoma: Markers of increased melanoma risk for affected persons and blood relatives. *J. Am. Acad. Dermatol.*, *22*, 69–75, 1990.
15. Donawho, C. and Wolf, P. Sunburn, sunscreen, and melanoma. *Curr. Opin. Oncol.*, *8*, 159–166, 1996.
16. Gallagher, R.P., McLean, D.I., Yang, C.P., Coldman, A.J., Silver, H.K., Spinelli, J.J., and Beagrie, M. Suntan, sunburn, and pigmentation factors and the frequency of acquired melanocytic nevi in children. Similarities to melanoma: The Vancouver Mole Study. *Arch. Dermatol.*, *126*, 770–776, 1990.
17. Ruiz-Maldonado, R. and Orozco-Covarrubias, M.L. Malignant melanoma in children. A review. *Arch. Dermatol.*, *133*, 363–371, 1997.
18. Gilchrest, B.A., Eller, M.S., Geller, A.C., and Yaar, M. The pathogenesis of melanoma induced by ultraviolet radiation. *N. Engl. J. Med.*, *340*, 1341–1348, 1999.
19. Sober, A.J., Lew, R.A., Koh, H.K., and Barnhill, R.L. Epidemiology of cutaneous melanoma. An update. *Dermatol. Clin.*, *9*, 617–629, 1991.
20. Pathak, M.A. Ultraviolet radiation and the development of non-melanoma and melanoma skin cancer: Clinical and experimental evidence. *Skin Pharmacol.*, *4* (Suppl.), 85–94, 1991.

21. Halder, R.M. and Bridgeman-Shah, S. Skin cancer in African Americans. *Cancer*, 75, 667–673, 1995.
22. Kricker, A., Armstrong, B.K., and English D.R. Sun exposure and non-melanocytic skin cancer. *Cancer Causes Control*, 5, 367–392, 1994.
23. Preston, D.S. and Stern, R.S. Nonmelanoma cancers of the skin. *N. Engl. J. Med.*, 327, 1649–1662, 1992.
24. *Forty-Five Years of Cancer Incidence in Connecticut: 1935–79. National Cancer Institute Monograph 70.* Government Printing Office: Washington, DC, 1986.
25. MacLennan, R., Green, A.C., McLeod, G.R.C., and Martin, N.G. Increasing incidence of cutaneous melanoma in Queensland, Australia. *J. Natl. Cancer Inst.*, 84, 1427–1432, 1992.
26. Pathak, M.A. and Fitzpatrick, T.B. The role of natural photoprotective agents in human skin. In *Sunlight and Man*, Fitzpatrick, T.B., Pathak, M.A., Haber, L.C., Seiji, M., and Kukita, A., Eds., University of Tokyo Press: Tokyo, 1974, pp. 725–750.
27. Kaidbey, K.H., Poh Agin, P., Sayre, R.M., and Kligman, A.M. Photoprotection by melanin — a comparison of black and Caucasian skin. *J. Am. Acad. Dermatol.*, 1, 249–260, 1979.
28. Pathak, M.A., Hori, Y., Szabo, G., and Fitzpatrick, T.B. The photobiology of melanin pigmentation in human skin. In *Biology of Normal and Abnormal Melanocytes*, Kawamura, T., Fitzpatrick, T.B., and Seiji, M., Eds., University Park Press, Baltimore, MD, 1971, pp. 149–169.
29. Kobayashi, N., Nakagawa, A., Muramatsu, T., Yamashina, Y., Shirai, T., Hashimoto, M.W., Ishigaki, Y., Ohnishi, T., and Mori, T. Supranuclear melanin caps reduce ultraviolet induced DNA photoproducts in human epidermis. *J. Invest. Dermatol.*, 110, 806–810, 1998.
30. Szabo, G. Racial differences in the fate of melanosomes in human epidermis. *Nature*, 222, 1081, 1969.
31. Thody, A.J., Higgins, E.H., Wakamatzu, K., Burchill, S.A., and Marks, J.M. Pheomelanin as well as eumelanin is present in human epidermis. *J. Invest. Dermatol.*, 97, 340–344, 1991.
32. Hunt, G., Kyne, S., Ito, S., Wakamatsu, K., Todd, C., and Thody, A.J. Eumelanin and pheomelanin contents of human epidermis and cultured melanocytes. *Pigment Cell Res.*, 8, 202–208, 1995.
33. Pathak, M.A., Jimbow, K., and Fitzpatrick, T. Photobiology of pigment cells. In *Phenotypic Expression in Pigment Cells*, Seiji, M., Ed., University of Tokyo Press, Tokyo, Japan, 1980, pp. 655–670.
34. Kadekaro, A.L., Kanto, H., Kavanagh, R.J., Schwemberger, S., Babcock, G., and Abdel-Malek, Z.A. A novel role for the paracrine factors endothelin-1 and alpha-melanocortin as survival factors for human melanocytes. *Pigment Cell Res.*, 16, 416, 2003.
35. Felix, C.C., Hyde, J.S., Sarna, T., and Sealy, R.C. Melanin photoreaction in aerated media: Electron spin resonance evidence of production of superoxide and hydrogen peroxide. *Biochem. Biophys. Res. Commun.*, 84, 335–341, 1978.
36. Bustamante, J., Bredeston, L., Malanga, G., and Mordoh, J. Role of melanin as a scavenger of active oxygen species. *Pigment Cell Res.*, 6, 348–353, 1993.
37. Chedekel, M.R., Smith, S.K., Post, P.W., Pokora, A., and Vessell, D.L. Photodestruction of pheomelanin: Role of oxygen. *Proc. Natl. Acad. Sci. USA*, 75, 5395–5399, 1978.
38. Menon, I.A., Persad, S., Ranadive, N.S., and Haberman, H.F. Photobiological effects of eumelanin and pheomelanin. In *Biological, Molecular and Clinical Aspects of Pigmentation*, Bagnara, J.T., Klaus, S.N., Paul, E., and Schartl, M., Eds., University of Tokyo Press: Tokyo, 1985, pp. 77–86.
39. Sarna, T. Properties and function of the ocular melanin — a photobiophysical view. *J. Photochem. Photobiol. B*, 12, 215–258, 1992.
40. Leonard, A. and Lauwerys, R.R. Carcinogenicity, teratogenicity and mutagenicity of arsenic. *Mutat. Res.*, 75, 49–62, 1980.
41. Pershagen, G. The carcinogenicity of arsenic. *Environ. Health Perspect.*, 40, 93–100, 1981.
42. Frost, D.V. Arsenicals in biology — retrospect and prospect. *Fed. Proc.*, 26, 194–208, 1967.
43. Shannon, R.L. and Strayer, D.S. Arsenic-induced skin toxicity. *Hum. Toxicol.*, 8, 99–104, 1989.
44. Fischer, A.B., Buchet, J.P., and Lauwerys, R.R. Arsenic uptake, cytotoxicity and detoxification studied in mammalian cells in culture. *Arch. Toxicol.*, 57, 168–172, 1985.
45. Tseng, W.P., Chu, H.M., How, S.W., Fong, J.M., Lin, C.S. and Yeh, S. Prevalence of skin cancer in an endemic area of chronic arsenicism in Taiwan. *J. Natl. Cancer Inst.*, 40, 453–463, 1968.
46. Yeh, S., How, S.W., and Lin, C.S. Arsenical cancer of skin. Histologic study with special reference to Bowen's disease. *Cancer*, 21, 312–339, 1968.

47. Cebrian, M., Albores, A., Aguilar, M., and Blakely, E. Chronic arsenic poisoning in the north of Mexico. *Hum. Toxicol.*, 2, 121–133, 1983.
48. Yeh, S. Skin cancer in chronic arsenicism. *Hum. Pathol.*, 4, 469–485, 1973.
49. Hsu, C.-H., Yang, S.-A., Wang, J.-Y., Yu, H.-.S, and Lin, S.-R. Mutational spectrum of p53 gene in arsenic-related skin cancers from the blackfoot disease endemic area of Taiwan. *Br. J. Cancer*, 80, 1080–1086, 1999.
50. Matsui, M., Nishigori, C., Toyokuni, S., Takada, J., Akaboshi, M., Ishikawa, M., Imamura, S., and Miyachi, Y. The role of oxidative DNA damage in human arsenic carcinogenesis: Detection of 8-hydroxy-2′-deoxyguanosine in arsenic-related Bowen's disease. *J. Invest. Dermatol.*, 113, 26–31, 1999.
51. Rossman, T.G., Uddin, A.N., Burns, F.J., and Bosland, M.C. Arsenite is a cocarcinogen with solar ultraviolet radiation for mouse skin: An animal model for arsenic carcinogenesis. *Toxicol. Appl. Pharmacol.*, 176, 64–71, 2001.
52. Hartwig, A., Groblinghoff, U.D., Beyersmann, D., Natarajan, A.T., Filon, R., and Mullenders, L.H.F. Interaction of arsenic(III) with nucleotide excision repair in UV-irradiated human fibroblasts. *Carcinogenesis*, 18, 399–405, 1997.
53. Andrew, A.S., Karagas, M.R., and Hamilton, J.W. Decreased DNA repair gene expression among individuals exposed to arsenic in United States drinking water. *Int. J. Cancer*, 104, 263–268, 2003.
54. Fisher, M.S. and Kripke, M.L. Systemic alteration induced in mice by ultraviolet light irradiation and its relationship to ultraviolet carcinogenesis. *Proc. Natl. Acad. Sci. USA*, 74, 1688–1692, 1977.
55. Noonan, F.P., Simon, J.D., Hart, P.H., Garssen, J., and De Fabo, E.C. Urocanic acid photobiology. In *Photobiology for the 21st Century*, Coohill, T.P. and Valenzeno, D.P., Eds., Valdenmar Publishing Company, Overland Park, KS, 2001, pp. 69–79.
56. Finlay-Jones, J.J. and Hart, P.H. Ultraviolet irradiation, systemic immunosuppression and skin cancer: Role of urocanic acid. *Australas. J. Dermatol.*, 38 (Suppl.), S7–S12, 1997.
57. Aubin, F. Mechanisms involved in ultraviolet light-induced immunosuppression. *Eur. J. Dermatol.*, 13, 515–523, 2003.
58. Meyer, T., Arndt, R., Nindl, I., Ulrich, C., Christophers, E., and Stockfleth, E. Association of human papillomavirus infections with cutaneous tumors in immunosuppressed patients. *Transpl. Int.*, 16, 146–153, 2003.
59. McCormack, C.J., Kelly, J.W., and Dorevitch, A.P. Differences in age and body site distribution of the histological subtypes of basal cell carcinoma. A possible indicator of differing causes. *Arch. Dermatol.*, 133, 593–596, 1997.
60. Bastiaens, M.T., Hoefnagel, J.J., Bruijn, J.A., Westendorp, R.G., Vermeer, B.J., and Bouwes Bavinck, J.N. Differences in age, site distribution, and sex between nodular and superficial basal cell carcinoma indicate different types of tumors. *J. Invest. Dermatol.*, 110, 880–884, 1998.
61. Mora, R.G., Perniciaro, C., and Lee, B. Cancer of the skin in blacks. III. A review of nineteen black patients with Bowen's disease. *J. Am. Acad. Dermatol.*, 11, 557–562, 1984.
62. Graham, J.H. and Helwig, E.B. Bowen's disease and its relationship to systemic cancer. *AMA Arch. Dermatol.*, 80, 133–159, 1959.
63. Glogau, R.G. The risk of progression to invasive disease. *J. Am. Acad. Dermatol.*, 42, 23–24, 2000.
64. Bentham, G. and Aase, A. Incidence of malignant melanoma of the skin in Norway, 1955–1989: Associations with solar ultraviolet radiation, income and holidays abroad. *Int. J. Epidemiol.*, 25, 1132–1138, 1996.
65. Nelemans, P.J., Groenendal, H., Kiemeney, L.A., Rampen, F.H., Ruiter, D.J., and Verbeek, A.L. Effect of intermittent exposure to sunlight on melanoma risk among indoor workers and sun-sensitive individuals. *Environ. Health Perspec.*, 101, 252–255, 1993.
66. Holman, C.D.J., Armstrong, B.K., and Heenan, P.J. Relationship of cutaneous malignant melanoma to individual sunlight-exposure habits. *J. Natl. Cancer Inst.*, 76, 403–414, 1986.
67. Weinstock, M.A. Controversies in the role of sunlight in the pathogenesis of cutaneous melanoma. *Photochem. Photobiol.*, 63, 406–410, 1996.
68. Setlow, R.B., Grist, E., Thompson, K., and Woodhead, A.D. Wavelengths effective in induction of malignant melanoma. *Proc. Natl. Acad. Sci. USA*, 90, 6666–6670, 1993.

69. Drobetsky, E.A., Turcotte, J., and Chateauneuf, A. A role for ultraviolet A in solar mutagenesis. *Proc. Natl. Acad. Sci. USA*, *92*, 2350–2354, 1995.
70. Dillman, A.M. Photobiology of skin pigmentation. In *Pigmentation and Pigmentary Disorders*, Levine, N., Ed., CRC Press, Boca Raton, FL, 1993, pp. 61–94.
71. Beitner, H. Clinical and experimental aspects of long-wave ultraviolet (UVA) irradiation of human skin. *Acta Derm. Venereol. Suppl. (Stockh)*, *123*, 1–56, 1986.
72. Gange, R.W. and Rosen, C.F. UVA effects on mammalian skin and cells. *Photochem. Photobiol.*, *43*, 701–705, 1986.
73. Kvam, E. and Tyrrell, R.M. Induction of oxidative DNA base damage in human skin cells by UV and near visible radiation. *Carcinogenesis*, *18*, 2379–2384, 1997.
74. Yohn, J.J., Norris, D.A., Yrastorza, D.G., Bruno, I.J., Leff, J.A., Hake, S.S., and Repine, J.E. Disparate antioxidant enzyme activities in cultured human cutaneous fibroblasts, keratinocytes, and melanocytes. *J. Invest. Dermatol.*, *97*, 405–409, 1991.
75. Setlow, R.B. and Woodhead, A.D. Temporal changes in the incidence of malignant melanoma: Explanation from action spectra. *Mutat. Res.*, *307*, 365–374, 1994.
76. Ley, R.D. Ultraviolet radiation A-induced precursors of cutaneous melanoma in *Monodelphis domestica*. *Cancer Res.*, *57*, 3682–3684, 1997.
77. Lubbe, J., Reichel, M., Burg, G., and Kleihues, P. Absence of p53 gene mutations in cutaneous melanoma. *J. Invest. Dermatol.*, *102*, 819–821, 1994.
78. Montano, X., Shamsher, M., Whitehead, P., Dawson, K., and Newton, J. Analysis of p53 in human cutaneous melanoma cell lines. *Oncogene*, *9*, 1455–1459, 1994.
79. Weinstock, M.A. Early detection of melanoma. *JAMA*, *284*, 886–889, 2000.
80. Kanzler, M.H. and Mraz-Gernhard, S. Primary cutaneous malignant melanoma and its precursor lesions: Diagnostic and therapeutic overview. *J. Am. Acad. Dermatol.*, *45*, 260–276, 2001.
81. Elwood, J.M. and Jopson, J. Melanoma and sun exposure: An overview of published studies. *Int. J. Cancer*, *73*, 198–203, 1997.
82. Cohen, L.M. Lentigo maligna and lentigo maligna melanoma. *J. Am. Acad. Dermatol.*, *33*, 923–936; quiz 937–940, 1995.
83. Yuspa, S.H. and Poirier, M.C. Chemical carcinogenesis: From animal models to molecular models in one decade. *Adv. Cancer Res.*, *50*, 25–70, 1988.
84. Yuspa, S.H. The pathogenesis of squamous cell cancer: Lessons learned from studies of skin carcinogenesis — thirty-third G.H.A. Clowes Memorial Award Lecture. *Cancer Res.*, *54*, 1178–1189, 1994.
85. Quintanilla, M., Brown, K., Ramsden, M., and Balmain, A. Carcinogen-specific mutation and amplification of Ha-ras during mouse skin carcinogenesis. *Nature*, *322*, 78–80, 1986.
86. Brown, K., Quintanilla, M., Ramsden, M., Kerr, I.B., Young, S., and Balmain, A. v-ras genes from Harvey and BALB murine sarcoma viruses can act as initiators of two-stage mouse skin carcinogenesis. *Cell*, *46*, 447–456, 1986.
87. Roop, D.R., Lowy, D.R., Tambourin, P.E., Strickland, J., Harper, J.R., Balaschak, M., Spangler, E.F., and Yuspa, S.H. An activated Harvey ras oncogene produces benign tumours on mouse epidermal tissue. *Nature*, *323*, 822–824, 1986.
88. van der Schroeff, J.G., Evers, L.M., Boot, A.J., and Bos, J.L. Ras oncogene mutations in basal cell carcinomas and squamous cell carcinomas of human skin. *J. Invest. Dermatol.*, *94*, 423–425, 1990.
89. Pierceall, W.E., Goldberg, L.H., Tainsky, M.A., Mukhopadhyay, T., and Ananthaswamy, H.N. Ras gene mutation and amplification in human nonmelanoma skin cancers. *Mol. Carcinog.*, *4*, 196–202, 1991.
90. Knudson, A.G., Jr. Genetics of human cancer. *Genetics*, *79* (Suppl.), 305–316, 1975.
91. Nowell, P.C. The clonal evolution of tumor cell populations. *Science*, *194*, 23–28, 1976.
92. Deamant, F.D. and Iannaccone, P.M. Clonal origin of chemically induced papillomas: Separate analysis of epidermal and dermal components. *J. Cell Sci.*, *88*, 305–312, 1987.
93. Iannaccone, P.M., Weinberg, W.C., and Deamant, F.D. On the clonal origin of tumors: A review of experimental models. *Int. J. Cancer*, *39*, 778–784, 1987.
94. Yuspa, S.H., Ben, T., Hennings, H., and Lichti, U. Phorbol ester tumor promoters induce epidermal transglutaminase activity. *Biochem. Biophys. Res. Commun.*, *97*, 700–708, 1980.
95. Hennings, H., Shores, R.A., Poirier, M.C., Reed, E., Tarone, R.E., and Yuspa, S.H. Enhanced malignant conversion of benign mouse skin tumors by cisplatin. *J. Natl. Cancer Inst.*, *82*, 836–840, 1990.

96. Tennenbaum, T., Weiner, A.K., Belanger, A.J., Glick, A.B., Hennings, H., and Yuspa, S.H. The suprabasal expression of alpha 6 beta 4 integrin is associated with a high risk for malignant progression in mouse skin carcinogenesis. *Cancer Res.*, *53*, 4803–4810, 1993.

97. Imamoto, A., Beltran, L.M., and DiGiovanni, J. Evidence for autocrine/paracrine growth stimulation by transforming growth factor-alpha during the process of skin tumor promotion. *Mol. Carcinog.*, *4*, 52–60, 1991.

98. Glick, A.B., Kulkarni, A.B., Tennenbaum, T., Hennings, H., Flanders, K.C., O'Reilly, M., Sporn, M.B., Karlsson, S., and Yuspa, S.H. Loss of expression of transforming growth factor beta in skin and skin tumors is associated with hyperproliferation and a high risk for malignant conversion. *Proc. Natl. Acad. Sci. USA*, *90*, 6076–6080, 1993.

99. Sawey, M.J., Goldschmidt, M.H., Risek, B., Gilula, N.B., and Lo, C.W. Perturbation in connexin 43 and connexin 26 gap-junction expression in mouse skin hyperplasia and neoplasia. *Mol. Carcinog.*, *17*, 49–61, 1996.

100. Gimenez-Conti, I., Aldaz, C.M., Bianchi, A.B., Roop, D.R., Slaga, T.J., and Conti, C.J. Early expression of type I K13 keratin in the progression of mouse skin papillomas. *Carcinogenesis*, *11*, 1995–1999, 1990.

101. Larcher, F., Bauluz, C., Diaz-Guerra, M., Quintanilla, M., Conti, C.J., Ballestin, C., and Jorcano, J.L. Aberrant expression of the simple epithelial type II keratin 8 by mouse skin carcinomas but not papillomas. *Mol. Carcinog.*, *6*, 112–121, 1992.

102. Bianchi, A.B., Fischer, S.M., Robles, A.I., Rinchik, E.M., and Conti, C.J. Overexpression of cyclin D1 in mouse skin carcinogenesis. *Oncogene*, *8*, 1127–1133, 1993.

103. Peto, R., Roe, F.J., Lee, P.N., Levy, L., and Clack, J. Cancer and ageing in mice and men. *Br. J. Cancer*, *32*, 411–426, 1975.

104. Loeb, L.A. Mutator phenotype may be required for multistage carcinogenesis. *Cancer Res.*, *51*, 3075–3079, 1991.

105. Weinberg, R.A. The molecular basis of oncogenes and tumor suppressor genes. *Ann. N. Y. Acad. Sci.*, *758*, 331–338, 1995.

106. Dumaz, N., van Kranen, H.J., de Vries, A., Berg, R.J.W., Wester, P.W., van Kreijl, C.F., Sarasin, A., Daya-Grosjean, L., and de Gruijl, F.R. The role of UV-B light in skin carcinogenesis through the analysis of {lp53} mutations in squamous cell carcinomas of hairless mice. *Carcinogenesis*, *18*, 897–904, 1997.

107. Jiang, W., Ananthaswamy, H.N., Muller, H.K., and Kripke, M.L. p53 protects against skin cancer induction by UV-B radiation. *Oncogene*, *18*, 4247–4253, 1999.

108. Kripke, M.L. Ultraviolet radiation and immunology: Something new under the sun–presidential address. *Cancer Res.*, *54*, 6102–6105, 1994.

109. Ziegler, A., Jonason, A.S., Leffell, D.J., Simon, J.A., Sharma, H.W., Kimmelman, J., Remington, L., Jacks, T., and Brash, D.E. Sunburn and p53 in the onset of skin cancer. *Nature*, *372*, 773–776, 1994.

110. Aubin, F., Donawho, C.K., and Kripke, M.L. Effect of psoralen plus ultraviolet A radiation on in vivo growth of melanoma cells. *Cancer Res.*, *51*, 5893–5897, 1991.

111. Jonason, A.S., Kunala, S., Price, G.J., Restifo, R.J., Spinelli, H.M., Persing, J.A., Leffell, D.J., Tarone, R.E., and Brash, D.E. Frequent clones of p53-mutated keratinocytes in normal human skin. *Proc. Natl. Acad. Sci. USA*, *93*, 14025–14029, 1996.

112. Kastan, M.B., Onyekwere, O., Sidransky, D., Vogelstein, B., and Craig, R.W. Participation of p53 protein in the cellular response to DNA damage. *Cancer Res.*, *51*, 6304–6311, 1991.

113. Levine, A.J. p53, the cellular gatekeeper for growth and division. *Cell*, *88*, 323–331, 1997.

114. Maltzman, W. and Czyzyk, L. UV irradiation stimulates levels of p53 cellular tumor antigen in nontransformed mouse cells. *Mol. Cell Biol.*, *4*, 1689–1694, 1984.

115. Barker, D., Dixon, K., Medrano, E.E., Smalara, D., Im, S., Mitchell, D., Babcock, G., and Abdel-Malek, Z.A. Comparison of the responses of human melanocytes with different melanin contents to ultraviolet B irradiation. *Cancer Res.*, *55*, 4041–4046, 1995.

116. Haupt, Y., Maya, R., Kazaz, A., and Oren, M. Mdm2 promotes the rapid degradation of p53. *Nature*, *387*, 296–299, 1997.

117. Kubbutat, M.H., Jones, S.N., and Vousden, K.H. Regulation of p53 stability by Mdm2. *Nature*, *387*, 299–303, 1997.

118. Sakaguchi, K., Herrera, J.E., Saito, S., Miki, T., Bustin, M., Vassilev, A., Anderson, C.W., and Appella, E. DNA damage activates p53 through a phosphorylation-acetylation cascade. *Genes Dev.*, *12*, 2831–2841, 1998.

119. Ashcroft, M., Kubbutat, M.H.G., and Vousden, K.H. Regulation of p53 function and stability by phosphorylation. *Mol. Cell Biol.*, *19*, 1751–1758, 1999.

120. Jimenez, G.S., Khan, S.H., Stommel, J.M., and Wahl, G.M. p53 regulation by post-translational modification and nuclear retention in response to diverse stresses. *Oncogene*, *18*, 7656–7665, 1999.

121. El-Deiry, W.S., Tokino, T., Velculescu, V.E., Levy, D.B., Parsons, R., Trent, J.M., Lin, D., Mercer, W.E., Kinzler, K.W., and Vogelstein, B. WAF1, a potential mediator of p53 tumor suppression. *Cell*, *75*, 817–825, 1993.

122. Smith, M.L., Chen, I.-T., Zhan, Q., O'Connor, P.M., and Fornace, A.J., Jr. Involvement of the p53 tumor suppressor in repair of UV-type DNA damage. *Oncogene*, *10*, 1053–1059, 1995.

123. Bates, S. and Vousden, K.H. p53 in signaling checkpoint arrest or apoptosis. *Curr. Opin. Genet. Dev.*, *6*, 12–19, 1996.

124. Hahn, H., Wicking, C., Zaphiropoulos, P.G., Gailani, M.R., Shanley, S., Chidambaram, A., Vorechovsky, I., Holmbeg, E., Unden, A.B., Gillies, S., Negus, K., Smyth, I., Pressman, C., Leffell, D.J., Gerrard, B., Goldstein, A.M., Dean, M., Toftgard, R., Chenevix-Trench, G., Wainwright, B., and Bale, A.E. Mutations of the human homolog of Drosophila *patched* in the nevoid basal cell carcinoma syndrome. *Cell*, *85*, 841–851, 1996.

125. Wicking, C., Shanley, S., Smyth, I., Gillies, S., Negus, K., Graham, S., Suthers, G., Haites, N., Edwards, M., Wainwright, B., and Chenevix-Trench, G. Most germ-line mutations in the nevoid basal cell carcinoma syndrome lead to a premature termination of the PATCHED protein, and no genotype–phenotype correlations are evident. *Am. J. Hum. Genet.*, *60*, 21–26, 1997.

126. Gailani, M.R., Stahle-Backdahl, M., Leffell, D.J., Glynn, M., Zaphiropoulos, P.G., Pressman, C., Unden, A.B., Dean, M., Brash, D.E., Bale, A.E., and Toftgard, R. The role of the human homologue of *Drosophila patched* in sporadic basal cell carcinomas. *Nat. Genet.*, *14*, 78–81, 1996.

127. Aszterbaum, M., Rothman, A., Johnson, R.L., Fisher, M., Xie, J., Bonifas, J.M., Zhang, X., Scott, M.P., and Epstein, E.H., Jr. Identification of mutations in the human *PATCHED* gene in sporadic basal cell carcinomas and in patients with basal cell nevus syndrome. *J. Invest. Dermatol.*, *110*, 885–888, 1998.

128. Ping, X.L., Ratner, D., Zhang, H., Wu, X.L., Zhang, M.J., Chen, F.F., Silvers, D.N., Peacocke, M., and Tsou, H.C. *PTCH* Mutations in squamous cell carcinoma of the skin. *J. Invest. Dermatol.*, *116*, 614–616, 2001.

129. Unden, A.B., Holmberg, E., Lundh-Rozell, B., Stahle-Backdahl, M., Zaphiropoulos, P.G., Toftgard, R., and Vorechovsky, I. Mutations in the human homologue of *Drosophila patched PTCH* in basal cell carcinomas and the Gorlin syndrome: different *in vivo* mechanisms of *PTCH* inactivation. *Cancer Res.*, *56*, 4562–4565, 1996.

130. Denef, N., Neubuser, D., Perez, L., and Cohen, S.M. Hedgehog induces opposite changes in turnover and subcellular localization of patched and smoothened. *Cell*, *102*, 521–531, 2000.

131. Bale, A.E. and Yu, K.-P. The hedgehog pathway and basal cell carcinomas. *Hum. Mol. Genet.*, *10*, 757–762, 2001.

132. Fan, H., Oro, A.E., Scott, M.P., and Khavari, P.A. Induction of basal cell carcinoma features in transgenic human skin expressing Sonic Hedgehog. *Nat. Med.*, *3*, 788–792, 1997.

133. Fan, H. and Khavari, P.A. Sonic hedgehog opposes epithelial cell cycle arrest. *J. Cell Biol.*, *147*, 71–76, 1999.

134. Piepkorn, M. Melanoma genetics: An update with focus on the CDKN2A(p16)/ARF tumor suppressors. *J. Am. Acad. Dermatol.*, *42*, 705–722, 2000.

135. Dracopoli, N.C. and Fountain, J.W. CDKN2 mutations in melanoma. *Cancer Surv.*, *26*, 115–132, 1996.

136. Pollock, P.M., Yu, F., Qiu, L., Parsons, P.G., and Hayward, N.K. Evidence for UV induction of *CDKN2* mutations in melanoma cell lines. *Oncogene*, *11*, 663–668, 1995.

137. Flores, J.F., Walker, G.J., Glendening, J.M., Haluska, F.G., Castresana, J.S., Rubio, M.P., Pastorfide, G.C., Boyer, L.A., Kao, W.H., Bulyk, M.L., Barnhill, R.L., Hayward, N.K., Housman, D.E., and Fountain, J.W. Loss of the p16INK4a and p15INK4b genes, as well as neighboring 9p21 markers, in sporadic melanoma. *Cancer Res.*, *56*, 5023–5032, 1996.

138. Stone, S., Jiang, P., Dayananth, P., Tavtigian, S.V., Katcher, H., Parry, D., Peters, G., and Kamb, A. Complex structure and regulation of the P16 (MTS1) locus. *Cancer Res.*, *55*, 2988–2994, 1995.

139. Mao, L., Merlo, A., Bedi, G., Shapiro, G.I., Edwards, C.D., Rollins, B.J., and Sidransky, D. A novel p16INK4A transcript. *Cancer Res.*, *55*, 2995–2997, 1995.

140. Hall, M., Bates, S., and Peters, G. Evidence for different modes of action of cyclin-dependent kinase inhibitors: p15 and p16 bind to kinases, p21 and p27 bind to cyclins. *Oncogene*, *11*, 1581–1588, 1995.

141. Coleman, K.G., Wautlet, B.S., Morrissey, D., Mulheron, J., Sedman, S.A., Brinkley, P., Price, S., and Webster, K.R. Identification of CDK4 sequences involved in cyclin D1 and p16 binding. *J. Biol. Chem.*, *272*, 18869–18874, 1997.

142. Pomerantz, J., Schreiber-Agus, N., Liegeois, N.J., Silverman, A., Alland, L., Chin, L., Potes, J., Chen, K., Orlow, I., Lee, H.W., Cordon-Cardo, C., and DePinho, R.A. The Ink4a tumor suppressor gene product, p19Arf, interacts with MDM2 and neutralizes MDM2's inhibition of p53. *Cell*, *92*, 713–723, 1998.

143. De Stanchina, E., McCurrach, M.E., Zindy, F., Shieh, S.Y., Ferbeyre, G., Samuelson, A.V., Prives, C., Roussel, M.F., Sherr, C.J., and Lowe, S.W. E1A signaling to p53 involves the p19(ARF) tumor suppressor. *Genes Dev.*, *12*, 2434–2442, 1998.

144. Box, N.F., Wyeth, J.R., O'Gorman, L.E., Martin, N.G., and Sturm, R.A. Characterization of melanocyte stimulating hormone receptor variant alleles in twins with red hair. *Hum. Mol. Genet.*, *6*, 1891–1897, 1997.

145. Smith, R., Healy, E., Siddiqui, S., Flanagan, N., Steijlen, P.M., Rosdahl, I., Jacques, J.P., Rogers, S., Turner, R., Jackson, I.J., Birch-Machin, M.A., and Rees, J.L. Melanocortin 1 receptor variants in Irish population. *J. Invest. Dermatol.*, *111*, 119–122, 1998.

146. Mountjoy, K.G., Robbins, L.S., Mortrud, M.T., and Cone, R.D. The cloning of a family of genes that encode the melanocortin receptors. *Science*, *257*, 1248–1251, 1992.

147. Suzuki, I., Cone, R., Im, S., Nordlund, J., and Abdel-Malek, Z. Binding capacity and activation of the MC1 receptors by melanotropic hormones correlate directly with their mitogenic and melanogenic effects on human melanocytes. *Endocrinology*, *137*, 1627–1633, 1996.

148. Hunt, G., Kyne, S., Wakamatsu, K., Ito, S., and Thody, A.J. Nle[4]DPhe[7] α-melanocyte-stimulating hormone increases the eumelanin: Phaeomelanin ratio in cultured human melanocytes. *J. Invest. Dermatol.*, *104*, 83–85, 1995.

149. Harding, R.M., Healy, E., Ray, A.J., Ellis, N.S., Flanagan, N., Todd, C., Dixon, C., Sajantila, A., Jackson, I.J., Birch-Machin, M.A., and Rees, J.L. Evidence for variable selective pressures at MC1R. *Am. J. Hum. Genet.*, *66*, 1351–1361, 2000.

150. Palmer, J.S., Duffy, D.L., Box, N.F., Aitken, J.F., O'Gorman, L.E., Green, A.C., Hayward, N.K., Martin, N.G., and Sturm, R.A. Melanocortin-1 receptor polymorphisms and risk of melanoma: Is the association explained solely by pigmentation phenotype? *Am. J. Hum. Genet.*, *66*, 176–186, 2000.

151. Bastiaens, M.T., ter Huurne, J.A.C., Kielich, C., Gruis, N.A., Westendorp, R.G.J., Vermeer, B.J., and Bouwes Bavinck, J.N. Melanocortin-1 receptor gene variants determine the risk of non-melanoma skin cancer independent of fair skin type and red hair. *Am. J. Hum. Genet.*, *68*, 884–894, 2001.

152. Scott, M.C., Wakamatsu, K., Ito, S., Kadekaro, A.L., Kobayashi, N., Groden, J., Kavanagh, R., Takakuwa, T., Virador, V., Hearing, V.J., and Abdel-Malek, Z.A. Human *melanocortin 1 receptor* variants, receptor function and melanocyte response to ultraviolet radiation. *J. Cell Sci.*, *115*, 2349–2355, 2002.

153. Box, N.F., Duffy, D.L., Chen, W., Stark, M., Martin, N.G., Sturm, R.A., and Hayward, N.K. MC1R genotype modifies risk of melanoma in families segregating CDKN2A mutations. *Am. J. Hum. Genet.*, *69*, 765–773, 2001.

154. van der Velden, P.A., Sandkuijl, L.A., Bergman, W., Pavel, S., van Mourik, L., Frants, R.R., and Gruis, N.A. Melanocortin-1 receptor variant R151C modifies melanoma risk in Dutch families with melanoma. *Am. J. Hum. Genet.*, *69*, 774–779, 2001.

155. Jafari, M., Papp, T., Kirchner, S., Diener, U., Henschler, D., Burg, G., and Schiffmann, D. Analysis of ras mutations in human melanocytic lesions: Activation of the ras gene seems to be associated with the nodular type of human malignant melanoma. *J. Cancer Res. Clin. Oncol.*, *121*, 23–30, 1995.

156. van Elsas, A., Zerp, S.F., van der Flier, S., Kruse, K.M., Aarnoudse, C., Hayward, N.K., Ruiter, D.J., and Schrier, P.I. Relevance of ultraviolet-induced N-ras oncogene point mutations in development of primary human cutaneous melanoma. *Am. J. Pathol.*, *149*, 883–893, 1996.

157. Funasaka, Y., Boulton, T., Cobb, M., Yarden, Y., Fan, B., Lyman, S.D., Williams, D.E., Anderson, D.M., Zakut, R., Mishima, Y., and Halaban, R. c-Kit-Kinase induces a cascade of protein tyrosine phosphorylation in normal human melanocytes in response to mast cell growth factor and stimulates mitogen-activated protein kinase but is down-regulated in melanomas. *Mol. Biol. Cell*, *3*, 197–209, 1992.

158. Medrano, E.E., Yang, F., Boissy, R.E., Farooqui, J.Z., Shah, V., Matsumoto, K., Nordlund, J.J., and Park, H.-Y. Terminal differentiation and senescence in the human melanocyte: Repression of tyrosine-phosphorylation of the extracellular-signal regulated kinase 2 (ERK2) selectively defines the two phenotypes. *Mol. Biol. Cell*, *5*, 497–509, 1994.

159. Bohm, M., Moellmann, G., Cheng, E., Alvarez-Franco, M., Wagner, S., Sassone-Corsi, P. and Halaban, R. Identification of p90^{RSK} as the probable CREB-Ser^{133} kinase in human melanocytes. *Cell Growth Differ.*, *6*, 291–302, 1995.

160. Davies, H., Bignell, G.R., Cox, C., Stephens, P., Edkins, S., Clegg, S., Teague, J., Woffendin, H., Garnett, M.J., Bottomley, W., Davis, N., Dicks, E., Ewing, R., Floyd, Y., Gray, K., Hall, S., Hawes, R., Hughes, J., Kosmidou, V., Menzles, A., Mould, C., Parker, A., Stevens, C., Watt, S., Hooper, S., Wilson, R., Jayatilake, H., Gusterson, B.A., Cooper, C., Shipley, J., Hargrave, D., Pritchard-Jones, K., Maitland, N., Chenevix-Trench, G., Riggins, G.J., Bigner, D.D., Palmieri, G., Cossu, A., Flanagan, A., Nicholson, A., Ho, J.W.C., Leung, S.Y., Yuen, S.T., Weber, B.L., Seigler, H.F., Darrow, T.L., Paterson, H., Marais, R., Marshall, C.J., Wooster, R., Stratton, M.R., and Futreal, P.A. Mutations of the *BRAF* gene in human cancer. *Nature*, *417*, 949–954, 2002.

161. Brose, M.S., Volpe, P., Feldman, M., Kumar, M., Rishi, I., Gerrero, R., Einhorn, E., Herlyn, M., Minna, J., Nicholson, A., Roth, J.A., Albelda, S.M., Davies, H., Cox, C., Brignell, G., Stephens, P., Futreal, P.A., Wooster, R., Stratton, M.R., and Weber, B.L. *BRAF* and *RAS* mutations in human lung cancer and melanoma. *Cancer Res.*, *62*, 6997–7000, 2002.

162. Pollock, P.M., Harper, U.L., Hansen, K.S., Yudt, L.M., Stark, M., Robbins, C.M., Moses, T.Y., Hostetter, G., Wagner, U., Kakareka, J., Salem, G., Pohida, T., Heenan, P., Duray, P., Kallioniemi, O., Hayward, N.K., Trent, J.M., and Meltzer, P.S. High frequency of *BRAF* mutations in nevi. *Nat. Genet.*, *33*, 19–20, 2003.

163. Bastian, B.C. Genomic analyses of melanocytic neoplasms: Insights into biology and opportunities for classification. *Pigment Cell Res.*, *16* (Abstract), 421, 2003.

164. Oro, A.E., Higgins, K.M., Hu, Z., Bonifas, J.M., Epstein, E.H., Jr., and Scott, M.P. Basal cell carcinomas in mice overexpressing Sonic hedgehog. *Science*, *276*, 817–821, 1997.

165. Aszterbaum, M., Epstein, J., Oro, A., Douglas, V., LeBoit, P.E., Scott, M.P., and Epstein, E.H., Jr. Ultraviolet and ionizing radiation enhance the growth of BCCs and trichoblastomas in patched heterozygous knockout mice. *Nat. Med.*, *5*, 1285–1291, 1999.

166. Chin, L. The genetics of malignant melanoma: Lessons from mouse and man. *Nat. Rev. Cancer*, *3*, 559–570, 2003.

167. Chin, L., Pomerantz, J., Polsky, D., Jacobson, M., Cohen, C., Cordon-Cardo, C., Horner, J.W., II, and DePinho, R.A. Cooperative effects of INK4a and ras in melanoma susceptibility in vivo. *Genes Dev.*, *11*, 2822–2834, 1997.

168. Sharpless, N.E. and Chin, L. The *INK4a/ARF* locus and melanoma. *Oncogene*, *22*, 3092–3098, 2003.

169. Bradl, M., Klein-Szanto, A., Porter, S., and Mintz, B. Malignant melanoma in transgenic mice. *Proc. Natl. Acad. Sci. USA*, *88*, 164–168, 1991.

170. Bardeesy, N., Bastian, B.C., Hezel, A., Pinkel, D., DePinho, R.A., and Chin, L. Dual inactivation of RB and p53 pathways in RAS-induced melanomas. *Mol. Cell Biol.*, *21*, 2144–2153, 2001.

171. Halaban, R., Rubin, J.S., Funasaka, Y., Cobb, M., Boulton, T., Faletto, D., Rosen, E., Chan, A., Yoko, K., White, W., Cook, C., and Moellmann, G. Met and hepatocyte growth factor/scatter factor signal transduction in normal melanocytes and melanoma cells. *Oncogene*, *7*, 2195–2206, 1992.

172. Natali, P.G., Nicotra, M.R., Di Renzo, M.F., Prat, M., Bigotti, A., Cavaliere, R., and Comoglio, P.M. Expression of the c-Met/HGF receptor in human melanocytic neoplasms: Demonstration of

the relationship to malignant melanoma tumour progression. *Br. J. Cancer*, *68*, 746–750, 1993.

173. Li, G., Schaider, H., Satyamoorthy, K., Hanakawa, Y., Hashimoto, K., and Herlyn, M. Downregulation of E-cadherin and Desmoglein 1 by autocrine hepatocyte growth factor during melanoma development. *Oncogene*, *20*, 8125–8135, 2001.

174. Noonan, F.P., Recio, J.A., Takayama, H., Duray, P., Anver, M.R., Rush, W.L., De Fabo, E.C., and Merlino, G. Neonatal sunburn and melanoma in mice. *Nature*, *413*, 271–272, 2001.

175. Noonan, F.P., Otsuka, T., Bang, S., Anver, M.R., and Merlino, G. Accelerated ultraviolet radiation-induced carcinogenesis in hepatocyte growth factor/scatter factor transgenic mice. *Cancer Res.*, *60*, 3738–3743, 2000.

19 Cell and Molecular Biology of Cancer of the Brain

Øle Didrik Laerum

CONTENTS

19.1 INTRODUCTION

Malignant tumors of the nervous system represent a special challenge in cancer research. The tumors are not easily accessible both due to their confinement to the intracranial space and also because, their host organ, the brain is very susceptible to all types of infringement, limiting the access by invasive diagnostic or experimental procedures. In addition, the malignant cell populations in these tumors largely represent the complexity of cell types and functions in the nervous system. Therefore, multiple directions of differentiation and functions including their combinations may be encountered.

In particular, the malignant process in the brain kills by local growth and spread. Brain tumors only rarely metastasize to other organs. Hence, their local invasive behavior is by far the most dangerous property, whereby recurrence often occurs even after extensive attempts of radical surgery.

403

In this chapter, a survey of primary brain tumors and their subtypes are given together with some notes on their causes, their occurrence, and common classification. Biological properties which can be of importance for their malignant behavior are emphasized. Their molecular genetics, including genetic alterations and gene expression, are addressed, followed by a discussion of invasiveness as a key factor for understanding their malignant behavior. Finally, their biology and molecular genetics as background for experimental therapy are discussed. At present, the clinical outcome of most tumors is dismal, and only minor improvements have been achieved in the recent years. However, basic research is at present contributing considerably to the understanding of this type of cancer. There is also a multitude of different experimental approaches for their therapy. This gives hope for improvement of the prognosis in a not too distant future.

19.2 EPIDEMIOLOGY AND ETIOLOGICAL FACTORS

19.2.1 EPIDEMIOLOGY

When compared to breast, prostatic, colon, and lung cancers, brain tumors are relatively rare. Still, they are responsible for a substantial number of deaths in many countries, not least because of the bad prognosis of the most common subtypes. Comparative studies on their occurrence have been done on a world-wide basis. This has had some limitations due to differences in diagnostic procedures, availability of neurosurgical services, and variable standardization by histopathology diagnostics. Still, a rough comparison of data can be done [1,2]. With a few exceptions, the incidence of most subtypes and especially of high-grade gliomas increases substantially with age. One in ten primary brain tumors occurs in children, of which the most frequent are astrocytomas, medulloblastomas, and ependymomas. More males than females develop brain cancer. The mortality and incidence rates for primary malignant brain tumors in the United States have been reported to be higher in whites than in blacks.

Brain cancer incidence and mortality rates have been modestly increasing for the last 25 years. Much of this increase has been interpreted as due to improvement in diagnostics and reporting of the tumors. Thus, numerous reports have concluded that the increase in brain tumor rates is largely artifactual and correlated with increases in diagnostic practices. Some portion of the increase may be real, especially in children [1,2]. Otherwise, incidence rates are higher in the upper social classes.

International variation in brain cancer mortality rates, which is of the magnitude about twofold, is considerably less than for other cancer types. Taken together, these data might indicate that environmental contributions to the brain cancer incidence are less than for other common types of human cancer.

The overall survival rate of brain tumor patients is poor. A relative survival estimate of 5 years for all types of malignant brain tumors in the United States has been estimated to be 27%. This is mainly a reflection of the short survival of the most frequent and malignant type, the glioblastoma multiforme. Otherwise, population based survival rates have been improved for some of the other brain tumor types for the last 25 years, reflecting improvements both in diagnosis and treatment of the patients.

19.2.2 ETIOLOGICAL FACTORS

There is an accumulating body of data showing that inherited susceptibility and some genetic polymorphisms are associated with increased risk of developing a malignant brain tumor. This includes defects in DNA repair and detoxification genes as well as mutagen sensitivity. In addition, there are data showing a causative relation to workplace, dietary, and other personal and residential exposures. Only a small proportion of cases has been related to proven causes such as therapeutic irradiation to the brain, some rare heritage syndromes, and immune suppression. The latter factor is only related to the development of brain lymphomas. For several years electromagnetic fields have been suspected as causative agents for brain tumor development, including excessive use of cellular phones. However, this has not been verified [1,2].

Inherited syndromes such as the Li–Fraumeni cancer family syndrome, neurofibromatosis type 1, and von Hippel–Lindau's disease as well as several others have been related to increased risk of different types of brain tumors. However, their total numbers are small.

In the 1950s and 1960s, it was shown that different chemical agents could cause specific types of brain tumors both by local and systemic injection, and in particular *N*-nitroso compounds. A high proportion of malignant brain tumors developed in the offspring of pregnant mothers who were treated with some of these compounds, showing that transplacental carcinogenesis might occur. This was related to alkylation of DNA and to the fact that in the immature brain the repair of such adducts was inefficient. Therefore, the risk for developing a tumor was increased manyfold as compared to the adult brain.

Human exposure to *N*-nitroso compounds is also known to occur from different types of industries and chemical laboratories. This leads to the fear that some chemical agents might be a substantial risk factor for developing brain tumors. This issue still remains open. There are several reports of occupational factors that may be related to the occurrence of brain tumor, including petrochemical and rubber industry, as well as in chemists and pathologists. However, the data obtained so far are equivocal, although they do not rule out a moderately increased risk. Some relation between physical trauma to the brain and the development of meningiomas has been reported, although the importance of this is still unclear. A female preponderance of meningiomas may indicate that hormonal factors play a role in etiology, and regional variations may suggest a role of cultural behavior, including diet. The observation of lower risk for developing gliomas when there is a history of allergy or some common infections may suggest a possible role of immune factors both in glioma development and progression [2–4]. Improved methods for biological and genetic classification of brain tumors and their relation to different causative factors may give a clearer picture of this issue (see also Reference 5).

19.3 CLASSIFICATION

Thanks to considerable international work and not least the development of classification systems through WHO, worldwide studies based on a uniformed histological classification have become important. Thereby, the incidence, the diagnostics, and outcome of treatment can be directly compared between different countries and hospitals. The WHO classification is summarized in Table 19.1, and the most important types are explained below.

Astrocytic tumors comprise neoplasms that differ in their location within the CNS and biological properties as well as tendency for progression in clinical course. A considerable body of evidence indicates that these differences reflect the type and sequence of genetic alterations during malignant transformation. The most important types are low-grade diffuse astrocytomas (WHO grade II), anaplastic astrocytomas (WHO grade III), and some rarer subtypes of astrocytomas.

The most common and most malignant astrocytic tumor is the glioblastoma multiforme. This is composed of poorly differentiated neoplastic astrocytes with areas of vascular proliferation and necrosis. The tumor typically affects adults and is often located in the cerebral hemisphere. Usually, there is a short clinical history and a bad prognosis after the diagnosis has been made in spite of intensive treatment. The mean survival is somewhat more than one year (Figure 19.1 and Figure 19.2[a]).

Oligodendrogliomas (Figure 19.2[b]) tumors are histologically well-defined and have their origin from differentiated oligodendrocytes or progenitor cells which are responsible for myelin formation in the nervous system. They can be mixed with other types of gliomas, mainly astrocytomas, and both the pure form and the mixed one can occur in different grades (WHO grade II and III).

Ependymal tumors arise from ependymal lining of the cerebral ventricles and from remnants of the central canal of the spinal cord. They mainly occur in children and young adults.

TABLE 19.1

Classification of Malignant Brain Tumors

WHO histology groupings

Tumors of neuroepithelial tissue
Diffuse astrocytoma (protoplasmic, fibrillary)
Anaplastic astrocytoma
Glioblastoma
Pilocytic astrocytoma
Unique astrocytoma variants
Oligodendroglioma
Anaplastic oligodendroglioma
Ependymoma/anaplastic ependymoma
Ependymoma variants
Mixed glioma
Astrocytoma, NOS[a]
Glioma, malignant, NOS[a]
Choroid plexus tumors
Neuroepithelial tumors
Benign neuronal/glial, neuronal and mixed
Pineal parenchymal tumors
Embryonal/primitive/medulloblastoma

Tumors of cranial and spinal nerves
Nerve sheath, benign and malignant

Tumors of the meninges
Meningioma
Other mesenchymal, benign and malignant
Hemangioblastoma

Lymphomas and hemopoietic neoplasms

Germ-cell tumors

Cysts and tumor-like lesions

Tumors of the sellar region
Craniopharyngioma
Chordoma/chondrosarcoma

Local extensions from regional tumors

Unclassified tumors

[a]NOS, not otherwise specified.
Source: From Bigner, D.D., McLendon, R., and Brunet, J.M., Eds., *Russel and Rubinstein's Pathology of Tumors of the Nervous System*, Vol. 1 of 2, Arnold Publishers, London, Sydney, Auckland, 1998. With permission.

The medulloblastomas belong to the embryonal tumors and mainly occur in children. It is a malignant, invasive tumor of the cerebellum, predominantly with neuronal differentiation and an inherent tendency to metastasize via the cerebrospinal fluid (Figure 19.2[c]).

Meningiomas are benign tumors of the brain coverings that comprise 15 to 20% of the primary intracranial tumors. They are composed of neoplastic meningial cells. The tumors may rarely develop into malignant growth. There are also a number of other brain tumors (see Table 19.1), which are more rarely seen and comprise specific cell types [6].

Several of the malignant brain tumors consist of mixed cell populations with different types of differentiation into various neural cells. Further subclassification is partly done on the basis of their biological markers, as can be visualized by immunohistochemistry. In addition, the tumor cell population can be characterized on the basis of their biological properties and molecular genetics. This is the matter for further discussion.

19.4 TUMORIGENESIS IN THE NERVOUS SYSTEM

As mentioned earlier, several chemical agents may cause the development of malignant tumors in the brain. This knowledge is mainly based on experimental animal models, although some insights have also been provided by the study of sequential alterations during the progression of human tumors.

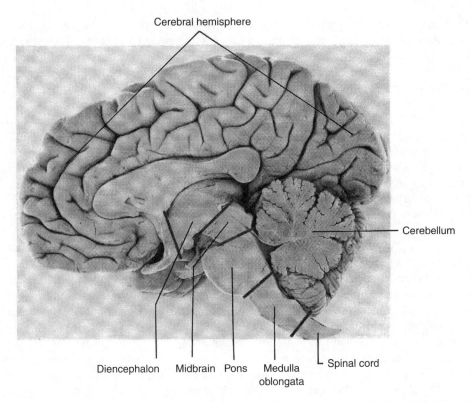

FIGURE 19.1   Regions in the adult human brain, longitudinal cross section through the middle (From Barr, 1974, size ×5/8).

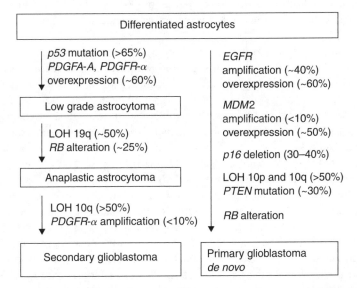

FIGURE 19.2   Some common types of primary brain tumors: (a) A lower-grade oligodendroglioma, (b) glioblastoma multiforme which is a high-grade tumor with short survival, and (c) neuroblastoma, a tumor with neuronal properties mainly arising in the cerebellum of children. Histological sections stained with hematoxylin and eosin.

For more than a generation animal models of tumors of the nervous system have been available (see e.g., References 7 and 8). This includes spontaneous tumors both in rodents and other animals, as well as other oncogenic agents. Tumor induction can be achieved by chemical substances, irradiation, viruses, and by genetic modification such as transgenic and knockout mice. Most of the common types of human brain cancer can be induced in experimental models, although the histological subtypes do not reflect the human situation entirely [9,10]. Different types of models using various rat strains have been especially well developed and refined, whereby detailed studies on tumor development have been possible. The combination of a relatively short life-span and induction time and reasonable size of the brain may have been contributing factors. In particular, the use of transplacental carcinogenic assays have been valuable and have given increased insight into the general process of neural carcinogenesis.

Mouse model systems have been fewer, although they have been available for nearly 50 years. The more recent development of transgenic and knockout procedures have been useful. In particular, the importance of different oncogenes, loss of suppressor functions, and not the least, the direct effects of autocrine mechanisms have been studied directly. Thus, brain tumors can be induced in mice using a recombinant platelet derived growth factor (*PDGF*) gene in a retrovirus [11]. In the 1970s, detailed morphological studies on carcinogen treated rat brains showed how the process started with small clusters of preneoplastic glial cells that later developed into microtumors and further to fully invasive brain tumors. The roles of structural DNA modifications, DNA repair, and neural target cell populations were also elucidated in a series of elegant studies (for review see Reference 12). With the availability of tissue culture systems whereby neural cells could be maintained for prolonged periods, it also became possible to follow directly the process of malignant transformation *in vitro*. This was pioneered by the explantation of transplacentally treated rat fetuses, whereby the neural cells with a specific carcinogenic damage *in vivo* were studied directly *in vitro*. The treated neural cells underwent a stepwise process whereby properties characteristic of malignant cells appeared in sequence (for review see Reference 13). Thus, the sequence of malignant transformation roughly followed the one observed in parallel *in vivo*, both with morphological alterations and in time. *In vitro*, stepwise alterations of differentiation and electrophysiological properties, DNA aberrations, and loss of anchorage dependence were seen until the neural cells were tumorigenic after 2 to 300 days *in vitro*, also reflecting the fully invasive phenotype. Later the sequence has been further elucidated by identification of a neu/erb2 mutant [14].

In the human tumors, some information on their development and progression has been obtained by careful radiographic analysis, although such studies are scarce for ethical reasons. During the last ten years major improvements have been made with regard to insights into the progression of human brain tumors and the genetic anomalies which lie behind them [15–18]. For example, the most aggressive tumors, the glioblastomas, show mutations of the cellular mechanisms that control cell cycling. Many specific cellular control mechanisms seem to be disturbed, and the more benign astrocytomas often show loss or mutations of the *TP53* gene. During the evolution of glioblastomas, at least two different pathways seem to operate. One is into the so-called *primary gliomas* which seem to develop *de novo* and frequently show over expression of the epidermal growth factor (EGF) receptor. The other main pathway is the progression from low grade or anaplastic astrocytoma to a so-called *secondary glioblastoma*. In these tumors TP53 mutations occur early and, in addition, other genetic aberrations with involvement of oncogenes and growth factors are characteristic, as discussed in the next section. Several other genetic aberrations occur during the development of malignant gliomas, and from year to year, the picture becomes more and more complicated. From a morphological point of view, there are several cases where biopsies are available from different stages of tumor progression, where an accumulation of genetic aberrations has been observed (see e.g., Reference 19). The loss of other suppressor functions, such as the *PTEN* gene, and over expression of autocrine signalling involving EGF and PDGF are also characteristic during this development [20].

During the development of human brain tumors, there are several reports of an associated infection with different viruses, including SV40. So far, this mainly seems to be a secondary event, for example, mediated by SV40-contaminated polio vaccine ([21], see also Reference 22).

Finally, there are several other cellular events that are associated with the development of brain neoplasms. This includes the induction of vascular supply with interactions between the malignant neural cells and endothelial cells [23,24]. In addition, the occurrence of large necrotic areas in glioblastomas may be a secondary event to the late stage of progression [23].

In conclusion, development of malignant brain tumors comprises a complexity of neural-cell populations and vascular endothelial cells. A stepwise process with acquisition of altered genetic and phenotypic properties and involving gradual loss of cellular control functions develops over periods up to large parts of the lifetime that ends with a fully invasive, rapidly progressive tumor. This cellular heterogeneity and complexity of their interactions are further discussed. So far, the combination of epidemiological studies with molecular genetics and clinical and histopathological examinations has been instrumental for the current insight into neuro-oncogenesis [25].

19.5 BIOLOGICAL PROPERTIES

Since directly observable malignant tumors in the nervous system often represent an intermediate or a late event in the process of neuro-oncogenesis, it is often difficult to know for certain the early cellular processes that are critical for malignancy. Phenotypic features of the malignant cell populations may be more or less important for the clinical outcome, but still be secondary to their malignant process. With this reservation, the different cell types and phenotypic properties in some of these tumors are discussed.

19.5.1 CELL TYPES

19.5.1.1 Glial Cells

As an estimate, glial cells outnumber the neuronal cells in the nervous system by a factor of nearly ten. It is therefore not surprising that most intracranial neoplasms have a glial phenotype. Generally, the tumors with the lowest malignant potential and the highest degree of differentiation also have a close morphological resemblance to the differentiated normal counterparts of glial cells. This mainly applies to low-grade astrocytomas and oligodendrogliomas (see References 1 and 6). With tumor progression and higher degree of malignancy, more heterogeneity in the tumor population may be seen. In addition, lower-grade tumors have a lower growth fraction and less proliferative activity that can be verified by histochemical markers (see Reference 26).

Oligodendrocytes, which are normally responsible for myelin formation in the nervous system, develop from the precursor cells designed O2A. In addition to their characteristic morphological appearance as oligodendrogliomas, these tumor cells also possess characteristic phenotypic and genotypic properties (see e.g. References 27 and 28). Thereby, they can be differentiated from low-grade astrocytomas. However, mixed oligodendrogliomas and astrocytomas (olegoastro-cytomas) are also known as an entity where both neoplastic astrocytes and oligodendrocytes are present simultaneously [29].

Once progression has occurred, several immunohistochemical markers may be indicative of the prognosis. This includes the increased proliferative activity, more domination by pleomorphic cells with increased proliferative activity, over expression and mutated growth factor receptors such as EGF-R, and the presence of different oncogenic products (see e.g., Reference 30). Increasing admixture of a special differentiated form of astrocytes called gemistocytes, is also characteristic [31]. These are cells with large eosinophilic cytoplasm, cell processes, and a marked cytoplasmic accumulation of glial fibrillary acidic protein (GFAP). In the so-called gemistocytic astrocytomas where these cells are more predominant, there is a considerably low survival rate of 5 years. They increase during progression and may also be common in glioblastomas. Their neoplastic nature has been verified by genetic analysis [31–33].

It should also be noted that phenotypic properties in neoplastic glial cells may also show overlap with other types, including neuronal cells. For example, the neuronal class III β-tubulin has been localized both in astrocytomas and in oligodendrogliomas [34,35]. The presence of this neuronal marker is also indicative of their higher malignant potential, showing that aberrant differentiation may be correlated to the neoplastic process. Likewise, the lack of markers, which are otherwise present in the differentiated glial cell counterpart, may also be indicative of neoplastic cells (see References 1 and 6).

19.5.1.2 Neuronal Cells

Several types of intracranial neoplasms may partly or mainly comprise cells with neuronal properties. This may apply to benign neuronal and mixed neuronal–glial tumors such as gangliocytomas, gangliogliomas, and other neuroepithelial tumors [6]. In addition, medulloblastomas as invasive embryonal tumors have predominantly neuronal differentiation [36]. In general, these are highly invasive both locally in cerebellum as well as in the intracranial space. It has also been known for a long time that medulloblastoma or neuroblastoma cells from childhood tumors can be induced to differentiate into neuronal cells by various stimuli in cell culture. Electrophysiological properties of neurons with emission of action potentials may also be seen. In recent years it has been shown that a high proportion of glioblastomas contain subpopulations of cells with neuronal properties, which can be directly identified by their electrophysiological properties *in vitro* [37,38]. The highly heterogeneous cell populations in these tumors could either arise from parallel neoplastic transformation of multiple cell types, or they are derived from a common precursor or stem cell. Data from neoplastic transformation of brain cells from the subependymal zone after administration of carcinogenic agents have long been interpreted in this direction (see e.g., Reference 39).

19.5.1.3 Other Cell Types

Apart from vascular endothelial cells, practically all known differentiated cell types of the nervous system have been associated with neoplastic development. This includes meningiomas, which are the second most common intracranial tumors. Although largely benign, they exhibit characteristic gene-expression profiles as well as genetic and differentiated aberrations (see e.g., References 6 and 40). Rarely, they may progress to sarcomas.

Similarly, ependymomas have phenotypic characteristics of ependymal cells with morphological resemblance to their normal counterparts. Schwannomas representing phenotypic properties of nerve sheath or Schwann cells are also seen [6]. In addition, children with inherited predisposition for neurofibromatosis also have increased frequency of WHO grade I pilocytic astrocytomas (see e.g., Reference 41).

19.5.1.4 Neural Stem Cells

The development of glioneuronal tumors of the CNS both in children and in adults strongly suggests a common precursor cell being the target of the neoplastic process (see References 41 and 42). In addition, it has been known for some years that multipotent populations of human neural progenitor cells cannot only be identified and isolated, but also propagated *in vitro* (see e.g., Reference 43). Upon differentiation they form neurons, astrocytes, and oligodendrocytes in defined culture media with addition of different growth factors. Such cells can be explanted from the human embryonic forebrain. However, there is considerable evidence that such progenitor cells are also present in the adult human brain. Thus, the intermediate filament protein nestin is not only expressed by primitive neuroepithelial cells during development, but also in the adult central nervous system. There, it is a marker for neural stem cells lining the ventricular wall and the central canal. Nestin and vimentin

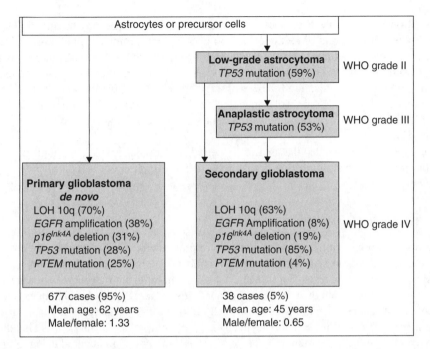

Astrocytes or precursor cells

Low-grade astrocytoma
TP53 mutation (59%)
WHO grade II

Anaplastic astrocytoma
TP53 mutation (53%)
WHO grade III

Primary glioblastoma
de novo
LOH 10q (70%)
EGFR amplification (38%)
p16^{Ink4A} deletion (31%)
TP53 mutation (28%)
PTEM mutation (25%)

Secondary glioblastoma

LOH 10q (63%)
EGFR Amplification (8%)
p16^{Ink4A} deletion (19%)
TP53 mutation (85%)
PTEM mutation (4%)
WHO grade IV

677 cases (95%)
Mean age: 62 years
Male/female: 1.33

38 cases (5%)
Mean age: 45 years
Male/female: 0.65

FIGURE 19.3   Main genetic pathways in the evolution of primary and secondary glioblastomas (From Ohgaki et al. *Cancer Research*, 64, 6892, 2004. With permission).

are co-expressed in neuroepithelial cells as well as in neuronal and glial progenitor cell in man [44–46]. It is therefore not surprising that nestin can be directly identified by immunohistochemistry in several pediatric brain tumors [47]. This also includes primitive neuroectodermal tumors (PNETs), and low-grade gliomas (Figure 19.3).

At present there is a large body of evidence pointing at primary brain cancer as a result of malignant transformation of common precursor or stem cells. Since the progression of the more malignant tumors also involves an increased renewal potential as well as altered differentiation with more primitive cells, it cannot be ruled out that this progression can also be the result of reversion to more primitive cell stages. The induction of telomerase activity in glioblastoma cells as a consequence of one of the several genetic alterations may point in this direction [48]. Further progress in molecular genetic studies may give a key to this question.

19.5.2   Cell Interactions

There is a large body of evidence about multiple cell interactions in primary tumors from the nervous system. This means that the complex cell populations comprising tumors, which have already been described, may have an influence on each other. Theoretically, this could both lead to the stimulation of certain subtypes as well as inhibition of other subtypes. In consequence, tumor progression may be the result of loss of suppressor functions not only as internal regulatory loops, but also from external control systems mediated by other cell types. The identification of several growth factors which are subject to autocrine or paracrine secretion in these neoplasms, such as EGF, hepatocyte growth factor (HGF), PDGF, and others, may point in this direction. The same applies to the amplification or over expression of their receptors (see e.g., References 49–51). Thereby, excessive growth and invasive potential of brain tumors may not only be due to intrinsic cellular factors, but also to their mutual stimulation and "cooperation." Cellular interactions in neural neoplasms are not only a matter of secretion of regulatory factors and the operation of their receptors, but also of direct communications. First, it has been shown that malignant glioma cells may be directly electrically coupled *in vitro* [52]. During brain tumor invasion, malignant glioma cells and normal brain cells may also form direct junctions [53,54]. In addition, the astrocytic gap junction protein, connexin43 has been directly identified in cell to cell contacts between cocultured glioma cells and astrocytes. Direct gap junction communication has been observed by extensive dye transfer. On the other hand, reduced expression of the P2 form of connexin43 has been described in malignant meningiomas, indicating altered cell to cell communication as a sign of more rapid growth and less favorable prognosis [55].

In total, it may be concluded that malignant neural cell populations do communicate by various means, but that their communication is disturbed or more or less out of control as part of their malignant potential.

19.5.3 TUMOR BIOLOGY

With the rapid development of methods for studies of proteomics of both the normal nervous system and malignant brain tumors, a correspondingly rapid progress in the understanding of biologic phenomena is expected. As discussed in the next session, multiple biological processes are altered in malignant brain tumors, both quantitatively and qualitatively, and this leads to complex interactions related to the malignant behavior. Studies on such multifactorial phenomena at a general systemic scale are present, but only in their infancy (see e.g., Reference 56). However, it is already a well-known phenomenon that alteration in one type of protein such as p53 may modulate multiple other proteins in the tumors, leading to a cascade of different phenotypic events (see e.g., Reference 57). As an example, HGF/scatter factor, may both protect human glioblastoma cells against cytotoxic death, as well as stimulate growth and overcome contact induced growth arrest both via upregulation and downregulation of different cellular factors [58,59].

An extensive review of all biological phenomena, which have been related to cancer of the human brain, would be outside the scope of this chapter. Another complicating factor is that alteration from so-called normal biological behavior of neural cells is not necessarily a part of the neoplastic process per se. They could equally well be secondary adapted processes or due to alterations in the cell composition and functional stage of the neoplastic cell population. However, some key factors are mentioned.

In principle, the identification of altered biological properties in tumors may be interpreted in two main directions. One is that the property is a primary event in the neoplastic process or progression and thus is partly responsible for the malignant behavior. The other possibility is that such alterations accompany the malignant process, by which a prognostic value is also implied. It must also be kept in mind that several such properties are related to each other, so that their value as independent prognostic markers is limited. Important information for the daily assessment of clinical course is obtained both by histopathology and immunohistochemistry (see e.g., Reference 60).

It is well known that the fractions of proliferative cells in the brain tumor tissue has a prognostic importance and is also related to the tumor grade [61,62]. The density of growth factor receptors, their mutational state, and also secretion of different ligands have already been mentioned. The direct response to growth factors in brain tumors may vary considerably from tumor to tumor within the same histological subtype [63]. This may indicate that subclasses of cells have a higher malignant potential than others in a heterogeneous tumor. The density of different growth-factor receptors may be of critical importance, as can be evidenced from the degree of receptor gene amplification [64]. More recently, a considerable interest has been focused on the insulin-like growth factors (IGFs) and their receptors which may play a pivotal role in brain tumor growth (for review see Reference 65). Since these growth factors also inhibit apoptosis, they may cause increased survival of the malignant cells. Specific IGF binding proteins may on the other hand inhibit tumor growth, and their expression correlates with tumor grading and patient survival. Similarly, interferons may intervene with malignant behavior in brain tumors. Thus, interferon alpha inhibits the cell cycling in glioblastomas, which again may be mediated, at least in part by the tumor suppressor gene product p21 [66]. Thus, complex interactions through activation of different signal transduction pathways may strongly modulate the tumor cell behavior. Recently, it has also been shown that spontaneous activation and signalling by over expressed growth factor receptors in glioblastoma cells may add to this mechanism [67].

Survival of the malignant cells has been the matter of increased focus. First of all, their direct survival mediated by escape from apoptosis is critical. On one side, the Fas ligand may be expressed in human gliomas, and this ligand may then cause apoptosis when the receptor is expressed [68]. On the other hand, it is known that malignant brain tumors, including in children may be resistant to apoptosis induction (see e.g., Reference 69). Different types of protein kinase C may likewise have either stimulating or inhibitory effects on both proliferation and apoptosis in glioma cells [70]. In this connection, malignancy in the nervous system is similar to that in other organs.

A considerable interest is centered around angiogenesis in brain tumors and its cascade of regulators (for recent review see Reference 71). A multitude of angiogenic inducers and endogeneous inhibitors have been identified which may be of importance for the more or less pathological vascular supply that is characteristic in higher grades of brain tumors. Leaky blood vessels may lead to capillary effusion and peritumoral edema. In addition, the formation of coiled, glomerulus-like capillary loops is frequently associated with glioblastomas. Since the most malignant type of glioblastomas, gliomatosis cerebri is largely characterized by diffuse infiltration of single cells in the brain tissue and with negligible vascular formation, the direct importance of angiogenesis as a central part of the malignant process remains to be established.

While programmed cell death by apoptosis may be restricted in these tumors, large areas of necrosis are characteristic [72]. Ischemic necroses are most frequent in primary and less in secondary glioblastomas. However, elevated Fas expression does not seem to be the consequence of this, but occurs by an independent mechanism, which is also unrelated to the TP53 status [73]. It has also been found that activation of the Akt signal transduction pathway may convert anaplastic astrocytoma cells to glioblastoma multiforme. Upon intracranial injection into nude mice these human tumor cells not only formed tumors with large areas of necrosis, but were also surrounded by extensive neovascularization. This indicates that activation of this pathway is sufficient to induce some of the hallmarks of glioblastomas [74]. Furthermore, a high secretion of glutamate from malignant gliomas may be responsible for some of the development of necrosis as well as extensive neurotoxicity in the vicinity, and this may be combined with increased growth of the malignant cells as such [75].

19.6 MOLECULAR GENETICS

It is generally known that neural cells may express wide varieties and a high proportion of the human genome, both due to the complexity of cell types and functions. The extent to which the malignant counterparts reflect the same genetic pattern, in consequence, may give a similar multitude of patterns. Although this field is still in its infancy, a large body of data is available on genetic alterations in malignant brain tumors. Some of these are mentioned together with some newer aspects of genetic aberrations and gene expression profiles.

19.6.1 CYTOGENETICS, ONCOGENES, AND SUPPRESSOR GENES

Although multiple different chromosome aberrations and mutational events have been described in the common types of brain tumors, some main patterns and pathways seem to exist (for review see References 6 and 78). While cytogenetic and allelotypic studies of low-grade adult astrocytomas have shown few deviations, comparative genomic hybridization and other methods have shown several abnormalities. With increasing degree of malignancy the gain of chromosome 7 and losses of all or parts of chromosome 10 occur in more than three fourth of all cases. In addition, a substantial amount of tumors contain losses of 9p,13,17p, and 22q. It is also known that more than half of childhood anaplastic astrocytomas and glioblastomas also exhibit normal stem lines. Since the known suppressor genes *CDKN2*, *Rb1*, and *TP53* are located on chromosome 9p, 13q, and 17p, and PTEN on chromosome 10, this may indicate that loss of suppressor functions are pivotal for development of glioblastomas.

In differentiated low-grade astrocytomas, TP53 mutation and over expression of PDGF receptor are common. With progression to anaplastic astrocytoma additional aberrations are seen, including loss of heterozygosity of the *PTEN* gene at chromosome 10, deleted in colorectal cancer (DCC) loss of expression, and PDGFR alpha amplification. In contrast, primary glioblastomas that occur *de novo* have a high proportion of amplification or over expression of EGF receptor, somewhat less of MDM2 and 80% PTEN deletion in addition to one third of cases with p16 deletion. However, other pathways of primary glioblastoma development with TP53 mutation seem to exist. Basically, and in line with other human tumors, TP53 mutations have important consequences both for escape from apoptosis as well as less efficient DNA repair of mutational events. Mutations of regulatory genes for cell-cycling and loss of heterozygosity for the retinoblastoma gene are important additional factors. In particular, loss of chromosome 10 seems to be important for release of suppressor functions.

Gross chromosomal aberrations may be directly visualized by altered DNA stem lines and several stem lines by use of DNA flow cytometry [77]. In addition, demonstration of intratumoral cytogenetic heterogeneity is a hallmark. This largely reflects genetic instability within a tumor and especially in glioblastomas that are some of the most malignant of tumors known [78].

In line with this it has been shown that allelic loss on chromosome 10 and 17p as well as *EGF-R* gene amplification relates to a less favorable prognosis in human malignant astrocytomas [79]. This reflects both the transition to higher tumor grades as well as older versus younger patients [80,81]. So far, it has been possible to identify at least four different genetic pathways that may lead to the development of anaplastic astrocytomas [81–86]. Such classification may also be used for subgrouping of heterogeneous tumors as glioblastoma multiforme with regard to prognosis and response to therapy [84]. This can, for example, be seen by substantial genotypical differences between short-term and long-term survivors of glioblastomas [87]. Differences in the proportions of PTEN mutations as well as alterations of cell-cycle regulatory genes point in the same direction [88,89].

Since the presence of genetic aberrations in the different histological types of brain tumors have been reviewed earlier, only some additional factors need to be mentioned. On one hand, germ-line mutations of the *TP53* gene lead to an increased risk of developing several different types of tumors in humans, including those of the brain. This applies to both lower- and higher-grades of astrocytomas, glioblastomas, and choroid plexus tumors [90]. However, once a tumor is established, it seems at least for glioblastomas that the presence of TP53 mutations as well as its location on the gene does not seem to effect prognosis [91]. The development of two primary glioblastomas has been described in the same patient with ten years difference. Both tumors had different types of genetic aberrations. Therefore, widely differing biological features may still comprise the same histological and clinical pattern [92] (Figure 19.4).

Several other aberrations, such as downregulation of the *BR-1* and *2* genes in low-grade human gliomas [93,94] identification of the human glioma associated growth factor gene granulin [95], as well as the mutational state of different suppressor and other genes have been described [96–98]. The findings indicate that such aberrant expression due to homozygous deletion or hypermethylation may be related to malignant progression [94–96].

With regard to other histological types of brain tumors, it has been found that molecular subtypes of oligodendrogliomas may preferentially arise in different areas of the brain and with differing patterns of growth. For example, allelic loss of chromosomes 1p and 19q were most frequently seen in the frontal lobes [99,100].

Gliosarcomas show a similar genetic profile to glioblastomas, except for the absence of EGF receptor amplification/over expression [101]. Anaplastic oligodendrogliomas also show a concurrent inactivation of RB1 and TP53 pathways, and together with other aberrations this may contribute

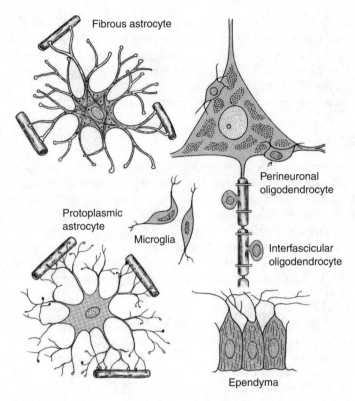

Fibrous astrocyte

Perineuronal
oligodendrocyte

Protoplasmic
astrocyte

Microglia

Interfascicular
oligodendrocyte

Ependyma

FIGURE 19.4   Schematic drawing of the main types of neuroglial cells of the central nervous system (From Barr, 1974).

substantially to their malignant phenotype [102]. Sporadic medulloblastomas seem to be associated with mutations of the adenomatosis polyposis coli (*APC*) gene [103], while supratentorial primitive neuroectodermal tumors (PNETs) of the childhood have a similar, but not identical pattern to pediatric glioblastomas [104]. These findings reflect the strong genetic heterogeneity of intracranial tumors.

It has been shown that mutation of PTEN and activation of EGF receptor also lead to regulatory alterations of mRNA expression for other genes such as the vascular endothelial growth factor [105]. Thus, mutational events may lead to a cascade of regulatory additions in accessary cell types, for example, neovascularization may promote further growth and progression.

19.6.2 GENE EXPRESSION

Apart from mutational events, deletions and over expressions/implifications of genes caused by the malignant process and as discussed earlier, more global alterations of gene-expression patterns may be seen in malignant brain tumors as in other types of tumors. A large number of recent reports have shown by use of real-time Polymerase Chain Reaction (PCR) analysis, microarray analysis, and related techniques that substantial quantitative changes may be seen in several cell functions of malignant brain tumors. This applies to both well-known genetic aberrations [106,107] and tumor markers, common with other types of tumors [108], as well as to expression of different extracellular matrix proteins that may be of importance both for neovascularization and tumor progression [109]. By use of microarray methodology, characteristic gene-expression profiles have been obtained both in low-grade astrocytomas [110], pleomorphic xanthoastrocytomas [111], and embryonal tumors of the nervous system [112]. Performing cluster analysis of multiple gene-expression patterns seems to be of increasing value both for subclassification and prognostic assessment of these tumors [113–115]. The combination of this with public databases for gene expression in human cancers (see e.g., Reference 116) gives promising prospects for an integrated approach to cancer genomics and proteomics [117]. Thereby, a better integration between biological properties, gene expression, and genetic aberrations in the tumor cells on one side, and their histological and clinical features on the other side, may greatly facilitate diagnostic work and therapy.

19.7 INVASION

Malignant brain tumors kill by their local growth and invasion into the surrounding normal tissue. They only rarely metastasize outside the intracranial space, and mainly when there has been some kind of surgical intervention, a shunt, or other mechanisms for opening the blood–brain barrier. Thus, invasiveness has become more and more recognized as the main malignant property. Extensive research during the last two decades has given a deeper insight into mechanisms related to brain tumor invasion. By use of specific model systems, this process can also be studied directly outside the body. In this section a general outline of invasiveness are first be given, followed by a description of available model systems. Finally, more recent insights into the mechanisms are described in short.

19.7.1 GENERAL ASPECTS

Basically, malignant invasion implies that tumor cells or tissue occupy another tissue. This process is progressive in time and will usually lead to destruction of the normal tissue (for reviews see References 73 and 118–120). The tumor cells are able to actively proliferate and migrate into the surrounding normal brain tissue and may spread along the different anatomical structures as single cells and partly as groups or solid nests of cells. Gradually, the surrounding brain tissue will be replaced by the tumor, and the malignant cells may also migrate to distant parts of the brain. By the time a malignant brain tumor is diagnosed, the tumor cells are, in many cases, already out of the reach of surgical intervention. Recurrence and further progression is therefore common.

The characteristic pattern of local spread in the brain was described three fourth of a century ago by the Belgian pathologist H.-J. Scherer. The term "primary structure" was used for the initial tumor with characteristic histological features, including pseudopalisading of nuclei adjacent to necrosis in glioblastomas, rosettes of medulloblastomas and ependymal channels, and rosettes in ependymomas. Upon the subsequent invasion into the normal brain tissue, the tumor cells tended to follow the existing anatomical structures and migrate around blood vessels and nerve cells. They also migrated along structural interfaces in the brain, such as grey and white substance and the ventricular lining. The invading tumor cells tended to become organized according to the neural tissue elements along which they were migrating, termed "secondary structure." As a consequence, the tumor invasion also led to impaired circulation, hemorrhage, and necrosis, which caused secondary reactions in the affected normal tissues. This was denoted as "tertiary structure" (see e.g., Reference 120).

At the periphery of the tumor, there is an undefined border between malignant invading cells and preexisting neural tissue together with reacting glial cells. From there the malignant cells form extensions of the tumor tissue into the surrounding, often edematous brain structures. They particularly tend to migrate along nerve fiber tracts and often spread into and within the Virchow–Robin and subarachnoid spaces. They also infiltrate around the blood vessels and make a circular "cuff" of tumor cells. Glioma cells often invade the superficial pia mater, but seldom proceed to the peripheral blood border of the perivascular spaces. The malignant cells also invade the cortex of the opposite convolution and accumulate along the leptomeninges, although these are not penetrated. They may be transported by the cerebrospinal fluid to distant sites, a process which in some cases can be very rapid. The migration through the brain tissue may occur in a few days in experimental tumors (for review see Reference 119). However, the different subtypes of primary brain tumors may show different patterns of local spread. A particular subtype of the glioblastoma, the already mentioned gliomatosis cerebri is characterized by diffuse infiltrative growth over large parts of the brain and with little tendency to solid tumor formation and induction of vascular supply (see also Reference 71).

As discussed later in this chapter, proteolysis is an important part of the invasion process, and this is also the case for brain metastases from malignant tumors in other organs [121]. Generally, cancer invasion has a similarity to tissue remodelling as can be seen during normal regeneration in different tissues. Thus, normal wound healing also implies proteolytic matrix degradation and extensive tissue remodelling together with induction of vascular supply, extensive cell migration, and an inflammatory reaction [122,123].

As pointed out in several studies during organogenesis, and not least in the nervous system, invasion is also exerted by normal cells. However, this is limited in time, although extensive migration and remodelling of tissues may occur. Hence, invasiveness per se is mediated by principally normal cell functions, although their regulation is totally disturbed and the reaction excessive when mediated by cancer cells. This is particularly the case with brain tumors.

19.7.2 Model Systems

Brain tumor invasiveness can be directly studied *in situ* on induced or transplanted tumors, although the limited access to the brain restricts the type of studies. A rapid progress in the use of *in vivo* model systems has occurred during the last decade, whereby cell lines as well as primary human tumors can be transplanted and studied in immunodeficient mice or rats (see e.g., References 71 and 119). This applies to both the formation of invasive localized tumors as well as diffuse infiltrative growth. In such model systems, detailed quantitative and qualitative data on glioma spread have been obtained [124,125].

A great advantage was given when *in vitro* model systems became available in the 1980s. Since monolayer culture in principle does not reflect a three-dimensional structure, it was not suited for studies of invasion. However, by the introduction of organoid tissue-culture models, this problem was overcome. Both fragments of solid brain tissue as well as reaggregating cultures forming a reconstituted brain tissue from rat fetuses were used. Such spheroids would rapidly be invaded when confronted with spheroids of malignant brain tumors. Both cell aggregates and precultured solid tissue pieces could be used, whereby an organoid structure of brain tumor would invade. Several of the key functions of invasiveness are mimicked by such coculture methods, including malignant cell migration, adhesion to the normal tissue, and gradual invasive growth and destruction. Also degradation of the normal extracellular matrix by proteolytic enzymes takes place [126–128]. For natural reasons, vascular and inflammatory reactions that develop in the normal brain are not seen in such confrontation cultures. It is also a limitation that fetal tissue has mainly been used as target, and this may have different properties from the adult brain tissue [129]. The progressive nature of invasion in such cocultures can be visualized and quantitated by the use of confocal laser microscopy (see e.g., References 126 and 128).

Also brain slices can be used as targets for glioma cells *in vitro* within a limited time frame, and several other modifications of organotypic models have later been designed (see e.g., Reference 130). In addition, different modifications of matrigel and collagen gels can be used as targets for invasive cells *in vitro*. Their advantage is that they are simple and easy to handle, but are limited by the available matrix as the only target and not living tissues. Therefore, considerable criticism has been made against the currently used model systems: the gel assay for being too simple and not reflecting the complexity of invaded tissue *in vivo*. Although the three-dimensional coculture systems to some extent reflect the complexity of the target tissues, they are still far from the *in vivo* situation. In particular, the extensive use of fetal brain tissue, which is easily invaded by the malignant cells, may be too permissive as compared to the adult brain tissue (see Reference 129). On the other hand, solid pieces of brain tumors forming spheroids *in vitro* seem to be better suited as the invading malignant tissue, although the *in vitro* situation may never reflect the complexity of the intact brain.

Such, spheroids *in vitro* may also reflect the central hypoxia and necrosis seen in gliomas *in vivo*, which may be important contributing factors to the tumor reaction [131].

19.7.3 MECHANISMS

An initial and key property of malignant cells is the ability to migrate. This is dependent on an intect locomotary system, including the cytoskeleton and seems to be more or less independent from other cell properties related to invasiveness (see e.g., Reference 128). The cells may use different cell-surface receptors for their migration, depending on the cell substrate that is available. Extracellular matrix proteins of high molecular weight, including laminin and fibronectin play an important role in this process. Both laminin and fibronectin stimulate migration, and the $\alpha_3\beta_1$ integrin receptor on the cell surface seems to be of particular importance [132,133]. Adhesion is an important part of this migration, which again can be blocked by antibodies to the matrix substrates [134–137]. In particular, CD44 is a cell-surface receptor for several extracellular matrix components, partly regulated by the Rho family of proteins. Blocking of CD44 leads to inhibition of glioma invasion (see e.g., Reference 138), and in addition tumor cell migration is dependent on cleavage of this surface protein by a membrane associated metalloproteinase [139]. Diffuse brain invasion by glioma cells in particular seems to require beta1 integrins [140]. However, invading glioma cells can also stimulate the production of extracellular matrix components in the normal brain, including fibronectin and collagen type IV as well as laminin [141,142].

A critical part of the migration and penetrance into the surrounding tissue is degradation of different parts of the complex extracellular matrix components in the brain (for review see Reference 143). This requires secretion of different types of proteases and in particular metalloproteinases, plasminogen activators, and cathepsinB (for review see Reference 144). These can be secreted into the surrounding tissue or be membrane bound to the tumor cells, acting at a short distance (see e.g., References 145 and 146).

Interactions between the malignant cells and the extracellular matrix together with enzymatic degradation are highly dynamic processes that vary in time and are regulated and influenced by a multitude of different regulatory factors. Thus, enzyme secretion can be greatly stimulated by different growth factors, such as EGF, Fibroblast Growth Factor (FGF), IL10, and others [147]. In addition over expression of cyclin D1 induces glioma invasion both by matrix metalloproteinase secretion and enhanced cell motility [148].

On the other hand, blockage of enzyme secretion or activity as well as of interferon gamma and growth factor receptors decreases glioma cell growth, migration, and invasion [149–151]. However, the stimulation of invasion by different growth factors is strongly variable, depending on the individual tumor and also the type of growth factor. Using the *in vitro* coculture assay, Pedersen et al. [152] found that greatest stimulation was exerted by EGF, followed by FGF and PDGF.

In general, growth factors, either autocrine, paracrine, or exogenous mainly seem to stimulate existing invasiveness, and blocking of their receptors may reduce invasion. Therefore, over expression and amplification of growth factor receptors on the tumor cells will have the same effect. In general, growth factor stimulation does not induce invasiveness in noninvasive cells, although both EGF and FGF have been related to malignant transformation as such. Conversely, blocking of growth factor receptors, including EGF-R can retard invasion [149,151]. To what extent it can also block invasion entirely *in vivo* is not yet known, although this is technically possible *in vitro*.

In vivo there seems to be a complex interplay between normally occurring growth factors and cytokines in the nervous system, and the malignant cell population invading into the normal brain tissue. This also applies to inflammatory cells in the invasion zone that secrete different messengers. Hence, invasion seems to be a complex interplay between normal and malignant tissue. Interference with signal transduction pathways in the malignant cells may also interfere with invasive capability. On the other hand, the fact that brain tumor invasion rate does not correlate with proliferative parameters, indicates that proliferation as such or lack of proliferation are not necessary for this process to occur (see e.g., References 119 and 153).

Invasiveness *in vitro* as well as *in vivo* may also be related to lytic capability of the malignant cells, and to a great extent this may be related to their enzyme secretion. In line with normal astrocytes, malignant glioma cells also possess phagocytic activity, which *in vitro* can be extensive [154].

At present, the genetic determinants of invasiveness are not completely understood. Although over expression and amplification of growth factors and their receptors as well as release of suppressor mechanisms may enhance the activity of an invasive phenotype, it is not clear which cellular processes are critical for this malignant property. Over expression of some genes, such as *P311*, coding for a protein product localized at focal adhesions has been suggested as an invasion gene in human glioblastoma cells [155]. Another candidate has been the *Sparc* gene (secreted protein acidic and rich in cystein), where increased expression promotes glioblastoma invasion *in vitro* [156]. However, since multiple cell properties are involved in malignant invasion, the current developments in large-scale analysis of gene aberrations and gene expression in invasive cells may greatly contribute to a deeper understanding of this process (see e.g., References 157).

19.8 EXPERIMENTAL THERAPY

The treatment of cancer of the brain in patients has so far had only limited success. Although the response to treatment by surgery, radiation, and chemotherapy, either singly or in combinations shows partly good results in lower-grade tumors, high-grade tumors and especially glioblastoma multiforme are associated with a short survival. A key question, therefore, is whether the extensive research on biology and molecular genetics of these tumors may lead to new treatment modalities. At present there are many reports in the literature about successful treatment *in vitro* or in animal models, and several clinical trials are on the way. Although the prospect of substantial prolongation of life seems to be far away, there are several experimental approaches that may be promising [158–164]. It should also be noted that the expression of multidrug resistance in human gliomas is a great clinical problem which is an urgent need to overcome for progress in chemotherapy (see e.g., Reference 165).

Specific gene therapy as well as antisense RNA therapy have been used for blocking critical cell functions in the malignant cells, including telomerase activity, which otherwise may immortalize the cells (see e.g., References 165–168). The other approach is to interfere directly with a cell function or surface components by use of specific antibodies (see e.g., References 161 and 162). This can for example be used for blocking vascular formation with antibodies against VEGF-R-2 [169]. An elegant approach is to deliver alginate-encapsulated producer cells directly to the tumor, by which, for example, antivascular and antitumor effects of human endostatin can be obtained *in vivo* ([170], see also Reference 71). As part of immunotherapy, vaccination of brain tumor patients with peptide pulsed dendritic cells has been shown to induce intracranial inflammatory reaction [171]. Moreover, inactivation of the p53 protein in tumor cells may sensitize them to chemotherapy [172].

Induction of apoptosis in the tumor cells by using the CD95 Fas ligand [173], by blockage of glutamate or different growth factor receptors [174,175], or by direct irradiation in model systems *in vitro* [176,177], have so far given promising results. Selective treatment with agents blocking epidermal growth factor receptor [177], cannabinoid receptor [178], cell adhesion molecules [179], or by use of various kinase inhibitors [180] have all shown deleterious effects on glioma cells in experimental systems. However, the antitumor activity of these actions as related to side effects from the delicate structures of the normal human brain remains to be established. Reversion of the neoplastic phenotype by different means, including connexin43 as structural component of the gap junctions [181], represents a particular challenge, although the most malignant cells may escape from such treatment.

A particular interest has been centered around the possibilities for antiinvasive therapy of malignant brain tumors. Using the general knowledge from mechanisms which are operating during invasion, a corresponding blockage of these functions should be ideal for limiting tumor dissemination. In addition, such treatment may effect several different cell functions at the same time, such as blocking of both cell motility and invasion by inhibition of phospholipaseC [182], blocking of cell surface molecules such as integrins [183], CD44 [184], as well as the use of metalloproteinase inhibitors [185,186]. A particular advantage may be offered by simultaneous inhibition of both glioma proliferation, invasion, and angiogenesis, which for example in some cases can be obtained by blocking growth factor receptors [187], or the use of a fragment derived from the autocatalytic digestion of matrix metalloproteinases [188]. Similar effects can be obtained by various synthetic matrix metalloproteinase inhibitors [189] or delivery of antisense gene to urokinase type plasminogen activator receptor [190], whereby both invasive capacity and proliferation are suppressed *in vitro*.

In conclusion, both manipulation with the extracellular matrix and its receptors, as well as growth factor and cytokine networks, proteinases, cytoskeletal components, and oncogenes or tumor suppressor genes, either as single factors or in combination may be promising avenues for further development of antiinvasive therapy (see also Reference 191). Further progress in the understanding of underlying biological mechanisms may greatly improve the selectivity of such therapy, although this is at present a severely limiting factor.

19.9   ARCHIVAL PATHOLOGY

The current research on cancer of the brain is mainly focused on prospective clinical or experimental studies, or the use of tissue material that has been collected over a short time period. This is a reflection of the need for obtaining fresh or even viable material for biological studies, including propagation of malignant cells *in vitro*. Use of only fresh, collected tissue may severely limit the access to research material, such as subgroups of tumors or rare tumor types as well as obtaining a wide range of biological variation. In the recent years, this situation has become changed by the improvement of storage of material over time, including larger biobanks. In addition, the possibility to do studies on molecular genetics, gene expression, and phenotypic expression by use of immunochemistry, has changed this picture. Especially, tissue that is embedded in paraffin for histological examination and thereafter stored, has shown to be of utmost importance for such research. For example, a great number of the reports cited here on molecular genetics in malignant gliomas are based on studies of DNA extracted from archival paraffin-embedded material which was collected nearly 30 years back in time. The use of such material greatly increases the access to research material, and with the rapid improvement in methodology studies for archival tissue, even up to one century back in time, opens a new avenue for studies on brain tumor biology. Especially when correlated to clinical data and epidemiological registries, biological and etiological factors can be studied over long time periods.

19.10   CONCLUDING REMARKS

Cancer of the brain, at present, represents one of the great challenges in current cancer research. The delicacy and difficult accessibility of the target tissue creates a particular challenge for discriminating between normal and malignant cells when designing the therapy. The complexity of tumors from the nervous system and involvement of many cell functions in invasiveness imply multiple cell processes and interactions. Insight into these processes given by cell biology, molecular genetics, and the design of different model systems have greatly facilitated progress in this field. Malignant gliomas may arise from neural stem cells or more differentiated cells that revert to an earlier developmental stage. Thereby, the cellular heterogeneity of the tumor population with both glial and neuronal elements may be explained. For further progress a deeper understanding of the invasive process into the brain may be of critical importance. Experimental therapy models indicate that multiple aberrant functions in the malignant cell population must be targets for treatment.

REFERENCES

1. Bigner, D.D., McLendon, R., and Brunet, J.M., Eds., *Russel and Rubinstein's Pathology of Tumors of the Nervous System*, Vol. 1 of 2, Arnold Publishers, London, Sydney, Auckland, 1998.
2. Wrensch, M., Minn, Y., Chew, T., Bondy, M., and Berger, M.S., Epidemiology of primary brain tumors: Current concepts and review of the literature, *Neuro-Oncology*, 4, 278, 2002.
3. Wikstrand, C.T., Ashley, D.M., Bigner, D.D., and Hole, L.P., Cellular immunology. Lymphocyte populations, cytokiner and target — effector systems, in *Russel and Rubinstein's Pathology of Tumors of the Nervous System*, Vol. 1, Bigner, D.D., McLendon, R., and Brunet, J.M., Eds., Arnold Publishers, London, Sydney, Auckland, 1998.
4. Debinski, W., Gibo, D., and Mintz, A., Epigenetics in high-grade astrocytomas: Opportunities for prevention and detection of brain tumors, *Ann. N.Y. Acad. Sci.*, 983, 232, 2003.
5. Wikstrand, C.J., Cokgor, I., Sampson, J.H., and Bigner, D.D., Monoclonal antibody therapy of human gliomas: Current status and future approaches, *Cancer Metastasis Rev.*, 18, 451, 1999.
6. Kleihues, P. and Cavenee, W.K., *Pathology and Genetics of Tumours of the Nervous System*, 2nd ed., IARC Press (International Agency for Research on Cancer), Lyon, 2000.
7. Laerum, O.D., Bigner, D.D., and Rajewsky, M.F., Eds., *Biology of Brain Tumors*, International Union against Cancer, Geneva, 1978.
8. Walker, V.E., Walker, D.M., and Swenberg, J.A., Neurocarcinogens and experimentally induced brain tumors, in *Russel and Rubinstein's Pathology of Tumors of the Nervous System*, Vol. 1, Bigner, D.D., McLendon, R., and Brunet, J.M., Eds., Arnold Publishers, London, Sydney, Auckland, 1998.
9. Sampson, J.H. and Bigner, D.D., Experimental tumors and the evaluation of neurocarcinogens, in *Russel and Rubinstein's Pathology of Tumors of the Nervous System*, Vol. 1, Bigner, D.D., McLendon, R., and Brunet, J.M., Eds., Arnold Publishers, London, Sydney, Auckland, 1998.
10. Holland, E.C., Brain tumor animal models: Importance and progress, *Curr. Opin. Oncol.*, 13, 143, 2001.
11. Uhrbom, L., Hesselager, G., Nistér, M., and Westermark, B., Induction of brain tumors in mice using a recombinant platelet-derived growth factor B-chain retrovirus, *Cancer Res.*, 58, 5275, 1998.
12. Kleihues, P. and Rajewsky, M.F., Chemical neurooncogenesis: Role of structural DNA modifications, DNA repair and neural target populations, *Prog. Exp. Tumor Res.*, 27, 1, 1984.
13. Laerum, O.D., Mørk, S.J., and De Ridder, L., The transformation process, *Prog. Exp. Tumor Res.*, 27, 17, 1984.
14. Kindler-Röhrborn, A., Kind, A.B., Koelsch, B.U., Fischer, C., and Rajewsky, M.F., Suppression of ethylnitrosourea-induced schwannoma development involves elimination of neu/erbB-2 mutant premalignant cells in the resistant BDIV rat strain, *Cancer Res.*, 60, 4756, 2000.
15. Aguzzi, A., Oncogenesis in the nervous system, in *Brain Tumor Invasion: Biological Clinical and Therapeutic Considerations*, Mikkelsen, T., Bjerkvig, R., Laerum, O.D., and Rosenblum, M.L., Eds., Wiley-Liss, Inc., New York, 1998, pp. 29–60.
16. Westermark, B. and Nistér, M., Growth factors and oncogenes in neuro-oncogenesis, in *Brain Tumor Invasion: Biological, Clinical and Therapeutic Considerations*, Mikkelsen, T., Bjerkvig, R., Laerum, O.D., and Rosenblum, M.L., Eds., Wiley-Liss, Inc., New York, 1998, pp. 61–70.
17. Collins, V.P., Progression as exemplified by human astrocytic tumors, *Sem. Cancer Biol.*, 9, 267, 1999.
18. Kleihues, P. and Ohgaki, H., Primary and secondary glioblastomas: From concept to clinical diagnosis, *Neuro-Oncology*, 9, 44, 1999.
19. Ohgaki, H., Watanabe, K., Peraud, A., Biernat, W., von Deimling, A., Yasargil, M.G., Yonekawa, Y., and Kleihues, P., A case history of glioma progression, *Acta Neuropathol.*, 97, 525, 1999.
20. Lokker, N.A., Sullivan, C.M., Hollenbach, S.J., Israel, M.A., and Giese, N.A., Platelet-derived growth factor (PDGF) autocrine signalling regulates survival and mitogenic pathways in glioblastoma cells: Evidence that the novel PDGF-C and PDGF-D ligands may play a role in the development of brain tumors, *Cancer Res.*, 62, 3729, 2002.
21. Ohgaki, H., Huang, H., Haltia, M., Vainio, H., and Kleihues, P., More about cell and molecular biology of simian virus 40: Implications for human infections and disease, *J. Natl. Cancer Inst.*, 92, 495, 2000.
22. Fuxe, J., Liu, L., Malin, S., Philipson, L., Collins, V.P., and Pettersson, R.F., Expression of the coxsackie and adenovirus receptor in human astrocytic tumors and zenografts, *Int. J. Cancer*, 103, 723, 2003.
23. Ohgaki, H., Reifenberger, G., Nomura, K., and Kleihues, P., Gliomas, in *Prognostic Factors in Cancer*, 2nd ed., Gospodarowicz, M.K., Ed., Wiley-Liss, Inc., 2001, p. 725.

24. Venetsanakos, E., Mirza, A., Fanton, C., Romanow, S.R., Tlsty, T., and McMahon, M., Induction of tubulogenesis in telomerase-immortalized human microvascular endothelial cells by glioblastoma cells, *Exp. Cell. Res.*, 1, 273, 2002.

25. Osborne, R.H., Houben, M.P., Tijssen, C.C., Coebergh, J.W., and van Duijn, C.M., The genetic epidemiology of glioma, *Neurology*, 57, 1751, 2001.

26. Lafuente, J.V., Alkiza, K., Garibi, J.M., Alvarez, A., Bilbao, J., Figols, J., and Cruz-Sanchez, F.F., Biologic parameters that correlate with the prognosis of human gliomas. *Neuropathology*, 20, 176–83, 2000.

27. Engelhard, H.H., Stelea, A., and Cochran, E.J., Oligodendroglioma: Pathology and molecular biology, *Surg. Neurol.*, 58, 111, 2002.

28. Hatanpaa, K.J., Burger, P.C., Eshleman, J.R., Murphy, K.M., and Berg, K.D., Molecular diagnosis of oligodendroglioma in paraffin sections, *Lab. Invest.*, 83, 419, 2003.

29. Watanabe, T., Nakamura, M., Kros, J.M., Burkhard, C., Yonekawa, Y., Kleihues, P., and Ohgaki, H., Phenotype versus genotype correlation in oligodendrogliomas and low-grade diffuse astrocytomas, *Acta Neuropathol.*, 103, 267, 2002.

30. Korshunov, A., Golanov, A., and Sycheva, R., Immunohistochemical markers for prognosis of cerebral glioblastomas, *J. Neuro-Oncol.*, 58, 217, 2002.

31. Reis, R.M., Hara, A., Kleihues, P., and Ohgaki, H., Genetic evidence of the neoplastic nature of gemistocytes in astrocytomas, *Acta Neuropathol.*, 102, 422, 2001.

32. Watanabe, K., Tachibana, O., Yonekawa, Y., Kleihues, P., and Ohgaki, H., Role of gemistrocytes in astrocytoma progression, *Lab. Invest.*, 76, 277, 1997.

33. Kösel, S., Scheithauer, B.W., and Graeber, M.B., Genotype-phenotype correlation in gemistocytic astrocytomas, *Neurosurgery*, 48, 187, 2001.

34. Katsetos, C.D., Del Valle, L., Geddes, J.F., Assimakopoulou, M., Legido, A., Boyd., J.C., Balin, B., Parikh, N., Maraziotis, T., de Chadarevian, J.-P., Varakis, J.N., Matsas, R., Spano, A., Frankfurter, A., Herman, M.M., and Khalili, K., Abberant localization of the neuronal class III β-tubulin in astrocytomas: A marker for anaplastic potential, *Arch. Pathol. Lab. Med.*, 125, 613, 2001.

35. Katsetos, C.D., Del Valle, L., Geddes, J.F., Aldape, K., Boyd, J.C., Legido, A., Khalili, K., Perentes, E., and Mörk, S.J., Localization of the neuronal class III β-tubulin in oligodendrogliomas. Comparison with Ki-67 proliferative index and 1p/19q status, *J. Neuropathol. Exp. Neurol.*, 61, 307, 2002.

36. Ellison, D., Classifying the medulloblastoma: Insights from morphology and molecular genetics, *Neuropathol. Appl. Neurobiol.*, 28, 257, 2002.

37. Patt, S., Labrakakis, C., Bernstein, M., Weydt, P., Cervos-Navarro, J., Nish, G., and Kettenmann, H., Neuron-like physiological properties of cells from human oligodendroglial tumors, *Neuroscience*, 71, 601, 1996.

38. Labrakakis, C., Patt, S., Hartmann, J., and Kettenmann, H., Functional GABA(A) receptors on human glioma cells, *Eur. J. Neurosci.*, 10, 231, 1998.

39. Rajewsky, M.F., Specificity of DNA damage in chemical carcinogenesis, in *Molecular and Cellular Aspects of Carcinogen Screening Tests*, Montesano, R., Bartsch, H., and Tomatis, L., Eds., IARC Scientific Publications, No. 27, 1980, pp. 41–54.

40. Evans, J.J., Lee, J.H., Hawthorne, L., Bondar, J., Harwalkar, J.A., Tchernova, O., Signorelli, K., Barnett, G.H., Cowell, J.K., and Golubic, M., Gene expression profiles in meningiomas, *Neuro-Oncology*, 3, 299, 2001.

41. Gutmann, D.H., Rasmussen, S.A., Wolkenstein, P., MacCollin, M.M., Guha, A., Inskip, P.D., North, K.N., Poyhonen, M., Birch, P.H., and Friedman, J.M., Gliomas presenting after age 10 in individuals with neurofibromatosis type 1 (NF1), *Neurology*, 59, 759, 2002.

42. McLendon, R.E. and Provenzale, J., Glioneuronal tumors of the central nervous system, *Brain Tumor Pathol.*, 19, 51, 2002.

43. Carpenter, M.K., Cui, X., Hu, Z.-Y., Jackson, J., Sherman, S., Seiger, Å., and Wahlberg, L.U., In vitro expansion of a multipotent population of human neural progenitor cells, *Exp. Neurol.*, 158, 265, 1999.

44. Dahlstrand, J., Zimmerman, L.B., McKay, R.D.G., and Lendahl, U., Characterization of the human nestin gene reveals a close evolutionary relationship to neurofilaments, *J. Cell Sci.*, 103, 589, 1992.

45. Dahlstrand, J., Lardelli, M., and Lendahl, U., Nestin in RNA expression correlates with the central nervous system progenitor cell state in many, but not all, regions of the developing central nervous system, *Dev. Brain Res.*, 84, 109, 1995.

46. Almqvist, P.M., Immunohistochemical detection of nestin in pediatric brain tumors, *J. Histochem. Cytochem.*, 50, 147, 2002.

47. Galli, R.G., Binda, E., Orfanelli, U., Cipelletti, B., Gritti, A., De Vitis, S., Fiocco, R., Foroni, C., Dimeco, F., and Vescovi, A., Isolation and characterization of tumorigenic, stem-like neural precursors from human glioblastoma, *Cancer Res.*, 64, 7011, 2004.

48. Leuraud, P., Aguirre-Cruz, L., Hoang-Xuan, K., Criniere, E., Tanguy, M.L., Golmard, J.L., Kujas, M., Delattre, J.Y., and Sanson, M., Telomerase reactivation in malignant gliomas and loss of heterozygosity on 10p15.1. *Neurology*, 60, 1820–2, 2003.

    Chang, G.H., Aldape, K.D., Hirose, Y., Berger, M.S., and Pieper, R.O., Genomic alterations in human glioblastoma multiforme occur independently of telomerase reactivation, *Neuro-Oncology*, 299, 2001.

49. Nistér, M. and Westermark, B., Mechanisms of altered growth control, in *Russel and Rubinstein's Pathology of Tumors of the Nervous System*, Vol. 1, Bigner, D.D., McLendon R., and Brunet, J.M., Eds., Arnold Publishers, London, Sydney, Auckland, 1998, pp. 83–116.

50. Laerum, O.D., The role of growth factors and oncogenes in tumor cell invasion, in *Brain Tumor Invasion*, Mikkelsen, T., Bjerkvig, R., Laerum, O.D., and Rosenblum, M.L., Eds., Wiley-Liss, Inc., New York, 1998, pp. 323–342.

51. Welch, W.C., Kornblith, P.L., Michalopoulos, G.K., Petersen, B.E., Beedle, A., Gollin, S.M., and Goldfarb, R.H., Hepatocyte growth factor (HGF) and receptor (c-met) in normal and malignant astrocytic cells, *Anticancer Res.*, 19, 1635, 1999.

52. Laerum, O.D., Hülser, F., and Rajewsky, M.F., Electrophysiological properties of ethylnitrosourea-induced, neoplastic neurogenic rat cell lines cultured in vitro and in vivo, *Cancer Res.*, 36, 2153, 1976.

53. Zhang, W., Couldwell, W.T., Simard, M.F., Song, H., Lin, J.H.-C., and Nedergaard, M., Direct gap junction communication between malignant glioma cells and astrocytes, *Cancer Res.*, 59, 1994, 1999.

54. Steinsvåg, S.K. and Laerum, O.D., Transmission electron microscopy of cocultures between normal rat brain tissue and rat glioma cells, *Anticancer Res.*, 5, 137, 1985.

55. Sato, K., Gratas, C., Lampe, J., Biernat, W., Kleihues, P., Yamasaki, H., and Ohgaki, H., Reduced expression of the P_2 form of the gap junction protein connexin43 in malignant meningiomas, *J. Neuropathol. Exp. Neurol.*, 56, 835, 1997.

56. Zheng, P.P., Kros J.M., Sillevis-Smitt, P.A., and Luider, P.T., Proteomics in primary brain tumors, *Front Biosci.*, 8, 451, 2003.

57. Hoi, H.S., Banaie, A., Rigby, L., Chen, J., and Meltzer, H., Alteration in p53 modulates glial proteins in human glial tumour cells, *J. Neuro-Oncol.*, 48, 191, 2000.

58. Bowers, D.C., Fan, S., Walter, K.A., Abounader, R., Williams, J.A., Rosen, E.M., and Laterra, J., Scatter factor/hepatocyte growth factor protects against cytotoxic death in human glioblastoma via phosphatidylinositol 3-kinase- and AKT-dependent pathways, *Cancer Res.*, 60, 4277, 2000.

59. Walter, K.A., Luddy, C., Goel, N., and Laterra, J., Scatter factor/hepatocyte growth factor stimulation of malignant glioma cell lines overcomes contact induced G1/G0 growth arrest via upregulation of C-myc and downregulation of p27, *Neuro-Oncology*, 3, 282, 2001.

60. Stemmer-Rachamimov, A.O. and Louis, D.N., Histopathologic and immunohistochemical prognostic factors in malignant gliomas, *Curr. Opin. Oncol.*, 9, 230, 1997.

61. Ellison, D.W., Steart P.V., Bateman, A.C., Pickering, R.M., Palmer, J.D., and Weller, R.O., Prognostic indicators in a range of astrocytic tumours: An immunohistochemical study with Ki-67 and p53 antibodies, *J. Neurol. Neurosurg. Psychiat.*, 59, 413, 1995.

62. Vaquero, J., Zurita, M., Morales, C., Oya, S., and Coca, S., Prognostic significance of endothelial surface score and MIB-1 labeling index in glioblastoma, *J. Neuro-Oncol.*, 46, 11, 2000.

63. Pedersen, P.H., Ness, G.O., Engebraaten, O., Bjerkvig, R., Lillehaug, J.R., and Laerum, O.D., Heterogeneous response to growth factors (EGF, PDGF (bb), TGFα, bFGF, IL-2) on glioma spheroid growth, migration and invasion, *Int. J. Cancer*, 56, 255, 1994.

64. Diedrich, U., Lucius, J., Baron, E., Behnke, J., Pabst, B., and Zoll, B., Distribution of epidermal factor receptor gene amplification in brain tumours and correlation to prognosis, *J. Neurol.*, 242, 683, 1995.

65. Zumkeller, W., IGFs and IGF-binding proteins as diagnostic markers and biological modulators in brain tumors, *Expert. Rev. Mol. Diagn.*, 2, 473, 2002.

66. Tanabe, T., Kominsky, S.L., Subramaniam, P.S., Johnson, H.M., and Torres, B.A., Inhibition of the glioblastoma cell cycle by type I IFNs occurs at both the G1 and S phases and correlates with the upregulation of p21$^{WAF1/CIP1}$, *J. Neuro-Oncol.*, 48, 225, 2000.

67. Thomas, C.Y., Chounard, M., Cox, M., Parsons, S., Stallings-Mann, M., Garcia, R., Jove, R., and Wharen, R., Spontaneous activation and signalling by overexpressed epidermal growth factor receptors in glioblastoma cells, *Int. J. Cancer*, 104, 19, 2003.

68. Gratas, C., Tohma, Y., Van Meir, E.G., Klein, M., Tenan, M., Ishii, N., Tachibana, O., Kleihues, P., and Ohgaki, H., Fas ligand expression in glioblastoma cell lines and primary astrocytic brain tumors, *Brain Pathol.*, 7, 863, 1997.

69. Riffkin, C.D., Gray, A.Z., Hawkins, C.J., Chow, C.W., and Ashley, D.M., Ex vivo pediatric brain tumors express Fas (CD95) and FasL (CD95L) and are resistant to apoptosis induction, *Neuro-Oncology*, 3, 229, 2001.

70. Mandil, R., Ashkenazi, E., Blass, M., Kronfeld, I., Kazimirsky, G., Rosenthal, G., Umansky, F., Lorenzo, P.S., Blumberg, P.M., and Brodie, C., Protein kinase α and protein kinase Cδ play opposite roles in the proliferation and apoptosis of glioma cells[1], *Cancer Res.*, 61, 4612, 201.

71. Visted, T., Enger, P.O., Lund-Johansen, M., and Bjerkvig, R., Mechanisms of tumor cell invasion and angiogenesis in the central nervous system, *Frontiers Biosci.*, 8, 289, 2003.

72. Raza, S.M., Lange, F.F., Aggarwal, B.B., Fuller, G.N., Wildrick, D.M., and Sawaya, R., Necrosis and glioblastoma: A friend or a foe? A review and a hypothesis, *Neurosurgery*, 51, 2, 2002.

73. Tohma, Y., Gratas, C., Van Meir, E.G., Desbaillets, I., Tenan, M., Tachibiana, O., Kleihues, P., and Ohgaki, H., Necrogenesis and Fas/APO-1 (CD95) expression in primary (de novo) and secondary glioblastomas, *J. Neuropathol. Exp. Neurol.*, 57, 239, 1998.

74. Sonoda, Y., Ozawa, T., Aldape, K.D., Deen, D.F., Berger, M.S., and Pieper, R., Akt pathway activation converts anaplastic astrocytoma to glioblastoma multiforme in human astrocyte model of glioma, *Cancer Res.*, 61, 6674, 2001.

75. Takano, T., Lin, J.H.-C., Arcuino, G., Gao, Q., Yang, J., and Nedergaard, M., Glutamate release promotes growth of malignant gliomas, *Nat. Med.*, 7, 1010, 2001.

76. Bigner, S.H., Batra, S.K., and Rasheed, B.K., Mechanisms of altered growth control. Cytogenetics, oncogenes and suppressor genes, in *Russel and Rubinstein's Pathology of Tumors of the Nervous System*, Vol. 1, Bigner, D.D., McLendon, R., and Brunet, J.M., Eds., Arnold Publishers, London, Sydney, Auckland, 1998, pp. 47–87.

77. Laerum, O.D., Flow cytometry and DNA ploidy measurements, in *Russel and Rubinstein's Pathology of Tumors of the Nervous System*, Vol. 1, Bigner, D.D., McLendon, R., and Brunet, J.M., Eds., Arnold Publishers, London, Sydney, Auckland, 1998, pp. 703–714.

78. Harada, K., Nishizaki, T., Ozaki, S., Kubota, H., Ito, H., and Sasaki, H., Intratumoral cytogenetic heterogeneity detected by comparative genomic hybridisation and laser scanning cytometry in human gliomas, *Cancer Res.*, 58, 4694, 1998.

79. Leenstra, S., Bijlsma, E.K., Troost, D., Osting, J., Westerveld, A., Bosch, D.A., and Hulsebos, T.J.M., Allele loss on chromosomes 10 and 17p and epidermal growth factor receptor gene amplification in human malignant astrocytoma related to prognosis, *Br. J. Cancer*, 70, 684, 1994.

80. Kunwar, S., Mohapatra,G., Bollen, A., Lamborn, K.R., Prados, M., and Feuerstein, B.G., Genetic subgroups of anaplastic astrocytomas correlate with patient age and survival, *Cancer Res.*, 61, 7683, 2001.

81. Ohgaki, H., Dessen, P., Jourde, B., Horstmann, S., Nishikawa, T., Di Patre, P.-L., Bukhard, C., Schüler, D., Probst-Hensch, N.M., Maiorka, P.C., Baeza, N., Pisani, P., Yonekawa, Y., Yasargil, M.G., Lütolf, U.M., and Kleihues, P., Genetic pathways to glioblastoma: A population-based study, *Cancer Res.*, 64, 6892, 2004.

82. Sonoda, Y., Ozawa, T., Hirose, Y., Aldape, K.D., McMahon, M., Berger, M.S., and Pieper, R.O., Formation of intracranial tumors by genetically modified human astrocytes defines four pathways critical in the development of human anaplastic astrocytoma, *Cancer Res.*, 61, 4956, 2001.

83. Sasaki, H., Betensky, R.A., Cairncross, G., and Louis, D.N., DMBT1 polymorphisms: Relationship to malignant glioma tumorigenesis, *Cancer Res.*, 62, 1790, 2002.

84. Wiltshire, R.N., Rasheed, B.K., Friedman, H.S., Friedman, A.H., and Bigner, S.H., Comparative genetic patterns of glioblastoma multiforme: Potential diagnostic tool for tumor classification, *Neuro-Oncology*, 2, 164, 2000.

85. Dyer, S., Prebble, E., Davison, V., Davies, P., Ramani, P., Ellison, D., and Grundy, R., Genomic imbalances in pediatric intracranial ependymomas define clinically relevant groups, *Am. J. Pathol.*, 161, 2133, 2002.

86. Fujisawa, H., Kurrer, M., Reis, R.M., Yonekawa, Y., Kleihues, P., and Ohgaki, H., Acquisition of the glioblastoma phenotype during astrocytoma progression is associated with loss of heterozygosity on 10q25-qter, *Am. J. Pathol.*, 155, 387, 1999.

87. Burton, E.C., Forsyth, P.A., Scott, J., Lamborn, K.R., Uyhara-Lock, J., Prados, M.D.J., Feuerstein, B.G., Berger, M.S., Uhm, J., O'Neill, B., Passe, S., Jenkins, R., and Aldape, K.D., Glioblastomas from patients who survive long-term are genotypically distinct from those found in typical survivors, *Neuro-Oncology*, 3, 293, 2001.

88. Tohma, Y., Gratas, C., Biernat, W., Peraud, A., Fukuda, M., Yonekawa, Y., Kleihues, P., and Ohgaki, H., PTEN (MMAC1) mutations are frequent in primary glioblastomas (de novo) but not in secondary glioblastomas, *J. Neuropathol. Exp. Neurol.*, 57, 684, 1998.

89. Biernat, W., Tohma, Y., Yonekawa, Y., Kleihues, P., and Ohgaki, H., Alterations of cell cycle regulatory genes in primary (de novo) and secondary glioblastomas, *Acta Neuropathol.*, 94, 303, 1997.

90. Vital, A., Bringuier, P.-P., Huang, H., San Galli, F., Rivel, J., Ansoborlo, S., Cazauran, J.-M., Taillandier, L., Kleihues, P., and Ohgaki, H., Astrocytomas and choroid plexus tumors in two families with identical p53 germline mutations, *J. Neuropathol. Exp. Neurol.*, 57, 1061, 1998.

91. Shiraishi, S., Tada, K., Nakamura, H., Makino, K., Kochi, M., Saya, H., Kuratsu, J., and Ushio, Y., Influence of p53 mutations on prognosis of patients with glioblastoma, *Cancer*, 95, 249, 2002.

92. Reis, R.M., Herva, R., Brandner, S., Koivukangas, J., Mironov, N., Bär, W., Kleihues, P., and Ohgaki, H., Second primary glioblastoma, *J. Neuropathol. Exp. Neurol.*, 60, 208, 2001.

93. Wei, K.C., Berger, M.S., and Sehgal, A., Molecular characterization of a novel BR-2 gene that is down-regulated in human low grade glioma tumors, *Anticancer Res.*, 22, 649, 2002.

94. Wei, K.C., Berger, S.M., and Sehgal, A., Cloning, sequencing and expression analysis of a novel gene BR-1 that is expressed in normal human brain tissue but not in glioma tumor samples, *Anticancer Res.*, 22, 745, 2002.

95. Liau, L.M., Lallone, R.L., Seitz, R.S., Buznikov, A., Gregg, J.P., and Kornblum, H.I., Identification of a human glioma-associated growth factor gene, granulin, using differential immuno-absorption, *Cancer Res.*, 60, 1353, 2000.

96. Nakamura, M., Yonekawa, Y., Kleihues, P., and Ohgaki, H., Promoter hypermethylation of the RB1 gene in glioblastomas, *Lab. Invest.*, 81, 77, 2001.

97. Nakamura, M., Watanabe, T., Klangby, U., Asker, C., Wiman, K., Yonekawa, Y., Kleihues, P., and Ohgaki, H., P14^{ARF} deletion and methylation in genetic pathways to glioblastomas, *Brain Pathol.*, 11, 159, 2001.

98. Nakamura, M., Watanabe, T., Yonekawa, Y., Kleihues, P., and Ohgaki, H., Promoter methylation of the DNA repair gene MGMT in astrocytomas is frequently associated with G:C → A:T mutations of the TP53 tumor suppressor gene, *Carcinogenesis*, 22, 1715, 2001.

99. Zlatescu, M.C., Yazdi, A.R., Sasaki, H., Megyesi, J.F., Betensky, R.A., Louis, D.N., and Cairncross, J.G., Tumor location and growth pattern correlate with genetic signature in oligodendroglial neoplasms, *Cancer Res.*, 61, 6713, 2001.

100. Ueki, K., Nishikawa, R., Nakazato, Y., Hirose, T., Hirato, J., Funada, N., Fujimaki, T., Hojo, S., Kubo, O., Ide, T., Usui, M., Ochiai, C., Ito, S., Takahashi, H., Mukasa, A., Asai, A., and Kirino, T., Correlation of histology and molecular genetic analysis of 1p, 19q, 10q, TP53, EGFR, CDK4, and CDKN2A in 91 astrocytic and oligodendroglial tumors, *Clin. Cancer*, 8, 196, 2002.

101. Reis, R.M., Könü-Leblebicioglu, D., Lopes, J.M., Kleihues, P., and Ohgaki, H., Genetic profile of gliosarcomas, *Am. J. Pathol.*, 156, 425, 2000.

102. Watanabe, T., Yokoo, H., Yokoo, M., Yonekawa, Y., Kleihues, P., and Ohgaki, H., Concurrent inactivation of RB1 and TP53 pathways in anaplastic oligodendrogliomas, *J. Neuropathol. Exp. Neurol.*, 60, 1181, 2001.

103. Huang, H., Mahler-Araujo, B.M., Sankila, A., Chimelli, L., Yonekawa, Y., Kleihues, P., and Ohgaki, H., APC mutations in sporadic medulloblastomas, *Am. J. Pathol.*, 156, 433, 2000.

104. Kraus, J.A., Felsberg, J., Tonn, J.C., Reifenberger, G., and Pietsch, T., Molecular genetic analysis of the TP53, PTEN, CDKN2A, EGFR, CDK4 and MDM2 tumour-associated genes in supratentorial

primitive neuroectodermal tumours and glioblastomas of childhood, *Neuropathol. Appl. Neurobiol.*, 28, 325, 2002.

105. Pore, N., Liu, S., Haas-Kogan, A., O'Rourke, D.M., and Maity, A., PTEN mutation and epidermal growth factor receptor activation regulate vascular endothelial growth factor (VEGF) mRNA expression in human glioblastoma cells by transactivating the proximal VEGF promoter, *Cancer Res.*, 63, 236, 2003.

106. Tchirkov, A., Rolhion, C., Kémény, J.-L., Irthum, B., Puget, S., Khalil, T., Chinot, O., Kwiatkowski, F., Périssel, B., Vago, P., and Verrelle, P., Clinical implications of quantiative real-time RT-PCR analysis of hTERT gene expression in human gliomas, *Br. J. Cancer*, 88, 516, 2003.

107. Ding, H., Roncari, L., Shannon, P., Xiaoli, W., Lau, N., Karaskova, J., Gutmann, D.H., Squire, J.A., Nagy, A., and Guha, A., Astrocyte-specific expression of activated p21-ras results in malignant astrocytoma formation in a transgenic mouse model of human gliomas, *Cancer Res.*, 61, 3826, 2001.

108. Riemenschneider, M.J., Büschges, R., Wolter, M., Reifenberger, J., Boström, J., Kraus, J.A., Schlegel, U., and Reifenberger, G., Amplification and overexpression of the MDM4 (MDMX) gene from 1q32 in a subset of malignant gliomas without TP53 mutation or MDM2 amplification, *Cancer Res.*, 59, 6091, 1999.

109. Ljubimova, J.Y., Lakhter, A.J., Loksh, A., Yong, W.H., Riedinger, M.S., Miner, J.H., Sorokin, L.M., Ljubimov, A.V., and Black, K.L., Overexpression of α4 chain-containing laminins in human glial tumors identified by gene microarray analysis, *Cancer Res.*, 61, 5601, 2001.

110. Huang, H., Colella, S., Kurrer, M., Yonekawa, Y., Kleihues, P., and Ohgaki, H., Gene expression profiling of low-grade diffuse astrocytomas by cDNA arrays, *Cancer Res.*, 60, 6868, 2000.

111. Kaulich, K., Blaschke, B., Numann, A., von Deimling, A., Wiestler, O.D., Weber, R.G., and Reifenberger, G., Genetic alterations commonly found in diffusely infiltrating cerebral gliomas are rare or absent in pleomorphic xanthoastrocytomas. *J. Neuropathol Exp. Neurol.*, 61, 1029–9, 2002.

112. Pomeroy, S.L., Tamayo, P., Gaasenbeek, M., Sturla, L.M., Angelo, M., McLaughlin, M.E. et al., Prediction of central nervous system embryonal tumour outcome based on gene expression, *Nature*, 415, 436, 2002.

113. Freije, W.A., Castro-Vargas, E.F., Fang, Z., Horvath, S., Cloughesy, T., Liau, L.M., Mischel, P.S., and Nelson, S.F., Gene expression profiling of gliomas strongly predicts survival, *Cancer Res.*, 64, 6503, 2004.

114. Caskey, L.S., Fuller, G.N., Bruner, J.M., Yung, W.K., Sawaya, R.E., Holland, E.C., and Zhang, W., Toward a molecular classification of the gliomas: Histopathology, molecular genetics, and gene expression profiling, *Histol. Histopathol.*, 15, 971, 2000.

115. Qi, Z.Y., Li, Y., Ying, K., Wu, C.Q., Tang, R., Zhou, Z.X., Chen, Z.P., Hui, G.Z., and Xie, Y., Isolation of novel differentially expressed genes related to human glioma using cDNA microarray and characterizations of two novel full-length genes, *J. Neuro-Oncol.*, 56, 197, 2002.

116. Lal, A., Lash, A.E., Altschul, S.F., Velculescu, V., Zhang, L., McLendon, R.E., Marra, M.A., Prange, C. et al., A public database for gene expression in human cancers, *Cancer Res.*, 59, 5403, 1999.

117. Hanash, S.M., Bobek, M.P., Rickman, D.S., Williams, T., Rouillard, J.M., Kuick, R., and Puravs, E., Integrating cancer genomics and proteomics in the post-genome era, *Proteomics*, 2, 69, 2002.

118. Laerum, O.D., Local spread of malignant neuroepithelial tumors, *Acta Neurochir.*, 139, 515, 1997.

119. Mikkelsen, T., Bjerkvig, R., Laerum, O.D., and Rosenblum, M.L., Eds., *Brain Tumor Invasion*, Wiley-Liss, Inc., New York, 1998.

120. Laerum, O.D. and Mørk, S.J., Mechanisms of altered growth control: Invasion and metastasis, in *Russel and Rubinstein's Pathology of Tumors of the Nervous System*, Vol. 1, Bigner, D.D., McLendon R., and Brunet, J.M., Eds., Arnold Publications, London, Sydney, Auckland, 1998, pp. 117–140.

121. Yamamoto, M., Ueno, Y., Hayashi, S., and Fukushima, T., The role of proteolysis in tumor invasiveness in glioblastoma and metastatic brain tumors, *Anticancer Res.*, 22, 4265, 2002.

122. Johnsen, M., Lund, L.R., Romer, J., Almholt, K., and Danø, K., Cancer invasion and tissue remodeling: Common themes in proteolytic matrix degradation, *Curr. Opin. Cell Biol.*, 10, 667, 1998.

123. Lund, L.R., Rømer, J., Bugge, T.H., Nielsen, B.S., Frandsen, T.L., Degen, J.L., Stephens, R.W., and Danø, K., Functional overlap between two classes of matrix-degrading proteases in wound healing, *EMBO J.*, 18, 4645, 1999.

124. Guillamo, J.S., Lisovoski, F., Christov, C., Le Guerinel, C., Defer, G.L., Peschanski, M., and Lefrancois, T., Migration pathways of human glioblastoma cells xenografted into the immunosuppressed rat brain, *J. Neuro-Oncol.*, 52, 205, 2001.

125. Mourad, P.D., Farrell, L., Stamps, L.D., Santiago, P., Fillmore, H.L., Broaddus, W.C., and Silbergeld, D.L., Quantitative assessment of glioblastoma invasion in vivo, *Cancer Lett.*, 20, 97, 2003.

126. Nygaard, S.J.T., Pedersen, P.-H., Mikkelsen, T., Terzis, A.J.A., Tysnes, O.-B., and Bjerkvig, R., Glioma cell invasion visualized by scanning confocal laser microscopy in an in vitro co-culture system, *Invasion Metastasis*, 15, 179, 1995.

127. Pedersen, P.-H., Nygaard, S.J.T., Baardsen, R., Tysnes, B.B., Engebraaten, O., and Bjerkvig, R., Glioma invasion studied in organotypic coculture models, in *Brain Tumor Invasion*, Mikkelsen, T., Bjerkvig, R., Laerum, O.D., and Rosenblum, M.L., Eds., Wiley-Liss, New York, 1998.

128. Nygaard, S.J.T., Haugland, H.K.R., Laerum, O.D., Lund-Johansen, M., Bjerkvig, R., and Tysnes, O.-B., Dynamic determination of human glioma invasion in vitro, *J. Neurosurg.*, 89, 441, 1998.

129. Laerum, O.D., Nygaard, S.J.T., Steine, S., Mørk, S.J., Engebraaten, O., Peraud, A., Kleihues, P., and Ohgaki, H., Invasiveness in vitro and biological markers in human primary glioblastomas, *J. Neuro-Oncol.*, 54, 1, 2001.

130. de Bouard, S., Christov, C., Guillamo, J.S., Kassar-Duchossoy, L., Palfi, S., Leguerinel, C., Masset, M., Cohen-Hagenauer, O., and Peschanski, M., Invasion of human glioma biopsy specimens in cultures of rodent brain slices: A quantitative analysis, *J. Neurosurg.*, 97, 169, 2002.

131. Franko, A.J., Parliament, M.B., Allalunis-Turner, M.J., and Wolokoff, B.G., Variable presence of hypoxia in M006 human glioma spheroids and in spheroids and xenografts of clonally derived sublines, *Br. J. Cancer*, 78, 1261, 1998.

132. Kaczarek, E., Zapf, S., Bouterfa, H., Tonn, J.C., Westphal, M., and Giese, A., Dissecting glioma invasion: Interrelation of adhesion, migration and intercellular contacts determine the invasive phenotype, *Int. J. Neurosci.*, 17, 625, 1999.

133. Mahesparan, R., Tysnes, B.B., and Read, T.-A., Extracellular matrix-induced cell migration from glioblastoma biopsy specimens in vitro, *Acta Neuropathol.*, 97, 231, 1999.

134. Mahesparan, R., Tysnes, B.B., Edvardsen, K., Haugland, H.K., Cabrera, I.G., Lund-Johansen, M., Engebraaten, O., and Bjerkvig, R., Role of high molecular weight extracellular matrix proteins in glioma cell migration, *Neuropathol. Appl. Neurobiol.*, 23, 102, 1997.

135. Tysnes, B.B., Larsen, L.F., Ness, G.O., Mahesparan, R., Edvardsen, K., Garcia-Cabera, I., and Bjerkvig, R., Stimulation of glioma-cell migration by laminin and inhibition by anti-$\alpha 3$ and anti-$\beta 1$ integrin antibodies, *Int. J. Cancer*, 67, 777, 1996.

136. Goldbrunner, R.H., Bernstein, J.J., and Tonn, J.C., Cell-extracellular matrix interaction in glioma invasion, *Act- Neurochir-Wien*, 141, 295, 1999.

137. Rooprai, H.K., Vanmeter, T., Panou, C., Schnull, S., Trillo-Pazos, G., Davies, D., and Pilkington, G.J., The role of integrin receptors in aspects of glioma invasion in vitro, *Int. J. Dev. Neurosci.*, 17, 613, 1999.

138. Wiranowska, M., Rojiani, A.M., Gottschall, P.E., Moscinski, L.C., Johnson, J., and Saporta, S., CD44 expression and MMP-2 secretion by mouse glioma cells: Effect of interferon and anti-CD44 antibody, *Anticancer Res.*, 20, 4301, 2000.

139. Isamu, O., Yoshiaki, K., Mitsuhiro, M.l.L., Moritaka, S., Kozo, K., Masayuki, A., and Hideyuki, S., Regulated CD44 cleavage under the control of protein kinase C, calcium influx, and the Rho family of small G proteins, *J. Biol. Chem.*, 274, 25525, 1999.

140. Paulus, W., Baur, I., Beutler, A.S., and Reeves, S.A., Diffuse brain invasion of glioma cells requires $\beta 1$ integrins, *Lab. Invest.*, 75, 819, 1996.

141. Knott, J.C., Mahesparan, R., Garcia-Cabrera, I., Tysnes, B.B., Edvardsen, K., Ness, G.O., Mørk, S.J., Lund-Johansen, M., and Bjerkvig, R., Stimulation of extracellular matrix components in the normal brain by invading glioma cells, *Int. J. Cancer*, 75, 864, 1998.

142. Tysnes, B.B., Mahesparan, F.A., Thorsen, F., Haugland, H.K., Porwols, T., Enger, P.Ø., Lund-Johansen, M., and Bjerkvig, R., Laminin expression by glial fibrillary acidic protein positive cells in human gliomas, *Int. J. Dev. Neurosci.*, 17, 531, 1999.

143. Chintala, S.K. and Rao, J.S., Invasion of human glioma: Role of extracellular matrix proteins, *Frontiers Biosci.*, 1, d324, 1996.

144. Binder, D.K. and Berger, M.S., Proteases and the biology of glioma invasion, *J. Neuro-Oncol.*, 56, 149, 2002.

145. Llano, E., Pendás, A.M., Freije, J.P., Nakano, A., Knäuper, V., Murphy, G., and Lópe-Otin, C., Identification and characterization of human MT5-MMP, a new membrane-bound activator of progelatinase A overexpressed in brain tumors, *Cancer Res.*, 59, 2570, 1999.

146. Demchik, L.L., Sameni, M., Nelson, K., Mikkelsen, T., and Sloane, B.F., Cathepsin B and glioma invasion, *Int. J. Dev. Neurosci.*, 17, 483, 1999.

147. Laerum, O.D., The role of growth factors and oncogenes in tumor cell invasion, in *Brain Tumor Invasion: Biological, Clinical and Therapeutic Considerations*, Mikkelsen, T., Bjerkvig, R., Laerum, O.D., and Rosenblum, M.L., Eds., Wiley-Liss, Inc., New York, 1998, pp. 323–341.

148. Wagner, S., Stegen, C., Bouterfa, H., Huettner, C., Kerkau, S., Roggendorf, W., Roosen, K., and Tonn, J.C., Expression of matrix metalloproteinases in human glioma cell lines in the presence of IL-10, *J. Neuro-Oncol.*, 40, 113, 1998.

149. Lund-Johansen, M., Bjerkvig, R., Humphrey, P.A., Bigner, S.H., Bigner, D.D., and Laerum, O.D., Effect of epidermal growth factor on glioma cell growth, migration, and invasion in vitro, *Cancer Res.*, 50, 6039, 1990.

150. Knüpfer, M.M., Knüfper, H., Jendrossek, V., van Gool, S., Wolff, J.E.A., and Keller, E., Interferon-gamma inhibits growth and migration of A172 human glioblastoma cells, *Anticancer Res.*, 21, 3989, 2001.

151. Penar, P.L., Khoshyomn, S., Bhushan, A., and Tritton, T.R., Inhibition of epidermal growth factor receptor-associated tyrosine kinase blocks glioblastoma invasion of the brain, *Neurosurgery*, 40, 141, 1997.

152. Pedersen, P.H., Ness, G.O., Engebraaten, O., Bjerkvig, R., Lillehaug, J.R., and Laerum, O.D., Heterogeneous response to the growth factors (EGF, PDGF (bb), TGF-α, bFGF, IL2) on glioma spheroid growth, migration and invasion, *Int. J. Cancer*, 56, 255, 1994.

153. Khoshyomn, S., Lew, S., DeMattia, J., Singer, E.B., and Penar, P.L., Brain tumor invasion rate measured in vitro does not correlate with Ki-67 expression, *J. Neuro-Oncol.*, 45, 111, 1999.

154. Bjerknes, R., Bjerkvig, R., and Laerum, O.D., Phagocytic capacity of normal and malignant rat glial cells in culture, *J. Natl. Cancer Inst.*, 78, 279, 1987.

155. Mariani, L., McDonough, W.S., Hoelzinger, D.B., Beaudry, C., Kaczmarek, E., Coons, S.W., Giese, A., Moghaddam, M., Seiler, R.W., and Berens, M.E., Identification and validation of P311 as a glioblastoma invasion gene using laser capture microdissection, *Cancer Res.*, 61, 4190, 2001.

156. Golembieski, W.A., Ge, S., Nelson, K., Mikkelsen, T., and Rempel, S.A., Increased SPARC expression promotes U87 glioblastoma invasion in vitro, *Int. J. Dev. Neurosci.*, 17, 463, 1999.

157. Fuller, G.N., Rhee, C.H., Hess, K.R., Caskey, L.S., Wang, R., Bruner, J.M., Yung, W.K.A., and Zhang, W., Reactivation of insulin-like growth factor binding protein 2 expression in glioblastoma multiforme: A revelation by parallel gene expression profiling, *Cancer Res.*, 59, 4228, 1999.

158. Tremont-Lukats, I.W. and Gilbert, M.R., Advances in molecular therapies in patients with brain tumors, *Cancer Control*, 10, 125, 2003.

159. Kuan, C.T., Wikstrand, C.J., and Bigner, D.D., EGFRvIII as a promising target for antibody-based brain tumor therapy, *Brain Tumor Pathol.*, 17, 71, 2000.

160. Kimura, T., Takeshima, H., Nomiyama, N., Nishi, T., Kino, T., Kochi, M., Kuratsu, J.I., and Ushio, Y., Expression of lymphocyte-specific chemokines in human malignant glioma: Essential role of LARC in cellular immunity of malignant glioma, *Int. J. Oncol.*, 21, 707, 2002.

161. Heimberger, A.B., Learn, C.A., Archer, G.E., McLendon, R.C., Chewning, T.A., Tuck, F.L., Pracyk, J.B., Friedman, A.H., Friedman, H.S., Bigner, D.D., and Sampson, J.H., Brain tumors in mice are susceptible to blockade of epidermal growth factor receptor (EGFR) with the oral, specific, EGFR-tyrosine kinase inhibitor ZD1839 (Iressa), *Clin. Cancer Res.*, 8, 3496, 2002.

162. Sampson, J.H., Reardon, D.A., Friedman, A.H., Friedman, H.S., Coleman, R.E., McLendon, R.E., Pastan, I., and Bigner, D.D., Sustained radiographic and clinical response in patient with bifrontal recurrent glioblastoma multiforme with intracerebral infusion of the recombinant targeted toxin TP-38: case study. *Neuro-oncol.*, 7(1), 90–6, 2005.

163. Yang, L., Ng, K.Y., and Lillehei, K.O., Cell-mediated immunotherapy: A new approach to the treatment of malignant glioma, *Cancer Control*, 10, 138, 2003.

164. Mischel, P.S. and Cloughesy, T.F., Targeted molecular therapy of GBM, *Brain Pathol.*, 13, 52, 2003.

165. Mohri, M., Nitta, H., and Yamashita, J., Expression of multidrug resistance-associated protein (MRP) in human gliomas, *J. Neuro-Oncol.*, 49, 105, 2000.

166. Spear, M.A., Gene therapy of gliomas: Receptor and transcriptional targeting, *Anticancer Res.*, 18, 3223, 1998.

167. Mukai, S., Kondo, Y., Koga, S., Komata, T., Barna, B.P., and Kondo, S., 2-5A antisense telomerase RNA therapy for intracranial malignant gliomas, *Cancer Res.*, 60, 4461, 2000.

168. Komata, T., Kondo, Y., Kanzawa, T., Hirohata, S., Koga, S., Sumiyoshi, H., Srinivasa, S.M., Barna, B.P., Germano, I.M., Takakura, M., Inoue, M., Alnemri, E.S., Shay, J.W., Kyo, S., and Kondo, S., Treatment of malignant glioma cells with the transfer of constitutively active caspase-6 using the human telomerase catalytic subunit (human telomerase reverse transcriptase) gene promoter, *Cancer Res.*, 61, 5796, 2001.

169. Kunkel, P., Ulbricht, U., Bohlen, P., Brockmann, M.A., Fillbrandt, R., Stavrou, D., Westphal, M., and Lamszus, K., Inhibition of glioma angiogenesis and growth in vivo by systemic treatment with a monoclonal antibody against vascular endothelial growth factor receptor-2[1], *Cancer Res.*, 61, 6624, 2001.

170. Read, T.-A., Farhadi, M., Bjerkvig, R., Olsen, B.R., Rokstad, A.M., Huszthy, P.C., and Vajkoczy, P., Intravital microscopy reveals novel antivascular and antitumor effects of endostatin delivered locally by alginate-encapsulated cells, *Cancer Res.*, 61, 6830, 2001.

171. Yu, J.S., Wheeler, C.J., Zeltzer, P.M., Ying, H., Finger, D.N., Lee, P.K., Yong, W.H., Incardona, F., Thompson, R.C., Riedinger, M.S., Zhang, W., Prins, R.M., and Black, K.L., Vaccination of malignant glioma patients with peptide-pulsed dendritic cells elicits systemic cytotoxicity and intracranial T-cell infiltration, *Cancer Res.*, 61, 842, 2001.

172. Yu, G.W., Nutt, C.L., Zlatescu, M.C., Keeney, M., Chin-Yee, I., and Cairncross, J.G., Inactivation of p53 sensitized U87MG glioma cells to 1,3-*bis*(2-chloroethyl)-1-nitrosourea, *Cancer Res.*, 61, 4155, 2001.

173. Weller, M., Kleihues, P., Dickgans, J., and Ohgaki, H., CD95 ligand: Lethal weapon against malignant glioma? *Brain Pathol.*, 8, 285, 1998.

174. Ishiuchi, S., Tsuzuki, K., Yoshida, Y., Yamada, N., Hagimura, N., Okado, H., Miwa, A., Kurihara, H., Nakazato, Y., Tamura, M., Sakaki, T., and Ozawa, S., Blockage of Ca^{2+}-permeable AMPA receptors suppresses migration and induces apoptosis in human glioblastoma cells, *Nat. Med.*, 8, 971, 2002.

175. Fukumoto, M., Takahashi, J.A., Murai, N., Ueba, T., Kono, K., and Nakatsu, S., Induction of apoptosis in glioma cells: An approach to control tumor growth by blocking basic fibroblast growth factor autocrine loop, *Anticancer Res.*, 20, 4905, 2000.

176. Bauman, G.S., Fisher, B.J., McDonald, W., Amberger, V.R., Moore, E., and Del-Maestro, R.F., Effects of radiation on a three-dimensional model of malignant glioma invasion, *Int. J. Dev.-Neurosci.*, 17, 643, 1999.

177. Krishnan, S., Rao, R.D., James, C.D., and Sarkaria, J.N., Combination of epidermal growth factor receptor targeted therapy with radiation therapy for malignant gliomas, *Front Biosci.*, 8, E1, 2003.

178. Sánchez, C., de Ceballos, M.L., del Pulgar, T.G., Rueda, D., Corbacho, C., Velasco, G., Galve-Roperh, I., Huffman, J.W., Ramóny Cajal, S., and Guzman, M., Inhibition of glioma growth in vivo by selective activation of the CB[2] Cannabinoid receptor[1], *Cancer Res.*, 61, 5784, 2001.

179. Sehgal, A., Ricks, S., Warrick, J., Boynton, A.L., and Murphy, G.P., Antisense human neuroglia related cell adhesion molecule hNr-CAM, reduces the tumorigenic properties of human glioblastoma cells, *Anticancer Res.*, 19, 4947, 1999.

180. Kilic, T, Alberta, J.A., Zdunek, P.R., Acar, M., Iannarelli, P., O'Reilly, T., Buchdunger, E., Black, P.M., and Stiles, C.D., Intracranial inhibition of platelet-derived growth factor-mediated glioblastoma cell growth by an orally active kinase inhibitor of the 2-phenylaminopyrimidine class, *Cancer Res.*, 60, 5143, 2000.

181. Huang, R.-P., Fan, Y., Hossain, M.Z., Peng, Ao., Zeng, Z.L., and Boynton, A.L., Reversion of the neoplastic phenotype of human glioblastoma cells by connexin 43 (cx43)[1], *Cancer Res.*, 58, 5089, 1998.

182. Khoshyomn, S., Penar, P.L., Rossi, J., Wells, A., Abramson, D.L., and Bhushan, B.A., Inhibition of phospolipase Cα1 activation blocks glioma and invasion fetal aggregates, *Neurosurgery*, 44, 568, 1999.

183. Bello, L., Lucini, V., Guissani, C., Carrabba, G., Pluderi, M., Scaglione, F., Tomei, G., Villani, R., Black, P.M., Bikfalvi, A., and Carroll, R.S., IS201, a specific $\alpha v \beta 3$ integrin inhibitor, reduces glioma growth in vivo, *Neurosurgery*, 52, 177, 2003.

184. Wiranowska, M., Tresser, N., and Saporta, S., The effect of interferon and anti-CD44 antibody on mouse glioma invasiveness in vitro, *Antiancer Res.*, 18, 3331, 1998.

185. Noha, M., Yoshida, D., Watanabe, K., and Teramoto, A., Suppression of cell invasion on human malignant glioma cell lines by a novel matrix-metalloproteinase inhibitor SI-27: In vitro study, *J. Neuro-Oncol.*, 48, 217, 2000.

186. Yoshida, D., Watanabe, K., Noha, M., Takahashi, H., Teramoto, A., and Sugisaki, Y., Anti-invasive effect of an anti-matrix metalloproteinase agent in a murine brain slice model using the serial monitoring of green fluorescent protein-labelled glioma cells, *Neurosurgery*, 52, 187, 2003.

187. Auguste, P., Gürsel, D.B., Lemière, S., Reimers, D., Cuevas, P., Carceller, F., Di Santo, J.P., and Bikfalvi, A., Inhibition of fibroblast growth factor/fibroblast growth factor receptor activity in glioma cells impedes tumor growth by both angiogenesis-dependent and -independent mechanisms, *Cancer Res.*, 61, 1717, 2001.

188. Bello, L., Lucini, V., Carrabba, G., Giussani, C., Machluf, M., Pluderi, M., Nikas, D., Zhang, J., Tomei, G., Villani, R.M., Carroll, V.R., Bikfalvi, A., and Black, P.M., Simultaneous inhibition of glioma angiogenesis, cell proliferation, and invasion by a naturally occurring fragment of human metalloproteinase-2[1], *Cancer Res.*, 61, 8730, 2001.

189. Tonn, J.C., Kerkau, S., Hanke, A., Bouterfa, H., Mueller, J.G., Wagner, S., Vince, G.H., and Roosen, K., Effect of synthetic matrix-metalloproteinase inhibitors in invasive capacity and proliferation of human malignant gliomas in vitro, *Int. J. Cancer*, 80, 764, 1999.

190. Mohan, P.M., Chintala, S.K., Mohanam, S., Gladson, C.L., Kim, E.S., Gokaslan, Z.L., Lakka, S.S., Roth, J.A., Fang, B., Sawaya R., Kyritsis, A.P., and Rao, J.S., Adenovirus-mediated delivery of antisense gene to urokinase-type plasminogen activator receptor suppressor glioma invasion and tumor growth, *Cancer Res.*, 59, 3369, 1999.

191. Tysnes, B.B. and Mahesparan, R., Biological mechanisms of glioma invasion and potential therapeutic targets, *J. Neuro-Oncol.*, 53, 192, 2001.

192. Ware, M.L., Berger, M.S., and Binder, D.K., Molecular biology of glioma tumorigenesis, *Histol. Histopathol.*, 18, 207, 2003.

193. Fulci, G., Ishii, N., Maurici, D., Hainaut, P., Gernert, K., and van Meir, E.G., Sequential mutations in TP53 alleles can induce progressive loss of p53 functions and drive malignant glial tumor progression, *Neuro-Oncol.*, 3, 300, 2001.

194. Sasaki, M., Nakahira, K., Kawano, Y., Katakura, H., Yoshimine, T., Shimizu, K., Kim, S.A., and Ikenaka, K., MAGE-E1, a new member of the melanoma-associated antigen gene family and its expression in human glioma, *Cancer Res.*, 61, 4809, 2001.

195. Watson, M.A., Perry, A., Budhjara, V., Hicks, C., Shannon, W.D., and Rich, K.M., Gene expression profiling with oligonucleotide microarrays distinguishes World Health Organization grade of oligodendrogliomas, *Cancer Res.*, 61, 1825, 2001.

196. Kitange, G., Shibata, S., Tokunaga, Y., Yagi, N., Yasunaga, A., Kishikawa, M., and Naito, S., Ets-1 transcription factor-mediated urokinase-type plasminogen activator expression and invasion in glioma cells stimulated by serum and basic fibroblast growth factors, *Lab. Invest.*, 79, 407, 1999.

197. Mori, T.R., Abe, T., Wakabayashi, Y., Hikawa, T., Matsuo, K., Yamada, Y., Kuwano, M., and Hori, S., Up-regulation of urokinase-type plasminogen activator and its receptor corelates with enhanced invasion activity of human glioma cells mediated by transforming growth factor-α or basic fibroblast growth factor, *J. Neuro-Oncol.*, 46, 115, 2000.

198. Arato-Ohshima, T. and Sawa, H., Over-expression of cyclin D1 induces glioma invasion by increasing matrix metalloproteinase activity and cell motility, *Int. J. Cancer*, 83, 387, 1999.

199. Barr, M.L., *The Human Nervous System. An Anatomical Viewpoint*, 2nd ed., Harper International Edition, Harper & Row Publishers, Hagerstown, Maryland, New York, Evanston, San Francisco, London, 1974.

20 Thyroid Cancer: Molecular Biology and Clinical Aspects

Johan R. Lillehaug, Øystein Fluge, and Jan Erik Varhaug

CONTENTS

20.1 INTRODUCTION

There are two main groups of thyroid carcinomas, (1) carcinomas arising from thyroid follicular cells, the thyroid cells "proper," and (2) carcinomas arising from the neuroendocrine C-cells (calcitonin producing cells). The difference between the main groups is of major biological importance and has consequences for diagnosis and follow-up, as described in this chapter. In the thyroid, like in other organs, cells from connective tissue, lymphoid cells, etc. may also give rise to malignant tumors (e.g., sarcomas, lymphomas), but those tumors are not described here.

The prognosis after treatment for thyroid carcinoma is usually very good. However, there is a wide spectrum of clinical behavior of various subtypes of thyroid carcinoma. One subtype, fortunately with rare occurrence, is among the most aggressive cancer types known in man.

20.1.1 EPIDEMIOLOGY

Thyroid carcinomas comprise about only 1% of new cancers per year and occur more frequently in females than in males. Annually, about 18,000 new cases are diagnosed in the United States [1]. The epidemiology of thyroid cancer in Northern Europe and much of the Western world can be illustrated by data from the population based Cancer Registry of Norway (www.kreftregisteret.no, English version). The incidence rates were (1996–2000) 4.1 per 100,000 women and 1.5 per 100,000 males per year. Thyroid carcinoma thus occurs relatively often among women. The incidence rate for women aged 25–29 years was 8.7 per 100,000. Because the long-time survival rate for thyroid carcinoma is generally very good, a relatively large group of patients (about 17 times the number of new cases per year) are under follow-up and observation after treatment for thyroid carcinoma. Thyroid carcinoma may recur many (>10–20) years after primary treatment and recurrences are often treatable.

20.1.2 ANATOMY

The thyroid gland is a soft, butterfly-shaped, red brownish gland consisting of a left- and a right-sided lobe, connected by a transverse bridge (isthmus). Usually there is also a slender midline extension (pyramidal lobe) extending toward the tongue, which is the region the thyroid gland stems from embryologically (Figure 20.1). The thyroid in adults weighs 15 to 20 g.

   The follicular cells are arranged in a pattern of follicles, that is, spheroids lined by a single layer of follicular cells on a basement membrane and surrounding a lumen filled with colloid, which is a gel-like substance containing thyreoglobulin. The follicles are 15 to 500 μm in diameter. Interspersed between the follicles there are (relatively few) parafollicular cells, C-cells producing the hormone calcitonin (Figure 20.1).

20.1.3 PHYSIOLOGY

The thyroid follicular cells produce the hormones thyroxin (T4) and triiodothyronine (T3) [2]. The thyroid hormones are stored in the follicular lumen until later uptake by the capillary blood vessels surrounding the follicles and distribution to the body at large. Thyroid hormones are essential to life

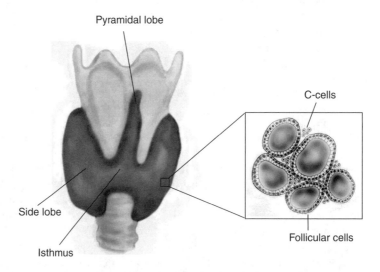

FIGURE 20.1 Anatomy of the thyroid gland. Section showing follicular cells surrounding the lumen filled with colloid containing thyreoglobulin. The follicles are 15 to 500 μm in diameter. Interspersed between the follicles there are a few parafollicular cells, C-cells, which produce the hormone calcitonin.

and influence metabolism of all types of cells. Iodine is essential for thyroid hormone production. A protein called sodium/iodide symporter (Na/I symporter, NIS) is essential in the process of trapping and concentration of iodine in the thyroid follicular cells [3]. The ability of high-iodine uptake and the production of thyreoglobulin are specific functions of thyroid follicular cells. These differentiated functions are often retained, although somewhat altered, in the thyroid carcinoma cells. This has implications for the follow-up and treatment of thyroid carcinoma.

The production and secretion of thyroid hormones from the thyroid gland are regulated in the form of a negative feedback by thyroid simulating hormone (TSH) produced by the pituitary gland [2]. If the thyroid hormone concentration in blood decreases, TSH production will increase and stimulate the follicular cells to increase their production of thyroid hormones, and vice versa (negative feedback). TSH also stimulates growth of thyroid cells and may, therefore, also stimulate carcinoma cells with functional TSH receptors [4].

20.2 CLINICAL PICTURE

Most thyroid carcinomas cause few symptoms. The first sign is often a symptomless visible node (tumor) in the lower anterior neck and the tumor usually moves by the action of swallowing. The level of thyroid hormones and metabolic rate are seldom affected. Enlarged lymph nodes in the mid- or lateral neck may be found. In some cases the tumor rapidly enlarges causing hoarseness and swallowing dysfunction, sometimes airway obstruction.

Clinical examination, ultrasonography examination of the neck, as well as needle aspiration biopsies for cytology examination, are the main diagnostic approaches [4].

20.2.1 ETIOLOGY

The etiology of thyroid carcinomas in man is not known, except the observation that exposure to irradiation increases the incidence of papillary thyroid carcinoma (PTC). There are many observations that irradiation is carcinogenic for the thyroid cells. A 100-fold increase in the incidence of PTC was observed in children who were heavily exposed to radioactive iodine from the Chernobyl accident in 1986 [5]. The fact that carcinomas from follicular thyroid cells are more common in women than in males, and especially so in younger women, indicates that endocrine and reproductive factors may influence the risk of developing thyroid carcinoma. Mechanisms for this are not known.

In geographic areas where the population has low iodine intake, the incidence of thyroid carcinoma is lower than in areas with adequate iodine intake. In iodine deficient areas there is a relative dominance of follicular and anaplastic carcinomas, whereas papillary carcinomas (PC) are more frequent in areas without iodine inefficiency (see Figure 20.2 and Table 20.1) [6].

Papillary and follicular carcinomas are called "differentiated" thyroid carcinomas and the prognosis after treatment is generally very good [4,7].

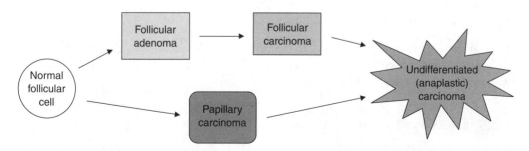

FIGURE 20.2 Model for thyroid cancer development. The model indicates possible routes for undifferentiated thyroid carcinoma.

TABLE 20.1

Types of Thyroid Carcinoma with Relative Frequencies

Thyroid carcinomas from follicular cells

Papillary carcinoma	~70%
Follicular carcinoma	~20%
Poorly differentiated carcinoma	<5%
Anaplastic carcinoma	<5%

Thyroid carcinomas from C-cells

Medullary carcinoma	~5%

20.2.2 PAPILLARY CARCINOMA

The papillary carcinomas may be solitary tumors, but more often there are multiple primary tumors and often also metastases to the neck lymph nodes. The long-time survival rate is about 80%. In younger persons, less than 45 years of age, the cancer-specific death rate is less than 5% [8]. Thus, biological aggressivity varies with age (more aggressive in young children and in old persons). Metastases to lungs, bones, and brain are rare. There seems to be a familiar hereditary type of PC and a genetic basis for this is not known.

20.2.3 FOLLICULAR CARCINOMA

The mean age of patients with follicular carcinoma (FC) is somewhat higher than for patients with PC. FC has a higher tendency than PTC for hematogenous metastases (lungs, skeleton). FC has a somewhat higher mortality rate than PTC, but among FCs too there are subtypes (e.g., low-grade carcinoma) with extremely good long-term prognosis.

20.2.4 ANAPLASTIC CARCINOMA

This is one of the most malignant tumors occurring in man. Mean survival time after diagnosis is less than one year. Anaplastic carcinoma may develop from long-standing differentiated carcinomas [9], but probably also develops as anaplastic carcinomas *de novo*. Local growth is infiltrative to the airways and oesophagus and the occurrence of distant metastasis is frequent. Anaplastic carcinoma cells have lost the specific qualities of thyroid cells like iodine accumulation and thyreoglobulin production.

20.2.5 POORLY DIFFERENTIATED CARCINOMA

This is an intermediate group between differentiated and anaplastic; oftentimes with both differentiated and less-differentiated areas within the tumor. Some of these patients may benefit from treatment for many years even though cure may not be feasible in all cases [10].

20.2.6 MEDULLARY THYROID CARCINOMA

Medullary thyroid carcinoma (MTC) may be "sporadic" (75%) or familial (hereditary) (25%). MTC develops from the neuroendocrine C-cells. Serum calcitonin is elevated and that is diagnostic for the disease. Hypercalcitoninemia gives no specific symptoms. MTC has a tendency for early spread to regional lymph nodes and long-time prognosis is less good than for differentiated follicular-cell thyroid carcinomas [11]. Familial medullary thyroid carcinoma (FMTC) and MTC as part of multiple endocrine neoplasia (MEN type 2) is caused by a germ-line mutation of the *RET* gene [12]. Patients with MTC as part of the MEN syndrome have a high risk of concomitant or asynchronous tumors of the adrenal medulla and the parathyroid glands.

20.2.7 TREATMENT

Surgery with total thyroidectomy and removal of areas with metastatic lymph nodes is the main core of treatment [4,13]. There are different opinions as to the need for total thyroidectomy in "all" cases, and less than total thyroidectomy may sometimes be sufficient, especially for small, solitary primary tumors without metastasis.

Because the cells of differentiated thyroid carcinomas often retain the capability of concentrating iodine, radioactive iodine can be used for detecting and treating tumor remnants or metastases [13]. When performing diagnostic or therapeutic procedures with radioactive iodine (^{131}I is the most commonly used isotope), a high level of s-TSH is necessary to promote iodine uptake. A high TSH level can be obtained endogenously, by withholding thyroxin medication for four to five weeks, or exogenously by using recombinant human TSH (rhTSH) [14]. When the thyroid gland and all tumor tissue have been removed, the level of serum thyroglobulin should be below "detection level." Serum thyroglobulin can be used as a marker for persistent or recurrent disease. Furthermore, thyroglobulin secretion is stimulated by TSH, and serum thyroglobulin, measured under endogenous or exogenous TSH stimulation, is the most sensitive marker for remnant or recurrent tumor disease [15].

Based on knowledge that has emerged from studies of thyroid cancer molecular biology, novel therapies have been explored and some have been tried in man, in cases with advanced disease without radioiodine uptake. Attempts to redifferentiate tumor cells, using retinoic acid to restore or increase the expression of sodium–iodine symporter, have resulted in reuptake of radioiodine [16]. Several other targeted therapies, among those, tyrosine kinase inhibitors, histone deacetylase inhibitors, DNA methyltransferase inhibitors, angiogenesis inhibitors, and gene therapy are in various preclinical stages [17].

Serum calcitonin is the main tumor marker for MTC [4]. Elevated basal serum calcitonin or a pathologic increase after provocation test with pentagastrin and calcium may be indicative of remaining or recurrent disease. In addition, the neuroendocrine tumor marker chromogranin A and s-carcinoembryonic antigen (CEA) are used as tumor markers for MTC. Somatostatin receptor scintigraphy can sometimes demonstrate MTC metastases.

20.3 MOLECULAR BIOLOGY

20.3.1 GERM-LINE MUTATIONS

Only about 5% of the differentiated carcinomas of the thyroid can be linked to germ-line mutations of genes often with a dominant inheritance. Several genes have been linked to familiar differentiated thyroid cancer but only in the case of FMTC and the multiple endocrine neoplasia type 2 (MEN2A and 2B) has a single gene been identified and causally linked to the disease. The MTCs as part of these syndromes carry an activating mutation in the *ret* gene encoding the Ret growth factor receptor (see Figure 20.3 and Figure 20.4). Even so, only about 25% of all medullary carcinomas are inherited. Several point mutations have been identified in MEN2A, MEN2B, and familial hereditary medullary carcinomas. One important type of mutation is when the *ret* gene is mutated in either of the two receptor domain cysteines that in the wild-type receptor participate in an intramolecular cys–cys-bridge formation. When one of the cysteines is replaced by another amino acid, the intramolecular cys–cys-bridge cannot form, and the remaining cysteines are free to participate in illegitimate receptor

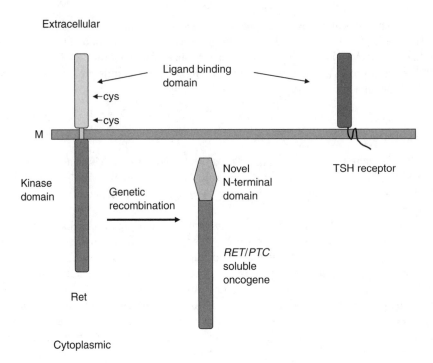

FIGURE 20.3 Schematic representation of the Ret and TSH receptors. M: Cell membrane, cys: cysteine amino acid. *Ret/PTC* oncogene protein: Result of gene recombination caused by for instance radioactive exposure.

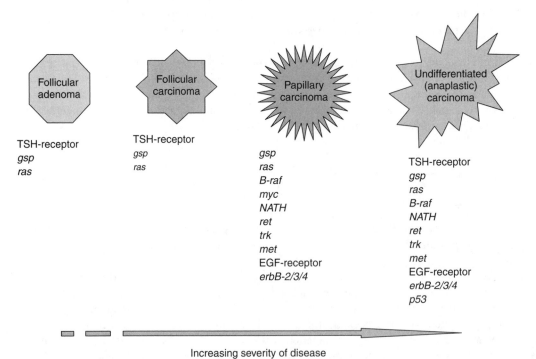

FIGURE 20.4 Overview of some of the known genes associated with thyroid tumors.

dimerization and signaling will occur independent of ligand and ligand-binding. Thus, in these cases, the mutated Ret receptor is continuously providing a growth-promoting signal via signal pathways such as the Ras–Raf–MAPK cascade. For a review of the Ras–Raf–MAPK signaling pathways see Reference 18 and references therein.

Genetic studies have attempted to identify a single genetic marker for inherited thyroid cancers without success since a number of different genes and genetic markers located on several different chromosomes seem to contribute to the development of the disease. It is therefore most likely that differentiated thyroid carcinomas represent a disease that is polygenic in nature [19]. A complete causative explanation for inherited differentiated thyroid cancers remains to be established.

20.3.2 RET

Transmembrane growth factor tyrosine kinase receptors are important proto-oncogenes in several different tumors. These transmembrane receptors have an extracellular domain participating in ligand-binding that activates its cytoplasmic tyrosine kinase domain by inducing receptor dimerization (Figure 20.3). In thyroid cancers, mutations in two such receptors, Ret, Trk, have been linked to PTCs (Ret, Trk) [20,21], MEN2A (Ret) [22], MEN2B (Ret) [23], while TSH receptor mutations have been linked to hyperfunctioning follicular adenomas and carcinoma (TSH receptor) [24,25]. Of significant interest is the translocations found involving the *ret* gene and one of several other partner genes, generating the chimeric oncogenes referred to as ret/PTCs [26]. In a high fraction of PCs (up to 40% in some studies), a translocation with the breakpoint located immediately near the C-terminal to the trans-membrane domain of the *ret* gene has occurred, generating a constitutively active Ret tyrosine kinase. These translocations are somatic mutations that appear to be almost unique to the thyroid gland. Since the rearrangements remove the extracellular ligand-binding domain with the signal peptide, the chimeric Ret/PTC protein cannot penetrate the plasmamembrane and is therefore located at the cytoplasm. The chimeric Ret/PTC protein, like the wild-type receptor, is dependent on dimerization for tyrosine kinase activation. It thus exists in the cytoplasm in a dimer configuration through dimerization of the Ret/PTC-partner protein. This active Ret/PTC tyrosine kinase is a strong oncogene. To date, at least eight different genes have been shown to recombine with *ret*, generating eight different *Ret/PTC* oncogenes [27]. The *ret* translocation event in humans has been closely linked to environmental radioactive exposure since Ret/PTCs are frequent in PCs in children exposed to radioactivity from the Chernobyl disaster in 1986 [28]. Interestingly, while Ret/PTC rearrangements due to radiation are frequent in humans, this event is not commonly found in animal studies. It was therefore assumed that radioactive irradiation would not be important as a cause of thyroid cancers in man. The difference in response observed between animal and man in this particular case is an important reminder that extrapolation from animal model studies to the human situation should be made with great caution.

Because early RNA expression studies did not detect wild-type Ret mRNA in thyroid tissue, it was accepted that only the Ret/PTC translocation proteins were expressed in the thyroid. However, recently, it was recognized that wild-type Ret is expressed in normal thyroid cells and at an elevated level in PCs [29]. The function of wild-type Ret in thyroid follicular cells is not fully understood but adds to the complexity of thyroid tumor biology.

20.3.3 TRK

The thyroid follicular cells express one of the nerve growth factor (NGF) receptors (TRK). TRK is a transmembrane tyrosine kinase resembling the Ret protein in structure. In concordance with the rearrangement mutations of *ret*, the *trk* gene undergoes rearrangement in a small fraction (few percent) of PCs [30]. Furthermore, in analogy with RET/PTC, the four identified rearranged TRK variants may function as oncogenes.

20.3.4 TSH Receptor and GS-α Protein

The TSH receptor is, like the Ret receptor, a transmembrane protein composed of an outside ligand-binding domain but lacks the cytoplasmic tyrosine kinase domain (Figure 20.3). Instead it is coupled to the G-proteins. The receptor is activated upon binding of the ligand, TSH, and the downstream signal is mediated by the G-protein cascades involving activated cAMP-dependent protein kinase A (PKA). Since TSH not only activates the thyroid-specific functions of the follicular cells but also generates a growth stimulus, mutations that generate a constitutively active TSH receptor will have profound effects on these cells. Mutated TSH receptors have been found in differentiated thyroid carcinomas but are most closely linked to hyperfunctioning thyroid follicular toxic adenomas [24,31]. The mutations are usually located on the extracellular domain next to the transmembrane domain. Immediately downstream of the TSH receptor, the G-protein, Gsα, encoded by the *gsp* gene, transmits the TSH signal by activating the cAMP-dependent PKA. The PKA activity is essential for follicular thyroid cell proliferation. Several studies have addressed the possibility of *gsp* mutations in thyroid cancer and mutations were found in about 15% of PC and 8% of FC in sporadic tumors [32]. The number of cases examined is, however, very low and these estimates of frequencies are therefore statistically inaccurate and the estimated frequency may be too high. Thus the role of *gsp* mutations in thyroid cancer is yet to be fully understood.

20.3.5 Ras

Mutated Ras is a strong oncogene in many different tissues and cell types acting through its activation of the Raf–MAPK pathway as well as activating PI 3-Kinase and Akt. Particularly, mutations in codons 12 and 13 that inactivate the GTPase activity of the Ras protein are of importance. In the history of thyroid cancer molecular biology investigations, *ras* mutations were early identified [33] and mutations have been found in all three *ras* genes (Ha-, Ki-, and N-*ras*) of human thyroid tumor biopsies. The frequency of *ras* mutations in these tumors varies broadly (from 6 to 45%) between different studies, for example, the United States, England, and France. The mutation frequency also varies between tumor types. In a report by Garcia-Rostan et al. [34] *ras* mutations were more frequent in poorly differentiated-compared to the more differentiated PCs. Such variations in occurrence are most often explained by genetic differences between the populations studied. However, genetic analyses have failed to verify this hypothesis [30,35] and one experimental study in rats indicates that *ras* mutations may be infrequent in x-ray-induced thyroid tumors [36]. On the other hand, rats treated with di-isopropanolnitrosamine developed cancerous lesions carrying Ki-*ras* mutations in codon 12 [37]. These findings indicate that *ras* mutations occurring in thyroid cancers may not be related to genetic differences between populations but rather to differences in environmental exposures. The latter phenomenon is well established for other forms of cancers. Unfortunately, due to the low number of human cases studied, safe conclusions about the role of mutated Ras in the initiation and progression of thyroid tumors is difficult to make at present.

The different Ras homologs may activate separate signaling pathways [38] but the best studied is the Ras–Raf–Mek–Erk signaling cascade. Three *raf* genes, c-*raf*, A-*raf*, and B-*raf* have been identified in humans. Recent studies have demonstrated high frequencies of point mutations in the B-*raf* gene in melanomas [39]. Other groups have subsequently demonstrated the importance of B-*raf* mutations in papillary thyroid cancers [40–43]. Some of these data show that *Ret/PTC*, *ras*, and B-*raf* mutations may be mutually exclusive in papillary thyroid cancers. However, unpublished data from our group indicate the presence of combinations of these three mutations in some of the patients. Nevertheless, the high mutation frequencies of *Ret/PTC* and B-*raf*, in addition to the *ras* mutations, provide strong evidence for a dominating role of the Ret/PTC, Ras, and B-Raf signaling pathway in oncogenic thyrocyte transformation.

20.3.6 Tumor Suppressor Genes

p53 is a transcription factor regulating the expression of several genes with key functions in the control of cell proliferation [44]. p53 is considered to be a tumor suppressor gene that is activated upon DNA damage to stop cell progression through the cell cycle, allowing DNA repair to take place before the cells are allowed to reenter the cell cycle. In cases of severe cell and DNA damage, p53 will activate programmed cell death, apoptosis. Thus, inactivating p53 mutations may lead to increased mutation frequencies and oncogenesis [45]. In thyroid cancers, p53 mutations are found in the poorly differentiated and anaplastic carcinomas at high frequencies (up to 70%) in some reports [46]. The fact that p53 mutations are not found in differentiated thyroid cancers, suggests that this is a late event in the progression of the disease. Furthermore, one may speculate that wild-type p53 could also play a role in maintaining a differentiated state of FCs and thyroid cancers.

p53 function and stability is closely linked to the activity of PTEN (phosphatase and tensin homolog deleted on chromosome 10) [47]. Germ-line mutations of PTEN are not only responsible for the Cowden syndrome, which carry an increased risk of breast cancer, but also for thyroid carcinomas. PTEN balances the phosphorylation state of protein kinases such as Akt. Since Akt activates mdm2 leading to degradation of p53, inactivating mutations of PTEN will allow high levels of phosphorylated (active) Akt and mdm2 in the cells and a high rate of p53 degradation [48]. This, in turn, may generate a situation of insufficient amounts of p53 in the cells. In this situation, the cell is no longer capable of responding properly to DNA damage or other insults. In thyroid tumors, only a few studies have reported on PTEN mutations and loss of heterozygosity (LOH) [49]. At present, PTEN mutations are very rare in sporadic thyroid cancers but recent data indicate that PTEN expression may be strongly decreased in anaplastic carcinomas [50]. The cause of downregulation is not understood but DNA methylation and epigenetic events like histone acetylation and methylation are likely factors that contribute to it. In the development of the malignant state, one can perhaps argue that loss of PTEN expression due to gene silencing will have the same consequence for the cell as an inactivating p53 mutation since absence of PTEN activity will allow the very rapid degradation of wild-type p53. Some of the best studied genes associated with thyroid cancer development are presented in Figure 20.4. It is noteworthy that an increasing number of genes are found with increasing severity of disease.

20.3.7 Differential Gene Expression

Several genes without known mutations are often found to play important roles in the biology of cancers. As is apparent from the present discussion, mutations in genes encoding components of signal transduction pathways may lead to constitutively activated protein products that are often oncogenic. On the other hand, wild-type genes involved in signal transduction and gene regulation are also known to contribute to the development of cancers by being constitutively over expressed or downregulated. Constitutively over-expressed genes may provide the cells with strong growth stimulatory signals and essentially generate the same biological effects as mutated oncogenes expressed at normal levels. For instance, high *myc* gene copy number and consequently, high Myc protein expression, is correlated to poor prognosis in childhood neuroblastomas.

Several studies have focused on the gene-expression levels in thyroid tumors. In particular, the family of ErbB receptors, epidermal growth factor (EGF) receptor, ErbB-2, -3, and -4, have all been shown to be over expressed in PCs [51] even though none of the genes encoding these receptors have been shown to be mutated or amplified in thyroid cancers. In particular, the expression of the EGF receptor is of interest since one of its ligands, TGFα, is strongly expressed in the same tumor cells [52]. This implies that PC cells can provide itself with a growth factor stimulus through an autocrine stimulatory mechanism, thus providing the tumor cells with a growth advantage. In addition, the EGF receptor is often found not only at its proper location, the plasma membrane, but also in the cytoplasm. (The cytoplasmic location of the EGF receptor was a negative prognostic factor.) Another group of ErbB tyrosine kinase ligands, the heregulins, are similarly over expressed in PCs [53]. It is of considerable interest that different heregulin splice forms are present in PC and in particular that the preheregulin form is located in the nucleus in these tumors but not in normal thyroid tissue. These observations are indicative of multiple functions of the same gene in the biology of these cells.

The vascular endothelial growth factors (VEGFs) and the platelet-derived growth factors (PDGFs) constitute another family of mitogens that are expressed in thyroid cancers [54]. PDGF-C, which is modestly over expressed in PTCs [55], has a molecular structure more similar to VEGFs than to PDGFs [56]. Normally, these growth factors play important roles in angiogenesis and cell proliferation. It is therefore important to note that the thyroid tumors are highly vascularized, a prerequisite for growth of most solid tumors [57]. The precise role in thyroid tumor biology of each of the different members of this group of growth factors has not been established and it is puzzling that growth factors with seemingly overlapping activities are expressed in the same tumor.

Activating mutations of the growth factor receptor Ret is frequent in certain types of thyroid cancers (discussed earlier in 20.3.2) but other growth factor receptors such as the ErbB family (EGF-receptor, ErbB2, -3, and -4) are not mutated. However, over expression of, in particular, the ErbB2 receptor is found in PC [58]. Since the ErbB2 receptor can heterodimerize with the other ErbB receptors upon ligand-binding, the over expression of ErbB2 facilitates efficient and enhanced signal transduction of extracellular growth factors specific for the ErbB family. The simultaneous presence of all four receptors in the same cells allows for a very diverse and complex signaling cascade that may give the tumor cells a specific growth or survival advantage.

The molecular mechanisms allowing over expression of certain genes in thyroid cancers have not been studied but mutations or changes in expression levels of transcription factors is likely to be of importance. p53 is a transcription factor, and the p53 mutations mentioned earlier are certainly of great significance for the development of undifferentiated carcinoma. Immunohistochemistry data [52] suggests that the transcription factor Myc is over expressed in PC. *In vitro* studies using thyroid cell cultures indicate the importance of members of the Ets transcription factor family in transformation. High Ets transcriptional activity was strongly associated with cell neoplastic transformation indicative of an important role of these transcription factors in thyroid cell transformation [59]. Furthermore, the role of epigenetic factors such as the high mobility group 1 (HMGA1) family of chromosomal proteins has recently come into focus. HMGA1 is over expressed in many human cancers, and rat thyroid cell culture experiments suggest that these proteins are necessary for sufficient expression of the *fra-1* gene, a member of the AP-1 transcription factor family [60]. These findings point to a possible role of AP-1 in thyroid cell transformation. AP-1 is one of the most studied transcriptions factor, a heterodimer consisting of the early identified oncogenes *Jun* and *Fos* [61].

20.3.8 RNA Splicing

In recent years, the importance of RNA splicing in providing phenotypic diversity from one gene, or primary RNA transcript, has been increasingly recognized. RNA splicing requires a large number of different proteins such as the SR-proteins organized in so-called spliceosomes, for proper splicing to take place. It appears that at least for some genes, the splicing process may be coupled to transcription [62]. It is conceivable that malfunctioning spliceosomes can lead to mRNAs encoding protein variants that may have oncogenic properties not found at the DNA level. For example, in the thyroid, incompletely spliced *ret* RNA transcripts are found in normal tissue and in PC but at significantly higher levels in PTC, however, only very low levels of spliced *ret* mRNA are found in normal tissue while spliced *ret* mRNA is significantly present in the tumor cells. In addition, several novel *ret* RNA splice forms are detected [29]. The exact role and importance of alternative RNA splicing in malignant cell transformation remains to be determined.

20.3.9 ONCOGENE COOPERATION AND TUMOR PROGRESSION

Based on the fact that several oncogenes are identified in the different types of thyroid cancers, and that only in a relatively low fraction of tumors, more than one oncogene is activated simultaneously must have implications as to how the mutation and expression data are interpreted with respect to tumor development and progression. It is generally expected that there exists a "linear" development of the thyroid tumors beginning with benign lesions that stepwise develop into the most malignant forms (Figure 20.2). If this is the case, the oncogenes identified so far, except the various *ret* mutations of MEN2A, MEN2B, and familial medullary carcinomas and translocations generating the RET/PTCs, appear not to be the "initiating" oncogene but are rather factors that facilitates the oncogenic development and possibly provide survival advantages such as rendering the tumor cells less susceptible to apoptosis. On the other hand, the different types of thyroid tumors may be envisioned to develop individually and change biology from a less to a more malignant form only occasionally. If that is the case, somatic gene lesions such as *p53* mutations could represent an early rather than a late event in anaplastic carcinoma development. The complexity of thyroid cancers is further underpinned by genetic evidence indicating differentiated thyroid carcinomas to be multigenic disorders [19].

20.4 CLINICAL CONSEQUENCES OF THYROID MOLECULAR BIOLOGY

Germ-line mutations of the *RET* gene predispose for MTC. The inheritance is autosomal dominant with high penetrance. Inheriting such a mutation often means 90% risk of developing a potentially lethal cancer of the thyroid. During recent years it has therefore become a practice to advice prophylactic thyroidectomy in children with inherited RET mutation of this type. There are differences between various mutations as to when in life the carcinoma develops and metastasizes [12].

Information from studies of the molecular biology of papillary and follicular thyroid carcinomas has so far had limited consequences for their diagnosis and treatment. One has especially searched for cellular markers to be able to distinguish benign from malignant follicular tumors. This important difference is usually not feasible to make without a surgical operation. Some marker genes (e.g., cytokeratine 19, galectin-3) have been shown to have elevated or altered expression in malignant as compared to benign tumours, however, without a complete confinement to malignant cells [63,64].

REFERENCES

1. Ries, L.A.G., Eisner, M.P., Kosary, C.L. et al. *SEER Cancer Statistics Review, 1993–1997*. NCI, Bethesda, 2000.
2. Braverman, L.E. and Utiger, R.D., Eds. *Werner and Ingbar's The Thyroid. A Fundamental and Clinical Text*, 8th ed. Lippingcott Williams & Wilkins, Philadelphia, 2000.
3. Filetti, S., Bidart, J.M., Arturi, F., Caillou, B., Russo, D., and Schlumberger, M. Sodium/iodide symporter: A key transport system in thyroid cancer cell metabolism. *Eur. J. Endocrinol.*, *141*: 443–457, 1999.
4. Sherman, S.I. Thyroid carcinoma. *Lancet*, *361*: 501–511, 2003.
5. Nagataki, S. and Nystrom, E. Epidemiology and primary prevention of thyroid cancer. *Thyroid*, *12*: 889–896, 2002.
6. Feldt-Rasmussen, U. Iodine and cancer. *Thyroid*, *11*: 483–486, 2001.
7. Mazzaferri, E.L. and Jhiang, S.M. Long-term impact of initial surgical and medical therapy on papillary and follicular thyroid cancer. *Am. J. Med.*, *97*: 418–428, 1994.
8. Mazzaferri, E.L. Long-term outcome of patients with differentiated thyroid carcinoma: Effect of therapy. *Endocr. Pract.*, *6*: 469–476, 2000.

9. Hunt, J.L., Tometsko, M., LiVolsi, V.A., Swalsky, P., Finkelstein, S.D., and Barnes, E.L. Molecular evidence of anaplastic transformation in coexisting well-differentiated and anaplastic carcinomas of the thyroid. *Am. J. Surg. Pathol.*, *27*: 1559–1564, 2003.

10. Nishida, T., Katayama, S., Tsujimoto, M., Nakamura, J., and Matsuda, H. Clinicopathological significance of poorly differentiated thyroid carcinoma. *Am. J. Surg. Pathol.*, *23*: 205–211, 1999.

11. Hyer, S.L., Vini, L., A'Hern, R., and Harmer, C. Medullary thyroid cancer: Multivariate analysis of prognostic factors influencing survival. *Eur. J. Surg. Oncol.*, *26*: 686–690, 2000.

12. Machens, A., Niccoli-Sire, P., Hoegel, J., Frank-Raue, K., van Vroonhoven, T.J., Roeher, H.D., Wahl, R.A., Lamesch, P., Raue, F., Conte-Devolx, B., and Dralle, H. Early malignant progression of hereditary medullary thyroid cancer. *N. Engl. J. Med.*, *349*: 1517–1525, 2003.

13. Vini, L. and Harmer, C. Management of thyroid cancer. *Lancet Oncol.*, *3*: 407–414, 2002.

14. Robbins, R.J. and Robbins, A.K. Clinical review 156: Recombinant human thyrotropin and thyroid cancer management. *J. Clin. Endocrinol. Metab.*, *88*: 1933–1938, 2003.

15. Mazzaferri, E.L., Robbins, R.J., Spencer, C.A., Braverman, L.E., Pacini, F., Wartofsky, L., Haugen, B.R., Sherman, S.I., Cooper, D.S., Braunstein, G.D., Lee, S., Davies, T.F., Arafah, B.M., Ladenson, P.W., and Pinchera, A. A consensus report of the role of serum thyroglobulin as a monitoring method for low-risk patients with papillary thyroid carcinoma. *J. Clin. Endocrinol. Metab.*, *88*: 1433–1441, 2003.

16. Schmutzler, C. and Kohrle, J. Retinoic acid redifferentiation therapy for thyroid cancer. *Thyroid*, *10*: 393–406, 2000.

17. Braga-Basaria, M. and Ringel, M.D. Clinical review 158: Beyond radioiodine: A review of potential new therapeutic approaches for thyroid cancer. *J. Clin. Endocrinol. Metab.*, *88*: 1947–1960, 2003.

18. McKenna, W.G., Muschel, R.J., Gupta, A.K., Hahn, S.M., and Bernhard, E.J. The RAS signal transduction pathway and its role in radiation sensitivity. *Oncogene*, *22*: 5866–5875, 2003.

19. Links, T.P., van Tol, K.M., Meerman, G.J., and de Vries, E.G. Differentiated thyroid carcinoma: A polygenic disease. *Thyroid*, *11*: 1135–1140, 2001.

20. Fusco, A., Berlingieri, M.T., Di Fiore, P.P., Portella, G., Grieco, M., and Vecchio, G. One- and two-step transformations of rat thyroid epithelial cells by retroviral oncogenes. *Mol. Cell Biol.*, *7*: 3365–3370, 1987.

21. Pierotti, M.A., Bongarzone, I., Borrello, M.G., Mariani, C., Miranda, C., Sozzi, G., and Greco, A. Rearrangements of TRK proto-oncogene in papillary thyroid carcinomas. *J. Endocrinol. Invest.*, *18*: 130–133, 1995.

22. Mulligan, L.M., Kwok, J.B., Healey, C.S., Elsdon, M.J., Eng, C., Gardner, E., Love, D.R., Mole, S.E., Moore, J.K., Papi, L. et al. Germ-line mutations of the RET proto-oncogene in multiple endocrine neoplasia type 2A. *Nature*, *363*: 458–460, 1993.

23. Eng, C., Smith, D.P., Mulligan, L.M., Nagai, M.A., Healey, C.S., Ponder, M.A., Gardner, E., Scheumann, G.F., Jackson, C.E., Tunnacliffe, A. et al. Point mutation within the tyrosine kinase domain of the RET proto-oncogene in multiple endocrine neoplasia type 2B and related sporadic tumours. *Hum. Mol. Genet.*, *3*: 237–241, 1994.

24. Parma, J., Duprez, L., Van Sande, J., Cochaux, P., Gervy, C., Mockel, J., Dumont, J., and Vassart, G. Somatic mutations in the thyrotropin receptor gene cause hyperfunctioning thyroid adenomas. *Nature*, *365*: 649–651, 1993.

25. Russo, D., Arturi, F., Schlumberger, M., Caillou, B., Monier, R., Filetti, S., and Suarez, H.G. Activating mutations of the TSH receptor in differentiated thyroid carcinomas. *Oncogene*, *11*: 1907–1911, 1995.

26. Fusco, A., Grieco, M., Santoro, M., Berlingieri, M.T., Pilotti, S., Pierotti, M.A., Della Porta, G., and Vecchio, G. A new oncogene in human thyroid papillary carcinomas and their lymph-nodal metastases. *Nature*, *328*: 170–172, 1987.

27. Fagin, J.A. Perspective: Lessons learned from molecular genetic studies of thyroid cancer — insights into pathogenesis and tumor-specific therapeutic targets. *Endocrinology*, *143*: 2025–2028, 2002.

28. Klugbauer, S. and Rabes, H.M. The transcription coactivator HTIF1 and a related protein are fused to the RET receptor tyrosine kinase in childhood papillary thyroid carcinomas. *Oncogene*, *18*: 4388–4393, 1999.

29. Fluge, O., Haugen, D.R., Akslen, L.A., Marstad, A., Santoro, M., Fusco, A., Varhaug, J.E., and Lillehaug, J.R. Expression and alternative splicing of c-ret RNA in papillary thyroid carcinomas. *Oncogene*, *20*: 885–892, 2001.

30. Suarez, H.G. Genetic alterations in human epithelial thyroid tumours. *Clin. Endocrinol. (Oxf)*, *48*: 531–546, 1998.

31. Van Sande, J., Parma, J., Tonacchera, M., Swillens, S., Dumont, J., and Vassart, G. Somatic and germline mutations of the TSH receptor gene in thyroid diseases. *J. Clin. Endocrinol. Metab.*, *80*: 2577–2585, 1995.

32. Suarez, H.G. Molecular basis of epithelial thyroid tumorigenesis. *C.R. Acad. Sci. III*, *323*: 519–528, 2000.

33. Lemoine, N.R., Mayall, E.S., Wyllie, F.S., Farr, C.J., Hughes, D., Padua, R.A., Thurston, V., Williams, E.D., and Wynford Thomas, D. Activated ras oncogenes in human thyroid cancers. *Cancer Res.*, *48*: 4459–4463, 1988.

34. Garcia-Rostan, G., Zhao, H., Camp, R.L., Pollan, M., Herrero, A., Pardo, J., Wu, R., Carcangiu, M.L., Costa, J., and Tallini, G. Ras mutations are associated with aggressive tumor phenotypes and poor prognosis in thyroid cancer. *J. Clin. Oncol.*, *21*: 3226–3235, 2003.

35. Manenti, G., Pilotti, S., Re, F.C., Della Porta, G., and Pierotti, M.A. Selective activation of ras oncogenes in follicular and undifferentiated thyroid carcinomas. *Eur. J. Cancer*, *30A*: 987–993, 1994.

36. Haugen, D.R., Lillehaug, J.R., and Akslen, L.A. Enhanced expression of EGF receptor and low frequency of ras mutations in x-ray-induced rat thyroid tumours. *Virchows Arch.*, *435*: 434–441, 1999.

37. Kobayashi, Y., Kawaoi, A., and Katoh, R. Mutation of ras oncogene in di-isopropanolnitrosamine-induced rat thyroid carcinogenesis. *Virchows Arch.*, *441*: 289–295, 2002.

38. Sheng, H., Shao, J., and DuBois, R.N. Akt/PKB activity is required for Ha-Ras-mediated transformation of intestinal epithelial cells. *J. Biol. Chem.*, *276*: 14498–14504, 2001.

39. Davies, H., Bignell, G.R., Cox, C., Stephens, P., Edkins, S., Clegg, S., Teague, J., Woffendin, H., Garnett, M.J., Bottomley, W., Davis, N., Dicks, E., Ewing, R., Floyd, Y., Gray, K., Hall, S., Hawes, R., Hughes, J., Kosmidou, V., Menzies, A., Mould, C., Parker, A., Stevens, C., Watt, S., Hooper, S., Wilson, R., Jayatilake, H., Gusterson, B.A., Cooper, C., Shipley, J., Hargrave, D., Pritchard-Jones, K., Maitland, N., Chenevix-Trench, G., Riggins, G.J., Bigner, D.D., Palmieri, G., Cossu, A., Flanagan, A., Nicholson, A., Ho, J.W., Leung, S.Y., Yuen, S.T., Weber, B.L., Seigler, H.F., Darrow, T.L., Paterson, H., Marais, R., Marshall, C.J., Wooster, R., Stratton, M.R., and Futreal, P.A. Mutations of the BRAF gene in human cancer. *Nature*, *417*: 949–954, 2002.

40. Kimura, E.T., Nikiforova, M.N., Zhu, Z., Knauf, J.A., Nikiforov, Y.E., and Fagin, J.A. High prevalence of BRAF mutations in thyroid cancer: Genetic evidence for constitutive activation of the RET/PTC-RAS-BRAF signaling pathway in papillary thyroid carcinoma. *Cancer Res.*, *63*: 1454–1457, 2003.

41. Soares, P., Trovisco, V., Rocha, A.S., Lima, J., Castro, P., Preto, A., Maximo, V., Botelho, T., Seruca, R., and Sobrinho-Simoes, M. BRAF mutations and RET/PTC rearrangements are alternative events in the etiopathogenesis of PTC. *Oncogene*, *22*: 4578–4580, 2003.

42. Xu, X., Quiros, R.M., Gattuso, P., Ain, K.B., and Prinz, R.A. High prevalence of BRAF gene mutation in papillary thyroid carcinomas and thyroid tumor cell lines. *Cancer Res.*, *63*: 4561–4567, 2003.

43. Fukushima, T., Suzuki, S., Mashiko, M., Ohtake, T., Endo, Y., Takebayashi, Y., Sekikawa, K., Hagiwara, K., and Takenoshita, S. BRAF mutations in papillary carcinomas of the thyroid. *Oncogene*, *22*: 6455–6457, 2003.

44. Sharpless, N.E. and DePinho, R.A. p53: Good cop/bad cop. *Cell*, *110*: 9–12, 2002.

45. Morgan, S.E. and Kastan, M.B. p53 and ATM: Cell cycle, cell death, and cancer. *Adv. Cancer Res.*, *71*: 1–25, 1997.

46. Donghi, R., Longoni, A., Pilotti, S., Michieli, P., Della Porta, G., and Pierotti, M.A. Gene p53 mutations are restricted to poorly differentiated and undifferentiated carcinomas of the thyroid gland. *J. Clin. Invest.*, *91*: 1753–1760, 1993.

47. Eng, C. PTEN: One gene, many syndromes. *Hum. Mutat.*, *22*: 183–198, 2003.

48. Stambolic, V., Suzuki, A., de la Pompa, J.L., Brothers, G.M., Mirtsos, C., Sasaki, T., Ruland, J., Penninger, J.M., Siderovski, D.P., and Mak, T.W. Negative regulation of PKB/Akt-dependent cell survival by the tumor suppressor PTEN. *Cell*, *95*: 29–39, 1998.

49. Gimm, O., Perren, A., Weng, L.P., Marsh, D.J., Yeh, J.J., Ziebold, U., Gil, E., Hinze, R., Delbridge, L., Lees, J.A., Mutter, G.L., Robinson, B.G., Komminoth, P., Dralle, H., and Eng, C. Differential nuclear and cytoplasmic expression of PTEN in normal thyroid tissue, and benign and malignant epithelial thyroid tumors. *Am. J. Pathol.*, *156*: 1693–1700, 2000.

50. Frisk, T., Foukakis, T., Dwight, T., Lundberg, J., Hoog, A., Wallin, G., Eng, C., Zedenius, J., and Larsson, C. Silencing of the PTEN tumor-suppressor gene in anaplastic thyroid cancer. Genes chromosomes. *Cancer, 35*: 74–80, 2002.

51. Haugen, D.R.F., Akslen, L.A., Varhaug, J.E., and Lillehaug, J.R. Expression of c-erbB-3 and c-erbB-4 proteins in papillary thyroid carcinomas. *Cancer Res., 56*: 1184–1188, 1996.

52. Haugen, D.R., Akslen, L.A., Varhaug, J.E., and Lillehaug, J.R. Demonstration of a TGF-alpha-EGF-receptor autocrine loop and c-myc protein over-expression in papillary thyroid carcinomas. *Int. J. Cancer, 55*: 37–43, 1993.

53. Fluge, O., Akslen, L.A., Haugen, D.R., Varhaug, J.E., and Lillehaug, J.R. Expression of heregulins and associations with the ErbB family of tyrosine kinase receptors in papillary thyroid carcinomas. *Int. J. Cancer, 87*: 763–770, 2000.

54. Gartner, R. Growth factors in thyroid cells. *Curr. Top. Pathol., 91*: 65–81, 1997.

55. Fluge, O., Bruland, O., Akslen, L.A., Varhaug, J.E., and Lillehaug, J.R. Identification of a novel growth factor homolog overexpressed in papillary thyroid carcinomas using a replica cDNA screening approach. In AACR 2000, San Francisco, 4/2000, 2000.

56. Reigstad, L.J., Sande, H.M., Fluge, O., Bruland, O., Muga, A., Varhaug, J.E., Martinez, A., and Lillehaug, J.R. Platelet-derived growth (PDGF)-C, a PDGF family member with a Vascular endothelial growth factor-like structure. *J. Biol. Chem., 278*(19), 1714–17120, 2003.

57. Akslen, L.A. and LiVolsi, V.A. Prognostic significance of histologic grading compared with subclassification of papillary thyroid carcinoma. *Cancer, 88*: 1902–1908, 2000.

58. Haugen, D.R., Akslen, L.A., Varhaug, J.E., and Lillehaug, J.R. Expression of c-erbB-2 protein in papillary thyroid carcinomas. *Br. J. Cancer, 65*: 832–837, 1992.

59. de Nigris, F., Mega, T., Berger, N., Barone, M.V., Santoro, M., Viglietto, G., Verde, P., and Fusco, A. Induction of ETS-1 and ETS-2 transcription factors is required for thyroid cell transformation. *Cancer Res., 61*: 2267–2275, 2001.

60. Berlingieri, M.T., Pierantoni, G.M., Giancotti, V., Santoro, M., and Fusco, A. Thyroid cell transformation requires the expression of the HMGA1 proteins. *Oncogene, 21*: 2971–2980, 2002.

61. Vogt, P.K., Aoki, M., Bottoli, I., Chang, H.W., Fu, S., Hecht, A., Iacovoni, J.S., Jiang, B.H., and Kruse, U. A random walk in oncogene space: The quest for targets. *Cell Growth Differ., 10*: 777–784, 1999.

62. Auboeuf, D., Honig, A., Berget, S.M., and O'Malley, B.W. Coordinate regulation of transcription and splicing by steroid receptor coregulators. *Science, 298*: 416–419, 2002.

63. Bartolazzi, A., Gasbarri, A., Papotti, M., Bussolati, G., Lucante, T., Khan, A., Inohara, H., Marandino, F., Orlandi, F., Nardi, F., Vecchione, A., Tecce, R., and Larsson, O. Application of an immunodiagnostic method for improving preoperative diagnosis of nodular thyroid lesions. *Lancet, 357*: 1644–1650, 2001.

64. Faggiano, A., Talbot, M., Lacroix, L., Bidart, J.M., Baudin, E., Schlumberger, M., and Caillou, B. Differential expression of galectin-3 in medullary thyroid carcinoma and C-cell hyperplasia. *Clin. Endocrinol. (Oxf), 57*: 813–819, 2002.

21 Chemoprevention of Chemical Carcinogenesis and Human Cancer

Mark A. Morse and Gary D. Stoner

CONTENTS

21.1 INTRODUCTION

Cancer chemoprevention is defined as prevention of cancer by the administration of one or more chemical entities, either as individual drugs or as dietary supplements [1]. If one includes the common forms of skin cancer (i.e., basal cell and squamous cell carcinomas), then nearly 1% of the U.S. population will be diagnosed with some form of cancer in the year 2002 [2]. The 5-year survival rate of cancer patients has approximately improved from 51 to 62% overall in the last 10 years or so; yet over half a million American citizens are expected to die of cancer in 2002 [2,3]. Despite considerable advances in terms of earlier detection and the eventual promise of improved therapeutics, the outlook for many solid tumors remains dismal. Several organ sites continue to yield exceptionally low 5-year survival rates, such as lung and bronchus (15% overall 5-year survival), liver (6% overall 5-year survival), stomach (22% overall 5-year survival), esophagus (14% overall 5-year survival), and pancreas (4% overall 5-year survival) [2]. Thus, the aggressive prevention of cancer remains a cornerstone of basic public health policy.

Primary cancer prevention involves the prevention of disease occurrence, typically by protecting individuals from exposure to etiologic agents. This can be accomplished by a number of means, including physical removal of the individual from the source of exposure or by the use of protective outerwear and other worn barriers in the presence of etiologic agents. A chemopreventive corollary to primary prevention would be the administration of chemopreventive chemicals that block internal exposure to etiologic agents *in vivo*. Secondary cancer prevention involves early detection and intervention in the carcinogenic process, ideally reversing or arresting the disease process prior to its clinical appearance [4]. Again, chemoprevention offers an alternative to conventional secondary prevention, since a number of chemopreventive agents may result in the reversal and arrest of the carcinogenic process. Thus, chemoprevention is considered a legitimate tool in the comprehensive cancer control armamentarium, and can be considered an adjunct or alternative to primary and secondary preventive practices in human populations.

21.2 TARGET POPULATIONS

The current target populations for cancer chemoprevention consist of high-risk groups, such as: (1) individuals who engage in risk-taking behaviors or lifestyles (e.g., smokers and heavy alcohol users), (2) individuals who have received occupational exposure to known carcinogens (e.g., asbestos workers, uranium miners, etc.), (3) individuals who are known to be genetically predisposed to the development of cancer (e.g., people with familial colonic polyposis [FAP], xeroderma pigmentosum, etc.), (4) individuals who possess premalignant lesions (e.g., rectal polyps in FAP patients, oral leukoplakia in snuff users, etc.), (5) survivors of primary cancers with a high degree of recurrence or a high tendency toward formation of second primary tumors (head and neck cancer, breast cancer, etc.), and (6) cancer survivors who had received significant chemotherapy and radiation therapy [3,5]. We do not foresee the administration of specific, potent, targeted chemopreventive agents in the general population. In all likelihood, chemoprevention in the general population will probably consist of the self-administration of simple dietary supplements.

21.3 CLASSES OF CHEMOPREVENTIVE AGENTS

The absolute classification of all known chemopreventive compounds is difficult since precise mechanisms of action are not known for some compounds, and a number of well-characterized compounds exhibit multiple mechanisms of action. The classification scheme that follows is an adaptation of a scheme that was originally developed by Wattenberg [6] and is based primarily on the stage of carcinogenesis in which a chemopreventive agent shows activity in animal models. Using this system as a basis, one can define at least three major categories of chemopreventive agents: (1) inhibitors of carcinogen exposure, (2) antiinitiating or blocking agents (Figure 21.1), and (3) antipromotion/ antiprogression agents or suppressing agents (Figure 21.2) [3]. All three categories are subject to further stratification on the basis of the specific mechanistic principles by which each individual agent acts (Tables 21.1–21.3).

21.3.1 Inhibitors of Carcinogen Exposure

Wattenberg originally defined a class of compounds that inhibited carcinogen formation; such compounds were primarily inhibitors of *in vivo* nitrosation that prevented the formation of nitrosamines

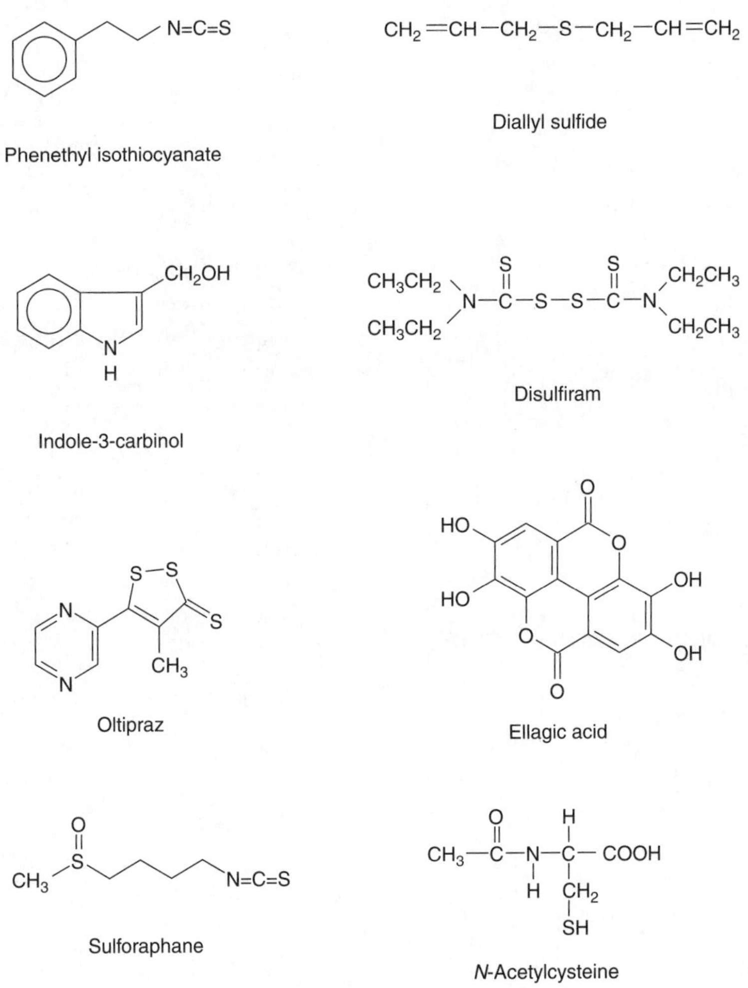

FIGURE 21.1   Representative anti-initiating agents.

from secondary amines and nitrite in an acidic environment. To this class, we have also added chemicals that prevent the absorption or uptake of carcinogenic stimuli (Table 21.1). Since both processes ultimately prevent internal exposure to carcinogenic influences, we consider all such compounds to be inhibitors of carcinogen exposure. A classical inhibitor of endogenous nitrosation, ascorbic acid can significantly decrease nitrosamine production when present in appreciable quantities [7]. Other compounds that inhibit nitrosamine formation include phenols such as caffeic acid and ferulic acid [8], as well as several sulfhydryl compounds [9]. Proline and thioproline scavenge nitrite by reacting with it to form nonmutagenic nitrosamines [10]. By binding to bile acids, calcium can inhibit their uptake by colon cells, thus reducing exposure to these tumor promoters [11]. Sodium thiosulfate can prevent the absorption of reactive electrophiles in the diet by reacting with such compounds, rendering them detoxified prior to uptake by the gastrointestinal tract [12]. Finally, the use of sunscreens that block carcinogenic UVB irradiation has been encouraged for years in Western countries [13].

21.3.2 ANTIINITIATING AGENTS

There are several means of chemical intervention in the initiation stage. With the exception of direct-acting carcinogens, most genotoxic carcinogens must first be metabolically activated to

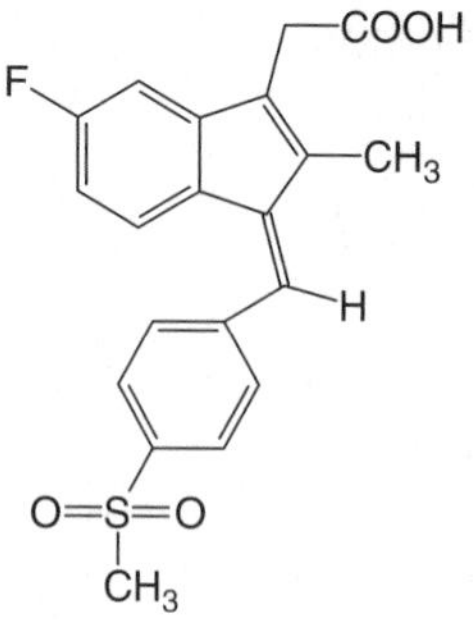

Exisulind

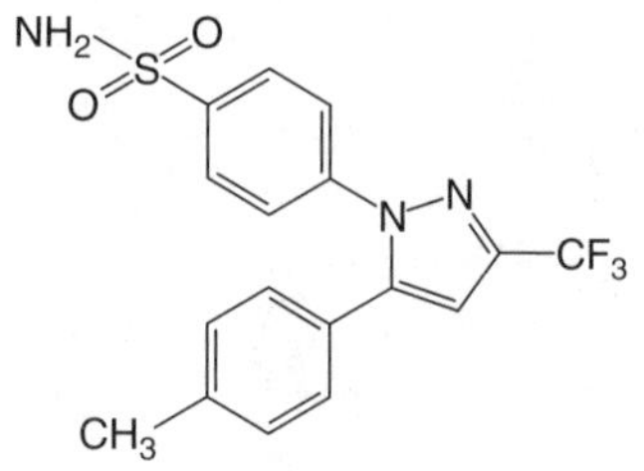

Celecoxib

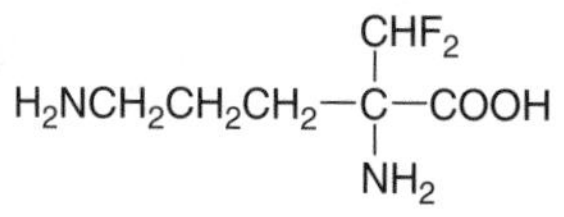

α-Difluoromethylornithine (DFMO)

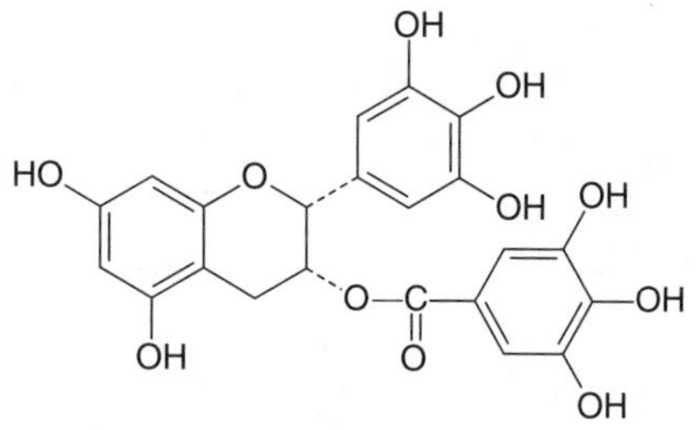

(–) Epigallocatechin-3-gallate (EGCG)

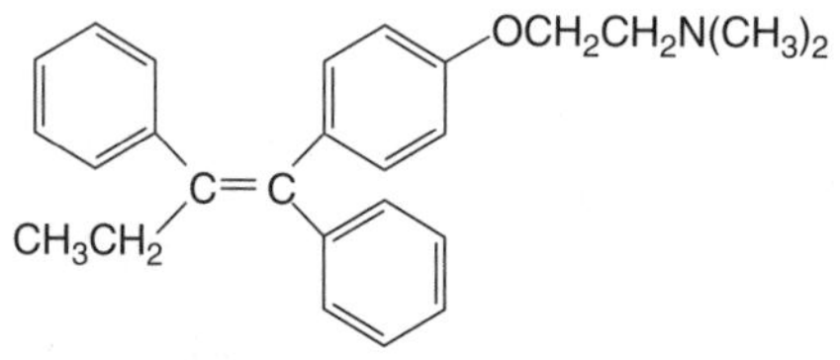

Tamoxifen

FIGURE 21.2 Representative antipromotion/antiprogressive agents.

electrophilic forms that can damage DNA. Although pathways of metabolic detoxification (often phase II pathways) can protect the cell by conjugating either procarcinogens or their reactive electrophilic metabolites, in many cases, at least a small proportion of carcinogen escapes detoxification. The electrophilic and ultimately carcinogenic form must then react with DNA, forming adducts that may induce mutations by resulting in direct mispairing, polymerase-induced errors, or by other mechanisms. Using the preceding as a basis of classification, the majority of blocking agents can be assigned to one or more of five major subclasses, namely: (1) inhibitors of cytochrome P450 enzymes; (2) inducers of cytochrome P450 enzymes; (3) inducers of phase II enzymes, such as glutathione *S*-transferase (GST); (4) alternative nucleophiles; and (5) inducers of DNA repair (Table 21.2).

21.3.2.1 Cytochrome P450 Inhibitors

A basic assumption inherent in the use of cytochrome P450 inhibitors as chemopreventive agents is that carcinogen activation presumably occurs largely in the target tissue. Thus, a selective blockade of cytochrome P450 enzymes, ideally restricted to a specific target tissue (and possibly to only

TABLE 21.1
Subclasses and Examples of Inhibitors of Carcinogen Exposure

Mechanism	Examples
Inhibitors of carcinogen formation (*inhibitors of nitrosation*)	Ascorbic acid α-Tocopherol Dietary phenols Proline
Inhibitors of carcinogen absorption/uptake	Calcium Sodium thiosulfate Sunscreens

Source: Adapted from Morse, M.A. and Stoner, G.D. *Carcinogenesis*, 14, 1737, 1993; Stoner, G.D., Morse, M.A., and Kelloff, G. *Environ. Health Perspect.,* 105, 945, 1997; Wattenberg, L.W. *Cancer Res.*, 45, 1, 1985. With permission.

TABLE 21.2
Subclasses and Examples of Antiinitiating Agents

Mechanism	Examples
Cytochrome P450 inhibitors	Phenethyl isothiocyanate Phenylpropyl isothiocyanate Diethyldithiocarbamate Diallyl sulfide Monoterpenes
Cytochrome P450 inducers	Indole-3-carbinol 5,6-Benzoflavone
Phase II enzyme inducers	Isothiocyanates Polyphenols Dithiolethiones
Alternative nucleophiles	Ellagic acid N-Acetylcysteine
DNA repair inducers	Vanillin N-Acetylcysteine

Source: Adapted from Morse, M.A. and Stoner, G.D. *Carcinogenesis*, 14, 1737, 1993; Stoner, G.D., Morse, M.A., and Kelloff, G. *Environ. Health Perspect.*, 105, 945, 1997; Wattenberg, L.W. *Cancer Res.*, 45, 1, 1985. With permission.

certain specific P450 isoforms), has the potential to decrease the incidence of initiating events in the target tissue without substantially altering P450 activity elsewhere. For such an approach to be truly successful, it is clear that there must be preexisting knowledge of the specific carcinogen(s) to which an individual is exposed along with an understanding of the specific P450 isoform(s) responsible for activation of the carcinogen(s). It is readily understood that a complete blockade of hepatic cytochrome P450 enzymes is not desirable, and would probably require the same restrictions in diet and medication followed in the management of end-stage hepatic disease (e.g., cirrhosis).

One of the first cytochrome P450 inhibitors shown to have chemopreventive activity was disulfiram (Figure 21.1), which inhibits the activation of dimethylhydrazine [14] and colon neoplasia induced by this compound [15]. Diallyl sulfide (Figure 21.1), a naturally occurring constituent of *Allium* vegetables, inhibits carcinogen activation [16] and tumorigenesis in a number of animal models [17–20]. The isothiocyanates are among the most potent chemopreventive agents known [21–30]. For example, dietary phenethyl isothiocyanate (Figure 21.1) at a concentration of 3 μmol/g diet can inhibit 4-(methylnitrosamino)-1-(3-pyridyl)-1-butanone (NNK)-induced lung tumors in F344 rats by about 50% [24] and completely inhibit N-nitrosomethylbenzylamine (NMBA)-induced esophageal tumors in F344 rats [28,29]. 6-Phenylhexyl isothiocyanate (PHITC) can inhibit NNK-induced lung tumorigenicity in strain A mice by more than 80% when administered at a dose 50-fold lower than NNK [27]. Unfortunately, increased length in the carbon chain separating the phenyl group from the isothiocyanate moiety does not always lead to decreased tumorigenesis; indeed, PHITC can enhance tumors induced by azoxymethane in the colon [31] and by NMBA in the esophagus [32]. Ellagic acid (Figure 21.1) inhibits NMBA metabolism *in vivo* and *in vitro* [33,34] and inhibits NMBA-induced esophageal tumors [35,36].

21.3.2.2 Cytochrome P450 Inducers

Inducers of cytochrome P450 can act as blocking agents as well, by increasing the production of activated metabolites at resistant nontarget tissues or by enhancing oxidative detoxification at any tissue site. Presumably, the ideal site to achieve induction is the liver, since hepatic cytochrome P450 activity is more readily inducible than in other tissues, and because orally ingested carcinogens must pass through the liver via the hepatic portal system prior to entering the systemic circulation. Indole-3-carbinol (I3C) (Figure 21.1) is a potent inducer of cytochrome P450 enzymes and has chemopreventive activity in a number of animal models [37–43]. In at least some of these models, the inductive abilities of I3C directly account for the observed inhibition of tumorigenicity, although it is clear that the gastric acid condensation products of I3C can directly inhibit cytochrome P450 enzymes [44]. However, the induction of P450 activity by I3C can lead to a shift in target organ through enhanced activation of carcinogen elsewhere [43,45]; this may, at least in part, account for the known ability of I3C to enhance tumorigenesis in many cases [46,47].

21.3.2.3 Phase II Enzyme Inducers

Among the most popular antiinitiating agents are phase II enzyme inducers, formerly referred to as monofunctional inducers based on their inability to simultaneously induce aryl hydrocarbon hydroxylase activity (largely due to CYP1A1 induction) [48]. Such phase II enzyme inducers have historically received preference over cytochrome P450 inducers as chemopreventive agents, since they may inhibit a greater range of target carcinogens and are regarded as less likely to enhance tumorigenesis themselves. However, it must be recognized that phase II enzymes *can* result in the production of reactive electrophiles. Examples include: (1) activation of alkyl dihalides following GST-catalyzed glutathione conjugation via spontaneous generation of an episulfonium ion [49], (2) production of acylating agents following UDP-glucuronosyltransferase-mediated conjugation of certain carboxylic acids [50], (3) spontaneous generation of carbonium ions from benzo[a]pyrene tetrols following sulfotransferase-mediated sulfate conjugation [51], and (4) the generation of reactive nitrenium ions from N-hydroxy arylamines following O-acetylation or O-sulfation catalyzed by N-acetyltransferases or sulfotransferases, respectively [52]. In addition, one should understand that the classic definition of monofunctional inducers was restricted to effects (or the lack thereof) on aryl hydrocarbon hydroxylase activity. Quite a few presumptive "monofunctional" inducers have been shown to induce other cytochrome P450 isoforms [53,54].

Much of the interest in phase II enzyme inducers has focused on inducers of those enzymes that can detoxify electrophiles, that is, GSTs and NAD(P)H quinone oxidoreductases (NQOs). (While some, notably Parkinson [55], regard NQO as a phase I enzyme because it largely catalyzes reduction reactions, many still regard NQO as a phase II enzyme.) However, such compounds normally induce UDP-glucuronosyltransferases as well. Benzyl isothiocyanate was one of the first chemopreventive agents whose mechanism was determined to be partially due to phase II enzyme induction, specifically induction of GST(s) [56]. Other isothiocyanates, such as phenethyl isothiocyanate [57] and sulforaphane [58] are also known inducers of GST, although in the case of phenethyl isothiocyanate, most of its inhibitory activity can be ascribed to cytochrome P450 inhibition and other mechanisms. Talalay and colleagues [58,59] initially identified sulforaphane as the chemical principally responsible for phase II enzyme induction in broccoli and later found that sulforaphane was a potent inhibitor of 7,12-dimethylbenz[a]anthracene (DMBA) induced mammary tumorigenesis in rats. The dithiolethione, oltipraz, is a potent inducer of GST and inhibits carcinogen-induced tumorigenesis in a number of animal models [60–66]. Kensler et al. [67] have demonstrated that dithiolthiones result in accumulation of the transcription factor *Nrf2*, which results in increased binding to antioxidant response elements (AREs) in the promoter regions of a number of phase II enzyme genes, resulting in transcriptional activation of these genes. It is likely that many phase II enzyme inducers act via similar mechanisms [68].

21.3.2.4 Alternative Nucleophiles

Alternative nucleophiles are chemical compounds that physically react with the activated (electrophilic) forms of carcinogens, and in so doing, protect DNA from electrophilic attack. This is analogous to the mechanism of action exhibited by many antioxidants, except that the damaging species here tend to be larger and exogenous in nature, as compared with the relatively smaller, endogenous reactive oxygen species (ROS). By acting as scavenging agents, alternative nucleophiles reduce DNA adduct formation and, thus, decrease the incidence of initiating events. In general, such compounds contain nucleophilic moieties such as hydroxyl groups, sulfhydryl functions, and so on. Examples of such compounds include ellagic acid and *N*-acetylcysteine (NAc) (Figure 21.1). Ellagic acid has been shown to react directly with the diol epoxide of benzo[a]pyrene to form a conjugate [69]; such activity may account for its inhibition of benzo[a]pyrene diol epoxide (BPDE)-induced mutagenicity and BPDE-induced carcinogenicity in mouse skin [70,71]. The sulfhydryl moiety of NAc can react with a number of electrophilic species, which may account for some of the antimutagenic and anticarcinogenic effects of NAc [72–74], although it is clear that NAc has other activities as well. A major disadvantage of such scavenging agents is that they must be maintained at relatively substantial concentrations in target tissues at all times during which carcinogens are present.

21.3.2.5 Inducers of DNA Repair

Alterations in DNA repair systems have profound effects on mutagenic or tumorigenic outcomes in prokaryotic and eukaryotic systems. Enhancement in the rate and fidelity of DNA repair in an individual exposed to genotoxic carcinogens should result in a decreased incidence of initiating events. Unfortunately, the majority of known inducers of DNA repair also cause DNA damage themselves, including UV irradiation, gamma-irradiation, and a host of electrophilic chemical carcinogens [75–80]. In a prior review, we cited vanillin as an example of a naturally occurring chemical that affects DNA repair [3]. Although vanillin shows significant inhibition of bacterial mutagenicity [81–83], chromosomal aberrations in *Drosophila melanogaster* [84], and mammalian cell mutagenicity [85–88], the utility of vanillin appears to be largely limited to inhibition of UVB and x-ray-induced mutagenesis. In the presence of alkylating agents, both vanillin and *o*-vanillin actually enhance the mutagenic response [88–91]. Thus, in many respects, the search for an otherwise innocuous chemical that can induce DNA repair represents the "holy grail" of antiinitiation research.

Kelloff [92] has listed NAc and protease inhibitors among enhancers of poly (ADP-ribose) polymerase (PARP). PARP activity normally increases significantly in the presence of DNA damage (i.e., DNA strand breaks), and PARP is believed to be important in DNA repair and the maintenance of genomic stability as well as other possible roles [93,94]. Chemical inhibition of PARP does often enhance tumorigenic responses in laboratory animals although it appears that PARP-1 knockout mice are not significantly more susceptible to tumor development [93]. However, as Burkle [95] points out, overproduction of PARP tends to sensitize cells to the cytotoxic effects of alkylating agents, suggesting that constitutive levels of PARP may be optimal for cell survival. More recently, cells pretreated with oltipraz experienced an enhanced rate of removal of total platinum-DNA adducts and interstrand cross-links [96].

Many investigators would maintain that the use of antiinitiating agents may be impractical in the human population, since individuals in high-risk groups have presumably received some exposure to initiating agents. As we have stated previously [3,5], the colorectal cancer model of Vogelstein and colleagues [97,98] demonstrates that human cancer is not adequately represented by the traditional initiation–promotion model; a more accurate depiction of the process involves an accumulation of several different genetic events. Thus, administration of blocking agents should be beneficial, since many individuals at high risk (e.g., smokers and the occupationally exposed) are continually exposed to genotoxic carcinogens. In addition, it is critical that individuals who are genetically susceptible to cancer avoid further mutations that might advance the carcinogenic process; such individuals are excellent candidates for prophylactic treatment with blocking agents. Also, the administration of inhibitors of promotion/progression should be beneficial in reducing the carcinogenic risk due to exposure to a wide range of carcinogens, regardless of the mechanistic model human carcinogenesis may follow. Our preferred strategy for cancer chemoprevention entails the use of combinations of antiinitiating and antipromotion/antiprogression agents.

21.3.3 ANTIPROMOTION/ANTIPROGRESSION AGENTS

The classification of these agents is more difficult since the precise sequence of events in the processes of promotion and progression is not known with certainty. There is also some difficulty in distinguishing which mechanistic effects are direct and which are indirect in nature. However, based on our previous reviews [3,5] and those of Kelloff [92] and De Flora and Ramel [99], one can classify most antipromotion/antiprogression agents as belonging to one or more of the following categories: (1) inducers of apoptosis, (2) inducers of differentiation, (3) inhibitors of angiogenesis, (4) inhibitors of arachidonic acid metabolism, (5) inhibitors of polyamine metabolism, (6) inhibitors of protease activity, (7) inhibitors of ROS, (8) modulators of hormonal activity, and (9) modulators of signal transduction (Table 21.3). This list of mechanistic categories should not be considered to be exhaustive, and as with the antiinitiating agents, it should be understood that virtually all of these agents possess multiple mechanisms of action. Thus, the listing of an agent in one category in Table 21.3 should not be taken as an indication that the mechanism indicated is the only or even necessarily the primary mechanism of action for that agent.

21.3.3.1 Inducers of Apoptosis

Apoptosis, or programmed cell death, is a means by which damaged cells can be removed when genomic DNA damage is excessive; tumor cells tend to lose their apoptotic responses [100]. Thus, restoration of the apoptotic response mechanisms in neoplastic cells represents a potential means of inhibiting cancer by selectively inducing cell death in premalignant lesions. Two primary apoptotic pathways are the death receptor pathway and the mitochondrial pathway, both of which lead to the activation of caspases, that is, cysteine proteases that cleave other proteins involved in the maintenance of nuclear integrity and cell cycle progression [100]. Because both pathways can be directly or indirectly influenced by a number of stimuli, a substantial number of chemopreventive

TABLE 21.3

Subclasses and Examples of Antipromotion/Antiprogression Agents

Mechanism	Examples
Inducers of apoptosis	Phenethyl isothiocyanate
	Exisulind
	EGCG
Inducers of differentiation	Retinoids
	Carotenoids
	Calcium
	Deltanoids
Inhibitors of angiogenesis	Linomide
	EGCG
Inhibitors of arachidonic acid metabolism	Celecoxib
	Piroxicam
	Indomethacin
	Aspirin
	Quercetin
	Curcumin
Inhibitors of polyamine metabolism	α-Difluoromethylornithine
Inhibitors of protease activity	Bowman–Birk protease inhibitor
Inhibitors of reactive oxygen species	EGCG
	N-Acetylcysteine
	Trolox
Modulators of hormonal activity	Tamoxifen
	LY353381
	Finasteride
Modulators of signal transduction	Staurosporine
	Monoterpenes
	R115777

Source: Adapted from Morse, M.A. and Stoner, G.D. *Carcinogenesis*, 14, 1737, 1993; Stoner, G.D., Morse, M.A., and Kelloff, G. *Environ. Health Perspect.*, 105, 945, 1997; Wattenberg, L.W. *Cancer Res.*, 45, 1, 1985. With permission.

agents likely induce apoptosis, albeit by indirect means in many cases. Besides its antiinitiating activities, phenethyl isothiocyanate at high doses can induce apoptosis *in vitro* and *in vivo* [101–103]. One of the most prominent apoptotic agents under current investigation is exisulind (Figure 21.2), a sulfone derivative of sulindac [104]. Mechanistically, exisulind activates and induces cyclic GMP-dependent protein kinase G, resulting in enhanced phosphorylation of β-catenin and apoptosis in tumor cells apparently via the mitochondrial pathway [105,106]. Initially, exisulind was found to be a potent inhibitor of human colon cancer cell growth along with a number of nonsteroidal antiinflammatory agents; however, it was clear that inhibition of prostaglandin synthesis was not involved, since exisulind is a poor inhibitor of cyclooxygenase (COX) [107]. Exisulind was found to inhibit methylnitrosourea-induced mammary cancer in rats by a mechanism that included the induction of apoptosis [108]. In a phase I clinical trial, exisulind appeared to induce apoptosis in the polyps of FAP patients [109]. Epigallocatechin gallate (EGCG), the green tea polyphenol, also appears to induce apoptosis by the mitochondrial pathway; EGCG specifically reduces expression of apoptotic inhibitors such as bcl-2 while increasing expression of the apoptotic enhancer bax, resulting in caspase-9 activation [110].

21.3.3.2 Inducers of Differentiation

Tumor cells are phenotypically dedifferentiated in nature [111]. The induction of terminal differentiation in a proliferating cell results in the production of a cell that is no longer capable of proliferation and that is more susceptible to apoptosis [111]. Hence, inducers of differentiation such as retinoids, caretonoids, calcium ion, and deltanoids, offer the possibility of arresting the growth of a neoplasm at the premalignant stage (Table 21.3). There appear to be a number of mechanisms that can induce terminal differentiation in cells, including up-regulation of cellular gap junctions [112] as well as inhibition of histone deacetylase [113]. Retinoids can be considered as classic inducers of differentiation, although it is clear that much of their chemopreventive activity can be ascribed to other mechanisms as well. Retinoids have long been known to induce the differentiation of epithelial cells *in vivo* and *in vitro* [114,115]. It is known that retinoids bind to nuclear receptors and regulate the expression of a number of genes [116]. The effects of retinoids on cell growth and on *in vitro* transformation are well correlated with their abilities to affect gap junctional communication; in this regard, carotenoids have similar effects *in vitro* [116–118]. Retinoids have been effective in inhibiting tumorigenesis in a number of animal model systems [119–128], and in the past, represented major agents evaluated in some of the early chemoprevention clinical trials [3,5,92,129]. As we noted earlier [3,5], many conventional retinoids display a number of undesirable attributes, such as their tendency to induce hepatotoxicity and teratogenicity [129], and there are difficulties in distinguishing between physiologically and pharmacologically appropriate doses [130,131]. Calcium and vitamin D agonists (deltanoids), such as retinoids, stimulate cell differentiation [132] and inhibit tumor promotion in animal models, particularly when high fat diets are employed [133–136].

21.3.3.3 Inhibitors of Angiogenesis

Like all other cells, tumor cells require access to adequate levels of nutrients and removal of cellular wastes in order to support their growth and proliferation. When solid tumors approach a size of approximately 2 to 3 mm in diameter, further growth tends to be inhibited unless the tumor is able to develop its own blood supply; in addition, tumors require access to blood vessels (or lymphatic vessels) in order to metastasize [137]. The process by which new blood vessels develop from existing vessels is termed angiogenesis [137]. Thus, by arresting (or possibly, reversing) tumor development in the premalignant stage or by inhibiting the metastasis of *in situ* neoplasias, inhibition of angiogenesis appears to be a viable chemopreventive mechanism. The immunomodulator linomide (a quinoline-3-carboxamide) has demonstrable antiangiogenic activity *in vitro* and *in vivo*, inhibits tumorigenesis at a number of different target organs, including mammary and prostate, and is considered to be an excellent prospective chemopreventive agent for prostate cancer [138–142]. EGCG inhibits endothelial cell proliferation as well as tumor vascularization [143,144]. Such antiangiogenic activities may make a significant contribution to the chemopreventive activity of EGCG. A number of other chemopreventive agents that are normally placed in other mechanistic categories also display antiangiogenic activities, at least *in vitro*. Such agents include, but are not limited to, certain nonsteroidal anti-inflammatory drugs (NSAIDs) such as aspirin and sulindac, classic retinoids such as all-*trans*-retinoic acid, 9-*cis*-retinoic acid, and 13-*cis*-retinoic acid, the Bowman-Birk inhibitor (BBI), thalidomide, tamoxifen, and α-difluoromethylornithine (DFMO) [145].

21.3.3.4 Inhibitors of Arachidonic Acid Metabolism

Among the many deleterious sequelae that occur during experimentally induced tumor promotion is an increased metabolism of arachidonic acid, which consequently contributes to an overall inflammatory response [3,5]. Two major enzymes controlling two distinct branches of this metabolic pathway are COX and lipoxygenase (LOX). The COX pathway converts arachidonic acid to prostaglandins, prostacyclins, and thromboxanes, while LOX converts arachidonic acid to leukotrienes and hydroxyeicosatetraenoic acids [146]. There are two distinct types of COX: COX-1, which is constitutively expressed in most tissues, and COX-2, which is normally induced by pro-inflammatory agents such as cytokines and growth factors [147]. Most untoward effects of NSAIDs, such as gastrointestinal bleeding and ulcer formation, appear to be largely due to inhibition of COX-1 activity, while COX-2 inhibition appears to correlate much better with thera-peutic effects, such as anti-inflammation, analgesia, and antipyresis [147]. Of these therapeutic effects, the anti-inflammatory effect of NSAIDs is believed to provide a substantial contribution to the chemopreventive mechanism of the agents, although other chemopreventive effects, such as antiangiogenesis and the induction of apoptosis, clearly contribute.

Most conventional NSAIDs tend to be nonspecific, whereas there is a considerable interest in the development of specific COX-2 inhibitors such as celecoxib (Figure 21.2) [148]. In the past, a variety of nonspecific COX inhibitors (i.e., piroxicam, indomethacin, ibuprofen) have been found to inhibit experimental colon cancer or cancers at other sites [149–152]. Celecoxib is an effective inhibitor of tumorigenesis in azoymethane-induced colon tumorigenesis in rats [153] and significantly decreases tumor burden in human FAP patients [154]. Because of its effects in humans, celecoxib has been approved by the Food and Drug Administration as an adjunctive treatment for FAP. Classical LOX inhibitors such as 3,4,2′,4′-tetrahydroxychalcone and quercetin inhibit ornithine decarboxylase (ODC) induction as well as the promotion of 7-bromomethylbenz[a]anthracene-initiated epidermal tumors [155]. Curcumin, the major pigment in mustard and turmeric, effectively inhibits LOX and COX activities [156] as well as 12-O-tetradecanolphorbol-13-acetone-induced promotion in mouse skin [157]. However, it has only recently been appreciated that there appear to be both procarcinogenic forms of LOX (5-LOX, 8-LOX, and 12-LOX) as well as anticarcinogenic forms of LOX (15-LOX-1 and 15-LOX-2) [158]. Thus, the use of a LOX inhibitor in human populations may be somewhat premature at this point.

21.3.3.5 Inhibitors of Polyamine Metabolism

Ornithine decarboxylase is the rate-limiting enzyme of polyamine biosynthesis [159]. The polyam-ines spermidine and spermine and their precursor putrescine play key roles in normal cell growth and differentiation, tumor promotion, and malignant transformation and are found in virtually all mam-malian tissues [160]. ODC activity is substantially increased in rapidly proliferating cells, including most tumor cells [161]. A number of tumor promoters induce ODC, including classical promoters such as 12-O-tetradecanoylphorbol-13-acetate, chrysarobin, and benzoyl peroxide, secondary bile acids, saccharin, and phenobarbital [161]. Typically, the degree of induction is dose-dependent and will be related to the degree of tumor promotion. A number of signal transduction pathways are known to affect polyamine biosynthetic pathways and thus, regulate polyamine levels [162]. Although the precise functions of polyamines are not known with complete certainty, the binding of polyamines to DNA and the ability of polyamines to perturb DNA-protein interactions appear to be crucial in accounting for the molecular effects of polyamines on cellular proliferation [162].

Inhibitors of polyamine metabolism may be best exemplified by the classical suicide inhibitor of ODC, DFMO (Figure 21.2) [163]. Since ODC elevation appears to be a virtually universal event in tumorigenesis, it is not surprising that DFMO inhibits tumorigenesis induced by a number of different carcinogens [164–171]. Preliminary clinical trials with DFMO indicate that it may be an effective inhibitor of colorectal cancer and other forms of cancer. Several clinical trials involving the use of this agent are in progress [172].

21.3.3.6 Inhibitors of Protease Activity

Protease inhibitors are active inhibitors of the early stages of promotion in mouse skin models [173], and they are also effective in the suppression of bladder, breast, colon, liver, and lung tumorigen-esis [174–177]. They have been shown to prevent transformation of NIH-3T3 cells by the H-*ras* oncogene [178] and inhibit the formation of promoter-induced oxygen radicals and hydrogen per-oxide [179]. Protease inhibitors that can inhibit chymotrypsin are more potent in the inhibition of oxygen radicals and appear to be better inhibitors of tumor promotion [180]. Historically, the untoward effects of protease inhibitors have included pancreatic enlargement in some experimental species as well as decreased rates of growth in young animals [181].

The most exhaustively examined protease inhibitor is the BBI. The BBI is a soybean-derived peptide protease inhibitor with a molecular weight of approximately 8000 Da [182–186]. The BBI is part of a larger family of polypeptides isolated from soybeans and other plants that are referred to collectively as "Bowman-Birk inhibitors" [182–184]. BBI is a potent inhibitor of the proteases trypsin and chymotrypsin [182]. Although BBI is a fairly large polypeptide, nearly 50% of an oral dose of BBI appears to be absorbed and distributed throughout the body, with elimination occurring primarily in the urine [182]. Besides its protease inhibitory activity, BBI possesses anti-inflammatory effects [184]. It is believed that BBI may also reverse the initiating event(s) in carcinogenesis, as well as affect the promotional stage of carcinogenesis [184]. Indeed, protease inhibitors are normally classified as suppressing agents. BBI and Bowman-Birk inhibitory concentrate (BBIC) decrease the expression of oncogenes such as *c-myc* and *c-fos* and inhibit certain types of proteolytic activities that are found to be elevated in carcinogen-treated tissues or transformed cells [184–186]. While BBI is known to inhibit apoptosis, this activity appears to be primarily due to the phospholipids that co-purify with BBI [187]. At least part of BBI's anti-inflammatory activity may be due to inhibition of oxygen radical formation; it is clear that BBI inhibits the generation of superoxide anion radicals in HL-60 cells although it does not act as a direct scavenger of free radicals [188].

In most experimental models of tumorigenesis, BBI and BBIC inhibit tumors, including 1,2-dimethylhydrazine-induced intestinal tumors in mice [189–191], 7,12-dimethylbenz[a]anthracene-induced cheek pouch tumors in hamsters [192,193], NMBA-induced esophageal tumorigenesis in rats [194], radiation-induced thymic lymphosarcoma in mice [195], murine pulmonary adenomas induced by the polycyclic aromatic hydrocarbon 3-methylcholanthrene [196], and spontaneous intestinal tumor incidence and multiplicity in the Min mouse, the mouse model for FAP [197]. The Food and Drug Administration granted BBIC Investigational New Drug status in April of 1992 [182]. BBIC has been employed in phase I and IIa trials conducted in patients with oral leukoplakia [172]. In the phase IIa study, BBIC elicited a complete or partial response in approximately one third of the individuals [198]. BBIC is still under evaluation in clinical trials.

21.3.3.7 Inhibitors of ROS

The generation of ROS and free radicals is facilitated by a number of promoters and complete carcinogens [199]. At low concentrations, ROS serve as regulatory molecules in various cellular signaling processes, while at high concentrations, these moieties promote cellular damage and inflammatory processes [200]. ROS are involved in initiation, the early stages of promotion, as well as in tumor progression. The production of superoxide, hydroxyl radicals, and hydrogen peroxide can result in the formation of DNA adducts, including 8-oxodeoxyguanosine, thymine glycol, and 5-hydroxymethyl uracil [199]. Indeed, the formation of such DNA adducts from ROS after treatment with promoters has forced some reconsideration of the use of the term, "genotoxic" and "epigenetic." Although we have chosen to place ROS inhibitors under antipromotion/antiprogression agents, it is clear that antioxidants that scavenge ROS can provide protection at all stages of carcinogenesis.

It appears that many of the effects of green tea or its semipurified polyphenol fraction can be ascribed to EGCG. Despite the fact that EGCG exhibits apoptotic effects, antiangiogenic effects, and other effects, many of the anticancer effects of EGCG are attributed to its antioxidant activities. In most *in vitro* test systems, EGCG exhibits considerably greater antioxidant potential compared with the other catechins [201]. The antioxidant effect is due to the metal ion chelating potential of the catechol moieties, the ability to scavenge superoxide and hydroxyl radicals, and the ability to quench lipid peroxidation chain reactions [201]. In some cases, EGCG can inhibit cytochrome P450 enzymes responsible for the activation of chemical carcinogens or scavenge the ultimately carcinogenic forms of carcinogens [201]. EGCG can also inhibit the expression of critical oncogenes and prime the immune system by promoting the proliferation of B-cells [202,203]. EGCG inhibits TPA-induced increases in H_2O_2 as well as the formation of oxidized DNA bases [199]. Under conditions in which EGCG does not affect NNK-induced DNA alkylation, EGCG reduces both 8-oxodeoxyguanosine formation and tumorigenesis in lung induced by NNK [204]. Inhibition of oxidative DNA damage may well be the mechanism by which a number of other compounds with antioxidant activity inhibit carcinogenesis. Administration of pure EGCG or in the form of green tea or green tea polyphenolic (GTP) fractions results in the inhibition of mouse skin tumorigenesis induced by UVB, 3-methylcholanthrene, benzo[a]pyrene, or 7,12-dimethylbenz[a]anthracene [205–213]. Green tea and GTP also inhibits the development of lung adenomas in mice induced by urethane, N-nitrosodiethylamine, NNK, and benzo[a]pyrene [214–217]. In most cases, the test agents inhibited both the initiation and promotion stages of lung tumorigenesis. Green tea inhibits murine forestomach tumorigenesis induced by either N-nitrosodiethylamine or benzo[a]pyrene [215,217]. Several studies have demonstrated the efficacy of green tea infusions in drinking water in the inhibition of rat esophageal tumorigenesis induced by the potent esophageal carcinogen NMBA or its precursors [218–220]. Additional studies show inhibitory effects on azoxymethane-induced colon tumors in rats, aflatoxin B1- and N-nitrosodiethylamine-induced liver tumors in rats, and N-nitroso-bis(2-oxopropyl)amine-induced pancreatic tumors in hamsters [201].

21.3.3.8 Modulators of Hormonal Activity

Hormones are implicated in the etiologies of breast cancer, endometrial cancer, ovarian cancer, prostate cancer, and other cancers [221–225]. For example, in the case of endometrial or breast cancer, the unopposed estrogen hypothesis has been offered as a major etiologic factor in these two types of cancer [221–223]. This hypothesis holds that exposure to estrogens unopposed by progesterone or synthetic progestins results in increased mitotic activity of target tissue cells, decreases in terminally differentiated cells, and increased number of DNA replication errors and somatic mutations, ultimately yielding malignancy [223]. Similar effects can be hypothesized for androgens in the case of prostate cancer, inasmuch as antiandrogen therapies can provide effective treatment of prostate cancer [226].

Tamoxifen (Figure 21.2), which possesses both estrogen agonist and estrogen antagonist activity, is the prototypical example of the class of compounds referred to as selective estrogen receptor modulators (SERMs) [227]. When used as an adjuvant in the therapy of breast cancer, it was found that tamoxifen resulted in a decreased risk of contralateral breast cancer [228,229]. As part of the National Surgical Adjuvant Breast and Bowel Project, a phase III clinical trial of tamoxifen was conducted in women at high risk for breast cancer [230]. In this trial, tamoxifen reduced both invasive and noninvasive breast cancer by approximately 50% [230]. In addition, women carrying breast cancer-predisposing germline mutations in the *BRCA2* gene experienced a 62% reduction in breast cancer incidence compared with *BRCA2* mutation carriers that took placebo [231]. As a consequence of these results, tamoxifen was formally approved by the Food and Drug Administration in 1998 as a chemopreventive agent for breast cancer [227]. However, tamoxifen does possess a number of potential liabilities. For example, tamoxifen has been found to be genotoxic [232], and it induces hepatocellular adenomas and carcinomas in rats [233–235]. In addition, tamoxifen has a number of adverse effects on endometrial tissue, including endometrial hyperplasia, cervical and endometrial polyps, endometrial adenocarcinoma, ovarian cysts, and uterine sarcoma [236]. The annual incidence of endometrial cancer for women taking tamoxifen is 0.2% and appears to be associated with the cumulative dose of tamoxifen administered [236]. As a consequence of these adverse side effects, there is increased interest in another SERM, raloxifene. Like tamoxifen, raloxifene appears to prevent breast cancer but in contrast to tamoxifen, does not appear to enhance endometrial cancer incidence [237]. Tamoxifen and raloxifene are undergoing comparison for their breast cancer chemopreventive abilities and their adverse side effects in a randomized phase III clinical trial [227]. LY353381 or arzoxifene, is a benzothiopene SERM that appears to be an improvement on raloxifene for some therapeutic effects [238]. However, recent experimental data indicate that it may be no better than tamoxifen in terms of its effects on endometrial tissue [239].

Soy isoflavones are naturally occurring constituents of soybeans and other plant-derived materials [240]. Among these isoflavones are genistein, daidzein, genistin, and biochanin A, although a number of studies have demonstrated that genistein and daidzein are the predominant isoflavones in soy products. Epidemiologic studies indicate that breast and prostate cancers are considerably reduced in Asian countries with a high consumption of soy products [241]. Soy isoflavones appear to exhibit both antiestrogenic and antiandrogenic properties. Rats treated neonatally with genistein have increased latency and reduced incidence and multiplicity of DMBA-induced mammary adenocarcinomas [242]. It appears that neonatal genistein treatment enhances the maturation of terminal ductal structures and alters the endocrine system to reduce cell proliferation in the mammary gland [242]. Rats treated prepubertally with genistein have a reduced incidence and multiplicity of DMBA-induced mammary adenocarcinomas [243]. The terminal end buds from mammary glands of female rats treated prepubertally with genistein have significantly fewer cells in S-phase of the cell cycle, indicating that genistein induces maturation of the glands via terminal differentiation of cells [243]. Methylnitrosourea-induced rat mammary carcinogenesis is significantly reduced in rats fed soybeans or administered injected isoflavones [244,245]. Soy protein isolates containing high levels of genistein and daidzein significantly reduce the incidence and increase the latency of prostatic adenocarcinomas in Lobund–Wistar rats [246]. Genistein also inhibits the growth of cultured human benign prostatic hyperplastic tissue and human prostate cancer explants *in vitro* [241]. Genistein produces a dose-dependent inhibition of human breast adenocarcinoma MCF-7 cell growth with an IC50 of 40 μM [247]. Genistein also induces apoptosis in MCF-7 cells [248,249]. In MCF-7 cells, genistein inactivates bcl-2, an inhibitor of apoptosis [249]. The previously mentioned studies conducted in rat mammary *in vivo* suggest estrogen antagonist activity by isoflavones [242,243].

21.3.3.9 Modulators of Signal Transduction

Early changes during the process of tumorigenesis lead to enhanced signaling in various cellular growth pathways [250]. Many signal transduction effectors behave as oncogenes when activated by mutation, overexpression, and so on, such that a constant growth signal is furnished to the cell. Thus, blocking or attenuating such signal transduction pathways by chemopreventive agents affords the cell some restoration of normal growth controls [251]. In particular, the tyrosine kinase receptor-ras-raf-MAPK (mitogen-activated protein kinase)-ERK (extracellular signal-regulated kinase) signal transduction pathway is highly active in many human tumors [252]. The *ras* genes code for 21 KDa small G-proteins that are found anchored on the inside of the cell membrane and function as important signal transducers. The *ras* p21 proteins serve as a common relay point for receptor tyrosine kinases such as the epidermal growth factor, fibroblast growth factor, platelet-derived growth factor (PDGF), and other growth factors [250]. In addition, the protein kinase C (PKC) family of serine-threonine kinases is an important mediator of signaling cascades that affect cell growth, proliferation, and death [253]. Thus, one possible approach is to use specific inhibitors of these receptor kinases.

Phase II clinical trials are currently being conducted with iressa (ZD1839), a highly specific inhibitor of the epidermal growth factor receptor tyrosine kinase as well as with imatinib mesylate (STI-571), a highly selective inhibitor of the PDGF receptor tyrosine kinase [254,255]. Such signal transduction disruptors have been initially developed as potential chemotherapeutic agents; ultimately, such agents may prove to be far more effective as chemopreventive agents.

Another possible target for chemopreventive intervention is PKC. PKC actually constitutes a family of several isozymes that are activated by diacylglycerol [256,257]. The potent tumor promoter, 12-*O*-tetradecanoylphorbol-13-acetate, binds to, and activates PKC, in competition with diacylglycerol [258]. Known chemopreventive agents that have some inhibitory activity toward PKC include glycyrrhetic acid and tamoxifen [259]; of course, much of the inhibitory effects of these compounds is due to other mechanisms of action. Staurosporine, an inhibitor of the catalytic site of PKC, and threo-dihydrosphingosine, an inhibitor of the regulatory site of PKC, were both considered to have potential as chemotherapeutic and possibly, chemopreventive agents [260,261]. Staurosporine, although potent *in vitro*, was found to be fairly nonselective [262].

One major signal transduction target that has been selected is the common growth effector molecule, ras p21. It has been known for some time that *ras* proto-oncogene activation occurs in a number of human and rodent tumors [263–269]. In addition, in most cases, *ras* activation seems to be an event that occurs relatively early in the carcinogenic process [270–272]. The protein product of the *ras* gene, p21, undergoes farnesylation [273], which appears to be critical for its location in the plasma membrane and its ability to transform cells [274]. Monoterpenes such as limonene and perillyl alcohol inhibit p21 farnesylation at high concentrations and are effective chemopreventive agents in many models [275–283]. R115777, a novel potent farnesyltransferase inhibitor, has been tested in chemotherapeutic phase I trials [284–286] and is currently under evaluation in a phase II chemoprevention trial in patients with neurofibromatosis (Table 21.4).

21.3.4 HUMAN CLINICAL TRIALS

A number of chemopreventive clinical trials have been conducted in several different countries. As mentioned previously, on the basis of such trials, tamoxifen and celecoxib have been approved by the Food and Drug Administration for use as chemopreventive agents. In the United States, the principal sponsor of such trials is the Division of Cancer Prevention of the National Cancer Institute. We shall not attempt to discuss all past and present trials, but rather, a select few, including one well-known past failure and a brief discussion of current trials underway. (For more comprehensive reviews of human clinical trials of chemopreventive agents, the reader is referred to organ-specific reviews such as References 287 and 288.)

21.3.4.1 β-Carotene: A Tragic Failure

Perhaps among the most interesting and most disappointing clinical trials were those trials conducted with β-carotene [289,290]. A large number of epidemiologic studies, both retrospective and prospective, strongly suggested that β-carotene might be an effective inhibitor of lung cancer [291]. Such studies demonstrated that individuals who ate more fruits and vegetables, which are rich in carotenoids, and people who had higher serum β-carotene levels had a lower risk of lung cancer. It was also known that β-carotene acted as an antioxidant in many *in vitro* systems, as well as a precursor of vitamin A [290]. Literature reviews indicated that β-carotene was neither genotoxic nor showed the reprotoxic or teratogenic effects of many retinoids. In addition, no signs of organ toxicity had been observed in subacute, subchronic, or chronic oral toxicity studies that had been conducted in experimental animals at doses of 1 g/day or less. When tested in Sprague–Dawley rats or in CD-1 mice, no evidence of carcinogenicity was observed for β-carotene. Last, β-carotene had been listed as a GRAS (generally regarded as safe) substance by the Food and Drug Administration. These data collectively suggested both chemopreventive promise and safety in the case of β-carotene.

On the basis of these data, several trials were designed that employed β-carotene. These trials included the ATBC Study (α-Tocopherol, β-Carotene Cancer Prevention Study), the CARET Trial (The Carotene and Retinol Efficacy Trial), and Physicians' Health Study. The ATBC Study, conducted in Finland, was a randomized trial of 29,133 male smokers [292]. The subjects in this trial were administered α-tocopherol (50 mg), β-carotene (20 mg), both, or neither daily for a period of 5 to 8 years. While a 19% reduction in lung cancer incidence was observed in the highest versus lowest quintile of serum α-tocopherol, β-carotene administration was associated with increased lung cancer incidence. The Caret Trial was a randomized trial of approximately 18,000 smokers in the Pacific Northwest, performed at the Fred Hutchinson Cancer Center in Seattle [293]. Subjects were randomized to treatment arms of 30 mg of β-carotene or 25,000 IU of retinyl palmitate daily. Unfortunately, smokers administered β-carotene suffered a 46% excess in lung cancer mortality

TABLE 21.4
Active Chemoprevention Clinical Trials

Agent	Type of study	Cancer targeted	Cohort	Endpoint	Site
Acitretin	Phase III	Non-melanoma skin cancer	Immunosuppressed organ transplant recipients with a history of basal cell carcinoma or squamous cell carcinoma	Tumor incidence and surrogate endpoint biomarkers	North Central Cancer Treatment Group
Buprion and smoking cessation	Phase III	Lung cancer	Current smokers with previously resected stage I or stage II non-small cell lung cancer	Second primaries, survival, behavioral measures	Cancer and Leukemia Group B, Eastern Cooperative Oncology Group, Southwest Oncology Group
Celecoxib	Phase I	Prostate cancer	Patients with localized prostate cancer	Modulation of prostaglandin levels following treatment by prostatectomy	Kimmel Cancer Center, Johns Hopkins, Baltimore, MD
Celecoxib	Phase II	Non-small cell lung cancer	Chronic smokers with evidence of airflow obstruction	Safety and long-term side effects	Jonsson Cancer Center, UCLA, Los Angeles, CA
Celecoxib	Phase II	Esophageal adenocarcinoma	Patients with low- or high-grade Barrett's dysplasia	Regression of Barrett's dysplasia	Kimmel Cancer Center, Johns Hopkins, Baltimore, MD
Celecoxib	Phase II	Non-melanoma skin cancer	Patients with Fitzpatrick type I–IV skin exposed to UV light	UV-induced erythema and surrogate biomarkers	Irving Cancer Center, New York, NY
Celecoxib	Phase II	Non-melanoma skin cancer	Patients with basal cell nevus syndrome	Development of basal cell carcinoma	UCSF Cancer Center, San Francisco, CA
Celecoxib	Phase II	Oral cancer	Patients with oral premalignant lesions	Reversion of premalignant lesions	Memorial Sloan-Kettering Cancer Center, New York, NY
Celecoxib	Phase II/III	Non-melanoma skin cancer	Patients with actinic keratosis	Inhibition of new actinic keratoses and reversal of existing actinic keratoses	University of Alabama-Birmingham Cancer Center, Birmingham, AL

(continued)

Agent	Phase	Cancer	Population	Endpoint	Institution
Celecoxib	Phase IIB/III	Bladder cancer	Patients with previously resected superficial transitional cell carcinoma	Tumor recurrence and surrogate endpoint biomarkers	MD Anderson Cancer Center, Houston, TX
Celecoxib	Phase III	Colorectal cancer	Patients with previously resected spontaneous adenomatous polyps	Occurrence of new polyps	Brigham and Women's Hospital, Boston, MA
DFMO (Eflornithine)	Phase II	Non-melanoma skin cancer	Patients with actinic keratosis	Reversal of actinic keratosis	Arizona Cancer Center, Tucson, AZ
DFMO	Phase IIB	Prostate cancer	Men at high risk for prostate cancer	Polyamine levels and gene expression	Chao Family Comprehensive Cancer Center, Orange, CA
DFMO	Phase IIB	Colorectal cancer	Patients with previously resected colorectal adenoma	Surrogate endpoint biomarkers	Chao Family Comprehensive Cancer Center, Orange, CA
DFMO and sulindac	Phase IIB	Colorectal cancer	Patients with previously resected colorectal adenoma	Surrogate endpoint biomarkers and adenomas	University of California, Irvine, CA
Dietary soy	Phase II	Prostate cancer	Patients with elevated prostate specific antigen levels	Prostate specific antigen levels and other biomarkers	Cancer and Leukemia Group B
Doxercalciferol	Phase II	Prostate cancer	Patients with localized prostate cancer	Intermediate endpoint biomarkers	University of Wisconsin Cancer Center, Madison, WI
Exisulind	Phase II/III	Colorectal cancer	Patients with familial adenomatous polyposis	Growth and development of duodenal adenomas	Huntsman Cancer Institute, Salt Lake City, UT
Fenretinide	Phase III	Bladder cancer	Patients with previously resected superficial bladder cancer	Tumor recurrence and surrogate endpoint biomarkers	MD Anderson Cancer Center, Houston, TX
Fenretinide	Phase III	Ovarian cancer	Women at high risk for ovarian cancer	Surrogate endpoint biomarkers	Gynecologic Oncology Group
Flutamide and Toremifene	Phase II	Prostate cancer	Patients with stage I or II prostatic adenocarcinoma	Percent of high-grade prostatic intraepithelial neoplasia and other biomarkers	University of Pittsburgh Cancer Institute, Pittsburgh, PA
Imiquimod	Phase II	Cervical cancer	Women with grade II or III cervical intraepithelial neoplasia (CIN) or persistent grade I–III CIN	Chemopreventive efficacy, toxicity, safety	North Central Cancer Treatment Group

**TABLE 21.4
(Continued)**

Agent	Type of study	Cancer targeted	Cohort	Endpoint	Site
Isoflavones	Phase III	Prostate cancer	Men with grade 1 or 2 prostate cancer	Tumor progression, surrogate endpoint biomarkers	Moffitt Cancer Center, Tampa, FL
Isotretinoin and vitamin E	Phase II	Lung cancer	Current smokers or former smokers at high risk for lung cancer	Surrogate endpoint biomarkers	University of Colorado Cancer Center, Denver, CO
Lutetium texaphyrin and photodynamic therapy	Phase I	Cervical cancer	Women with grade II or III CIN	Toxicity, optimal dose, and treatment sensitivity	University of Pittsburgh, Pittsburgh, PA
LY353381 hydrochloride	Phase II	Breast cancer	Women at high risk for breast cancer and with cytologic evidence of hyperplasia	Baseline cytology and potential surrogate endpoint biomarkers	University of Kansas Medical Center, Kansas City, KS
Provera and depo-provera	Phase III	Endometrial cancer	Women with atypical endometrial hyperplasia	Histological response	Gynecologic Oncology Group
R115777	Phase II	Neurofibromatosis	Pediatric patients with neurofibromatosis 1 and progressive plexiform neurofibromas	Progression of disease	NCI Center for Cancer Research, Bethesda, MD
Raloxifene	Phase II	Breast cancer	Women at high risk for breast cancer	Surrogate endpoint biomarkers	NCI Center for Cancer Research, Bethesda, MD
Raloxifene and exemestane	Phase II	Breast cancer	Women with stage 0, I, II, or III breast cancer after completion of adjuvant therapy	Surrogate endpoint biomarkers	Memorial Sloan-Kettering Cancer Center, New York, NY
Raloxifene and goserelin	Phase III	Breast cancer	Women at high genetic risk for breast cancer	Surrogate endpoint biomarkers and tumor occurrence	United Kingdom Coordinating Committee on Cancer Research, U.K.
Rofecoxib	Phase III	Colorectal cancer	Patients with previously resected stage II or III carcinoma	Survival and disease-free survival	Cancer Research Campaign Trials Unit-Birmingham (CRCTU), Birmingham, England, U.K.

(continued)

Agent	Phase	Cancer	Population	Endpoint	Group
Selenium	Phase III	Lung cancer	Previously resected stage I non-small cell lung cancer	Incidence of second primary tumors	American College of Surgeons Oncology Group, Cancer and Leukemia Group B, Eastern Cooperative Oncology Group, NCIC-Clinical Trials Group, North Central Cancer Treatment Group, Southwest Oncology Group
Selenium	Phase III	Prostate cancer	Men with high-grade prostatic intraepithelial neoplasia	Prostate cancer incidence and surrogate endpoint biomarkers	Cancer and Leukemia Group B, Eastern Cooperative Oncology Group, Southwest Oncology Group
Selenium and Vitamin E	Phase III	Prostate cancer	Males over the age of 50 to 55 years	Incidence of prostate cancer and other cancers	Cancer and Leukemia Group B, Eastern Cooperative Oncology Group, NCIC-Clinical Trials Group, North Central Cancer Treatment Group, Radiation Therapy Oncology Group, Southwest Oncology Group
Tamoxifen and LY353381 hydrochloride	Phase IB	Breast cancer	Postmenopausal women with newly diagnosed breast cancer	Expression of tissue biomarkers	University of Kansas Medical Center, Kansas City, KS
Tamoxifen and Raloxifene	Phase III	Breast cancer	Women at high risk for breast cancer	Incidence of breast cancer	National Surgical Adjuvant Breast and Bowel Project
Tretinoin	Phase III	Cervical cancer	Women with CIN II or III	Human papilloma virus-related surrogate endpoint biomarkers	University of Michigan Cancer Center, Ann Arbor, MI

Source: NCI Clinical Trials web page (http://www.nci.nih.gov/search/clinical_trials/).

compared with controls. The Physicians' Health Study included β-carotene supplementation (50 mg every other day) in the primary prevention of cancer in a group of 22,071 U.S. male physicians, which included very few current smokers [294]. In this trial, no decrease in overall cancer incidence or in lung cancer incidence was observed with β-carotene administration. Thus, in these three trials, an increased risk of lung cancer incidence or mortality occurred in two studies, with no evidence of a protective effect in the third study.

There are a number of factors that may account for the failure of these trials. Firstly, β-carotene had never been shown to be an effective chemopreventive agent against lung cancer in animals. In fairness, many consider the utility of animals as models for carotenoid testing to be of dubious value, since most animals do not absorb or metabolize carotenoids in a manner similar to humans. However, the current paradigm for drug development (in the United States) then and now, includes a requirement of the demonstration of preclinical efficacy, which was waived in the case of β-carotene. In addition, the potential limitations of the epidemiologic data were not fully appreciated. Although epidemiologic studies remain an important means of determining the beneficial or deleterious effects of various exposures, one must remember that statistically significant associations in such studies cannot prove causality. Also, one must remember that fruits and vegetables are complex mixtures of hundreds of different chemicals. It may be that in the prospective epidemiologic studies, which virtually uniformly showed protective effects of higher levels of serum β-carotene, that β-carotene simply served as a biomarker for high fruit and vegetable consumption, and that the truly chemopreventive chemicals were not identified or examined. It is also known that many antioxidants can behave as potent pro-oxidants under the right conditions. Under certain conditions (such as high oxygen partial pressures), β-carotene actually displays a dose-dependent pro-oxidant effect [295]; such an effect could have occurred in active smokers in the ATBC and CARET trials.

While the results of these β-carotene trials are disheartening, they do provide an abject lesson in how *not* to develop chemopreventive agents. For prospective chemopreventive agents, it seems prudent to insist on the following minimal benchmarks: (1) a clear demonstration of preclinical efficacy, (2) a thorough understanding of the mechanisms of action of the prospective agent, and (3) the initiation of phase III trials only after the successful completion of phase I trials and short-term phase II trials utilizing reliable surrogate endpoint biomarkers even in the case of a GRAS substance.

21.3.5 CURRENT HUMAN TRIALS

Over 35 human clinical trials (phases I–III) supported by NCI are currently in progress (Table 21.4). The majority of these studies are phase II or phase III clinical trials. The major targets of these studies include prostate cancer (8 trials), breast cancer (6 trials), colorectal cancer (5 trials), non-melanoma skin cancer (5 trials), lung cancer (4 trials), cervical cancer (3 trials), bladder cancer (2 trials), as well as other sites (5 trials total). Celecoxib, the selective COX-2 inhibitor, is currently under evaluation in over 20% (9 of 38) of the total trials listed, with a total of 6 different specific sites (colon, skin, esophagus, lung, oral cavity, and bladder). The next most frequent chemopreventive agents are: tamoxifen/raloxifene (5 trials), retinoids (5 trials), DFMO (4 trials), and selenium (3 trials). None of the agents under evaluation are purely antiinitiating agents, and the vast majority of agents are more properly classified as antipromotion/antiprogression agents. In addition, the vast majority of these trials are being conducted with single agents. In time, it is expected that combinations of different chemopreventive agents will become more pervasive.

21.3.6 CONCLUSIONS

Chemoprevention continues to merit consideration as an effective means of controlling cancer incidence. As we have already pointed out, a large number of identifiable groups at high risk for the development of cancer stand to benefit from chemopreventive approaches. A large number of chemopreventive compounds are currently being examined in preclinical experiments, with a correspondingly lower number of agents under consideration in current human clinical trials. As our understanding of the molecular mechanisms of action of these agents increases, we should expect to develop more rational combinations of such agents, for it is apparent that the most fruitful approach will be to employ such combinations, rather than individual agents. Finally, since a disproportionate share of the burden of funding chemoprevention efforts falls upon the public sector and since chemoprevention is relatively underfunded in comparison with other means of cancer control and treatment, it is incumbent upon us to step up our efforts to increase government funding for chemopreventive efforts, and to attempt to enlist greater support from the private sector as well.

REFERENCES

1. *Cancer Facts and Figures*, American Cancer Society, 2002, pp. 1–48.
2. *Cancer Facts and Figures*, American Cancer Society, 1992, pp. 1–3.
3. Morse, M.A. and Stoner, G.D. Cancer chemoprevention: Principles and prospects, *Carcinogenesis*, 14, 1737, 1993.
4. De Flora, S., Izzotti, A., D'Agostini, F., Balansky, R.M., Noonan, D., and Albini, A. Multiple points of intervention in the prevention of cancer and other mutation-related diseases, *Mutat. Res.*, 9, 480–481, 2001.
5. Stoner, G.D., Morse, M.A., and Kelloff, G. Perspectives in cancer chemoprevention, *Environ. Health Perspect.*, 105, 945, 1997.
6. Wattenberg, L.W. Chemoprevention of cancer, *Cancer Res.*, 45, 1, 1985.
7. Mirvish, S.S. Ascorbic acid inhibition of *N*-nitrosocompound formation in chemical, food and biological systems, in *Inhibition of Tumor Induction and Development*, Zedeck, M.S. and Lipkin, M., Eds., Plenum Publishing, New York, 1981, p. 101.
8. Kuenzig, W., Chau, J., Norkus, E., Holowaschenko, H., Newmark, H., Mergens, W., and Conney, A.H. Caffeic acid and ferulic acid as blockers of nitrosamine formation, *Carcinogenesis*, 5, 309, 1984.
9. Shenoy, N.R. and Choughuley, A.S.U. Inhibitory effect of diet related sulfhydryl compounds on the formation of carcinogenic nitrosamines, *Cancer Lett.*, 65, 227, 1992.
10. Wakabayashi, K., Nagao, M., and Sugimura, T. Heterocyclic amines, lipophilic ascorbic acid, and thioproline: Ubiquitous carcinogens and practical anticarcinogenic substances, in *Cancer Chemoprevention*, Wattenberg, L., Lipkin, M., Boone, C.W., and Kelloff, G.J., Eds., CRC Press, Boca Raton, FL, 1992, p. 311.
11. van der Veere, C.N., Schoemaker, B., van der Meer, R., Groen, A.K., Jansen, P.L., and Oude Elferink, R.P. Rapid association of unconjugated bilirubin with amorphous calcium phosphate, *J. Lipid Res.*, 36, 1697, 1995.
12. Wattenberg, L.W., Hochalter, J.B., and Galbraith, A.R. Inhibition of β-propiolactone-induced mutagenesis and neoplasia by sodium thiosulfate, *Cancer Res.*, 47, 4351, 1987.
13. Scherschun, L. and Lim, H.W. Photoprotection by sunscreens, *Am. J. Clin. Dermatol.*, 2, 131, 2001.
14. Fiala, E.S., Bobotas, G., Kulakis, C., Wattenberg, L.W., and Weisburger, J.H. The effects of disulfiram and related compounds on the *in vivo* metabolism of the colon carcinogen 1,2-dimethylhydrazine, *Biochem. Pharmacol.*, 26, 1763, 1977.
15. Wattenberg, L.W. Inhibitors of chemical carcinogenesis, in *Environmental Carcinogenesis*, Emmelot, P. and Kriek, E., Eds., Elsevier/North Holland Biomedical Press, Amsterdam, 1979, p. 241.
16. Hong, J.-Y., Smith, T., Lee, M.-J., Li, W., Ma, B.-L., Ning, S.M., Brady, J.F., Thomas, P.E., and Yang, C.S. Metabolism of carcinogenic nitrosamines by rat nasal mucosa and the effect of diallyl sulfide, *Cancer Res.*, 51, 1509, 1991.
17. Wargovich, M.J. Diallyl sulfide, a flavor component of garlic (*Allium sativum*), inhibits dimethylhydrazine-induced colon cancer, *Carcinogenesis*, 8, 487, 1987.
18. Sparnins, V.L., Barany, G., and Wattenberg, L.W. Effects of organosulfur compounds from garlic and onions on benzo(a)pyrene-induced neoplasia and glutathione *S*-transferase activity in the mouse, *Carcinogenesis*, 9, 131, 1988.
19. Wargovich, M.J., Woods, C., Eng, V.W.S., Stephens, L.C., and Gray, K. Chemoprevention of *N*-nitrosomethyl-benzylamine-induced esophageal cancer in rats by the naturally-occurring thioether, diallyl sulfide, *Cancer Res.*, 48, 6872, 1988.

20. Tadi, P.P., Teel, R.W., and Lau, B.H.S. Organosulfur compounds of garlic modulate mutagenesis, metabolism, and DNA binding of aflatoxin B_1, *Nutr. Cancer*, 15, 87, 1991.
21. Wattenberg, L.W. Inhibition of carcinogenic effects of polycyclic hydrocarbons by benzyl isothiocyanate and related compounds, *J. Natl. Cancer Inst.*, 58, 395, 1977.
22. Wattenberg, L.W. Inhibition of carcinogen-induced neoplasia by sodium cyanate, *tert*-butyl isocyanate, and benzyl isothiocyanate administered subsequent to carcinogen exposure, *Cancer Res.*, 41, 2992, 1981.
23. Wattenberg, L.W. Inhibitory effects of benzyl isothiocyanate administered shortly before diethylnitrosamine or benzo(a)pyrene on pulmonary and forestomach neoplasia in A/J mice, *Carcinogenesis*, 8, 1971, 1987.
24. Morse, M.A., Wang, C.-X., Stoner, G.D., Mandal, S., Conran, P.B., Amin, S.G., Hecht, S.S., and Chung, F.-L. Inhibition of 4-(methylnitrosamino)-1-(3-pyridyl)-1-butanone-induced DNA adduct formation and tumorigenicity in the lung of F344 rats by dietary phenethyl isothiocyanate, *Cancer Res.*, 49, 549, 1989.
25. Morse, M.A., Amin, S.G., Hecht, S.S., and Chung, F.-L. Effects of aromatic isothiocyanates on tumorigenicity, O^6-methylguanine formation, and metabolism of the tobacco-specific nitrosamine 4-(methylnitrosamino)-1-(3-pyridyl)-1-butanone in A/J mouse lung, *Cancer Res.*, 49, 2894, 1989.
26. Morse, M.A., Eklind, K.I., Amin, S.G., Hecht, S.S., and Chung, F.-L. Effects of alkyl chain length on the inhibition of NNK-induced lung neoplasia in A/J mice by arylalkyl isothiocyanates, *Carcinogenesis*, 10, 1757, 1989.
27. Morse, M.A., Eklind, K.I., Hecht, S.S., Jordan, K.G., Choi, C.-I., Desai, D.H., Amin, S.G., and Chung, F.-L. Structure-activity relationships for inhibition of 4-(methylnitrosamino)-1-(3-pyridyl)-1-butanone (NNK) lung tumorigenesis by arylalkyl isothiocyanates in A/J mice, *Cancer Res.*, 51, 1846, 1991.
28. Stoner, G.D., Morrissey, D.T., Heur, Y.-H., Daniel, E.M., Galati, A.J., and Wagner, S.W. Inhibitory effects of phenethyl isothiocyanate on *N*-nitrosobenzylmethylamine carcinogenesis in the rat esophagus, *Cancer Res.*, 51, 2063, 1991.
29. Wilkinson, J.T., Morse, M.A., Kresty, L.A., and Stoner, G.D. Effect of alkyl chain length on inhibition of *N*-nitrosomethylbenzylamine-induced esophageal tumorigenesis and DNA methylation by isothiocyanates, *Carcinogenesis*, 16, 1011, 1995.
30. Morse, M.A., Eklind, K.I., Amin, S.G., and Chung, F.-L. Effect of frequency of isothiocyanate administration on inhibition of 4-(methylnitrosamino)-1-(3-pyridyl)-1-butanone-induced pulmonary adenoma formation in A/J mice, *Cancer Lett.*, 62, 77, 1992.
31. Rao, C.V., Rivenson, A., Simi, B., Zang, E., Hamid, R., Kelloff, G.J., Steele, V., and Reddy, B.S. Enhancement of experimental colon carcinogenesis by dietary 6-phenylhexyl isothiocyanate, *Cancer Res.*, 55, 4311, 1995.
32. Stoner, G.D., Siglin, J.C., Morse, M.A., Desai, D.H., Amin, S.G., Kresty, L.A., Toburen, A.L., Heffner, E.M., and Francis, D.J. Enhancement of esophageal carcinogenesis in male F344 rats by dietary phenylhexyl isothiocyanate, *Carcinogenesis*, 16, 2473, 1995.
33. Mandal, S., Shivapurkar, N.M., Galati, A.J., and Stoner, G.D. Inhibition of *N*-nitrosobenzylmethylamine metabolism and DNA binding in cultured rat esophagus by ellagic acid, *Carcinogenesis*, 9, 1313, 1988.
34. Barch, D.H. and Fox, C.C. Dietary ellagic acid reduces the esophageal microsomal metabolism of methylbenzyl-nitrosamine, *Cancer Lett.*, 44, 39, 1989.
35. Mandal, S. and Stoner, G.D. Inhibition of *N*-nitrosobenzylmethylamine-induced esophageal tumorigenesis in rats by ellagic acid, *Carcinogenesis*, 11, 55, 1990.
36. Daniel, E.M. and Stoner, G.D. The effects of ellagic acid and 13-*cis*-retinoic acid on *N*-nitrosobenzylmethylamine-induced esophageal tumorigenesis in rats, *Cancer Lett.*, 56, 117, 1991.
37. Wattenberg, L.W. and Loub, W.D. Inhibition of polycyclic aromatic hydrocarbon-induced neoplasia by naturally-occurring indoles, *Cancer Res.*, 38, 1410, 1978.
38. Nixon, J.E., Hendricks, J.D., Pawlowski, N.E., Pereira, C.B., Sinnhuber, R.O., and Bailey, G.S. Inhibition of aflatoxin B_1 carcinogenesis in rainbow trout by flavone and indole compounds, *Carcinogenesis*, 5, 615, 1984.
39. Goeger, D.E., Shelton, D.W., Hendricks, J.D., and Bailey, G.S. Mechanisms of anti-carcinogenesis by indole-3-carbinol: Effect on the distribution and metabolism of aflatoxin B_1 in rainbow trout, *Carcinogenesis*, 7, 2025, 1986.

40. Dashwood, R.H., Arbogast, D.N., Fond, A.T., Hendricks, J.D., and Bailey, G.S. Mechanisms of anti-carcinogenesis by indole-3-carbinol: Detailed *in vivo* DNA binding dose-response studies after dietary administration with aflatoxin B_1, *Carcinogenesis*, 9, 427, 1988.

41. Dashwood, R.H., Arbogast, D.N., Fong, A.T., Pereira, C., Hendricks, J.D., and Bailey, G.S. Quantitative inter-relationships between aflatoxin B_1 carcinogen dose, indole-3-carbinol anti carcinogen dose, target organ DNA adduction and final tumor response, *Carcinogenesis*, 10, 175–181, 1989.

42. Tanaka, T., Mori, Y., Morishita, Y., Hara, A., Ohno, T., Kojima, T., and Mori, H. Inhibitory effect of sinigrin and indole-3-carbinol on diethylnitrosamine-induced hepatocarcinogenesis in male ACI/N rats, *Carcinogenesis*, 11, 1403, 1990.

43. Morse, M.A., LaGreca, S.D., Amin, S.G., and Chung, F.-L. Effects of indole-3-carbinol on lung tumorigenesis and DNA methylation induced by 4-(methylnitrosamino)-1-(3-pyridyl)-1-butanone (NNK), and on the metabolism and disposition of NNK in A/J mice, *Cancer Res.*, 50, 2613, 1990.

44. Takahashi, N., Dashwood, R.H., Bjeldanes, L.F., Williams, D.E., and Bailey, G.S. Mechanisms of indole-3-carbinol (I3C) anticarcinogenesis: Inhibition of aflatoxin B1-DNA adduction and mutagenesis by I3C acid condensation products, *Food Chem. Toxicol.*, 33, 851, 1995.

45. Morse, M.A., Wang, C.-X., Amin, S.G., Hecht, S.S., and Chung, F.-L. Effects of dietary sinigrin or indole-3-carbinol on O^6-methylguanine-DNA-transmethylase activity and 4-(methylnitrosamino)-1-(3-pyridyl)-1-butanone-induced DNA methylation and tumorigenicity in F344 rats, *Carcinogenesis*, 9, 1891, 1988.

46. Pence, B.C., Buddingh, F., and Yang, S.P. Multiple dietary factors in the enhancement of dimethylhydrazine carcinogenesis: Main effect of indole-3-carbinol, *J. Natl. Cancer Inst.*, 77, 269, 1986.

47. Bailey, G.S., Hendricks, J.D., Shelton, D.W., Nixon, J.E., and Pawlowski, N.E. Enhancement of carcinogenesis by the natural anticarcinogen indole-3-carbinol, *J. Natl. Cancer Inst.*, 78, 931, 1987.

48. Prochaska, H.J. and Talalay, P. Regulatory mechanisms of monofunctional and bifunctional anticarcinogenic enzyme inducers in murine liver, *Cancer Res.*, 48, 4776, 1988.

49. Guengerich, F.P., Peterson, L.A., Cmarik, J.L., Koga, N., and Inskeep, P.B. Activation of dihaloalkanes by glutathione conjugation and formation of DNA adducts, *Environ. Health Perspect.*, 76, 15, 1987.

50. Sallustio, B.C., Sabordo, L., Evans, A.M., and Nation, R.L. Hepatic disposition of electrophilic acyl glucuronide conjugates, *Curr. Drug Metab.*, 1, 163, 2000.

51. Glatt, H. Sulfation and sulfotransferases 4: Bioactivation of mutagens via sulfation, *FASEB J.*, 11, 314, 1997.

52. Windmill, K.F., McKinnon, R.A., Zhu, X., Gaedigk, A., Grant, D.M., and McManus, M.E. The role of xenobiotic metabolizing enzymes in arylamine toxicity and carcinogenesis: Functional and localization studies, *Mutat. Res.*, 376, 153, 1997.

53. Guo, Z., Smith, T.J., Wang, E., Sadrieh, N., Ma, Q., Thomas, P.E., and Yang, C.S. Effects of phenethyl isothiocyanate, a carcinogenesis inhibitor, on xenobiotic-metabolizing enzymes and nitrosamine metabolism in rats, *Carcinogenesis*, 13, 2205, 1992.

54. Debersac, P., Heydel, J.M., Amiot, M.J., Goudonnet, H., Artur, Y., Suschetet, M., and Siess, M.H. Induction of cytochrome P450 and/or detoxication enzymes by various extracts of rosemary: Description of specific patterns, *Food Chem. Toxicol.*, 39, 907, 2001.

55. Parkinson, A. Biotransformation of xenobiotics, in *Casarett and Doull's Toxicology*, 6th ed., Klaasen, C.D., Ed., McGraw-Hill, New York, 2001, p. 133.

56. Sparnins, V.L. and Wattenberg, L.W. Enhancement of glutathione *S*-transferase activity of the mouse forestomach by inhibitors of benzo[a]pyrene-induced neoplasia of the forestomach, *J. Natl. Cancer Inst.*, 66, 769, 1981.

57. Guo, Z., Smith, T.J., Wang, E., Sadrieh, N., Ma, Q., Thomas, P.E., and Yang, C.S. Effects of phenethyl isothiocyanate, a carcinogenesis inhibitor, on xenobiotic-metabolizing enzymes and nitrosamine metabolism in rats, *Carcinogenesis*, 13, 2205, 1992.

58. Zhang, Y., Talalay, P., Cho, C.G., and Posner, G. A major inducer of anticarcinogenic protective enzymes from broccoli: Isolation and elucidation of structure, *Proc. Natl. Acad. Sci. USA*, 89, 2399, 1992.

59. Zhang, Y., Kensler, T.W., Cho, C.G., Posner, G.H., and Talalay, P. Anticarcinogenic activities of sulforaphane and structurally related synthetic norbornyl isothiocyanates, *Proc. Natl. Acad. Sci. USA*, 91, 3147, 1994.

60. Wattenberg, L.W. and Bueding, E. Inhibitory effects of 5-(2-pyrazinyl)-4-methyl-1,2-dithiol-3-thione (oltipraz) on carcinogenesis induced by benzo(a)pyrene, diethylnitrosamine, and uracil mustard, *Carcinogenesis*, 7, 1379, 1986.

61. Kensler, T.W., Egner, P.A., Trush, M.A., Bueding, E., and Groopman, J.D. Modification of aflatoxin B_1 binding to DNA *in vivo* in rats fed phenolic antioxidants, ethoxyquin and a dithiolthione, *Carcinogenesis*, 6, 759, 1985.

62. Kensler, T.W., Egner, P.A., Dola, P.M., Groopman, J.D., and Roebuck, B.D. Mechanism of protection against aflatoxin tumorigenicity in rats fed 5-(2-pyrazinyl)-4-methyl-1,2-dithiol-3-thione (oltipraz) and related 1,2-dithiol-3-thiones and 1,2-dithiol-3-ones, *Cancer Res.*, 47, 4271, 1987.

63. Liu, Y.-L., Roebuck, B.D., Yager, J.D., Groopman, J.D., and Kensler, T.W. Protection by 5-(2-pyrazinyl)-4-methyl-1,2-dithiol-3-thione (oltipraz) against the hepatotoxicity of aflatoxin B_1 in the rat, *Toxicol. Appl. Pharmacol.*, 93, 442, 1988.

64. Davidson, N.E., Egner, P.A., and Kensler, T.W. Transcriptional control of glutathione *S*-transferase gene expression by the chemoprotective agent 5-(2-pyrazinyl)-4-methyl-1,2-dithiol-3-thione (oltipraz) in rat liver, *Cancer Res.*, 50, 2251, 1990.

65. Roebuck, B.D., Liu, Y.-L., Rogers, A.E., Groopman, J.D., and Kensler, T.W. Protection against aflatoxin B_1-induced hepatocarcinogenesis in F344 rats by 5-(2-pyrazinyl)-4-methyl-1,2-dithiol-3-thione (oltipraz): Predictive role for molecular dosimetry, *Cancer Res.*, 51, 5501, 1991.

66. Rao, C.V., Tokomo, K., Kelloff, G., and Reddy, B.S. Inhibition by dietary oltipraz of experimental intestinal carcinogenesis induced by azoxymethane in male F344 rats, *Carcinogenesis*, 12, 1051, 1991.

67. Kensler, T.W., Curphey, T.J., Maxiutenko, Y., and Roebuck, B.D. Chemoprotection by organosulfur inducers of phase 2 enzymes: Dithiolethiones and dithiins, *Drug Metab. Drug Interact.*, 17, 3, 2000.

68. Kwak, M.K., Itoh, K., Yamamoto, M., Sutter, T.R., and Kensler, T.W. Role of transcription factor Nrf2 in the induction of hepatic phase 2 and antioxidative enzymes in vivo by the cancer chemoprotective agent, 3H-1, 2-dimethiole-3-thione, *Mol. Med.*, 7, 135, 2001.

69. Sayer, J.M., Yagi, H., Wood, A.W., Conney, A.H., and Jerina, D.M. Extremely facile reaction between the ultimate carcinogen benzo[a]pyrene-7,8-diol 9,10-epoxide and ellagic acid, *J. Am. Chem. Soc.*, 104, 5562, 1982.

70. Wood, A.W., Huang, M.-T., Chang, R.L., Newmark, N.L., Lehr, R.E., Yagi, H., Sayer, J.M., Jerina, D.M., and Conney, A.H. Inhibition of the mutagenicity of bay-region diol-epoxides of polycyclic aromatic hydrocarbons by naturally occurring plant phenols: Exceptional activity of ellagic acid, *Proc. Natl. Acad. Sci. USA*, 79, 5513, 1982.

71. Chang, R.L., Huang, M.-T., Wood, A.W., Wong, C.-Q., Newmark, H.L., Yagi, H., Sayer, J.M., Jerina, D.M., and Conney, A.H. Effect of ellagic acid and hydroxylated flavonoids on the tumorigenicity of benzo(a)pyrene and $(\pm)$-7β,8α-dihydroxy-9α,10α-epoxy-7,8,9,10-tetrahydrobenzo(a)pyrene on mouse skin and in the newborn mouse, *Carcinogenesis*, 6, 1127, 1986.

72. DeFlora, S., Bennicelli, C., Zannachi, P., Camoirano, A., Morelli, A., and DeFlora, A. *In vitro* effects of *N*-acetylcysteine on the mutagenicity of direct-acting compounds and procarcinogens, *Carcinogenesis*, 5, 505, 1984.

73. DeFlora, S., Bennicelli, C., Camoirano, A., Serra, D., Romano, M., Rossi, G.A., Morelli, A., and DeFlora, A. *In vivo* effects of *N*-acetylcysteine on glutathione metabolism and on the biotransformation of carcinogenic and/or mutagenic compounds, *Carcinogenesis*, 5, 1735, 1985.

74. DeFlora, S., Astengo, M., Serra, D., and Bennicelli, C. Inhibition of urethan-induced lung tumors in mice by dietary *N*-acetylcysteine, *Cancer Lett.*, 32, 235, 1986.

75. McGregor, W.G. DNA repair, DNA replication, and UV mutagenesis, *J. Invest. Dermatol. Symp. Proc.*, 4, 1, 1999.

76. Elkind, M.M. Repair processes in radiation biology, *Radiat. Res.*, 100, 425–449, 1984.

77. Cooper, D.P., O'Connor, P.J., and Margison, G.P. Effect of acute doses of 2-acetylaminofluorene on the capacity of rat liver to repair methylated purines in DNA *in vivo* and *in vitro*, *Cancer Res.*, 42, 4203–4209, 1982.

78. Den Engelse, L., Floot, B.G.J., Menkveld, G.J., and Tates, A.D. Enhanced repair of O^6-methylguanine in liver DNA of rats pretreated with phenobarbital, 2,3,7,8-tetrachlorodibenzo-*p*-dioxin, ethionine, or *N*-alkyl-*N*-nitrosoureas, *Carcinogenesis*, 7, 1941, 1986.

79. Lefebvre, P. and Laval, F. Enhancement of O^6-methylguanine-DNA-methyltransferase activity induced by various treatments in mammalian cells, *Cancer Res.*, 46, 5701, 1986.

80. Chan, C.L., Wu, Z., Eastman, A., and Bresnick, E. Induction and purification of O^6-methylguanine-DNA-methyltransferase from rat liver, *Carcinogenesis*, 11, 1217, 1990.

81. Sasaki, Y.F., Imanishi, H., Ohta, T., and Shirasu, Y. Effects of antimutagenic flavourings on SCEs induced by chemical mutagens in cultured Chinese hamster cells, *Mutat. Res.*, 189, 313, 1987.

82. Takahashi, K., Sekiguchi, M., and Kawazoe, Y. Effects of vanillin and *o*-vanillin on induction of DNA-repair networks: Modulation of mutagenesis in *Escherichia coli*, *Mutat. Res.*, 230, 127, 1990.

83. Sato, T., Chikazawa, K., Yamamori, H., Ose, Y., Nagase, H., and Kito, H. Evaluation of the SOS chromotest for the detection of antimutagens, *Environ. Mol. Mutagen.*, 17, 258, 1991.

84. de Andrade, H.H., Santos, J.H., Gimmler-Luz, M.C., Correa, M.J., Lehmann, M., and Reguly, M.L. Suppressing effect of vanillin on chromosome aberrations that occur spontaneously or are induced by mitomycin C in the germ cell line of *Drosophila melanogaster*, *Mutat. Res.*, 279, 281, 1992.

85. Sasaki, Y.F., Imanishi, H., Ohta, T., and Shirasu, Y. Effects of vanillin on sister-chromatid exchanges and chromosome aberrations induced by mitomycin C in cultured Chinese hamster ovary cells, *Mutat. Res.*, 191, 193, 1987.

86. Sasaki, Y.F., Imanishi, H., Watanabe, M., Ohta, T., and Shirasu, Y. Suppressing effect of antimutagenic flavorings on chromosome aberrations induced by UV-light or x-rays in cultured Chinese hamster cells, *Mutat. Res.*, 229, 1, 1990.

87. Imanishi, H., Sasaki, Y.F., Matsumoto, K., Watanabe, M., Ohta, T., Shirasu, Y., and Tutikawa, K. Suppression of 6-TG-resistant mutations in V79 cells and recessive spot formations in mice by vanillin, *Mutat. Res.*, 243, 151, 1990.

88. Sasaki, Y.F., Ohta, T., Imanishi, H., Watanabe, M., Matsumoto, K., Kato, T., and Shirasu, Y. Suppressing effects of vanillin, cinnamaldehyde, and anisaldehyde on chromosome aberrations induced by x-rays in mice, *Mutat. Res.*, 243, 299, 1990.

89. Tamai, K., Tezuka, H., and Kuroda, Y. Different modifications by vanillin in cytotoxicity and genetic changes induced by EMS and H_2O_2 in cultured Chinese hamster cells, *Mutat. Res.*, 268, 231, 1992.

90. Takahashi, K., Sekiguchi, M., and Kawazoe, Y. Effects of vanillin and *o*-vanillin on induction of DNA-repair networks: Modulation of mutagenesis in *Escherichia coli*, *Mutat. Res.*, 230, 127, 1990.

91. Matsumura, H., Watanabe, K., and Ohta, T. *o*-Vanillin enhances chromosome aberrations induced by alkylating agents in cultured Chinese hamster cells, *Mutat. Res.*, 298, 163, 1993.

92. Kelloff, G.J. Perspectives on cancer chemoprevention research and drug development, *Adv. Cancer Res.*, 78, 199, 2000.

93. Tong, W.M., Cortes, U., and Wang, Z.Q. Poly(ADP-ribose) polymerase: A guardian angel protecting the genome and suppressing tumorigenesis, *Biochim. Biophys. Acta*, 1552, 27, 2001.

94. Herceg, Z. and Wang, Z.Q. Functions of poly(ADP-ribose) polymerase (PARP) in DNA repair, genomic integrity and cell death, *Mutat. Res.*, 477, 97, 2001.

95. Burkle, A. Poly(APD-ribosyl)ation, a DNA damage-driven protein modification and regulator of genomic instability, *Cancer Lett.*, 163, 1, 2001.

96. O'Dwyer, P.J., Johnson, S.W., Khater, C., Krueger, A., Matsumoto, Y., Hamilton, T.C., and Yao, K.S. The chemopreventive agent oltipraz stimulates repair of damaged DNA, *Cancer Res.*, 57, 1050, 1997.

97. Vogelstein, B., Fearon, E.R., Hamilton, S.R., Kern, S.E., Preisinger, A.C., Leppert, M., Nakamura, Y., White, R., Smits, A.M., and Bos, J.L. Genetic alterations during colorectal tumor development, *N. Engl. J. Med.*, 319, 52, 1988.

98. Fearon, E.R. and Vogelstein, B. A genetic model for colorectal tumorigenesis, *Cell*, 61, 759, 1990.

99. DeFlora, S. and Ramel, C. Classification of mechanisms of inhibitors of mutagenesis and carcinogenesis, *Basic Life Sci.*, 52, 461, 1990.

100. Gupta, S. Molecular steps of death receptor and mitochondrial pathways of apoptosis, *Life Sci.*, 69, 2957, 2001.

101. Kong, A.N., Yu, R., Chen, C., Mandlekar, S., and Primiano, T. Signal transduction events elicited by natural products: Role of MAPK and caspase pathways in homeostatic response and induction of apoptosis, *Arch. Pharm. Res.*, 23, 1, 2000.

102. Bonnesen, C., Eggleston, I.M., and Hayes, J.D. Dietary indoles and isothiocyanates that are generated from cruciferous vegetables can both stimulate apoptosis and confer protection against DNA damage in human colon cell lines, *Cancer Res.*, 61, 6120, 2001.

103. Yang, Y.M., Conaway, C.C., Chiao, J.W., Wang, C.X., Amin, S., Whysner, J., Dai, W., Reinhardt, J., and Chung, F.-L. Inhibition of benzo(a)pyrene-induced lung tumorigenesis in A/J mice by dietary

N-acetylcysteine conjugates of benzyl and phenethyl isothiocyanates during the postinitiation phase is associated with activation of mitogen-activated protein kinases and p53 activity and induction of apoptosis, *Cancer Res.*, 62, 2, 2002.

104. Goluboff, E.T. Exisulind, a selective apoptotic antineoplastic drug, *Expert Opin. Investig. Drugs*, 10, 1875, 2001.

105. Liu, L., Li, H., Underwood, T., Lloyd, M., David, M., Sperl, G., Pamukcu, R., and Thompson, W.J. Cyclic GMP-dependent protein kinase activation and induction by exisulind and CP461 in colon tumor cells, *J. Pharmacol. Exp. Ther.*, 299, 583, 2001.

106. Takuma, K., Phuagphong, P., Lee, E., Mori, K., Baba, A., and Matsuda, T. Anti-apoptotic effect of cGMP in cultured astrocytes: Inhibition by cGMP-dependent protein kinase of mitochondrial permeable transition pore, *J. Biol. Chem.*, 276, 48093, 2001.

107. Hixson, L.J., Alberts, D.S., Krutzsch, M., Einsphar, J., Brendel, K., Gross, P.H., Paranka, N.S., Baier, M., Emerson, S., Pamukcu, R. et al. Antiproliferative effect of nonsteroidal antiinflammatory drugs against human colon cancer cells, *Cancer Epidemiol. Biomarkers Prev.*, 3, 433, 1994.

108. Thompson, H.J., Jiang, C., Lu, J., Mehta, R.G., Piazza, G.A., Paranka, N.S., Pamukcu, R., and Ahnen, D.J. Sulfone metabolite of sulindac inhibits mammary carcinogenesis, *Cancer Res.*, 57, 267, 1997.

109. van Stolk, R., Stoner, G., Hayton, W.L., Chan, K., DeYoung, B., Kresty, L., Kemmenoe, B.H., Elson, P., Rybicki, L., Church, J., Provencher, K., McLain, D., Hawk, E., Fryer, B., Kelloff, G., Ganapathi, R., and Budd, G.T. Phase I trial of exisulind (sulindac sulfone, FGN-1) as a chemopreventive agent in patients with familial adenomatous polyposis, *Clin. Cancer Res.*, 6, 78, 2000.

110. Masuda, M., Suzui, M., and Weinstein, I.B. Effects of epigallocatechin-3-gallate on growth, epidermal growth factor receptor signaling pathways, gene expression, and chemosensitivity in human head and neck squamous cell carcinoma cell lines, *Clin. Cancer Res.*, 7, 4220, 2001.

111. Leszczyniecka, M., Roberts, T., Dent, P., Grant, S., and Fisher, P.B. Differentiation therapy of human cancer: Basic science and clinical applications, *Pharmacol. Ther.*, 90, 105, 2001.

112. Trosko, J.E. and Chang, C.C. Modulation of cell–cell communication in the cause and chemoprevention/chemotherapy of cancer, *Biofactors*, 12, 259, 2000.

113. Jung, M. Inhibitors of histone deacetylase as new anticancer agents, *Curr. Med. Chem.*, 8, 1505, 2001.

114. Moon, R.C., Rao, K.V.N., Detrisac, C.J., and Kelloff, G.J. Retinoid chemoprevention of lung cancer, in *Cancer Chemoprevention*, Wattenberg, L., Lipkin, M., Boone, C.W., and Kelloff, G.J., Eds., CRC Press, Boca Raton, FL, 1992, p. 83.

115. Lotan, R. Evaluation of the results of clinical trials with retinoids in relation to their basic mechanism of action, in *Cancer Chemoprevention*, Wattenberg, L., Lipkin, M., Boone, C.W., and Kelloff, G.J., Eds., CRC Press, Boca Raton, FL, 1992, p. 71.

116. Mehta, P.P., Bertram, J.S., and Lowenstein, W.T. The actions of retinoids on cellular growth correlate with their actions on gap-junctional communication, *J. Cell Biol.*, 108, 1053, 1989.

117. Hossain, M.Z., Wilkens, L.R., Mehta, P.P., Lowenstein, W., and Bertram, J.S. Enhancement of gap junctional communication by retinoids correlates with their ability to inhibit neoplastic transformation, *Carcinogenesis*, 10, 1743, 1989.

118. Bertram, J.S. Role of gap junctional cell/cell communication in the control of proliferation and neoplastic transformation, *Radiat. Res.*, 123, 252, 1990.

119. Chu, E.W. and Malmgren, R.A. An inhibitory effect of vitamin A on the induction of tumors of forestomach and cervix in the Syrian hamster by carcinogenic hydrocarbons, *Cancer Res.*, 25, 884, 1965.

120. Sporn, M.B., Squire, R.A., Brown, C.C., Smith, J.M., Wenk, M.L., and Springer, S. 13-*cis*-Retinoic acid: Inhibition of bladder carcinogenesis in the rat, *Science*, 195, 487, 1977.

121. Moon, R.C., Thompson, H.J., Becci, P.J., Grubbs, C.J., Gander, R.J., Newton, D.L., Smith, J.M., Phillips, S.R., Henderson, W.R., Mullen, L.T., Brown, C.C., and Sporn, M.B. *N*-(4-hydroxyphenyl)retinamide, a new retinoid for prevention of breast cancer in the rat, *Cancer Res.*, 39, 1339, 1979.

122. Becci, P.J., Thompson, H.J., Grubbs, C.J., Brown, C.C., and Moon, R.C. Effect of delay in administration of 13-*cis*-retinoic acid on the inhibition of urinary bladder carcinogenesis in the rat, *Cancer Res.*, 39, 3141, 1979.

123. Maiorana, A. and Gullino, P. Effect of retinyl acetate on the incidence of mammary carcinomas and hepatomas in mice, *J. Natl. Cancer Inst.*, 64, 655, 1980.
124. Ip, C. and Ip, M.M. Chemoprevention of mammary tumorigenesis by a combined regimen of selenium and vitamin A, *Carcinogenesis*, 2, 915, 1981.
125. Thompson, H.J., Meeker, L.D., and Becci, P.J. Effect of combined selenium and retinyl acetate treatment on mammary carcinogenesis, *Cancer Res.*, 41, 1413, 1981.
126. McCormick, D.L., Mehta, R.G., Thompson, C.A., Dinger, N., Caldwell, J.A., and Moon, R.C. Enhanced inhibition of mammary carcinogenesis by combination N-(4-hydroxyphenyl)retinamide and ovariectomy, *Cancer Res.*, 42, 509, 1982.
127. McCormick, D.L. and Moon, R.C. Retinoid–tamoxifen interaction in mammary cancer chemoprevention, *Carcinogenesis*, 7, 193, 1986.
128. Mawson, M.I., Chao, W.-R., and Helmes, C.T. Inhibition by retinoids of anthralin-induced mouse epidermal ornithine decarboxylase activity and anthralin-promoted skin tumor formation, *Cancer Res.*, 47, 6210, 1987.
129. Costa, A., Veronesi, U., De Palo, G., Chiesa, F., Formelli, F., Marubini, E., Del Vecchio, M., and Nava, M. Chemoprevention of cancer with the synthetic retinoid fenretinamide: Clinical trials in progress at the Milan Cancer Institute, in *Cancer Chemoprevention*, Wattenberg, L., Lipkin, M., Boone, C.W., and Kelloff, G.J., Eds., CRC Press, Boca Raton, FL, 1992, p. 95.
130. De Luca, L.M., Shores, R.L., Spangler, E.F., and Wenk, M.L. Inhibition of initiator–promoter-induced skin tumorigenesis in female SENCAR mice fed a vitamin A-deficient diet and reappearance of tumors in mice fed a diet adequate in retinoid or β-carotene, *Cancer Res.*, 49, 5400, 1989.
131. De Luca, L.M., Sly, L., Jones, C.S., and Chen, L.-C. Effects of dietary retinoic acid on skin papilloma and carcinoma formation in female SENCAR mice, *Carcinogenesis*, 14, 539, 1993.
132. Whitfield, J.F. Calcium switches, cell cycles, differentiation, and death, in *Calcium, Vitamin D, and Prevention of Colon Cancer*, Lipkin, M., Newmark, H.L., and Kelloff, G.J., Eds., CRC Press, Boca Raton, FL, 1991, p. 31.
133. Pence, B.C. Calcium and vitamin D effects of tumor promotion in the colon and mouse skin, in *Calcium, Vitamin D, and Prevention of Colon Cancer*, Lipkin, M., Newmark, H.L., and Kelloff, G.J., Eds., CRC Press, Boca Raton, FL, 1991, p. 191.
134. Carroll, K.K., Eckel, L.A., Fraher, L.J., Frei, J.V., and Newmark, H.L. Dietary calcium, phosphate, and vitamin D in relation to mammary carcinogenesis, in *Calcium, Vitamin D, and Prevention of Colon Cancer*, Lipkin, M., Newmark, H.L., and Kelloff, G.J., Eds., CRC Press, Boca Raton, FL, 1991, p. 229.
135. Skrypec, D.J. Effect of dietary calcium on azoxymethane-induced intestinal carcinogenesis in male F344 rats fed high-fat diets, in *Calcium, Vitamin D, and Prevention of Colon Cancer*, Lipkin, M., Newmark, H.L., and Kelloff, G.J., Eds., CRC Press, Boca Raton, FL, 1991, p. 241.
136. Benner, S.E., Hong, W.K., and Lippman, S.M. Treatment of field cancerization, in *Cancer Chemoprevention*, Wattenberg, L., Lipkin, M., Boone, C.W., and Kelloff, G.J., Eds., CRC Press, Boca Raton, FL, 1992, p. 113.
137. Hasina, R. and Lingen, M.W. Angiogenesis in oral cancer, *J. Dent. Educ.*, 65, 1282, 2001.
138. Khan, S.R., Mhaka, A., Pili, R., and Isaacs, J.T. Modified synthesis and antiangiogenic activity of linomide, *Bioorg. Med. Chem. Lett.*, 11, 451, 2001.
139. Gross, D.J., Reibstein, I., Weiss, L., Slavin, S., Stein, I., Neeman, M., Abramovitch, R., and Benjamin, L.E. The antiangiogenic agent linomide inhibits the growth rate of von Hippel–Lindau paraganglioma xenografts to mice, *Clin. Cancer Res.*, 5, 3669, 1999.
140. Joseph, I.B. and Isaacs, J.T. Macrophage role in the anti-prostate cancer response to one class of antiangiogenic agents, *J. Natl. Cancer Inst.*, 90, 1648, 1998.
141. Shaw, M., Ratanawong, S., Chou, P., Ray, V., Mirochnik, Y., Slobodskoy, L., Rubenstein, M., and Guinan, P. Paclitaxel, bropirimine and linomide: Effect on growth inhibition in a murine prostate cancer model by different growth regulatory mechanisms, *Methods Find. Exp. Clin. Pharmacol.*, 20, 111, 1998.
142. Kelloff, G.J., Lieberman, R., Steele, V.E., Boone, C.W., Lubet, R.A., Kopelovich, L., Malone, W.A., Crowell, J.A., Higley, H.R., and Sigman, C.C. Agents, biomarkers, and cohorts for chemopreventive agent development in prostate cancer, *Urology*, 57, 46, 2001.
143. Cao, Y. and Cao, R. Angiogenesis inhibited by drinking tea, *Nature*, 398, 381, 1999.

144. Jung, Y.D. and Ellis, L.M. Inhibition of tumour invasion and angiogenesis by epigallocatechin gallate (EGCG), a major component of green tea, *Int. J. Exp. Pathol.*, 82, 309, 2001.

145. Sharma, S., Ghoddoussi, M., Gao, P., Kelloff, G.J., Steele, V.E., and Kopelovich, L. A quantitative angiogenesis model for efficacy testing of chemopreventive agents, *Anticancer Res.*, 21, 3829, 2001.

146. Moncada, S., Flower, R.J., and Vane, J.R. Prostaglandins, prostacyclin, and thromboxane A$_2$, in *The Pharmacological Basis of Therapeutics*, 6th ed., Gilman, A.G., Goodman, L.S., and Gilman, A., Eds., McMillan Publishing Co., New York, 1980, p. 668.

147. Hinz, B. and Brune, K. Cyclooxygenase-2–10 years later, *J. Pharmacol. Exp. Ther.*, 300, 367, 2002.

148. Levine, L. Stimulated release of arachidonic acid from rat liver cells by celecoxib and indomethacin, *Prostaglandins Leukot. Essent. Fatty Acids*, 65, 31, 2001.

149. Reddy, B.S., Maruyama, H., and Kelloff, G. Dose-related inhibition of colon carcinogenesis by dietary piroxicam, a nonsteroidal antiinflammatory drug, during different stages of rat colon tumor development, *Cancer Res.*, 47, 5340, 1987.

150. Kudo, T., Narisawa, T., and Abo, S. Antitumor activity of indomethacin on methylazoxymethanol-induced large bowel tumors in rats, *Gann*, 71, 260, 1980.

151. Narisawa, T., Sato, M., Tani, M., Kudo, T., Takahashi, T., and Goto, A. Inhibition of development of methylnitrosourea-induced rat colon tumors by indomethacin, *Cancer Res.*, 41, 1954, 1981.

152. Boone, C.W., Steele, V.E., and Kelloff, G.J. Screening for chemopreventive (anticarcinogenic) compounds in rodents, *Mutat. Res.*, 267, 251, 1992.

153. Reddy, B.S., Hirose, Y., Lubet, R., Steele, V., Kelloff, G., Paulson, S., Seibert, K., and Rao, C.V. Chemoprevention of colon cancer by specific cyclooxygenase-2 inhibitor, celecoxib, administered during different stages of carcinogenesis, *Cancer Res.*, 60, 293, 2000.

154. Lynch, P.M. COX-2 inhibition in clinical cancer prevention, *Oncology*, 15, 21, 2001.

155. Nakadate, T., Yamamoto, S., Aizu, E., and Kato, R. Inhibition by lipoxygenase inhibitors of 7-bromomethylbenz(a)anthracene-caused epidermal ornithine decarboxylase induction and skin tumor promotion in mice, *Carcinogenesis*, 10, 2053, 1989.

156. Huang, M.-T., Lysz, T., Ferraro, T., Abidi, T.F., Laskin, J.D., and Conney, A.H. Inhibitory effects of curcumin on *in vitro* lipoxygenase and cyclooxygenase activities in mouse epidermis, *Cancer Res.*, 51, 813, 1991.

157. Huang, M.-T., Smart, R.C., Wong, C.-Q., and Conney, A.H. Inhibitory effect of curcumin, chlorogenic acid, caffeic acid, and ferulic acid on tumor promotion in mouse skin by 12-*O*-tetradecanoylphorbol-13-acetate, *Cancer Res.*, 48, 5941, 1988.

158. Shureiqi, I. and Lippman, S.M. Lipoxygenase modulation to reverse carcinogenesis, *Cancer Res.*, 61, 6307, 2001.

159. Pegg, A.E. Polyamine metabolism and its importance in neoplastic growth and as a target for chemotherapy, *Cancer Res.*, 48, 759, 1988.

160. Tabor, C.W. and Tabor, H. Polyamines, *Ann. Rev. Biochem.*, 53, 749, 1984.

161. Slaga, T.J. Overview of tumor promotion in animals, *Environ. Health Perspect.*, 50, 3, 1983.

162. Thomas, T. and Thomas, T.J. Polyamines in cell growth and cell death: Molecular mechanisms and therapeutic applications, *Cell Mol. Life Sci.*, 58, 244, 2001.

163. Metcalf, B.W., Bey, P., Danzin, C., Jung, M.J., Casara, P., and Vevert, J.P. Catalytic irreversible inhibition of mammalian ornithine decarboxylase (E.C. 4.1.1.17.) by substrate and product analogues, *J. Am. Chem. Soc.*, 100, 2551, 1978.

164. Fozard, J.R. and Prakash, N.J. Effects of D,L-α-difluoromethylornithine, an irreversible inhibitor of ornithine decarboxylase, on the rat mammary tumor induced by 7,12-dimethylbenz(a)anthracene, *Arch. Pharm.*, 28, 1, 1982.

165. Weeks, C.E., Herrmann, A.L., Nelson, F.R., and Slaga, T.J. α-Difluoromethylornithine, an irreversible inhibitor of ornithine decarboxylase, inhibits tumor-promoter-induced polyamine accumulation and carcinogenesis in mouse skin, *Proc. Natl. Acad. Sci. USA*, 79, 6028, 1982.

166. Thompson, H.J. and Ronan, A.M. Inhibition of 1-methyl-1-nitrosourea induced mammary tumorigenesis by α-difluoromethylornithine and retinyl acetate, *Proc. Am. Assoc. Cancer Res.*, 24, 86, 1983.

167. Kingsnorth, A.N., King, W.W.K., Diekema, K.A., McCann, P.P., Ross, J.S., and Malt, R.A. Inhibition of ornithine decarboxylase with 2-difluoromethylornithine: Reduced incidence of dimethylhydrazine-induced colon tumors in mice, *Cancer Research*, 43, 2545, 1983.

168. Takigawa, M., Verma, A.J., Simsiman, R.C., and Boutwell, R.K. Inhibition of mouse skin tumor promotion and of promoter-stimulated epidermal polyamine biosynthesis by α-difluoromethylornithine, *Cancer Res.*, 43, 3732, 1983.

169. Homma, Y., Kakizoe, T., Samma, S., and Oyasu, R. Inhibition of *N*-butyl-*N*-(4-hydroxybutyl) nitrosamine-induced rat urinary bladder carcinogenesis by α-difluoromethylornithine, *Cancer Res.*, 47, 6176, 1987.

170. Reddy, B.S., Nayini, J., Tokumo, K., Rigotty, J., Zang, E., and Kelloff, G. Chemoprevention of colon carcinogenesis by concurrent administration of piroxicam, a nonsteroidal antiinflammatory drug with DL-α-difluoromethylornithine, an ornithine decarboxylase inhibitor, in diet, *Cancer Res.*, 50, 2562, 1990.

171. Tanaka, T., Kojima, T., Hara, A., Sawada, H., and Mori, H. Chemoprevention of oral carcinogenesis by DL-α-difluoromethylornithine, an ornithine decarboxylase inhibitor: Dose-dependent reduction in 4-nitroquinoline 1-oxide-induced tongue neoplasms in rats, *Cancer Res.*, 53, 772, 1993.

172. Meyskens, F.L. Jr. Development of difluoromethylornithine and Bowman-Birk inhibitor as chemopreventive agents by assessment of relevant biomarker modulation: Some lessons learned, *IARC Sci. Publ.*, 154, 49, 2001.

173. Troll, W., Klassen, A., and Janoff, A. Tumorigenesis in mouse skin: Inhibition by specific inhibitors of proteases, *Science*, 169, 1211, 1970.

174. Troll, W., Garte, S., and Frenkel, K. Suppression of tumor promotion by inhibitors of poly(ADP)ribose formation, *Basic Life Sci.*, 52, 225, 1990.

175. Hozumi, M., Ogawa, M., Sugimura, T., Takeuchi, T., and Umezawa, H. Inhibition of tumorigenesis in mouse skin by leupeptin, a protease inhibitor from *Actinomycetes*, *Cancer Res.*, 32, 1725, 1972.

176. Nomura, T., Hata, S., Enomoto, T., Tanaka, H., and Shibata, K. Inhibiting effects of antipain on urethane-induced lung neoplasia in mice, *Br. J. Cancer*, 42, 624, 1980.

177. Yavelow, J., Finlay, T.H., Kennedy, A.R., and Troll, W. Bowman-Birk soybean protease inhibitor as an anticarcinogen, *Cancer Res.*, 43 (Suppl.), 2454s, 1983.

178. Troll, W., Lim, J.S., and Belman, S. Protease inhibitors suppress carcinogenesis *in vivo* and *in vitro*, in *Cancer Chemoprevention*, Wattenberg, L., Lipkin, M., Boone, C.W., and Kelloff, G.J., Eds., CRC Press, Boca Raton, FL, 1992, p. 503.

179. Frenkel, K., Chrzan, K., Ryan, C.A., Wiesner, R., and Troll, W. Chymotrypsin-specific protease inhibitors decrease H_2O_2 formation by activated human polymorphonuclear leukocytes, *Carcinogenesis*, 8, 1207, 1987.

180. Troll, W. Protease inhibitors interfere with the necessary factors of carcinogenesis, *Environ. Health Perspect.*, 81, 59, 1989.

181. Workshop Report from the Division of Cancer Etiology, National Cancer Institute, National Institutes of Health. Protease inhibitors as cancer chemopreventive agents, *Cancer Res.*, 49, 499, 1989.

182. Kennedy, A.R. The Bowman-Birk inhibitor from soybeans as an anticarcinogenic agent, *Am. J. Clin. Nutr.*, 68 (Suppl.), 1406s, 1998.

183. Kennedy, A.R. Chemopreventive agents: Protease inhibitors, *Pharmacol. Ther.*, 78, 167, 1998.

184. Kennedy, A.R. Prevention of carcinogenesis by protease inhibitors, *Cancer Res.*, 54 (Suppl.), 1999s, 1994.

185. Kennedy, A.R., Szuhaj, B.F., Newberne, P.M., and Billings, P.C. Preparation and production of a cancer chemopreventive agent, Bowman-Birk inhibitor concentrate, *Nutr. Cancer*, 19, 281, 1993.

186. Gladysheva, I.P., Balabushevich, N.G., Moroz, N.A., and Larionova, N.I. Isolation and characterization of soybean Bowman-Birk inhibitor from different sources, *Biochemistry (Mosc)*, 65, 198, 2000.

187. Foehr, M.W., Tomei, L.D., Goddard, J.G., Pemberton, P.A., and Bathurst, I.C. Antiapoptotic activity of the Bowman-Birk inhibitor can be attributed to copurified phospholipids, *Nutr. Cancer*, 34, 199, 1999.

188. Ware, J.H., Wan, X.S., and Kennedy, A.R. Bowman-Birk inhibitor suppresses production of superoxide anion radicals in differentiated HL-60 cells, *Nutr. Cancer*, 33, 174, 1999.

189. Weed, H., McGandy, R.B., and Kennedy, A.R. Protection against dimethylhydrazine induced adenomatous tumors of the mouse colon by the dietary addition of an extract of soybeans containing the Bowman-Birk protease inhibitor, *Carcinogenesis*, 6, 1239, 1985.

190. St. Clair, W., Billings, P., Carew, J., Keiler-McGandy, C., Newberne, P., and Kennedy, A.R. Suppression of DMH-induced carcinogenesis in mice by dietary addition of the Bowman-Birk protease inhibitor, *Cancer Res.*, 50, 580, 1990.

191. Billings, P.C., Newberne, P., and Kennedy, A.R. Protease inhibitor suppression of colon and anal gland carcinogenesis induced by dimethylhydrazine, *Carcinogenesis*, 11, 1083, 1990.

192. Messadi, P.V., Billings, P., Shklar, G., and Kennedy, A.R. Inhibition of oral carcinogenesis by a protease inhibitor, *J. Natl. Cancer Inst.*, 76, 447, 1986.

193. Kennedy, A.R., Billings, P.C., Maki, P.A., and Newberne, P. Effects of various protease inhibitor preparations on oral carcinogenesis in hamsters induced by 7,12-dimethylbenz[a]anthracene, *Nutr. Cancer*, 19, 191, 1995.

194. von Hofe, E., Newberne, P.M., and Kennedy, A.R. Inhibition of N-nitrosomethylbenzylamine induced esophageal neoplasms by the Bowman-Birk protease inhibitor, *Carcinogenesis*, 12, 2147, 1991.

195. Evans, S.M., Szuhaj, B.F., Van Winkle, T., Michel, K., and Kennedy, A.R. Protection against radiation induced thymic lymphosarcoma in C57BI/6NCriBR mice by an autoclave resistant factor present in soybeans, *Radiat. Res.*, 132, 259, 1992.

196. Witschi, H. and Kennedy, A.R. Modulation of lung tumor development in mice with the soybean-derived Bowman-Birk protease inhibitor, *Carcinogenesis*, 10, 2275, 1989.

197. Kennedy, A.R., Beazer-Barclay, Y., Kinzler, K.W., and Newberne, P.M. Suppression of carcinogenesis in the intestines of min mice by the soybean-derived Bowman-Birk inhibitor, *Cancer Res.*, 56, 679, 1998.

198. Armstrong, W.B., Kennedy, A.R., Wan, X.S., Taylor, T.H., Nguyen, Q.A., Jensen, J., Thompson, W., Lagerberg, W., and Meyskens, F.L. Jr. Clinical modulation of oral leukoplakia and protease activity by Bowman-Birk inhibitor concentrate in a phase IIa chemoprevention trial, *Clin. Cancer Res.*, 6, 4684, 2000.

199. Frenkel, K. Carcinogen-mediated oxidant formation and oxidative DNA damage, *Pharmacol. Ther.*, 53, 127, 1992.

200. Droge, W. Free radicals in the physiological control of cell function, *Physiol. Rev.*, 82, 47, 2002.

201. Yang, C.S. and Wang, Z.Y. Tea and cancer, *J. Natl. Cancer Inst.*, 85, 1038, 1993.

202. Hu, G., Han, C., and Chen, J. Inhibition of oncogene expression by green tea and (−)-epigallocatechin-O-gallate in mice, *Nutrition and Cancer*, 24, 203, 1995.

203. Hu, Z.Q., Toda, M., Okubo, S., Hara, Y., and Shimamura, T. Mitogenic activity of (−)-epigallocatechin gallate on B-cells and investigation of its structure–function relationship, *Ind. J. Immunopharmacol.*, 14, 1399, 1992.

204. Xu, Y., Ho, C.-T., Amin, S.G., Han, C., and Chung, F.-L. Inhibition of tobacco-specific nitrosamine-induced lung tumorigenesis in A/J mice by green tea and its major polyphenol as antioxidants, *Cancer Res.*, 52, 3875, 1992.

205. Wang, Z.Y., Agarwal, R., Bickers, D.R. et al. Protection against ultraviolet B radiation-induced photocarcinogenesis in hairless mice by green tea polyphenols, *Carcinogenesis*, 12, 1527, 1991.

206. Khan, W.A., Wang, Z.Y., Athar, M. et al. Inhibition of the skin tumorigenicity of (±)7β, 8α-dihydroxy-9α, 10α-epoxy-7, 8, 9, 10-tetrahydrobenzo(a)pyrene by tannic acid, green tea polyphenols and quercetin in Sencar mice, *Cancer Lett.*, 42, 7, 1988.

207. Wang, Z.Y., Khan, W.A., Bickers, D.R. et al. Protection against polycyclic aromatic hydrocarbon-induced skin tumor initiation in mice by green tea polyphenols, *Carcinogenesis*, 10, 411, 1989.

208. Katiyar, S.K., Agarwal, R., Wang, Z.Y. et al. (-)-Epigallocatechin-3-gallate in *Camellia sinensis* leaves from Himalayan region of Sikkim: Inhibitory effects against biochemical events and tumor initiation in Sencar mouse skin, *Nutr. Cancer*, 18, 73, 1992.

209. Huang, M.T., Ho, C.T., Wang, Z.Y. et al. Inhibitory effect of topical application of a green tea polyphenol fraction on tumor initiation and promotion in mouse skin, *Carcinogenesis*, 13, 947, 1992.

210. Wang, Z.Y., Zhou, Z.C., Bickers, D.R. et al. Inhibition of chemical and photocarcinogenesis in murine skin by green tea polyphenols, *Proc. Am. Assoc. Cancer Res.*, 31, 159, 1990.

211. Wang, Z.Y., Huang, M.T., Ferraro, T., Wong, C.Q., Lou, Y.R., Reuhl, K., Iatropoulos, M., Yang, C.S., and Conney, A.H. Inhibitory effect of green tea in the drinking water on tumorigenesis by ultraviolet light and 12-O-tetradecanoylphorbol-13-acetate in the skin of SKH-1 mice, *Cancer Res.*, 52, 1162, 1992.

212. Wang, Z.Y., Bickers, D.R., and Mukhtar, H. Protection against Experimental Skin Chemical Carcinogenesis in Mice by Green Tea Polyphenols, Paper presented at the International Tea-Quality-Human Health Symposium (China), 1987, p. 112.

213. Yoshizawa, S., Horiuchi, T., Fujiki, H. et al. Antitumor promoting activity of (−)-epigallocatechin gallate, the main constituent of "tannin" in green tea, *Phytother. Res.*, 1, 44, 1987.
214. Wu, R.R., Lin, Y.P., and Chen, H.Y. Effect of Fujian Oolong Tea, Jasmine Tea, Green Tea and Tea Standing Overnight on Urethan Induced Lung Neoplasia in Mice, Paper presented at the International Tea-Quality-Human Health Symposium (China), 1987, p. 118.
215. Wang, Z.Y., Agarwal, R., Khan, W.A. et al. Protection against benzo(a)pyrene and *N*-nitrosodiethylamine-induced lung and fore-stomach tumorigenesis in A/J mice by water extracts of green tea and licorice, *Carcinogenesis*, 13, 1491, 1992.
216. Chung, F.-L., Morse, M.A., Eklind, K.I., and Xu, Y. Inhibition of the tobacco-specific nitrosamine-induced lung tumorigenesis by compounds derived from cruciferous vegetables and green tea, in *Tobacco Smoking and Nutrition: Influence of Nutrition on Tobacco-Associated Health Risks*, Diana, J.N. and Pryor, W.A., Eds., New York Academy of Sciences, 1993, p. 186.
217. Wang, Z.Y., Hong, J.Y., Huang, M.T. et al. Inhibition of *N*-nitrosodiethylamine- and 4-(methylnitrosamino)-1-(3-pyridyl)-1-butanone-induced tumorigenesis in A/J mice by green tea and black tea, *Cancer Res.*, 52, 1943, 1992.
218. Han, C. and Xu, Y. The effect of Chinese tea on the occurrence of esophageal tumor induced by *N*-nitrosomethylbenzylamine in rats, *Biomed. Environ. Sci.*, 3, 35, 1990.
219. Xu, Y. and Han, C. The effect of Chinese tea on the occurrence of esophageal tumors induced by *N*-nitrosomethylbenzylamine formed *in vivo*, *Biomed. Environ. Sci.*, 3, 406, 1990.
220. Gag, G.D., Zhou, L.F., Qi, G. et al. Initial study of antitumorigenesis of green tea: Animal test and flow cytometry, *Tumor*, 10, 42, 1990.
221. Hulka, B.S. and Moorman, P.G. Breast cancer: Hormones and other risk factors, *Maturitas*, 38, 103, 2001.
222. Colditz, G.A. Hormones and breast cancer: Evidence and implications for consideration of risks and benefits of hormone replacement therapy, *J. Womens Health*, 8, 347, 1999.
223. Akhmedkhanov, A., Zeleniuch-Jacquotte, A., and Toniolo, P. Role of exogenous and endogenous hormones in endometrial cancer: Review of the evidence and research perspectives, *Ann. N.Y. Acad. Sci.*, 943, 296, 2001.
224. Roa, B.R. and Slotman, B.J. Action and counter-action of hormones in human ovarian cancer, *Anticancer Res.*, 9, 1005, 1989.
225. Hsing, A.W. Hormones and prostate cancer: What's next? *Epidemiol. Rev.*, 23, 42, 2001.
226. Wilding, G. The importance of steroid hormones in prostate cancer, *Cancer Surv.*, 14, 113, 1992.
227. Dalton, R.R. and Kallab, A.M. Chemoprevention of breast cancer, *Southern Med. J.*, 94,7, 2001.
228. Fornander, T., Rutqvist, L.E., Cedermark , B., Glas, U., Mattsson, A., Silfversward, C., Skoog, L., Somell, A., Theve, T., Wilking, N. et al. Adjuvant tamoxifen in early breast cancer: Occurrence of new primary cancers, *Lancet*, 8630, 117, 1989.
229. Jordan, V.C. Long-term adjuvant tamoxifen therapy for breast cancer, *Breast Cancer Res. Treat*, 15, 125, 1990.
230. Wolmark, N. and Dunn, B.K. The role of tamoxifen in breast cancer prevention: Issues sparked by the NSABP Breast Cancer Prevention Trial (P-1), *Ann. N.Y. Acad. Sci.*, 949, 99, 2001.
231. King, M.C., Wieand, S., Hale, K., Lee, M., Walsh, T., Owens, K., Tait, J., Ford, L., Dunn, B.K., Costantino, J., Wickerham, L., Wolmark, N., and Fisher, B. Tamoxifen and breast cancer incidence among women with inherited mutations in BRCA1 and BRCA2: National Surgical Adjuvant Breast and Bowel Project (NSABP-P1) Breast Cancer Prevention Trial, *JAMA*, 286, 2251, 2001.
232. White, I.N.H., de Matteis, F., Davies, A., Smith, L.L., Crofton-Sleigh, C., Venitt, S., Hewer, A., and Phillips, D.H. Genotoxic potential of tamoxifen and analogues in female Fischer F344/n rats, DBA/2 and C57BL/6 mice and in human MCL-5 cells, *Carcinogenesis*, 13, 2197, 1992.
233. Rattel, B., Löser, R., Dahme, E.G., Liehn, H.D., and Siebel, K. Comparative toxicology of droloxifene (3-OH tamoxifen) and tamoxifen: Hepatocellular carcinomas induced by tamoxifen, *Biennial International Breast Cancer Research Conference*, Abstract F18, 1987.
234. Williams, G.M., Iatropoulos, M.J., and Hard, G.C. Long-term prophylactic use of tamoxifen: Is it safe? *Eur. J. Cancer Prev.*, 1, 386, 1992.
235. Williams, G.M., Iatropoulos, M.J., Djordjevic, M.V., and Kaltenberg, O.P. The triphenylethylene drug tamoxifen is a strong liver carcinogen in the rat, *Carcinogenesis*, 14, 315, 1993.
236. Eltabbakh, G.H. and Mount, S.L. Tamoxifen and the female reproductive tract, *Expert Opin. Pharmacother.*, 2, 1399, 2001.

237. O'Regan, R.M. and Jordan, V.C. Tamoxifen to raloxifene and beyond, *Semin. Oncol.*, 28, 260, 2001.

238. Sato, M., Turner, C.H., Wang, T., Adrian, M.D., Rowley, E., and Bryant, H.U. LY353381.HCl: A novel raloxifene analog with improved SERM potency and efficacy in vivo, *J. Pharmacol. Exp. Ther.*, 287, 1, 1998.

239. Dardes, R.C., Bentrem, D., O'Regan, R.M., Schafer, J.M., and Jordan, V.C. Effects of the new selective estrogen receptor modulator LY353381.HCl (Arzoxifene) on human endometrial cancer growth in athymic mice, *Clin. Cancer Res.*, 7, 4149, 2001.

240. Barnes, S., Sfakianos, J., Coward, L., and Kirk, M. Soy isoflavonoids and cancer prevention. Underlying biochemical and pharmacological issues, *Adv. Exp. Med. Biol.*, 401, 87, 1996.

241. Fritz, W.A., Coward, L., Wang, J., and Lamartiniere, C.A. Dietary genistein: Perinatal mammary cancer prevention, bioavailability and toxicity testing in the rat, *Carcinogenesis*, 19, 2151, 1998.

242. Lamartiniere, C.A., Moore, J.B., Brown, N.M., Thompson, R., Hardin, M.J., and Barnes, S. Genistein suppresses mammary cancer in rats, *Carcinogenesis*, 16, 2833, 1995.

243. Murrill, W.B., Brown, N.M., Zhang, J.X., Manzolillo, P.A., Barnes, S., and Lamartiniere, C.A. Prepubertal genistein exposure suppresses mammary cancer and enhances gland differentiation in rats, *Carcinogenesis*, 17, 1451, 1996.

244. Gotoh, T., Yamada, K., Yin, H., Ito, A., Kataoka, T., and Dohi, K. Chemoprevention of *N*-nitroso-*N*-methylurea-induced rat mammary carcinogenesis by soy foods or biochanin A, *Jpn. J. Cancer Res.*, 89, 137, 1998.

245. Constantinou, A.I., Mehta, R.G., and Vaughan, A. Inhibition of *N*-methyl-*N*-nitrosourea-induced mammary tumors in rats by the soybean isoflavones, *Anticancer Res.*, 16, 3293, 1996.

246. Pollard, M. and Luckert, P.H. Influence of isoflavones in soy protein isolates on development of induced prostate-related cancers in L-W rats, *Nutr. Cancer*, 28, 41, 1997.

247. Geller, J., Sionit, L., Partido, C., Li, L., Tan, X., Youngkin, T., Nachtsheim, D., and Hoffman, R.M. Genistein inhibits the growth of human-patient BPH and prostate cancer in histoculture, *Prostate*, 34, 75, 1998.

248. Pagliacci, M.C., Smacchia, M., Migliorati, G., Grignani, F., Riccardi, C., and Nicoletti, I. Growth-inhibitory effects of the natural phyto-oestrogen genistein in MCF-7 human breast cancer cells, *Eur. J. Cancer*, 30A, 1675, 1994.

249. Constantinou, A.I., Kamath, N., and Murley, J.S. Genistein inactivates bcl-2, delays the G2/M phase of the cell cycle, and induces apoptosis of human breast adenocarcinoma MCF-7 cells, *Eur. J. Cancer*, 34, 1927, 1998.

250. Marshall, M.S. Ras target proteins in eukaryotic cells, *FASEB J.*, 9, 1311, 1995.

251. Porter, A.C. and Vaillancourt, R.R. Tyrosine kinase receptor-activated signal transduction pathways which lead to oncogenesis, *Oncogene*, 16, 1343, 1998.

252. Leirdal, M. and Sioud, M. Tyrosine kinase receptor-ras-ERK signal transduction pathway as therapeutic target in cancer, *Tidsskr Nor Laegeforen*, 122, 178, 2002.

253. Barry, O.P. and Kazanietz, M.G. Protein kinase C isozymes, novel phorbol ester receptors and cancer chemotherapy, *Curr. Pharm. Res.*, 7, 1725, 2001.

254. Nowak, A.K., Lake, R.A., Kindler, H.L., and Robinson, B.W. New approaches for mesothelioma: Biologics, vaccines, gene therapy, and other novel agents, *Semin. Oncol.*, 29, 82, 2002.

255. George, D. Platelet-derived growth factor receptors: A therapeutic target in solid tumors, *Semin. Oncol.*, 28, 27, 2001.

256. Housey, G.M., O'Brian, C.A., Johnson, M.D., Kirschmeier, P., and Weinstein, I.B. Isolation of cDNA clones encoding protein kinase C: Evidence for a protein kinase C-related gene family, *Proc. Natl. Acad. Sci. USA*, 84, 1065, 1987.

257. Knopf, J.L., Lee, M.H., Sultzman, L.A., Kritz, R.W., Loomis, C.R., Hewick, R.M., and Bell, R.M. Cloning and expression of multiple protein kinase C cDNAs, *Cell*, 46, 491, 1986.

258. Blumberg, P.M. Protein kinase C as the receptor for the phorbol ester tumor promoters: Sixth Rhoads memorial award lecture, *Cancer Res.*, 48, 1, 1988.

259. O'Brian, C.A., Ward, N.E., Ioannides, C.G., and Dong, Z. Potential strategies of chemoprevention through modulation of protein kinase C activity, in *Cellular and Molecular Targets for Chemoprevention*, Steele, V.E., Stoner, G.D., Boone, C.W., and Kelloff, G.J., Eds., CRC Press, Boca Raton, FL, 1992, p. 161.

260. Nakadate, T., Jeng, A.Y., and Blumberg, P.M. Comparison of protein kinase C functional assays to clarify mechanisms of inhibitor action, *Biochem. Pharmacol.*, 37, 1541, 1988.

261. Schwartz, G.K., Jiang, J., Kelsen, D., and Albino, A.P. Protein kinase C: A novel target for inhibiting gastric cancer cell invasion, *J. Natl. Cancer Inst.*, 85, 402, 1993.
262. Caponigro, F., French, R.C., and Kaye, S.B. Protein kinase C: A worthwhile target for anticancer drugs? *Anticancer Drugs*, 8, 26, 1997.
263. Barbacid, M. *ras* Genes, *Ann. Rev. Biochem.*, 56, 779, 1987.
264. Barbacid, M. Oncogenes and human cancer: Cause or consequence? *Carcinogenesis*, 7, 1037, 1986.
265. Stowers, S., Glover, P., Reynolds, S.H., Boone, L., Maronpot, R.R., and Anderson, M.W. Activation of the K-*ras* protooncogene in lung tumors from mice and rats chronically exposed to tetranitromethane, *Cancer Res.*, 47, 3212, 1987.
266. You, M., Candrian, U., Maronpot, R.R., Stoner, G.D., and Anderson, M.W. Activation of the K-*ras* protooncogene in spontaneously-occurring and chemically-induced lung tumors of the strain A mouse, *Proc. Natl. Acad. Sci. USA*, 86, 3070, 1989.
267. Wang, Y., You, M., Reynolds, S.H., Stoner, G.D., and Anderson, M.W. Mutational activity of the cellular Harvey *ras* oncogene in rat esophageal papillomas induced by methylbenzyl-nitrosamine, *Cancer Res.*, 50, 1591, 1990.
268. Reynolds, S.H., Anna, C.K., Brown, K.C., Wiest, J.S., Beattie, E.J., Pero, R.W., Iglehart, J.D., and Anderson, M.W. Activated protooncogenes in human lung tumors from smokers, *Proc. Natl. Acad. Sci. USA*, 88, 1085, 1991.
269. Slebos, R.J.C., Hruban, R.H., Dalesio, O., Mooi, W.J., Offerhaus, J.A., and Rodenhuis, S. Relationship between K-*ras* activation and smoking in adenocarcinoma of the human lung, *J. Natl. Cancer Inst.*, 83, 1024, 1991.
270. Nelson, M.A., Futscher, B.W., Kinsella, T., Wymer, J., and Bowden, G.T. Detection of mutant Ha-*ras* genes in chemically initiated mouse skin epidermis before the development of benign tumors, *Proc. Natl. Acad. Sci. USA*, 89, 6398, 1992.
271. Nelson, M.A., Bowden, G.T., Kinsella, T., Wymer, J., Cordero, G., and Rosenberg, R.K. Mutant K-*ras* in the lung: A potential marker of bronchial epithelial premalignancy and early lung cancer, *Proc. Am. Assoc. Cancer Res.*, 33, 104, 1992.
272. Cerny, W.L., Mangold, K.A., and Scarpelli, D.G. K-*ras* mutation is an early event in pancreatic duct carcinogenesis in the Syrian golden hamster, *Cancer Res.*, 52, 4507, 1992.
273. Casey, P.J., Solski, P.A., Der, C.J., and Buss, J.E. p21 *ras* is modified by a farnesyl isoprenoid, *Proc. Natl. Acad. Sci. USA*, 86, 8323, 1989.
274. Kato, K., Cox, A.D., Hisaka, M.M., Graham, S.M., Buss, J.E., and Der, C.J. Isoprenoid addition to *ras* protein is the critical modification for its membrane association and transforming activity, *Proc. Natl. Acad. Sci. USA*, 89, 6403, 1992.
275. Crowell, P.L., Chang, R.R., Ren, Z., Elson, C.E., and Gould, M.N. Selective inhibition of isoprenylation of 21–26 kDa proteins by the anticarcinogen d-limonene and its metabolites, *J. Biol. Chem.*, 266, 17679, 1991.
276. Wattenberg, L.W. Inhibition of neoplasia by minor dietary constituents, *Cancer Res.*, 43 (Suppl.), 2448s, 1983.
277. Elegbede, J.A., Elson, C.E., Qureshi, A., Tanner, M.A., and Gould, M.N. Inhibition of DMBA-induced mammary cancer by the monoterpene d-limonene, *Carcinogenesis*, 5, 661, 1984.
278. Elegbede, J.A., Elson, C.E., Tanner, M.A., Qureshi, A., and Gould, M.N. Regression of rat primary mammary tumors following dietary d-limonene, *J. Natl. Cancer Inst.*, 76, 323, 1986.
279. Elson, C.E., Maltzman, T.H., Boston, J.L., Tanner, M.A., and Gould, M.N. Anti-carcinogenic activity of d-limonene during the initiation and promotion/progression stages of DMBA-induced rat mammary carcinogenesis, *Carcinogenesis*, 9, 331, 1988.
280. Maltzman, T.H., Hurt, L.M., Elson, C.E., Tanner, M.A., and Gould, M.N. The prevention of nitrosomethylurea-induced mammary tumors by d-limonene and orange oil, *Carcinogenesis*, 10, 781, 1989.
281. Russin, W.A., Hoesly, J.D., Elson, C.E., Tanner, M.A., and Gould, M.N. Inhibition of rat mammary carcinogenesis by monoterpenoids, *Carcinogenesis*, 10, 2161, 1989.
282. Wattenberg, L.W., Sparnins, V.L., and Barany, G. Inhibition of *N*-nitrosodiethylamine carcinogenesis in mice by naturally occurring organosulfur compounds and monoterpenes, *Cancer Res.*, 49, 2689, 1989.

283. Wattenberg, L.W. and Coccia, J.B. Inhibition of 4-(methylnitrosamino)-1-(3-pyridyl)-1-butanone carcinogenesis in mice by d-limonene and citrus fruit oils, *Carcinogenesis*, 12, 115, 1991.

284. Zujewski, J., Horak, I.D., Bol, C.J., Woestenborghs, R., Bowden, C., End, D.W., Piotrovsky, V.K., Chiao, J., Belly, R.T., Todd, A., Kopp, W.C., Kohler, D.R., Chow, C., Noone, M., Hakim, F.T., Larkin, G., Gress, R.E., Nussenblatt, R.B., Kremer, A.B., and Cowan, K.H. Phase I and pharmacokinetic study of farnesyl protein transferase inhibitor R115777 in advanced cancer, *J. Clin. Oncol.*, 18, 927, 2000.

285. Punt, C.J., van Maanen, L., Bol, C.J., Seifert, W.F., and Wagener, D.J. Phase I and pharmacokinetic study of the orally administered farnesyl transferase inhibitor R115777 in patients with advanced solid tumors, *Anticancer Drugs*, 12, 193, 2001.

286. Karp, J.E., Lancet, J.E., Kaufmann, S.H., End, D.W., Wright, J.J., Bol, K., Horak, I., Tidwell, M.L., Liesveld, J., Kottke, T.J., Ange, D., Buddharaju, L., Gojo, I., Highsmith, W.E., Belly, R.T., Hohl, R.J., Rybak, M.E., Thibault, A., and Rosenblatt, J. Clinical and biologic activity of the farnesyltransferase inhibitor R115777 in adults with refractory and relapsed acute leukemias: A phase 1 clinical-laboratory correlative trial, *Blood*, 97, 3361, 2001.

287. Gwyn, K. and Sinicrope, F.A. Chemoprevention of colorectal cancer, *Am. J. Gastroenterol.*, 97, 13, 2002.

288. Lippman, S.M. and Spitz, M.R. Lung cancer chemoprevention: An integrated approach, *J. Clin. Oncol.* 19, 74s, 2001.

289. Vainio, H. Chemoprevention of cancer: A controversial and instructive story, *Br. Med. Bull.*, 55, 593, 1999.

290. Pryor, W.A., Stahl, W., and Rock, C.L. Beta carotene: From biochemistry to clinical trials, *Nutr. Rev.*, 58, 39, 2000.

291. van Poppel, G. Carotenoids and cancer: An update with emphasis on human intervention studies, *Eur. J. Cancer*, 29A, 1335, 1993.

292. Albanes, D., Heinonen, O.P., Huttunen, J.K., Taylor, P.R., Virtamo, J., Edwards, B.K., Haapakoski, J., Rautalahti, M., Hartman, A.M., Palmgren, J. et al. Effects of alpha-tocopherol and beta-carotene supplements on cancer incidence in the alpha-tocopherol beta-carotene cancer prevention study, *Am. J. Clin. Nutr.*, 62, 1427s, 1995.

293. Goodman, G.E. Prevention of lung cancer, *Crit. Rev. Oncol. Hematol.*, 33, 187, 2000.

294. Cook, N.R., Le, I.M., Manson, J.E., Buring, J.E., and Hennekens, C.H. Effects of beta-carotene supplementation on cancer incidence by baseline characteristics in the physicians' health study (United States), *Cancer Causes Control*, 11, 617, 2000.

295. Yeh, S. and Hu, M. Antioxidant and pro-oxidant effects of lycopene in comparison with beta-carotene on oxidant-induced damage in Hs68 cells, *J. Nutr. Biochem.*, 11, 548, 2000.

22 Carcinogen Biomarkers and Exposure Assessment

Glenn Talaska

CONTENTS

22.1 INTRODUCTION

There is no area of occupational or environmental health that is more daunting to practitioners than estimating the exposure and effects of carcinogens. Factors such as multiple routes of potential exposure, the severity of the disease, the lag time between initial exposure and manifest effects, and multiplicity of mechanisms contribute to the problem. In addition, regulatory agencies have yet to provide substantial guidance to help practitioners estimate carcinogen exposures and place them into perspective. The United States Environmental Protection Agency (USEPA) does not have a formal policy on carcinogen exposure assessment, although carcinogen risk assessment is an important part of their efforts [1]. The USEPA home page does not list "carcinogens" or specific carcinogenic compounds [1]. The National Institute for Occupational Safety and Health (NIOSH) that "has not identified thresholds for carcinogens that will protect 100% of the population . . . , recommends that **occupational exposures to carcinogens be limited to the lowest feasible concentration**" (emphasis in original) [2]. NIOSH goes on to describe appropriate respiratory protection without mentioning dermal exposure potential. The Occupational Safety and Health Administration (OSHA) lists 45 compounds or mixtures as known human carcinogens, but neither does it detail specific guidelines or tools for estimating exposure, nor provide an overall carcinogen exposure assessment philosophy [3]. Proscribing exposure across the board does not help the occupational health professional in the absence of specific tools for assessing exposure and dose. Estimating exposure to carcinogens is where "the rubber meets the road" for intervention in the carcinogenic process and cancer prevention. The focus of this chapter is to provide a brief overview of air monitoring methods for carcinogens. This is followed by a detailed description of methods of carcinogen biomonitoring and biomarkers, particularly for polycyclic aromatic hydrocarbons (PAHs) and aromatic amines.

22.1.1 AIR SAMPLING FOR CARCINOGENS

For those carcinogenic compounds that have significant vapor pressure and a low potential for dermal absorption, air sampling is a reasonable way of assessing exposure. For example, ethylene oxide is a gas at room temperature and the vapor pressure of *bis*-chloromethyl ether is 30 mm Hg at room temperature (equilibrium concentration of about 40,000 ppm), so highly significant airborne exposures are possible with these, and similar, compounds. NIOSH has recommended air sampling methods for carcinogens with these properties. A list of NIOSH carcinogens and recommended air sampling methods (where they exist) is given as Table 22.1 [4]. However, note that there are many compounds that possess a "skin" notation from the American Conference of Governmental Industrial Hygienists (ACGIH) [5]. In fact, more than 50% of all the compounds or mixtures listed as A1 or A2 carcinogens by the ACGIH and NIOSH and 48 of the 84 compounds in Table 22.1 carry this notation. The skin notation indicates that there is "potential significant contribution to the overall exposure by the cutaneous route . . . ." However, it should be noted that having the skin notation does not mean that the skin will be the major route of entry. An industrial hygiene evaluation is needed to determine the actual exposure routes in each circumstance. Nevertheless, a skin notation indicates that dermal uptake has been demonstrated. A significant number of other compounds might have the skin notation if testing information was available. Tsuruta [6] examined the dermal absorption of a series of halogenated hydrocarbons and found that those with the most dermal uptake did not always carry a skin notation and there was no other published work regarding dermal uptake. Air sampling will underestimate uptake if the assumptions implicit in the guidelines are violated. These include exposure by other routes, skin and gastrointestinal, higher ventilation rates caused by exertion or stress, or higher uptake from the lungs than anticipated due to distribution to body fat or high blood lipid levels.

For these reasons, biological monitoring has been proposed to estimate uptake by the individual. There are BEIs (Biological Exposure Indices) established by the ACGIH for several of the compounds in Table 22.1. The Documentation for the Threshold Limit Values (TLVs) and BEIs [5] should be consulted for specific sampling methods and applications. Of the 860 compounds with TLVs, 196 carry the skin notation, but BEIs are available for only 15 of compounds with that notation; there are a large number of compounds that carry skin notation for which BEIs are not available. The process for developing these advisory standards is driven by the committee's perceived need for such advice to the occupational health community and reasonable amount of literature supporting a value of a marker for a given exposure.

22.1.2 OVERVIEW OF BIOLOGICAL MONITORING

Biomarkers are measurements of the changes that occur in the individual in response to an exposure. Biomarkers can include an increase in the internal levels of the material itself, in one of its metabolites, or in the measurement of a cellular or biochemical change that the material causes. Biomarkers can be specific for exposure, as when the parent compound or a unique metabolite is measured. Other biomarkers are nonspecific. There are two general types of nonspecific biomarkers: 2 compounds may share a measured metabolite, such as the several chlorinated hydrocarbons (e.g., methyl chloroform, trichloroethylene) that are metabolized to trichloroacetic acid [5]. Additionally, the same biomarker of effect may be caused by many different compounds. Acetylcholinesterase inhibition by a large number of organophosphate pesticides is an example.

Surrogate markers are also commonly used in biomonitoring. Surrogate markers include those that are measured in an easily available matrix, such as blood or urine, and respond similarly to the marker that is affected to produce the effect and is not available for sampling. Acetylcholinesterase inhibition is also an example of such a surrogate marker [5]. A related type of surrogate marker

TABLE 22.1
Air Sampling Methods for Selected NIOSH Carcinogens[a]

Compound	ACGI HBEI[b]	ACGIH skin notation[b]	NIOSH analytical method number[c]
Acetaldehyde	No	No	2538
Acrylamide	No	Yes	21
Acrylonitrile	No	Yes	1604
Aldrin	No	Yes	5502
4-Aminobiphenyl	No	Yes	269
Amitrole	No	No	0500
Aniline	Yes	Yes	2002, BEI***
Arsenic	Yes	No	7900, BEI
Arsine	No	No	6001
Asbestos	NA	No	7400
Asphalt Fumes	No[4]	No	See specific NIOSH Criteria Document
Benzene	Yes	Yes	1500, 3700, BEI
Benzidine	No	Yes	5509
Beryllium	No	No	7102
1,3-Butadiene	No[5]	No	1024
t-butyl chromate	No	Yes	7604
Cadmium	Yes	No	7048
Carbon tetrachloride	No	Yes	1003
Chlordane	No	Yes	5510
Chlorinated champhene	No	Yes	S67
Chlorodiphenyl	No	Yes	5503
Chloroform	No	No	1003
Bis-chloromethyl ether	No	No	10
Chloromethyl methyl ether	No	No	PCAM 220
β-chloroprene	No	Yes	1002
Chromates	Yes	No	7600, 7024
Coal tar volatiles, coking	No[4]	No	5023
DDT	No	No	S274
2,4-diaminoanisole	No	N/A	None available
o-Dianisidine	No	N/A	5013
Dibromochloropropane	No	Yes	None available
Dichloroacetylene	No	No	None available
p-dichlorobenzene	No	No	1003
3,3'-dichlorobenzidine	No	Yes	5509
Dichloroethyl ether	No	Yes	1004
1,3-dichloropropene	No	Yes	None available
Diglycidyl ether	No	No	None available
4-dimethylaminoazobenzene	No	N/A	PCAM 284
Dimethylcarbamoyl chloride	No	No	None available
1,1-dimehtylhydrazine	No	Yes	S143
Dimethyl sulfate	No	Yes	2524
Dinitrotoluene	Yes	Yes	44, BEI
di-sec-octyl phthalate	No	N/A	5020
Dioxane	No	Yes	1602

(continued)

**TABLE 22.1
(Continued)**

Compound	ACGIH BEI[2]	ACGIH skin notation[2]	NIOSH analytical method number[3]
Epichlorohydrin	No	Yes	1010
Ethyl acrylate	No	Yes	1450
Ethylene dibromide	No	Yes	1008
Ethylene dichloride	No	No	1003
Ethyleneimine	No	Yes	2507
Ethylene oxide	No	No	1614
Ethylene thiourea	No	N/A	5011
Formaldehyde	No	No	3500
Heptachlor	No	Yes	S287
Hexachlorobutadiene	No	Yes	PCAM 307
Hexachloroethane	No	Yes	1003
Hexamethylphosphoramide	No	Yes	None available
Hydrazine	No	Yes	3503
Kepone	No	N/A	5508
Malonaldehyde	No		None available
Methoxychlor	No	No	S371
Methyl bromide	No	Yes	2520
Methyl chloride	No	Yes	1001
4,4′-methylene-*bis*-2-chloroaniline	Yes	Yes	None available, BEI
4-4′methylenedianiline	No	Yes	5029
Methyl hydrazine	No	Yes	3510
Methyl iodine	No	Yes	1014
β-naphthylamine	No	Yes	5518
Nickel carbonyl	No	No	6007
Nickel	No	No	7300
4-nitrobiphenyl	No	Yes	PCAM273
p-nitrochlorobenzene	No	Yes	2005
2-nitronaphthalene	No	N/A	None available
Nitrosodimethylamine	No	Yes	2522
Phenyl glycidyl ether	No	Yes	S74
Phenylhydrazine	No	Yes	S160
N-phenyl-2-naphthylamine	No	N/A	None available
Propane sultone	No	No	None available
β-propiolactone	No	No	None Available
Propylene dichloride	No	No	1013
Propylene imine	No	Yes	None available
Propylene oxide	No	No	1612
Silica, crystalline	NA	No	7500
Tetrachlorodibenzo-*p*-dioxin	No	N/A	None available
1,1,2,2-tetrachloroethane	No	Yes	1019
Tetrachloroethylene	Yes	No	1003, BEI
o-toluidine	No	Yes	2202, BEI

(continued)

TABLE 22.1
(Continued)

Compound	ACGIH BEI[2]	ACGIH skin notation[2]	NIOSH analytical method number[3]
Toluenediamine	No	N/A	5516
1,1,2-trichloroethane	No	Yes	1003
Trichloroethylene	Yes	No	1022, BEI
1,2,3-trichloropropane	No	Yes	1003
Vinyl bromide	No	No	1009
Vinyl chloride	No	No	1007
Vinyl cyclohexene dioxide	No	Yes	None available
Vinylidine chloride	No	No	1015
Wood dust	No	No	0500

[a] See Reference 4.
[b] See Reference 5 "Documentation of the TLVs and BEIs" for specifics.
[c] Details can be found at http://www.cdc.gov/niosh/nmam/nmampub.html.
[d] Information is being gathered for a BEI on polycyclic aromatic hydrocarbon containing compounds.
[e] Information is being gathered by the AIHA Biological Monitoring Committee for an advisory biological standard on this material.

estimates mutations in genes that are easily measured, but are not related to the process of carcinogenesis. Measurement of mutations in the red blood cell (RBC) surface marker, glycophorin A is an example [7]. Another type of surrogate is one that estimates exposure for classes of materials that are present in either levels that are too low or parts of mixtures too complex to measure individually. Urinary 1-hydroxypyrene is a surrogate marker that has proven useful for exposure to complex PAH mixtures [8,9].

There are BEIs available for eight of the NIOSH carcinogens listed in Table 22.1. Previous chapters have discussed specifics of many carcinogen exposure and effect biomarkers. The purpose of this chapter is to put them into the context of humans and human exposure conditions.

22.2 RELATIONSHIPS BETWEEN BIOMARKERS AND HEALTH OUTCOME

As our understanding of the initial phases of cancer biology has grown, we have hoped to measure the biochemical and physiological changes that occur following exposure predictors of events that may follow. The current view is that cancer is a disease initiated by specific mutations in multiple, specific genes. These genes that must be mutated may also vary in different tissues. Within many of these genes only mutations in certain base pairs lead to activation. For example, only about 12 basepairs out of 450 lead to activation of the *ras* oncogenes [10]. The targets are larger with tumor suppressor genes such as *p53*, in which any inactivating mutation can increase the oncogenic potential of the cell. In addition, more than one oncogene must be activated for a cell to be transformed and the number and temporal sequence of these activations also appears to be tissue specific. Although there may be multiple pathways to the same (or equivalent) endpoint in each tissue, it is obvious that the number of critical targets in a given tissue is very small relative to the total number of DNA bases in the genome, say, 10^{10}. The error rate of DNA polymerase is estimated to be on the order of $1/10^8$ to $1/10^9$

bases. Thus, if exposure were a rare event, the probabilities that any cell would acquire the specific mutations needed to activate specific oncogenes and initiate the tumor process would also be very small. If this were the only factor, the disease rate would be higher in more rapidly proliferating cell types and vice versa. However, people do develop cancer, sometimes without known external exposure. And, the sites of the most common, "spontaneous" cancers are not necessarily in the most rapidly dividing tissues (e.g., breast and prostate). Balancing the small number of targets is the relatively large number of projectiles (i.e., DNA modifying agents) randomly assaulting the genome and the cell, in general.

It has been estimated that there are 10,000 to 50,000 bases adducted by active oxygen species per cell, per day from endogeneous sources alone [11]. On the other hand, efficient repair processes appear to have evolved to deal with the majority of oxidative DNA damage. Occupational exposure can contribute significantly to the genotoxic exposure burden. For example, the ACGIH TLV for coal tar pitch volatiles (CTPV) is 0.2 mg/m^3. Benzo[a]pyrene (BAP) comprises approximately 0.4% of CTPV, so that a person exposed at the TLV would be absorbing on the order of 2×10^{16} BAP molecules per day, or approximately 10,000 molecules per cell, assuming equal distribution. The other PAHs in the mixture would add to the load. And although this is a large burden, it is counterbalanced by detoxifying metabolism, distribution, elimination, repair, and by the extremely small number of critical targets on DNA. Cancer latency, by this model is defined as the time when DNA is acquiring hits. There are at least two important implications for carcinogen biomonitoring. The first is that it will be difficult, if not impossible, to use biomarkers of carcinogen exposure, and most markers of carcinogen effect to predict health outcome for an individual, that is, diagnostically. To do so would require that one would have to know, *a priori*, which cells were to be transformed and also have a method of selecting them. These markers can only be interpreted in probabilistic terms. A person with ten times the level of a carcinogen biomarker relative to another person might be reasonably expected to be at greater risk, but one cannot say with surety if they will develop the disease. Cancer is an all or nothing, dichotomous disease: one cannot have a "touch of cancer" any more than one can be a "little pregnant," although the disease is clearly different at various stages. The second important implication of cancer latency and biomarkers is that if the levels of the exposure biomarker decreases prior to the frank transformation of a cell, then the probability of disease should decrease as well. This suggests that if exposure is terminated prior to a certain, as yet unknown, critical step, the disease may be prevented. These temporal effects are discussed later in this chapter.

So, what are the important considerations for cancer and the individual? Genetics clearly plays a role, particularly for the relatively small number of people predisposed to the disease by inheriting genes already possessing some of the critical mutations. For most people exposure is the most important factor in disease. In the general environment there is no question that a PAH-containing mixture, tobacco smoke, is the major cause of cancer of the lung, urinary bladder, esophagus, throat, and other sites. In the United States, 70–80% of lung cancer is due to this exposure [12]. Occupational exposure accounts for a fairly low percentage of the cancer cases in the general population. On the other hand, in certain occupations, exposures and disease rates were and are horrendous. For example, Mitch Zavon recently communicated to this author that 70% of the men in U.S. dye works have developed urinary bladder cancer form their exposure to benzidine, this is 50–100 times higher than expected in that population [13]. In a more recent biomarker study of benzidine exposure in India, individual workers had benzidine-DNA adduct levels from 8 to almost 250 times higher than the control population [14]. Diet, nutritional status, and genetic polymorphisms for phase 1 and 2 metabolic enzymes may prove to have a modifying effect, but the exposure is clearly the most important factor in the diseases seen [15]. The final factor in cancer for the individual following exposure is plain, dumb luck, or lack of it; cancer mechanisms are probabilistic, if nothing else.

In this light, one approach to understanding biomarkers and biological monitoring for carcinogens is a linear continuum between exposure and effects to disease (Figure 22.1); intervening stages that

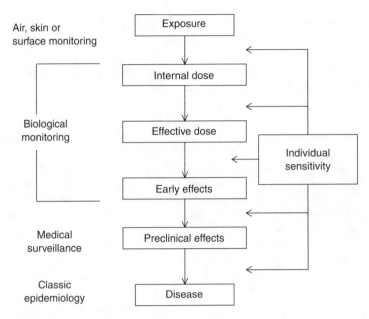

FIGURE 22.1 Continuum between exposure and effects with increasingly specific and predictive levels of biomarkers prior to disease.

can be measured as biomarkers. The scheme is shown as a linear continuum, although in reality there are many side paths and dead ends. The figure indicates that there are four very different ways that exposure and effects can be determined: monitoring the external exposure, biological monitoring, medical surveillance, and classical epidemiology. Air sampling for monitoring external exposure to carcinogens has been discussed earlier.

Biological monitoring and biomarkers encompass the next three levels of measurements. Biological monitoring is an industrial hygiene discipline concerned with estimating exposure. The markers in the next three classes can, for various reasons, be considered exposure markers and not disease markers. Markers of internal dose are used to estimate the amount of material entering the person and thus control for absorption routes as well as individual differences in absorption [16,17]. These biomarkers include measurement of urinary metabolites, elimination of the parent volatile compounds in exhaled air, or measurement of the parent compound in blood. In the case where the parent compound (e.g., a solvent like toluene) is the active agent, the internal dose is the effective dose. Urinary levels of 1-hydroxypyrene or 3-hydroxybenzo[a]pyrene are examples of internal dose markers for PAHs [8]. These markers are estimates of carcinogen load or body burden and give information as to the amount of PAH that has entered the body. They do not provide any information regarding the fraction of the absorbed dose interacting with the genetic material.

Effective dose biomarkers measure the amount of material that reaches the critical target in a form capable of causing damage. These markers are more solidly linked to effect than are internal dose markers and imply that a compound is in a biologically active form by either its own structure or because of metabolism. Exposure to n-hexane and methyl isobutyl ketone is associated with development of a peripheral neuropathy that has been traced to 2,5-hexanedione, one of the major metabolites of these materials [18]. Measurement of urinary 2,5-hexanedione estimates the effective dose of this material reaching the peripheral nerves, taking into account that only a fraction of the absorbed parent compounds will be metabolized to the specific toxin. Thus, effective dose biomarkers integrate for individual differences in exposure and metabolism. Measurement of hemoglobin adducts of PAH carcinogens would be an example of a surrogate effective dose marker. Formation of 4-aminobiphenyl (4-ABP) hemoglobin adducts requires the presence of N-hydroxy-4-ABP in the blood [19,20]. This is the same metabolite that reacts with DNA and has been found in the urinary bladder of human smokers. In this case, hemoglobin is a protective molecule acting as a trap for the reactive metabolite. Carcinogen-DNA adducts can also be considered effective dose markers. DNA adducts are the ultimate mutagens, but the likelihood that any measured DNA adduct would be responsible for a cancer is extremely small, as discussed above. Rather, they can be viewed as estimates of insult to the genetic material [21]. Because target organs are often not available, DNA adduct measurements are made in surrogate tissues such as white blood cells or blood lymphocytes. Effective dose biomarkers, in general, account for differences in metabolism and distribution responsible for the formation and delivery, respectively, of the active material to the target.

The next level includes biomarkers of early effects. These are measurements of important, reversible effects related to the disease process caused by the activated agent. Carboxyhemoglobin and inhibition of RBC acetylcholinesterase activity are commonly used examples of this type of marker. Carboxyhemoglobin levels measure directly the binding of carbon monoxide to its critical target, hemoglobin. This marker is truly a smoking gun for effect. RBC acetylcholinesterase is, on the other hand, a surrogate measurement for the inhibition of the central nervous system (CNS) isozyme affected acutely by exposure to organophosphate and carbamate pesticides. The delayed or long-term effects of these pesticides are poorly predicted by the RBC surrogate, so the search continues for a surrogate biomarker available in peripheral tissues that will predict the chronic CNS effects. Cytogenetic markers, such as micronuclei and chromosome aberrations and mutations markers such as glycophorin A and lymphocyte HGPRT mutation rates, also fit into the category of markers of early effects. In each case it is understood that the marker measured has nothing to do with the disease process but is an estimate of the average rate of mutation and damage in the genome that can be inferred to include damage to critical targets such as the *p53*, *ras*, and *myc* oncogenes [22]. Although not reversible or repairable, these biomarkers are surrogates for the types of damage, mutations, and chromosome translocations, respectively, which have been associated with the activation of oncogenes.

Preclinical effects are detected through medical surveillance. The markers involved include biological effects such as pleural plaques indicative of asbestosis, the restrictive pattern of airway pulmonary function changes associated with silicosis, the excretion of β_2-microglobin as a result of chronic cadmium exposure, and phenotypically altered cells detected during sputum or urinary cytology. Clinical markers such as circulating levels of hepatic enzymes (e.g., SGPT, SGOT) also fall into this general class. New molecular approaches within this paradigm include the detection of tumor-related growth factors and mutated oncogenes and oncoproteins [22]. These biomarkers share a higher positive predictive value than biomarkers of exposure, but have a lesser role in disease prevention.

Finally classical epidemiology, as discussed above, enumerates the cases of frank disease and attempts to associate the effects with an exposure. Increasingly epidemiologists have been looking to employ biomarkers as a confirmatory test for exposure and effects for reasons that will be discussed below.

The selection of a biomarker at a specific level should be guided by the question that must be addressed. If the level of current exposure is the key question, then a biomarker of internal dose should be used. If there is a need to take metabolic variability into consideration, then an effective dose biomarker should be selected and so on.

Individual differences have the potential to significantly alter the transition between each biomarker level. Factors such as exertion and respiratory rate, body fat, and site of dermal absorption significantly influence how much of an external dose actually enters the body. There is now a literature providing evidence that physical factors such as dermatitis and skin abrasions and chemical agents such as kerosene can enhance the dermal absorption of occupational carcinogens [23,24]. Metabolism (both phase 1 and 2) and distribution determine the fraction of the absorbed dose that gets converted to the effective dose at the target organ, and these have variability in the population [25]. Factors involved in the transition between effective dose early effects are less well understood, but presumably involve differences in damage repair. Even less well understood are the factors involved in the transitions to preclinical effects and disease. For noncarcinogenic effects, these factors may include differences in tissue redundancy, making some individuals better able to absorb chronic renal, brain, or peripheral nerve damage.

Clearly, when a measurement is made closer to the disease process as is the case for preclinical effects, the greater the positive predictive value; finding this positive marker will coincide with the disease. For this reason preclinical effect markers can be useful in surveillance programs, but only for diseases amenable to treatment after early detection. On the other hand, by measuring a marker close to the disease itself, the possibilities for intervention may be limited, or nonexistent. Those diseases for which early detection has little impact upon the ultimate disease outcome, for example, lung cancer, early detection will have little impact on the disease rate, although it may increase the time between detection of disease and death (According to the National Cancer Institute the 5-year survival rate for lung cancer has increased from 12.5% to 15% in the United States during the period 1974 to 1999. Worldwide survival rates are not available for the identical period, but the latest estimates appear to be about 13% 5-year survival.). Detection of early silicotic changes in the lung using a restrictive pattern of pulmonary function testing, often makes little difference as the disease is frequently progressive at this point. For preventive purposes, it is best to use biomarkers that identify that exposure is occurring and respond to exposure interventions. Positive measurements of internal dose could be used to accomplish this; however, it is not often practical to completely eliminate many occupational or environmental exposures. In these cases, it may be better to determine which persons have the potential for increased risk of disease due to their exposure and to determine metabolic factors that may predispose them to adverse outcomes. In this regard, an ideal biomarker would be one that measures a change associated mechanistically with the disease, but not inevitably so. A practical balance has to be decided upon in each case.

22.3 BIOMARKERS OF CARCINOGENESIS

Biomarkers for carcinogens have been developed for each of the levels given in Figure 22.1. These include measurement of metabolites of the carcinogens as internal dose markers, hemoglobin or DNA adducts as effective dose markers, mutation assays in surrogate genes, chromosomal aberrations, and micronuclei would be examples of early effect markers and mutations in oncogenes would be examples of preclinical effects. Other markers could be included for certain exposures. For example, the levels of sputum or urinary cotinine can be used to estimate the internal exposure to tobacco smoke [26].

Major issues to be considered when dealing with biomarkers of carcinogens are: the exposure window the biomarker opens; the relationship of the biomarker to disease; relationships between biomarkers at different levels; biomarker specificity; and the use of surrogate markers and tissues for target organ exposure.

22.3.1 CANCER BIOMARKERS AND THEIR EXPOSURE WINDOW

Cancer is a disease with a latency period, that is, it occurs generally 5 to 50 years after the first significant exposure; it very often occurs after exposure has ceased. The investigator who would examine the effects of an exposure that occurred long before the onset of a disease faces several potentially serious problems. On one hand, most exposure and effect biomarkers do not persist indefinitely when exposure stops. Persons who are no longer exposed may have background levels of biomarkers, but have a residual increased risk due to damage that accrued during the exposure and was exacerbated by other unmonitored events. For example, ex-tobacco smokers have about two times more the risk of urinary bladder cancer than do never-smokers, yet the levels of DNA adducts in urinary bladder biopsy samples of ex-smokers was no higher than that of never-smokers when cessation was started 5 years before the sampling [27]. It also appears now that lung cancer is being increasingly diagnosed in ex-smokers as well. To interpret the data, investigators must know the character of the exposure because the window on the disease provided by the biomarker is a function of the biomarker's biological half-life ($t_{1/2}$) and the type of exposure. On the other hand, the critical mutations that do persist are in cells that probably have similar behavior to the "sea" of normal cells in a tissue and thus are very difficult to detect. After all, except for relatively few mutations, these cells share the same genome as the other cells and the same phenotype for most tissue traits. This presents a particularly unprickly "needle in the haystack" problem.

With a model having but one level of exposure, and no biomarkers, prior to disease, it was impossible to define precisely the latency period, except to say that it is a time when the cancer lies somehow hidden within a tissue. However, DNA adduction and oncogene mutation studies have indicated that the latency period is the time when the potential tumor cells acquire DNA damage and mutational changes necessary for initiation to occur (vide supra). Considering the pharmacokinetic factors that limit the distribution of activated forms of carcinogens to target sites along with the bullet–target probabilities discussed above, it would appear that most of the latency period is required to attain the fully initiated phenotype. Latency is shortened if the individual is heterozygous for a putative oncogene as in the case of retinoblastoma or Wilm's tumor [28], or in individuals exposed to very high doses of the toxicant. However, the accumulation of mutations in cells may not be strictly independent: cells with one mutation in an oncogene might be more responsive to a promoting stimulus and these cells may have growth advantage and be more likely to be mutated a second time. In this way initiation and promotion might interact and increase the probabilities of the second and subsequent hits. It would seem that as long as individuals are exposed, their tissues will become increasingly mutagenized, thus increasing the probability of neoplasia.

If what is said above is true, one would expect that within an exposed tissue there would be clones of cells that have one, two, or more different mutations. The concept of "field cancerization," which is based upon the notion that as exposure increases, many, or most of the cells in a target tissue amass DNA damage and mutations, is supported by observations of multiple independent tumors seen at autopsy, and by recurrences of independent tumors following successful treatment. Finally, it has been recently shown that mutations are indeed widespread in normal appearing tissues of lung cancer patients [29]. Although cancer is an all or nothing disease, it is most likely to advance in discrete stages as mutations accrue. Field cancerization provides the most important impetus for the use of biomarkers in primary, that is, exposure prevention.

The biological half-life ($t_{1/2}$) of a marker must be considered from both the accumulation of the marker levels during the initial phases of exposure and the decline following exposure cessation. The $t_{1/2}$ values of biomarkers can vary greatly. The parent compounds generally persist as such for periods of several minutes to a few hours after the end of an exposure. Exceptions include compounds that have very high affinity for adipose tissue (e.g., TCDD, PERC). Metabolites being more water soluble generally have $t_{1/2}$s on the order of hours [8]. Xenobiotic metals tend to persist longer, particularly when they mimic essential elements, as lead does with calcium, for example [30]. The $t_{1/2}$ of compounds covalently bound to protein or DNA can be considerably longer than that of the parent compound or unbound, excretable metabolites, often being identical to the half-life of the protein or the tissue in which the binding occurs. Albumin has a $t_{1/2}$ of approximately 20–25 days, hemoglobin adducts may persist following the kinetics of hemoglobin, about 100–120 days. There appear to be two or three compartments for DNA adducts in tissues such as skin. Some adducts are rapidly cleared as if they were being repaired, while others appear to be repair resistant and persist as long as the cell survives. With chronic exposure, the slowly removed compartment becomes predominant. For example, we have recently shown that the $t_{1/2}$ of BAP-DNA adducts in the skin is eight times longer following a chronic exposure than with a single dose [31,32]. The $t_{1/2}$ of DNA adducts in nondividing tissues may be extremely long and contribute to the accumulation of very high levels of damage. Randerath and coworkers [33,34] reported that the highest levels of carcinogen-DNA adducts in human postmortem samples were in the heart and aortas. The biological

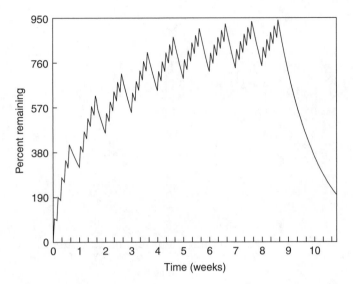

FIGURE 22.2   Elimination kinetics of a marker with a half-life of 24 h. The model assumes that exposure is uniform during the day for 8 h per day, 5 days per week.

relevance of these adducts is not known. Since heart muscle does not divide, mutations cannot be fixed; and, the rate of adult heart neoplasms is very low. On the other hand, there is some suggestion that other forms of heart disease may be related to the level of DNA damage [34,35].

Biomarker persistence should be considered in every study. The first consideration should be for the biomarker accumulation phase, to be certain that a representative level is being reported. Metabolites with a biological half-life of 12 h or greater may accumulate during the course of the workweek to levels higher than after a single dose. Figure 22.2 shows the excretion kinetics of a metabolite with a 24-h half-life during three workweeks, assuming five uniform, daily exposures per workweek. An end of shift sample on Thursday or Friday would have almost twice the level of the biomarker as a similar sample taken on Monday. Also note that there is no week-to-week accumulation of the marker after the first week. Biomarkers with longer half-lives have significant week-to-week accumulation. Figure 22.3 indicates that with a similar uniform exposure, the urinary levels of a biomarker with a $t_{1/2}$ of 1 week would not reach steady-state levels until 8 weeks of exposure assuming uniform exposure 8 h per day, 5 days per week. At steady state, the level of the biomarker would be about 900% higher than after the first exposure. Once steady-state has been reached with biomarkers having a long half-life, day-to-day variations in exposure have a lesser impact on biomarker levels than with biomarkers having a shorter $t_{1/2}$; for a marker with a 12-h half-life Monday post-shift levels will be 88% of those post-shift Friday compared with 64% for a marker with a 24-h $t_{1/2}$. Biomarkers with very long $t_{1/2}$s may take years of exposure to reach steady state. Using the general rule that steady state is reached after nine $t_{1/2}$s under these exposure conditions, it will take 36 months of exposure to reach a plateau. On the other hand, the effects of vacations and other variations in exposure will have smaller effects when $t_{1/2}$ is that long; biomarker levels will be 92% of their pre-vacation values upon return from a 2-week hiatus, for example.

22.3.2   VALIDATION OF CANCER BIOMARKERS

The validation of biomarkers for use in human exposure is performed on several levels. The most basic is the demonstration of a dose response in animals and humans. The next level of study design involves the use of multiple biomarkers as well as a traditional exposure assessment. In human studies, exposures are measured by questionnaire (e.g., how many cigarettes did you smoke each

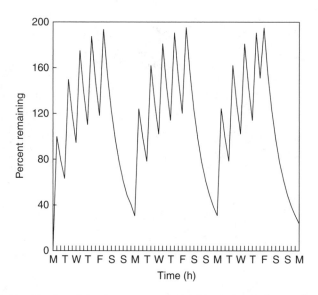

FIGURE 22.3 Elimination kinetics of a marker with a half-life of 196 h. Similar assumptions regarding exposure are made as in Figure 22.2.

day?), air sampling, or using validated biomarkers of internal dose. These latter studies are becoming increasingly common and more informative. Ultimately, the final validation is the relationship between the marker and the disease in populations without interventions. These are discussed first.

Most scientists would agree that markers such as DNA adducts are informative of risk [21]. The same may not be true at the legal and managerial levels of companies and agencies that manufacture or regulate carcinogens [1,2]. A first level of proof answers questions like "are the biomarkers positive and negative where they might be anticipated to be?" Cigarette smokers who are at increased risk for lung and urinary bladder cancer should have higher levels of carcinogen-DNA adducts in these tissues than do nonsmokers. This has been demonstrated in several studies [27,36]; adduct levels are significantly higher in the target organs of smokers and the difference between these groups is proportional to the increased risk of cancer. However, unambiguous proof of this type of relationship in humans requires a prospective model and a dose response to answer questions like "are people with higher levels of adducts more likely to get cancer?" Groopman and coworkers [37] have gone a long way to answer this question in their prospective study of excreted aflatoxin B_1 deoxyguanine in urine samples of persons with and without liver cancer in an area of endemic exposure to this compound. They also uncovered a striking synergy among the DNA adducts, hepatitis B infection, and liver cancer.

Animal studies have also clearly validated several carcinogen biomarkers. Adducts and chromosomal aberrations were correlated in target organs of treated mice [38]. Similar findings were reported for mutation frequency and adducts [39,40]. Taken separately, these data would not be sufficient to support the notion that DNA adducts are related to carcinogenesis. As a mass of evidence, they are much more powerful reinforcement of the notion.

Biomarkers should also be internally consistent, they should respond to changes in dose and be correlated with each other if there is reason for interaction. We conducted a study several years ago that was the first application of the analysis of carcinogen-DNA adducts in exfoliated urothelial cells [15]. Specific DNA adduct levels were higher in smokers than nonsmokers, but the differences were not statistically significant. However, when only the smokers were analyzed, there was a significant correlation between the number of cigarettes per day and the levels of three adducts. One of these adducts co-chromatographed with N-(deoxyguanosin-8-yl)-4-aminobiphenyl. This adduct was also well correlated ($r = 0.601, p = 0.01$) with the levels of 4-ABP hemoglobin adducts measured in each smoker. In addition, the levels of this adduct and of total adducts were significantly correlated with the measured urinary mutagenicity. These data suggested that there was an association between smoking and the adduct level in exfoliated urothelial cells and gave this assay some promise for use in human studies. The results also suggested that the problem we had was likely in the levels of background activity in the nonsmokers' samples. We have since focused on decreasing the background levels. We conducted another study more recently measuring DNA adduct levels in the exfoliated urothelial cells of workers exposed to benzidine and benzidine-based dyes.

An epidemiological study by Bi et al. [41] found that on average similarly exposed workers in China were from 5 to over 160 times more likely to develop urinary bladder cancer than were controls, depending on exposure category. On average these workers were 25 times more likely to develop urinary bladder cancer. The initial reading of the data from our biomarker study indicated that there was a 12-fold increase in the specific, acetylated benzidine-DNA adduct in the exfoliated urothelial cells of workers exposed to benzidine or benzidine-based dyes. However, when the exposed workers were analyzed on the basis of the fact that they either worked with benzidine itself or benzidine-based dyes and no benzidine, a more dramatic effect was seen. Adduct levels were twofold higher, marginally, but significantly different in the workers exposed to benzidine-based dyes and the controls. However, the benzidine-exposed workers had levels of the specific benzidine adduct that were on average 24 times higher than the controls. When the individual data points were examined, the range of values indicated that there were individuals who had only slightly elevated adduct levels, some who were in the middle, and others with adduct levels from 50 to 200 times that of the average in the controls. Thus, the average and the range of values we saw in the carcinogen-DNA adduct study were very similar to that seen in the earlier epidemiological study of workers with similar exposure. With such a range of results the possibility that individuals could be misclassified as to the extent of their exposure and potential effects existed if they were only categorized as "exposed" or "nonexposed." However, we found that there was a significant correlation between level of excreted benzidine urinary metabolites and the levels of the specific BZ-DNA adduct ($r = 0.68$, $p = 0.00001$). In this group of workers, recent exposure (measured by the relatively transient metabolites) predicted effect, suggesting that in these persons the daily exposure was likely fairly regular. And, because the adduct levels and urinary metabolite levels were congruent, these data also indicated that the distribution of the adduct levels we saw in the exposed group was due to differences in exposure. Again, it is also very interesting that not only the mean values, but also the range of adduct values corresponded to the excess risk of urinary bladder cancer reported by Bi et al. [41] in other workers with similar exposures. DeMarini et al. [42] measured the levels of excreted mutagens in the urine of the same workers and reported that urine mutagenicity was significantly elevated in the exposed workers and correlated with both the level of BZ-DNA adducts ($r = 0.59$, $p = 0.0006$) and excreted urinary metabolites ($r = 0.88$, $p = 0.0001$). The fact that the urinary mutagenicity is better correlated with the metabolite levels than with the DNA adduct levels is not surprising because the former are biomarkers with the same $t_{1/2}$ and in fact may be the same material, reflecting more recent exposure, where the DNA adduct biomarker has a much longer $t_{1/2}$ and the reported levels are the result of cumulative exposure to the urothelium and include factors such as DNA repair. That the measurement of the urinary metabolites and mutagenicity are closely related is reflected in the very similar correlations that each has with the levels of adducts (0.68 and 0.59, respectively). Thus, it would seem from these data that on average current exposure measured by either of these two biomarkers accounts for about 40% of the variability on the longer term biomarker, the specific DNA adducts. The fact that about 60% of the variance of the adduct data is not explained by the measurement of the urinary metabolites indicates that adduct measurements are necessary to predict effects.

Van Schooten and his group have taken a similar approach in the validation DNA adduct levels in induced sputum as an noninvasive biomarker of exposure to lung carcinogens. They reported that levels of diagonal zone DNA adducts were related in the induced sputum and in peripheral blood lymphocytes of smokers. The levels in the sputum-derived cells responded more quickly to changes in exposure cessation [43]. Response differences were thought due to differences in life span of the cells in the induced sputum sample. About 70% of the total cells in the sample are relatively short-lived macrophages, as compared with the peripheral blood leukocyte (PBL), which includes a subpopulation of long-lived cells (see below). Validation of DNA adduct biomarkers with mutation rates was reported by Hou et al. [44] (1999) who saw a significant correlation between DNA adduct levels and hgprt mutations in peripheral blood lymphocytes.

Useful markers also provide insight into the biology of exposure and effect. It has been largely through the utilization of biomarkers that the importance of the skin as a major route of PAH exposure was documented [8,45]. Biomarkers have also been used to show how physical and chemical factors such as body mass index and urinary pH may be as important as genetic variation in modifying the response to an exposure [46,47].

22.3.3 Specificity of Cancer Biomarkers

Biomarker specificity is often critical in data interpretation. Biomarker specificity to the disease process (the epidemiological sense) was discussed earlier. The chemical specificity of a biomarker is whether it reflects effects of one or a group of agents; that is, are there multiple pathways leading to the biomarker? As an example, cancer is a multifactorial disease caused by radiation, asbestos, wood dust, aromatic amines, PAHs, and a variety of other chemical and physical agents. Other biological process such as chromosome breaks, mutations, and inhibited or induced enzymes are also multifactorial and their measurement is nonspecific. Chemical biomarkers are more specific if they offer satisfying proof of chemical identity. Specific techniques or the potential for artifacts increase the potential for ambiguity. For example, the technique used by Skipper, Tannenbaum, and coworkers for the analysis of 4-ABP hemoglobin adducts involves the hydrolysis of ABP from the Hb molecule followed by derivatization and analysis by GC-MS [48]. Among their findings were puzzling increases in the adduct levels in humans and animals thought to be unexposed to the compound [49]. These workers realized that although the MS data was sufficient to demonstrate ABP, the compound could come from ABP-Hb or from any ABP contamination during the procedure. Realizing this, they undertook heroic measures to preclude contamination for their findings. In some cases absolute chemical identity is or was simply not possible using existing analytical methods and the weight of best possible evidence must be used. For example, using ^{32}P-postlabelling, we described the presence of a specific 4-ABP-DNA adduct in the bladders of tobacco smokers [27]. Being a thin layer chromatography based system, ^{32}P-postlabelling does not have a high chemical specificity as other adducts could conceivably share chromatographic identity in the systems. An authentic standard was available for this adduct and we were able to show that the putative adduct co-chromatographed with the standard in several TLC systems. In addition, the putative adduct was eluted from the thin layer and injected into an HPLC. Again, co-chromatography with the standard was demonstrated. Nonetheless, some uncertainty remained as to whether a 4-ABP adduct was specified. However, findings of 4-ABP adducts in the human urothelium has been corroborated by MS and immunohistochemical techniques, which were at the point of the original publication too insensitive to be used on these samples [50]. Combining analytical approaches increased the specificity of the ^{32}P-postlabelling method and the weight of evidence now indicates that the findings were valid.

Specificity is also an issue regarding surrogate tissues that are used to estimate effects in the target. It is well- recognized that the target for acute poisoning with organophosphate pesticides are the acetylcholinesterases of the CNS [5]. The biomarker that is measured to estimate the extent of exposure is the inhibition of an acetylcholinesterase isozyme in the RBC. The fact that obtaining blood samples is relatively easy and noninvasive favors the use of a surrogate tissue. There is now a large accumulation of data that generally supports the association between the marker, and acute effects for many pesticides supports the use of the surrogate tissue in this case for this effect. Conversely, the ability of RBC acetylcholinesterase inhibition to predict chronic or cumulative effects of these exposures appears limited [5]. Apparently inhibition of another isozyme results in the chronic effects and the two inhibitions are less closely tied. Surrogate tissues are often used for cancer biomarkers for the same reasons. Samples from target tissues such as lung, liver or urinary bladder are thought to be difficult to obtain without invasive procedures. Blood is plentiful and relatively easy to obtain and contains RBCs, which contain hemoglobin, which has been shown to covalently react with carcinogens and white blood cells that contain DNA. These tissues have been used in a variety of studies of environmental and occupational exposures.

Because cancer biomarkers are often not measured in the target organ affected by environmental agents, another level of specificity with regard to how the biomarker measured in a surrogate tissue or material relates to changes in the target organ. Hemoglobin and hemoglobin adducts are surrogates for DNA and DNA adducts for example. Hemoglobin contains nucleophilic sites such as sulfur, oxygen, and nitrogen, which are capable of reacting with electrophilic carcinogens. It has been shown that certain cysteinyl residues on hemoglobin are very reactive to certain carcinogens and hemoglobin itself participates in the activation of 4-ABP [48]. Hemoglobin is also abundant; it circulates in the central compartment and is relatively easy to obtain by venipuncture. There are disadvantages to using hemoglobin as a surrogate tissue. Hemoglobin is not the genetic material, so the significance of measuring this "effect" does not have the same biological support as measuring binding to DNA. The nucleophilic sites on hemoglobin have different reactivities as those on DNA, and electrophiles are likely to react with them differently than they do with the nucleophilic sites on DNA.

Peripheral blood leukocytes have also been used as surrogate tissues for monitoring target organ effects. They resolve the initial disadvantages of measuring hemoglobin adducts in that they contain DNA. They are also plentiful and relatively easy to obtain. Potential disadvantages for the use of leukocytes include the fact that leukocyte metabolism may be quite different than that of the target organ. Leukocytes also include several populations of cells with greatly differing $t_{1/2}$. The majority of the granulocytic fraction of blood leukocytes have a $t_{1/2}$ of 1 or 2 days, while the $t_{1/2}$ of some T-lymphocytes is several years. Granulocytes may comprise 40 to 70% of the total leukocytes in blood. When the exposure rate is constant the effect of the different $t_{1/2}$ is minimal; biomarkers resulting from daily habits such as tobacco smoking would not be greatly impacted. The effects of occupational exposures would be more dramatic with the level of adducts in total leukocytes expected to decline by 50% from Friday night to Monday morning assuming a granulocytic fraction of 60% and a $t_{1/2}$ of 24 h. Savela and Hemminki [51] and Weinike et al. [36] have shown that there are significant differences in the adduct levels in these cell types.

Whether or not a particular surrogate cell or molecule can be used to monitor an exposure depends directly on the proportionality between the levels of the biomarker in the surrogate and target tissues. Unfortunately very few studies have investigated this relationship. Already mentioned are the results of our study of carcinogen-DNA and hemoglobin adducts in tobacco smokers; we saw a significant relationship between the specific 4-ABP adducts in samples from each person. We have also examined the leukocyte and exfoliated urothelial cell-DNA adduct levels in our study of benzidine-exposed workers [52]. We saw a striking correlation between the two markers in each person ($r = 0.798$, $p = 0.001$). These data suggest that leukocytes may be useful as a surrogate tissue for the urinary bladder for benzidine exposure. This finding was quite surprising because we and others earlier reported a low correspondence between DNA adduct levels on leukocytes and the target tissues for 4-ABP. The findings were also surprising because it appears that the urinary pH level plays a significant role in the level of DNA adduct in the urinary bladders of the workers [47]. It is difficult to imagine how similar factors could come into play in leukocytes. On the other hand, these data suggest that for a 6-day per week, 8 to 10 h per day exposure of these workers, the total leukocyte levels were sufficient to predict the target organ effect and that no leukocyte fractionation is necessary.

Godschalk et al. [53] studied the relationship between the DNA adduct levels in PBL and various target organs in rats treated with benzo[a]pyrene by different routes. They reported significant correlations between the levels in many cases regardless of the route of exposure. Thus, it does appear that DNA adduct levels in PBL are predictive of target organ adduct levels, for these two compounds, at least.

One emergent and most controversial area of genetic toxicology is the potential impact of individual susceptibility markers on effect biomarkers and, ultimately, cancer risk. Some research has indicated that there may be a 20 to 60-fold variability in the human population for some metabolic activities [54]. One of the first-reported susceptibility markers is the acetylator phenotype and its impact upon the toxicity of several drugs and the carcinogenicity of aromatic amines [55]. In the literature there is a consistent increase in the rate of urinary bladder cancer in slow acetylators [55]. Although there are exceptions and extreme values reported in certain cases, it appears that there is about a twofold excess urinary bladder cancer risk associated with the slow phenotype. We have conducted two studies in which we looked for the effect of this phenotype on the biomarker results. Slow acetylators were about two times as likely to be positive for the 4-ABP DNA adduct in their exfoliated urothelial cells as were fast acetylators, and, the levels of 4-ABP hemoglobin adducts were significantly elevated in slow acetylators [56]. These data are consistent with what is known for the activation of 4-ABP; acetylation is a detoxification pathway that reduces the amount of N-oxidation of 4-ABP [57]. We also examined acetylator phenotype (and genotype) in the study of benzidine-exposed workers. We found that there was no difference in the levels of BZ-DNA adducts in the exfoliated urothelial cells by acetylator phenotype. These data were also consistent with what is known about the activation of BZ and the epidemiology of specific BZ exposure. Although some reports suggested that slow acetylators exposed to benzidine might be at increased risk, there was strong evidence that exposures may have been mixed with other monofunctional aromatic amines. Hayes et al. [58] reported that there was no increase in urinary bladder cancer in workers exposed exclusively to benzidine. Benzidine metabolism (and perhaps that of other di-amines) is unusual in that acetylation appears to be required for N-oxidation to take place [59]. Our work confirmed the importance of acetylation in the activation of this compound because only acetylated benzidine adducts were detected. These findings together make it appear that the impact of the slow acetylator phenotype is substrate, that is, exposure, specific. Thus, no gross generalizations of risk can be made without taking into account the complete exposure picture.

We recently worked with the laboratory of Dr. Daniel W. Nebert to define more precisely the role of specific CYP enzymes in activation of model compounds by knocking the *CYP 1A1* and *CYP 1A2* genes in mice exposed to benzo[a]pyrene and 4-aminobiphenyl, respectively. These enzymes were shown, *in vitro*, to have the greatest activity with these respective substrates. In the scientific culture, they became virtually synonymous with activation of these materials. In studying these materials in the respective knockout mice, we were surprised to learn that ablating these genes [60] had very little effect on the DNA adduct levels [61,62].

These data further underscore the need to reevaluate the impact of metabolic "susceptibility" factors. A reevaluation was also recently suggested by Garte [63].

22.4 SURROGATE BIOMARKERS FOR CARCINOGENS IN COMPLEX MIXTURES

22.4.1 1-HYDROXYPYRENE AND BAP AND PYRENE IN COMPLEX CARCINOGENIC MIXTURES

The basic premise of this chapter has been that there is a need for tools to help occupational health professionals to control human exposures to PAHs in complex mixtures containing them. The research community favors biomarkers that may be more specific for the disease process (see Figure 22.1), and provide more information relative to genetic variability. Unfortunately, the status of markers such as DNA and protein adducts is still judged to be experimental and these tests are often very expensive [64]. On the other hand, the occupational health practitioner may be better served by a biomarker that simply indicates exposure intensity. This is predicated the guiding principle of industrial hygiene, namely, "protect the health of the worker" and prevent occupational disease. A useful tool for a practitioner is one that indicates uptake and exposure routes. A marker of internal dose fits this need. The BEIs committee of the ACGIH recently proposed that 1-hydroypyrene (1HP) be used to help understand and control exposure to PAH mixtures [9]. The committee did not find sufficient data to justify a numeric value for the BEI based on the TLV for coal tar pitch, or on health effects. However, it did determine that end-of workweek, post-shift levels of 1 HP $> 1\mu g/l$ urine were evidence of occupational exposure, based upon the distribution of 1HP levels in the nonoccupationally exposed population.

The selection of pyrene as the compound to monitor is interesting because pyrene is not carcinogenic. Rather 1HP levels are considered surrogates used to estimate the uptake of total PAH during an exposure. 1HP was selected as a surrogate for several practical reasons. First, pyrene is a relatively abundant component of almost all PAH-containing mixtures. Second, a very sensitive and robust analytical method based on fluorescence HPLC was developed by Jongeneelen et al. [8] with relatively minor modifications by several others [65]. The method is sensitive to at least one order of magnitude below 1 $\mu g/l$. Finally, there is an extensive literature relating 1HP levels to occupational exposure and potential confounders such as tobacco smoking, diet, and air pollution. The interpretation of the 1HP levels >1 $\mu g/l$ would also be fairly straightforward for the occupational health practitioner: "exposure is occurring, it is greater than the background in the general population, investigate possible sources and routes." The documentation of the BEI contains extensive information detailing methodology and how 1HP levels can be used in a program to differentiate between occupational, environmental, and dietary exposures, and between current and past occupational exposures, and how to adjust the recommended level based upon the relative composition of the mixture.

Probably the most important finding that has come out of the studies of 1HP levels in exposed workers has been the apparent impact of the dermal exposure route on total dose. Prior to the application of biomarkers to the study of uptake, it was generally assumed that the respiratory route was a predominant route of exposure in occupations such as coke oven workers, roofers, pavers, and electrode paste workers (for aluminum reduction electrodes), for example. However, more recent data using 1HP as the exposure gauge suggest that the skin may contribute at least as much to these exposures as does the respiratory route [45,66,67]. For example, VanRooij et al. [45] estimated that approximately 70% of the total pyrene dose in coke oven workers was via the dermal route. Subsequently, this group reported that affecting dermal exposure by such simple means as new gloves, laundered clothes, and washing of face and hands before and after breaks, reduced the 1HP levels by 37% in coke oven workers [66]. The same research group also studied creosote workers and estimated that simply changing the overalls worn by the workers reduced exposure by 35% [67]. Elovaara et al . [68] reported that the amount of 1HP in the urine was as much as 50 times greater than the amount of pyrene in breathing zone samples of these workers. Quinlan et al. [69] reported very similar findings in workers involved in coal liquication; clean clothes reduced exposure by about 50% and contamination on a skin pad under the clothes by 80%. LaFontaine et al. [70] reported that about 45 to 50% of the total absorbed dose in aluminum reduction plant workers was via the dermal route.

The major advantage of knowing that the dermal route makes such a significant contribution to total dose is that it is relatively easy to control PAH exposure by this route. Clothing that is discolored is obviously contaminated. One might wonder why workers would tolerate these exposures and work with contaminated clothing. The most likely explanation is they were probably never made aware that this was a problem. The low acute toxicity of PAH often belies the potential for chronic effects. The use of 1HP as a biomarker for exposure has another advantage of giving both the workers and the occupational hygienist a feedback regarding the efficacy of interventions. Quinlan et al. [69] proposed the use of 1HP in this fashion in devising strategies to control exposure in their coal liquification plant.

The question of risk invariably enters the picture when discussing biomonitoring for carcinogens. 1HP measures nicely skirt the issue; levels of the detoxified metabolite of a noncarcinogen are clearly related to exposure, not to risk. The water:octanol partition coefficients and other physicochemical properties of the PAHs indicate that carcinogens are being absorbed at the same time as pyrene. In addition, the level of 1 μg/l will be higher than found in 99% of the smokers without occupational exposure and it is well known that the highest consumers of tobacco products are at more than a tenfold excess risk of lung and other cancers [5]. However, it is also clear that tobacco smoke is different than most occupational PAH exposures in that the former involves co-exposure to a plethora of other toxicants that are likely to exacerbate the effects of the PAH in the lung. These important considerations were not lost on the BEI committee and the documentation also contains a similar discussion.

Biological monitoring will be an important component of an occupational or environmental health program to control exposure to PAHs. That multiple routes of exposure, diverse mixture composition, differences in dietary and other environmental exposures can be controlled by the measurement of an internal dose biomarker will be strong impetus to employ the marker in studies and occupational health practice. In addition, as more studies are done the relationship between the internal dose and effective dose and early effect markers and disease itself could well be established and improve out ability to prevent disease from exposure to these compounds.

ACKNOWLEDGMENT

This chapter is dedicated to the memory of a first rate man and cancer researcher, Roy E. Albert.

REFERENCES

1. http://www.epa.gov/.
2. http://www.cdc.gov/niosh/homepage.html.
3. http://www.osha.gov/.
4. National Institute for Occupational Safety and Health (NIOSH), NIOSH Pocket Guide to Chemical Hazards, U.S. Department of Health and Human Services, U.S. GPO, 1996.
5. American Conference of Governmental Industrial Hygienists, Documentation of the threshold limit values, 8th ed., ACGIH, Cincinnati, OH, 2002.
6. Tsuruta, H. (1990) Dermal absorption. In Biological Monitoring of exposure to industrial Chemicals, *American Conference of Governmental Industrial Hygienists*, Fiserova-Bergerova, V. and Ogata, M., Eds., Cincinnati, OH, USA, pp. 131–136.
7. Lee, K.-H., Lee, J., Ha, M., Choi, J.-W., Cho, S.-H., Hwang, E.-S., Park, C.-G., Strickland, P.T., Hirvonen, A., and Kang, D. (2002) Influence of polymorphism of GSTM1 gene on association between glycophorin a mutant frequency and urinary PAH metabolites in incineration workers. *J. Toxicol. Environ. Health*, 65, 355–363.
8. Jongeneelen, F.J., Anzion, R.B.M., and Henderson, P.Th. (1987) Determination of hydroxylated metabolites of polycyclic aromatic hydrocarbons in urine. *J. Chromatog.*, 413, 227–232.
9. American Conference of Governmental Industrial Hygienists, Documentation of the threshold limit values, Notice of intended changes, 2003, ACGIH, Cincinnati, OH. http://www.acgih.org/tlv/NIClist.htm.
10. Lundberg, A.S., Hahn, W.C., Gupta, P., and Weinberg, R.A. (2000) Genes involved in senescence and immortalization. *Curr. Opin. Cell Biol.*, 12, 705–709.
11. Beckman, K.B. and Ames, B.N. (1999) Endogeneous oxidative damage of mtDNA. *Mutat. Res.*, 424, 51–58.
12. American Cancer Society. *Cancer Facts and Figures, 2001*, American Cancer Society, Atlanta, GA, 1995.
13. Zavon, M., Hoegg, U., and Bingham, E. (1973) Benzidine exposure as a cause of bladder tumors. *Arch. Environ. Health*, 27, 1–7.
14. Rothman, N., Bhatnagar, V.K., Hayes, R.B., Kashyap, R., Parikh, D.J., Kashyap, S.K., Schulte, P.A., Butler, M.A., Jaeger, M., and Talaska, G. (1996) The impact of interindividial variability in *N*-acetyltransferase activity on benzidine urinary metabolites and urothelial DNA adducts in exposed workers, *Proc. Natl. Acad. Sci. USA*, 93, 5084–5089.

15. Bhatnagar, V.K. and Talaska, G. (1999) Carcinogen exposure and effect biomarkers (Invited Review). *Toxicol. Lett.*, 108, 107–116.

16. Opdam, J.J.G. and Smolders, J.F.J. (1986) Alveolar sampling and fast kinetics of tetrachloroethene in man. *Br. J. Ind. Med.*, 43, 814–824.

17. Lauwerys, R.R., Kivits, A., Lhoir, M., Rigolet, P., Houbeau, D., Buchet, J.P., and Roels, H.A. (1980) Biological surveillance of workers exposed to dimethylformamide and the influence of skin protection on its percutaneous absorption. *Int. Arch. Occup. Environ. Health*, 45, 189–203.

18. DeCaprio, A.P., Olajos, N.P., and Weber, P. (1982) Covalent binding of a neurotoxic n-hexane metabolite: conversion of primary amines to substituted pyrrole adducts by 2,5-hexanedione. *Toxicol. Appl. Pharmacol.*, 65, 440–450.

19. Maclure, M., Katz, R.B.-A., Bryant, M.S., Skipper, P.L., and Tannenbaum, S.R. (1989) Elevated blood levels of carcinogens in passive smokers. *Am. J. Publ. Health*, 79, 1381–1384.

20. Skipper, P.L. and Tannenbaum, S.R. (1994) Molecular dosimetry of aromatic amines in human populations. *Environ. Health Perspect.*, 102, 17–21.

21. Hemminki, K. (1993) DNA adducts, mutation and cancer. *Carcinogenesis*, 14, 2007–2012.

22. Brant-Rauf, P.W. and Pincus, M.R. (1998) Molecular markers of carcinogenesis. *Pharmacol. Ther.*, 77, 135–148.

23. Lee, J.-H., Roh, J.H., Burks, D., Warshawsky, D., and Talaska, G. (2000) Skin cleaning with kerosene facilitates passage of carcinogens to the lungs of animals treated with used gasoline engine oil. *Appl. Occup. Environ. Hyg.*, 15, 362–369.

24. Vermeulen, R., Kromhout, H., Bruynzeel, D.P., de Boer, E.M., and Brunekreef, B. (2001) Dermal exposure, hand washing and hand dermatitis in the rubber manufacturing industry. *Epidemiol.*, 12, 350–354.

25. Nebert, D.W., McKinnon, R.A., and Puga, A. (1996) Human drug metabolizing enzymes polymorphisms: effects on risk of toxicity and cancer. *DNA Cell Biol.*, 15, 273–280.

26. Etzel, R. (1990) A review of the use of saliva cotinine as a marker of tobacco smoke exposure. *Prev. Med.*, 19, 190–197.

27. Talaska, G., Al-Juburi, A.Z.S.S., and Kadlubar, F.F. (1991) Smoking-related carcinogen-DNA adducts in biopsy samples of human urinary bladder: Identification of N-(deoxyguanosin-8-yl)-4-aminobiphenyl as a major adduct. *Proc. Natl. Acad. Sci. USA*, 88, 5350–5354.

28. Yunis, J.J. and Ramsay, N. (1979) Retinoblastoma and deletion of chromosome 13. *Am. J. Dis. Child*, 132, 161–163.

29. Sozzi, G., Miozzo, M., Pastorino, U., Pilotti, S., Donghi, R., Giarola, M., DeGregorio, L., Manenti, G., Radice, P., Minoletti, F., Della Porta, G., and Pierotti, M.A. (1995) Genetic evidence for an independent origin of multiple preneoplastic and neoplastic lung tissues. *Cancer Res.*, 55, 135–140.

30. Mahaffey, K.R. (1977) Quantities of lead producing health effects in humans: Sources and bioavailability. *Environ. Health Perspect.*, 19, 285–298.

31. Talaska, G., Jaeger, M., Reilman, R., Collins, T., and Warshawsky, D. (1996) Chronic, topical exposure to benzo[a]pyrene induces relatively high steady state levels of DNA adducts in target tissues and alters kinetics of adduct loss. *Proc. Natl. Acad. Sci. USA*, 93, 7789–7793.

32. Albert, R., Miller, M., Talaska, G., Underwood, P., Cody, T., and Andringa, S. (1995) Epidermal cytokinetics, DNA adducts and dermal inflammation in the mouse in response to repeated benzo(a)pyrene exposures. *Toxicol. Appl. Pharmacol.*, 136, 67–74.

33. Randerath, E., Miller, R.H., Mittal, D., Avitts, T.A., Dunsford, H.A., and Randerath, K., (1989) Covalent DNA damage in tissues of cigarette smokers as determined by ^{32}P-postlabelling assay. *J. Natl. Cancer Inst.*, 81, 341–346.

34. Randerath, E., Mittal, D., and Randerath, K. (1988) Tissue distribution of covalent DNA damage in mice treated dermally with cigarette 'tar': preference for lung and heart DNA. *Carcinogenesis*, 9, 75–80.

35. Albert, R.E., Vanderlaan, M., Burns, F.J., and Nishizumi, M. (1977) Effects of carcinogens on chicken atherosclerosis. *Cancer Res.*, 37, 2232–2235.

36. Wiencke, J.K., Kelsey, K.T., Varkonyi, A., Semey, K., Wain, J., Eugene, M., and Christiani, D.C. (1995) Correlation of DNA adducts in blood mononuclear cells with tobacco carcinogen-induced damage in human lung. *Cancer Res.*, 55, 4910–4914.

37. Ross, R.K., Yuan, J.-M., Yu, M.C., Wogan, G.N., Qian, G.-S., Tu, J.-T., Groopman, J.D., Gao, Y.-T., and Henderson, B.E. (1992) Urinary aflatoxin biomarkers and risk of hepatocellular carcinoma. *Lancet*, 339, 943–946.

38. Talaska, G., Au, W.W., Ward, J.B. Jr., Randerath, K., and Legator, M.S. (1987) The correlation between DNA adducts and chromosomal aberrations in the target organ of benzidine exposed, partially-hepatectomized mice. *Carcinogenesis*, 8, 1899–1905.

39. Perera, F.P., Dickey, C., Santella, R., O'Neill, J.P., Albertini, R.J., Ottman, R., Tsai, W.Y., Mooney, L.A., Savela, K., and Hemminki, K. (1994) Carcinogen-DNA adducts and gene mutation in foundry workers with low level exposure to polycyclic aromatic hydrocarbons. *Carcinogenesis*, 15, 2905–2910.

40. Nesnow, S., Ross, J.A., Nelson, G., Wilson, K., Roop, R.C., Jeffers, A., Galati, A.J., Stoner, G.D., Sangaiah, R., Gold, A., and Mass, M. (1994) Cyclopenta[cd] pyrene induced tumorigenicity, Ki-*ras*, codon 12 mutations and DNA adducts in strain A/J mouse lung. *Carcinogenesis*, 15, 601–606.

41. Bi, W., Hayes, R.B., Feng, P., Qi, Y., You, X., Zhen, J., Zheng, M., Qu, B., Fu, Z., Chen, M., Chein, H.T., and Blot, W.J. (1992) Mortality and incidence of urinary bladder cancer in benzidine exposed workers in China. *Am. J. Ind. Med.*, 21, 481–489.

42. DeMarini, D.M., Brooks, L., Bhatnagar, V.K., Hayes, R.B., Eischen, B.T., Shelton, M.L., Zenser, T.V., Talaska, G., Kashyap, S.K., Dosemeci, M., Kashyap, R., Parikh, D.J., Lakshmi, V., Hsu, F., Davis, B.B., Jaeger, M., and Rothman, N. (1997) Urinary mutagenicity as a biomarker in benzidine-exposed workers: correlation with urinary metabolites and urothelial DNA adducts. *Carcinogenesis*, 18, 981–988.

43. Besarati Nea, A., Maas, L.M., Brouwer, E.M.C., Kleinjans, J.C.S., and Van Schooten, F.J. (2000) Comparison between smoking-related DNA adduct analysis in induced sputum and peripheral blood lymphocytes. *Carcinogenesis*, 21, 1335–1340.

44. Hou, S.-M., Yang, K., Nyberg, F., Hemminki, K., Pershagen, G., and Lambert, B. (1999) Hprt mutant frequency and aromatic DNA adduct level in non-smoking and smoking lung cancer patients and population controls. *Carcinogenesis*, 20, 437–444.

45. Van Rooij, J.G.M., Bodelier-Bade, M.M., Hopmans, P.M.J., and Jongeneelen, F.J. (1994) Reduction of urinary 1-hydroxypyrene excretion in coke oven workers exposed to polycyclic aromatic hydrocarbons due to improved hygienic skin protective measures. *Ann. Occup. Hyg.*, 38, 247–256.

46. Godschalk, R.W.L., Feldker, D.E.M., Borm, P.J.A., Wonters, E.F.M., and Van Schooten, F.J. (2002) Body mass index modulates aromatic amine DNA adduct levels and their persistence in smokers. *Cancer Epidemiol. Biomarkers Prev.*, 11, 790–793.

47. Rothman, N., Zenser, T., and Talaska, G. (1997) Acidic urine pH is associated with elevated levels of free urinary benzidine and *N*-acetylbenzidine and urothelial cell DNA adducts in exposed workers. *Cancer Epidemiol. Biomarkers Prev.*, 6, 1039–1042.

48. Skipper, P.L. and Tannenbaum, S.R. (1990) Protein adducts in the molecular dosimetry of chemical carcinogens. *Carcinogenesis*, 11, 507–518.

49. Skipper, P.L., Bryant, M.S., Tannenbaum, S.R., and Groopman, J.D. (1986) Analytical methods for assessing exposure to 4-aminobiphenyl based on protein adduct formation. *J. Occup. Med.*, 28, 643–646.

50. Hsu, T.-M., Zhang, Y.-J., and Santella, R.M. (1997) Immunoperoxidase quantitation of 4-aminobiphenyl and polycyclic aromatic hydrocarbon DNA adducts in exfoliated oral and urothelial cells of smokers and non-smokers. *Cancer Epidemiol. Biomarkers Prev.*, 6, 193–199.

51. Savela, K. and Hemminki, K. (1991) DNA adducts in leukocytes and granulocytes of smokers and nonsmokers detected by the ^{32}P-postlabelling assay. *Carcinogenesis*, 12, 503–508.

52. Zhou, Q., Talaska, G., Jaeger, M., Bhatnagar, V.K., Hayes, R.B., Zenser, T.V., Kashyap, S.K., Lakshmi, V.M., Kashyap, R., Dosemeci, M., Hsu, F.F., Parikh, D.J., Davis, B.B., and Rothman, N. (1997) Benzidine-DNA adducts levels in human peripheral white blood cells correlate significantly with levels in exfoliated urothelial cells. *Mutat. Res.*, 393, 199–205.

53. Godschalk, R.W.L., Moonen, E.J.C., Schilderman, P.A.E.L., Broekmans, W.M.R., Kleinjans, J.C.S., and Van Schooten, F.J. (2000) Exposure route dependent DNA adduct formation by polycyclic aromatic hydrocarbons. *Carcinogenesis*, 11, 87–92.

54. Autrup, H. (1990) Carcinogen metabolism in cultured human tissues and cells. *Carcinogenesis*, 11, 707–712.

55. Hein, D.W., Doll, M.A., Fretland, A.J., Leff, M.A., Webb, S.J., Xiao, G.H., Devanaboyina, U.-S., Nangju, N.A. and Feng, Y. (2000) Molecular genetics and epidemiology of the NAT1 and NAT2 acetylation polymorphisms. *Cancer Epidemiol. Biomarkers Prev.*, 9, 29–42.

56. Vineis, P., Caporaso, N., Tannenbaum, S., Skipper, P., Glogowski, J., Bartsch, H., Coda, M., Talaska, G., and Kadlubar, F. (1990) Acetylation phenotype, carcinogen-hemoglobin adducts, and cigarette-smoking. *Cancer Res.*, 50, 3002–3004.

57. Kadlubar, F.F. and Badawi, A.F. (1995) Genetic susceptibility and DNA adduct formation in human urinary bladder carcinogenesis. *Toxicol. Lett.*, 82, 627–632.

58. Hayes, R.B., Bi, W., Rothman, N., Broly, F., Caporaso, N., Feng, P., You, T., Yin, S., Woolsley, R., and Meyer, U. (1993) A phenotypic and genotypic analysis of *N*-acetylation and bladder cancer in benzidine-exposed workers. *Carcinogenesis*, 14, 675–678.

59. Frederick, C.B., Weiss, C.C., Flammang, T.J., Martin, C.N., and Kadlubar, F.F. (1985) Hepatic *N*-oxidation, acetyltransfer and DNA binding of the acetylated metabolites of the carcinogen, benzidine. *Carcinogenesis*, 6, 959–965.

60. Dalton, T.P., Dieter, M.Z., Matlib, R.S., Childs, N.L., Shertzer, H.G., Genter, M.B., and Nebert, D.W. (2000) Targeted knockout of Cyp1a1 gene does not alter hepatic constitutive expression of other genes in the mouse[Ah] battery. *Biochem. Biophys. Res. Commun.*, 267, 184–189.

61. Uno, S., Dalton, T.P., Shertzer, H.G., Genter, M.B., Warshawsky, D., Talaska, G., and Nebert, D.W. (2001) Benzo[a]pyrene-induced toxicity: Paradoxical protection in Cyp1a1($-/-$) knockout mice having increased hepatic BaP-DNA adduct levels. *Biochem. Biophys. Res. Commun.*, 289, 1049–1056.

62. Tsuneoka, Y., Dalton, T.P., Miller, M.M., Clay, C., Shertzer, H.G., Talaska, G., and Nebert, D.W. (2003) Liver and urinary bladder toxicity, oxidative stress, and adduct formation induced by 4-aminobiphenyl in *Cyp1a2*($-/-$) and *Cyp1a2*($+/+$) mice. JNCI 95, 1227–1237.

63. Garte, S. (2001) Metabolic susceptibility genes as cancer risk factors: Time for reassessment. *Cancer Epidemiol. Biomarkers Prev.*, 10, 1233–1238.

64. Dor, F., Dab, W., Empereur-Bissonnet, and Zmirou, D. (1999) Validity of biomarkers in environmental health studies: The case of PAH and benzene. *Crit. Rev. Toxicol.*, 29, 129–168.

65. Hansen, A.M., Christensen, J.M., and Sherson, D. (1995) Estimation of reference values for urinary 1-hydroxypyrene and alpha-naphthol in Danish workers. *Sci. Tot. Environ.*, 163, 211–219.

66. Boogaard, P.J. and van Sittert, N.J. (1994) Exposure to polycyclic aromatic hydrocarbons in petrochemical industries by measurement of urinary 1-hydroxypyrene. *Occup. Environ. Med.*, 51, 250–258.

67. Van Rooij, J.G., van Lieshout, E.M.A., and Bodelier-Bade, M.M. (1993) Effect of the reduction of skin contamination on the internal dose of creosote workers exposed to polycyclic aromatic hydrocarbons. *Scand. J. Work Environ. Health*, 19, 200–207.

68. Elovaara, E., Heikkila, P., Pyy, L., Mutanen, P., and Riihimaki, V. (1995) Significance of dermal and respiratory uptake in creosote workers: exposure to polycyclic aromatic hydrocarbons and urinary excretion of 1-hydroxypyrene. *Occup. Environ. Med.*, 52, 196–203.

69. Quinlan, R., Kowalczyk, G., Gardiner, K., and Calvert, I. (1995) Exposure to polycyclic aromatic hydrocarbons in coal liquefaction workers: impact of a work wear policy on excretion of urinary-hydroxypyrene. *Occup. Environ. Med.*, 52, 600–605.

70. LaFontaine, M.L., Gendre, C., Morele, Y., and Laffite-Rigaud, G. (2002) Excretion of urinary 1-hydroxypyrene in relation to the penetration routes of polycyclic aromatic hydrocarbons. *Polycyclic Aromat. Compd.*, 22, 579–588.

23 Risk Assessment for Chemical Carcinogens

Andrew G. Salmon

CONTENTS

23.1 INTRODUCTION

23.1.1 Risk Assessment Objectives: What is Risk Assessment?

Risk assessment is the use of scientific methods and data to predict the probability of a particular outcome (usually, one perceived as adverse) as a result of a particular set of measured or hypothesized starting circumstances. This concept has been applied to a wide range of problems in fields as diverse as engineering and economics. The present discussion is concerned with the use of risk assessment in assessing the impact of exposures to toxic chemicals, in which the ultimate purpose is to assist in identifying and mitigating impacts on public health.

23.1.2 Risk Assessment Methodology

The use of risk assessment techniques in considering the possible impact of exposures to carcinogens is now standard practice; the basic techniques have been widely discussed and available for two or three decades, and they have been routinely used by regulatory authorities such as the U.S. Environmental Protection Agency (US EPA) since the 1970s (Anderson et al., 1983). However, the scientific basis for risk assessment, particularly for carcinogens, is subject to substantial uncertainties, in both the underlying models and assumptions and in the actual values used for input parameters. Continuing scientific studies have reduced some of these uncertainties substantially, but they remain considerable (NRC, 1994). Even under the best possible circumstances, risk estimates should be regarded as "order of magnitude" estimates rather than precise numerical predictions.

Risk assessment is used as an objective method of arbitration between fundamentally opposed and competing interests, involving financial interests, at all scales from individual to macroeconomic on the one hand, and the lives and health of individuals, communities, and regional or global populations on the other. It is, therefore, essential to use formal risk assessment procedures, which have received scientific peer review and public comment (National Research Council [NRC], 1983, 1994). In the United States, regulatory authorities at both State (Office of Environmental Health Hazard Assessment [OEHHA], 1999b) and Federal (US EPA, 1986, 2003a) level have published cancer risk assessment guidelines to provide a common basis for the public debates and legal proceedings that commonly ensue. Some other jurisdictions have chosen a less open process with greater reliance on deliberation by expert committees in closed sessions. Such an approach often has the appearance of greater efficiency, but clearly depends on the willingness of the various parties to the social contract to agree on the composition and powers of the expert committees.

23.1.3 STATISTICAL ESTIMATES OF RISK: UNCERTAINTY AND VARIABILITY

The methods used are designed to make predictions of probability and address the outcome for a population, rather than an individual. It is almost impossible to say with certainty that a certain chemical exposure (or any other event) "caused" an adverse health outcome in an individual, such as the appearance of a tumor, or to make a specific prediction of the outcome of such an exposure for a particular individual. However, the epidemiologist uses statistical methods to examine the frequencies and associations of exposures and health outcomes and can draw plausible inferences as to the causal relationships between exposure or other initiating events and conditions, and the appearance of disease. The well-known criteria defined by Bradford Hill (1965, 1971) are used to evaluate the likelihood that an association observed in an epidemiology study represents a causal relationship. Similarly, when considering a population rather than an individual, the risk assessor can make statistical predictions as to the probability of certain outcomes, given a particular set of initiating circumstances.

Retrospective analysis by the epidemiologist and prediction of future events by the risk assessor are both prone to some degree of error. The outcome in any individual case is subject to both uncertainty (such as inaccurate measurement of initial and final states and lack of knowledge about the actual mechanisms connecting the two) and variability (when a particular individual responds to a given event or circumstance in a slightly different way to other members of the population) (Bailar and Bailer, 1999).

Both uncertainty and variability can be characterized to some degree in either a retrospective analysis or a predictive risk assessment (NRC, 1996). The variation of outcomes due to both uncertainty and variability can be described by a mathematical distribution. This is often used to determine whether the population studied is homogeneous with regard to its sensitivity to a cause of disease, or whether there are sensitive subpopulations within the overall population with greater or lesser sensitivity. This might be the result of having a variant form of the gene for a particular enzyme, for instance (e.g., see Ginsberg et al., 2002). It is important for the risk assessor to know whether the target population of the assessment includes either a high degree of variability overall or specific subpopulations with markedly different responses to the exposure. Predictions of risk assessments are often associated with a considerable degree of both variability and uncertainty. Some description of these is therefore usually included in the final assessment. Risks to public health are often discussed in terms of an upper confidence bound (usually a 95th percentile) on the risk estimate, as well as or in place of the mean, median, or maximum likelihood estimates. There are two distinct reasons for considering an upper confidence bound estimate. First, even if a very serious outcome (such as a high incidence of cancer) were not the most likely prediction, it would be prudent to know if it is at all plausible when devising strategies to protect public health against possible adverse consequences (Bailar and Bailer, 1999). Second, the statistical properties of some prediction methods result in some measures of central tendency in the risk estimate distribution being unstable, that is, relatively small changes in the input data can result in very large changes in this measure of the estimated risk (Crump et al., 1977). The upper 95% confidence level provides a more stable estimate and is therefore a better way of describing the overall character of the risk estimate, and it is a better basis for comparison of different materials or other elements of the risk assessment. This property applies to the dose–response estimation methodology conventionally used for carcinogens, as described later in this chapter.

The usual scientific approach to a problem involves a predictive or hypothesis-development phase and an experimental phase in which measurement and observation validate or negate the hypothesis. It is rare for this complementarity to appear in a risk assessment, because generally the objective of the process is either to establish that no significant adverse outcomes are expected, or to provide a justification for legal or remedial action to prevent the exposure. A complicating and ironic feature of the discipline is that the most successful risk assessments are those that can never be tested. Conversely, it is very important to take advantage of any available opportunities to validate a prediction, such as when the causative agent is a general environmental pollutant with significant population exposure. Many of the important methods and insights used in risk assessment derive from this type of investigation.

23.1.4 KEY STEPS IN RISK ASSESSMENT

The National Academy of Sciences/National Research Council (NRC, 1983) presented their analysis "Risk Assessment in the Federal Government: Managing the Process" in 1983. Although risk assessment methodology has developed in various ways since (NRC, 1994, 1996), that publication remains a critical landmark in the development and formalization of risk assessment procedures in the United States. The broad principles and procedures laid out in this publication remain the basis of most risk assessment activity in the United States at the Federal and State level.

Drawing on various earlier academic analyses, and on experience with risk assessment particularly within the US EPA and the U.S. Food and Drug Administration (FDA), this document provides a convenient summary of the key steps in risk assessment. Risk assessment is considered to be a stage in the overall process of evaluation and regulatory control of a hazardous material or situation. It is preceded by a phase of research and measurement to provide data as input to the assessment process and is followed by a risk management phase in which technical options for mitigation or elimination of the risk are considered along with other social and economic factors, leading to eventual practical and legal action. The key steps of the risk assessment process are hazard identification, exposure assessment, dose–response assessment, and risk characterization. These key steps are illustrated in Figure 23.1, and described in detail below, with particular reference to risk assessment for carcinogens.

23.1.5 RISK ASSESSMENT AND RISK MANAGEMENT

A key principle embraced by NRC (1983) is the concept of separation between risk assessment and risk management. This insulates risk assessment conclusions from influence by risk management issues such as cost, political feasibility, or pressure from special interests involved in the outcome. A risk manager can only make optimal decisions if provided with unbiased input data. Thus, it is often important to provide analyses such as technical feasibility and cost–benefit, but these should not affect the risk predictions provided by the risk assessment. Most environmental and public health regulatory programs respect this distinction by clearly separating the risk assessment and risk management roles. More recent regulatory experience has suggested that this separation has sometimes been taken too far, resulting in risk assessments that do not address the information needs of the risk managers charged with implementing their conclusions. A second report (NRC, 1994) was critical of then-current practice in this regard, and urged that the risk managers and risk assessors should be more closely related. In particular, the risk assessor needs to understand the needs and perspectives of the risk manager, and to thoroughly explain the basis and conclusions of the risk assessment. Recent risk assessment guidance documents have emphasized the importance of a thorough explanation of both the risk predictions, and the likely uncertainties therein, as part of the risk characterization step (US EPA, 2003a).

23.1.6 SCOPE OF A RISK ASSESSMENT PROJECT

A complete risk assessment includes characterization of the chemical agents, their toxicological properties, the exposed population, and the routes and extents of exposure. Usually however an individual project will not be so comprehensive. Most commonly a risk assessor working as a consultant for interested parties (exposing or exposed), or for the risk management arm of a regulatory agency, will conduct a site-specific risk assessment in which the emphasis is on identification of chemical agents, characterization of exposed individuals or populations, and measurement or modeling of exposure.

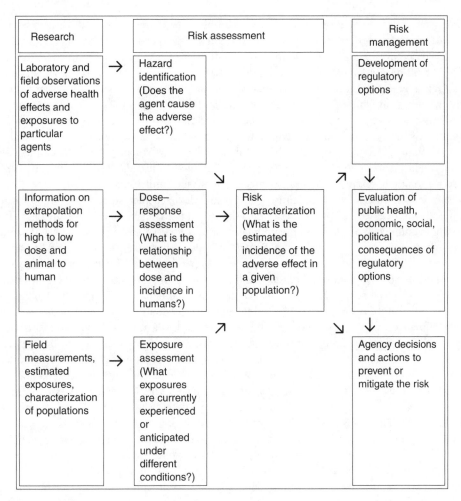

FIGURE 23.1 Elements of risk assessment and risk management (Adapted from National Research Council [1983] Risk Assessment in the Federal Government Managing the Process National Academies Press, Washington, DC, 191 pp.).

A risk assessor working for a regulatory standard-setting program, or as one of the academic or public interest advisors to such programs, may undertake a compound-specific risk assessment in which the emphasis is on understanding the toxicological impact and dose–response characteristics of a particular chemical, or commonly encountered mixture. In this case the exposure component of the assessment may be reduced to a generic default or a simple hypothetical model, with the emphasis on providing input on hazard and dose–response that can be incorporated by reference into site-specific assessments of actual exposure situations.

23.2 HAZARD IDENTIFICATION FOR CARCINOGENS

23.2.1 DATA SOURCES

23.2.1.1 Human Studies: Epidemiology, Ecological Studies and Case Reports

A risk assessment project aims to determine impact on human health. Therefore, the ideal hazard identification relies on studies in humans by which the nature and extent of the hazard may be demonstrated. In a few situations, such as clinical trials of experimental drugs that may benefit the subjects of the trial more than the current standard treatments, experiments in humans can be ethically justified. These have yielded valuable information on human response to pharmacological agents, including toxic responses as well as the intended effects. However, other experimental studies of toxic effects in human subjects are rarely undertaken and seldom justifiable. Purely pharmacokinetic volunteer studies, in which the doses used are well below the threshold for any toxic effect, are well established. Apart from this type of situation, the risk assessor will rely on epidemiological studies of existing occupational or environmental exposures. It is not intended in this chapter to provide a detailed account of epidemiological methods, but the student of risk assessment should refer to standard texts on epidemiology (e.g., Lilienfeld and Lilienfeld, 1980; Rothman, 1986) for an introduction to these methods and an appreciation of their limitations.

It is important when using epidemiological data to follow accepted standards of rigor in drawing inferences from observations of morbidity or mortality. It is easier to meet these standards in situations where both the exposures and the population are substantial and well defined, and accessible to direct measurement rather than anecdotal reports or recall. Thus, many important findings of carcinogenicity to humans are based on analysis of occupational exposures. Examples include benzidine (Meigs et al., 1986), vinyl chloride (Creech and Johnson, 1974), inhaled chromium VI (Mancuso and Hueper, 1951; Mancuso, 1975), coke oven emissions (Redmond, 1983), and diesel exhaust (e.g., Garshick et al., 1988). A common problem in occupational situations is simultaneous exposure to several different known or suspected carcinogens. Often quantification of the exposures is vague or anecdotal; this is certainly complicates the later task of dose–response assessment, but may also be important for hazard identification when trying to resolve the role of multiple agents in a workplace. Other confounding exposures such as active or passive cigarette smoking may also complicate the interpretation. A further difficulty in using occupational data for public health risk assessment is that workers in chemical-using industries tend to be healthy adults. The historical database of occupational studies also has a bias in favor of male Caucasians. Thus, the hazard analysis of these studies may not accurately characterize effects on women, infants, children, the elderly, or on members of other ethnic groups. Despite these difficulties, the analysis of occupational epidemiological studies has proved an important source for unequivocal identification of human carcinogens.

Other important sources of epidemiological evidence are situations in which a large but clearly identifiable segment of a general population has some unusual exposure, due to, for instance, the presence of the material of interest in drinking water or food sources. This may occur either naturally, or because of human activity (industry, mining, etc.). Unusual food preferences, cooking and storage methods, and cultural practices may also create populations having greater or lesser exposure than elsewhere. Rigorous cohort and case–control studies may sometimes be possible, in which exposed individuals are identified, their exposure and morbidity or mortality evaluated, and compared with less exposed but otherwise similar controls. More often at least the initial investigation is an "ecological" study, in which prevalence of exposures and outcomes is compared in the control and exposed populations. Such studies are less definitive than cohort and case–control studies, but are important sources of information nevertheless, and can often also justify more costly and labor-intensive follow-up studies. The investigation of exposures via drinking water has been important especially for naturally occurring toxic minerals, such as the classic and continuing investigations of carcinogenesis and other toxic effects caused by arsenic in Taiwan, Chile, Argentina, and Bangladesh. Important results have also been obtained in the study of microbial toxins (aflatoxins, fumonisins) in food. In some cases, the agent responsible is unknown except as a component of a unique food material, such as the "salted fish" carcinogen.

The clinical medical literature contains many "case reports" where some interesting or unusual medical outcome is reported along with unusual exposures or situations that might have contributed to the occurrence. These reports typically involve a single patient or a small group, and have no statistical significance. They are, however, of interest in hazard identification as indications of possible associations that deserve follow-up using epidemiological methods, and as supporting evidence, addressing the plausibility of associations measured in larger studies. Many thorough and definitive investigations were initiated in response to such reports; the risk assessor is most interested in the mathematical rigor of these follow-up studies.

23.2.1.2 Animal Studies

Although the observation of human disease in an exposed population is the definitive hazard identification, this analysis represents a failure of the public health protection system, not a success. Ideally we would have identified the hazard earlier, and avoided the exposure before human sickness or death resulted. Usually, risk estimates have been based on studies in experimental animals, and extrapolation of these results to predict human toxicity. The animals used are mostly rodents, typically the common laboratory strains of rat and the mouse. Laboratory rodents are relatively inexpensive, survive and breed well in captivity without requiring extraordinary diet, housing conditions, or handling, and live for a convenient span of about 2 years — manageable for lifetime studies. They also include genetically well-defined inbred strains with constant and well-characterized anatomy, physiology, and biochemistry.

Rats and mice are similar to humans in many ways. Physiology and biochemistry are similar for all mammals, especially at the fundamental levels of xenobiotic metabolism, DNA replication, and DNA repair that we are most concerned with in identifying carcinogens. More complex functions (e.g., behavior, endocrine function, and reproduction) also share many common fundamental themes, although there are well-characterized differences. The fact that rodents are omnivores makes them more similar to humans in biochemical terms than some other mammals that are phylogenetically closer to humans, but have a purely herbivorous or carnivorous diet.

In spite of the important similarities, in interpreting the results of animal studies the risk assessor should also consider the known and specific differences between rodents and humans. When interpreting carcinogenicity data, it must be remembered that rodents, with a short lifespan, have differences in cell growth regulation compared with longer-lived species such as humans. For instance, although laboratory investigations have suggested that two regulatory gene mutations (e.g., *H-ras* and *p-53*) are sometimes sufficient to convert a rodent cell to a tumorigenic state, many human cancers observed clinically have seven or eight such mutations. The use of genomics to study chemical carcinogenesis is relatively new, but the differences at present appear to be a matter of degree rather than kind.

Differences in regulation of cell division are one possible reason for the variation between species that is often observed in the site of action of a carcinogen, or its potency at a particular site. A positive finding of carcinogenesis in the mouse liver, for instance, is a reasonably good indicator of potential for carcinogenesis at some site in the human, but in the majority of cases not in human liver. The mouse liver (and to a lesser extent that of the rat) is a common site of spontaneous tumors. It is also relatively sensitive to chemical carcinogenesis. The human liver is apparently more resistant to carcinogenesis; human liver tumors are unusual except when associated with additional predisposing disease, such as hepatitis B or alcoholic cirrhosis. Conversely, other tumor sites are more sensitive in the human. The other major reason for interspecies variation in site and sensitivity to carcinogenesis is differences in pharmacokinetics and metabolism, especially for carcinogens in which metabolic activation or detoxification is important. This variability may cause important differences in sensitivity between individuals in a diverse population such as humans. Variability between individuals in both susceptibility and pharmacokinetics or metabolism is probably less in experimental animal strains that are bred for genetic homogeneity.

The original design of animal carcinogenesis studies was intended to maximize the chances of detecting a positive effect, and thus did not mimic realistic human exposure scenarios. Much early work on the carcinogenic effects of polycyclic aromatic hydrocarbons (PAHs) involved dermal application, a known and important route of human exposure to these materials, but frequently also involved promotion with agents such croton oil (or its active ingredient, phorbol ester). Although promotion by injury or recurrent irritation does occur in human dermal carcinogenesis, the typical initiation–promotion study in mouse skin is an extreme application of this phenomenon. Similarly, in unpromoted studies by the oral or inhalation routes, doses are very large compared with those commonly encountered in the environment or workplace. This is designed to counter the limitation in statistical power caused by the relatively small size of an animal experiment. Whereas the exposed population of an epidemiological study might number in the hundreds (or, for a large ecological study, even the thousands or millions), the typical animal study might have 50 individuals per exposure group. With this group size any phenomenon with an incidence of less than about 5% is likely to be undetectable unless it is an extremely rare tumor.

Early animal studies were often even smaller, but this was not necessarily a problem when studying extremely potent carcinogens. Statistically significant results were obtained even with groups as small as ten animals per dose group, when incidence of a tumor that is rare in the controls approached 100% in a treated group. However, agents as potent as this are relatively unusual. Some studies have also used routes such as subcutaneous or intraperitoneal injection. These may be useful in specific circumstances as investigational tools. They are less helpful for hazard identification because of the frequent appearance of tumors at the site of injection and the bypassing of important metabolic processes to which carcinogens are exposed during uptake by routes more relevant to human exposure (e.g., oral or inhalation).

The consensus experimental design for animal carcinogenesis studies, which has evolved over the last 50 years of investigation, is represented by the protocol used by the U.S. National Toxicology Program for studies using oral routes (diet, gavage, or drinking water) or inhalation. The first step is to establish a likely maximum tolerated chronic dose by means of a small subchronic study. "Maximum tolerated dose" (MTD) is defined as the highest dose that causes no mortality or other severe noncancerous pathology and that has a marginal impact (about 10% ideally) on body weight gain. If there is no impact at all on body weight gain, it may be suspected that the MTD was not achieved. Conversely, if the impact at the highest dose level was greater than 10 to 15%, or if there is treatment-related mortality or morbidity (other than related to tumors, of course), the MTD may have been exceeded. For the main carcinogenesis study, groups of 50 animals of a single sex and species are used, with a control group, and at least two dose groups, one receiving the MTD and one receiving one half that dose. More recent designs have emphasized the desirability of more dose levels. The MTD is still the target for the top dose level, but the study now typically includes at least three levels covering a decade with "logarithmic" spacing (i.e., MTD, 1/2 MTD or 1/3 MTD, and 1/10 MTD). This extended design is aimed at providing better dose–response information, and it may contribute important additional information, such as mechanistic insights, for the hazard identification phase.

These carcinogenesis bioassays usually involve control and exposed groups for both sexes of an experimental species and most often two species. The National Toxicology Program has standardized the use of the C57BlxC3H F_1 hybrid mouse, and the Fischer 344 rat as the standard test species, which offer an acceptable compromise between low background rates of common tumors and sensitivity to chemically induced tumors. These strains are also well characterized and stable. There is now an extensive database of background tumor incidences, normal physiology, biochemistry, histology, and anatomy, which aid the interpretation of pathological changes observed in experiments. Although this background information is helpful in interpretation, it is generally regarded as essential that the controls and experimental groups in a single experiment be conducted concurrently. There is, for instance, enough variation in background rates of common tumors that the use of "historical control" data in hazard identification or dose–response assessment is undesirable, except as a last resort in analyzing poorly conducted experiments, or to indicate anomalous outcomes in concurrent controls.

23.2.1.3 Supporting Evidence: Genetic Toxicity, Mechanistic Studies

Animal studies of the size and complexity described in the previous section are expensive and time-consuming, and even when conducted appropriately may yield equivocal results. Investigators have therefore developed additional data sources that can support or modify the conclusions of animal carcinogenesis bioassays, and provide information on mechanisms of action of agents suspected of being carcinogenic based on epidemiological studies or animal bioassays.

A substantial body of information has been obtained in the last 30 years on chemicals that cause genetic damage in exposed organisms. This damage includes both gene mutations (point or frameshift), and larger scale effects such as deletions, gene amplification, sister-chromatid exchanges, translocations, and loss or duplication of segments or whole chromosomes. These genetic effects of chemical exposures are deleterious in their own right. In addition, since carcinogenesis results from somatic mutations and similar genetic alterations, agents that cause genetic damage generally have carcinogenic potential. Conversely, many known carcinogens are also known to be genotoxic, although there is also a significant class of carcinogens that are not directly genotoxic according to the usual tests. These latter agents presumably work by some other mechanism, although recent genetic studies have shown that even tumors induced by these agents show mutations, deletions, or amplification of growth regulatory genes.

Genetic toxicity has been measured by a number of experimental procedures. These may involve exposure of intact animals, and examination of genetic changes in, for example, bone marrow cells (or cells descended from these, for example, the micronucleus test, which detects remnants of chromosomal fragments in immature erythrocytes), mutations in *Drosophila* flies, or appearance of color spots in the coat of mice. However, the majority of genetic tests have employed single celled organisms or mammalian cells in culture, since these tests are generally simpler, faster, and cheaper to undertake. The best known of these tests is the *Salmonella* reverse mutation assay, popularly known as the Ames test after its inventor. This is representative of a larger class of tests for mutagenic activity in prokaryotic organisms (bacteria), which necessarily only look at gene-level mutations. Similar tests in eukaryotic microorganisms (yeasts, *Aspergillus*) and cultured mammalian cells also detect chromosomal effects. A further class of tests in mammalian cells examines the induction of cell transformation, that is, impaired growth regulation, where cells pile up instead of remaining as a monolayer (morphological transformation), or gain the ability to divide while suspended in soft agar (anchorage-independent growth transformation). These changes are seen as analogues of the conversion of a normal tissue cell into a tumorigenic cell.

These different classes of genetic test contribute slightly different information, which together may be used to amplify and confirm conclusions drawn from human studies or animal bioassays, or to draw tentative conclusions in the absence of epidemiological or bioassay data. In the latter case they have been an important source of data in prioritizing agents for more complete evaluation by means of bioassays, or conversely for identifying chemical structures that would not be worth further investigation as a possible drug or pesticide.

23.2.2 INFERENCE GUIDELINES AND CLASSIFICATIONS OF CARCINOGENS

Risk assessors providing input to public regulatory processes and arbitration or litigation value consistency and transparency, so that any interested party has access to all the input data and can follow the logic by which the conclusions are reached. An important contributor to this goal is the use of standard inference guidelines. These are generally peer reviewed to ensure their acceptance by the broad scientific community and may also receive regulatory endorsement. The US EPA originally published a guidance document for carcinogen risk assessment in 1986 (see US EPA, 1986). An update (US EPA, 2003a) is presently still in draft form after several revisions, but is considered generally useful and authoritative nevertheless. Issues within this framework have also received extensive treatment in the scientific literature, at conferences organized by the US EPA and other regulatory agencies, and by scientific societies.

In the specific area of hazard identification, the International Agency for Research on Cancer (IARC), a component agency of the World Health Organization (WHO) based in Lyon, France, is often seen as the ultimate arbiter of both methodology and conclusions. The IARC Monograph series provides a description of the data inputs and inference guidelines used in a preamble, followed by detailed evaluations of the carcinogenicity of individual substances or commonly occurring mixtures. The inference guidelines used are similar to those used by regulatory authorities, although IARC is a scientific, rather than a regulatory, organization.

The key inputs to hazard identification for carcinogens are human epidemiological studies and animal bioassays. The considerations of data quality and problems of interpretation of these have been described. IARC evaluations determine the quality of evidence for both these data classes as falling into one of four categories: sufficient evidence of carcinogenicity, limited evidence of carcinogenicity, inadequate evidence of carcinogenicity, and evidence suggesting lack of carcinogenicity. Stringent requirements for data quality are imposed. Entry of a chemical into the fourth category requires concrete evidence to support the conclusion, for instance repeated negative test results and mechanistic data. Mere lack of data or studies which are "nonpositive" due to lack of statistical power are classified as inadequate.

A conclusion of "sufficient" evidence from human studies requires that a causal relationship between the exposure and an increase in malignant tumors be established. This depends on statistically significant results in properly designed studies and appropriate control of confounding carcinogenic influences such as tobacco smoking. In clear-cut cases, a single such study might suffice. However, due to the complexity and insensitivity of epidemiological investigations, a series of such studies are usually needed, possibly supported by other human data such as case reports. Mechanistic evidence supporting the plausibility of the causal relationship is another important contributor to the conclusion. Evidence of carcinogenicity that falls short of this standard but has an element of credibility is considered "limited." Studies that are equivocal, contradictory, or where the causative agent has not been clearly identified are deemed "inadequate."

For animal studies a conclusion of sufficient evidence normally requires positive results either in two different species of animals or in two different and independent experiments in a single species. IARC and other evaluators normally consider experiments as independent only if conducted at different times or in different laboratories. Positive results in both sexes of a single species may be regarded as sufficient evidence by some evaluators. This depends on supporting considerations, such as convincing statistics, substantial tumor incidence, and appearance of tumors that are rare in controls. Exceptionally, a single result might be considered sufficient, depending on the supporting evidence. IARC utilizes the evaluations of animal and human data, along with supporting evidence such as genotoxicity, structure–activity relationships (see below), and identified mechanisms to reach an overall evaluation of the potential for carcinogenicity in humans. This is expressed as a numerical grouping, as shown in Table 23.1.

Other scientific and regulatory agencies typically use evaluation schemes like those developed by IARC, and reach similar conclusions, although inevitably there is disagreement about agents that fall on the borderline between categories. Some agents are also differently classified because new and influential data appeared after one evaluation that was available to another organization. In this situation, it may be important to consider the later evaluation even if it conflicts with a source such as IARC, which is usually considered authoritative.

In their original guidelines (US EPA, 1986), the US EPA recommended an overall classification scheme based on that used by IARC, but with letters instead of numbers for the categories (to support their sense of individuality). These categories are also shown in Table 23.1. Groups A, B, and E are equivalent to IARC's groups 1, 2, and 4, respectively. Some descriptions subdivide Group B into B1 and B2, corresponding to IARC's groups 2A and 2B, but in the 1986 guidelines US EPA made no particular distinction between these. For the majority of agents in Group B there is sufficient evidence in animals, with or without limited evidence in humans. Although the two agencies showed some differences with regard to their interpretation of certain issues (particularly the occurrence of benign tumors), these did not result in major differences of opinion in many cases. In fact, with the development of US EPA's evaluation process and IARC's guidance on interpretation of bioassay and

TABLE 23.1

Categories for Evaluation of Carcinogenic Potential by IARC and US EPA

IARC category	Description (IARC)	US EPA category
Group 1 The agent is carcinogenic to humans.	This category is used when there is sufficient evidence of carcinogenicity in humans.	*Group A* Carcinogenic to humans.
Group 2A The agent is probably carcinogenic to humans.	There is limited evidence of carcinogenicity in humans, and sufficient evidence of carcinogenicity in animals. Exceptionally, agents may be placed in this group with limited evidence in humans or sufficient evidence in animals, but not both, if there is other strong supporting evidence.	*Group B* Probably carcinogenic to humans.
Group 2B The agent is possibly carcinogenic to humans.	There is limited evidence in humans and limited or inadequate evidence in animals, or sufficient evidence in animals but only inadequate evidence or no data in humans. A few agents are placed here with only limited evidence in animals, and inadequate evidence or no data in humans, on the basis of strong supporting evidence.	
Group 3 The agent is not classifiable as to its carcinogenicity to humans.	Agents not covered by any other group, due to limited evidence or no data in animals, and inadequate evidence or no data in humans. This broad category includes agents that are a "near miss" for Group 2B, as well as those that nearly qualified for Group 4, along with many for which there are no interpretable data.	*Group C* Possibly carcinogenic to humans. *Group D* Not classifiable as to human carcinogenicity.
Group 4 The agent is probably not carcinogenic to humans.	There is evidence suggesting lack of carcinogenicity in both animals and humans. Agents with such evidence in animals but inadequate evidence or no data in humans might be included if there is strong supporting evidence suggesting lack of carcinogenicity. A relatively small number of agents qualify for this classification.	*Group E* Evidence of noncarcinogenicity for humans.

supporting data since that time, they have converged. However, the US EPA guidelines achieved an important clarification by their separation of the Group C agents, for which there is some evidence of carcinogenicity (but not enough to qualify for Group 2B/B), from Group D, for which there is insufficient evidence to make any substantial decision.

In contrast to the earlier categorical classification, US EPA's recent draft guidelines (US EPA, 2003a) emphasize the provision of a narrative covering the nature of the contributing evidence and a more detailed summary of the conclusion, including a description of the attendant uncertainties. However, responding to the needs of risk managers and others with limited time for weighing the details, they recommend the use of the following summary phrases or "Descriptors for Summarizing

Weight of Evidence":

"Carcinogenic to Humans"
"Likely to be Carcinogenic to Humans"
"Suggestive Evidence of Carcinogenicity, but Not Sufficient to Assess Human Carcinogenic Potential"
"Data Are Inadequate for an Assessment of Human Carcinogenic Potential"
"Not likely to be Carcinogenic to Humans"

Although these descriptors correspond to the earlier categories A through E (but with improved clarity both in the titles and their supporting explanations), the newer draft guideline places greater emphasis on the importance of supporting data and mechanistic information as a component of the overall evaluation.

These evaluation guidelines are important in providing an agreed and consistent basis for hazard identification of carcinogens. However, their use (particularly where narrative is emphasized rather than merely categorization) produces a complex result, which the risk manager ultimately has to reduce to a simple dichotomous decision: whether or not the agent should be regulated as a carcinogen (often involving an outright ban, or very stringent control or remediation measures). The final decisions made by the risk manager involve a number of nonhealth related factors, such as the legal, practical, and financial consequences of a given choice. Because of these differing contextual factors, it is possible and reasonable for risk managers to choose differently for different programs or purposes, based on the same risk assessment input. However, the risk assessor should provide clear guidance to the risk manager from the health risk viewpoint, and explore the consequences of alternative decisions. Where dietary, environmental, or occupational exposures occur to agents in IARC categories 1, 2A, and 2B or bearing one of the first two US EPA descriptors (i.e., potential human carcinogens), these are generally regulated stringently. Regulatory agencies in the United States at Federal and State levels generally regard agents in the US EPA category C (suggestive evidence of carcinogenicity, but not sufficient to assess human carcinogenic potential) with caution. Some of these agents are regulated as carcinogens, especially in programs addressing widespread exposures to the general public. Others may be regulated based on their noncancer toxicity, alone or with additional uncertainty factors reflecting the suspicion of carcinogenicity. Category D (insufficient data) agents are not generally regarded as carcinogens, but registration as a drug, food additive, or pesticide may be denied until data gaps are filled.

23.2.3 STRUCTURE–ACTIVITY RELATIONSHIPS IN HAZARD IDENTIFICATION

Chemical carcinogens cover a very broad range of different chemical structures, reflecting the enormous diversity of possible pathways for metabolism to reactive intermediates, sources of direct reactivity to DNA, and the other possible mechanisms leading to carcinogenic effects. However, certain classes of chemical structure are especially associated with chemical carcinogenesis. In several cases this has been connected to specific structural features, which are susceptible to a metabolic transformation that results in a reactive metabolite. Other classes of carcinogens may share structural features associated with binding to a receptor that has impacts on cell growth regulation. In many such cases, the structural resemblance between members of a group known to be carcinogenic provides an indicator of possible carcinogenicity for similar compounds for which epidemiological or animal bioassay evidence is not available. Such structural analogies are rarely used in isolation as a basis for actual regulation, but they are supporting evidence for evaluating data that might otherwise be considered equivocal. Structural analogies have also been helpful in identifying compounds requiring further study, including candidates for the National Toxicology Program's bioassays.

23.2.3.1 The "Usual Suspects": Major Classes of Chemical Carcinogens

One of the best-known chemical carcinogen classes, and the one most studied in practice, is that of the PAHs. These ubiquitous products of incomplete combustion occur widely as occupational and environmental contaminants. They are also found in some cooked foods, due to pyrolysis of food components and contamination by smoke. Many soots, tars, and similar products have been identified as carcinogenic, starting with the observation of carcinogenic effects of soot in the scrotum of chimney sweeps by Percival Pott (1775), which is generally cited as the earliest systematic description of chemical carcinogenesis. PAHs and mixtures containing them have been found to be carcinogenic in many organs, including lung and other parts of the respiratory tract, skin, the alimentary tract, liver, and mammary gland. Organ selectivity depends on several factors, including the metabolic capacity of the tissue for both activation and deactivation of the compound in question. Unsubstituted PAHs are pro-carcinogens, that is, they require activation to a reactive species by metabolism before causing the genotoxic effect that underlies their carcinogenicity. Route of exposure and resulting toxicokinetic processes are also important in determining the site and extent of the carcinogenicity of PAHs.

The PAH class consists of compounds having at least two fused, conjugated aromatic rings of carbon atoms. Compounds with alkyl or other hydrocarbon substituents on these rings are included. (A larger class of chemicals, including the PAHs and related derivatives, is "polycyclic organic matter" (POM), which is defined as a hazardous air pollutant by US EPA and a toxic air contaminant by California. This includes all PAHs, and derivatives of these such as nitro-substituted PAHs, multiring aromatic heterocycles, quinones, and multiring but unfused aromatic structures such as biphenyl ether, diphenyl ether, and their derivatives.) Theoretical and experimental study of the PAHs has shown the importance of epoxidation as a metabolic route leading to reactive intermediates, and eventually (after Phase II conjugation reactions, especially with glutathione) to nonreactive and excretable water-soluble metabolites. Many potent carcinogenic PAHs are metabolized by a two-step epoxidation process, producing dihydrodiol–epoxide intermediates, which react with DNA and are relatively resistant to the epoxide hydrolase enzyme that usually detoxifies reactive epoxide intermediates. Structural features promoting this type of metabolism such as the so-called bay region have been identified. The art of predicting carcinogenicity and relative potency at this finer level of structural analysis is incomplete. Other mechanisms of carcinogenesis, including reactions with DNA by other intermediate metabolites, such as simple epoxides and semiquinones, are sometimes important. The risk assessor usually regards any hydrocarbon, or related compound, with the characteristic molecular structure of the PAHs as a potential carcinogen.

The aromatic amines form another important class of carcinogens, of which 2-naphthylamine is the archetype. Many of these compounds were formerly used in the dyestuffs and rubber industries, and were found to be associated with substantially elevated incidences of bladder cancer. The mechanism of this effect involves the acetylation or glucuronidation of the amino group in the liver. The resulting conjugate appears in the urine, but is subject to further metabolism in the bladder epithelium to the unstable N-hydroxy intermediate. This rearranges to a reactive nitrenium ion, which is responsible for DNA damage and cytotoxicity, both of which contribute to bladder carcinogenesis. A number of carcinogens act by this mechanism. They are characterized by an aromatic amino group (or a substituted form, such as an amide) in a location accessible to the various oxidases and transferases involved. Aromatic nitro groups are also a characteristic marker for this effect, since they are readily reduced to aromatic amino or hydroxylamino groups.

Some of these materials are active liver carcinogens, although at higher doses than in the bladder. Animal studies have readily demonstrated the liver carcinogenicity in rodents, but it is difficult to produce the bladder tumors characteristic of humans exposed to aromatic amines. This is because humans and other primates typically retain their urine in the bladder for extended periods. Rodents on the other hand do not, instead maintaining a more or less continuous output. To demonstrate bladder carcinogenesis by an aromatic amine, a bioassay typically requires either very high dose levels, or use of a test species, such as the dog or the ferret, which stores its urine. Identification of the mechanism of action for several compounds of this type has greatly refined the understanding of detailed structure–activity relationships for various subclasses of aromatic amines. To benefit from this, the risk assessor needs actual biochemical or bioassay data on the compound of interest. However, any molecule containing an aromatic amino group arouses suspicion of carcinogenicity, although not all such compounds are in fact carcinogens.

Carcinogenic nitrosamines occur as food or beverage contaminants. A common traditional method of preserving meats is to add nitrite. This reacts with secondary amines and cyclic imino groups in the food, including certain amino acids, to form nitrosamines. Nitrosamines are also generated in combustion processes, being present in tobacco smoke, and beer made from malt dried in open-hearth kilns, *inter alia*. Some nitrosamines are industrial chemicals and intermediates, and certain natural products such as antibiotics and alkaloids contain related structures. However, many nitrosamine carcinogens are only laboratory curiosities. These were synthesized so that their chemical and biological properties could be studied, but were not found naturally or resulting from industrial activities. This research has demonstrated dependence of both the strength and the site of action of the carcinogenic effect on the structural characteristics of individual nitrosamines. The mechanism of action of nitrosamine carcinogens has been studied extensively. For simple alkyl nitrosamines, the first step is metabolic activation by cytochrome P-450 mediated hydroxylation at the α carbon atom of one alkyl group. The molecule thus formed rearranges to a positively charged carbonium/nitrenium ion, which is highly electrophilic and reacts with DNA. Related chemical structures such as nitrosamides are susceptible to a similar rearrangement reaction at an appropriate pH without metabolic activation. Nitrosamines have been used extensively in research on carcinogenesis because of their specificity for certain tissues, and because many are efficient carcinogenic initiators but have little promoting activity. This has enabled the study of various chemicals and procedures that act as promoters.

Risk assessors are likely to encounter carcinogenic metals when dealing with environmental or occupational exposures. For instance, carcinogenic nickel, chromium VI, cadmium, arsenic, and lead compounds are widely distributed as a result of industrial activities. It is difficult to discern structure–activity relationships for the carcinogenic metals. They appear to act by different mechanisms, several of which are described elsewhere in this volume. Often the carcinogenic action is route dependent: exposure by inhalation is associated with a high potency for nickel, chromium VI, and arsenic-containing compounds. Carcinogenicity is often associated with a particular valence state, or certain compounds. These differences appear to relate to chemical properties, absorption, metabolism, or toxicokinetics of the individual metals and compounds.

23.2.3.2 Systematic and Computerized Methodologies

Attempts have been made to relate carcinogenic activity to identifiable features in molecular structure such as the well-known specific groups noted above. These have often involved statistical analysis of structural data, combined with either actual carcinogenicity data or prediction rules developed by panels of experts. In some cases these analyses have been made commercially available as computer software packages that develop some kind of "score" to characterize the inferred likelihood and degree of carcinogenic activity for any given structure. It is unlikely that hazard identification would be based exclusively on such analyses, but they do allow systematic inclusion of structural information in the overall weight of evidence, for prioritization, or to assist in resolution of difficult or doubtful cases.

23.2.4 AUTHORITATIVE RESOURCES FOR HAZARD IDENTIFICATION

The risk assessor dealing with specific sites or situations is seldom dealing with previously unevaluated chemicals, and will often rely for hazard identification on an existing assessment. This hazard identification may be exactly specified by certain regulatory programs. In dealing with air pollutants, for example, both U.S. and California regulations identify a list of over 100 agents as "Hazardous Air Pollutants" or "Toxic Air Contaminants" respectively. Other programs such as State and U.S. drinking water standards, solid waste and transportation regulations maintain lists of agents of concern, and they specify the nature and extent of the hazard associated with these materials. California's Safe Drinking Water and Toxic Enforcement Act ("Proposition 65") specifies lists of identified carcinogens and developmental or reproductive toxicants, for which warnings are required where significant exposures may occur.

In other cases, applicable regulatory programs may not specify the specific chemical hazards to be considered, or there may be additional exposures besides those listed. In such cases the risk assessor may use general sources such as the US EPA listing of carcinogens in the IRIS database (US EPA, 2003b, compiled from evaluations from all the individual regulatory programs, with further review to establish consensus). For occupational situations, evaluations by National Institute for Occupational Safety and Health (NIOSH) are particularly relevant. The European Union has a similar centralized listing (ECDIN) of toxicological data including carcinogenic evaluations. In addition nonregulatory hazard evaluations such as those provided by IARC and the U.S. National Toxicology Program will be useful. Risk assessors developing the lists for regulatory programs often cite these as "authoritative." Where one of these agencies provides a scientific evaluation, the regulatory process may take this as the basis for listing by administrative procedure, rather than conduct a full and independent evaluation for that specific program.

23.3 EXPOSURE ASSESSMENT

The second step needed in the risk assessment process is exposure assessment. The hazard identification phase will have established the nature of the potential risk from a carcinogen or other toxicant, but this will only be realized as an actual risk if, and to the extent that, exposure to the causative agent occurs.

In determining the exposure, it is important to identify the characteristics of the exposed individuals, the routes of exposure, and the intensity and duration of the exposure. The intensity of exposure may be further divided into the concentration in the external media, the bioavailability (i.e., extent to which absorption from these media occurs), and finally the toxicokinetics that determine the delivery of the material or its toxic metabolites to the sensitive tissues, and their eventual detoxification and removal from the body. Although these different processes form a continuum, it is conventional in risk assessment to consider concentrations in, and exposure to, the external media, and the bioavailability, under the heading of exposure assessment. Toxicokinetics are generally considered under the heading of dose–response assessment.

23.3.1 THE EXPOSED INDIVIDUAL

A first requirement is to identify the exposed subject for whom the risk must be calculated. Depending on circumstances, this may be an actual individual whose exposure may be measured by personal sampling, estimation of biomarkers such as urinary metabolites, or measurements of the environment in which they are located, coupled with measurements of uptake parameters such as breathing rates, water consumption, and so on.

Often, the "exposed individual" is instead a model of what might occur under particular hypothetical circumstances. The risk assessment thereby identifies precautionary requirements to avoid significant harm occurring to the local or general population. The exposure analysis often defines a hypothetical "maximally exposed individual" using health-conservative assumptions that are unlikely to be exceeded. Several exposure estimates may be derived, representing both the maximally exposed individual, and others such as medians for the whole population and for specified subgroups. Often a "tiered" approach is used, starting with an estimate using simple methodology and assumptions for a maximally exposed individual. If the risk prediction for this case is insignificant, further analysis is unnecessary. However, if this simple but extreme analysis predicts a significant risk it may be necessary to employ a more complex and realistic analysis to better estimate the exposure.

Individual behavior has important effects on exposure. Some aspects of this will be covered in the chosen exposure model, but activity levels must be considered for the target individual. High levels of physical activity substantially increase water consumption and breathing rate. Children behave differently from adults showing higher physical activity levels and greater propensity to absorb toxicants orally from soil, household dust, and similar sources by hand-to-mouth transfer. Behavioral and cultural factors also greatly influence the amount and types of food consumed. This has been extensively studied with regard to toxicants in fish. Most North Americans and Europeans eat modest amounts of commercial fish and little or no personally caught "sport" fish, but some individuals undertake subsistence fishing and obtain a substantial proportion of their total protein intake from this source.

Individuals vary markedly in physiological parameters, including body weight, and in factors such as age and health status that may also affect their exposure to toxicants. Uptake parameters such as breathing rate, food and water consumption are often assumed as "standard" values in simple models, but are more fully described by distributions based on values measured for sample populations. The standard values may represent a hypothetical individual (often a healthy adult male), or they may be chosen as points on a population distribution. Points of interest usually include the 95th percentile (desirable in assessments designed to provide protection to the majority of the exposed population) and the mean or median. The whole distribution may be incorporated in a complex analysis; its shape is often assumed normal or lognormal, and defined by mean and standard deviation estimates. Monte Carlo methods are often used to combine distributions for multiple parameters, to avoid the overprediction found in simpler models in which several 95th percentile estimates are incorporated. Special distributions may be chosen to represent subgroups of the population, such as children or infants in certain age ranges, pregnant women, or members of a cultural or occupational group.

23.3.2 MEDIA AND ROUTES OF EXPOSURE

The analysis needs to consider the media to which an individual may be exposed. Media such as air or drinking water directly convey the toxicant from the source to the exposed individual, but there are also more complex routes such as contaminated food grown in soil containing the toxicant, or inhalation (and dermal) exposure to volatile materials present in the domestic water supply during showering and other water-handling activities. Exposure assessment generally requires determination of concentrations and often time-dependent functions such as flows, chemical reactions, and biodegradation, in both the media into which the toxicant is first emitted, and the media to which the individual of interest is exposed. This situation is frequently addressed using computer models (see below).

The nature of the toxicant and the media containing it to which the individual is exposed determine the routes by which the material is absorbed into the body. The route of exposure is often an important determinant of the toxic effect. Some toxicants such as reactive irritants produce their major impacts on tissues at or near their first point of contact. For instance, formaldehyde is strongly reactive with proteins, nucleic acids, and other components of tissues, and this reaction occurs similarly whether the point of contact is the skin, the eye, the respiratory system, or the gastrointestinal tract. Likely consequences include immediate tissue damage, with necrosis in severe cases, and perhaps effects such as carcinogenesis due to reaction with DNA. However, the sensitivity of tissues to these effects varies considerably. The skin has an outer layer of dead keratinized cells, which present a barrier protecting the living tissue beneath. It also has various defense and repair mechanisms to avoid or limit the damage sustained. Conversely, the epithelia that line the respiratory tract expose live cells directly at the surface. The gastrointestinal tract has varying sensitivity, depending on the structure of the lining in the various regions. The uptake rate and toxic effects are also modified by the presence of food and secretions that may react with, or preferentially dissolve, the toxicant. Thus formaldehyde is an acute irritant, and an allergic sensitizer by any route of application, although the effective dose is substantially higher for dermal or oral exposures. Since it is reactive to DNA, formaldehyde is potentially carcinogenic by any route, but this result is only seen after inhalation exposure, as has been observed in particular in the upper respiratory tract of rodents. Oral exposures to formaldehyde are generally not observed to be carcinogenic. This example shows the importance of route in determining the eventual consequences of exposure, since the carcinogenic potency and other measures of toxicity vary substantially from route to route.

Carcinogens and other toxicants often act at target organs remote from the point of first contact, following distribution via the blood. One would not expect different potencies for different routes of exposure when the dose is expressed in terms of the amount actually absorbed. However, the uptake by different routes may vary greatly.

For inhaled materials, the concentration in air and the solubility of the material in blood are important determinants of the extent of uptake. Where there is equilibrium between a vapor phase in the lung alveoli and dissolved material in the blood, a partition coefficient can be defined. This, along with the blood flow through the lung, determines how much material is absorbed. Where equilibrium is not achieved, the uptake may be better described by the diffusion coefficient through the alveolar surface. Pharmacokinetic models are often used to describe these uptake processes, especially at the high exposure concentrations observed in some occupational situations. At low doses such as may be found in the general environment, one may use simplifying assumptions. For relatively water-soluble gases and vapors, and for particles having an aerodynamic diameter of 10 μm or less, which enter the respiratory system freely, it may be sufficient to assume that 100% of the material present in inspired air is absorbed. For less soluble gases and vapors it may be more appropriate to assume 70% absorption.

Thus,

$$D = \frac{CVu}{\text{bw}}$$

where D is the dose (μg/kg/day), C is the air concentration (μg/m^3), V is the volume of air breathed per day (default 20 m^3), u is the fraction absorbed (assume 1.0 or 0.7, or from model), and bw is the body weight (default 70 kg).

The volume of air inspired in a given period depends on the individual, being affected by variables such as body weight, age, sex, and activity level. It may be sufficient to use the default value of 20 m^3 per day assumed for an adult male, who is also assigned a default body weight of 70 kg. More sophisticated analyses may use population breathing rate distributions (such as those published for the California Air Toxics Hot Spots risk assessment program), and consideration of different "standard" individuals, such as an infant, child, generic adult female, or a worker with an especially high physical activity level.

Uptake from food or drinking water depends on the amount of food or water consumed. Although national and regional databases listing consumption of food commodities may be useful, one may need to consider more local and appropriate cases based on actual data, especially for cases where special cultural or socioeconomic factors apply, for example, subsistence fishers. Drinking water consumption may also be estimated using population distribution, real data, or default values. A default consumption of 2 l/day is often assumed, but this will increase for high levels of physical activity and hot climates.

In considering consumption of drinking water, this figure includes actual drinking and an allowance for water consumption as a result of food preparation and cooking. An additional factor in household situations is that piped water is used in various other domestic activities such as showering, flushing toilets, washing clothes and dishes, and so on. These will result in additional exposures to volatile contaminants in drinking water via inhalation since the materials will equilibrate between the supplied water and the indoor air of the residence. Dermal exposure to water contaminants will also occur during bathing or showering. The extent of these additional exposure factors varies according to the vapor pressure, water solubility, and dermal permeability of the compounds: in certain cases the effective consumption may be as much as 2 to 3 times the direct consumption as drinking water. The relevant compound-specific physical properties are conventionally estimated by relating them to physical constants that may be measured or calculated such as the Henry's Law coefficient and log P (octanol/water partition coefficient). The CalTox program provides a convenient default approach to estimating the effective overall consumption of piped water, given the physical characteristics of the compound of interest.

Both food and water contaminants are subject to partial uptake, such as air contaminants. The proportion taken up by the oral route is referred to as a bioavailability factor. There are extensive experimental studies of bioavailability for pharmaceuticals and most of the expertise in measurement and modeling derives from this area. Data are limited for most environmental contaminants, especially at low doses. A default of 100% bioavailability is often assumed in the absence of data, to protect public health. Data for either water or dietary sources of organic toxicants is very less, but some metals, particularly lead, and those for which an essential dietary role has been identified, have been studied more extensively. In the case of lead, the bioavailability via the oral route is low, but there are important interactions with dietary metals such as iron and calcium, and also substantial variation with age. Infants and children are especially sensitive to the neurotoxic action of lead, and they also show higher bioavailability of oral lead exposures. This illustrates the need for caution in interpreting experimental bioavailability data, especially when relying on studies in adults with good nutritional status. US EPA (1999) has developed guidelines for dealing with oral uptake of lead from contaminated sites, as part of an integrated lead uptake model, but for most other toxicants and situations 100% bioavailability will be assumed.

For volatile materials, absorption from the gastrointestinal tract may be effectively 100%, but there may be loss of unchanged material through the lungs by equilibration with alveolar air, especially at higher doses. If the toxicant is metabolically activated, as is often the case for carcinogens, a pharmacokinetic model may be useful. This can estimate the amount of material metabolized, which is often a better measure of the effective dose for such chemicals than the applied dose or exposure concentration. In occupational epidemiology and experimental animal studies, in which exposures are relatively high, the metabolized dose is often substantially less than the applied dose. At the lower exposures typical of drinking water, food, or environmental contamination situations, this difference may be unimportant. The risk assessor may then choose cautious default assumptions such as 100% metabolism or, for highly volatile materials with low solubility and slow metabolism, 70%, as was described for inhalation.

In a few cases, the issue of "bioavailability" by the oral route is complicated by reaction of the toxicant with food or digestive secretions, resulting in greatly reduced uptake of the actual toxic material or even, effectively, none at all. The minimal carcinogenicity of some reactive substances such as formaldehyde by the oral route may be considered to reflect this situation, among other contributing factors. A number of other reactive irritants are highly potent in causing tumors of the stomach or, in the rodent, the forestomach, so this protective factor cannot be assumed as a panacea. Another example of reactivity of the toxic agent is hexavalent chromium. The carcinogenicity of Cr^{VI} by inhalation is well known, whereas Cr^{III} is generally considered inactive. Some experimental studies have found carcinogenic effects of Cr^{VI} by the oral route, while others have been negative, and this is generally thought to result from reduction of Cr^{VI} to Cr^{III} in the stomach, a process that is favored by the low pH and high organic substrate content in that region. There is a currently unresolved debate about extent to which this reduces the oral bioavailability of Cr^{VI}, and thus lowers the effective carcinogenic potency. Other reactions in the stomach, such as formation of nitrosamines from nitrate or nitrite and amino-substituted dietary constituents (or contaminants, additives, and pharmaceuticals), may result in creation of carcinogenic materials. This case also illustrates how the route, with associated factors such as dietary status and concurrent exposures to other materials, is important in determining the effective exposure to carcinogens and other toxicants.

23.3.3 DURATION OF EXPOSURE

In the simple case of a more or less constant exposure to an environmental contaminant the question of duration of exposure is generally addressed in risk assessment for carcinogens by assuming that the exposure will continue at a constant level for the standard human lifetime, that is, 70 years. Exposure assessment for carcinogens becomes much more complex when shorter periods are involved, because both the underlying animal or human data and the risk model used in dose–response assessment are based on chronic exposures that continue for a substantial fraction of the lifetime or an exposed human or experimental animal. It is necessary to determine as a matter of policy what exact measure of risk is desired.

Consider the case of a toxic waste site adjacent to a residential development, resulting in exposure of the residents to a constant level of a carcinogen. The nature and quantity of material in the site is such that depletion or degradation is not a significant issue over at least 70 years. What should be the assumed duration of this exposure? The "average" resident stays in their own house for a period of around 7 years for many U.S. populations. However, this might be significantly longer elsewhere or for specific populations (e.g., those who have difficulty selling a house located next to a toxic waste site). So perhaps a "plausible upper bound" residence time should be used: the US EPA has often used 35 years (0.5 lifetime) as a plausible upper bound on the residence time when calculating the risk to a maximally exposed individual. However, unless the house is of unusually shoddy construction, it is likely to remain in use for at least 70 years; even if it is demolished, it is most likely to be replaced with another residential structure. Although the first resident may move out after 7 (or 35) years, somebody else will move in to take his or her place. The risk to the population as a whole is therefore best represented by an estimate based on the full standard lifetime exposure period of 70 years. The risk assessor must choose a scenario that reflects the purpose and objectives of the assessment being undertaken, but will generally take a cautious approach to defining the effective duration of exposure to a chronic toxicant such as a carcinogen.

Other situations may provide a more obviously limited duration of exposure, such as when an emission is limited by the nature of the process to a short period, or a known quantity of toxicant results in a predictably decreasing exposure due to depletion or degradation of the source material. In such cases, it may be appropriate to consider the predicted duration of the exposure in risk assessment. The assessor should consider whether an exposure episode is likely to be repeated, and if so how often, since the effects of exposure to carcinogens may be presumed to be cumulative. The assessor should also consider the reliability of any predictions of changes in exposure. For instance, is there statutory authority to force an emission to cease after a defined period, and will it be used? Does the activity have a naturally limited duration or is the prediction relying on some sort of average figure? If the emission, or exposure to it, is intermittent, is the frequency controlled, and if so to what extent? The risk assessor charged with protecting public health will use effective duration estimates that allow for such uncertainties.

A final problem to be considered is that the risk model used in evaluating carcinogenic exposures does not apply to short-term acute, or highly episodic, exposures. Extensive data in the scientific literature show that single exposures to carcinogens are effective in causing tumors, but ways of using these data to provide quantitative risk estimates have not been developed. This has sometimes led to the risks from single short-term exposures being ignored, as often happens when considering the effects of accidental releases of nonpersistent carcinogens. Some argue that the risks of such exposures are small, since if a peak exposure lasts a day or two the resulting "chronic" exposure, calculated as a time-weighted average over 70 years, is very small. However, we do not know whether this accurately reflects the risk. It is plausible, but unproven, that this situation might result in higher risks than those predicted by the time-weighted average.

For many risk assessment programs, where the effective period of an exposure is limited to a total period of less than 70 years, it is appropriate to calculate a time-weighted average over 70 years and use this as the input to the dose–response equation. However, theoretical considerations require that the effective exposure duration not be reduced below the minimum that could be considered a "chronic" exposure, that is, one occurring for a substantial percentage of lifetime, since both the experimental data and the risk model used are based on chronic exposures. There are experimental data in animals and anecdotal reports in humans that report significant tumor incidences after a single brief exposure to a carcinogen, but it has not generally been possible to quantitatively relate the potency of such exposures to the potency derived from lifetime exposure experiments. For carcinogenic exposures, the California Air Toxics programs have sometimes defined a period such as 8 years as the minimum duration of exposure that can be considered in a risk assessment.

23.3.4 COMPUTER MODELING IN EXPOSURE ASSESSMENT

Development of exposure models for the variety of situations where exposure to carcinogens and other toxicants is of interest is a large and highly complex endeavor, which is often assisted by computerized models. These include models describing the fate and transport of toxicants emitted into the environment (including in some cases transformations such as those caused by photochemical, biological, or hydrolytic reactions). They may simulate features such as dispersion of plumes in the atmosphere, or movement of groundwater. Movement of materials between different media may be described in detail using rate constants and partial differential equations, or more simply by "transfer factors" describing the portion of material that moves from one medium to another at a presumed equilibrium state. Computerized models are available to estimate individual exposure and uptake from specific media, using either measured data on the compound of interest or general chemical data such as vapor pressure, water solubility, or log P (a measure of hydrophobic or hydrophilic character). These are all specific to particular situations, and a detailed description is beyond the scope of this chapter. However, various environmental regulatory programs offer detailed exposure assessment guidance including, in many cases, recommended models and computer programs, sample input parameters and so on. The reader is referred to the US EPA's Risk Assessment Guidance for Superfund (RAGS) for solid waste and contaminated site modeling. California EPA provides a spreadsheet modeling tool ("CalTox"), which is also useful for dealing with a number of multimedia exposure assessment problems. For assessment of local area or point sources of air pollution, the reader may consult the Technical Support Document on exposure assessment for Air Toxics Hot Spots risk assessment published by California EPA.

23.4 DOSE–RESPONSE ASSESSMENT

At this point, we have identified that a certain exposure has the potential for causing an effect on biological systems and quantified this exposure by measurement or modeling. The next step is to determine the relationship between the dose received by the target system (usually a human, or a wild animal species of interest in the case of ecological investigations), and the magnitude of the effect produced. The measurement or prediction of this relationship is referred to as the dose–response assessment phase of risk assessment. Although exposure to a number of different toxic compounds simultaneously is the norm in the real world, one can only study the dose–response relationship for mixtures having a defined, constant, and repeatable composition. Usually this relationship is determined for the individual compounds one at a time. As discussed later, the integration or summation of the predicted effects presents significant problems and uncertainties. For carcinogens, it is usually assumed that the effects of different compounds are independent of one another, although this is not always the case.

23.4.1 ANIMAL AND HUMAN DATA

The effect on exposed humans is of interest to the risk assessor, so it is naturally simpler and more accurate to rely on human data to define the dose–response relationship. This avoids the difficulties and uncertainties of extrapolating from an experimental animal species to the human, and also (in the case where population-based epidemiological studies are used as the source data) may provide information on the variability in response that may be expected in the target population. Conversely, if the source data are an occupational cohort, the subjects are likely to be healthy adults, often exclusively male, and less diverse than the population at large with regard to age, ethnicity, and socioeconomic status. In this latter situation, the risk assessor makes allowance for the greater diversity of the population as a whole, if the general population, including children, the aged, and other potentially susceptible subgroups, is the target for the risk assessment.

The quantitative problems in determining the extent of an effect in an epidemiological study are well known. Unless carefully controlled for, confounding factors may either mask the effect of interest, or, if exposure to the compound of interest and the confounding exposures are correlated, may suggest a spuriously large effect at a given dose. The most serious confounding factor for studies of carcinogenesis is undoubtedly cigarette smoking, which is associated with a high rate of lung and other cancers. Well-designed studies either measure smoking behavior of exposed and reference subjects and analyze the data with smoking as an independent variable, or at least attempt to choose groups in which the exposed and control subjects have similar smoking rates. This is particularly important when the exposure of interest is a dietary or lifestyle related agent, since smoking behavior may be correlated with diet, beverage consumption, and so on.

In occupational cohorts, it may be possible to control adequately for smoking by choosing exposed and nonexposed workers from similar geographical and socioeconomic groups (especially workers at the same job site). However, with this type of study other confounding exposures to toxic chemicals in the workplace may occur. Unless these can be measured accurately, and unless groups of workers can be identified where these are not perfectly correlated with the exposure of interest, it may not be possible to use the data to provide an unequivocal dose–response relationship. Since multiple exposures are common in the workplace, the best that can be done is sometimes to provide a maximum estimate of risk, assuming that all the observed effect is due to the compound of interest.

Confounding exposures may result in an excessively high prediction of the extent of risk from a given exposure. This may be acceptable from the public health point of view, where the most important priority is to protect the population at large from harm. However, an overly pessimistic prediction may impose unnecessary costs on the party generating the exposure, or responsible for cleanup of a toxic site. Equally, this may divert effort and resources away from other exposures, which, if all were accurately assessed, would be seen as a more urgent priority.

Another common failing of epidemiological data, misclassification, generally results in under-prediction of risk. This occurs when subjects identified as exposed are mistakenly identified as unexposed (or low instead of high exposure if several exposure levels are identified), or vice versa. These errors generally dilute the response observed, since the disease incidence in the controls is increased by the addition of some exposed and affected subjects, whereas the incidence in the exposed group is decreased by the inclusion of unexposed and unaffected subjects.

Some types of epidemiological study such as those using the case–control design are intrinsically only capable of reporting the incidence of an event as a relative risk, that is the ratio of the incidence of the disease (or other event) in the exposed population to the incidence in a matching control population. In addition to these, most authors of cohort studies (which may be in principle capable of reporting risks as absolute numbers rather than ratios) choose to use the relative risk model in their calculations. This allows simpler correction for variations in background incidence with age, date of exposure, geographical location, and other possibly influential variables within the control and exposed populations. The relative risk model assumes that the observed incidence of an effect in the exposed population is a function (usually, although not necessarily, linear) of the intensity of the exposure, and of the background incidence in unexposed individuals having the same characteristics as the exposed population. For diseases with a significant incidence in unexposed individuals, this is a reasonable model assumption. It breaks down when the natural background incidence is zero, or too low to be measured accurately even in a large population. In this case, the epidemiologist has often expressed the incidence as a relative risk compared with some other more inclusive category, for example, mesothelioma compared with all thoracic cancers. This may or may not be reasonable.

In using these studies for risk assessment, risks are conventionally presented in terms of absolute numbers rather than ratios (this being how the controlling legislation is usually drafted, apart from any other considerations). For analyses derived from case–control studies or cohort studies with relatively common endpoints, the risk assessor may obtain a numerical risk estimate by multiplying the relative risk by a standard incidence value from a reference source, or, preferably if available, actual incidence data relating to the target population of the risk assessment. For rare tumors where the true background incidence is unknown or zero it may actually be preferable to attempt a reanalysis of the study using an additive rather than relative risk model.

In practice, it is difficult to find epidemiological data suitable for dose–response assessment of carcinogenic effects. Many studies that are useful in hazard identification are unsuitable for dose–response assessment because the exposure measures used are vague and general (e.g., "high," "medium," and "low," without further quantification), and because of confounding or misclassification problems that may affect the strength of the observed relationship. Also, epidemiological studies are typically insensitive unless extremely large study populations are available, and the dose levels are extremely high relative to the potency of the carcinogen (or other toxic agent). Thus in the majority of cases, the risk assessor will attempt to predict the incidence of tumors in humans exposed to a given dose of a chemical on the basis of data obtained in controlled experiments on animals.

This attempt is subject to uncertainties, and it is important to recognize the source of these, and to reflect this knowledge in the eventual risk predictions offered to the risk manager or the interested public. One major source of uncertainty is the necessity of extrapolating from the test species, usually a rodent, to humans. Rodents are small, short-lived animals with a number of important physiological, genetic, and biochemical differences between them and humans. In particular, there are suggestions that the underlying genetic processes of cell regulation in the tissues of large, long-lived animals such as humans are more complex and stringent than those of rodents. This must in fact be so since the cells of large, long-lived animals must divide many more times, and function for much longer, than those of small, short-lived animals. In spite of years of lively debate among cell biologists, biochemists, and geneticists about how it is that whales do not contract cancer long before reaching their adult size, it does not appear that there is at present a full answer to this conundrum. Background cancer rates on a per-lifetime basis are somewhat comparable between species of widely differing sizes. On the other hand, when there are significant differences between these rates the variation seems to depend more on individual species-specific factors than on any consistent variation with size or longevity. This has led to the custom of calculating tumor incidence rates on a per-lifetime basis in experimental animal studies and to using this per-lifetime rate (rather than a per-year rate, for instance) as the starting point for extrapolation to other species.

23.4.2 STANDARD DEFAULT ASSUMPTIONS

To deal with the many uncertainties in the generation of risk estimates, the usual approach has been to address data gaps in the relationships between the source data and the desired estimate by using default assumptions. These are assumed values for parameters or relationships that are missing in the specific data available for the case under consideration, which may be based either on "reasonable" values obtained from generic situations or measured values from situations assumed to be similar. The extrapolation of cancer rates from rats with short lifetimes to humans with long lifetimes was considered previously. To define an appropriate basis for this extrapolation, it is necessary to determine the lifetime of rats and humans. In a few cases these two durations may be determined as part of the data available, but usually this is not the case, and in any event it may be desirable to standardize these values to provide comparability between different assessments. The "standard" lifetime of a rat is generally assumed to be 2 years (104 weeks). This duration is assumed in the standard carcinogenesis bioassay design: indeed, it is usually imposed by terminating the experiment at this point, killing all surviving animals and conducting the required pathology studies on both exposed and control animals. A parallel assumption is made that the standard human lifetime is 70 years, although based on recent demographics this is somewhat shorter than the true value. Similarly, in defining both exposure parameters (as noted previously) and in choosing values to insert into dose–response models, it may be useful to use "standard" values for physiological parameters such as body weight. In the consideration of exposures, it was noted that sometimes with the data either for the subject population of an epidemiological study or for a group of experimental animals, it is possible to define a range or distribution for some of these physiological parameters. However, this had not been generally undertaken for dose–response assessment. There are two possible reasons for this. First, the mathematical complexities of fitting assumed dose–response relationships to data are already sufficiently severe without adding additional variables, and second, it is not clear that the inter-individual variation of these parameters within a species is actually relevant to the way members of that species respond in, for example, a cancer dose–response model. Many of these extrapolation models are based on general observations of what happens in "typical" members of the species. These models have never been related to distributions of physiological parameters within either the model or the target species.

23.4.3 EMPIRICAL FITS VERSUS MECHANISTIC MODELS

A mathematician analyzing a data set may develop a model that fits the data in some sense, for example, defining a smooth curve through all the points. The features of the model are assumed to be arbitrary and to have no necessary connection with the phenomena represented by the data points. This process has its uses, for instance allowing interpolation but not, with any reliability, extrapolation. Usually the analyst has the objective of learning more about the data, and this requires something more than a purely empirical fit. Statisticians fit models to data and describe the properties of the data in terms of how well the data fit these assumed models. They also seek to determine whether there are any systematic patterns in the deviation of the data from these assumed models. They may take a more pragmatic look at the data and try to display them in such a way that any interesting relationships between variables or categories are apparent. Normally, simple models are used, such as an assumed linear relationship between two variables, or the assumption that variability follows a distribution such as normal or lognormal. These assumed models are very generally applicable, and were often originally derived from first principles of mathematics. They can be seen as a kind of "how the world works" model rather than saying anything very special about the phenomena underlying the data.

Conversely, a biologist is interested in the biological processes of the system. It is not so much whether the data follow the expected statistical models and simple relationships that is of interest, as the identification of statistically significant deviations from these simple generic patterns, which indicate that biologically interesting things are happening. Often the biologist will develop a mechanistic hypothesis, and then use this hypothesis to develop a mathematical model that predicts how the data should appear. The questions then are, do the data fit the hypothetical model (allowing for the expected statistically described variability, measurement error, and so on), and also does this hypothesis provide a unique explanation of the data or could other hypotheses (which might lead to the same model of the data, or a subtly different one) provide an equally plausible fit? If the hypothesis provides a unique and satisfying fit to the data, this is seen as endorsing the usefulness and potential correctness of the hypothesis. We may then even contemplate using the hypothetical model to allow for extrapolation, to predict outcomes outside the range of currently observed data, which as noted earlier cannot be justified for purely empirical models. Such extrapolation depends on the quality and correctness of the underlying hypothesis, as well as the range and quality of the data. If we do have the opportunity to make such predictions by extrapolation and then test them experimentally this is an important way to further validate the hypothesis, and perhaps eventually elevate it to the exalted status of a theory.

Because one of the key objectives of risk assessment is to predict outcomes that we cannot observe directly, or would prefer not be realized, extrapolation is a fundamental tool on which the risk assessor has depended on for the development of models. These are both statistical and, where possible, based on biological mechanisms and measured parameters, to support the necessary extrapolation. An early use of mechanistic analysis to support risk assessment was the development of the Armitage–Doll multistage model of dose–response for carcinogenesis. This model was developed into the so-called linearized multistage model, which has been extensively used for carcinogen risk assessment. It leads to a number of partially verifiable predictions, including linearity of the dose–response relationship at low doses, which is observed for many genotoxic carcinogens. It also predicts the form of the dose–response relationship at higher doses, which generally follow a polynomial form (subject to sampling and background corrections) except where other identifiable factors such as pharmacokinetics intervene. The multistage description was initially developed on theoretical grounds, using mathematical curve fitting as the primary tool. Subsequent discovery of the molecular biology of proto-oncogenes has provided a basis for explaining the model in terms of actual biological events and systems (Barrett and Wiseman, 1987). More recently, it has been felt that the simple linearized form of the multistage model has limitations, which detract from its usefulness and generality.

Various experimental investigators showed that cell proliferation is an important process in the progression of cancer. For a few carcinogens it may actually be the primary mechanism of action, as opposed to the direct modification of DNA by the carcinogen or a metabolite, which is the "classical" mechanism assumed in the multistage description. Moolgavkar and colleagues developed a cell proliferation model ((Moolgavkar and Knudson, 1981) that not only retains many features of the original multistage model, but also considers proliferation, death, or terminal differentiation of both normal and "initiated" (genetically transformed) cells. This model and its subsequent derivatives are generally considered to be better descriptions of carcinogenesis from a biological standpoint than the original multistage model. However, they have been little used in risk assessment because a complete model description requires a large number of parameters that are difficult to define and measure (such as proliferation and death rates for various classes of cell). If these cannot be accurately determined, the model has so many free parameters that it can be made to predict virtually any continuous curve shape the analyst chooses, and it is not helpful in defining extrapolated values for risk assessment purposes. This highlights a general problem in using mechanistic models in carcinogen risk assessment, which is that the quality and quantity of the carcinogenesis data themselves are generally insufficient to fully define the dose–response curve shape. The analysis is therefore supplemented with policy-based assumptions (such as the expectation of linearity a low doses) and, wherever possible, additional experimental measurements in order to make meaningful prediction of risk from environmental exposures to humans. Most recently, in US EPA draft guidelines (US EPA, 2003) and in some risk assessments by California regulatory authorities, a less overtly mechanistic approach has been advocated. This approach combines benchmark dose methodology (described below) with explicit policy-based choice of the method for low-dose extrapolation, either assuming low-dose linearity or, for certain carcinogens where data indicate that this is appropriate, a "margin of exposure" or safety/uncertainty factor-based approach.

23.4.4 LINEARIZED MULTISTAGE MODEL

23.4.4.1 Quantal Analyses

For regulatory purposes, the "Multistage" polynomial (Anderson et al., 1983; US EPA, 1986) is often used to model the lifetime risk of cancer. This describes the relationship between dose and risk

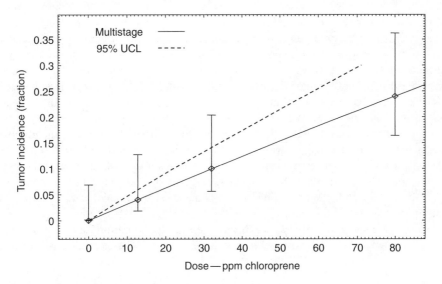

FIGURE 23.2   Multistage model fit to typical tumor incidence data.

by a polynomial. It is usually used for analysis of animal bioassay data, although related approaches are occasionally used in epidemiology. In mathematical terms, the probability of dying with a tumor (P) induced by an average daily dose (d) is:

$$P(d) = 1 - \exp[-(q_0 + q_1 d + q_2 d^2 + \cdots + q_i d^i)]$$

with constraints

$$q_i \geq 0 \quad \text{for all } i.$$

The q_i model parameters are constants that can be estimated by fitting the polynomial to the data from the bioassay, that is, the number of tumor bearing animals (as a fraction of the total at risk) at each dose level, including the controls. The fit is optimized using the maximum likelihood methodology. The maximum number of terms allowed is one less than the number of points in the data set (i.e., the number of degrees of freedom). All the coefficients of the terms are constrained to be zero or positive, so the curve is required to be straight or upward curving, with no maxima, minima, or other points of inflection.

Figure 23.2 shows an example of such a curve fit to tumor incidence data in a bioassay. The data were obtained from an NTP inhalation bioassay on chloroprene (TR 467; NTP, 1998). The data shown are the incidences of squamous cell papilloma or carcinoma of the oral cavity in groups of 50 male rats that were exposed to 12.8, 32, or 80 ppm of chloroprene for 6 h per day, 5 days per week for 105 weeks. The polynomial calculated included nonzero values for q_0, q_1, and q_2. The parameter q_0 represents the background lifetime incidence of the tumor. The slope factor q_1, or its upper bound, is often called the cancer potency. For small doses, it is the ratio of excess lifetime cancer risk to the average daily dose received. In addition to the maximum likelihood estimates of the parameters, the program calculates upper 95% confidence bounds on these parameters, which yield the confidence limit curve also shown in the figure. Details of the estimation procedure are given in Crump (1981) and Crump et al. (1977).

The cancer potency is generally defined as q_1^*, the upper 95% confidence bound on q_1. In addition to making appropriate allowance for uncertainty and variability in the data, the use of an upper confidence bound on the estimate of q_1 provides a stable estimate of this parameter under all conditions. For some data sets, small changes in the input data can result in changes in the order of the best-fit polynomial, or large shifts in the balance between the different terms. The maximum likelihood estimate of q_1 is therefore unstable. It may also erratically have a zero value, even when the data imply a significant positive dose–response relationship. The upper bound q_1^* provides a more stable estimate that is not subject to these mathematical peculiarities. When dose is expressed in units mg/kg-day, the parameters q_1 and q_1^* are given in units $(mg/kg\text{-}day)^{-1}$. For use in risk estimation, the slope factor value in this experiment needs to be corrected for the intermittent exposure. In the absence of a pharmacokinetic model, the potency would ordinarily be based on the time-weighted average exposure during the dosing period. This is in addition to the correction for exposure duration that would be necessary if the study had not lasted for 105 weeks, and the interspecies correction, both of which are described below. Risk calculations using this potency value estimate the cancer risk at low doses only, with the higher order terms of the fitted polynomial being ignored since their contribution is negligible at low dose.

To estimate potency in animals (q_{animal}) from experiments of duration T_e, rather than the natural lifespan of the animals (T), it is assumed that lifetime incidence of cancer increases with the third power of age:

$$q_{animal} = q_1^*(T/T_e)^3$$

The experimental justification for this assumption is shown below, in the section on time-dependent analyses. Usually, a natural lifespan of 2 years is assumed for mice and rats (Anderson et al., 1983; Gold et al., 1984), so for experiments lasting T_e weeks in these rodents

$$q_{animal} = q_1^*(104/T_e)^3$$

The interspecies extrapolation methods described below are then used as the basis for an estimate of the potency for humans exposed to the carcinogen at low doses. The human potency is multiplied by average daily dose. The risk estimate obtained is referred to by the US EPA (Anderson et al., 1983) as "extra risk," and is equivalent to that obtained by using the Abbott (1925) correction for background incidence.

23.4.4.2 Time-Dependent Analyses

The standard quantal analysis of grouped tumor incidence data uses average values for the survival of the animals, and thus the effective duration of the exposure to the carcinogen. Not all experiments conform accurately to this assumption. Survival of animals is poor in some bioassays, or differs among dosed groups. Also, dosing may have occurred for significantly less than the lifetime of the animals, or may have been discontinuous or at variable levels. Under these circumstances, results obtained by fitting the simple version of the multistage model may not accurately reflect the true dose–response relationship for tumor incidence. These problems can be addressed most effectively using an extended version of the multistage model and fitting this to survival and tumor incidence data for each individual animal. This level of detail is not always available in published bioassay reports, but it is provided by the technical report series describing the NTP bioassays.

For experiments with significant intercurrent mortality, performed with constant dosing schedules, an extended model form can be used that retains the multistage polynomial in dose, and adds a Weibull exponential term to describe the time dependence of tumor incidence. In this extended form of the model, the probability of tumor ($P(t,d)$) by time t is:

$$P(t,d) = 1 - \exp[-(q_0 + q_1d + \cdots + q_kd^k)(t - t_0)^m]$$

with

$$q_i \geq 0 \quad \text{for all } i, \text{ and } 0 \leq t_0 < t \geq t$$

For this analysis, either the tumor in question is assumed to have caused the death of the animal (the "lethal" assumption), or it is assumed that the animal died of a cause other than the particular tumor of interest (the "incidental" assumption). The term t is usually interpreted as the latency period, and m is the age exponent (Krewski et al., 1983). The carcinogenic potency at the end of the experiment (T) is given by the upper 95% confidence bound on Q, where

$$Q = q_1(T - t_0)^m$$

A fit to time-to-tumor data is more difficult to visualize graphically, since there are more dimensions of variation (time, dose, variance) than can be conveniently shown in two dimensions. Some aspects of the model, in particular the increase in tumor yield with time in both control and dose groups can be seen in Figure 23.3. This is based on liver tumor yield in female mice an inhalation study of tetrachloroethylene by NTP (1986).

For regulatory purposes, and in order to provide comparison between results from experiments of different durations, experimental cancer potency is normalized to lifetime cancer potency at $T = 104$ weeks, the natural lifespan typically assumed for rats and mice (Anderson et al., 1983). It has been observed for a large number of analyses of this type that the age exponent (m) generally has a value between 1 and 6, with an approximate intermediate value of 3 being the most usual. This (along with similar observations from epidemiological data) is the basis for the assumption that cumulative cancer incidence increases with the third power of age (Anderson et al., 1983), which is used in correcting for the duration of experiments with grouped incidence data that last less than the standard lifetime.

The linearized multistage model has also been modified (Crouch, 1983; Crump and Howe, 1984) to allow analysis of data sets with variable dosing over time. This can also be useful in the analysis of experiments in which the dosing period was much shorter than the nominal lifetime of the test animals, or the overall observation period of the experiment.

As discussed by Crouch (1983), if the probability per unit time of the stage transformation depends linearly on dose rate ($d(t)$), and the carcinogen only affects a single "stage," the probability

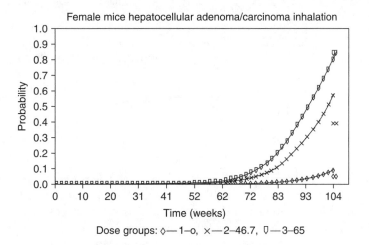

FIGURE 23.3 Time-to-tumor data: perchloroethylene.

of tumor by time T_e under Armitage and Doll (1954) becomes

$$P(T_e) = 1 - \exp[-(A + BD)]$$

with

$$D = \beta(m,j)(1/T)^m \int_0^{T_e} (T_e - t)^{m-j} t^{j-1} \, \mathrm{d}(t) \, \mathrm{d}t$$

where T_e is the time to observation, and $\beta(m,j)$ is Euler's beta function. The natural lifetime of the test animal, T, is assumed to be 2 years for rats and mice (Anderson et al., 1983). The integer m (the number of "stages") specifies the rate of increase in incidence with time and j is the "stage" affected by the carcinogen. In analysis of lifetime incidence data by this method, a linear relationship between applied dose and the probability that a "stage transition" has occurred is assumed. The stage affected by the carcinogen and the number of "stages" is also assumed.

This analysis can be combined with time-dependent analysis of individual animal data, which allows determination of the number of stages (i.e., the value of m) to be measured. However, one must still in this case assume that a particular stage (j) is affected by the carcinogen. In the absence of contrary information, it is often assumed the first stage is primarily impacted by the applied dose ($j = 0$) and a value of 3 is assumed for m. The potency in animals, q_{animal}, then, is given by the upper 95% confidence bound on B. This calculation allows for abbreviated or variable dosing schedules and for observation periods less than the nominal lifetime of the test animals. The effective dose may be calculated using a correction factor multiplied by the applied dose for each period of constant dose rate during the experiment. A spreadsheet to accomplish this calculation is represented in the following table, with actual cell references and formulae shown on one line and some sample calculations on the following two lines:

	A	B	C	D	E	F	G	H
1	Armitage–Doll dose correction							
2	Age-related exponent		3					
3	Nominal lifespan		728	(days)	*(enter data in column C if different)*			
4	Time of observation		500	(days)	*(enter data in column C)*			
5								
6	Dose per treatment (mg/kg)	Start of treatment (days)	Treatment days per week	Weeks of treatment	End of treatment (days)	Dose (mg/kg-day)	Correction factor	Corrected dose (mg/kg-day)
7	*enter data*	*enter data*	*enter data*	*enter data*	= B8 + D8*7	= (A8 * (C8/7))* D8 * 7)/(E8-B8)	=(1/\$C\$4^\$C\$3) * ((\$C\$5-B8)^\$C\$3- (\$C\$5-E8)^\$C\$3)	=F8 * G8
8	1	30	5	6	72	0.714285714	0.065884863	0.047060616
9	1	72	5	6	114	0.714285714	0.054144015	0.038674296
10							Total Corrected dose (mg/kg-day)	=SUM(H8: H9)
11								0.085734913

The default ($m = 3$, $j = 0$) is reasonable for carcinogenic initiators (e.g., potent genotoxic carcinogens). It implies that exposures early in life have a greater effect than those later on, which is an important consideration in determining risk of exposures to children. Early exposure to a carcinogenic initiator produces mutations that are available for "promotion" (according to this model, a sequence of later mutations that occur independently of further exposure) for longer than if it occurred late in life (the so-called shelf life effect). Mechanistic data may suggest other assumptions. For a carcinogen with greater promoting effects, the age-dependence may not be so large; if j is near or equal to m (a late-stage carcinogen), the tumor-inducing effect of early exposures will not be much different from that of late exposures.

23.4.4.3 Calculation Methods for Linearized Multistage Models

Evidently, the mathematical operations necessary to fit a linearized multistage model to a set of experimental data are complex. The necessary procedures have been incorporated into several computer programs. Early work needed mainframe computing resources, but advances in computer hardware and software allow the programs now in use to run easily on desktop computers. Crump and coworkers developed the original GLOBAL program (with US EPA sponsorship) to fit a linearized multistage model to bioassay data, and versions (ADOLL and WEIBULL) to fit time-dependent data to the variable-dose-rate model and the Weibull time-to-tumor model, respectively. These were originally mainframe FORTRAN programs, subsequently recompiled for MS-DOS desktop systems, and their origins are apparent in a strict batch-processing style of control and very inconvenient user interfaces.

More recently, Crump and colleagues developed the Tox_Risk program, which provides fits of the linearized multistage model and various other mathematical models (many of which are also useful in considering noncancer data) to both quantal and time-to-tumor data. This program also runs on MS-DOS (and later, Windows)-based desktop systems, with a user interface which, if not exactly state-of-the-art, is nevertheless a vast improvement on the earlier programs. It also more flexibly accommodates variable-dose-rate data in analysis of time-to-tumor data.

Other software packages have also been developed to fit linearized multistage models to bioassay data. Crouch developed the MSTAGE program to fit quantal data as a mathematically more efficient alternative to GLOBAL, and this program ran rapidly even on hardware that was barely adequate to support GLOBAL86 (the most recent iteration of that program). It also has the virtue of being shared freely by the author with academic collaborators.

More recently the US EPA has sponsored the development of the BMDS system, which is primarily designed to implement benchmark dose methodology for cancer and noncancer risk assessment. However, this includes the algorithms required to fit a multistage polynomial to quantal data, and thus may be used to derive a linearized multistage result, the cancer potency being provided by the upper bound on the linear term of the fitted polynomial (q_1^*). This program also has the advantage of being freely distributed, and it is available from the US EPA web site.

23.4.5 BENCHMARK DOSE METHODOLOGIES

The use of benchmark dose methodology has been explored by Gaylor and Kodell, Crump and other investigators, as a tool for dose–response extrapolation for both carcinogenic (US EPA, 2003) and noncarcinogenic (US EPA, 1995) endpoints. The basic approach is to fit an arbitrary function to the observed incidence data, and to select a "point of departure" or benchmark dose *within the range of the observed data*. From this a low-dose risk estimate or assumed safe level may be obtained by extrapolation, using an assumed function (usually linear) or by application of uncertainty factors. The critical issue here is that no assumptions are made about the nature of the underlying process in fitting the data. The assumptions about the shape of the dose–response curve (linear, threshold, etc.) are explicitly confined to the second step of the estimation process, and are chosen on the basis of policy, mechanistic evidence, or other supporting considerations. The point of departure chosen is a point at the low end of the dose–response curve but one at which effects are clearly observable. Usually an ED_{05} or ED_{10} (dose at which the incidence of the effect is 5% or 10% respectively) is chosen for animal studies, although lower effect levels may be appropriate for large epidemiological data sets. Because real experimental data include variability in the response of individual subjects, and measurement errors, maximum likelihood methodology is usually applied in fitting the data. A lower confidence bound (usually 95%) of the effective dose, rather than its maximum likelihood estimate, is used as the benchmark dose. This properly reflects the uncertainty in the estimate, taking a more cautious interpretation of highly variable or error-prone data.

For cancer dose–response estimation using the benchmark dose method, either animal bioassay data or epidemiological data provide a suitable basis. The model used to fit the data can be chosen arbitrarily, but in practice, the multistage polynomial fit developed for the linearized multistage model works well. Here it is being used merely as a mathematical curve-fitting tool, without making assumptions about its validity as a biological model of carcinogenesis.

Suitable polynomial fits and estimates of the benchmark (usually LED_{10} — the 95% lower confidence bound on the dose producing 10% tumor incidence) may be obtained using the same software — Tox_Risk or US EPA's BMDS — as was used for the linearized multistage model. The benchmark required is the dose value at which the polynomial predicts 10% incidence. This value and its confidence bounds are reported by Tox_Risk, but the earlier GLOBAL program reports the equation parameters, without a solution of the equation for a specified risk. With this program, the dose for a specified level of risk is usually calculated from the linear term only. This is reasonable for low risk levels such as 10^{-5} or 10^{-6}, for which this program was intended, and is consistent with the "linearized" feature in this use of the Armitage–Doll cancer model. However, it is inappropriate in calculating an LED_{10} value as a benchmark dose. The GLOBAL program and its relatives are therefore less suitable as curve-fitting tools for benchmark dose analysis.

It is usually assumed in cancer risk estimation that the low-dose–response relationship is linear, so risk estimates and a potency value (slope factor) may be obtained by linear extrapolation from an appropriate benchmark dose. Dose levels at defined risk levels, or risk levels at defined doses, may be read off from the straight line between the LED_{10} and the origin (zero dose, zero risk). The potency is the slope of that line ($0.1/LED_{10}$). The low-dose linearity assumption is a general default for any carcinogen, and it is unlikely to be altered for genotoxic carcinogens.

A calculation using the standard modeling approach (a polynomial model with exponents restricted to zero or positive values) and linear extrapolation from the LED_{10} to obtain a potency estimate is shown in Figure 23.4 (the figure was generated by the US EPA's BMDS program). This is based on tumor incidence data from an actual experiment with vinyl bromide in rats (Benya et al., 1982), with metabolized dose calculated by means of a pharmacokinetic model (Salmon et al., 1992). The value

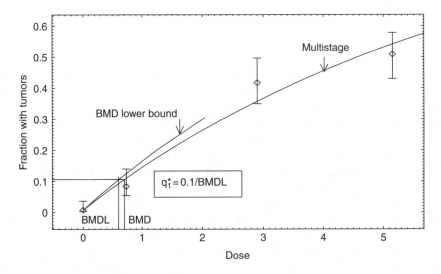

FIGURE 23.4   Multistage model with 0.95 confidence.

of q_1* obtained by this calculation would then be corrected for the duration of the experiment if it had lasted for less than the standard rat lifetime, and for bodyweight and route-specific pharmacokinetic factors as described above.

The benchmark dose methodology allows for the use of more complex functions where these are justified by experimental data, such as the determination of cell proliferation rates and incorporation of these rates into a Moolgavkar type of model. If such a mechanistic model were available and validated, it might also be advantageous to use this model to fit the actual incidence data in determining the benchmark dose, although the implementation of a maximum likelihood fit of such a model to data is mathematically challenging.

In a few cases there is sufficient mechanistic evidence available to support the conclusion that a carcinogen acts by a "nongenotoxic" mechanism, and that moreover this mechanism involves a threshold dose below which it will have no carcinogenic effect. Only a few such cases (e.g., saccharin as a bladder carcinogen in rodents) have been sufficiently clearly established at this time for this assumption to be acceptable in public health risk assessment. In such cases, an uncertainty factor (UF) approach is used to predict a safe level. The benchmark dose is divided by appropriate factors that allow for the expected shape of the curve in the threshold range, the uncertainty associated with interspecies extrapolation and inter-individual variability in the human population, and the need for caution in protecting public health from a severe effect such as cancer. In view of these considerations, risk assessors have generally opted for substantial UFs such as 10^3 below the benchmark.

The application of the linear extrapolation procedure, and of the UF approach (referred to by US EPA as a "margin of exposure" method — a term used with slightly different implications in other contexts), and the criteria for choosing the UF approach rather than the default linear extrapolation, are further discussed in the US EPA's current draft cancer risk assessment guidelines. Comparisons of the two approaches are also shown in some recent risk assessments, such as those undertaken by California EPA for methyl tertiary butyl ether in drinking water (OEHHA, 1999a).

23.4.6 TOXICODYNAMIC MODELS

23.4.6.1 Cell Proliferation Models

The cell proliferation model proposed by Moolgavkar and others (Moolgavkar and Knudson, 1981) is a more biologically plausible alternative to the original Armitage–Doll model of cancer dose–response and time course (Portier et al., 1993). It has been used with some success to describe the observed tumor incidence patterns for certain carcinogens for which induction of cell proliferation is important in the observed tumor induction, rather than mutational events alone. Thus, the incidence of bladder tumors in rats exposed from an early age to large doses of sodium saccharin has been successfully modeled by assuming that sustained cell proliferation (caused by microcrystals appearing in the urine) interacting with the background mutation rate observed in this tissue leads to the tumors. This is consistent with the absence of significant metabolism of saccharin in the rat, and its lack of mutagenic activity either *in vivo* or *in vitro*. Another interesting series of studies undertaken by Connelly and others on formaldehyde (CIIT, 1999) successfully described the observed dose–response in terms of the interaction between mutagenic effects and cytotoxic effects that result in cell proliferation.

Use of these modeling techniques for actual risk assessment, as opposed to analysis of research investigations, has been limited. Extrapolation beyond the range of observed data with the models can be very uncertain. The models contain a large number of variables, far more than can be determined by fitting the cancer incidence data. Values of these variables must be obtained independently from experimentally determined rates of mutation, cell division, and cell death. There are few carcinogens, or target tissues, for which these studies have been performed. Even where these results are available, the actual parameters required by the model must generally be deduced indirectly from what can be measured experimentally, with attendant large uncertainty ranges. If the required values are not closely determined from external data, but merely guessed at or used as free parameters in the model fit, the resulting equations can fit virtually response curves of any shape and not provide any useful limits on extrapolated risk levels. However, although their use in risk assessment is not a routine or reliable choice at present, an increasing number of recent studies have provided measurements of proliferation and mutation rates in various tissues, so the cell proliferation models may be of greater use in future.

23.4.6.2 Nonlinear and Threshold Models

For a few carcinogens, sufficient mechanistic data are available to indicate that their mode of action involves interaction with cellular control mechanisms, such as binding to a receptor, rather than direct mutation. One of the best-known and most widely accepted examples is 2,3,7,8-tetrachlorodibenzodioxin (TCDD) and the other "dioxin-like" chlorinated dioxins, dibenzofurans, and polychlorinated biphenyls (PCBs). In these cases, in principle, one could calculate a dose–response curve based on the binding constant between the carcinogen and the receptor on which it acts. If the binding follows simple mass-action principles, the dose–response would be linear at low doses, with indications of saturation at higher doses. This is similar to the Michaelis–Menten description of enzyme kinetics (see Medinsky and Valentine [2001] or any basic biochemistry text). Pharmacologists often assume that binding of an agonist to a receptor is cooperative, and they use the Hill equation (originally used to describe the cooperative binding of oxygen to hemoglobin [Hill, 1925]) or similar relationships to describe the concentration — saturation relationship. These curves often show a marked inflection point or "pseudo-threshold," below which little binding occurs.

Such equations have been proposed to fit the dose–response curve of some receptor-binding (or so-called hormonal) carcinogens. This is feasible for the dioxins, since the properties of the Ah receptor and its affinity for dioxin-like compounds are fairly well understood (US EPA, 2000b, in particular Part II, Chapter 8). However, the result is less interesting than it might be from the risk assessment standpoint, since the implied pseudo-threshold is below the background dioxin level in the general human population. Therefore, any additional exposures considered in a risk assessment would be additive to this background and result in a real increment in risk, even though the mode of action is often described as implying a threshold type of dose–response.

Other modes of action implying possible threshold effects have been proposed. If a stress on some biochemical system can be tolerated or adapted to up to a certain level, and when this level exceeded cytotoxicity, cell necrosis or apoptosis and response proliferation occur. As predicted by the Moolgavkar model, this may lead eventually to increased tumor incidence. This cytotoxicity model is accepted by pharmacologists to describe the production of tissue damage by toxic chemicals. It has been proposed to explain some carcinogenic actions, notably those of carbon tetrachloride and trihalomethanes in rodent liver. However, to date most of these descriptions have not been fully quantitative. The assumption of a definite threshold is usually regarded as controversial, and not protective of public health, unless supported by extensive studies establishing both the plausibility of this hypothesis, and the lack of other competing or contributing processes. The debate over the mode of action for trihalomethanes has been marked by agreement that cytotoxicity is important at least at higher dose levels. However, it is uncertain whether mutagenic reactive metabolites (free radicals, carbenes, carbonyl halides, etc.) are formed, and if they are, whether they contribute to a major extent, or perhaps only a minor degree, to the observed carcinogenesis. Such uncertainties can have large impacts on the predicted risk at low-dose levels, which are not observable in the laboratory, but may be highly relevant to human environmental exposures. Because of these difficulties and controversies, not many risk assessments use an actual mechanistic model of a threshold process to predict dose–response. In those few cases in which evidence of a threshold is sufficiently convincing, the benchmark dose method is usually used to analyze the observed cancer dose–response, coupled with uncertainty factors to provide a sufficient margin of exposure in setting acceptable exposure levels to protect public health.

23.4.7 PHARMACOKINETIC MODELS

Except in the case of carcinogens that act at the site where they first contact the body, the extent of absorption, and the subsequent distribution of the compound have an important impact on the dose received by those tissues that are most sensitive to the carcinogen action. Tissue specificity may result from selective uptake of the carcinogen by those tissues. It is generally the case that metabolism of carcinogens is important both in activating the applied material to an active form, and in converting it to inactive substances that are then excreted. The rate and route of excretion are important determinants of the concentration and duration of the exposure received by sensitive tissues. These factors are described by the field of pharmacokinetics (or "toxicokinetics" as it may be called for compounds having only adverse effects as opposed to pharmaceuticals with allegedly beneficial effects as well). There is an extensive literature describing efforts to measure and model the kinetics of carcinogens and other toxic compounds, in addition to the even larger published literature and regulatory studies on kinetics of drugs. The risk assessor should be aware of the utility of this approach in refining the analysis of carcinogen dose–response curves, and in supporting the comparison of results between animal species or their extrapolation to humans.

An important additional application has been the investigation of inter-individual variability in the human population, including the identification of specific subpopulations who may be more or less sensitive to a particular carcinogen due to their individual metabolic capabilities. These subpopulations may include groups distinguished by age (especially infants, children, and the elderly), by ethnicity, sex, medical status, and so on, which may be important to consider as special targets in a risk assessment in which members of the general population may be exposed.

Analyses of animal dose–response data have traditionally used the applied dose as the key exposure parameter, but where pharmacokinetic data are available, these are often included. If a specific target organ is remote from the site of application of the external dose, the risk assessor may examine the distribution of the compound to predict the dose (concentration and time course) to which that organ is exposed. Where there are data on the routes of metabolism and the enzymes involved, it may be possible to determine enzyme kinetic parameters and thus model saturation of certain pathways at high doses. This phenomenon is often responsible for applied dose–response curves that show leveling off at high doses. These are hard to explain or fit with the model equations usually considered appropriate for cancer dose–response analysis. Application of a relatively simple kinetic description, such as the Michaelis–Menten equation, allows estimation of the amount of material metabolized, or the concentration × time integral for a reactive intermediate, in the target tissue. This tissue dose estimate often allows a better fit to the tumor data with the usual multistage polynomial. In the case of vinyl chloride, pharmacokinetic analysis improved the quality of fit (US EPA, 2000a), and thus the risk prediction, in analysis of the animal bioassay data. See Medinsky and Valentine (2001) and Leung (1999) for recent discussions the application of pharmacokinetic modeling techniques in toxicology.

23.4.8 INTERSPECIES EXTRAPOLATION

The problem of relating tumor incidence in species of greatly different longevities has been addressed using the "per-lifetime" basis for comparison, although the reasoning behind this assumption has not really been well defined. This basis generally works for background rates of cancer, apart from notable differences at particular sites (both between short-lived animals such as rodents and long-lived animals such as humans, and between different species of similar longevity such as rats, mice, and hamsters). However, the sensitivity of these species to induction of tumors by externally applied chemicals does not follow this simple relationship. This issue is important when extrapolating the results of animal bioassays to potential human cancer risk. One of the few sets of data in which this relationship can be explored is that for the anticancer drugs. Many of these act by damaging DNA, and thus have a selective action in killing rapidly dividing cells such as those of a tumor. On the other hand, these drugs are in fact mutagenic, and some cause a measurable incidence of subsequent cancer in treated patients, and also tumors in experimental animals. The mutagenic, carcinogenic, and cytotoxic effects of these materials are assumed to result from the same basic mechanism of action. Their relative effectiveness has been assessed by measuring cytotoxicity (including acute lethality, which is much studied in animals, and not unknown in human subjects since the therapeutic ratio for these drugs is often very small). In comparing the MTD, or the LD_{50}, of these drugs in humans and in various experimental animals, it is observed that humans are usually more sensitive than smaller animals. This sensitivity follows a power function of body weight similar to that observed for basal metabolic rate. Not surprisingly, given the relationship between surface area and heat loss, metabolic rate follows the relationship between body surface area and body weight, that is,

$$\text{Basal metabolic rate} \propto \text{Surface area} \propto \text{Body weight}^{2/3}$$

Using this analogy, it was assumed that cancer potencies would follow the same relationship, so the cancer potency estimate for humans was related to that observed in an animal study by a factor derived from the ratio of the default body weights of the two species:

$$q_{human} = q_{animal} \times (bw_{human}/bw_{animal})^{1/3}$$

where the q values are the two potency estimates and the bw values are the body weights for the two species. Sometimes a true average body weight for the experimental animals in the study was used, although this is complicated when considering a lifetime study in rodents since these animals grow continuously throughout life (although at a lower rate when adult than when young), unlike humans who cease somatic growth after adolescence. More often, "standard" values either generic to the species, or specific to the particular strain used in the study, are used, and these are often chosen as a mean body weight during the adult lifespan. US EPA (1988) provides tables of standard values, which are widely used as a reference source, especially when detailed body weight data are not reported in the experimental studies. The source data on which this assumed relationship is based are generic to entire species, with little power even to discern the impact of variation between strains, let alone individuals. The justification for adjustments to allow for the characteristics of individual experimental groups, or even specific strains, is therefore questionable.

This default basis for interspecies extrapolation of cancer potencies has been modified recently. In general, the maximal rates of metabolism of foreign compounds (e.g., carcinogens) do not exactly follow the 2/3-power law seen for basal metabolic rates, but rather correlate with body weight to the 3/4-power. Detailed comparisons of the relative toxicity of cytotoxic drugs in animals and humans have been largely inconclusive. The relationships for individual drugs vary from powers of 0.5 to about 1.5 of body weight. Although the mean exponent is near 2/3 or 3/4, there is insufficient consistency or statistical power to distinguish the two cases. Nevertheless, the consensus among risk assessors now favors the 3/4-power relationship. Using "generic" body weights for mice, rats, and humans (respectively, 30 g, 350 g, and 70 kg), the power relationships default to fixed ratios, which may be sufficient for simpler risk assessments:

Species	Body weight ratio	q_{human}/q_{animal}: 2/3-power	q_{human}/q_{animal}: 3/4-power
Rat	200	5.85	3.76
Mouse	2333	13.3	6.95

The assumed 3/4-power relationship reflects the known importance of metabolism for many carcinogens. However, if a pharmacokinetic model of carcinogen metabolism in two species is examined, the variation of important output values, such as the peak concentration or area-under-curve for reactive intermediates, does not follow the 3/4-power relationship assumed for individual metabolic capabilities of Phase I and Phase II enzymes. The observed difference in sensitivity of humans and small experimental species probably reflects both pharmacokinetic differences and intrinsic sensitivity to carcinogenesis, DNA damage, or mutagenesis. The assumed 2/3- or 3/4-power relationship is a policy choice, which is consistent with the data available, rather than an exact conclusion based on knowledge of mechanisms.

It is important to resolve how the interspecies adjustment should be handled when a physiologically based pharmacokinetic (PBPK) model has been developed and parameterized for both the experimental species and humans. This should permit interspecies extrapolation of the potency factor without arbitrary assumptions such as the 3/4-power relationship, to the extent that the different responses of the test species and humans relate to differences in metabolism of the carcinogen. However, the differences in sensitivity of these organisms to carcinogenesis are believed to reflect both toxicokinetic and toxicodynamic differences. The simplistic approach of eliminating the body weight relationship and relying on the PBPK model alone neglects an important source of the variation. This may lead to an underestimation of the carcinogenic potency of the agent in humans — a significant error with adverse implications for public health. It is also inconsistent to ignore the improved information on extrapolation of the metabolic events provided by the PBPK model, by continuing to apply the 3/4-power relationship as if no model were available.

This debate has been conducted with some vigor in recent years in the arena of noncancer toxic effects, where uncertainty factors are used to extrapolate from a no adverse effect level in an experimental species to a level presumed safe for humans. In the absence of additional information, the long-standing default assumption has been to apply a tenfold uncertainty factor to results in rodents. Where pharmacokinetic information is available in the test species and in humans, this factor is now usually divided into two parts ($\sqrt{10}$, or 3.16, approximating to threefold in each case) representing uncertainty due to toxicokinetic factors on one hand, and toxicodynamic factors (i.e., differences in sensitivity to the critical effect) on the other. There are only a few cases where the toxicodynamic uncertainty factor can be replaced by objective data, but it has become increasingly usual to replace the toxicokinetics factor, either with an actual PBPK model, or (for inhaled materials) with the "Human Equivalent Concentration" methodology developed by US EPA (1994).

Similar division of the assumed interspecies extrapolation factor in cancer risk assessment might be proposed. This would probably not cause much dispute in the few cases where a margin-of-exposure approach is applied to the LED_{10} for tumor induction. Usually, however, risk is predicted by linear extrapolation, and the power relationship to body weight is customarily used for interspecies adjustment. So far, there is no clear agreement as to how this adjustment (which in practical terms functions as a type of uncertainty factor) should be subdivided, if at all. Solution of the dilemma in a particular case remains the responsibility of the individual risk assessor.

23.4.9 STRUCTURE–ACTIVITY RELATIONSHIPS — TEFs AND PEFs

In general, risk assessment relies only on cancer potencies derived from bioassay data with the specific compound of interest. Theoretically, structure–activity relationships might be used to estimate cancer potencies for compounds for which no suitable bioassay data are available. They may also assist in route-to-route extrapolation, where data are only available for experimental routes of exposure that are not relevant to the human situation of interest. Only two cases are currently accepted in which potencies are estimated using structure–activity relationships. The first is for the various congeners of the chlorinated dioxins, dibenzofurans, and coplanar PCBs (often referred to as the "dioxin-like compounds"). The second is for certain PAHs that are known to be carcinogenic from studies using oral, dermal, or intraperitoneal exposures, but for which there are no experimental data by inhalation. This route is important for many PAHs, which are regularly encountered as air pollutants produced by incomplete combustion.

For the dioxin-like compounds, bioassay data for individual congeners are only available for (TCDD) and 2,3,4,7,8,9-hexachlorodibenzodioxin. These environmental contaminants (and the PCBs which were formerly produced industrially) are usually encountered as mixtures of many of the possible congeners. There is substantial evidence that many or all of these components contribute to the toxicological effects, including carcinogenesis, of the mixtures. Estimates of the potency of all individual congeners are therefore required to estimate the effects of mixtures, which vary in composition due to differences in their sources, and as a result of differential bioconcentration and degradation in the environment.

The mode of action of the dioxin-like compounds has been shown to involve binding with high affinity to the Ah receptor, which is responsible for control of expression of various cytochrome P-450 enzymes as well as a range of growth regulatory events. Toxic effects observed include carcinogenesis, immunotoxicity, teratogenicity, various adverse endocrine impacts, and dermal toxicity. All of these are believed to relate directly or indirectly to continuous and inappropriate stimulation of the Ah receptor. Standard bioassays on all the possible individual compounds would be enormously expensive and time-consuming, and have not been performed. In contrast, a number of biochemical measurements *in vitro* and short-term toxicity experiments are feasible and have been conducted for most or all of the dioxin-like compounds. Results from these experiments provide a ranking of all the congeners with regard to their ability to bind to the Ah receptor. This is assumed to represent the relative toxicity for all the related endpoints, including carcinogenicity. The ranking is provided as a table of factors (Toxic Equivalency Factors or TEFs), relative to the index compound TCDD, for which a carcinogenic potency is available. The factors are rounded to the nearest order of magnitude. The effective dose of a mixture is calculated in terms of an equivalent dose of TCDD (Toxic equivalents or TEQ). This scheme has been used, in various internationally agreed iterations, by many agencies including the WHO, the US EPA, and the State of California. Its latest version (currently used in draft reports from US EPA and California's Toxic Air Contaminants program) appears in reports from the WHO expert committee (van Leeuwen, 1997; van den Berg et al., 1998).

A similar approach was developed by the State of California to estimate the carcinogenicity of mixtures of PAHs by inhalation. For a range of PAHs identified as carcinogenic in animals by IARC, a ranking of potency by inhalation was predicted by comparing the carcinogenic potency by other routes such as oral, intraperitoneal, or in the commonly used skin carcinogenesis assay with promotion by phorbol ester (Collins et al., 1998). The potency ranking (similar to the dioxin TEFs) was described in a table of factors relative to the index compound benzo[a]pyrene (BaP), in this case referred to as Potency Equivalency Factors (PEFs). Estimated factors are rounded to the nearest order of magnitude, although more exact factors, for compounds having inhalation potencies based on bioassay data, are included in the calculation of total BaP dose equivalents. This scheme is used by California's Toxic Air Contaminants program.

23.4.10 SOURCES FOR SLOPE FACTORS AND SIMILAR STANDARD VALUES

Estimation of carcinogenic potency requires substantial mathematical resources and the application of scientific judgment and experience. Often, an individual risk assessor dealing with a site-specific situation will not generate these estimates *de novo*. Instead, the values used are either mandated by the regulatory program(s) addressed by the assessment or obtained from a source seen as reliable and widely accepted. Where the regulatory program specifies the potencies, no further discussion is needed unless the mandated values are seen as unreasonable or outdated. However, where the situation requires selection of suitable values, an appropriate source must be chosen. For hazard identification, the international preeminence of IARC has been noted, but this organization does not perform quantitative risk assessment or recommended potency values, and no other single organization of equivalent stature does so. Within the United States a general default assumption has been to use values calculated by the US EPA for pesticides and environmental contaminants. Most slope factors approved for general use by US EPA programs appear in the IRIS database (US EPA, 2003). Other federal agencies such as the FDA and OSHA (advised by NIOSH) also provide values for compounds within the scope of their interests. The State of California also provides potency values for many carcinogens (available on the OEHHA Web site), expressed either as potencies (slope factors or unit risks; see OEHHA, 1999b), or as risk-specific intake levels such as the No Significant Risk Levels provided by the Proposition 65 program and the Public Health Goal levels developed for the California drinking water program. In a situation unconstrained by regulatory requirements, the risk assessor may wish to examine all these sources to find a suitable value. In case of disagreement between agencies, the preferred choice will be the value using the most recent data or otherwise appearing most suitable to the assessment being undertaken. When using carcinogenic potency values from any of these sources, the risk assessor is encouraged to review the assessment documentation provided with the number, especially the risk characterization section as described below, rather than simply taking values from a likely looking table without further consideration.

23.5   RISK CHARACTERIZATION

23.5.1   PURPOSE

The risk characterization section of a risk assessment document integrates the conclusions of the earlier sections (hazard identification, exposure assessment, and dose–response assessment), to provide an analysis of the problem that prompted the risk assessment (NRC, 1996). It generally takes the form of a narrative summarizing the conclusions of the earlier sections and, where necessary, additional calculations to combine these.

23.5.2   CALCULATION OF INDICATOR VALUES

In an assessment of the risk experienced by a maximally exposed individual at a specific site, the exposure analysis provides estimated exposures to those carcinogens identified as posing a potentially significant hazard. The dose–response analysis will present the carcinogenic potencies of these agents (derived or obtained from reference sources). In the simple case, the predicted doses are multiplied by the potencies to obtain the estimated lifetime risk from these exposures. A risk assessment for a regulatory program might include calculations in the risk characterization section to provide proposals for safe levels of exposure to the material of interest. These are expressed in terms of carcinogenic potencies, "unit risks" (risk per unit of concentration in an environmental medium), maximum acceptable daily intake, or other values required by the program for which the assessment was prepared. These calculations use the potency, and the assumed or modeled uptake rates, bioavailability, or metabolic conversions detailed in the earlier exposure assessment and dose–response sections.

For instance, the intake level (I, in mg/day) associated with a specified lifetime excess individual cancer risk R, from exposure to a carcinogen is

$$I = \frac{R\mathrm{bw_h}}{q_1^*}$$

where $\mathrm{bw_h}$ is the body weight (commonly 70 kg for adult humans), and q_1^* the cancer potency estimate for the carcinogen, in $(\mathrm{mg/kg\text{-}day})^{-1}$. Similarly, the public health protective concentration (C) for a chemical in an environmental medium such as air or drinking water may be calculated as follows:

$$C = \frac{R\mathrm{bw_h}}{q_1^*\mathrm{Vol}}$$

where Vol is the volume of the medium to which the individual is exposed per day. This might be the daily volume of water consumed (a default of 2 l/day for an adult, plus any other volume in liter equivalents/day that may be needed to account for additional inhalation and dermal exposures from household use of drinking water). Similarly for air exposures, Vol would be the volume of air inhaled per day, by default 20 m^3. The concentration might be expressed in mg/l or μg/m^3 depending on whether q_1^* is in mg or μg per kg-day, and the volume in liters or cubic meters.

The risk manager (rather than the risk assessor) determines what constitutes a reasonable exposure level for a carcinogen or other toxic agent. However, for some programs the acceptable level of risk is defined in advance for any carcinogen. Where this definition exists, it is appropriate to include in the risk characterization a comparison of the calculated risk to the predefined standard. This applies when the program defines a specific numerical risk standard, rather than a nebulous and unreachable standard such as "zero risk." In other cases, it may be helpful to the risk managers and other readers of the assessment to include for comparison commonly accepted lifetime risk standards such as 10^{-6} (the most widely used cutoff level for "negligible" risk), 10^{-5} (a commonly used action level for public notification programs such as California's Safe Drinking Water and Toxics Enforcement Act, a.k.a. Proposition 65), and 10^{-4} (a level used as a trigger for immediate remediation in many programs at State and Federal level in the United States). The risk assessor may present calculations of doses required to reach these risk levels, or concentrations in relevant media assuming constant lifetime exposure or modeled briefer or intermittent exposures. A risk assessment using emission and exposure modeling from a facility may compare the risks calculated for various production or operation assumptions to the standard values.

23.5.3 Additivity of Carcinogenic Risks

The effects of simultaneous exposures to multiple genotoxic carcinogens are generally assumed to be independent, so the combined risk is the sum of the risks estimated for the individual exposures. In a few cases, synergistic or antagonistic interactions may be expected, usually due to induction of enzymes involved in activation or detoxification. The influence of such interactions may be predictable by pharmacokinetic models, but more often, the necessary information is not available so the risk assessor has to use the default assumption of additivity. Carcinogens acting by "hormonal" or receptor-based mechanisms, or others having important effects on cell proliferation as well as or instead of genotoxic effects, may have much more complex interactions in the case of multiple exposures, and there is no generic treatment of these cases. The risk assessor will have to consider the specific biological processes involved in each specific mechanism, and either (in the rare case where the necessary data are available) use these as the basis of a combined risk estimate, or (more usually) rely on a suitably cautious default assumption. Even in these cases, the assumption of additivity may be reasonable. For instance, this is implicitly assumed in the TEQ calculations for dioxin-like compounds, and is supported by data, at least for short-term effects in animals.

The effects of complex carcinogenic mixtures are difficult to predict from the sum of the effects of identified components. Mixtures such as cigarette smoke and diesel exhaust show greater carcinogenic activity than that predicted by adding up the potencies of the components (weighted by mass fraction of the total mixture). This may partly result from the presence of additional unidentified carcinogenic materials in the mixture, and partly from synergistic interactions between components, including both complete carcinogens and promoters. Fortunately, several of these complex and incompletely defined mixtures have been studied by animal bioassays or by epidemiology. Independent carcinogenic potency estimates are thus available for the complete mixtures (e.g., for diesel exhaust, Dawson and Alexeeff, 2001), which should generally be used in preference to sums of component values.

23.5.4 Uncertainty Estimates

A key component of a risk characterization is to present the degree of confidence in a particular estimate. Risk assessments have for many years included standardized statements emphasizing that the potency estimates derived by standard methodology are confidence bound estimates (usually 95%

upper bound), and that the "true" risk value is likely to be less than this estimate. With the development of stochastic exposure estimation methodology, it has become possible to more precisely estimate the plausible range for risks in specified circumstances, at least as far as uncertainties in exposure are concerned. Quantitative descriptions of the uncertainty associated with a potency estimate have been harder to provide. When using the linearized multistage model, instability of the model means that a comparison between the maximum likelihood estimate (MLE) and 95% upper confidence bound estimates is not necessarily a reliable guide to the limits or degree of confidence that may be placed on the estimate (although if this difference is very large it may indicate problems with the data or model fit). Statistical treatment of the benchmark dose method, such as comparisons between LED_{10} and ED_{10}, may be somewhat more acceptable, especially where the benchmark is within the range of the observed data. However, in any potency analysis, the "model uncertainty" is normally substantially larger than the uncertainty due to variability, measurement error, or other sources that are susceptible to statistical measurement. There is intrinsic uncertainty in extrapolating from doses in an animal experiment or an occupational exposure setting to substantially lower environmental exposures, and similarly in extrapolating from animal species to humans. No amount of statistical manipulation will reduce this uncertainty. It can be reduced by mechanistic studies and comparison of multiple data sources, but nevertheless remains large and difficult to quantify. Where there are important model uncertainties the uncertainty of the risk estimate may be very large, as for instance when one proposed mechanism of action suggests a linear extrapolation to low doses whereas another implies a threshold.

It is important that these uncertainties are described in the risk characterization, but not exaggerated or used to deflect legitimate concerns. The extent to which the results of bioassays at high dose levels in animals (especially the $B6C3F_1$ mouse) reflect risk to humans under plausible exposure scenarios has been extensively debated. Doubts about positive findings in such experiments have been supported in some cases. For instance, an isolated finding of tumors in male mouse liver, but at no other sites and in no other species or sex, is often discounted as an indicator of risk to humans. On the other hand, tumor findings in mice overall are substantially concordant with positive findings in other species, and they are usually thought to indicate plausible risk to humans. Recent judgments of IARC on trichloroethylene and tetrachloroethylene, which are based partly on human epidemiological data, have substantially upgraded their evaluation of the human carcinogenic potential of these compounds. This is in spite of the fact that these compounds were previously at the center of the debate about the relevance or otherwise of the positive tumor findings in mouse liver.

23.5.5 TARGET AUDIENCES

A risk characterization section needs to reflect the audience(s) for whom the risk assessment is being prepared. For the analytical sections (hazard identification, exposure, and dose–response assessment) the standard of writing is the same as for any scholarly scientific publication. Sufficient rigor and detail should be provided so that another expert in the field could understand, repeat, and fully evaluate the calculations and conclusions presented. Although the risk characterization section still needs to meet these standards, it also needs to be comprehensible to its target audience. This may include others such as risk managers (who may be highly qualified in other disciplines such as engineering, but not in cancer biology or statistics), economists, various interested parties at the site or situation being considered, or even members of the general public. This section would be written differently for different audiences, for example:

- A risk assessment presenting improved methodology in a scientific publication, intended for technical risk assessment specialists.
- The regulatory assessment of a chemical, whose primary audience is risk managers in regulatory enforcement programs.
- A site assessment to guide engineers in designing cleanup strategies, which also needs to respond to substantial local public interest.

In fact, the needs of the various interested groups may be so diverse that different versions of the risk characterization may need to be written to address their requirements.

23.6 CONCLUSIONS

23.6.1 Do Current Risk Assessment Methods Serve Their Stated Objectives?

Current knowledge and risk assessment methodology are only partly successful in meeting the objectives of those using the techniques or depending on their results. This is inevitable given the uncertainty that is part of the risk assessment paradigm. If one could be certain of the outcome in given circumstances, it would not be necessary to extrapolate from known to unknown situations, or to use probabilistic concepts such as likelihood. Concerns about risk assessment as a scientific discipline are expressed by those who want, or are used to dealing with, conclusions where the confidence in the prediction can be described in everyday terms as certainty. (Theoretical physicists describe the physical universe in terms of probability functions and uncertainty distributions, but the average person in the street does not use these concepts except when gambling.) The desire for a "certain" or "just" conclusion with no permissible doubts or alternatives is one reason for the long-running conflict of viewpoint between scientific risk assessors and the law. It may be necessary in social or economic terms to have the legal system make hard-and-fast decisions on a particular issue, but such legal decisions appear arbitrary to the scientist, and are often strongly influenced by the individuals involved in the legal decision-making process and their particular interests, abilities, and knowledge.

Some of the often-expressed dissatisfaction with scientific risk assessment is therefore an inevitable consequence of the nature of the activity and the context in which it operates. On the other hand, scientific developments such as improved understanding of toxicological mechanisms, and improved assessment methodology, may significantly reduce the uncertainties and possibility for error. Methods of analysis should accurately quantify the range of possible outcomes in terms of uncertainty (caused by lack of knowledge about the situation), variability (actual variation in the values of quantities that are important in determining the outcomes — such as individual differences in exposure or sensitivity), and measurement error. Recent developments in risk assessment methodology emphasize the importance of these concepts, and individual assessments are now required to include such descriptions. This should in turn assist the risk manager to make well-informed and well-balanced decisions and also improve the acceptability of the eventual regulatory and practical outcomes to those affected by them.

23.6.2 Current Issues and Problems

23.6.2.1 Uncertainty and Variability

The concepts of variability and uncertainty have been discussed already. This is an area of active development in the methods used in risk assessment. The description of variability is well understood by statisticians, and methods are available for measuring it and predicting its effect on observations. This knowledge has been increasingly applied in risk assessment, especially in the area of exposure measurement and modeling. It is also used when considering inter-individual variability, both in data analysis (of epidemiological studies and animal experiments) and in risk estimation, where the characteristics of an exposed population are considered. Estimation of experimental error in measurements is also a well-known and extensively studied concept, especially in the physical sciences; thus any exposure measurement description requires appropriate discussion of accuracy, precision, detection limits, and so on. However, the risk assessor often faces an additional problem, usually described as "model uncertainty" — where the underlying mechanisms of carcinogenesis are not fully understood, and plausible alternatives predict widely different risks. This uncertainty can be large compared with variability and measurement error, yet it is hard to quantify. Some such efforts have been described in the recent risk assessment literature, with varying degrees of success. Any success is an improvement over the previous way of handling this issue, which was to ignore it. Usually attempts to characterize model uncertainty have started with the exploration of the implications of different possible mechanistic models or interpretations of the experimental data, and at least presented the risk implications of these alternatives. Some attempts have been made to present a systematic and unbiased choice between the alternatives, or to provide a quantitative weighting of the risk implications of the different possibilities, but there is no standard methodology at present for doing this.

23.6.2.2 Children's Health Issues

The concept of variability in the target population has been explored in adults having differences in physiological, biochemical, or other properties that contribute to susceptibility, and different circumstances or behaviors that affect exposure. These concerns apply with even greater force to the susceptibility of children to environmental pollutants. The original assumption that children are basically small adults has been strongly criticized by pediatricians and developmental physiologists. It is now understood that many important physical characteristics and behavior patterns distinguish different human life stages. Size, metabolic capability and needs, and types and levels of activity change drastically, especially during infancy and early childhood. These are anticipated to affect exposure and sensitivity of children to environmental toxic agents. Furthermore, structural and functional development of various major organ systems (particularly the respiratory, immune, nervous, and endocrine/reproductive systems) continues to a very substantial degree during childhood and into adolescence, in addition to the overall growth in size and cell number of all organs. This may not only change sensitivity to toxic endpoint observed in adults, but also offer whole new possibilities for toxic agents to modify developmental processes, which have no precedent in adults. Risk assessors have only recently begun to address these issues systematically, at least in part as a result of public concern about demonstrated effects such as the increase in childhood asthma, and the impact of childhood lead exposure on intelligence. There are also concerns about less well-characterized possibilities, such as the impact of endocrine disruptors. Although experimental science has examined effects *in utero* for many years (ever since the Thalidomide effect was recognized), the study of postnatal juvenile-specific toxicity is a newer field, and so far, the assessment of risks for children is significantly hampered by lack of needed data. There are some indications of differential sensitivity to carcinogenesis in children and young animals, over and above the "shelf life" effect, implied by the Armitage–Doll model, such as those seen for vinyl chloride (Cogliano et al., 1996). However, analysts have not so far agreed on a general method for including these concerns in cancer risk assessments. California risk assessors recently presented a review of possible special impacts on children from both carcinogens and other toxic substances that are found as air pollutants (OEHHA, 2001). The current draft of the revised US EPA carcinogen risk assessment guidelines (US EPA, 2003) is accompanied by a supporting document that examines ways of accommodating the special sensitivities of children in assessments of carcinogenic risk. An adjustment factor of ten for exposures in the first 2 years of life, or three for exposures at ages of 3 to 15, is suggested for risk assessments using an underlying "adult" potency value obtained from epidemiological studies or lifetime animal bioassays. However, these proposals are limited in scope (addressing only "genotoxic" carcinogens), and rely primarily on proposed policy defaults rather than actual data or validated models. This issue is currently being studied actively both by laboratory and epidemiological research investigators and by regulatory risk assessors.

23.6.2.3 "Incorporating the Latest Scientific Knowledge" Versus the Precautionary Principle

A debate along these lines occurs frequently in relation to regulatory risk assessment of major industrial chemicals and pesticides, where stringent regulations have already been set in order to protect public health and these regulations are perceived by users or manufacturers of the chemicals to have a large economic impact on their activities. Interested parties will often fund substantial scientific investigations in the hope of generating data that reduces the uncertainty involved in the regulatory risk assessment, and convincing the regulators and other guardians of public health that a particular chemical is not as carcinogenic as they previously thought. (Less frequently, research is undertaken to investigate newly appreciated hazards or to determine whether regulators have not previously been overoptimistic in their assessment of a health impact.) Reduction of uncertainty is important from a scientific standpoint. If new data are available that indicate with good confidence that a previous risk estimate (which may have had to allow for some adverse possibilities that can now be discounted) can be revised downward, then clearly the new data should be used and the assessment updated. Often however, new data may suggest the possibility of a lower risk estimate, but fall short of establishing the new proposal as the true interpretation. Perhaps a new metabolic pathway is identified that has low risk implications, but is the earlier one discounted or do the two operate in parallel, and if so in what ratio? The risk assessor is then confronted with model uncertainty, and will not adjust the risk estimate downward. Some interested parties and experimental scientists are accustomed to espousing a new theory "on balance" that is if it is considered to have a probability of correctness >50%. On the other hand, the risk assessor needs to allow for highly adverse implications of a theory with (e.g.) only 10% probability of correctness. This is often described as the "precautionary principle" — possible adverse consequences to public health need to be avoided even when the probability of their occurrence is considered low, especially if the potential for harm is great. While it has been argued that this principle has been abused or applied overzealously at times, it remains a core principle of public health risk assessment.

REFERENCES

Abbott, W.S. (1925). A method of computing the effectiveness of an insecticide. *J. Econ Entomol.*, 18: 265–267.

Anderson, E.L. and the U.S. Environmental Protection Agency Carcinogen Assessment Group (1983) Quantitative approaches in use to assess cancer risk. *Risk Anal.*, 3, 277–295.

Armitage, P. and Doll, R. (1954) The age distribution of cancer and a multistage theory of carcinogenesis. *Br. J. Cancer*, 8, 1–12.

Bailar, J.C. and Bailer, A.J. (1999) Risk assessment — the mother of all uncertainties. Disciplinary perspectives on uncertainty in risk assessment. *Ann. N.Y. Acad. Sci.*, 895, 273–285.

Barrett, J.C. and Wiseman, R.W. (1987) Cellular and molecular mechanisms of multistep carcinogenesis: relevance to carcinogen risk assessment. *Environ. Health. Perspect.*, 78, 65–70.

Benya, T.J., Busey, W.M., Dorato, M.A., and Berteau, P.E. (1982) Inhalation carcinogenicity bioassay of vinyl bromide in rats. *Toxicol. Appl. Pharmacol.*, 64, 367–379.

Chemical Industry Institute of Toxicology (CIIT, 1999) *Formaldehyde: Hazard Characterization and Dose–Response Assessment for Carcinogenicity by the Route of Inhalation*, revised ed. Chemical Industries Institute of Toxicology, September 28, 1999, Research Triangle Park, NC. (Unfortunately, the part of these studies relating to the cell proliferation model has not yet been published in the open literature, but this report is currently available on request from CIIT.)

Cogliano, V.J., Hiatt, G.F.S., and Den, A. (1996) Quantitative cancer assessment for vinyl chloride: Indications of early-life sensitivity. *Toxicology*, 111, 21–28.

Collins, J.F., Brown, J.P., Alexeeff, G.V., and Salmon, A.G. (1998) Potency equivalency factors for some polycyclic aromatic hydrocarbons and polycyclic aromatic hydrocarbon derivatives. *Regul. Toxicol. Pharmacol.*, 28, 45–54.

Creech, J.L., Jr., and Johnson, M.N. (1974) Angiosarcoma of liver in the manufacture of polyvinyl chloride. *J. Occup. Med.*, 16, 150–151.

Crouch, E. (1983). Uncertainties in interspecies extrapolations of carcinogenicity. *Environ. Health Perspect*, 5: 321–327.

Crump, K.S. (1981) An improved procedure for low-dose carcinogenic risk assessment from animal data. *J. Environ. Pathol. Toxicol.*, 52, 675–684.

Crump, K.S. and Howe, R.B. (1984) The multistage model with a time-dependent dose pattern: Applications to carcinogenic risk assessment. *Risk. Anal.*, 4, 163–176.

Crump, K.S., Guess, H.A., and Deal, L.L. (1977) Confidence intervals and test of hypotheses concerning dose–response relations inferred from animal carcinogenicity data. *Biometrics*, 33, 437–451.

Dawson, S.V. and Alexeeff, G.V. (2001) Multi-stage model estimates of lung cancer risk from exposure to diesel exhaust, based on a U.S. railroad worker cohort. *Risk Anal.*, 21, 1–18.

Garshick, E., Schenker, M., Munoz, A., Segal, M., Smith, T., Woskie, S., Hammond, S., and Speizer, F. (1988) A retrospective cohort study of lung cancer and diesel exhaust exposure in railroad workers. *Am. Rev. Respir. Dis.*, 137, 820–825.

Ginsberg, G., Smolenski, S., Hattis, D., and Sonawane, B. (2002) Population distribution of aldehyde dehydrogenase-2 genetic polymorphism: Implications for risk assessment. *Regul. Toxicol. Pharmacol.*, 36, 297–309.

Gold, L.S., Sawyer, C.B., Magaw, R., Backman, G., deVeciana, M., Levinson, R., Hooper, N.K., Havender, W.R., Bernstein, L., Peto, R., Pike, M., Ames, B.N. (1984). A carcinogenic potency database of the standardized results of animal bioassays. *Environ. Health Perspect.*, 58: 9–319.

Hill, A.B. (1965) The environment and disease: Association or causation? *Proc. Roy. Soc. Med.*, 58, 295–300.

Hill, A.B. (1971) Statistical evidence and inference. In *Principles of Medical Statistics.*, (A.B. Hill, Ed.) Oxford University Press, Oxford, UK/New York, pp. 309–323.

Hill, R. (1925) *Proc. R. Soc.*, B100, 419. For a more accessible description of the Hill equation and its modern interpretations and implications, see any standard biochemistry textbook.

International Agency for Research on Cancer (IARC, ongoing series) *IARC Monographs on the Evaluation of Carcinogenic Risks to Humans*, IARC, Lyon. See also http://monographs.iarc.fr/.

Krewski, D., Crump, K.S., Farmer, J., Gaylor, G.W., Howe, R., Portier, C., Salsburg, D., Sielkin, R.L., van Ryzin, J. (1983). A comparison of statistical methods for low dose extrapolation utilizing tie-to-tumor data. *Fund Appl Toxicol* 3: 140–160.

Leung, H.-W. (1999) Chapter 7: Physiologically based pharmacokinetic modeling. In *General and Applied Toxicology*, Ballantyne, B., Marrs, T.C., Syversen, T., Eds. McMillan, London, UK, 1999 and Groves' Dictionaries, New York, 1999, chapter 7.

Lilienfeld, A.M. and Lilienfeld, D.E. (1980) *Foundations of Epidemiology.* Oxford University Press, Oxford, UK and New York, 375 pp.

Mancuso, T. (1975) Consideration of chromium as an industrial carcinogen. In *Proceedings of the International Conference on Heavy Metals in the Environment*. Toronto, Ontario, Canada, pp. 343–356.

Mancuso, T. and Hueper, W. (1951) Occupational cancer and other health hazards in a chromate plant: A medical appraisal. I. Lung cancers in chromate workers. *Indust. Med. Surg.*, 20, 358–363.

Medinsky, M.A. and Valentine, J.L. (2001) Toxicokinetics. In *Casarett and Doull's Toxicology*, 6th ed. Klaassen, C.D., Ed. McGraw-Hill, New York, 2001, Chapter 7.

Meigs, J.W., Marrett, L.D., Ulrich, F.U., and Flannery, J.T. (1986) Bladder tumor incidence among workers exposed to benzidine: A thirty-year follow-up. *J. Natl. Cancer Inst.*, 76, 1–8.

Moolgavkar, S.H. and Knudson, A.G. (1981) Mutation and cancer: A model for human carcinogenesis. *J. Natl. Cancer Inst.*, 66, 1037–1052.

National Research Council (NRC, 1983) Risk Assessment in the Federal Government: Managing the Process. (Committee on the Institutional Means for Assessment of Risks to Public Health, Commission on Life Sciences, National Research Council) National Academies Press, Washington, DC, 191 pp.

National Research Council (NRC, 1994) Science and Judgment in Risk Assessment. (Committee on Risk Assessment of Hazardous Air Pollutants, National Research Council) National Academies Press, Washington, DC, 1983, 672 pp.

National Research Council (NRC, 1996) Understanding Risk: Informing Decisions in a Democratic Society. (Stern, P.C. and Fineberg, H.V., Eds., Committee on Risk Characterization, National Research Council.) National Academies Press, Washington, DC, 264 pp.

National Toxicology Program (NTP, 1986) Toxicology and Carcinogenesis Studies of Tetrachloroethylene (Perchloroethylene) in F344/N Rats and B6C3F$_1$ Mice (Inhalation Studies). TR 311, NIH publication no. 86-2567. National Toxicology Program, Research Triangle Park, NC.

National Toxicology Program (NTP, 1998) Toxicology and Carcinogenesis Studies on Chloroprene in F344/N Rats and B6C3F$_1$ Mice. TR 467, NIH publication no. 98-3957. National Toxicology Program, Research Triangle Park, NC.

Office of Environmental Health Hazard Assessment, California Environmental Protection Agency (OEHHA, 1999a) Public Health Goal for Methyl Tertiary Butyl Ether (MTBE) in Drinking Water. Available from http://oehha.ca.gov/water/phg/pdf/mtbe_f.pdf.

Office of Environmental Health Hazard Assessment, California Environmental Protection Agency (OEHHA, 1999b) Air Toxics Hot Spots Program: Risk Assessment Guidelines, Part II. Technical Support Document for Describing Available Cancer Potency Factors. Available from the OEHHA Web site: http://www.oehha.ca.gov/air/cancer_guide/hsca2.html#download.

Office of Environmental Health Hazard Assessment, California Environmental Protection Agency (2000) Air Toxics Hot Spots Program: Risk Assessment Guidelines, Part IV. Exposure Assessment and Stochastic Analysis Technical Support Document.

Office of Environmental Health Hazard Assessment, California Environmental Protection Agency (OEHHA, 2001) Prioritization of Toxic Air Contaminants — Children's Environmental Health Protection Act — Final Report. Available from the OEHHA website: http://www.oehha.ca.gov/air/toxic_contaminants/SB25finalreport.htm.

Poiley, S.M. (1972) Growth tables for 66 strains and stocks of laboratory animals. *Lab. Animal Sci.*, 22, 759–779.

Portier, C.J., Kopp-Schneider, A., and Sherman, C.D. (1993) Using cell replication data in mathematical modeling in carcinogenesis. *Environ. Health Perspect.*, 101 (Suppl. 5), 79–86. (This volume presents the proceedings of a conference on cell proliferation and carcinogenesis, and contains a number of review articles relevant to this topic.)

Pott, P. (1775) *Chirurgical Observations Relative to the Cataract, the Polypus of the Nose, the Cancer of the Scrotum, the Different Kinds of Ruptures, and the Mortifications of the Toes and Feet.* London, UK, Hawse, Clark and Collins.

Redmond, C.K. (1983) Cancer mortality among coke oven workers. *Environ. Health Perspect.*, 52, 67–73.

Rothman (1986) *Modern Epidemiology.* Little, Brown & Co., Boston, M.A. and Toronto, Canada, 358 pp.

Salmon, A.G., Monserrat, L., and Brown, J.P. (1992) Use of a pharmacokinetic model in cancer risk assessment for vinyl bromide. Presented at the Society of Toxicology Annual Meeting, Seattle, WA, February 1992. Abstract: *Toxicologist*, 12, 96.

US EPA (1986) Guidelines for Carcinogen Risk Assessment. EPA/630/R-00/004. Risk Assessment Forum, U.S. Environmental Protection Agency, Washington, DC, September 1986. Published: Federal Register 51, 33992–34003, September 24, 1986.

See the US EPA's website: http://www.epa.gov/ncea/raf/car2sab/guidelines_1986.pdf.

US EPA (1994) U.S. Environmental Protection Agency Methods for Derivation of Inhalation Reference Concentrations and Application of Inhalation Dosimetry. EPA/600/8-90/066F. Office of Research and Development. Washington, DC.

US EPA (1995) Use of the Benchmark Dose Approach in Health Risk Assessment. EPA/630/R-94/007. 01 Feb. 1995. U.S. Environmental Protection Agency, Risk Assessment Forum, Washington, DC, 100 pp.

US EPA (1988). Recommendations for and documentation of biological values for use in risk assessment. EPA/600/6-87-008. Office of Health and Environmental Assessment, Cincinnati, OH, 1988.

US EPA (1999) IEUBK Model Bioavailability Variable. EPA #540-F-00-006 OSWER #9285.7-32 October 1999. Office of Solid Waste and Emergency Response, U.S. Environmental Protection Agency, Washington, DC. See http://www.epa.gov/superfund/programs/lead/products/sspbbioc.pdf.

US EPA (2000a) Toxicological Review of Vinyl Chloride in Support of Summary Information on the Integrated Risk Information System (IRIS). U.S. Environmental Protection Agency, Washington, DC, May 2000.

US EPA (2000b). Draft Exposure and Human Health Reassessment of 2,3,7,8-Tetrachlorodibenzo-p-Dioxin (TCDD) and Related Compounds. U.S. Environmental Protection Agency, Washington, DC, September 2000. On the US EPA's website: http://cfpub.epa.gov/ncea/cfm/dioxreass.cfm.

US EPA (2003a) Draft Final Guidelines for Carcinogen Risk Assessment (External review draft, February 2003). NCEA-F-0644A. 03 March 2003. U.S. Environmental Protection Agency, Risk Assessment Forum, Washington, DC, 125 pp. (1999). See the US EPA's website (NCEA/ Risk assessment forum): http://cfpub.epa.gov/ncea/raf/recordisplay.cfm?deid=55868 for this document, prior versions and supporting materials. Note: On March 29, 2005 the US EPA released a final version of their Guidelines for Carcinogen Risk Assessment, which is available at http://cfpub.epa.gov/ncea/cfm/recorddisplay.cfm?deid=11628. The basic elements of the guidelines are not different from those in the most recent draft version published in 2003, although there are some updates and revisions at the detail level.

US EPA (2003b) Integrated Risk Information System (IRIS). On the US EPA's website: http://www.epa.gov/iris/.

van den Berg, M., Birnbaum, L., Bosveld, A.T.C., Brunstrom, B., Cook, P., Feeley, M., Giesy, J.P., Hanberg, A., Hasegawa, R., Kennedy, S.W., Kubiak, T., Larsen, J.C., Van Leeuwen, F.X.R., Liem, A.K.D., Nolt, C., Peterson, R.E., Poellinger, L., Safe, S., Schrenk, D., Tillitt, D., Tysklind, M., Younes, M., Waern, F., and Zacharewski, T. (1998) Toxic equivalency factors (TEFs) for PCBs, PCDDs, PCDFs for humans and wildlife. *Environ. Health Perspect.*, 106, 775–792.

van Leeuwen, F.X.R. (1997) Derivation of toxic equivalency factors (TEFs) for dioxin-like compounds in humans and wildlife. *Organohalogen Compounds*, 34, 237.

24 Prevention and Regulation Approaches to Carcinogens

Eula Bingham and Jon Reid

CONTENTS

Currently, in general, the public assumes carcinogens are regulated by standards and requirements that do not allow these agents to enter their lives through the food supply, consumer products, workplace, or through environmental media such as water or air. In fact, many carcinogens are not regulated by the United States Federal Government. The eleventh United States Public Health Service (USPHS) Report on Carcinogens [1] lists the carcinogens and also whether or not they are regulated.

This chapter presents the evolution of preventive measures and regulations to reduce exposure through more than a century of recognition that certain chemical and physical agents cause cancer. Most of the recognition of cancer-causing chemicals and physical agents came from studies of worker populations. These early discoveries are presented in Table 24.1. As can be seen, these early findings alerted the medical community to the hazardous nature of certain agents. Examination of the steps taken by the medical community and the various regulatory authorities reveal the approaches to preventing cancer induced by chemical and physical agents.

24.1 EARLY HISTORY

The efforts in the United States to prevent cancer arising from exposures to chemical and physical agents lagged behind Europe. This lag is probably because of the advanced development of occupational medicine in Europe in contrast to the United States. Dr. Alice Hamilton was a physician in the United States and one of the very few who were interested in diseases of workers. Dr. Hamilton wrote in her autobiography that when she attended meetings in Europe on occupational diseases before World War I, she was chided because the United States did not know about occupational diseases [2].

TABLE 24.1
Early Industrial Carcinogens

Date	First Reported by	Reported Agent or Process	Site
1775	P. Pott	Soot	Scrotum
1822	J.A. Paris	Arsenic	Skin
1875	R. Volkmann	Crude wax from coal	Skin
1876	B. Bell	Shale oil	Skin
1879	F.H. Härting and W. Hesse	Ionizing radiation	Lung
1894	P.G. Unna	Ultraviolet radiation	Skin
1895	L. Rehn	Aromatic amine	Bladder
1898	S. Mackenzie	Creosote	Skin
1917	Leymann	Crude anthracene	Skin
1928	Delore	Benzene	Leukemia
1929	H.S. Martland	Radium	Bone
1935	K.M. Lynch and W.A. Smith	Asbestos production	Lung
1952	C.S. Weil et al.	Isopropyl alcohol manufacture	Sinuses and larynx

The history of environmental and occupational science tracks most closely with preventive measures and specific regulations of carcinogens. It has been necessary in some instances for specific cancer studies to be performed several times before there could be preventive measures or regulatory actions. Certainly, this was the case for complex mixtures derived from combustion of fossil fuels and aromatic amines in the last century and continues to date.

In the case of complex mixtures, it was only after numerous cases of cancer were reported to occur among workers in Britain exposed to pitch, tar, or tarry compounds that those carcinogens were added to the third schedule of the British Workmen's Compensation Act in 1907. That year, the definition, "scrotal epithelioma occurring in chimney sweeps and epitheliomatous cancer of ulceration of the skin occurring in handling or use of pitch, tar, or tarry compounds," was included [3]. In 1914, "bitumen, mineral oil or paraffin or any compound or products or residue of any of these substances," were added [3]. Finally, in 1920, a list of diseases was provided that required notification to the Chief Inspector of Factories under Section 73 of the existing Factory & Warehouse Act (now Section 66 of the Factories Act, 1937). This included the definition, "Epitheliomatous ulceration or cancer of the skin due to pitch, tar, bitumen, paraffin or mineral oil or any other compound or residue of any of these substances or any product there of contracted in a factory or workshop" [3].

One of the earliest and most important regulatory agencies is the Office of Inspector of Factories in Great Britain. A good example of the role of the Inspector of Factories is that of S.A. Henry, a Federal Inspector, who collected information on cancer among workers exposed to mineral oils. Descriptions of the numerous trades or types of manufacturing where tumors from these complex mixtures occurred were presented in a classic report from S.A. Henry, formerly H. M. Medical Inspector of Factories [3]. The use of personal protective clothing and improved personal hygiene (washing) was important in reducing exposures among these workers and as a result, cancer was reduced. This type of action remains central to prevention today among industrial hygienists and occupational physicians in reducing skin cancer and preventing the absorption of toxic agents into the systemic circulation.

Another example of the role of the European medical community coupled with the affected industry was the efforts to prevent bladder cancer caused by aromatic amines. Both Germany and England were in the forefront of the scientific investigations [4]. These efforts led to preventive measures and legislation to regulate beta-naphthylamine, benzidine, and other aromatic amines.

These measures were taken in the 1920s and 1930s, whereas the United States did little to regulate these compounds until the 1970s. Several states had regulations to prevent exposures that were developed in the 1940s (Pennsylvania, Massachusetts).

The first regulation of a carcinogenic agent in the United States is probably the decision made by the Food and Drug Administration (FDA) to keep 2-acetylaminofluorene out of the food supply in the late 1930s. This compound is an insecticide and was intended to replace such materials as arsenic and lead. When testing was performed in a chronic feeding study, it was found to be a potent carcinogen, and the FDA did not approve it. Not only were consumers protected, but also were the workers who would have manufactured 2-acetylaminofluorene in the 1930s and 1940s. We, now know how potent this carcinogen is, based on the many scientific publications of its toxicity and mechanism of action. Experimental studies with this compound have provided much of the modern understanding of electrophilic compounds and cancer induction [5]. Even under the most modern methods of manufacture it would be difficult to control exposure to this compound and would certainly expose pesticide applicators in the performance of their job.

Another landmark era in the history of regulating carcinogens was the late 1950s when Congress passed the Delaney Clause. In 1950, the United States began a comprehensive review of food additives, with a view toward remedying deficiencies in the basic 1938 Act, which covered a food additive only when a food product caused injury to the consumer. Hearings by the "Committee to Investigate the Use of Chemicals in Food" that became known as the Delaney Committee (chaired by Congressman Delaney) were held off until 1958, when the food additive amendments were passed. In 1958, the hearings included two days of testimony from a panel of scientists and experts selected by the National Academy of Sciences by request of the Subcommittee, and representatives from the Department of Health, Education, and Welfare [6]. The House Committee on Commerce passed a bill to which became attached an amendment that said: "Provided, that no additive shall be deemed to be safe if it is found to induce cancer when ingested by man or animal, or if it is found, after tests which are appropriate for the evaluation of the safety of food additives, to induce cancer in man or animals" [7]. The bill was passed by the Senate on September 6, 1958 wherein it became Public Law 85-929. This clause became known as the Delaney Clause.

24.2 THE ERA OF ENVIRONMENTAL REGULATION FOR CARCINOGENS

24.2.1 OCCUPATIONAL SAFETY AND HEALTH ADMINISTRATION (OSHA), 1970

At that time, in 1971, OSHA regulated carcinogens using the 6(a) authority of the Occupational Safety and Health Act. The Agency adopted Permissible Exposure Limits (PELs) for about 400 chemicals used in workplaces, of which many were recognized in the medical literature as potential carcinogens either by standard consensus organizations such as the American Conference of Governmental Industrial Hygienists (ACGIH) or American National Standards Institute (ANSI). The first carcinogen regulated by OSHA was asbestos, using the 6(b) rulemaking authority that says,

> The Secretary, in promulgating standards dealing with toxic materials or harmful physical agents under this subsection, shall set the standard which most adequately assures, to the extent feasible, on the basis of the best available evidence, that no employee will suffer material impairment of health or functional capacity even if such employee has regular exposure to the hazard dealt with by such standard for the period of his working life. Development of standards under this subsection shall be based upon research, demonstrations, experiments, and such other information as may be appropriate. In addition to the attainment of the highest degree of health And safety protection for the employee, other considerations shall be the latest available scientific data in the field, the feasibility of the standards, and experience gained under this and other health and safety laws. Whenever practicable, the standard promulgated shall be expressed in terms of objective criteria and of the performance desired [8].

After the passage of the Act, the standard proposed was finalized in 1972. While this specific chemical, asbestos, was regulated, no specific emphasis was placed on the cancer-causing properties of asbestos.

The next effort was to specifically identify and regulate carcinogens, and was instigated by a petition from the Health Research Group (HRG) and the Chemical and Atomic Workers (OCAW) in 1973. They argued that an Emergency Temporary Standard (ETS) should be issued for 14 carcinogens, which had been listed by the ACGIH, a private consensus standard organization mentioned earlier. This group of cancer-causing agents had been listed in the appendix of the ACGIH's Threshold Limit Values (TLVs) in 1968. The standard for these chemicals (often called the "14 carcinogens standard") in this first formal governmental attempt to reduce cancer in the workplace, became official in 1974. These 14 included *bis*(chloromethyl) ether, chloromethyl methyl ether, 2-naphthylamine, 1-naphthylamine, 4-aminobiphenyl, 2-acetylaminofluorene, 4-nitrobiphenyl, benzidine, 3,3'-dichlorobenzidine, B-propiolactone, 4-(dimethylamino)azobenzene, *N*-nitrosodimethylamine, 4,4'methylene *bis*(2-chlorobenzenamine), and ethylenimine.

The process that followed as a result of this petition was to establish an Advisory Committee that had 270 days to issue a report and to issue an ETS. Several significant recommendations were made by the Advisory Committee, most of which were adopted by OSHA over the subsequent months in the so-called 14 carcinogen standards (this became 13 after a procedural legal challenge). Some of the significant recommendations are as follows:

1. *No open vessels in carcinogen manufacturing*: While the ETS described precautions for handling of carcinogens in open vessels, the Advisory Committee was unanimous in recommending "no *open* vessels for carcinogens." This was a landmark decision in protecting workers. Up until that time, many vat operations were performed in the open air; carcinogens were literally shoveled and scrapped off filter presses.
2. *Carcinogen areas*: Controversial at the time, but now generally accepted throughout the industry, is the demarcation of areas where carcinogens are in use.
3. *Laboratory or bench scale operations regulation*: It was recognized that many small operations where carcinogens are used, particularly on a bench- or laboratory scale, should be performed in hoods with "appropriate" ventilation. It is interesting to note that . . . ."
4. *Signs of warning for carcinogens and a cancer hazard were required in workplaces*: Using the word cancer was argued by many to result in scaring workers so they would not perform the jobs. Today, it is generally recognized that specific warnings are necessary and legal.

It should be noted that the atmosphere of the Advisory Committee deliberations was highly charged and two persons resigned about midway through the 270-day period. They were replaced and on the last day a report was delivered to the Labor Department with the above recommendations.

Since that early rulemaking numerous other carcinogens have been regulated and are listed in Table 24.2.

Generally, a PEL (permissible exposure limit) is provided for the carcinogens based on a risk assessment and other requirements included in the OSHA Act, such as their technical and economic feasibility. It should be noted that at many of the listed PELs, the risk for developing cancer is substantial. For example, at the current benzene standard (1 ppm), estimates are that for the one specific type of cancer used in the risk assessment, acute myelogenous leukemia (AML) there could be three cancers of AML per thousand workers. Evidence now exists for other cancers of benzene-exposed workers, such as non-Hodgkins lymphoma, but it is not used in the risk estimates. The notion that the OSHA standards represent "safe levels" is not correct. They represent levels that are set based on criteria in addition to health, such as feasibility and are created at a certain point in time. Standards are updated infrequently.

The Hazard Communication Standard (HCS) regulates chemicals that pose health hazards. This standard requires chemical manufacturers or importers to evaluate chemicals based on criteria listed

TABLE 24.2
OSHA Regulated Carcinogens

Compound	OSHA PEL
2-Acetylaminofluorene	CFR 1910,10014[a]
Acrylonitrile	2 ppm C 10 ppm CFR 1910.1045
Arsenic, inorganic	0.01 mg/m^3 CFR 1910.1018
Asbestos	CFR 1910.1001
All forms	CFR 1910.1101
Amosite	0.2 fiber/cm^3
Chrysotile	0.2 fiber/cm^3
Crocidolite	0.2 fiber/cm^3
Other forms	2.0 fiber/cm^3
Tremolite	2.0 fibers/cm^3
Benzene	1 ppm, STEL 5 ppm CFR 910.1028
Coke-oven emissions	0.15 CFR 1910.1029
Benzidine	CFR 910.1010[a]
Benzo[a]pyrene as coal tar, pitch volatiles	0.2 mg/m^3
Cadmium, dust, and Salts, as Cd	0.2, C 0.6
Bis(Chloromethyl) ether	CFR 910.1008[a]
Chloromethyl methyl ether (also methyl chloromethyl ether)	CFR 1910.1006[a]
1,2-Dibromo-3-chloropropane (DBCP)	0.001 ppm CFR 1910.1044
3,3′-Dichlorobenzidine (and its salts)	CFR 910.1007[a]
4-(Dimethylamino) azobenzene	CFR 910.1015[a]
Ethylene oxide	1.0 ppm, STEL 9 ppm CFR 1910.1047
Formaldehyde	1.0 ppm, C 2.5 ppm CFR 1910.1048
1-Naphtylamine (α-napthylamine)	CFR 1910.1004[a]
2-Naphthylamine (β-napthylamine)	CFR 1910.1009[a]
4-nitrodiphenyl (also 4-nitrobiphenyl)	CFR 1910.1003[a]
B-Propiolactone	CFR 1910.1013[a]
Vinyl chloride	1.0 ppm, C 5 ppm CFR1910.1017

[a] one of original carcinogens — lowest feasible level.

in Appendix 13 of the standard for carcinogenicity. This appendix specifies that the determination of whether a chemical is a carcinogen or a potential carcinogen be based on findings by the National Toxicology Program (NTP), the International Agency for Research on Cancer (IARC), and OSHA. For positive determination to be a "hazardous material" such as a carcinogen, the following information is required in a material safety data sheet (MSDS):

1. *Identity of the material*: The identity on the MSDS must match the identity on the label. The MSDS must also provide the chemical and common names of the hazardous chemical. In the case of a mixture, the chemical and common names of the ingredients must be listed. If the mixture has been tested as a whole to determine its hazards, the chemical and common names of any ingredient contributing to the known hazards must be listed. If the hazardous chemical has not been tested as a whole, the chemical and common names of all hazardous ingredients that comprise 1% or greater of the composition must be listed, except for carcinogens, which must be listed if they comprise 0.1% or greater in the composition. In all cases, the chemical and common names of ingredients that present a physical hazard when present in the mixture must be listed. In some states, the Central Abstract Services (CAS) number is also required for each component. The standard does include provisions for protecting trade secrets of chemical manufacturers. The specific identity of hazardous chemicals can be withheld from the MSDS if this information

regarding the properties is a trade secret; however, all other information regarding the properties and effects of the chemical must be provided.

2. Physical and chemical characteristics.
3. *Physical hazards*: This includes information regarding potential for fire, explosion, and reactivity.
4. *Health hazards*: Health hazard information must include signs and symptoms of exposure, and medical conditions aggravated by exposure to the chemical.
5. *Primary routes of entry into the body*: Inhalation, ingestion, and skin.
6. *Exposure limits*: These limits include one OSHA PEL, the ACGIH TLV, or other exposure limit used or recommended.
7. *Carcinogenicity*: The MSDS must state whether the chemical has been determined to be a carcinogen or potential carcinogen by the NTP Annual Report on Carcinogens, IARC Monographs, or OSHA.
8. Precautions for safe handling and use.
9. *Control measures*: Including engineering controls, work practices, or personal protective equipment.
10. Emergency and first aid procedures.
11. Date of preparation.
12. *Name, address, and telephone number*: Of the chemical manufacture, importer, employer, or other responsible party preparing or distributing the MSDS.

The description of the programs for defining and classifying agents as carcinogens are provided in the following sections on the NTP and the IARC because these classifications, while not regulations in themselves, may trigger certain regulatory actions (Hazard Communications Standard and OSHA as previously indicated).

24.3 NATIONAL TOXICOLOGY PROGRAM

The Public Health Service act, Section 301(b)(4) as amended, requires the Department of Health and Human Services (DHHS) to publish a report containing a list of substances, which either are *known to be human carcinogens* or may *reasonably be anticipated to be human carcinogens*, and to which a significant number of persons in the United States are exposed. These reports are prepared by the NTP, and published as the *Report on Carcinogens* (*RoC*). It is an informational, scientific, and public health document that identifies and discusses agents, substances, mixtures, or exposure circumstances that may pose a carcinogenic hazard to human health. This report is a compilation of data on the carcinogenicity, genotoxicity, and potential biologic mechanisms of action for the listed substances (human and animal). It also includes the potential for exposure to these substances, and finally, the regulations promulgated by Federal agencies to limit exposures.

An agent, substance, mixture, or exposure circumstance can be listed in the *RoC* either as *known to be a human carcinogen* or as *reasonably anticipated to be a human carcinogen*. The *known* category is reserved for those substances for which there is sufficient evidence of carcinogencity from studies in humans that indicates a cause and effect relationship between exposure and human cancer. The *reasonably anticipated* category includes those substances for which there are limited evidences of carcinogenicity in humans and sufficient evidence of carcinogencity in experimental animals. Conclusions regarding carcinogenicity in humans and sufficient evidence of carcinogenicity in experimental animals are based on expert, scientific judgment, with consideration given to all relevant information [1].

The 11th Report on Carcinogens, Vol. II, provides a summary of the regulations for each chemical listed as a carcinogen. There is a summary of the regulations for each chemical and then a table

TABLE 24.3
Cadmium and Cadmium Compounds

First Listed in the *First Annual Report on Carcinogens as Reasonably Anticipated to be Human Carcinogens* updated to *Known to be Human Carcinogens* in the *eleventh Report on Carcinogens.*

Regulations

Under the Clean Water Act (CWA), the water quality criteria published by EPA for cadmium and its compounds for the protection of human health are identical to Safe Drinking Water Act (SDWA) standards of 10 μg/l. EPA's Carcinogen Assessment Group includes cadmium oxide, cadmium sulfide, and cadmium sulfate on its list of potential carcinogens. Under the Comprehensive Environmental Response, Compensation, and Liability Act (CERCLA) reportable quantities (RQs) have been established for cadmium and cadmium chloride. EPA issued a Rebuttable Presumption Against Registration (RPAR) for cadmium-containing pesticides under FIFRA. Also under FIFRA, there are labeling and reporting requirements. By 1997, all cadmium pesticides had undergone voluntary cancellation (OPP, 1997).

Cadmium and cadmium compounds are also regulated under the Resource Conservation and Recovery Act (RCRA) and Superfund Amendments and Reauthorization Act (SARA). Both RACA and SARA subject cadmium and its compounds to reporting requirements. FDA, under FD&CA, has set a maximum concentration level of 0.005 mg Cd/l in bottled water and limits the amount of cadmium in color additives and direct food additives. In 1984, NIOSH recommended that exposures to cadmium be reduced to the lowest possible level (NIOSH, 1996). OSHA adopted permissible exposure limits (PELs) for toxic effects other than cancer for cadmium: 0.1 mg/m^3 as a ceiling for fumes. 0.2 mg/m^3 as an 8-h TWA for dust, and 0.6 mg/m^3 as a ceiling for dust: the standards were adopted by OSHA. OSHA regulates cadmium and certain cadmium compounds under the Hazard Communication Standard and as chemical hazards in laboratories. Regulations are summarized in Table 24.5.

follows listing the exact Federal Register notice or Code of Federal Regulations citation. Cadmium compounds illustrate the kinds of regulatory information presented (see Table 24.3 and Table 24.4).

24.4 INTERNATIONAL AGENCY FOR RESEARCH ON CANCER

In 1969, the IARC initiated a program to evaluate the carcinogenic risk of chemicals to humans and to produce monographs on individual chemicals. The monograph's program now includes assessment of the carcinogenic potential of various complex mixtures of chemicals, radiations, and viruses.

With the help of international working groups of scientific experts, the monographs present a critical review and evaluation of evidence on the carcinogenicity of a wide range of human exposures. The evaluations of IARC working groups are scientific, qualitative judgments about the evidence for or against carcinogenicity provided by the available data. These evaluations may represent a part of the body of information that regulatory agencies use. The categorization of an agent, mixture, or exposure circumstances is a matter of scientific judgment, reflecting the strength of the evidence derived from studies in humans and in experimental animals and from other relevant data.

Group 1 — The agent (mixture) is carcinogenic to humans. The exposure circumstance entail exposures that are carcinogenic to humans. This category is used when there is sufficient evidence of carcinogenicity in humans.

Group 2A — The agent (mixture) is probably carcinogenic to humans. The exposure circumstances entail exposures that are probably carcinogenic to humans. This category is used when there is limited evidence of carcinogenicity in experimental animals.

Group 2B — The agent (mixture) is possibly carcinogenic to humans. The exposure circumstances entail exposures that are possibly carcinogenic to humans. This category is used for

TABLE 24.4
Regulations for Cadmium and Cadmium Compounds

Regulatory action	Effect of regulation/other compounds
CPSC regulations	
45 FR 51631. Published 8/5/80. CPSC 10: Commission denied a petition for a rule declaring sewage sludge products to be banned as hazardous substances.	The commission noted that sludges may contain high levels of cadmium and cadmium compounds, but declined to act based on anticipated EPA action on this matter. (Some municipal sewage sludges have been sold as fertilizers for home use.)
EPA regulations	
40 CFR 60 — Part — Standards of Performance for New Stationary Sources. Promulgated: 36 FR24877, 12/23/71. The provisions of this part apply to cadmium, cadmium chloride, cadmium sulfate, and cadmium sulfide.	
40 CFR 60.740 — Subpart VVV — Standards of Performance for Polymeric Coating of Supporting Substrates Facilities. The provisions of this part apply to cadmium, cadmium chloride, cadmium sulfate, and cadmium sulfide.	
40 CFR 61 — PART 61 — National Emission Standards for Hazardous Air Pollutants. Promulgated: 38 FR 8826, 04/06/73. U.S. Code: 7401, 7412, 7414, 7416, 7601. The provisions of this part apply to cadmium.	Part 61 lists substances that, pursuant to section 112 of the CAA, have been designated as hazardous air pollutants, and applies to the owner or operator of any stationary sources for which a standard is prescribed under this part.
40 CFR 63 — Part 63 — National Emission Standards for Hazardous Air Pollutants for Source Categories. Promulgated: 57FR 61922, 12/29/92. U.S. Code: 7401 et seq.; CAA. The Provisions of this part apply to cadmium and cadmium compounds (not otherwise specified).	Standards that regulate specific categories of stationary sources that emit (or have potential to emit) one or more hazardous air pollutants are listed in this part pursuant to section 112(b) of the CAA.

agents, mixtures, and exposure circumstances for which there is limited evidence of carcinogenicity in humans and less than sufficient evidence of carcinogenicity in experimental animals.

Group 3 — The agent (mixture or exposure circumstances) is not classifiable as to its carcinogenicity in humans. This category is used most commonly for agents, mixtures, and exposure circumstances for which the evidence of carcinogenicity is inadequate or limited in experimental animals.

Group 4 — The agent (mixture) is probably not carcinogenic to humans. This category is used for agents or mixtures for which there is evidence that suggests lack of carcinogenicity in humans and in experimental animals [9].

24.5 THE US EPA CARCINOGEN POLICY (MOST RECENT, 1996)

24.5.1 HISTORY

The present "proposed" carcinogen policy (1996) follows a long history of application of risk assessment procedures initiated in various segments of the U.S. government. Before these "risk assessment" concepts were popular in the vernacular, toxicologists were using "No Observed Adverse Effects Levels (NOAELs)" and "Acceptable Daily Intakes (ADIs)" as a safety-factor method for evaluating the hazard associated with chemicals. The method of "risk assessment" began as a spin-off from the field of radiation control [10]. The EPA Carcinogen Assessment Group developed a quantitative risk assessment of environmental exposures to carcinogenic agents in 1976 as the first guidelines on the subject [11]. In 1983, the National Academy of Sciences (NAS)/National Research Council published "Risk Assessment in the Federal Government: Managing the Process [12]," where they recommended establishment of "inference guidelines" to ensure consistency and technical quality in risk assessment. The 1986 cancer guidelines were proposed in 1984 and issued on September 24, 1986. The 1986 document was developed in response to a 1983 National Research Council report [12].

The 1996 Proposed Guidelines suggested principles and procedures to guide EPA scientists in the conduct of Agency cancer risk assessments. Numerous interagency EPA sessions and external work groups were conducted prior to publishing these guidelines. The following is a brief review of the changes published by EPA in the 1996 document, "Proposed Guidelines for Carcinogen Risk Assessment," which was offered for public comment in April 1996 [13]. To date, the 1996 document is still under review by the EPA. However, much of the guidance given is currently being applied to carcinogen risk assessments.

24.5.2 Significant Aspects of the 1996-Guidance

24.5.2.1 Weighing Hazard Evidence

A major recommended change is to broaden considerations from the original emphasis on tumor finding in human and animal studies to take into account modes of action at cellular and subcellular levels, toxicokinetic, and metabolic factors. The weighing of evidence is expanded from simply a finding of carcinogenic potential to a more detailed characterization, with dimensions of the carcinogenic potential including information for determining a dose–response assessment.

24.5.2.2 Classification Descriptors

Previously, summary rankings for human and animal cancer studies ranged from A to E (A, human carcinogen; B, Probable Human Carcinogen; C, Possible Human Carcinogen; D, Undetermined; and E, Not Human Carcinogen).

In the new approach, no classification of tumor findings occurs. It is replaced with weighing of evidence in a single step with descriptors of conclusions rather than letter designations divided into 3 categories: (1) "known/likely," (2) "cannot be determined," or (3) "not likely," and this is related to the route of exposure, for example, "by inhalation."

The narrative is intended to describe issues/strengths/limitations and mode of action. The category "known/likely" describes the situation when the available tumor effects and other key data are adequate to convincingly demonstrate carcinogenic potential for humans, including: (1) agents known to be carcinogenic in humans based on either epidemiologic evidence or a combination of epidemiologic and experimental evidence, demonstrating causality between human exposure and cancer; (2) agents that should be treated as if they were known human carcinogens, based on a combination of epidemiologic data showing a plausible causal association (not demonstrating it definitively) and strong experimental evidence; (3) agents that are likely to produce cancer in humans due to the production or anticipated production of tumors by modes of action that are relevant or assumed to be revelant to human carcinogenicity.

Modifying descriptors for particularly high or low ranking in the "known/likely" group can be applied based on scientific judgment and experience and are as follows:

- Agents that are *likely* to produce cancer in humans based on data that are at the *high end* of the weights of evidence typical of this group.

- Agents that are *likely* to produce cancer in humans based on data that are at the *low end* of the weights of evidence typical of this group.

Regarding the category *cannot be determined*, it is appropriate when available tumor effects or other key data are suggestive or conflicting or limited in quantity and, thus, are not adequate to convincingly demonstrate carcinogenic potential in humans. In general, further agent-specific and generic research and testing are needed to be able to describe human carcinogenic potential.

The descriptor *cannot be determined* is used with a subdescriptor that captures the rationale:

- Agents whose carcinogenic potential *cannot be determined*, but for which there is *suggestive* evidence that raises concern for carcinogenic effects.
- Agents whose carcinogenic potential *cannot be determined* because the existing evidence is composed of *conflicting data* (e.g., some evidence is suggestive of carcinogenic effects, but other equally pertinent evidence does not confirm any concern).
- Agents whose carcinogenic potential *cannot be determined* because there are *inadequate data* to perform an assessment.
- Agents whose carcinogenic potential *cannot be determined* because *no data* are available to perform an assessment.

The descriptor *not likely* is appropriate when experimental evidence is satisfactory for deciding that there is no basis for human hazard concern.

The document provides case studies for examples of assignments into these categories.

24.5.2.3 Dose–Response Assessment

The most far-reaching change is the recommendation that conclusions with regard to hazard assessment relate to the potential modes of action. When animal studies are the basis, the estimate of human equivalent toxicokinetic data may be used or a default procedure. For oral dose, the default is to scale daily-applied doses for a lifetime in proportion to body weight to the 0.75 power. For inhalation, the default methodology estimates respiratory deposition of particles and gases and utilizes estimates of internal doses with different absorption characteristics. This is a change from the previous guidance, which was to scale body weight to the 0.66 power. In addition, nontumor data are analyzed in addition to tumor incidence.

Very significantly, the new guidance proposes that a case-specific dose–response model be developed from the range of empirical data if possible. Alternatively, as a default, the lower 95% confidence limit on a dose associated with an estimated 10% increased tumor or relevant nontumor response (LED10) is taken as a point of departure for extrapolating the relationship to environmental exposure levels of interest, usually significantly lower levels. It is also possible to use other points of departure, if shown to be appropriate.

The next step, extrapolation to low-dose levels, is based on a biologically specific model if supported by the data. Otherwise, the default is applied appropriate to the mode of action of the agent. The possibilities are linearity or nonlinearity or both.

24.5.2.3.1 Linearity
For linearity, a straight line is extended to the zero dose. This is a change from the former guidance that utilized a "linearized multistage" procedure. The results are not significantly different.

24.5.2.3.2 Nonlinearity
When the linear extrapolation is clearly inappropriate, a margin of exposure approach is utilized (as for a clearly nongenotoxic carcinogen). This is simply the ratio of the LED10 to the environment exposures of interest. Key to decision making is the magnitude of the margin of exposure. If human

TABLE 24.5
Extrapolation Models

Data for support of bio/based model	Yes	No	No	No	No
Linearity		x		x	
Nonlinearity			x	x	
Extrapolation model	1[a]	2[b]	3[c]	4[d]	2[b]

[a] Model.
[b] Default linear.
[c] Default nonlinear.
[d] Default linear and nonlinear.

variability cannot be estimated, at least a 10-fold factor is considered acceptable. If comparisons of species sensitivities cannot be estimated, then humans can be considered to be tenfold more sensitive. Alternately, if humans are less sensitive than animals, a factor no smaller than one tenth may be assumed.

24.5.2.3.3 Both
Note that if either linear or nonlinear approach could be appropriate, both are presented in the assessment. Table 24.5 summarizes the above information.

24.5.2.4 Miscellaneous

The document provides considerable discussion on the use and nonuse of default assumptions. Emphasis on the mechanism of action leads to considerations of *similarity or lack of similarity from animals to humans.* One example given dealt with the finding of male rat kidney neoplasia involving alpha-2u globulin. The key question was whether this mechanism could be present in humans. The EPA in its 1991 policy guidance made the conclusion that this *mechanism is not relevant to humans* and therefore not indicative of human hazard.

Otherwise, animal data is still considered generally relevant to human cancer. It is assumed that positive effects in animal cancer studies indicate carcinogenic potential in humans. If animal data alone available then they are the basis for assessing carcinogenic hazard to humans. This is supported by the fact that nearly all agents known to cause cancer in humans are also carcinogenic in animals [14,15,16]. Almost one third of human carcinogens were identified subsequent to animal testing [17]. Finally, the carcinogenic mechanism, in general is remarkably similar among species and highly conserved in evolution.

Regarding high doses, which are necessary in some animal experiments, if the conclusion is that the effect is specifically related to *excessive toxicity rather than carcinogenicity*, then this dose can be considered as not relevant to the human situation.

On the other hand, arsenic, which shows only minimal to no effect in animals, is positive in humans. It is recognized that animal studies and epidemiologic studies have low power to detect cancer effects (10% is suggested — with the exception of rare tumor types — for example, angiosarcoma caused by vinyl chloride).

Concordance of the site of animal target organ is not required for decisions with regard to cancer in humans, as a wealth of data suggests differences in target organs between animals and humans. An exception to this is with regard to pharmacokinetic modeling where there should be concordance. The guidance indicates the inclusion of benign tumors if they have the capacity to progress to malignancies.

In the final step, "Risk Characterization," an Integrative Analysis is recommended, which combines the exposure assessment, dose response, and hazard characterization. This analysis is utilized to prepare a Risk Characterization Summary, which is written in general terms so that one not versed in the process may understand.

Detailed descriptions regarding the use of animal and human (epidemiological) data is presented in more detail than the predecessor document, including a short description of "meta analysis," which is recommended for comparing and synthesizing a number of studies.

There is a short discussion on "Criteria for Causality" including seven aspects for consideration: that is: Temporal Relationship (Latency Period), Magnitude of the Association, Consistency, Biological Gradient (dose response appropriate increase), Specificity of the Association, Biological Plausibility, and Coherence (re: natural history of the disease).

With regard to animal studies and the high dose required for a valid carcinogenic study, such a dose is considered to be one that causes no more than a 10% reduction in body weight (assuming that there is no specific organ toxicity or perturbation of function, behavior or clinical chemistry, organ weight or morphology, or significant changes).

The document provides that the following observations add significance to the tumor findings: uncommon tumor types, tumors at multiple sites, tumors by more than one route of administration, tumors in multiple species, strains, or both sexes, progression of lesions from preneoplastic to benign to malignant, reduced latency of neoplastic lesions, metastases, unusual magnitude of tumor response, proportion of malignant tumors, and dose-related increases.

REFERENCES

1. Eleventh Report on Carcinogens, U.S.P.H.S., 2005.
2. Hamilton, A. *Exploring the Dangerous Trades; The Autobiography of Alice Hamilton, M.D.*, Little, Brown, and Company, Boston, MA, 1943.
3. Henry, S.A. Occupational cutaneous cancer attributable to certain chemicals in industry, *Br. Med. Bull.*, 4, 389–401, 1947.
4. Hueper, W.C., Wiley, F.H., and Wolfe, H.D. Experimental production of bladder tumors in dogs by administration of beta-naphthylamine, *J. Industl. Hyg. Toxicol.*, 20, 46–84, 1938.
5. Wilson, R.H., DeEds, F., and Cox, A.J., Jr. The toxicity and carcinogenic activity of 2-acetamino-fluorene, *Cancer Res.*, I, 595, 1941.
6. Hearings on Food Additives Before the Subcommittee on Health and Science of the House Committee on Interstate and Foreign Commerce, 85th Congress, 1st and 2nd Sessions, 1957–1958.
7. 100 Cong. Rec. H. 16015 daily ed. August 13, 1958.
8. Public Law 91-596, 91st Congress, S. 2193, December 29, 1970.
9. International Agency for Research on Cancer (IARC), *Monographs on the Evaluation of Carcinogenic Risks to Humans*, 71, Part I, Lyon, France, 1999.
10. An Historical Perspective on Risk Assessment in the Federal Government, Center for Risk Analysis, Harvard School of Public Health, March 1994.
11. U.S. EPA, Interim Procedures and Guidelines for Health Risks and Economic Impact Statements of Suspected Carcinogens, Federal Register Vol. 41, May 25, 21402, 1976.
12. Risk Assessment in the Federal Government: Managing the Process, National Research Council (NRC), Committee on the International Means for Assessment of Risks to Public Health, Commission on Life Sciences, NCR, Washington, DC, National Academy Press, 1983.
13. U.S. EPA, *Proposed Guidelines for Carcinogen Risk Assessment*, 1996.
14. IARC, Some industrial chemicals, *Monographs on the Evaluation of Carcinogenic Risks to Humans*, 60, 13–33, 1994.
15. Tomatis, L., Aitio, A., Wilbourn, J., and Shuker, I. Human carcinogens so far identified, *Jpn. J. Cancer Res.*, 80, 795–807, 1989.
16. Huff, J.E. Chemicals causally associated with cancers in humans and laboratory animals. A perfect concordance, in *Carcinogenesis*, Waalkes, M.P., and Ward, J.M., Eds., Raven Press, 1994, pp. 25–37.
17. Huff, J.E. Chemicals and cancer in humans: First evidence in experimental animals, *Environ. Health Perspect.*, 80, 201–210, 1993.

Index